AF357324

The Role of Natural Products on Diabetes Mellitus Treatment

The Role of Natural Products on Diabetes Mellitus Treatment

Editors

Diana Marcela Aragon Novoa
Fátima Regina Mena Barreto Silva

Basel • Beijing • Wuhan • Barcelona • Belgrade • Novi Sad • Cluj • Manchester

Editors

Diana Marcela Aragon Novoa
Pharmacy
Universidad Nacional de
Colombia
Bogotá D.C.
Colombia

Fátima Regina Mena Barreto Silva
Biochemistry
Universidade Federal de
Santa Catarina
Florianópolis
Brazil

Editorial Office
MDPI
St. Alban-Anlage 66
4052 Basel, Switzerland

This is a reprint of articles from the Special Issue published online in the open access journal *Pharmaceutics* (ISSN 1999-4923) (available at: https://www.mdpi.com/journal/pharmaceutics/special_issues/diabetes_mellitus_treatment).

For citation purposes, cite each article independently as indicated on the article page online and as indicated below:

Lastname, A.A.; Lastname, B.B. Article Title. *Journal Name* **Year**, *Volume Number*, Page Range.

ISBN 978-3-7258-0010-0 (Hbk)
ISBN 978-3-7258-0009-4 (PDF)
doi.org/10.3390/books978-3-7258-0009-4

Cover image courtesy of Diana Marcela Aragon Novoa

Contents

About the Editors

Diana Marcela Aragon Novoa

Diana Marcela Aragon Novoa is a pharmacist with a PhD in Pharmaceutical Science and a full professor at the National University of Colombia. She has experience in the research and development of new drug delivery systems, extract standardization, and the study of the pharmacokinetic and biopharmaceutical properties of different active pharmaceutical agents. She has supervised more than 20 Master and PhD theses and has published more than 50 scientific papers, seven scientific book chapters, and one book about the role of women in science. Her PhD thesis was awarded as a "Meritorious Thesis". She has been awarded for her performance as a researcher at the National University of Colombia and is a senior researcher, according to the Ministry of Science of Colombia. She has led research projects funded by different organizations with Colombian and foreign participants and has been an invited researcher at different foreign universities. She is a member of the Ibero-American Pharmacometrics Network (RedIF) and of the Organization for Women in Science for the Developing World (OWSD-Colombia).

Fátima Regina Mena Barreto Silva

Fátima Regina Mena Barreto Silva is an undergraduate in Chemistry License with a Master and PhD degrees in Biochemistry and a Post-doc at the Université de CAEN, France. She is a full professor at the Federal University of Santa Catarina, Brazil, and a supervisor of Master and PhD programs in Pharmacy, Biochemistry, and Biological Sciences. She is the coordinator of the following: International Cooperation Program among UNICAEN and UFSC; International Project at the University of Guelph, Canada, and UFSC for Special Invited Professors supported by CNPq/Brazil; Project for Undergraduate and Graduation Exchange Students among UNICAEN and UFSC; Project of International Collaboration at the Sorbonne Universités, Paris and UFSC; Project of International Collaboration in Biochemistry, UFSC; Project of International Collaboration among University of California, Riverside, and UFSC; Project of International Collaboration among Universidad Nacional de Colombia and UFSC; and Project of International Collaboration among De Montfort University, Leicester and UFSC. She is a fellowship student of the Japan International Cooperation Agency, Chiba, Japan; a visiting scientist at UCR/USA and at UNICAEN; and an invited professor at UNICAEN/France in the Biology undergraduate and graduation courses. She presented at international scientific meetings (Japan, Uruguay, the USA, India, France, Poland, China, Greece, and Canada) and the International Scientific Committee (Ciencia Y Tecnologia para el Desarrollo/CYTED). She is an Associate Editor for *Frontiers in Physiology* and a Lead Editor Guest for *Life Sciences*, *CTMC*, *Pharmacologia*, *Frontiers in Endocrinology*, *BioMed Res. Int.*, and *CDT*. She has more than 135 international publications and 9 book chapters. She supervised 24 Masters, 16 PhDs, 12 Post-docs, and more than 70 undergraduate students for Scientific Initiation. Her research experience includes studies in Pharmacognosy and Medicinal Chemistry, as well as the mechanisms of action of hormones and natural and synthetic compounds related to diabetes and male fertility in the field of Biochemistry.

Preface

It is with great pleasure that we introduce this Special Issue of *Pharmaceutics*, dedicated to exploring the multifaceted realm of diabetes mellitus (DM) treatment through the lens of natural products. The global burden of DM necessitates innovative approaches, and within these pages, we embark on a journey through the promising avenues offered by natural compounds for managing this complex metabolic disorder.

This Special Issue serves as a comprehensive repository of recent advancements, encapsulating the latest patents, the profound effects of characterized extracts, and the intricate cellular mechanisms underpinning isolated secondary metabolites in DM therapy. Our endeavor is to shed light on the burgeoning research that delves into the diverse array of natural products, uncovering their potential as effective tools in the fight against DM.

The inclusion of clinical trials within this Special Issue underscores the practical implications and translational aspects of these findings. It is our fervent hope that these contributions will not only expand the scientific discourse but also catalyze further exploration and development of novel therapeutic interventions for DM.

We extend our heartfelt gratitude to the esteemed contributors whose dedication and scholarly pursuits have enriched this collection. Their commitment to advancing our understanding of natural product-based treatments for DM has been instrumental in shaping this compendium.

We invite you, our readers and fellow enthusiasts in the field, to immerse yourselves in this assemblage of knowledge and discoveries. May this Special Issue serve as a catalyst for continued innovation, inspiring new avenues in the pursuit of more effective, accessible, and holistic strategies in diabetes mellitus management.

Diana Marcela Aragon Novoa and Fátima Regina Mena Barreto Silva

Editors

 *pharmaceutics*

Review

Natural Products as Outstanding Alternatives in Diabetes Mellitus: A Patent Review

Ingrid Andrea Rodríguez [1], Mairim Serafini [2], Izabel Almeida Alves [3], Karen Luise Lang [4], Fátima Regina Mena Barreto Silva [5] and Diana Marcela Aragón [1,*]

[1] Departamento de Farmacia, Facultad de Ciencias, Universidad Nacional de Colombia, Bogotá 110321, D.C., Colombia

[2] Departamento de Farmácia, Universidade Federal de Sergipe, Sao Cristovao 49100-000, SE, Brazil

[3] Department of Medicines, Faculty of Pharmacy, Universidade Federal da Bahia, Salvador 40170-115, BA, Brazil

[4] Departamento de Farmácia, Campus Governador Valadares, Universidade Federal de Juiz de Fora, Governador Valadares, Juiz de Fora 36038-330, MG, Brazil

[5] Departamento de Bioquímica—Centro de Ciências Biológicas, Universidade Federal de Santa Catarina, Rua João Pio Duarte Silva, Florianópolis 88037-000, SC, Brazil

* Correspondence: dmaragonn@unal.edu.co

Abstract: Diabetes mellitus (DM) is a metabolic syndrome that can be considered a growing health problem in the world. High blood glucose levels are one of the most notable clinical signs. Currently, new therapeutic alternatives have been tackled from clinicians' and scientists' points of view. Natural products are considered a promising source, due to the huge diversity of metabolites with pharmaceutical applications. Therefore, this review aimed to uncover the latest advances in this field as a potential alternative to the current therapeutic strategies for the treatment of DM. This purpose is achieved after a patent review, using the Espacenet database of the European Patent Office (EPO) (2016–2022). Final screening allowed us to investigate 19 patents, their components, and several technology strategies in DM. Plants, seaweeds, fungi, and minerals were used as raw materials in the patents. Additionally, metabolites such as tannins, organic acids, polyphenols, terpenes, and flavonoids were found to be related to the potential activity in DM. Moreover, the cellular transportation of active ingredients and solid forms with special drug delivery profiles is also considered a pharmaceutical technology strategy that can improve their safety and efficacy. From this perspective, natural products can be a promissory source to obtain new drugs for DM therapy.

Keywords: diabetes mellitus; natural products; flavonoids; polyphenols; patent

Citation: Rodríguez, I.A.; Serafini, M.; Alves, I.A.; Lang, K.L.; Silva, F.R.M.B.; Aragón, D.M. Natural Products as Outstanding Alternatives in Diabetes Mellitus: A Patent Review. *Pharmaceutics* **2023**, *15*, 85. https://doi.org/10.3390/pharmaceutics15010085

Academic Editor: Crispin R. Dass

Received: 23 November 2022
Revised: 16 December 2022
Accepted: 22 December 2022
Published: 27 December 2022

1. Introduction

Diabetes mellitus is a metabolic disbalance among carbohydrates, lipids, and proteins that are associated with hyperglycemia, cardiovascular events, kidney, and ocular perturbances, peripheral neuropathy, ulcers, and even lower limb amputations [1–3]. A defect in insulin secretion and/or action is identified as one of the pathogeny causes [4]. Type 2 diabetes is the most reported form [5]. It occurs when the insulin levels decrease progressively, and the insulin resistance increases. Furthermore, type 1 diabetes is defined as a chronic condition in which the pancreas produces low levels of insulin or almost none [6].

The International Diabetes Federation reports estimate that about 537 million people live with DM worldwide. The countries with the highest incidence levels are by China, India, the United States, Pakistan, and Brazil [7]. In addition, low- and middle-income countries have the largest number of DM cases, and 1.5 million deaths are directly caused by diabetes each year. In recent decades, new case numbers and prevalence indicators in DM have been increasing. This has triggered sanitary alerts in international health systems, stimulating the research and development of new therapeutic strategies.

Currently, there are different treatments for DM type II, including changes in lifestyle, the use of oral hypoglycemic drugs or biotechnology products, gene therapy, pancreatic transplants, and precision medicine, to personalize diabetes prevention and management [8–12]. However, it has recently been shown that natural products and isolated substances play an important role in DM [13–17].

Galega officinalis L. (Fabaceae) was one of the first medicinal plants associated with antidiabetic activity prescribed in the Middle Ages [18]. Galegine metabolite was isolated from this plant, which is also known as galega or goat's rue. This substance corresponds to a guanidine derivative with high toxicity levels [19]. However, this important fact helped to establish a precursor prototype to develop metformin the first hypoglycemic drug of the biguanide type. Their structures are shown in Figure 1.

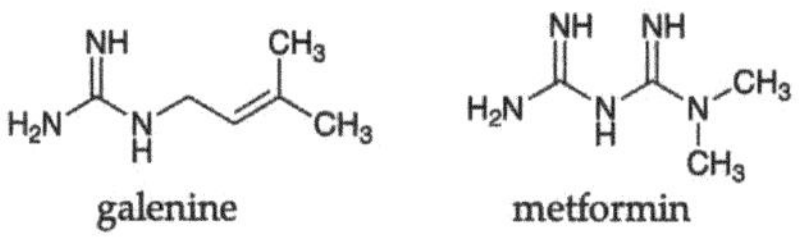

Figure 1. Chemical structure of galegine and metformin.

Pycnogenol (acquose extract derived from *Pinus pinaster* (Pinaceae)), acarbose, miglitol, and voglibose (α-glucosidase inhibitors isolated from microorganisms) are other examples of natural products applied in DM treatment [20,21]. Ethnopharmacology criteria represent a valuable tool that allows advancing systematic analysis of natural products as innovative drugs in the industrial application field. A systematic patent review was carried out in the 2016–2022 period (Table 1) in which research with therapeutic DM potential based on natural products was selected. *Curcuma longa, Trigonella foenum-graecum*, and *Panax ginseng* were the main assessed as well as seaweed, fungi, and several minerals' derivatives. Among the therapeutic targets, the inhibition of α-amylase, effects on glucose uptake, effects on glucose transporters, improvement in insulin secretion, the proliferation of pancreatic β-cells, and inhibition of proteins such as tyrosine phosphatase were found [15,22–24].

Furthermore, organic acids, polyphenols, saponins, terpenes, and flavonoids were identified as the responsible metabolites for the potential activity. Mainly, improved insulin signaling, effects on lipids, reduction in fat accumulation, inhibition of α-glycosidase activity, an influence on glucose metabolism, antioxidant activity, and influence on the intestinal microbiota were the effects derived from their use. Likewise, some pharmacotechnical strategies such as solid forms (tablets and granules), liquid forms (syrups and solutions), heterodisperse systems (suspensions), and transmucosal adhesive forms were also discussed. Finally, safety assessments and toxicity tests (in vitro and in vivo), and clinical efficacy reports were included.

Table 1. Pharmaceutical product patents with natural origin published by EPO from 2016 to 2022.

Year	Country	Registration Number	Species	Part of the Plant	Secondary Metabolites	Mechanism of Action	Formulation	Safety Assessments	Efficacy Tests	Ref
2020	WO	WO2020115767 (A1)	Premna herbacea	Leaves	isoverbascoside	Improved insulin signaling via the AET/AM PK pathway	ND	A single dose of 2000 mg/kg of *P. herbacea* methanol extract was reported as safe after acute toxicity evaluation in Swiss ICR male mice.	In vitro: [a] Uptake of glucose in L6 cells [b] Total accumulation of lipids in HepG2 cell lines by staining with oil red 0 In vivo: ND Clinical studies: ND	[25]
2020	US	US11007237 (B2); US2020147160 (A1)	*Pinus* spp.	Resin	ND	α-glucosidase inhibition	Syrup, tablets, capsules, granules, and powders.	ND	In vitro: [a] Suppression of α-glucosidase activity [b] Ability to uptake glucose Clinical studies: ND	[26]
2020	WO	WO2020012299 (A1)	*Curcuma longa, Phylanthus emblica*	ND	ND	Insulin sensitization and antihyperglycemic	ND	Efficacy studies were carried out with non-cytotoxic doses after evaluation of cytotoxicity and PPAR γ expression in fibroblasts using the MTT assay.	In vitro: [a] Evaluation of PPAR γ expression in fibroblasts cell culture. In vivo: Evaluation of streptozotocin-induced insulin resistance in rats. Clinical studies: ND	[27]
2019	SG	SG11201908574T (A)	*Curcuma longa, Emblica officinalis, Vernonia anthelmintica, Tinospora cordifolia, Trigonella foenum-graecum, Ixora coccinea, Syzygium cumini*	ND	Tannins, organic acids, curcuminoids	ND	Powders, pastes, granules, capsules, tablets, emulsions, suspensions, syrup, elixir, oral drops, and nutraceuticals	Acute toxicity test in male and female Swiss ICR mice, at doses. It was reported as safe.	In vitro: [a] Evaluation of glucose uptake potential in L6 cells [rat skeletal muscle cells] In vivo: ND Clinical studies: ND	[28]

Table 1. *Cont.*

Year	Country	Registration Number	Species	Part of the Plant	Secondary Metabolites	Mechanism of Action	Formulation	Safety Assessments	Efficacy Tests	Ref
2019	WO	WO2019205943 (A1)	*Polygonatum* spp., *Poria cocos, Lycium barbarum, Pueraria Panax ginseng, Shiitake, cordyceps, Ganoderma lucidum, Hericium erinaceus, Tremella Dendrobium officinale*	ND	Flavonoids and oligosaccharides (inulin, fructooligosaccharides, oligosaccharides)	Reduction in the proportion of Firmicutes/Bacteroides in the gastrointestinal tract.	ND	The description of the methodology was not detailed, but it was concluded that the preparation is safe.	In vitro: AND In vivo: Oral glucose tolerance test in different groups of mice. Clinical studies: ND	[29]
2019	WO	WO2019186579 (A1)	*Trigonella foenum-graecum, Solanum lycopersicum, Dudhi bhopla, Lagenaria siceraria, Vitis vinifera, Medicago sativa*	Roots, stems, leaves, buds, flowers, fruits, and seeds.	Osmotin	Adiponectin agonists/mimetics. Activation of the AMPK pathway	Tablets, syrup, granules, and film for transmucosal administration.	ND	In vitro: ND In vivo: ND Clinical studies: ND	[30]
2019	US	US10967025 (B2) US2019209634 (A1)	*Moringa Oleifera, Origanum vulgare, Shilajit, Blue-Green Algea, Phyllanthus emblica, Piper nigrum, Salvia rosmarinus, Punica granatum, Trigonella foenum-graecum, Curcuma longa.*	Leaves, fruits, and seeds.	Polyphenols, carvacrol, phycocyanin	Antioxidant and anti-inflammatory—via COX-1 inhibition	Capsules	ND	In vitro: [a] ROS absorption capacity [b] DPPH radical scavenging activity [c] Oxidation of LDL in a cell-free system [d] Insulin release in islets of rats [e] Effect on cholesterol absorption in Caco-2 cells [f] Effect on COX-1 Enzyme activity. In vivo: ND Clinical studies: ND	[31]

Table 1. *Cont.*

Year	Country	Registration Number	Species	Part of the Plant	Secondary Metabolites	Mechanism of Action	Formulation	Safety Assessments	Efficacy Tests	Ref
2019	WO	WO2019088958 (A2); WO2019088958 (A3)	*Paiiurus spina-christi*	Fruits	Naringenin	Effect on ALT and AST regulation enzyme levels, antioxidant activity, increased insulin secretion, regulates blood glucose, and increased magnesium absorption	ND	ND.	In vitro: ND In vivo: Rats with streptozotocin-induced diabetes Clinical studies: ND	[32]
2019	MX	MX2018004489 (A)	*Agavaceae* spp.	Leaves, rhizomes, and mead	Steroidal saponins and sapogenins (kammogenin, manogenin, gentrogenin, and hecogenin).	Lipid and glucose metabolism improvements, energy expenditure, gut microbiota health, muscle oxidative capacity, and thermogenesis, stimulating PGC-1Q and UCP1, and activating AMPK	Pills, capsules.	ND	In vitro: ND In vivo: ND Clinical studies: ND	[33]
2019	US	US10640480 (B2); US2019031635 (A1)	*Gnetum gnemon*	Fruits	Gnetin C	Antioxidant, antibacterial, inhibition of lipase enzymes, α-glucosidase, and α-amylase.	Powders, granules, tablets, capsules.	ND	In vitro: ND In vivo: ND Clinical studies: ND	[34]
2019	AU	AU2018278958 (A1); AU2018278958 (B2)	*Cichorium endivia, Latifolium; Lactuca sativa, Longifolia; Lactuca sativa.* Plantas das famílias *Asteraceae* spp., *Lamiaceae* spp., *Brassicaceae* spp., *Amaranthaceae* spp.	Fruits, leaves, stems, and roots	ND	IRS2 branch activation of the insulin-mediated signal transduction cascade.	N.D.	ND	In vitro: ND In vivo: ND Clinical studies: ND	[35]

Table 1. *Cont.*

Year	Country	Registration Number	Species	Part of the Plant	Secondary Metabolites	Mechanism of Action	Formulation	Safety Assessments	Efficacy Tests	Ref
2019	US	US10576117 (B2); US2019125816 (A1)	*Salacia chinensis, Gymnema sylvestre, Emblica officinalis, Eugenia jambolana, Curcuma longa, Commiphora mukul, Tinospora cordifolia. Ithania somnifera, Terminalia chebula, Andrographis paniculata, Boerhavia diffusa, Azhadirachta indica, Aristolochia indica, Aegle marmelos, Cyperus rotundus, Hemedesmus indicus, Trichosanthes dioica, Santalum alba, Terminalia arjuna, Woodfordia fruiticosa, Glycerrhiza glabra, Mucuna pruriens, Myrica nagi, Plumbago rosea, Inula racemosa, Zingiber officinalis, Piper longum y Piper nigrum*	Roots, fruits, bark, leaves, heart-wood, flowers, seeds, rhizomes, gum resin, and stems.	ND	Lipids/cholesterol reduction, tissue phosphatases, and tissue transaminases reduction, reversal of glycogen depletion, pancreatic islets cell regeneration, hypoglycemic, hypolipidemic, cytoprotective, and immunomodulatory action. FBS, PPBS, HbA1C reduction	Tablets, granules, capsules, solution, emulsion, suspension	According to OECD guidelines in rodent tests 423, 407, 408, and 452. It was reported as safe.	In vitro: ND In vivo: Model of diabetes in Wistar rats diabetic by alloxan. Clinical studies: A prospective, open-label, non-randomized, phase III clinical trial conducted in outpatients at Muniyal Ayurvedic Hospital and Research Centre, Manipal, India.	[36]
2017	US	US10668122 (B2); US2017252393 (A1)	*Toona sinensis*	leaves	Monoterpenes, diterpenes, triterpenes, sesquiterpenes, saturated and unsaturated fatty acids	Glucose absorption improvements, lipid degradation, inhibition of large lipid clusters, and inhibition of metabolic syndrome	ND	ND	In vitro: 24 h glucose consumption of 3T3-L1 adipocyte culture medium. 2- Adipogenesis blocked by TS-SCF in adipocytes 3- Antidiabetic effect of high to medium/high polar components of TSL, except TSL-SCF In vivo: ND Clinical studies: ND	[37]

Table 1. *Cont.*

Year	Country	Registration Number	Species	Part of the Plant	Secondary Metabolites	Mechanism of Action	Formulation	Safety Assessments	Efficacy Tests	Ref
2017	US	US10111923 (B2); US2017173103 (A1)	*Cyclocarya paliurus, Puerariae lobatae Radix, Polygonati odorati Rhizoma. L*	ND	ND	ND	Pills, capsules, granules, powders, or tea formulation	ND	In vitro: ND In vivo: Rats with streptozotocin-induced diabetes Clinical studies: ND	[38]
2017	NZ	NZ630125 (A)	*Calophyllum inophyllum*	Cortex	ND	Inhibition of DGAT-1 and SCD-1 enzyme activity	Tablets, coated tablets, capsules, powders, granules, elixir, and syrup.	ND	In vitro: 1- Assay of DGAT-1. 2- Inhibition of hDGAT-1 3- SCD-1 Assay 4- Triglyceride synthesis assay in HepG2 cells In vivo: Streptozotocin-induced diabetic rats fed a high-fat diet. Clinical studies: ND	[39]
2016	US	US2016361341 (A1); US9828442 (B2)	*Hirsutella sinensis*	ND	Polysaccharides	Hyperglycemia reduction and improvement of insulin sensitivity in humans and animals.	ND	ND	In vitro: ND In vivo: Streptozotocin-induced diabetic rats fed a high-fat diet. Clinical studies: ND	[40]

Table 1. *Cont.*

Year	Country	Registration Number	Species	Part of the Plant	Secondary Metabolites	Mechanism of Action	Formulation	Safety Assessments	Efficacy Tests	Ref
2016	US	US10028930 (B2); US2016228400 (A1)	*Ephedra sinica and* brown algae concentrate	ND	Fucoxanthin, caffeine, alkaloids such as ephedrine	Thermogenesis potentiator	Extended-release tablets, transdermal patches	ND	In vitro: ND In vivo: ND Clinical studies: ND	[41]
2016	US	US10799547 (B2); US2016067294 (A1)	*Suaeda japonica*	Leaves and seeds	ND	The cytoprotective activity of pancreatic β-cells, hypoglycemic action, insulin secretion increase, improvement of glucose tolerance, and increase of adiponectin content in the blood.	Pills, powders, granules, capsules.	ND	In vitro: ND In vivo: Overfeeding diabetic OLETF mice. ND clinical studies	[42]
2016	UA	UA104161 (U)	*Equiseti arvensis, Sambucus* spp., *Hypericum perforatum, Tiliae* spp., *Polygonum aviculare, Urticae* spp. *Inula helenium* e *Mirtilli* spp.	Flowers, leaves. Additional plant material: rhizomes and roots	Ascorbic acid, tannins, organic acids, flavonoids, hydroxycinnamic acids	Hypoglycemic action due to exposure to glucagon-like peptide 1 [GLP-1]. Increased insulin levels, improves carbohydrate and lipid metabolism.	ND	ND	In vitro: ND In vivo: Alloxan-diabetic female rats. Clinical studies: ND	[43]

Abbreviations: WO: World Intellectual Property Organization (WIPO). xUS: United States. SG: Singapore. MX: Mexico, AU: Australia. NZ: New Zealand.UA: Ukraine, AKT/AM PK: Protein kinase B or AKT / AMP-activated protein kinase. ICR: Cancer Research Institute. PPAR γ: Peroxisome proliferator-activated receptor. COX-1: Cyclooxygenase-1. ROS: reactive oxygen species. DPPH: 2,2-Diphenyl-1-picrylhydrazyl. LDL: low-density lipoproteins. ALT: alanine aminotransferase. AST: Aspartate aminotransferase. PGC-1Q: Peroxisome proliferator-activated gamma recipe coactivator. UCP1: Uncoupling protein 1 or thermogenin. IRS2: insulin receptor substrate 2. FBS: Fasting blood sugar. PPBS: Postprandial blood sugar test. HbA1C: glycosylated hemoglobin. TS-SCF: T. sinensis- supercritical carbon dioxide fluid. TSL-SCF: Hojas de T. sinensis- supercritical carbon dioxide fluid. DGAT-1: Diacylglycerol O-acyltransferase 1. Human hDGAT-1: Diacylglycerol Acyltransferase -1. SCD-1: stearoyl-CoA desaturates.

2. Materials and Methods

Espacenet developed by the European patent office (EPO) was chosen as a database repository. The keyword diabetes in the title or abstract and the code a61k36/00 were chosen as the inclusion criteria. According to the International Patent Classification (IPC), the code a61k36/00 is associated with medicinal preparations of the indeterminate constitution, which contain algae material, lichens, fungi or plants, or their derivatives, and traditional herbal medicines. The review process was carried out between February to April 2022.

By Espacenet, 183 patents satisfied the first criteria. Subsequently, 5 cases were duplicated, and they were rejected. The publication date was considered as the second criterion, and then the patents published before 2016 were also discarded ($n = 144$). Additionally, the titles and abstracts were read, and 3 of them were due to a lack of relation to DM.

Then, 31 patents were fully analyzed, and 12 documents were out of the scope of this review. Finally, 19 documents were accomplished with all the criteria established. This allows the identification of natural products with potential applications in the pharmaceutical industry related to DM treatment. The screening scheme is represented in Figure 2.

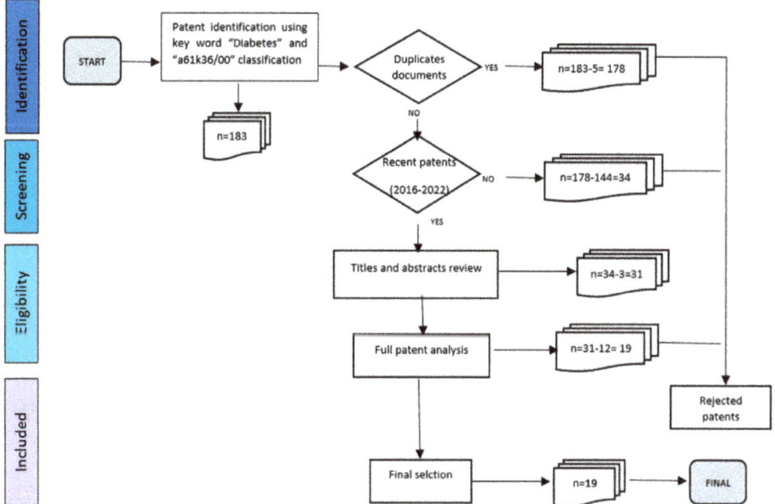

Figure 2. A systematic approach to the patent review with inclusion and exclusion criteria.

3. Results

The United States, Mexico, Ukraine, Singapore, Australia, and New Zealand were the countries of origin of the reviewed patents. This can be observed in Figure 3.

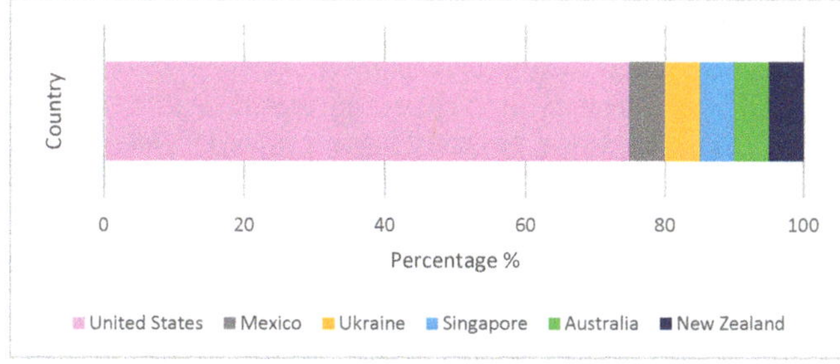

Figure 3. Patents for DM with A61K36/00 IPC classification that were published between 2016 and 2022. (100% represents the total number of 19 patents).

The number of patents published and the scientific articles found in PubMed with keywords Diabetes (AND) herbal and Diabetes (AND) algae was also compared, as shown in Figure 4.

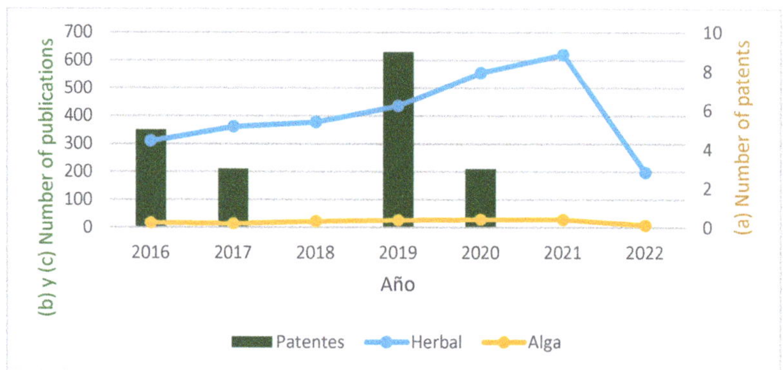

Figure 4. (a) Number of patents with IPC classification A61K36/00 published between period 01/2016 and 04/2022. (b) Number of articles published in PubMed between 01/2016 and 04/2022 associated with diabetes and herbal preparations. (c) Number of articles published in PubMed between 01/2016 and 04/2022 associated with diabetes and algae-based preparations.

The worldwide impact of diabetes has increased in the latest years. Ongoing research work needs to be rigorous and constant to establish alternative therapeutic options. This context aims to analyze natural products with diverse and promissory mechanisms of action and their possible clinical applications.

This framework led to an in-depth investigation of the documents shown in Table 1. Plants, fungi, and seaweed extraction methods and some mineral compounds of traditional uses are compiled from patents in the field. Qualitative formula descriptions, pharmaceutical dosage form strategies derived from different extracts, and multi-component preparation (association of several natural derivatives materials) are also included.

4. Discussion

4.1. Main Hypoglycemic Mechanisms of Action

The multi-component systems are described in Table 1. Often, these formulations are made up of plants or a group of them. Therefore, some metabolites' roles are highlighted in the same product. This is the case for SG11201908574T (A) patent, in which the ellagitannins (Figure 5) are the predominant substances with high maltase-inhibitory activity and energy expenditure (EE) increase by enhancing thermogenesis in brown adipose tissue [44,45]. Other activities derived from ellagitannins such as beneficial effects against insulin resistance, oxidative stress control in the polyol flow pathway, the activation of protein kinase C and the suppression of α-glucosidase, α-amylase, and lipase activities have been described [46–48].

Curcuminoids (Figure 5) are present in SG11201908574T (A) pharmaceutical preparation. Especially, bisdemethoxycurcumin [44] present in *Curcuma longa* has been related to the inhibitory activity of α-glucosidase derived from in vitro tests. Furthermore, some cellular and molecular effects have been attributed to these metabolites. For instance, a decrease in the sterol regulatory element-binding protein 1 (SREBP1c), and carbohydrate-responsive element-binding protein (ChREBP) and an increase in the carnitine palmitoyltransferase (CPT1) and the acyl-coenzyme A:cholesterol acyltransferase (ACAT) levels can allow the regulation of lipid metabolism [49]. An increase in PPAR-γ via AMPK activation and a decrease in lipid peroxidation effects are also associated with antioxidant activity with curcuminoids [50,51]. Finally, a decrease in monocyte chemoattractant protein-1 (MCP-1), IL-1β, TNF-α, IL-6, and COX-2, can be resulted in an anti-inflammatory effect [52–54]. Furthermore, an influence on the intestinal microbiota is also suggested, owing to a possible mediation in the metabolic syndrome process [55,56].

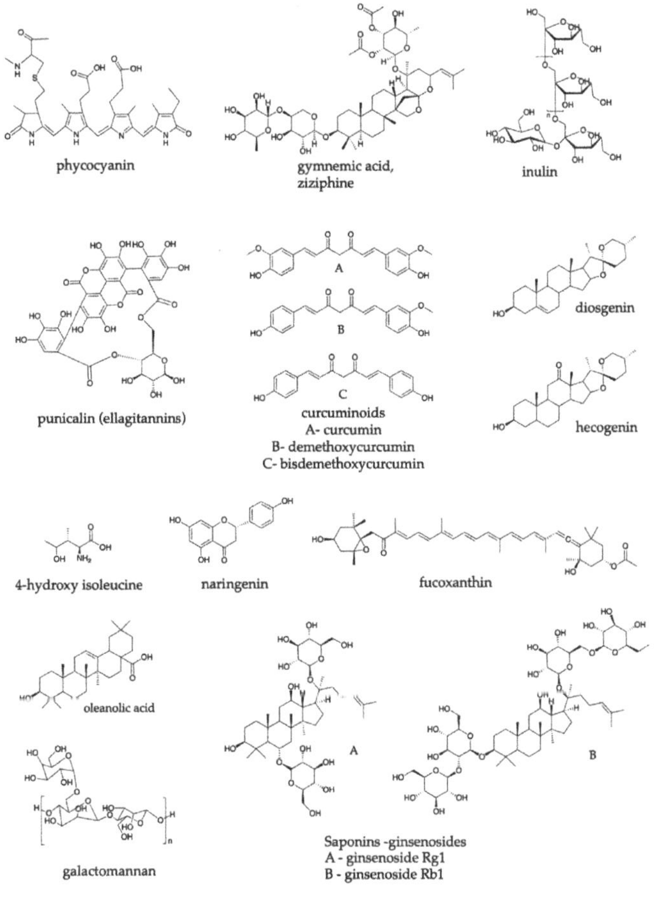

Figure 5. Principal metabolites reported by patents from natural products.

Meanwhile, different terpenes also have promising effects in DM treatment. Oleanolic acid is present in the species *Sambucus* spp. [57] has demonstrated an inhibitory activity on α-glucosidase, α-amylase, and protein tyrosine phosphatase 1B (PTP 1B) [58,59]. In this context, saponins or triterpenoids are also thought to be responsible for the beneficial effect in DM [60]. It was found that saponins such as kammogenin, manogenin, gentrogenin, and hecogenin (Figure 5) exhibited an α-glucosidase-inhibitory activity. They are included in the composition of the patent product MX2018004489 (A) [33]. A powerful hypoglycemic activity in an oral glucose tolerance assay in rats is also exhibited by these saponins [44].

SG11201908574T (A) patent refers to the presence of organic acids, which describe a potent antioxidant and hypoglycemic activity that can be evaluated in animal models of hyperglycemia [28,61]. The above also suggests stimulating effects on insulin secretion [44]. Likewise, it attenuates diabetic nephropathy through anti-inflammatory and antifibrotic effects, which are found in flavonoids such as the flavonone naringenin (Figure 5); this was identified in WO2019088958 (A2) and WO2019088958 (A3) patents [9,32]. These metabolites are important components of the human diet. and several studies have demonstrated a broad range of beneficial effects in DM, such as the balancing of glucose metabolism [14].

Some polysaccharides were described in the composition of the products in the US2016361341 (A1) and US9828442 (B2) patents. Decreased glucose absorption, increased serum insulin concentration, reduced postprandial hyperglycemia, and modulation of liver enzymes are the proposed mechanisms of action of these metabolites [40]. Inulin (Figure 5) is an example of these substances that can be applied to DM [9,44]. Additionally, it is known that oligosaccharides such as non-digestible carbohydrates, a subtype of this group of compounds, also play a role in modulating the intestinal bacterial population after dysbiosis [55,62].

Different mechanisms of action derived from secondary metabolites are mentioned above in the reviewed patents. Nevertheless, an important number of patents do not mention the complete chemical composition. Only the species of plants, fungi, or seaweed and their parts such as roots, fruits, bark, leaves, heartwood, flowers, seeds, rhizomes, gum resin, or stems are detailed. The lack of information on the chemical identity in the preparation can be considered a limitation of the study. The poor description affects the monitoring of a particular pharmacological marker as being responsible for anti-DM effects. Even so, seven categories of various mechanisms of action that could contribute to the treatment of DM have been proposed (Figure 6) based on the authors' annotations.

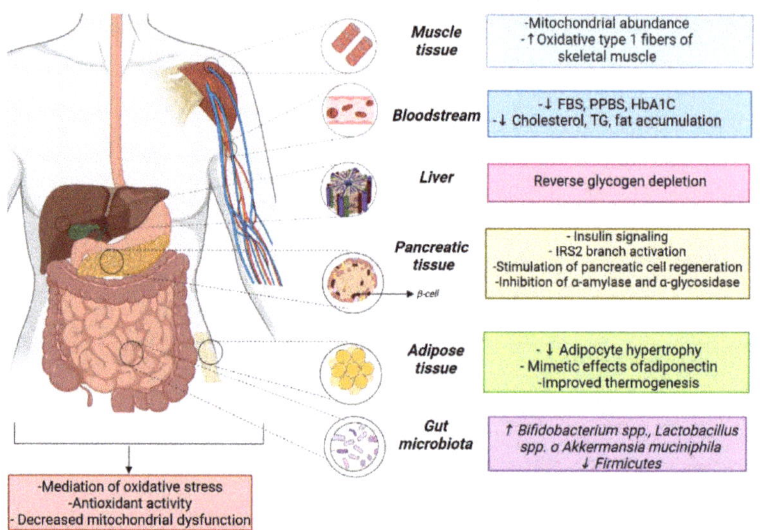

Figure 6. Mechanisms of action in DM related to natural products derived from the patent reviewed. (↑ increase ↓ decrease).

4.1.1. Insulin Signaling

Anabolic processes comprise a series of harmonized steps that are oriented toward the construction of complex biomolecules. The procedure is carried out after the generation of covalent molecular bonds. As a consequence of these reactions, the reserve energy is available to be released according to the body's demand [63].

Insulin is synthesized in the pancreas, a glandular organ located in the abdomen [64]. This hormone contributes to the anabolic process, which is mediated through glucose uptake in myofibrils and adipocytes [64]. DM cases are derived from the total or partial absence of insulin, which triggers a metabolic imbalance. As a solution, agents that promote or improve insulin signaling are considered a useful strategy in this field [65].

About 40% of the reviewed patent analyzed the improvement in insulin signaling as a benchmark derived from their preparations. For instance, an increase in insulin secretion and regeneration of pancreatic cells is mentioned in the US10576117 (B2) and US2019125816

(A1) patents [22]. This scenario is attributed to the use of *Gymnema sylvestre* extracts, whose major components are gymnemic acid, saponins, and gurmarin [22,66–68].

On the other hand, *Herbaceous Premna* and *Paiiurus spina-christi Mill* species are present in WO2020115767 (A1), WO2019088958 (A2), and WO2019088958 (A3) patents, in which the authors suggest an improvement insulin signaling via the AKT/AM PK pathway [25,69].

4.1.2. Effects on Lipids

Lipidic perturbances are also characteristics of DM. As a result, a strong relationship between dyslipidemia, diabetes, and cardiovascular disease has been documented. It shows the metabolic interdependence between carbohydrate metabolism and lipids [70].

Higher levels of specific cholesterol subtypes and very low lipoproteins (VLDL), low-density lipoprotein (LDL), and triglycerides increase the likelihood of obstruction in vessels and capillaries, which promotes vascular changes. As result, this target can allow better lipid regulation and avoid complications resulting from DM [71].

In the WO2020115767 patent, a significant reduction in lipids was obtained. The decrease was induced by palmitate in a dose-dependent manner in HepG2 cells after *P. herbacea* administration. Then, it suggests the applicability of the herbal preparation to complications associated with hepatic insulin resistance in DM2 [25].

In the same way, the reduction in obesity, overweight, accumulation of visceral adipose fat and central obesity, adipocyte hypertrophy, intracytoplasmic accumulation of hepatic triglycerides, dyslipidemia, and LDL and the improvement in lipid degradation were also studied in the MX2018004489 (A), US10668122 (B2), and US2017252393 (A1) patents [33,37].

4.1.3. Glucose Metabolism

Many natural products are useful in the control of DM due to their effects on glucose metabolism. The main ones are the action on degrading enzymes, glucose re-uptake, and agonist effects on specific receptors. During the digestion process, enzymes help break down macromolecules to make them available for absorption. One of them, α-glucosidase, helps in the cleaving processes of the starch molecules, allowing the absorption of released monosaccharides [72]. This fact explains the glucose bioavailability levels. The higher the α-glucosidase levels, the higher the probability of free glucose in blood circulation, which is critical in DM patients.

Digestive enzyme regulation can be considered a target of DM treatment. The possible delay of carbohydrate absorption in the intestine and the subsequent decrease in the post-prandial insulin level [57] derived from enzyme control are highlighted by the US11007237 and (B2) US10640480 (B2) patents.

Additionally, the control of glucose levels may be mediated by peripheral glucose reuptake. Therefore, the glucose transporter (GLU), peroxisome proliferator-activated receptors (PPARs), and adipocytokines [72] have been described as targets of interest.

For instance, adiponectin (a type of adipocytokine) represents a protein involved in insulin sensitivity, glucose uptake, and lipid metabolism. Chronic inflammatory processes can be expressed in DM due to low serum levels of this substance [73]. In addition, osmotin is described as a structural homolog of the β-barrel domain of adiponectin in WO2019186579 (A1). This enzyme can bind and activate the adiponectin receptor (ADI-POR1), which led carrying out kinase phosphorylation through AMPK activation ($3'\ 5'$-adenosine-monophosphate-kinase) [74]. The previous process takes place on muscle cells and has a beneficial effect on DM treatment.

Nonetheless, natural products can strengthen the pancreatic tissue and promote insulin secretion and decrease the intestinal absorption of glucose [9].

Trigonella foenum-graecum species present in two patents contain metabolites such as diosgenin, galactomannan, and 4-hydroxy-isoleucine (Figure 5) that have been associated with lowering blood glucose levels and insulin sensitivity [75].

Ginsenosides (Figure 5) from *Panax ginseng* have preclinical and clinical reports reflecting the improvement in glucose metabolism and renal function [22,75]. Moreover, the European Medicines Agency (EMA) established its inhibitory effect on the plasma glucose-lowering action of Rh2 by opioid μ-receptor blockers [76].

4.1.4. Influence on the Intestinal Microbiota

The relationship between the gut microbiota, the onset of insulin resistance, and diabetes has been investigated for years. Intestinal permeability, endotoxemia, interaction with bile acids, and changes in the proportion of adipose tissue are possible mechanisms of interconnection [77]. An increase in the abundance of Firmicutes phylum bacteria and a reduction in Bacteroidetes phylum bacteria have been reported in people with obesity and DM [78]. However, natural products can ameliorate the control and balance of microorganisms' growth and strengthen intestinal health.

Some polyphenols of natural origin have probiotic activity and can influence the intestinal microbiota, being benefic for obesity and metabolism. That is the case of curcumin (Figure 5), a representative metabolite of *Curcuma longa*. Intestinal barrier function, reduced circulating LPS levels, and improves glucose tolerance in LDLR 1 were found after the oral administration of curcumin in mice [55].

A possible reduction in the imbalance of the Firmicutes/Bacteroidetes ratio in the gastrointestinal tract was researched in WO2019205943 (A1), which is derived from various plants, fungi, or their association [29,79]. Likewise, several beneficial effects such as a decrease in Firmicutes and an increase in Bacteroidetes, *Lactobacillus* spp., *Bifidobacterium* spp., and *Akkermansia muciniphila* are also discussed in MX2018004489 (A). The microbiota perturbances are related to mucin degradation, which represents a protective glycoprotein in the intestinal mucus layer and has been associated with better insulin sensitivity and decreased fat gain [33,36].

4.1.5. Oxidative Stress

High levels of glucose in DM may be associated with the presence of reactive oxygen species (ROS) due to the degradation of LDL lipids [75]. The presence of ROS triggers chemical reactions binding to lipid and protein membranes. The disruption of physiological conditions exacerbates oxidative damage and modifies the cellular environment [72]. The conditions mentioned result in an insulin resistance scenario. The antioxidant activity is discussed in several patents. Limiting the propagation of radicals and interrupting membrane damage are characteristics of these processes.

The improvement in pancreatic β cell function and regulation of insulin tolerance has been explained as an antioxidant and antidiabetic effect from curcumin in the patent review [75]. Meanwhile, polyphenols in *Vitis vinifera* (patent WO2019186579 (A1)) were associated with antioxidant capacity and insulin resistance decrease [30,75]. *Moringa oleifera* and *Zingiber officinale* also contribute to the mitigation of oxidative stress in animal models [22]. In addition, cyanobacterial-derived metabolites, such as phycocyanin (Figure 5), act against oxidative damage and contribute to preventing chronic inflammation [31].

4.1.6. Other Mechanisms

Thermogenic effects are also found as a pharmacological strategy favorable in DM cases. It is linked to the body's ability to generate heat after metabolic reactions. Free fatty acid concentration and insulin sensitivity are associated with an improvement in metabolic parameters through the activation of brown fat [80]. Thermogenic effect is mentioned on MX2018004489 (A), in which the inhibition of phosphodiesterase activity is carried out [33]. In addition, stimulating PGC-1Q and UCP1 and activating AMPK [41] mechanisms are shown in US10028930 (B2) and US2016228400 (A1) as a result of thermogenic effects.

Other documents, such as US10576117 (B2) and US2019125816 (A1), describe the reversing processes of glycogen depletion and proteins that are generally observed in the tissues of diabetic subjects [36].

4.2. Perspectives for Pharmaceutical Dosage Forms

Conventional pharmacotherapeutic schemes in DM include the daily administration and multiple doses of several drugs [81]. These facts can transform into disadvantages such as adverse effects and difficulties in adherence to therapy [75]. Consequently, providing adequate and effective clinical strategies is a constant challenge in pharmaceutical research.

In this sense, the application of controlled release systems is thought to be a promising approach to reducing the number of drugs in DM treatment [82]. This technology requires the critical choice of excipients, which will support the drug release according to the physiological conditions. As result, dose–response curves and retention times of the drugs can be adjusted to contribute to the pharmacokinetic parameters and reduce drug ingestion.

For instance, a protective bilayer dosage form is designed in the S10028930 (B2) and A US2016228400 (A1) patents. In this case, thermogenic substances such as fucoxanthin and caffeine are released with an hour of difference, which can increase energy expenditure and lipid oxidation [83]. It even suggests a dual-acting anti-diabetic and anti-hypertensive derived from this type of metabolite [84]. Another method is related to the administration of active salt forms that exert their action according to the pH of the medium. Therefore, drugs are often chemically made into their salt forms to enhance how the drug dissolves, boost its absorption into the bloodstream, and increase its effectiveness [85]. Finally, prodrugs are also included in the patent review. They can undergo chemical pH-dependent transformations, which allow extended time activation according to the medium.

A transdermal patch with a prodrug system was discussed in the US10028930 (B2) and A US2016228400 (A1) patents. The prodrug-to-drug conversion was performed due to an enzyme or chemical reagent included in the matrix [41]. Additionally, tablets or pills containing the extract of *Calophyllum inophyllum* were also proposed by NZ630125 (A). The invention performs a modified coating that prolongs the action of metabolites in DM treatment.

Otherwise, several inventions do not detail the delivery systems in which extracts are incorporated. For example, US10799547 (B2) and US2016067294 (A1) patent presents a wide range of forms, including granules, powders, syrup, solutions, suspension, tablets, injectables, poultices, and capsules to transport *Suaeda japonica* extract. These dosage forms realize different release forms such as rapid, continuous, or delayed [42].

In WO2019186579 (A1) patent, structural homology between osmotin and adiponectin domain, provides a possibility of a controlled release system through different pharmaceutical forms, such as tablets, granules, syrups, and films for transmucosal administration [86].

On the other hand, mucosal films have been tackled in patents. As part of improving the dissolution drug in the oral cavity, a transmucosal film of metformin and linagliptin was carried out. The omission of first-pass metabolism, a shorter time, and a faster-acting response compared to conventional pharmaceutical forms were the main advantages of this technology [87,88].

Other findings in the patent review are oriented to the optimization of extraction processes to maximize yields and achieve a better loading of the pharmaceutical vehicle. An extraction solvent based on an emulsifier, cyclodextrin, acid, water, and/or an organic solvent was related to *Gnetum gnemon* extract by the US10640480 (B2) and US2019031635 (A1) patents. This solution boosts the gnetin C glycoside content due to the solvolysis reaction [34].

4.3. Safety Test Assessments

Alternative therapies for DM from natural origins must follow the non-maleficence principle, in which the new substance administration does not represent a health risk. Safety in this field is based on the correct tracking of metabolites, their quantification, and the correct doses administrated. The current open science idea increases the safety background to divulge key data from diverse natural products. This fact mitigates the premature use of raw materials and unreliable data in natural products [72]. In this sense, acute and chronic toxicity studies are considered valuable tools to analyze the possible adverse effects derived from natural products.

The results obtained in these tests can guide the next steps in the research perspective. For instance, particular findings were obtained in the in vivo acute toxicity test performed in SG11201908574T (A). Inventors mentioned behavioral changes from 30 min until the 15th day after oral administration, characterized by tremors, sedation, convulsions, catatonia, state alertness, muscle spasms, cyanosis, writhing, irritability, and diarrhea.

On the other hand, lethal doses (LD) are also a parameter studied in the safety stage. For instance, greater LD values of 5000 mg/kg in rats were demonstrated in a safety assessment (short and long-term) in the US10576117 (B2) product.

Nevertheless, this information must be tackled carefully. Curcuminoids (present in several patents) have been identified as hepatotoxic in rodents [89,90]. The toxicity was linked to the metabolic pathways, especially 3,4-epoxidation predominant in rats. However, in humans, the metabolism occurs through the 7-hydroxylation pathway, which indicates that the hepatotoxic metabolites are not formed [72,91,92]. In contrast, these human metabolites are related to minor symptoms, such as itching, constipation, vertigo, and diarrhea [52].

Other safety warnings in natural products are focused on terpenes. Some terpenes have high molecular weights, and their permeability properties can be limited in physiological conditions. Although some tests suggest their accumulation in the liver, which can be dangerous, other researchers do not indicate adverse effects after acute and chronic ingestion [57]. Furthermore, a co-administration therapeutic regimen can also be reviewed. The concomitant use of traditional drugs and natural products may result in complications. An example can be the water-soluble fraction of okra (*Abelmoschus officinalis*) fruits, which modified the metformin absorption, reducing its serum concentration [72].

4.4. Pharmacological Assessment

To confirm the benefits of DM therapy, the preparation must be challenged to guarantee its effectiveness. This ensures that the mentioned therapeutic targets are achieved.

Despite the number of biochemical or cellular assays to test the target activity, the (AU2018278958 (A1) and AU2018278958 (B2) patents describe an alternative assay. The applicants have created a system in which a cell-specific protein-based target can identify activators of the IRS2 branch in the insulin-mediated signal transduction cascade. To support the procedure, control and test cells were derived from the 32D myeloid progenitor cell line [35,93].

However, the in vivo models are still predominant. To model DM, the use of streptozotocin (STZ) and alloxan (ALX) were mentioned by several patents and literature reviews [94,95]. These substances easily accumulate in pancreatic β-cells through the glucose transporter GLUT2. The STZ model is characterized by the presence of methylnitrosourea. This metabolite modifies biological macromolecules, fragments DNA, and destroys β-cells. Meanwhile, the ALX model has an oxidative stress capacity mediated by dialuric acid. In this case, the metabolite activates ROS cascades, impacting β-cell dysfunction [96].

Other findings in in vivo models indicate the use of OLETF (Otsuka Long-Evans Tokushima Fatty) mice, which are mentioned in the US10799547 (B2) and US2016067294 (A1) documents. The DM model is performed with a hypercaloric diet. When the DM model is stabilized, the extract is administered. Then, the blood and plasma samples are tracked to determine their content in the body.

Once the in vitro and in vivo stages are optimized, the next step is to carry out clinical trial assessments [52]. Only 15% of patients reported clinical studies annotations.

The influence of Ayurvedic medicine was observed in the clinical tests of DM in the patent review. The product administration was carried out in a personalized way. According to the Ayurvedic standard, dose adjustment and regimen timelines of administration led to the clinical criteria for practitioners to use the inventions [36].

To illustrate this practice, the US10028930 (B2) and US2019125816 (A1) patent inventors performed a prospective, open-label, non-randomized, phase III clinical trial. The procedure took place in the Muniyal Ayurvedic Hospital and Research Center in Manipal, India, and the patients were individuals from 30 to 60 years old [36]. London protocol elements such as inclusion and exclusion criteria or voluntary withdrawal were also documented. Despite the clinical trial details not being shown, the aim of the practice was clear, namely to verify the effectiveness [41].

Another example of clinical trials in this review refers to the use of *Trigonella foenum-graecum* in DM. Nevertheless, a lack of harmonization in the administration schedule was identified. This framework suggests an improvement opportunity for variable control in further assessment to reach a convergence protocol [9].

Although clinical assessments must be carried out within the transparency standards (under compliance agreements), the status of their execution must be critically monitored. Government entities such as the National Institute of Health (NIH) record the execution and status of these studies. The information is saved in an online database (https://clinicaltrials.gov/) and available to the general public. There are about 50 active trials related to DM and herbal components, predominantly located in China, India, and the United States [97].

4.5. Other Natural Resources with Anti-DM Potential

To achieve a therapeutic effect, WO2019205943 (A1), US10028930 (B2), US2016228400 (A1), US2016361341 (A1), and US9828442 (B2) describe the use of extracts and metabolites from algae, fungi, and metals or their derivatives.

Fucoxanthin is a carotenoid extracted from brown algae. Their effects in DM are observed in US10028930 (B2) and A US2016228400 (A1) [41]. Previous reports of this algae-derived metabolite have highlighted its role in inducing UCP1 in the white adipose tissue mitochondria. Fatty acid oxidation, heat production, insulin resistance improvements, and decrease blood glucose levels are the main effects of their use [98].

Additionally, the *spirulina cyanobacterium* metabolite phycocyanin has powerful antioxidant, anti-inflammatory, adipogenesis reducing and thermogenic-activating properties that can be considered in this field [99,100]. Thus, their presence increases the body's ability to prevent the development of chronic inflammation, which is frequent in DM patients [31].

Similarly, polysaccharides isolated from *Hirsutella sinensis* (fungi) improved insulin sensitivity in rodents in in vivo studies [40]. Increased expression of protein markers of thermogenesis, intestinal integrity improvements, anti-inflammatory effects, and lipid metabolism regulation were found for this substance [101,102].

Regarding Ayurvedic medicine influence, the shilajit (asphaltum black) a complex derived from the decomposition of plant and mineral materials was deepened by US 10967025 (B2) and US2019209634 (A1) [103]. The complex contains more than 85 minerals, as well as humic acid, dibenzo-α-pyrones, and fulvic acid. Antioxidant and cytoprotective properties are associated with it, which can be exploited in the prevention of DM disease [31].

In the same way, the use of minerals such as mica, pyrite, tin, lead, zinc, coral, iron, and copper and bhasmas or calcined preparations such as Swarna Makshika Bhasma, Abhraka Bhasma, Loha Bhasma, Vanga Bhasma, Yashada Bhasma and Pravala Bhasma derived from Ayurvedic medicine were reported in US10576117 (B2); US2019125816 (A1) Nevertheless, the literature available in online databases is focused on the synthesis and physicochemical characterization of these Herbo-Mineral preparations [36,104–107]. Therefore, further research on the subject is required.

5. Conclusions

This review shows different methods of preparation and extraction of natural products from plants, algae derivatives, fungi, and some minerals. The most reported and studied secondary metabolites in patents are tannins, organic acids, polyphenols, saponins, terpenes, and flavonoids. Among the mechanisms of action, the effects on the circulatory system, intestines, pancreas, adipose tissue, and muscles have been identified.

Additionally, some pharmacotechnical strategies such as solid forms with a modified release, bilayer systems, and transmucosal films are reported. Finally, some key safety and efficacy points are discussed, which provide a critical analysis of these substances in the current context.

The above demonstrates the importance of research using medicinal plants in the design and development of new drugs for the treatment of DM.

Author Contributions: Conceptualization, D.M.A. and F.R.M.B.S.; methodology, M.S. and I.A.A.; formal analysis, I.A.R.; investigation, I.A.R.; data curation I.A.R. and K.L.L.; writing—original draft preparation, I.A.R.; writing—review and editing, D.M.A., K.L.L. and F.R.M.B.S.; supervision, D.M.A.; All authors have read and agreed to the published version of the manuscript.

Funding: This research received no external funding.

Institutional Review Board Statement: Not applicable.

Informed Consent Statement: Not applicable.

Data Availability Statement: Not applicable.

Conflicts of Interest: The authors declare no conflict of interest.

References

1. Association Diabetes American Diagnosis and classification of diabetes mellitus. *Diabetes Care* **2009**, *32*, S62–S67. [CrossRef]
2. Maida, C.D.; Daidone, M.; Pacinella, G.; Norrito, R.L.; Pinto, A.; Tuttolomondo, A. Diabetes and Ischemic Stroke: An Old and New Relationship an Overview of the Close Interaction between These Diseases. *Int. J. Mol. Sci.* **2022**, *23*, 2397. [CrossRef] [PubMed]
3. WHO. Diagnosis and Management of Type 2 Diabetes (HEARTS-D). Geneva. (WHO/UCN/NCD/20.1). Available online: https://www.who.int/publications/i/item/who-ucn-ncd-20.1 (accessed on 17 June 2022).
4. National Institutes of Health. What is Diabetes? NIDDK. Available online: https://www.niddk.nih.gov/health-information/diabetes/overview/what-is-diabetes (accessed on 16 April 2022).
5. National Diabetes Statistics Report. 2020. Available online: https://www.cdc.gov/diabetes/pdfs/data/statistics/national-diabetes-statistics-report.pdf (accessed on 16 April 2022).
6. Association American Diabetes 2. Classification and Diagnosis of Diabetes: Standards of Medical Care in Diabetes-2021. *Am. Diabetes Assoc.* **2021**, *44*, S15–S33.
7. International Diabetes Federation. Five questions on the IDF Diabetes Atlas. *Diabetes Res. Clin. Pract.* **2013**, *102*, 147–148. [CrossRef] [PubMed]
8. Tan, S.Y.; Mei Wong, J.L.; Sim, Y.J.; Wong, S.S.; Mohamed Elhassan, S.A.; Tan, S.H.; Ling Lim, G.P.; Rong Tay, N.W.; Annan, N.C.; Bhattamisra, S.K.; et al. Type 1 and 2 diabetes mellitus: A review on current treatment approach and gene therapy as potential intervention. *Diabetes Metab. Syndr. Clin. Res. Rev.* **2019**, *13*, 367–372. [CrossRef]
9. Salehi, B.; Ata, A.; Kumar, N.V.A.; Sharopov, F.; Ramírez-Alarcón, K.; Ruiz-Ortega, A.; Ayatollahi, S.A.; Fokou, P.V.T.; Kobarfard, F.; Zakaria, Z.A.; et al. Antidiabetic potential of medicinal plants and their active components. *Biomolecules* **2019**, *9*, 551. [CrossRef]
10. Rubio-Almanza, M.; Cámara-Gómez, R.; Merino-Torres, J.F. Obesity and type 2 diabetes: Also linked in therapeutic options. *Endocrinol. Diabetes y Nutr.* **2019**, *66*, 140–149. [CrossRef]
11. Serván, P.R. Diet recomendations in diabetes and obesity. *Nutr. Hosp.* **2018**, *35*, 109–115. [CrossRef]
12. Xie, F.; Chan, J.C.N.; Ma, R.C.W. Precision medicine in diabetes prevention, classification and management. *J. Diabetes Investig.* **2018**, *9*, 998–1015. [CrossRef]
13. Rey, D.; Ospina, L.; Aragón, D. Inhibitory effects of an extract of fruits of Physalis peruviana on some intestinal carbohydrases. *Rev. Colomb. Cienc. Químico-Farm.* **2015**, *44*, 72–89. [CrossRef]
14. Rey, D.; Miranda Sulis, P.; Alves Fernandes, T.; Gonçalves, R.; Silva Frederico, M.J.; Costa, G.M.; Aragon, M.; Ospina, L.F.; Mena Barreto Silva, F.R. Astragalin augments basal calcium influx and insulin secretion in rat pancreatic islets. *Cell Calcium* **2019**, *80*, 56–62. [CrossRef] [PubMed]
15. Monzón Daza, G.; Meneses Macías, C.; Forero, A.M.; Rodríguez, J.; Aragón, M.; Jiménez, C.; Ramos, F.A.; Castellanos, L. Identification of α-Amylase and α-Glucosidase Inhibitors and Ligularoside A, a New Triterpenoid Saponin from Passiflora ligularis Juss (Sweet Granadilla) Leaves, by a Nuclear Magnetic Resonance-Based Metabolomic Study. *J. Agric. Food Chem.* **2021**, *69*, 2919–2931. [CrossRef] [PubMed]
16. Bernal, C.A.; Castellanos, L.; Aragón, D.M.; Martínez-Matamoros, D.; Jiménez, C.; Baena, Y.; Ramos, F.A. Peruvioses A to F, sucrose esters from the exudate of Physalis peruviana fruit as α-amylase inhibitors. *Carbohydr. Res.* **2018**, *461*, 4–10. [CrossRef] [PubMed]
17. Durazzo, A.; Lucarini, M.; Santini, A. Molecular Sciences Plants and Diabetes: Description, Role, Comprehension and Exploitation. *Int. J. Mol. Sci.* **2021**, *22*, 3938. [CrossRef] [PubMed]

18. Ríos, J.L.; Francini, F.; Schinella, G.R. Natural Products for the Treatment of Type 2 Diabetes Mellitus. *Planta Med.* **2015**, *81*, 975–994. [CrossRef] [PubMed]
19. Wang, G.S.; Hoyte, C. Review of Biguanide (Metformin) Toxicity. *J. Intensive Care Med.* **2019**, *34*, 863–876. [CrossRef] [PubMed]
20. Liu, X.; Wei, J.; Tan, F.; Zhou, S.; Würthwein, G.; Rohdewald, P. Antidiabetic effect of Pycnogenol® French maritime pine bark extract in patients with diabetes type II. *Life Sci.* **2004**, *75*, 2505–2513. [CrossRef]
21. Vitetta, L.; Butcher, B.; Dal Forno, S.; Vitetta, G.; Nikov, T.; Hall, S.; Steels, E. A Double-Blind Randomized Placebo-Controlled Study Assessing the Safety, Tolerability and Efficacy of a Herbal Medicine Containing Pycnogenol Combined with Papain and Aloe vera in the Prevention and Management of Pre-Diabetes. *Medicines* **2020**, *7*, 22. [CrossRef]
22. Governa, P.; Baini, G.; Borgonetti, V.; Cettolin, G.; Giachetti, D.; Magnano, A.R.; Miraldi, E.; Biagi, M. Phytotherapy in the management of diabetes: A review. *Molecules* **2018**, *23*, 105. [CrossRef]
23. Artasensi, A.; Pedretti, A.; Vistoli, G.; Fumagalli, L. Type 2 diabetes mellitus: A review of multi-target drugs. *Molecules* **2020**, *25*, 1987. [CrossRef]
24. AragónNovoa, D. *Passifloraligularis Juss. (Granadilla): Farmacológicos de una Estudios Químicos y Plantacon Potencial Terapéutico*; Universidad Nacional de Colombia—Sede Bogotá: Bogotá, Colombia, 2021.
25. Narayan, T.; Sanjay, B.; Bhaswati, K.; Simanta, B.; Sagar, B.; Barsha, D.; Yunus, S.; Seydur, R.; Aparajita, G.; Pratim, D.P.; et al. A Herbal Composition from Premna herbacea, Useful for Prevention of Obesity and Type 2 Diabetes and a Method for Its Extraction. WO Patent 2020115767 (A1), 11 June 2020. Available online: https://worldwide.espacenet.com/publicationDetails/biblio?DB=EPODOC&adjacent=true&locale=en_EP&FT=D&date=20200611&CC=WO&NR=2020115767A1&KC=A1 (accessed on 29 March 2022).
26. Saya, O.; Reiko, T.; Takuya, Y.; Norio, I. Agent for Suppressing Carbohydrate Breakdown and Absorption. U.S. Patent 2020147160 [A1], 14 May 2020. Available online: https://worldwide.espacenet.com/publicationDetails/biblio?DB=EPODOC&adjacent=true&locale=en_EP&FT=D&date=20200514&CC=US&NR=2020147160A1&KC=A1 (accessed on 29 March 2022).
27. Lal, H. A Novel Synergic Herbal Formulation for the Prevention and Treatment of Pre-Diabetes, Diabetes and Other Insulin Resistance Cases. WO Patent 2020012299 [A1], 16 January 2020. Available online: https://worldwide.espacenet.com/publicationDetails/biblio?DB=EPODOC&adjacent=true&locale=en_EP&FT=D&date=20200116&CC=WO&NR=2020012299A1&KC=A1 (accessed on 29 March 2022).
28. Patanjali, S.; Sundaram, C.; Jayashree, M.; Anilkumar, K.; Sarala, S.; Bishwajit, N.; Ramesh, V.; Nitish, N.; Jayarajan, K. Herbal Composition. SG Patent 11201908574T [A], 30 October 2019. Available online: https://worldwide.espacenet.com/publicationDetails/biblio?DB=EPODOC&adjacent=true&locale=en_EP&FT=D&date=20191030&CC=SG&NR=11201908574TA&KC=A (accessed on 29 March 2022).
29. Hexiao, S.; Guolong, L. Composite Preparation for Improving Insulin Resistance, Preparation Method and Application. WO Patent 2019205943 [A1], 31 October 2019. Available online: https://worldwide.espacenet.com/publicationDetails/biblio?DB=EPODOC&adjacent=true&locale=en_EP&FT=D&date=20191031&CC=WO&NR=2019205943A1&KC=A1 (accessed on 29 March 2022).
30. Vasant, S.C.; Anantrao, S.V. Plant Fractions Having Anti-Pathogenesis Properties. WO Patent 2019186579 [A1], 3 October 2019. Available online: https://worldwide.espacenet.com/publicationDetails/biblio?DB=EPODOC&adjacent=true&locale=en_EP&FT=D&date=20191003&CC=WO&NR=2019186579A1&KC=A1 (accessed on 29 March 2022).
31. Vieira, K. Herbal Nutraceutical Formulation to Reduce Oxidative Stress, Viral and Microbial Infections, and Inflammation. U.S. Patent 2019209634 [A1], 11 July 2019. Available online: https://worldwide.espacenet.com/publicationDetails/biblio?DB=EPODOC&adjacent=true&locale=en_EP&FT=D&date=20190711&CC=US&NR=2019209634A1&KC=A1 (accessed on 29 March 2022).
32. Kasim, T. The Use of Jerusalem Thorn Fruits as Herbal Tea For Diabetes Treatment. WO Patent 2019088958 [A2], 9 May 2019. Available online: https://worldwide.espacenet.com/publicationDetails/biblio?DB=EPODOC&adjacent=true&locale=en_EP&FT=D&date=20190509&CC=WO&NR=2019088958A2&KC=A2 (accessed on 29 March 2022).
33. Diaz, A.M.L.; Uribe, J.A.G.; Torres, N.T.Y.; Palacio, A.R.T.; Lopez, L.G.N. Agavaceae Extract Comprising Steroidal Saponins to Treat or Prevent Metabolic Disorder Related Pathologies. MX Patent 2018004489 [A], 21 January 2019. Available online: https://worldwide.espacenet.com/publicationDetails/biblio?DB=EPODOC&adjacent=true&locale=en_EP&FT=D&date=20190121&CC=MX&NR=2018004489A&KC=A (accessed on 29 March 2022).
34. Eishin, K.; Shinya, H. Gnetin c-Rich Melinjo Extract and Production Method Thereof. U.S. Patent 2019031635 [A1], 31 January 2019. Available online: https://worldwide.espacenet.com/publicationDetails/biblio?DB=EPODOC&adjacent=true&locale=en_EP&FT=D&date=20190131&CC=US&NR=2019031635A1&KC=A1 (accessed on 29 March 2022).
35. Housey, G.; Balash, M.E. Plant Extracts with Anti-Diabetic and Other Useful Activities. AU Patent 2018278958 [A1], 17 January 2019. Available online: https://worldwide.espacenet.com/publicationDetails/biblio?DB=EPODOC&adjacent=true&locale=en_EP&FT=D&date=20190117&CC=AU&NR=2018278958A1&KC=A1 (accessed on 29 March 2022).
36. Vijayabhanu, S. Herbo-Mineral Formulation for Prevention, Treatment and Management of Diabetes and Method of Preparation Thereof. U.S. Patent 2019125816 [A1], 2 May 2019. Available online: https://worldwide.espacenet.com/publicationDetails/biblio?DB=EPODOC&adjacent=true&locale=en_EP&FT=D&date=20190502&CC=US&NR=2019125816A1&KC=A1 (accessed on 29 March 2022).

37. Lien, P.-J.; Sun, C.-C.; Kuo, T.-C.; Yang, S.-C.; Wu, Y.-C.; Chang, F.-R.; Hsieh, T.-J.; Tsai, Y.-H.; Du, Y.-C. Extract of Toona Sinensis from Supercritical Fluid Extraction for Treating Diabetes and Metabolic Disease, the Preparation Method and the Use Thereof. U.S. Patent 2017252393 [A1], 7 September 2017. Available online: https://worldwide.espacenet.com/publicationDetails/biblio?DB=EPODOC&adjacent=true&locale=en_EP&FT=D&date=20170907&CC=US&NR=2017252393A1&KC=A1 (accessed on 29 March 2022).

38. Wah, M.C.; Zhen, L.; Xia, Z.; Xiaolei, G. Compositions Comprising Cyclocarya Paliurus Extract and Preparation Method and Uses Thereof. U.S. Patent 2017173103 [A1], 22 June 2017. Available online: https://worldwide.espacenet.com/publicationDetails/biblio?DB=EPODOC&adjacent=true&locale=en_EP&FT=D&date=20170622&CC=US&NR=2017173103A1&KC=A1 (accessed on 29 March 2022).

39. Arvind, S.; Parikshit, G.; Aslam, B.; Somesh, S. Herbal Composition for the Treatment of Metabolic Disorders. NZ Patent 630125 [A], 27 January 2017. Available online: https://worldwide.espacenet.com/publicationDetails/biblio?DB=EPODOC&adjacent=true&locale=en_EP&FT=D&date=20170127&CC=NZ&NR=630125A&KC=A (accessed on 29 March 2022).

40. Ko, Y.-F.; Jan, M.; Liau, J.-C.; Chang, I.-T.; Lee, C.-S.; Wang, W.-C.; Chiu, W.-C.; Chang, C.Y.; Lin, C.-S.; Wu, T.-R.; et al. Method to Prepare Hirsutella Sinensis Polysaccharides Possessing Insulin-Sensitizing Properties and Applications Thereof. U.S. Patent 2016361341 [A1], 15 December 2016. Available online: https://worldwide.espacenet.com/publicationDetails/biblio?DB=EPODOC&adjacent=true&locale=en_EP&FT=D&date=20161215&CC=US&NR=2016361341A1&KC=A1 (accessed on 29 March 2022).

41. Horn, G.; Trimbo, S. Compositions Capable of Enhancing Thermogenesis and Uses Thereof. U.S. Patent 2016228400 [A1], 11 August 2016. Available online: https://worldwide.espacenet.com/publicationDetails/biblio?DB=EPODOC&adjacent=true&locale=en_EP&FT=D&date=20160811&CC=US&NR=2016228400A1&KC=A1 (accessed on 29 March 2022).

42. Sik, H.K.; Yong, C.J.; Young, P.S.; Zhangjun, H. Composition Comprising Suaeda Japonica for Preventing or Alleviating Diabetes. U.S. Patent 2016067294 [A1], 10 March 2016. Available online: https://worldwide.espacenet.com/publicationDetails/biblio?DB=EPODOC&adjacent=true&locale=en_EP&FT=D&date=20160310&CC=US&NR=2016067294A1&KC=A1 (accessed on 29 March 2022).

43. Mihaylivna, M.S.; Oleksandrivna, S.A.; Yurivna, K.O. Herbal Species for Treatment of Diabetes Mellitus Type ll. UA Patent 104161 [U], 12 January 2016. Available online: https://worldwide.espacenet.com/publicationDetails/biblio?DB=EPODOC&adjacent=true&locale=en_EP&FT=D&date=20160112&CC=UA&NR=104161U&KC=U (accessed on 29 March 2022).

44. Gaikwad, S.B.; Krishna Mohan, G.; Rani, M.S. Phytochemicals for Diabetes Management. *Pharm. Crop.* **2014**, *5*, 983–992. [CrossRef]

45. Xia, B.; Shi, X.C.; Xie, B.C.; Zhu, M.Q.; Chen, Y.; Chu, X.Y.; Cai, G.H.; Liu, M.; Yang, S.Z.; Mitchell, G.A.; et al. Urolithin A exerts antiobesity effects through enhancing adipose tissue thermogenesis in mice. *PLoS Biol.* **2020**, *18*, e3000688. [CrossRef] [PubMed]

46. Laddha, A.P.; Kulkarni, Y.A. Tannins and vascular complications of Diabetes: An update. *Phytomedicine* **2019**, *56*, 229–245. [CrossRef] [PubMed]

47. Wu, S.; Tian, L. A new flavone glucoside together with known ellagitannins and flavones with anti-diabetic and anti-obesity activities from the flowers of pomegranate (Punica granatum). *Nat. Prod. Res.* **2019**, *33*, 252–257. [CrossRef] [PubMed]

48. Yuan, T.; Ferreira, D.; Seeram, N. Punicatannins A and B: α-glucosidase inhibitory ellagitannins from pomegranate (Punica granatum) flowers. *Planta Med.* **2012**, *78*, PI154. [CrossRef]

49. Lee, M.; Nam, S.-H.; Yoon, H.-G.; Kim, S.; You, Y.; Choi, K.-C.; Lee, Y.-H.; Lee, J.; Park, J.; Jun, W. Fermented Curcuma longa L. Prevents Alcoholic Fatty Liver Disease in Mice by Regulating CYP2E1, SREBP-1c, and PPAR-α. *J. Med. Food* **2022**, *25*, 456–463. [CrossRef]

50. Pujimulyani, D.; Yulianto, W.A.; Setyowati, A.; Arumwardana, S.; Kusuma, H.S.W.; Sholihah, I.A.; Rizal, R.; Widowati, W.; Maruf, A. Hypoglycemic activity of curcuma mangga val. Extract via modulation of GLUT4 and ppar-γ mrna expression in 3T3-L1 adipocytes. *J. Exp. Pharmacol.* **2020**, *12*, 363–369. [CrossRef]

51. Mohammadi, E.; Behnam, B.; Mohammadinejad, R.; Guest, P.C.; Simental-Mendía, L.E.; Sahebkar, A. Antidiabetic Properties of Curcumin: Insights on New Mechanisms. In *Advances in Experimental Medicine and Biology*; Springer: Cham, Switzerland, 2021; Volume 1291.

52. Pivari, F.; Mingione, A.; Brasacchio, C.; Soldati, L. Curcumin and type 2 diabetes mellitus: Prevention and treatment. *Nutrients* **2019**, *11*, 1837. [CrossRef]

53. Kotha, R.R.; Luthria, D.L. Curcumin: Biological, pharmaceutical, nutraceutical, and analytical aspects. *Molecules* **2019**, *24*, 2930. [CrossRef]

54. Zeng, L.; Yu, G.; Hao, W.; Yang, K.; Chen, H. The efficacy and safety of Curcuma longa extract and curcumin supplements on osteoarthritis: A systematic review and meta-analysis. *Biosci. Rep.* **2021**, *41*, 6. [CrossRef]

55. Jin, T.; Song, Z.; Weng, J.; Fantus, I.G. Curcumin and other dietary polyphenols: Potential mechanisms of metabolic actions and therapy for diabetes and obesity. *Am. J. Physiol.—Endocrinol. Metab.* **2018**, *314*, E201–E205. [CrossRef] [PubMed]

56. Scazzocchio, B.; Minghetti, L.; D'archivio, M. Interaction between gut microbiota and curcumin: A new key of understanding for the health effects of curcumin. *Nutrients* **2020**, *12*, 2499. [CrossRef] [PubMed]

57. Nazaruk, J.; Borzym-Kluczyk, M. The role of triterpenes in the management of diabetes mellitus and its complications. *Phytochem. Rev.* **2015**, *14*, 675–690. [CrossRef] [PubMed]

58. Zhong, Y.Y.; Chen, H.S.; Wu, P.P.; Zhang, B.J.; Yang, Y.; Zhu, Q.Y.; Zhang, C.G.; Zhao, S.Q. Synthesis and biological evaluation of novel oleanolic acid analogues as potential α-glucosidase inhibitors. *Eur. J. Med. Chem.* **2019**, *164*, 706–716. [CrossRef]
59. Ding, H.; Hu, X.; Xu, X.; Zhang, G.; Gong, D. Inhibitory mechanism of two allosteric inhibitors, oleanolic acid and ursolic acid on α-glucosidase. *Int. J. Biol. Macromol.* **2018**, *107*, 1844–1855. [CrossRef]
60. Chen, W.; Balan, P.; Popovich, D.G. Review of ginseng anti-diabetic studies. *Molecules* **2019**, *24*, 4501. [CrossRef]
61. Sun, Y.; Gao, H.Y.; Fan, Z.Y.; He, Y.; Yan, Y.X. Metabolomics signatures in type 2 diabetes: A systematic review and integrative analysis. *J. Clin. Endocrinol. Metab.* **2020**, 105. [CrossRef]
62. Kim, Y.A.; Keogh, J.B.; Clifton, P.M. Probiotics, prebiotics, synbiotics and insulin sensitivity. *Nutr. Res. Rev.* **2018**, *31*, 35–51. [CrossRef]
63. Judge, A.; Dodd, M.S. Metabolism. *Essays Biochem.* **2020**, *64*, 607–647. [CrossRef]
64. Tokarz, V.L.; MacDonald, P.E.; Klip, A. The cell biology of systemic insulin function. *J. Cell Biol.* **2018**, *217*, 2273–2289. [CrossRef]
65. Sims, E.K.; Carr, A.L.J.; Oram, R.A.; DiMeglio, L.A.; Evans-Molina, C. 100 years of insulin: Celebrating the past, present and future of diabetes therapy. *Nat. Med.* **2021**, *27*, 1154–1164. [CrossRef] [PubMed]
66. Devangan, S.; Varghese, B.; Johny, E.; Gurram, S.; Adela, R. The effect of Gymnema sylvestre supplementation on glycemic control in type 2 diabetes patients: A systematic review and meta-analysis. *Phyther. Res.* **2021**, *35*, 356–359. [CrossRef] [PubMed]
67. Sandech, N.; Jangchart, R.; Komolkriengkrai, M.; Boonyoung, P.; Khimmaktong, W. Efficiency of Gymnema sylvestre-derived gymnemic acid on the restoration and improvement of brain vascular characteristics in diabetic rats. *Exp. Ther. Med.* **2021**, *22*, 1420. [CrossRef] [PubMed]
68. Al-Romaiyan, A.; Liu, B.; Persaud, S.; Jones, P. A novel Gymnema sylvestre extract protects pancreatic beta-cells from cytokine-induced apoptosis. *Phyther. Res.* **2020**, *34*, 161–172. [CrossRef]
69. Abdulrahman, M.D.; Zakariya, A.M.; Hama, H.A.; Hamad, S.W.; Al-Rawi, S.S.; Bradosty, S.W.; Ibrahim, A.H. Ethnopharmacology, Biological Evaluation, and Chemical Composition of Ziziphus spina- christi (L.) Desf.: A Review. *Adv Pharmacol. Pharm. Sci.* **2022**, *2022*, 4495688. [CrossRef]
70. Kane, J.P.; Pullinger, C.R.; Goldfine, I.D.; Malloy, M.J. Dyslipidemia and diabetes mellitus: Role of lipoprotein species and interrelated pathways of lipid metabolism in diabetes mellitus. *Curr. Opin. Pharmacol.* **2021**, *61*, 32–37. [CrossRef]
71. Athyros, V.G.; Michael Doumas Konstantinos P Imprialos; Stavropoulos, K.; Georgianou, E.; Katsimardou, A.; Karagiannis, A. Diabetes and lipid metabolism. *Hormones* **2018**, *17*, 61–67. [CrossRef]
72. Manukumar, H.M.; Shiva Kumar, J.; Chandrasekhar, B.; Raghava, S.; Umesha, S. Evidences for diabetes and insulin mimetic activity of medicinal plants: Present status and future prospects. *Crit. Rev. Food Sci. Nutr.* **2017**, *57*, 2712–2729. [CrossRef]
73. Choi, H.M.; Doss, H.M.; Kim, K.S. Multifaceted physiological roles of adiponectin in inflammation and diseases. *Int. J. Mol. Sci.* **2020**, *21*, 1219. [CrossRef]
74. Fang, H.; Judd, R.L. Adiponectin regulation and function. *Compr. Physiol.* **2018**, *8*, 1031–1063. [CrossRef]
75. Unuofin, J.O.; Lebelo, S.L. Antioxidant Effects and Mechanisms of Medicinal Plants and Their Bioactive Compounds for the Prevention and Treatment of Type 2 Diabetes: An Updated Review. *Oxid. Med. Cell. Longev.* **2020**, *2020*, 1356893. [CrossRef] [PubMed]
76. EMA. Assessment Report on Panax Ginseng C.A. Meyer, Radix 2014. Available online: https://www.ema.europa.eu/en/documents/herbal-report/final-assessment-report-panax-ginseng-ca-meyer-radix_en.pdf (accessed on 19 September 2022).
77. Muñoz-Garach, A.; Diaz-Perdigones, C.; Tinahones, F.J. Gut microbiota and type 2 diabetes mellitus. *Endocrinol. Nutr.* **2016**, *63*, 560–568. [CrossRef] [PubMed]
78. Hills, R.D.; Pontefract, B.A.; Mishcon, H.R.; Black, C.A.; Sutton, S.C.; Theberge, C.R. Gut microbiome: Profound implications for diet and disease. *Nutrients* **2019**, *11*, 1613. [CrossRef] [PubMed]
79. Magne, F.; Gotteland, M.; Gauthier, L.; Zazueta, A.; Pesoa, S.; Navarrete, P.; Balamurugan, R. The firmicutes/bacteroidetes ratio: A relevant marker of gut dysbiosis in obese patients? *Nutrients* **2020**, *12*, 1474. [CrossRef]
80. Betz, M.J.; Enerbäck, S. Targeting thermogenesis in brown fat and muscle to treat obesity and metabolic disease. *Nat. Rev. Endocrinol.* **2018**, *14*, 77–87. [CrossRef]
81. Pfeiffer, A.F.H.; Klein, H.H. Therapie des diabetes mellitus typ 2. *Dtsch. Arztebl. Int.* **2014**, *111*, 69–82. [CrossRef]
82. Abrilla, A.A.; Pajes, A.N.N.I.; Jimeno, C.A. Metformin extended-release versus metformin immediate-release for adults with type 2 diabetes mellitus: A systematic review and meta-analysis of randomized controlled trials. *Diabetes Res. Clin. Pract.* **2021**, *178*, 108824. [CrossRef]
83. Li, H.; Qi, J.; Li, L. Phytochemicals as potential candidates to combat obesity via adipose non-shivering thermogenesis. *Pharmacol. Res.* **2019**, *147*, 104393. [CrossRef]
84. Chukwuma, C.I.; Matsabisa, M.G.; Ibrahim, M.A.; Erukainure, O.L.; Chabalala, M.H.; Islam, M.S. Medicinal plants with concomitant anti-diabetic and anti-hypertensive effects as potential sources of dual acting therapies against diabetes and hypertension: A review. *J. Ethnopharmacol.* **2019**, *235*, 329–360. [CrossRef]
85. Anderson, L.A. Drug Names and Their Pharmaceutical Salts—Clearing Up the Confusion. Drugs.com. 2022. Available online: https://www.drugs.com/article/pharmaceutical-salts.html (accessed on 14 October 2022).
86. de Freitas, C.D.T.; Nishi, B.C.; do Nascimento, C.T.M.; Silva, M.Z.R.; Bezerra, E.H.S.; Rocha, B.A.M.; Grangeiro, T.B.; de Oliveira, J.P.B.; Souza, P.F.N.; Ramos, M.V. Characterization of Three Osmotin-Like Proteins from Plumeria rubra and Prospection for Adiponectin Peptidomimetics. *Protein Pept. Lett.* **2020**, *27*, 593–603. [CrossRef]

87. Sushma, M.; Raju, Y.; Sundaresan, C.R.; Vandana, K.R.; Kumar, N.; Chowdary, V. Transmucosal Delivery of Metformin- A Comprehensive Study. *Curr. Drug Deliv.* **2014**, *11*, 172–178. [CrossRef] [PubMed]

88. Modgill, V.; Garg, T.; Goyal, A.; Rath, G. Transmucosal Delivery of Linagliptin for the Treatment of Type- 2 Diabetes Mellitus by Ultra-Thin Nanofibers. *Curr. Drug Deliv.* **2015**, *12*, 323–332. [CrossRef] [PubMed]

89. Huang, F.J.; Lan, K.C.; Kang, H.Y.; Liu, Y.C.; Der Hsuuw, Y.; Chan, W.H.; Huang, K.E. Effect of curcumin on in vitro early post-implantation stages of mouse embryo development. *Eur. J. Obstet. Gynecol. Reprod. Biol.* **2013**, *166*, R713–R715. [CrossRef] [PubMed]

90. Felter, S.P.; Vassallo, J.D.; Carlton, B.D.; Daston, G.P. A safety assessment of coumarin taking into account species-specificity of toxicokinetics. *Food Chem. Toxicol.* **2006**, *44*, 462–475. [CrossRef]

91. Miura, T.; Uehara, S.; Shimizu, M.; Murayama, N.; Suemizu, H.; Yamazaki, H. Roles of human cytochrome P450 1A2 in coumarin 3,4-epoxidation mediated by untreated hepatocytes and by those metabolically inactivated with furafylline in previously transplanted chimeric mice. *J. Toxicol. Sci.* **2021**, *46*, 525–530. [CrossRef]

92. Hsieh, C.Y.J.; Sun, M.; Osborne, G.; Ricker, K.; Tsai, F.C.; Li, K.; Tomar, R.; Phuong, J.; Schmitz, R.; Sandy, M.S. Cancer Hazard Identification Integrating Human Variability: The Case of Coumarin. *Int. J. Toxicol.* **2019**, *38*, 455–468. [CrossRef]

93. Zjablovskaja, P.; Danek, P.; Kardosova, M.; Alberich-Jorda, M. Proliferation and differentiation of murine myeloid precursor 32d/g-csf-r cells. *J. Vis. Exp.* **2018**, *2018*, e57033. [CrossRef]

94. Pandey, S.; Dvorakova, M.C. Future Perspective of Diabetic Animal Models. *Endocr. Metab. Immune Disord.—Drug Targets* **2019**, *20*. [CrossRef]

95. Kottaisamy, C.P.D.; Raj, D.S.; Prasanth Kumar, V.; Sankaran, U. Experimental animal models for diabetes and its related complications—A review. *Lab. Anim. Res.* **2021**, *37*, 23. [CrossRef]

96. Lenzen, S. The mechanisms of alloxan- and streptozotocin-induced diabetes. *Diabetologia* **2008**, *51*, 536–546. [CrossRef]

97. NIH. Search of: Herbal. Diabetes Mellitus. Adult—List Results—ClinicalTrials.gov 2022. Available online: https://www.clinicaltrials.gov/ct2/results?term=herbal&cond=Diabetes+Mellitus&age=1# (accessed on 21 September 2022).

98. Gammone, M.A.; D'Orazio, N. Anti-obesity activity of the marine carotenoid fucoxanthin. *Mar. Drugs* **2015**, *13*, 2196–2214. [CrossRef] [PubMed]

99. Bannu, S.M.; Lomada, D.; Gulla, S.; Chandrasekhar, T.; Reddanna, P.; Reddy, M.C. Potential Therapeutic Applications of C-Phycocyanin. *Curr. Drug Metab.* **2019**, *20*, 967–976. [CrossRef] [PubMed]

100. Seo, Y.J.; Kim, K.J.; Choi, J.; Koh, E.J.; Lee, B.Y. Spirulina maxima extract reduces obesity through suppression of adipogenesis and activation of browning in 3T3-L1 cells and high-fat diet-induced obese mice. *Nutrients* **2018**, *10*, 712. [CrossRef] [PubMed]

101. Wu, T.R.; Lin, C.S.; Chang, C.J.; Lin, T.L.; Martel, J.; Ko, Y.F.; Ojcius, D.M.; Lu, C.C.; Young, J.D.; Lai, H.C. Gut commensal Parabacteroides goldsteinii plays a predominant role in the anti-obesity effects of polysaccharides isolated from Hirsutella sinensis. *Gut* **2019**, *68*, 248–262. [CrossRef]

102. Lu, Z.; Li, S.; Sun, R.; Jia, X.; Xu, C.; Aa, J.; Wang, G. Hirsutella sinensis Treatment Shows Protective Effects on Renal Injury and Metabolic Modulation in db/db Mice. *Evid.-Based Complement. Altern. Med.* **2019**, *2019*, 4273290. [CrossRef]

103. Jafari, M.; Forootanfar, H.; Ameri, A.; Foroutanfar, A.; Adeli-Sardou, M.; Rahimi, H.R.; Najafi, A.; Zangiabadi, N.; Shakibaie, M. Antioxidant, cytotoxic and hyperalgesia-suppressing activity of a native Shilajit obtained from Bahr Aseman mountains. *Pak. J. Pharm. Sci.* **2019**, *32*, 2167–2173.

104. Pathiraja, P.M.Y.S.; Ranatunga, Y.M.M.K.; Herapathdeniya, S.K.M.K.; Gunawardena, S.H.P. Swarna Makshika Bhasma preparation using an alternative heating method to traditional Varaha Puta. *J. Ayurveda Integr. Med.* **2020**, *11*, 529–530. [CrossRef]

105. Kantak, S.; Rajurkar, N.; Adhyapak, P. Synthesis and characterization of Abhraka (mica) bhasma by two different methods. *J. Ayurveda Integr. Med.* **2020**, *11*, 236–242. [CrossRef]

106. Singh, A.; Ota, S.; Srikanth, N.; Galib, R.; Bojja, S.; Dhiman, K.S. Application of Spectroscopic and Chromatographic Methods for Chemical Characterization of an Ayurvedic Herbo-Mineral Preparation: Maha Yograja Guggulu. *J. Evidence-Based Integr. Med.* **2018**, *23*. [CrossRef]

107. Pareek, A.; Bhatnagar, N. Physico-chemical characterization of traditionally prepared Yashada bhasma. *J. Ayurveda Integr. Med.* **2020**, *11*, 228–235. [CrossRef]

pharmaceutics

Review

Evidence-Based Anti-Diabetic Properties of Plant from the Occitan Valleys of the Piedmont Alps

Valentina Boscaro [1], Matteo Rivoira [2,3], Barbara Sgorbini [1], Valentina Bordano [1], Francesca Dadone [1], Margherita Gallicchio [1], Aline Pons [2], Elisa Benetti [1,*,†] and Arianna Carolina Rosa [1,*,†]

[1] Dipartimento di Scienza e Tecnologia del Farmaco, University of Turin, Via Pietro Giuria 9, 10125 Turin, Italy
[2] Dipartimento di Studi Umanistici, University of Turin, Via Sant'Ottavio 20, 10124 Turin, Italy
[3] Atlante Linguistico Italiano (ALI), Via Sant'Ottavio 20, 10124 Turin, Italy
[*] Correspondence: elisa.benetti@unito.it (E.B.); ariannacarolina.rosa@unito.it (A.C.R.);
Tel.: +39-0116707137 (E.B.); +39-0116707152 (A.C.R.)
[†] These authors contributed equally to this work.

Abstract: Data on urban and rural diabetes prevalence ratios show a significantly lower presence of diabetes in rural areas. Several bioactive compounds of plant origin are known to exert anti-diabetic properties. Interestingly, most of them naturally occur in different plants present in mountainous areas and are linked to traditions of herbal use. This review will aim to evaluate the last 10 years of evidence-based data on the potential anti-diabetic properties of 9 plants used in the Piedmont Alps (North-Western Italy) and identified through an ethnobotanical approach, based on the Occitan language minority of the Cuneo province (*Sambucus nigra* L., *Achillea millefolium* L., *Cornus mas* L., *Vaccinium myrtillus* L., *Fragaria vesca* L., *Rosa canina* L., *Rubus idaeus* L., *Rubus fruticosus/ulmifolius* L., *Urtica dioica* L.), where there is a long history of herbal remedies. The mechanism underlying the anti-hyperglycemic effects and the clinical evidence available are discussed. Overall, this review points to the possible use of these plants as preventive or add-on therapy in treating diabetes. However, studies of a single variety grown in the geographical area, with strict standardization and titration of all the active ingredients, are warranted before applying the WHO strategy 2014–2023.

Keywords: diabetes; elderberry; yarrow; cornelian cherry; bilberry; wild strawberry; rosehip; raspberry; blackberry; stinging nettle

Citation: Boscaro, V.; Rivoira, M.; Sgorbini, B.; Bordano, V.; Dadone, F.; Gallicchio, M.; Pons, A.; Benetti, E.; Rosa, A.C. Evidence-Based Anti-Diabetic Properties of Plant from the Occitan Valleys of the Piedmont Alps. *Pharmaceutics* **2022**, *14*, 2371. https://doi.org/10.3390/pharmaceutics14112371

Academic Editor: Kathryn J. Steadman

Received: 27 July 2022
Accepted: 28 October 2022
Published: 3 November 2022

Publisher's Note: MDPI stays neutral with regard to jurisdictional claims in published maps and institutional affiliations.

1. Introduction

Diabetes mellitus is a chronic metabolic disorder characterized by persistent hyperglycemia [1]. The increased level of glucose in the blood damages body tissues over time, thus inducing the development of microvascular (including blindness, nephropathy, neuropathy and diabetic foot) and macrovascular (cardiovascular and stroke) complications that could be disabling and life-threatening. Unfortunately, diabetes prevalence is increasing globally, reaching epidemic proportions and represents a leading cause of death worldwide. Currently, about 538 million adult people are living with diabetes and by 2045 it is estimated that the number will increase to 783 million [2]. Nowadays, the economic burden associated with diabetes is very worrying, accounting for 12% of global healthcare expenditure [2,3] and these costs usually increase over time and with disease severity. Moreover, considering the expected increase in diabetes morbidity, the healthcare system expenditure on diabetes and its complications is bound to increase. Therefore, early investments into prevention and disease management are particularly worthwhile.

Accounting for over 90–95% of all cases, type 2 diabetes mellitus (T2DM) is the most common disease, thus representing the primary focus of diabetes research. The progressive impairment of insulin sensitivity results in insulin resistance, defined as an increased insulin requirement to maintain glucose homeostasis. This results from a reduced insulin ability to activate the insulin signaling pathway in the hormone-responsive tissue. Another diagnostic determinant of T2DM is pancreatic β-cell dysfunction. In the early stages of the disease, peripheral tissue responsiveness to circulating insulin is reduced and pancreatic β-cells increase insulin secretion to compensate for this insulin resistance. Over time, the ability of pancreatic β-cells to release sufficient insulin declines, resulting in impaired fasting glucose and impaired glucose tolerance associated with prediabetes. Further disease progression is characterized by continued β-cell deterioration and chronically elevated blood glucose concentrations [1,4]. Recognizing that the combinations of several unhealthy lifestyle factors play a crucial role in the pathogenesis of this disease, strategies focused on promoting an active lifestyle and a healthy diet are a cornerstone for preventing T2DM. One dietary factor of particular interest is the consumption of natural products. Several natural products, mostly plant-specialized (secondary) metabolites, have shown either insulin-mimetic or secretagogue properties, with a beneficial effect on glucose metabolism, thus suggesting that the consumption of these plants could lower the risk of T2DM [3,5,6]. These beneficial effects are mainly attributed to the content of different active ingredients, including alkaloids and terpenoids [7]. So far, in the literature, more than 400 plant species have been described as anti-hyperglycemic [8]. However, this definition raises several doubts and could appear quite superficial, because defining a substance as anti-hyperglycemic is a complex process requiring not only in vitro and in vivo evaluations, but above all clinical studies. Altogether, preclinical and clinical evaluations represent a two-step approach with the first suggesting and the second confirming the anti-diabetic effect. In vitro assays aim to assess the improvement in glucose uptake from insulin-dependent tissues, the modulation of insulin secretion and/or the possible inhibition of digestive enzymes such as α-glucosidase and α-amylase. In vivo measurements include fasting glycaemia, oral glucose tolerance test (OGTT), homeostatic model assessment for insulin resistance (HOMA-IR) and insulinemia. Positive results from these preliminary studies should be confirmed by a clinical evaluation based on the standard assays for the diagnosis of diabetes, such as glycosylated hemoglobin (HbA1c; $\geq$6.5%), fasting plasma glycaemia ($\geq$126 mg/dL), 2 h OGTT ($\geq$200 mg/dL) or random glycaemia ($\geq$200 mg/dL) [9].

Focusing on natural remedies the experimental studies are not conclusive, mainly replicating previous results, without providing added value to the potential efficacy of the clinical application. Clinical trials are usually lacking and the few available include a relatively small number of subjects and/or have a relatively short-term endpoint. Therefore, the bench-to-bedside translation is very difficult and the contribution of natural remedies in real life is still an unresolved challenge. Despite the limits on the knowledge of natural remedies' efficacy, reliance on them by the public is usually high as demonstrated by the wide use of this kind of product worldwide. This led to the investigation of the potential use of plants for different applications, including the most urgent therapeutic need area, as is ongoing with, for example, the proposed in vitro efficacy of *Sambucus nigra* L. [10,11], *Urtica dioica* L. [10,12], *Rasa canina* L. [13], *Rubus fructicosus* L. [14] or *Achillea millefolium* L. [15] against COVID-19 infection.

Consistently, promoting a culture of conscious use of these remedies is strictly urgent to avoid false efficacy expectations or, even worse, the onset of health issues. Diabetic patients have been reported to use herbal medicines [16] with an increase in their consumption of 380% in the US [17]. Therefore, a periodic review of the most recent literature on the anti-diabetic properties of herbal remedies allows for the addition of new pieces of evidence towards the conscious use of herbal remedies for the prevention/treatment of diabetes.

Interestingly, data comparing urban-to-rural diabetes prevalence ratios show a significantly lower presence of diabetes in rural areas [2] in Europe more than in other continents. According to epidemiological data in the Piedmont Region (North-Western Italy), the province of Cuneo (South-Western part of the Piedmont Region) is the area with the lowest prevalence of diabetes in the whole region, with a prevalence of diabetes in 2018 of 4.7% [18]. On the contrary, Turin (the capital city of Piedmont) registered in 2016–2018 a cumulative incidence of 11.5×1000 new diabetic patients with a prevalence of 5.9% [19]. Cuneo, the 19th Italian province by population, is covered more than half by mountains.

The geographical-naturalistic heritage of Cuneo province, based on rural-artisan environments and small city settlements [20], explains the rural lifestyle of residents. Interestingly, a cross-sectional study in the mountainous region of Nepal demonstrated that diabetes mellitus prevalence is lower among residents at a higher altitude compared to the general population [21]. Therefore, it is possible to speculate that people living at a higher altitude in the Alps of Cuneo have a reduced risk of developing diabetes compared to people living in the same region's plains. Moreover, residents could benefit from the specific Alpine biodiversity and ethnobotanical resources. The Piedmont Alps in Cuneo province show a peculiar heritage, as they belong to the Occitan-speaking valleys of Piedmont, a language minority of Alpine communities living 750 m above sea level (msl), with a solid herbal use tradition and agrarian economy tradition, beginning to be lost after the Second World War but recovering in the last decades [22], with the rediscovery of local traditions in favor of a return to agriculture flanked by tourism development. In the past years, the traditional knowledge of the plants of the Piedmont Alps was investigated through an ethnobotanical approach based on the concept of language and ethnic minority groups. Two papers, in particular, focused on the presence of several wild plants as famine foods and medicines in this region: the first [23] recorded 88 botanical taxa in the upper Val Varaita valley (Piedmont Alps, Cuneo Province), a quarter of them also known in other surrounding Occitan valleys, while the second [24] reported 92 plants belonging to 40 different families in three selected linguistic sites in the Alpine Occitan communities (Grana and Gesso valleys, both in the Piedmont Alps in Cuneo Province) [24]. These papers highlighted how different local folk uses of plants go far beyond their nutritional properties.

In the more general attempt to relocate wild alpine plants as functional foods (foods that have beneficial effects on one or more functions beyond their nutritional properties [25]) based on traditional knowledge, this review aims to analyze the spontaneous (not exclusive) presence of plants in the Piedmont Alps (Italy) used by the Occitan language minority in the Cuneo province, for which the literature showed in vitro and in vivo anti-diabetic properties and/or are subject to ongoing clinical evaluation for pre-diabetes and diabetes (Table 1).

Table 1. Plants used in the Occitan valleys of the Piedmont Alps for which anti-diabetic properties have been suggested by the scientific literature published between 2011–2021.

Family	Botanical Name	Plant	Altitude (msl)
Adoxaceae	*Sambucus nigra* L.	Black elderberry	0–1400
Asteraceae	*Achillea millefolium* L.	Yarrow	0–2500
Cornaceae	*Cornus mas* L..	Cornelian cherry or dogwoods	0–1500
Ericaceae	*Vaccinium myrtillus* L.	Bilberry	1200–2000 (rarely 300–2800)
Rosaceae	*Fragaria vesca* L.	Wild strawberry	200–1900 (rarely up to 2400)
Rosaceae	*Rosa canina* L.	Rosehip or dog rose	0–1900
Rosaceae	*Rubus fruticosus* L., *Rubus ulmifolius* L.	Blackberry	0–1400
Rosaceae	*Rubus idaeus* L.	Raspberry	200–2000
Urticaceae	*Urtica dioica* L.	Stinging nettle	0–1800

msl = meters above sea level.

Included in this task, following an ethnobotanical analysis of the Piedmont Alps, a literature search of papers published between 2011–2021 was conducted using keywords including the botanical name of each plant, diabetes, T2DM, hyperglycemia and insulin resistance. Reviews, experimental and clinical studies were all considered. Finally, a search of clinical trials was performed on the ClinicalTrial.gov database. The review is structured in four sections: the first identifies and describes the selected plants in ethnobotanical terms; the second describes the main active ingredients with presumed anti-diabetic activity; the third, divided into sub-sections by single plant, provides a description of the experimental studies published in the 2011–2021-time frame; the fourth reports the results obtained from the most recent clinical studies.

2. Plants from the Occitan Valleys of the Piedmont Alps and Their Traditional Use

Plants reported in Table 1 are all well known to the Piedmont Alpine Communities and their different local folk uses, far beyond their nutritional properties, are described below.

Still, no reference to anti-diabetic properties has been found in the traditional knowledge of the Occitan valleys of the Piedmont Alps.

Among plants belonging to the Rosaceae Family, different well-known species have been identified (Table 1); most of them are used for their delicious fruits as food but also present exciting folk uses in the Piedmont region. The Rubus genus is the largest genus of the Rosaceae family, distributed all over the world except Antarctica [26,27]. It comprises more than 750 species and 12 subgenera [26]. The species of this genus have been cultivated for centuries for their fruits, which are consumed fresh or as processed products [28,29].

2.1. Rubus idaeus L. (Raspberry)

Rubus idaeus L., or raspberry, belongs to the ampula type: according to some, the name is based on a pre-Indo-European base amp-; according to others, it is instead a continuer of a Latin ampulla form due to the shape of the fruit once detached from the peduncle which has a deep cavity inside [30–32]. Indeed, the pendulous fruits are aggregates of different berries, first green and then red when ripe (drupetum). Raspberry is a perennial shrub characterized by a creeping rhizome from which branch green or reddish-brown stems up to 2 m high. Stems are woody, pruinose, glabrous, sometimes with simple hairs and small patented or inclined spines, sometimes violet-colored. The flowers are hermaphroditic and grouped in inflorescences of up to 10 each; the calyx is composed of five persistent, triangular, reflexed, whitish-grey sepals. Petals are equal in number to the sepals, deciduous, white and shorter than the sepals. The leaves are deciduous, characterized by an upper light green page and a lower whitish, a petiole and a spiny-tomentose rachis. They are pinnate with three to seven ovate-lanceolate leaflets sprinkled with very short simple or star-shaped hairs, very variable in size and shape, with serrated margins [33]. EMA recognizes the traditional use of raspberry leaf, whose minimum content of tannins should be 3% according to the European Pharmacopoeia [34,35], for minor spasms associated with menstrual periods, for the symptomatic treatment of mild inflammation in the mouth or throat and the symptomatic treatment of mild diarrhea. The fruits of red raspberry can be consumed in raw or processed forms (frozen, dried, juiced, powdered) or extracted [36]. The information on the use of its fruits in the Occitan valleys is sparse but covers a wide range, from jam to syrup or liqueur [37,38]. In eastern Europe, raspberry fruits have also been used to treat the common cold, fever and flu-like infections [27]. These uses can be ascribed to the presence of several phenolic compounds, the predominant being anthocyanins (around 92.1 ± 19.7 mg anthocyanins/100 g of fresh fruit, although variability between fruit variety, season, developmental stage and methods to quantify the compounds have been reported [39]) and ellagitannins such as sanguiin H-6 (from 139.2 ± 14.4 mg/100 g dry weight to 633.1 ± 65.6 mg/100 g dry weight according to the different variety) and ellagic acid (from 26.1 ± 2.6 mg/100 g dry weight to 106.8 ± 10.9 mg/100 g dry weight according to the different varieties) [40]. Sanguiin H-6 activities have been tested in different in vitro experimental settings and extensively reviewed [41]. According to the values reported by the authors, the anti-inflammatory and anti-oxidant effects have been achieved in a range of concentrations between 2.5 and 250 µM, depending on the specific assay. Moreover, sanguiin H-6 showed a minimum inhibitory concentration (MIC) between 0.06 mg/mL (against *Clostridium sporogenes*) and 0.5 mg/mL (for *Streptococcus A, Pneumoniae, Bacillus subtilis, Moraxella catarrhalis*); on the contrary, the MIC calculated for *Candida albicans* is 5 mg/mL [41].

2.2. Rubus fructicosus L./Rubus ulmifolius (Blackberry)

Rubus fruticosus L. and *Rubus ulmifolius* Schott (blackberry) are other widely used mountain plants as folk remedies. The lexical type prevalent in the western Alps is *moura* with explicit reference to the dark color of the fruits [30], well known for being a functional food [29]. *Rubus ulmifolius* Schott is a perennial sarmentose shrub growing in almost all of Europe, North Africa and Southern Asia; it has also been introduced in America and Oceania. Shrubs present aerial stems characterized by a pentagonal section, up to 6 m long, provided with curved thorns. Leaves are dark green, quite persistent, ternate to palmate with stipules filiform to linear; leaflets are three to five, with a base rounded to cuneate, margins finely to moderately serrate, apex acute or acuminate. Flowers are white or pink with five petals and five sepals. They are grouped in racemes with 10 to 60 bisexual flowers forming oblong or pyramidal inflorescences. The edible fruit is composed of numerous small drupes (generally 10 to 40), green at first, then red and finally blackish when ripe, each arising from separate carpels but forming part of the same gynoecium [42]. Blackberries have a high content of carbohydrates, vitamins (ascorbic acid in particular), minerals (potassium included) and dietary fiber suggesting their use in diet-based therapies for improving human health. They are also rich in specialized bioactive metabolites, mainly phenolic compounds and ellagitannins, which may contribute to the anti-oxidant and anti-inflammatory properties of these berries [26,28,43,44]. In Italian folk medicine, blackberry treats ulcers, intestinal inflammations, diarrhea, abscesses, furuncles, red eyes, vaginal disorders and hemorrhoids [45]; it is also used as an anti-pyretic and carminative agent [26]. A crude methanolic extract and flavonoid fractions obtained from dried materials of aerial parts of blackberry showed significant anti-oxidant and anti-pyretic activity in rats with pyrexia [26]. The effect on the gastrointestinal tract can be due to the biological activity of polyphenols, with blackberry improving gastrointestinal digestion and protecting against ethyl carbamate and acrylamide-induced oxidative stress and cytotoxicity [46]. However, the primary folk use in the Occitan valleys recalls mostly the food consumption of blackberry fruits, with the preparation of jams. As a folk remedy, some evidence focused on its use for sore throats: a syrup prepared from the blackberries cooked, passed through a sieve to remove the small grains and cooked again with sugar and a gargling solution after infusion of spring jets, has been described [37,38]. Blackberry's use for sore throats or whooping cough stems from its anti-microbial, anti-oxidant and anti-inflammatory effects. Suit and leaves have a lower anti-microbial activity compared to root or stem [47]. The leaves are also rich in tannins and used to counteract diarrhea or, when used externally, for wound healing [48].

2.3. Fragaria vesca L. (Wild Strawberry)

Fragaria vesca L. (wild strawberry) always belongs to the Rosaceae family. The rugged woodland strawberry is native to Europe and Asia. It is the most widely distributed species of *Fragaria* occurring throughout Europe, Northern Asia, North America and Northern Africa. It is an herbaceous perennial plant characterized by an oblique and branched rhizome and short, slender, weakly woody stems. It reaches a height ranging from 5 to 20 cm. The trifoliate leaves are as long as the scape, with a long, pubescent petiole; the upper page is dark green, while the lower is whitish, slightly hairy, especially on the nerves; the margin is roughly toothed. The scape is characterized by one to three erect flowers, becoming pendulous at the fruiting stage; flowers have a corolla composed of five white petals with overlapping margins, triangular sepals and 20 stamens. Fruits are small, black achenes scattered on the swollen red receptacle, easily detached from the calyx [33,49]. Wild strawberry is reported to have two basic lexical types in the western Alpine valleys: *frola-frùi* and *maiòla-maiùso*, sometimes with an alternative suffix [30]. EMA recognizes in wild strawberry leaf its traditional use to increase urine production in patients with minor urinary tract problems and to relieve symptoms of mild diarrhea [35]. Moreover, its leaves were reported to have healing properties when applied to the top of cuts or wounds [50]. These effects can be due to all the active ingredients present in wild strawberry leaves, including phenolic compounds (mainly ellagitannins, flavonoids, proanthocyanins, in particular catechins, phenolic acids), terpenes characterizing volatile oils (e.g., borneol), methyl salicylate and trace amounts of alkaloids [51,52]. The plant is rich in vitamin C and folate [5]. Therefore, its extract-based hydrogel at 2% has been recently studied for topical cosmetic applications [53].

2.4. Rosa canina L. (Rosehip)

The last plant in the Rosaceae family approached in this review is *Rosa canina* L. (rosehip or dog rose, Table 1). In the western Occitan, there are three main lexical types of rosehips. The first, also in Piedmont, is *'grattaculo'*, no longer diversifying between the plant's name and the fruit. In this case, the etymological process proposed goes back to the name *crataeugus*, reinterpreted as *grattaculo*. The second is *agoulenchìer*, widespread in the high valleys and deriving from *aquilentum* crossed with *aculeus*, concerning the thorniness of the shrub. The third type collects those names that refer to the appearance of the plant (bush, presence of thorns). In some cases, the fruit is distinguished from the shrub *bossou*, a name elsewhere referring to the wild blackthorn [30]. It is a shrub-like species growing spontaneously throughout Europe. A woody, bushy and thorny deciduous shrub, its deep roots and pendulous branches covered with strong, bent, or hooked thorns, mainly with a swollen base, characterize it. The leaves are imparipinnate, with 2 lanceolate stipules at the base; leaflets are soft green, oval, or ovate-elliptical, usually glabrous, or slightly pubescent on the rachis, without glandular hairs. The flowers are solitary or in groups of two to three, delicately scented; corolla has five white or pinkish petals and five green concrescent sepals that fall off when the fruit ripens. The false (or pseudo) fruits, called cinorrods, ripen in autumn and are pyriform, fleshy and glabrous, usually borne by 10–20 mm long peduncles, bright red when ripe. The cinorrods contain many hard achenes covered with short, stiff hairs [33,49]. The cinorrods and petals are mainly used to flavor grappa, produce a sour jam or a sauce to accompany meats, or as a sort of bread by grinding the internal seeds. Moreover, harvested and dried cinorrods are used in herbal tea for their vitamin and diuretic properties. The volume *Piante e Fiori delle nostre Montagne* [54] reports that "the leaves have contrasting properties: both anti-diarrheal and mild laxative. The petals are corrective, refreshing, laxative, anti-diarrheal and flavoring and used externally in eye drops, mouthwashes, or washing; the rosehips are used as corrective, refreshing vitamins due to the high concentration of vitamin C, anti-inflammatory for the genitourinary system, diuretics, or astringents. The external use of pseudo-fruits includes toning, lightening and smoothing of the skin, or for their anti-hemorrhoidal properties. They are also prepared in preserves and jams. Some indicate the seeds as relief for mild symptoms of mental stress and as wormers against roundworms. Galls […] have diuretic and anti-perspiration activity; they are astringent and decongestant for external use. Galls were considered possible substitutes for ergot as regulators of uterine contractions". Consistently, different medical uses can be identified in folk medicine: decoctions for the treatment of bronchitis and cough, for liver stones and to regulate blood pressure [31]; the pseudo-fruit was used as food in times of famine [55]; the herbal tea was used to fight colds, or to treat inflammation of the gums, especially in children, as "pink honey" (prepared cooking rosehip petals with apples). Finally, the use of flower petals to make healing compresses for the eyes should be noted. All these effects reflect the identification of numerous bioactive compounds, including vitamin C, flavonoids, tocopherols, carotenoids, organic acids, sugars and essential fatty acids [56,57]. Rosehip, made up of the receptacle and the remains of the dried sepals of rosehip, should have a minimum vitamin C content of 0.3% (dried drug) according to the European Pharmacopoeia [34]. The anti-inflammatory properties have been generally attributed to linoleic acid, α-linolenic acid and, more importantly, the galactolipid (2S)-1,2-di-O-[(9Z,12Z,15Z)-octadeca-9-12-15-trienoyl]-3-O-β-d-galactopyranosyl glycerol (GOPO).

This latter active ingredient, isolated from seeds and fruits, reduced the chemotaxis of peripheral blood polymorphonuclear leukocytes, neutrophils, monocytes and C-reactive protein levels [58]. Consistently, preliminary clinical data showed its efficacy in reducing the complaints of patients with osteoarthritis, thus improving their quality of life [59].

2.5. Vaccinum myrtillus L. (Bilberry)

Well known to Alpine communities, *Vaccinium myrtillus* L. (bilberry) is a species belonging to the *Vaccinium* genus and the Ericaceae family. It belongs to a large genus: blueberry (*Vaccinium corymbosum*) and cranberry (*Vaccinium macrocarpon*). Several names are lexical types that refer to bilberry, also known as the European blueberry, whortleberry, huckleberry and blueberry. In the Western Alps, the most common types generally refer to the color of the berries, e.g., Latin *ater* 'dark' *aidre, arëzze, aize* (types of the Occitan valleys; for *areze/erze*, other proposals have also been made: *acinus + ater, acris* or *prel.* * *alisa*) or their shape, e.g., *prelatin * brom-* 'rounded object' *ambruna* (type of the Franco-Provençal valleys) [31]. Bilberry is a low-growing (between 10 and 40 cm) perennial shrub in North Europe, North America and Asia. Bilberry has small alternate leaves, green on both sides, characterized by a thin oval or elliptical lamina and short petioles. The flowers are hermaphrodite, actinomorphic with 4–7 mm reflexed pedicels, usually reddened. The calyx is gamosepalous and divided into five very short obtuse lobes. The corolla is greenish-white or pinkish with 5 small, revolute lobes. Fruits are small berries (5–9 mm in diameter), blue-violet or blackish, covered by bloom and slightly flattened at the apex. The inner part is paler and contains numerous brown seeds [33]. The flavor is acidulous and pleasant. Therefore, fruits have been used in various culinary preparations, fresh or jam, for a long time. Nevertheless, thanks to the high content in anthocyanins, mainly in the glycosylated forms (cyanidin-, delphinidin- and malvidin-3-O-glycosides), vitamins, catechins, phenolic acid derivatives and minerals [60], several beneficial effects have been reported for bilberry. Fresh fruits have a laxative effect, while dried or as a juice they have an astringent effect. The leaves have anti-inflammatory properties and are used for oral, skin, or ocular inflammation, thanks to anthocyanins. Its anti-thrombotic, anti-hypertensive and cardioprotective properties are well known [60], for which not only fruits but also leaves were used [31]. Notably, relieving discomfort and heaviness of legs related to minor venous circulatory disturbances or symptoms of bleeding into the skin from tiny blood vessels are traditionally recognized by EMA for fresh fruit. As for dried fruit, EMA recognizes traditional use for treating mild diarrhea and minor inflammations of the oral mucosa [35]. Both fresh and dried fruits are described in the European Pharmacopoeia, with a minimum content of 0.30% of anthocyanins, expressed as cyanidin 3-O-glucoside chloride and 0.8% of tannins, defined as pyrogallol, respectively [34]. Among other plant uses, boiled leaves were administered to newly relieved cows, facilitating the expulsion of the placenta [55].

2.6. Sambucus nigra L. (Black Elderberry)

Sambucus nigra L. (black elderberry or elder, elderberry, European elder, European elderberry and European black elderberry) is a species belonging to the Adoxaceae family, known in the Western Italian valleys as sambuca according to the primary Italian lexical type, which continues the Latin *Sambucus*. Also from Latin, *Sambucus nigra* L. is also referred to as *seuic*. It is a small tree or shrub not exceeding 5 m in height, growing native in Europe, northern Africa and southwestern Asia. It is characterized by a gray bark and numerous protruding lenticels. The leaves, bright green in color, are opposite, pinnate with 5–7 leaflets, oval or oblong, acuminate and with irregularly serrated margins. The odorous tiny white flowers are grouped in large corymb inflorescences; they are actinomorphic and are characterized by a rounded corolla composed of five ivory white, sometimes reddish, oval petals. The fruits are small globular drupes, first green, then purple-blackish, shiny and juicy when ripe, containing two to five oval, brown seeds. Fruits are grouped in pendulous infructescence on reddish peduncles [49]. Historically the whole plant was used, from the wood to the flowers and the fruits, thanks to the high content of carbohydrates, proteins, fats, fatty acids, organic acids, minerals, vitamins, essential oils and anti-oxidants such as polyphenols [61]. The anti-oxidant activity explains the use of elderberry fruits and flowers in different therapeutic contexts, including toothache, colds and udder diseases. It is used as a refreshing and purifying agent and as an anti-hypertensive remedy, also in animals. The leaves were used against rheumatism [31]. "In medicine, the flowers of the plant are prepared as compresses, attenuate swellings of various origins and, in herbal tea, calm toothache" [55]. Moreover, infusion of dried flowers was used against colds and bronchitis; the fruits, in sweetened juice or jams, as a mild laxative and against neuralgia [37,38]. EMA recognizes the traditional use of elderberry flowers to relieve early symptoms of common cold [35]. The minimum content of flavonoids should be 0.80%, expressed as isoquercitroside, according to the European Pharmacopoeia [34].

2.7. Achillea millefolium L. (Yarrow)

Much more widely known and used in the mountain areas than in the municipalities of the plains [37,38] is *Achillea millefolium* L. (yarrow), a species belonging to the Compositae family. It is an aromatic perennial herbaceous plant, rhizomatous and slightly suffruticose, covered with hair or down, 4–90 cm high. Leaves are alternate, oblong-lanceolate, green or greyish green, slightly pubescent on the upper surface and more pubescent on the lower surface, two–three pinnatisects divided into linear hairy segments. The flowering tops containing essential oil, also described in the European Pharmacopoeia, are the most active part of the plant [62,63]. Flower heads are white or pinkish and are arranged in corymbs at the end of the stem. Each flower head consists of a slightly convex receptacle, 4–5 ligulate ray florets and 3–20 tubular disk florets. Fruits are 2-mm shiny achenes with no pappus [33]. The lexical type prevalent in the Western Alps is cut grass, *taiouira* or *mal tagliata* "badly cut," due to its hemostatic properties and perhaps also in connection with mowing practices. Disinfectant and anti-inflammatory uses are also known. With local specific preparations and indications, it was used as a sedative in poultices for sprains and dislocations; decoctions were used against hemorrhoids and abdominal pain; in packs and infusions for hemorrhoids, flowering tops were collected to prepare digestive, calming and pain-relieving effects, in oil with camphor for muscle massage or whole plant to make compresses for toothache [31]. In other parts of Italy, yarrow is used for a variety of conditions, in particular gastro-intestinally [64,65]. Its use is also reported in Peru for gastritis, diabetes, cholesterol and mainly for skin infections [66]. Consistently with these local uses, EMA considers herbs and flowers as traditional herbal medicinal products for temporary loss of appetite and for the symptomatic treatment of mild, spasmodic gastro-intestinal complaints, including bloating flatulence or minor spasms associated with menstrual periods [35]. Moreover, they can be used as a traditional herbal medicine to treat minor superficial wounds (a leaf pack was also used as wound healing [31]). It can be found in herbal pharmacies as tinctures and capsules containing dry flowers or aerial parts [67]. Apart from the described folk uses, pleiotropic effects, including anti-diabetic, anti-tumor, anti-fungal, anti-septic and liver protective effects, have been suggested. These effects have been mostly proven in vitro, on isolated organs, or in vivo, but to the best of our knowledge not in the clinic; thus, their transferability to the real world is far from being demonstrated. The reason for this plethora of potential effects stems from the several active ingredients of yarrow including phenols (such as choline, caffeoylquinic acids (DCCAs) or salicylic acid), flavonoids (including apigenin, artemetin, luteolin-7-O-β-D-glucuronide, dihydro-dehydro-diconiferyl alcohol 9-O-β-D-glucopyranoside) and many monoterpenes (such as borneol, camphor, eucalyptol, α- and β-pinene, α-terpineol and sesquiterpenes (chamazulene, azulene, included), with monoterpenes constituting 90% of the essential oils compared to sesquiterpenes [63,67–74]. Actually, essential oil is the only feature reported by the European Pharmacopoeia, according to which its minimum content in whole or cut, flowering tops should be 2 mL/kg dried drug [34]. The essential oil is the formulation mostly related to the folk uses described above. Other uses have been suggested for different herbal preparations, such as decoction, hydroalcoholic, methanolic and aqueous extract [75], in which specific active ingredients can be recognized. For instance, a choleretic activity has been ascribed to teas or tinctures, for their content in luteolin-7-O-β-D-glucuronide and DCCAs, whose activity was demonstrated in isolated organ experiments [67,76]. The hepatoprotective effect was tested in vivo with ethanol [77], aqueous [78], or aqueous–methanol [79] extract from the aerial parts of the plant. The active ingredient detrimental to this effect has not yet been clearly established, although some evidence points out at the 5-hydroxy 3, 4′, 6, 7-tetramethoxy flavone [78]. The chloroform extract, in which flavonoids (apigenin, luteolin, centaureidin, casticin and artemetin) and sesquiterpenoids (paulitin, isopaulitin, psilostachyin C, desacetylmatricarin and sintenin) are present, showed in vitro antiproliferative activities [67].

2.8. Urtica dioica L. (Stinging Nettle)

Urtica dioica L. (stinging nettle) is a well-known plant with the same lexotype (lat. *Urtica*, meaning "to burn," attributed to its stinging hair) with different phonetic results [30]. The specific name *dioica* is related to the fact that it is a dioecious plant, having unisexual flowers (male or female) on different plants. Nettle is an herbaceous perennial plant growing worldwide. It is characterized by long stoloniferous rhizomes branched just below the surface, from which numerous robust, erect and striped quadrangular stems arise. Stems are covered with unicellular stinging hairs. The leaves are opposite, with a petiole shorter than the leaf blade (not reaching its midpoint); they are much longer than wide, ovate-lanceolate, with a heart-shaped base and acute apex; the margin is toothed. In addition, leaves are covered by long stinging hair together with short, simple hairs. The small flowers are grouped in the glomerulus, greenish-yellow or reddish, inserted at the leaf's axil. The fruit is an ovoid-elliptic, olive-brown diclesium, with a tuft of hairs at the apex, enclosed in the enlarged petals [33]. Stinging hairs contain several chemicals, including serotonin, histamine, acetylcholine, moroidin, leukotrienes and formic acid, that provoke pain, wheals, or a stinging sensation. However, the irritant power is lost during cooking, thus making this plant usable for culinary purposes, consistently with its high content in nutrients (up to 3.7% proteins, 0.6% fat, 2.1% ash, 6.4% dietary fiber and 7.1% carbohydrates in harvested upgrowths [80], high concentrations of vitamins and metals in fresh leaves [81] and about 30% proteins, 4% fats, 40% non-nitrogen compounds, 10% fiber and 15% ash in leaf powders [82]), it is appreciated as food in general: stinging nettle tips are the first to be harvested in spring. They are used in omelets or herb soups and risotto. Besides several nutrients, stinging nettle is rich in biologically-active compounds such as terpenoids [83], carotenoids, fatty acids, polyphenols [81,84–86], tannins, carbohydrates, sterols, polysaccharides, isolectins and minerals, the most important of which is iron [82]. The distribution of phytochemicals is uneven: leaves are rich in flavonoids (mainly quercetin, kaempferol, rutin and their 3-rutinosides and 3-glycosides, catechin, epicatechin and epigallocatechin-gallate); root contains lectins, polysaccharides, phytosterols, lignans, coumarins (scopoletin) and high amounts of fatty acids; nettles have abundant quantitative of flavonoids, in particular, anthocyanins; flowers contain high amounts of β-sitosterol and 7-flavonoid glycosides [87–89]. The European Pharmacopoeia describes only leaves and radix [34]. EMA recognizes the traditional use of the leaves and herbs to relieve minor articular pain and increase the amount of urine and as an adjuvant in minor urinary complaints. Herbs could also be used for seborrheic skin conditions [35]. The radix was traditionally used to relieve lower urinary tract symptoms related to benign prostatic hyperplasia. These uses are like stinging nettles by local populations in Piedmont, where also veterinary uses are described. In humans, it was applied as anti-anemic due to its content in iron, a digestive and purifying agent, against bruises and hair loss. Stinging nettle broth was used for ulcers and cosmetic applications, especially against acne or greasy hair. Finally, folk uses included the treatment of sciatica and rheumatism, as well as diuretic or metabolic effects. In cows, stinging nettles were used to treat diarrhea and in hens to promote the production of eggs (it is considered "warming"). The other best-known use of stinging nettle leaf macerate is against plant lice, possibly leaving it to macerate with tobacco [50].

2.9. Cornus mas L. (Cornelian Cherry)

Less used because not always known to local communities is *Cornus mas* L., cornelian cherry or dogwoods. It is a sporadic species in the Western valleys belonging to the Cornaceae family. The primary lexical type is etymological to the Italian: *cournal*. Cornelian cherry is a deciduous shrub that can grow up to 6–8 m in height. The trunk is erect, often twisted, highly branched at the top, gray with reddish cracks, peeling bark and short erect-patent twigs. The leaves, with a short hairy petiole, are oval, opposite and acuminate. The small yellow flowers bloom before the leaves and give off a faint honey smell; they are grouped in axillary umbels, surrounded by four acuminate greenish bracts. The calyx is composed of four acute greenish sepals, while the corolla by four acute, glabrous, golden yellow petals. The fruits are ovoid edible dark red drupes [33]. Records of its use come from mountain or foothills villages. Healing decoctions and sour jam were prepared from the fruits, characterized by local anti-inflammatory properties, due to the content of anti-oxidants such as phenolic metabolites, concerning anthocyanins [31,90].

3. Active Ingredients of Plants from the Occitan Valleys of the Piedmont Alps Recognized to Have Anti-Diabetic Properties: An Overview

The Alpine plants identified in our study are characterized by common active ingredients justifying the use of these plants as functional foods or nutraceuticals. However, each plant's percentage of active ingredients varies according to its species and genus. Therefore, despite the coexistence of different active ingredients pointing out a possible synergism, the anti-diabetic effects can be mostly attributed to some active ingredient classes such as polyphenols, anthocyanins in particular and oligosaccharides (Table 2). Their relevance relates to their direct effect on signaling pathways involved in glucose homeostasis.

Table 2. Plants used in the Occitan valleys of the Fiedmont Alps: potential direct anti-diabetic mechanisms (experimental evidence between 2011 and 2021).

Plant	Part of the Plant Tested		Mechanisms (Experimental Evidence)	References
	Formulation	**Identified Active Ingredients**		
Urtica dioica L.	Leaves		insulin secretion (++) insulin sensitivity (++) α-glucosidase/α-amylase inhibition (+)	Ahangarpour et al., 2012 [91] Dar et al., 2013 [92] Kadan et al., 2013 [93] Qujeq et al., 2013 [94] Rahimzadeh et al., 2014 [95] Ranjbari et al., 2016 [96] Obanda et al., 2016 [97] Obanda et al., 2016 [98] Gohari et al., 2018 [99] Abedinzade et al., 2019 [100] Fan et al., 2020 [101] Salim et al., 2020 [102]
	H_2O extract [92,94–96]	polyphenols (phenolic acids, flavonoids and anthocyanins)		
	Hexane extract [92]	terpenes, fatty acids [92]		
	Chloroform extract [92]			
	Ethyl-acetate [92]			
	MeOH [92]			
	EtOH extracts [94]			
	Hydroalcoholic extract [91]			
	Leaves and stem			
	Hydroalcoholic extract [93]			
	Aerial parts			
	Hydroalcoholic extract [100]			
	powder [101]			
	distillate [99]			
	H_2O extract [97,98]			
Achillea millefolium L.	Aerial parts		insulin secretion (++) insulin sensitivity (++) α-glucosidase inhibition (+)	Mustafa et al., 2012 [103] Ramirez et al., 2012 [104] Zolghadri et al., 2014 [77] Chávez-Silva et al., 2018 [105]
	H_2O extract [103]	Tannins, glycosides, terpenoids, flavonoids and phenolics [103]		
	MeOH extract [103]			
	Hydroalcoholic extract [77,104,105]			
Vaccinium myrtillus L.	Leaves with stems		insulin secretion (+/-) α-glucosidase inhibition (+) α-amylase inhibition (+/-) reduction of glucose absorption (+)	Brader et al., 2013 [106] Kim et al., 2015 [107] Asgary et al., 2016 [108] Buchholz & Melzig, 2016 [109] Bljajić et al., 2017 [110] Karcheva-Bahchevanska et al., 2017 [111] Xiao et al., 2017 [112] Schreck & Melzig, 2021 [113]
	H_2O extract [110]	Polyphenols: phenolic acids > flavonoids [110]		
	Hydroalcoholic extract [110]	Polyphenols: flavonoids > phenolic acids [110]		
	Fruits			
	MeOH:H_2O:trifluoroacetic acid extract [106]	Polyphenols (anthocyanins, phenolic acids, flavonols) [106]		
	Hydroalcoholic extract [107,112]	Polyphenols (anthocyanins, flavonoids and phenolic acids) [112] Anthocyanins [107]		

Table 2. Plants used in the Occitan valleys of the Piedmont Alps: potential direct anti-diabetic mechanisms (experimental evidence between 2011 and 2021).

Plant	Part of the Plant Tested		Mechanisms (Experimental Evidence)	References
	Formulation	**Identified Active Ingredients**		
	MeOH:H_2O: HCl [111]	Polyphenols [111]		
	Acetone: H_2O: HCl [111]			
	H_2O extract [109,111,113]			
	MeOH extract [109,113]			
	Powder [108]			
	Fruits		insulin sensitivity (+)	Zhu et al., 2018 [114]
Rubus idaeus L.	Powder [114,116,117]	Polyphenols [114,116]	α-glucosidases/α-amylase inhibition (+)	Xiong et al., 2018 [115]
	MeOH extract [118]	Polyphenols (anthocyanins) [118]		Zhao et al., 2018 [116]
				Xing et al., 2018 [117]
				Gutierrez-Albanchez et al., 2019 [118]
	Fruits			
Rubus fruticosus L.,	MeOH extract [119]		insulin sensitivity (+/-)	Bispo et al., 2015 [119]
Rubus ulmifolius L.	EtOH extract [46]			Gowd et al., 2018 [46]
	Leaves		α-glucosidase/α-amylase inhibition (+)	Takács et al., 2020 [120]
Fragaria vesca L.	H_2O extract [120]	Flavonoids [120]		
	Flowers			
	MeOH extract [121,122]			
	DCM extract [122,126]	Polyphenols (phenolic acids, flavonoids) [122]		Schrader et al., 2012 [121]
	EtOH extract [126]	Polyphenols [126]		Bhattacharya et al., 2013 [122]
	Hydroalcoholic extract [126]			Farrell et al., 2015 [123]
	H_2O extract [126]		insulin sensitivity (+/-)	Salvador et al., 2016 [124]
	Fruits		α-glucosidase/α-amylase inhibition (+)	Ho et al., 2017 [125]
Sambucus nigra L.	Anthocyanin-rich extract [123]	Anthocyanins [123]		Ho et al., 2017 [126]
	DCM extract [125]			Zielinska-Wasielica et al., 2019 [127]
	EtOH extract [125]			
	Hydroalcoholic extract [125]	Polyphenols (anthocyanins and procyanidins) [125]		
	H_2O extract [125]			
	pressed juice [125]			
	H_2O extract [127]	Polyphenols (anthocyanins, flavonoids, phenolic acids) [127]		

Table 2. Plants used in the Occitan valleys of the Piedmont Alps: potential direct anti-diabetic mechanisms (experimental evidence between 2011 and 2021).

Plant	Part of the Plant Tested		Mechanisms (Experimental Evidence)	References
	Formulation	Identified Active Ingredients		
Rose canina L.	Pseudo fruits and flowers		insulin secretion (++) insulin sensitivity (+/-) α-amylase inhibition (+)	Taghizadeh et al., 2016 [128] Fattahi et al., 2017 [129] Jemaa et al., 2017 [130] Chen et al., 2017 [131] Bahrami et al., 2020 [132] Rahimi et al., 2020 [133]
	MeOH extracts [130]	Flavonoids [130]		
	Pseudo fruits			
	Oligosaccharide fraction of extract [132,133]	Oligosaccharides [132,133]		
	H$_2$O extract [129,131]			
	Hydroethanolic extract [128]			
Cornus mas L.	Fruits		insulin sensitivity (+/-) α-glucosidase/α-amylase inhibition (+)	Capcarova et al., 2019 [134] Dzydzan et al., 2019 [135] Dzydzan et al., 2020 [136] Blagojevic et al., 2021 [137]
	Pressed juice [25,136]	Polyphenols (anthocyanins, phenolic acids, flavonols) and iridoids (loganic acid) [135] Iridoids [136]		
	Hydroalcoholic extract [137]	Iridoids and anthocyanins [137]		
	Homogenized [134]			

DCM: dichloromethane; H$_2$O: water; MeOH: methanol; EtOH: ethanol; ++ = in vitro and in vivo consistent results; + = in vivo or in vitro data; +/- = studies with contrasting results.

3.1. Poliphenols

The most commonly represented active ingredients in the plants considered are polyphenols. Polyphenols represent a large chemical class of natural compounds, with one or more hydroxyl groups attached to the aromatic ring [138], that share the ability to act as anti-oxidants, thus protecting against the damage caused by oxygen-free radicals to DNA and several cellular components [139]. The polyphenols could be classified into different groups by the phenol ring number and the structural elements connected to the ring [140]. Usually, a first distinction is between flavonoids (divided into anthocyanins, flavonols and flavanols based on the oxidation state of the heterocyclic pyran ring) and non-flavonoids (such as phenolic acid, including benzoic acids and hydroxycinnamic acids, stilbenes and its derivative, including gallotannins, ellagitannins and lignins) [141].

It is challenging to dissect each polyphenol's relative contribution to glucose metabolism, but polyphenols are usually evaluated for their anti-diabetic potential. Therefore, several mechanisms could explain their role in glucose metabolism: they reduce glucose absorption, increase glucose metabolism and pancreatic insulin secretion and enhance glucose use by muscle and adipocytes. Polyphenols such as flavonoids and tannins inhibit critical enzymes involved in the digestion of carbohydrates to glucose, such as α-glucosidase and α-amylase [142,143]. Polyphenols can regulate postprandial glycemia by a facilitated insulin response and the release of the glucose-dependent insulinotropic polypeptide (GIP) and glucagon-like polypeptide-1 (GLP-1) [144,145]. Glycolysis, glycogenesis and gluconeogenesis, generally impaired in diabetes, could be controlled by polyphenols; for example, epigallocatechin gallate, naringenin and hesperidin increased the expression of glucokinase (GK) at hepatic level [146,147] and naringenin decreases the expression of glucose-6-phosphatase. Moreover, polyphenols such as resveratrol, quercetin and epigallocatechin gallate can enhance the translocation of glucose transporter (GLUT)4 to the plasma membrane, thus increasing glucose uptake in muscle and adipocytes by the activation of the AMP-activated protein kinase (AMPK) pathway [148], which is a sensor of the cell's energy that plays a vital role in obesity. Polyphenols could suppress β-cell damage induced by prolonged exposure to high glucose concentrations, for example, by decreasing oxidative stress. Anthocyanins could protect β-cells from oxidative stress affecting phosphoinositide 3-kinases (PI3K)/protein kinase B (AKT) and extracellular signal-regulated kinase (ERK)1/2 pathways [148]. Other polyphenols could upregulate essential genes involved in β-cell functions, such as *Glut2*, *insulin* (*ins*)*1*, glucokinase gene (*Gck*) *and sirt1* [149]. Catechins increase insulin release upregulating the expression of the nuclear factor erythroid 2–related factor (*Nrf*)*2 and 1* gene [150]. Moreover, polyphenols can block inflammatory pathways by inhibiting several protein kinases overexpressed in diabetes [139].

Among polyphenols, anthocyanins are the largest group of water-soluble vacuolar polyphenolic compounds [151–153] in the plant kingdom with a positive charge in acidic solution; they are responsible for the characteristic red to blue color of plants [154] and are primarily present in fruits and flowers. These molecules consist of an anthocyanidin "core" with a sugar moiety (glucose, galactose, xylose, arabinose, or rhamnose) attached at various positions [151,155,156]. Hydroxyl and methoxyl groups can be attached to the core anthocyanin in different numbers and positions. These compounds are cyanidin, delphinidin, malvidin, pelargonidin, peonidin and petunidin. The primary sources of anthocyanins, apart from vegetables such as red cabbage and radish, are dark fruits such as blackberry, blueberry, cranberries, black and red currants, red grapes and raspberries. Accordingly, plants rich in anthocyanins, bilberry, blackberry and raspberry. are collected in natural environments in the Alps. Their content in anthocyanins varies across species and depends on environmental factors, such as solar radiation, temperature, soil content of nitrogen and phosphorus and the cultivation technique. Bilberry plants located at high latitudes or altitude have been demonstrated to produce berries with higher contents of anthocyanins [157]. Several studies demonstrated that anthocyanin amount could be particularly affected by processing. Frozen strawberries, blueberries and raspberries showed an average 42% decrease in anthocyanin content in their raw forms; the juice of berries retains only 22% of anthocyanins contents compared to their raw whole fruit form [36]. Storage temperature and heat treatment may degrade anthocyanins more than other flavonoids [158], while freeze-drying seems the most effective way to preserve anthocyanins in strawberries [159]. Anthocyanins have anti-oxidant effects causing the increased synthesis of endogenous anti-oxidant enzymes such as glutathione peroxidase (GPx), catalase (CAT) and superoxide dismutase (SOD) and exert different properties, such as DNA stabilization and adipocyte gene expression modification; they also have anti-apoptotic, anti-inflammatory and anti-bacterial effects. However, like other anti-oxidants, anthocyanins can produce a prooxidant effect in very high doses, not practically achievable via the diet, so the dietary intake of anthocyanins poses no concern. Numerous epidemiological studies demonstrated that a higher intake of total flavonoids, especially anthocyanins, could be associated with a lower risk of hypertension, myocardial infarction, stroke and T2DM [160–163]. The amelioration of insulin sensitivity could be explained through different mechanisms, such as the upregulation of GLUT4 gene expression in white adipose tissue and skeletal muscles [164], the activation of AMPK, which stimulates glucose uptake and insulin secretion by pancreatic β-cells. AMPK activation could also downregulate gluconeogenesis in the liver [165,166]. Anthocyanins could also regulate postprandial glycemia. Berries can reduce glucose absorption from a meal and inhibit α-glucosidase and pancreatic α-amylase activity, thus decreasing glycemia after starch-rich meals [167,168]. The mechanism of α-glucosidase inhibition action by anthocyanins is not fully understood. Still, one can assume that acarbose is a competitive inhibitor whose activity is due to the structural similarity between normal substrate maltose and the glucosyl groups β-linked to the anthocyanin. Therefore, anthocyanins bind to the enzyme's active site without undergoing hydrolysis [169]. Moreover, anthocyanins significantly reduced the blood glucose level in diabetic rats, increasing the expression and translocation of GLUT4 and enhancing the activation of the insulin receptor phosphorylation, thereby increasing the uptake and utilization of glucose by cells [160]. Anthocyanins also improve insulin secretion, which could be significant for people with T2DM whose pancreatic activity is damaged.

3.2. Oligosaccharides

Another class of active ingredients represented in the selected plants is oligosaccharides. Oligosaccharides, physiologically active fragments of plant cell wall polysaccharides, are low molecular weight polymers usually composed of 2 to 10 monosaccharides linked by glycosidic bonds non-susceptible to human digestive enzymes [170]. Of both natural and chemical, physical, or biochemical origin, the plethora of oligosaccharides includes functional compounds such as fosfo-fructo-oligosaccharides (FOSs), galacto-oligosaccharides (GOSs), mannan oligosaccharides (MOSs), xylo-oligosaccharides (XOS), neo-agaro-oligosaccharides, iso-malto-oligosaccharides and chito-oligosaccharides (COSs), used as prebiotic [171]. All oligosaccharides have the potential for prebiotic action. So far, the first hypothesis for their role in improving glycemic control and reducing the risk of developing diabetes could be based on their ability to improve the intestinal microbiota, which in turn stimulates insulin signal [172], for instance, increasing gut hormones including GLP-1 and peptide YY (PYY) [173].

Moreover, prebiotics have been demonstrated to inhibit intestinal α-glucosidase, α-amylase and glucose transporters, improve insulin resistance, glycogen synthesis and gluconeogenesis, affect the AMPK pathway in the liver and reduce the inflammatory status associated with diabetes and reduce circulating lipopolysaccharide (LPS), interleukin-6 (IL-6), monocyte chemoattractant protein-1 (MCP-1) and tumor necrosis factor-α (TNF-α). All these events have been long studied and extensively reported for the functional oligosaccharides, which means depolymerized or synthetic polymers used as prebiotics [173]. Compared to functional oligosaccharides, natural oligosaccharides occurring in plants were less studied, although at the mechanistic level they are supposed to exert several effects, including immune-stimulation and anti-oxidant effects, also by specific linkage of a glycan sequence and cell receptors [174].

4. Mechanistic Interpretation of the Anti-Diabetic Effects of Plants from the Occitan Valleys of the Piedmont Alps: Evidence from Experimental Studies

As the identified Alpine plants have many similarities in their active ingredient composition, they can share many anti-diabetic mechanisms. However, each mechanism could be more or less evident based on the percentage of each active ingredient. Table 2 summarizes the anti-diabetic mechanism(s) proposed for each plant according to the experimental studies published between 2011 and 2021.

Three events are triggered: (1) insulin secretion, (2) insulin sensitivity and (3) intestinal glucose absorption (Figure 1).

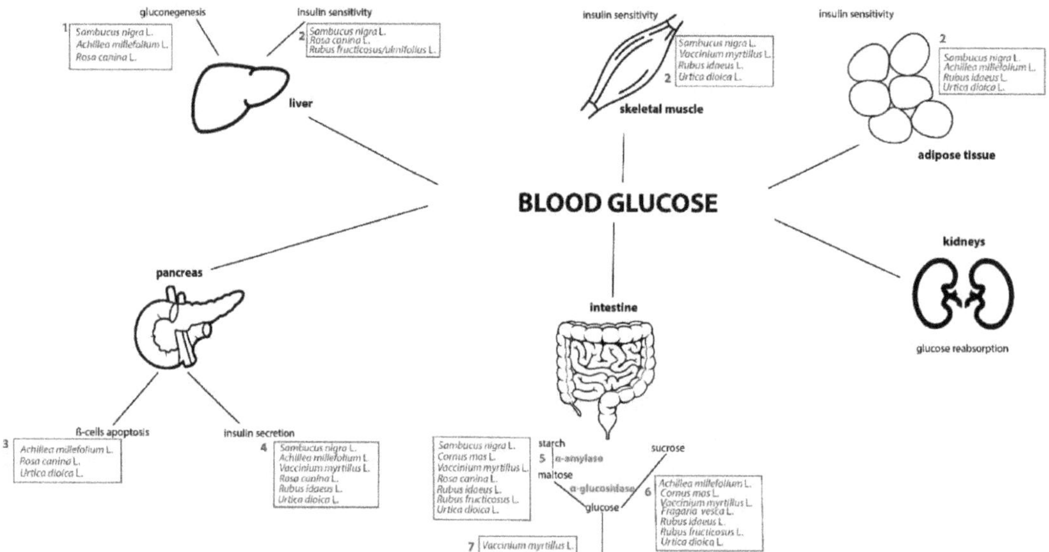

Figure 1. Plants used in the Occitan valleys of the Piedmont Alps: target tissues of their anti-diabetic effects. The selected plants have been suggested to reduce blood glucose levels by acting on various organs and tissues involved in glucose homeostasis. According to the experimental evidence published between 2011 and 2021 they can be grouped into plants promoting: 1 = gluconeogenesis downregulation; 2 = insulin sensitivity increase in liver, skeletal muscle and adipose tissue; 3 = islet atrophy inhibition and/or pancreatic β-cells regeneration promotion (associated with a reduction of inflammatory disease and increased proliferation); 4 = insulin secretion from pancreatic β-cells; 5 = α-amylase inhibition; 6 = α-glucosidase inhibition; 7 = reduction of glucose uptake from the intestinal epithelium.

The evidence of these mechanisms for each plant relies on a different number of studies, with stinging nettle one of the most extensively studied.

4.1. Stinging Nettle

Stinging nettle's hypoglycemic properties have been known for centuries, since the time of Avicenna [175]. In recent decades, several studies have investigated its anti-diabetic effects and the related mechanism of action. All the three events mentioned above have been demonstrated. These effects are ascribed to hydroxycinnamic acid derivatives and flavonoids, quercetin, in particular, being the central bioactive molecule of the aqueous extract [176]. Interestingly, the aqueous extract is the most effective in counteracting hyperglycemia compared to the hexane, chloroform, ethyl acetate, or methanol extracts [92].

The inhibition of intestinal glucose absorption is consistent with the observation of a reduction in glycaemia 1 h after glucose loading paralleled by a significant reduction of glucose absorbed in situ at the jejunum level by an aqueous extract of nettle (250 mg/kg by oral gavage). However, the same extract failed to reduce hyperglycemia in diabetic rats in which diabetes was induced by alloxan [177]. Specific inhibitory activity of the stinging nettle aqueous extracts on α-glucosidase and α-amylase enzymes was investigated and confirmed [95,178]. More recently, seco-isolariciresinol was identified as the phenolic compound responsible for the α-amylase and β-glucosidase inhibition [102]. Focusing on the increase of insulin secretion from the islets of Langerhans already reported in 2003 [179], compelling evidence comes from studies published between 2011–2021 and reporting the ability of stinging nettle to inhibit islet atrophy and promote regeneration of pancreatic β-cells. Indeed, recently, using the model of streptozotocin (STZ)-induced diabetes in rats, it was demonstrated that stinging nettle distillate (12.5 mL/kg/day for 4 weeks by oral gavage) was able to reduce hyperglycemia, increasing serum insulin levels, possibly correlated with the observed increase in β-cells number, an effect that was not shared by glibenclamide (0.6 mg/kg/day) or insulin (3 units/bid of protamine insulin) used as controls [99]. A positive effect of stinging nettle on pancreas proliferation was already observed in 2013 in STZ-induced diabetic rats: stinging nettle dried alcoholic and aqueous extracts induced an appropriate repair of pancreatic tissue [94]. Moreover, the administration of the stinging nettle aqueous extract to diabetic rats undergoing physical activity (swimming exercise) induced pancreatic β-cells proliferation, as demonstrated by hypercellularity revealed by histological analysis. These in vivo observations are in keeping with in vitro observations on RIN-5F pancreatic cells (an insulinoma cell line used as a β-cell model), which displayed a marked increase in glucose-stimulated insulin secretion with stinging nettle aqueous extract at both 1.5 mg/mL and 3 mg/mL [96]. Interestingly, the same study also provides evidence for the insulin sensitivity effect of stinging nettle. Indeed, rats treated with stinging nettle showed a significantly reduced HOMA-IR. Consistently, the extract stimulated in a concentration-dependent manner (1.5 mg/mL and 4 mg/mL) the glucose uptake by immortalized rat skeletal (L6) myoblast cells differentiated into a myotube muscle cell, phenotype naturally expressing GLUT4 [96]. GLUT4 translocation to the plasma membrane in both basal and insulin stimulated state was demonstrated in L6-GLUT4myc cells, a rat muscle cells line, exposed to 125 and 250 µg/mL of nettle extracts [93]. Beneficial effects on glucose uptake were also observed in C2C12 myotubes. Stinging nettle extract at 5 µg/mL counteracted the AKT dephosphorylation induced by free fatty acids and has improved glycogen synthesis. The ex-vivo analysis of gastrocnemius muscle from obesity-induced insulin resistance mice confirmed the effect of stinging nettle on AKT phosphorylation [97], which could be related to decreased activity of cytosolic protein phosphatase 2A (PP2A) [98], a serine/threonine phosphatase activated by ceramides which affect insulin signaling [180,181] inactivating AKT [182]. The activation of the AKT pathway is known to deactivate the glycogen synthase kinase (GSK)-3β, a regulator of glycogen metabolism [183]. Consistently, the stinging nettle extract reduces blood GSK-3β levels, possibly through the activation of the Kirsten rat sarcoma (KRAS) pathway, which in turn activates several downstream signaling molecules, such as PI3K [100] and consequently AKT (Figure 2).

Apart from AKT phosphorylation, the extract (1–10 µg/mL) also improved the expression of adiponectin and the activation of ceramidases, thus decreasing ceramide accumulation, a predictor of T2DM development [184], in 3T3-L1 adipocytes exposed to free fatty acid [97]. Ex vivo experiments demonstrated in a model of high-fat diet (HFD) induced-insulin resistance that stinging nettle increased the expression of genes involved in adipogenesis (peroxisome proliferator-activated receptor (PPAR)-γ, CEBP-α and CD36) [101] and decreased the expression of all markers of lipogenesis and triglyceride synthesis (diacylglycerol O-acyltransferase 1 and fatty acid binding protein 4). In the liver, stinging nettle increases the expression of markers of fatty acid oxidation like PPAR-α, fork-head box (FOX)O1 and carnitine palmitoyl-transferase (CPT)-1a [101]. All these described molecular effects were paralleled by a reduction in weight gain, fasting plasma glucose, insulin level and the HOMA-IR or glucose tolerance, both when stinging nettle supplementation was used as a preventive approach (supplementation started at the beginning of the study) [98] and when used as a treatment approach (after 6 weeks of HFD) [101]. Similar results were obtained administering an hydroalcoholic extract of stinging nettle to fructose-induced insulin resistance rats [91].

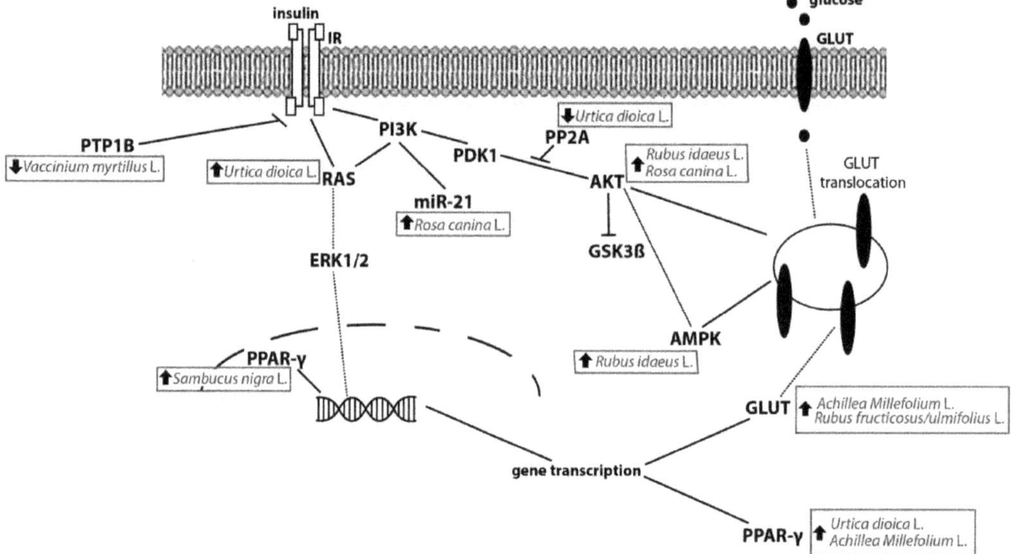

Figure 2. Plants used in the Occitan valleys of the Piedmont Alps: activation of the insulin signaling. According to the experimental evidence published between 2011 and 2021, the selected plants stimulate insulin signaling at different molecular levels. Their principal molecular targets are schematically represented. IR: insulin receptor; GLUT: glucose transporter; PTP1B: protein tyrosine phosphatase 1B; RAS: rat sarcoma; ERK1/2: extracellular signal-regulated kinase 1/2; PI3K: phosphoinositide 3-kinases; PDK1: phosphoinositide-dependent kinase-1; PP2A: protein phosphatase 2A; AKT: protein kinase B; GSK-3β: glycogen synthase kinase-3β; PPAR-γ: peroxisome proliferator-activated receptor-γ; AMPK: AMP-activated protein kinase.

Another exciting approach pursued in evaluating the anti-diabetic effect of stinging nettle deals with the ability of its extract to counteract the different complications associated with diabetes, from memory functions affected by alteration in the muscarinic cholinergic system in the hippocampus [95] to depressive behavior and cognitive dysfunction after dexamethasone-induced hyperglycemia [185], deterioration of seminiferous tubules [186] and even diabetic nephropathy [187]. This latter effect was investigated in vivo with the model of STZ-induced diabetic nephropathy in STZ-diabetic mice, a well-known and reproducible model of diabetes in which renal injury shares similarities to human diabetic nephropathy [188]. The combination of stinging nettle hydroalcoholic extract with pioglitazone showed a synergic effect on the reduction of blood glucose and prevention of renal dysfunction, measured in terms of serum urea and creatinine. The mechanism suggested for these effects was related to the anti-oxidant properties of the plant: stinging nettle significantly reduced reactive oxygen species (ROS) formation in renal supernatant, levels of malondialdehyde (MDA) and protein carbonyl and oxidation of glutathione (GSH) [187]. Interestingly, the stinging nettle extract administered to STZ rats attenuated the diabetes-related cardiac dysfunction and improved the mitochondrial function in the cardiac muscle by restoring NRF2 and the peroxisome proliferator-activated receptor-gamma coactivator (PGC)-1α deficits. These results suggested that stinging nettle could have a protective effect on cardiac tissue, thus minimizing the incidence of diabetic cardiomyopathy [189].

4.2. Yarrow

Well-characterized both in vitro and in vivo is also yarrow, for which different molecular pathways were investigated. Yarrow, thanks to its multiple active ingredients, ranging from monoterpenes and sesquiterpenes and phenolic compounds [67–72] to alkaloids, [73,74] showed different mechanisms of action that can contribute to the anti-diabetic effect demonstrated in several in vivo studies. The first common effect was β-cells protection, possibly associated with a reduction of inflammatory mediator release. Using the T1DM model of alloxan-induced diabetes in rats, it has been demonstrated that both an aqueous and a methanolic extract of yarrow at 500 mg/kg i.p. partially restored the normal cellular population and enlarged size of diabetic β-cells, with comparable effects despite the different yield of the two extracts (2.6% and 7.3% for aqueous and methanolic extracts, respectively) [103]. Interestingly, the effects shown after 14 days of treatment were comparable to those of glibenclamide (30 mg/kg i.p.). The protective effect on the pancreas was paralleled by the significant increase in glucose tolerance and reduced fasting glycemia, where the methanolic extract seemed to be the most effective [103]. These beneficial effects could be related to the reduction of interleukin (IL)-1β and inducible nitric oxide synthase (iNOS) mRNA elicited by yarrow [77]. Indeed, both IL-1β and iNOS can be considered effectors of β-cell destruction in T1DM, which are thought to act as early inflammatory signals [190]. A hydroalcoholic extract (80% of ethanol) of aerial parts of yarrow 100 mg/kg/day i.p. for 14 days caused a statistically significant reduction of STZ-induced hyperglycemia, with insulin serum levels restored to negative controls [77]. The authors suggest that the effects could be due to the protection of β-cells promoted by the downregulation of IL-1β and iNOS. Moreover, the pancreatic protection could be due also to the over-expression of PPARγ. Indeed, PPARγ activation has been reported to decrease the expression of inflammatory mediators, such as IL-1β and iNOS [191]. The hypothesis of PPARγ contribution comes from the observation that the hydroalcoholic extract induces the over-expression of PPARγ and that of the glucose transporter GLUT4 (Figure 2) in the preadipocytes cell line 3T3-L1. These events could drive the beneficial effect registered for insulin sensitivity in vivo, where this extract at 100 mg/kg i.p. was able to improve the oral glucose and sucrose tolerance tests in a model of chemically induced T2DM (nicotinamide 20 mg/kg i.p. followed by STZ 120 mg/kg i.p.) in CD1 mice [105]. These data highlight the potential insulin-sensitizing action of this plant. Simultaneously, using RINm5F cells, the same authors demonstrated that the hydroethanolic (70% of ethanol) extract of aerial parts of yarrow at 200 µg/mL induced an increase in intracellular Ca^{2+} concentration and accordingly of insulin secretion comparable to that of glibenclamide [105]. Consistently, the same study demonstrated, using the OGTT, that yarrow elicits a hypoglycemic response in normoglycemic CD1 mice, thus supporting that it could act as an insulin secretagogue. However, the possibility of a direct effect on the α-glucosidases was also evaluated in vitro on homogenate of Sprague Dawley rats' intestinal brush border: the yarrow extract inhibited the α-glucosidases activity by 55% [105], confirming previous data [104]. However, the anti-diabetic effects of yarrow could also be ascribed to a more general anti-oxidant potential of the plant observed when exploring the protective effect of a hydroalcoholic extract of yarrow (300 g of plant in 70% ethanol) on lipid profile, blood glucose levels, body weight and serum liver enzymes in a model of T1DM, STZ-induced, in rats [192]. The oral gavage with the yarrow at 50 mg/kg/day or 100 mg/kg/day for 28 days increased the body weight of both non-diabetic and diabetic rats [192]. Similar effects were already reported [103]. These data should be interpreted according to the model used: STZ-model causes a significant reduction in the body weight of diabetic animals, one of the typical T1DM features [193–195]. In the same model, yarrow supplementation at 100 mg/kg/day reduced blood glucose levels to a similar extent of metformin at 250mg/kg/day [192]. Comparable results were obtained in an HFD-induced insulin resistance model using a CO_2 supercritical fluid extract from yarrow. The extract at 800 mg/Kg, five times per week for 3 months, improved fasting glycaemia, oral glucose tolerance, insulin levels and HOMA-IR [196]. The beneficial effects on hyperglycemia and hepatic enzymes were paralleled by ameliorating lipid profile [103,192,196–198]. These effects were, at least in part, ascribed by the authors to the inhibition of fatty acids synthesis. However, the plant's accurate and complete mechanism of action still needs to be fully elucidated.

4.3. Bilberry

There is much evidence of the anti-diabetic properties of plants rich in anthocyanins, such as bilberry, blackberry and raspberry.

The use of bilberry leaf for diabetes was widespread at the beginning of the 20th century, before the discovery of insulin, was one of the most frequently used herbal anti-diabetic remedies [110]. Many bilberry leaf extracts are available on the market and bilberry leaf tea is still recommended and sold by more than 40% of herbalists to manage diabetes [110]. Its efficacy has been confirmed in vivo [108]. In this study, the dietary supplementation with bilberry fruit powder (2 g/day by oral gavage) for 4 weeks significantly reduced glycemia in diabetic rats. The effect on hyperglycemia was comparable to that of glibenclamide at 0.6 mg/kg i.p. Still, bilberry was also effective in improving the lipid profile [108]. Similar beneficial effects were observed for the bilberry leaf extract (2% for 50 days) in a model of diabetes induced by STZ in combination with a high-carbohydrate diet in Wistar rats with diabetes [199]. The effect could be due to the anthocyanin components, as suggested by using the model of Zucker fatty rats (ZDF) [200], which spontaneously develop obesity and insulin resistance, progressing into hyperglycemia with impaired β-cell function and decreased responsiveness to insulin and glucose [201]. In the same experimental model, bilberry was demonstrated to be effective only on the total and LDL-cholesterol levels, not on the glycemic profile [106]. These contrasting results could be due to the experimental protocol adopted with bilberry administered as a 15% substitute for the standard diet [106], or as a non-acilated anthocyanin extract from bilberry at 25–50 mg/kg/day [200].

Notably, the beneficial effects of bilberry on diabetes could be far beyond the anti-hyperglycemic: the supplementation with bilberry extract (100 mg/kg) orally administered to STZ-induced diabetic rats for 6 weeks could decrease markers of diabetic retinopathy, such as retinal vascular endothelial growth factor expression, even though it did not affect blood glucose levels and body weight [107]. The data reported is in keeping with a protective effect on retina [202]. The retina protection observed is not surprising for bilberry. At the same time, pilot data on the beneficial effect on hippocampal neurons/nerve fibers plasticity [203], if confirmed, could represent an important milestone in determining the functional role of bilberry supplementation in diabetes. On the contrary, using a model of chemically induced T2DM (the STZ+nicotinamide non-insulin-dependent model), no protective effects were observed on diabetic gastroenteropathy [204].

The direct anti-diabetic activity could be correlated to the ability of bilberry extracts to inhibit the activity of carbohydrate-hydrolyzing enzymes such as α-glucosidase and α-amylase, thus reducing postprandial hyperglycemia. This effect, which may be related to flavonoids [112], was demonstrated in different studies. The inhibitory activity of α-glucosidase of phenol-rich extracts obtained from bilberry fruits from different Bulgarian mountains have been tested [111]. The comparative evaluation of water (total polyphenolic content in extracts expressed as gallic acid equivalents, gallic acid equivalent (GAE)/100 mg, ranging from 290.83 ± 8.72 to 453.63 ± 17.23), acetone (GAE/100 mg ranging from 510.88 ± 16.10 to 636.46 ± 22.90) and methanol (GAE/100 mg, ranging from 548.30 ± 16.75 to 1299.60 ± 32.46) extract highlighted that the methanol extract was the most efficient in inhibiting the α-glucosidase activity, probably because of the highest amount in gallic acid, in accord with a previous report [169]. Leaf extracts showed good anti-oxidant power, too and the hydroethanolic one significantly inhibited α-glucosidase activity, with an IC_{50} value equal to that of acarbose [110]. On the other hand, the activity of leaf extract on α-amylase was not straightforward. Both aqueous and hydroethanolic extracts were unable to inhibit α-amylase activity [110]. However, bilberry fruits showed a positive response in previous screening [109]. In addition, recent evidence demonstrated that methanol or aqueous extract of bilberry fruits could inhibit 80% of the glucose uptake in Caco-2 cells [113].

The anthocyanin component, more than the flavonoid or phenolic, contributes to the beneficial effects of bilberry, with its potent anti-oxidant activity. Indeed, a comparative study on the anti-oxidant activity of hydroethanolic (75% of ethanol) extracts from blueberries, bilberries, cranberries and mulberries [112] demonstrated that the anti-oxidant activity follows the anthocyanin content: blueberry > bilberry > mulberry > cranberry. A parallel hierarchy was also found for the inhibitory effect on protein tyrosine phosphatase 1B (PTP1B), a negative regulator of the leptin and insulin signaling pathways [205] involved in insulin resistance [206]. Therefore, it is possible to speculate that bilberry could improve glucose homeostasis by inhibiting PTP1B (Figure 2), as muscle-specific PTP1B$^{-/-}$ mice exhibit improved muscle glucose uptake, systemic insulin sensitivity and glucose tolerance [207]. On the contrary, mulberry was the most effective in α-glucosidase activity [112].

4.4. Raspberry

Apart from bilberry, anthocyanins are particularly abundant in plants belonging to the Rosaceae Family, *Rubus* genus. Looking at raspberry, the anthocyanins are responsible for the berry's brilliant red color (also known as red raspberry). Red raspberry is rich in dietary fiber (6 g in 1 cup of frozen whole fruit) that may provide glycemic control benefits [208]. However, the plant also contains different micronutrients, such as folic acid and vitamins C and K, magnesium, potassium, calcium and iron. Still, it is poor in carbohydrates and consistently has a low glycemic index that supports its possible health-promoting benefits in diabetes. Alongside anthocyanins, other polyphenols, such as flavanols, flavonols and phenolic acid, such as ellagitannins [209], are the main bioactive compounds found in red raspberry. Altogether, these active ingredients may participate in the anti-diabetic effect of red raspberry. The proposed mechanism for the modulation of postprandial glycemia following consumption of dietary berries or their extracts is the inhibition of the enzymes α-amylase, even higher than realized by other berries, and α-glucosidase to prevent glucose absorption in the intestines [115,118]. Similar effects were registered in vitro for another species of the *Rubus* genus, *Rubus grandifolius* Lowe, an endemic species from the Madeira archipelago, whose leaf (composed of 44–49% of ellagitannins, followed by hydroxycinnamic acids, flavonols and flavones) and berry (anthocyanins 54%, ellagitannins 37–39%, flavonols 8–9% and hydroxycinnamic acids 0.7–0.9%) extracts strongly inhibited yeast α-glucosidase, showing lower IC$_{50}$ values than acarbose, with leaves being the most active samples [210]. The microbial metabolite of ellagitannins/ellagic acid, urolithins, were shown to play a role in glucose homeostasis [211]. However, anthocyanins are particularly involved in the beneficial effects of red raspberry and other *Rubus* species, including improved insulin response, glucose and lipid metabolism, and anti-oxidant and anti-inflammatory properties. Anthocyanins may enhance insulin secretion from pancreatic β-cells, thus improving insulin sensitivity [110]. A key role in these events is ascribed to AMPK, highlighted using knockout mice for the catalytic subunit of AMPK (AMPKα1$^{-/-}$ mice) [116]. An HFD supplemented freeze-dried raspberry (5% of dry fruits, containing polyphenols at 11 g gallic acid equivalent/kg of dry weight) for 10 weeks had beneficial effects on body weight and expression of pro-inflammatory cytokines, such as TNF-α, IL-1β, IL-6 and IL-18, only in wild-type animals and not in the AMPKα1$^{-/-}$ counterpart. Moreover, raspberry supplementation reduced lipid accumulation in skeletal muscle and increased GLUT4 mRNA and proteins (Figure 2). These effects were observed only in wild-type mice, in which raspberry induced a significant increase of phospho/total-AKT ratio [116], a well-known downstream effector of insulin signaling. Raspberry supplementation (5% freeze-dried fruits for 12 weeks) in HFD C57BL/6J obese mice activates the AMPK/Sirt1 pathway and increases the FNDC5/irisin contents, thus decreasing white adipose tissue hypertrophy, suppressing pro-inflammatory cytokines expression and macrophage infiltration in white adipose and improving insulin sensitivity [117]. Consistently, using a similar animal model, the consumption of the equivalent of a single daily serving of either raspberry puree concentrate or raspberry juice concentrate for 10 weeks in C57BL/6J mice fed an HFD was demonstrated to ameliorate the symptoms of metabolic syndrome, by inducing, among other effects, a significant reduction of glucose and insulin levels in comparison to HFD animals [212]. Interestingly, the preventive supplementation of raspberry improved insulin resistance and metabolic dysfunction in diet-induced obesity. These effects were associated with the suppression of the nucleotide-binding oligomerization domain, leucine-rich repeat and pyrin domain-containing (NLRP)3 inflammasome elicited by HFD [114]. The ability of raspberry to counteract diet-induced obesity and metabolic dysfunction, particularly to inhibit NLRP3 inflammasome, was investigated by comparing the effects of polyphenol fractions from pulp, whole fruit and seed [213]. Their results suggested that supplementation with raspberry polyphenols from pulp and, to a lesser extent, from entire fruit effectively prevents HFD-induced weight gain, insulin resistance and energy expenditure. The data is consistent with the inhibitory effect on NLRP3 in adipose tissue macrophages of epididymal white adipose tissue, which was higher with the pulp polyphenol fraction and less extensive for polyphenols from whole fruits. Notably, anthocyanins are the primary constituents of pulp polyphenols, while the seed fraction contains the highest amount of ellagic acid and epicatechin. Unexpectedly, the supplementation with seed polyphenol fractions failed to exert metabolic benefits. However, these results highlight anthocyanins' properties and crucial role in this extract [213].

4.5. Blackberry

Similar protective effects against hyperglycemia from anthocyanins were also recorded for blackberry. The extracts from its berries were tested as anti-hyperglycemic agents on diabetic rats (using the T1DM model by STZ) after anthocyanins extraction with 80% acidified ethanol in ultrasonic conditions at room temperature [214]. The extract (administered for 5 weeks in drinking water to obtain a concentration expressed as cyanidin-3-glucoside of 300 mg/L) determined a significant decrease in glucose level from 360 to about 270 mg/dL [214]. Previous findings suggested that the hypoglycemic effect of the blackberry is independent of the elevation of insulin secretion but is likely mediated by extra-pancreatic actions, such as the increase of glucose uptake from peripheral tissues [215]. This hypothesis was supported by the results obtained administering a *Rubus* extract (blackberry Loch Ness variety, obtained by lyophilization and extraction with 80% methanol in 0.1% HCl) in rats fed a standard diet or cafeteria diet [119], a hypercaloric diet model based on the recapitulation of oro-sensory properties and palatability of food promoting its overconsumption [216]. Interestingly, the authors showed that blackberry extract, in rats given a standard diet, decreased glycemia and increased insulin sensitivity, as demonstrated by the reduction in the HOMA-IR, with a sex-dependent response greater in females than in males. In parallel, animals showed decreased plasma glucose and insulin and increased plasma non-esterified fatty acids, glycerol and 3-hydroxybutyrate and triacylglycerols concentrations. However, no effect was seen in rats fed the cafeteria diet, probably due to their high insulin resistance [119].

In parallel, blackberry Schott variety was tested in vitro on the human hepatoma cells line HepG2. The glucose consumption in HepG2 cells was increased in a concentration-dependent manner by the different blackberry extracts: blackberry ethanol extract, gastrointestinal digested blackberry and gut metabolites of blackberry collected at 24 and 48 h fermentation. A similar significant anti-oxidant activity, most probably due to the phenolic metabolites, was observed in cells exposed to oxidative stress induced with high glucose (30 mM) plus palmitic acid (0.2 mM) for 24 h. Indeed, all the extracts decreased the ROS content and restored the cellular GSH and mitochondrial membrane potential [46].

4.6. Wild Strawberry

Still belonging to Rosaceae, a plant rich in anthocyanins is wild strawberry, which also contains high amounts of vitamin C, folate and phenolic constituents [5]. The anti-tumor, anti-oxidant and anti-diabetic properties of its fruits have been investigated. Standard mechanisms are based on their anti-oxidant and anti-inflammatory properties. A hydroal-coholic extract from wild strawberry (160 µg/mL) exerted on the macrophage cell line Raw 264.7 a direct nitric oxide scavenging activity. Moreover, the proteasome activity was inhibited and the conversion of the microtubule-associated protein light chain LC3-I to LC3-II, suggesting autophagy, was promoted [51]. The literature on the diabetic effect of strawberries is based on different plants belonging to the *Fragaria* genus. For instance, one of the most studied is *Fragaria* x *ananassa*, which differs from *Fragaria vesca* L. for the composition of volatile esters (acetate in *Fragaria vesca* L. and ethyl hexanoate in *Fragaria* x *ananassa*) [217,218]. The freeze-dried strawberry powder obtained from *Fragaria* x *ananassa* was tested as a supplement (700 mg/kg/day for 45 days) in diabetic rats (T1DM according to the alloxan model) [219]. The extract reverted hyperglycemia and insulin levels to near normal in that study.

Although the mechanism underlying these effects remains to be completely clarified, evaluating the strawberry extract's ability to inhibit metabolic syndrome-related enzymes in vitro showed positive results. In particular, the extract inhibited the α-glucosidase and α-amylase enzymes, thus supporting that this mechanism could underlie, at least in part, its anti-diabetic effect [118,120]. However, another study failed to demonstrate any decrease in hyperglycemia in a model of HFD rats following supplementation with wild strawberry fruit ethanolic extract up to 500 mg/kg/day for 30 days [220].

4.7. Elderberry

Elderberry represents again a plant with fascinating anti-diabetic properties, whose mechanism has been previously compared to that of sulfonylureas, having similar insulin-releasing action counteracted by diazoxide [221]. The beneficial effects of elderberry are related to phenols and anthocyanins, whose amount in fruits is even higher than in other plants [61]. Notably, these active principles are heat-stable, acetone-insoluble and unaffected by altered pH environment [222]. Apart from promoting insulin secretion from pancreatic cells, the aqueous extract of elderberry has insulin-sensitizing properties. It is well characterized in the muscles, where it can directly stimulate glucose uptake and glucose metabolism [222,223]. These observations have been confirmed in in vitro study on pig primary myotube cells [122]. The authors showed that the dichloromethane and the methanol elderflower extracts are both dose-dependent (6−1000 µg/mL) and able to increase glucose uptake, with the dichloromethane extract being more effective. Similarly, it was demonstrated that the 96% ethanol and the acidic methanol extracts of elderberry at 50 µg/mL increase glucose uptake in human skeletal muscle cells by 37.4 ± 4.8 and $42.6 \pm 6.7\%$, respectively. Notably, both extracts were strong α-glucosidase and α-amylase inhibitors, maybe to their high content of polyphenols. The further evaluation of the anti-oxidant activity of these extracts evidenced that the acidic methanolic extract (IC_{50} 95.9 ± 3.9 µg/mL) showed the highest radical scavenging activity, followed by pressed juice and the 96% ethanol extract (IC_{50} 128.3 ± 4.3 µg/mL) [125]. The same authors achieved similar results on HepG2 cells [126]. Muscle and liver are not exclusive targets of the elderberry, which also affects adipocytes. demonstrated that elderberry extracts increases glucose uptake in mature, fully differentiated insulin-resistant 3T3-L1 adipocytes more efficiently than rosiglitazone. In parallel, the authors observed a reduction in ROS and an improvement in adipokine balance: leptin is down-regulated and adiponectin is up-regulated [127]. This could promote a non-obese phenotype, supporting the observation of a reduction of fat accumulation in Caenorhabditis elegans treated with elderflower extract [122]. Together, these observations strengthen the hypothesis that this plant could exert beneficial effects on glucose utilization. Nevertheless, according to the above mentioned results [122], the possible induction of cellular stress as the mainstream mechanism of elderberry extracts on glucose uptake can be discharged. Evaluating the effects of elderflower extracts on the transcription regulation of heme oxygenase 1 (HMOX1, a marker of oxidative stress response) and heat shock protein 70 (HSP70, whose expression is induced by different kinds of cellular stress conditions), no indication of cellular stress on the myotubes were observed. In contrast, dichloromethane extract showed significant anti-oxidative properties by reducing the amount of intracellular ROS. Interestingly, PPAR-γ has been implicated in the insulin sensitization effects exerted by elderberry: the extract activated PPAR-γ (Figure 2) in COS-1 cells (fibroblast-like cells isolated from the kidney of an African green monkey), probably because of the presence of α-linolenic acid, linoleic acid and polyphenols such as naringenin [121]. However, in vivo contrasting results were reported. Female db/db-mice supplemented for 4 weeks with a methanol extract of elderflowers (0.1 and 1 g/kg/die), placebo, or rosiglitazone (0.02 g/kg/die) had no benefit from elderflower extract in terms of fasting plasma glucose, adiponectin and cholesterol levels [121].

On the contrary, the supplementation for 4 weeks with a lipophilic and polar extract (190 and 350 mg/kg/die, respectively) decreased fasting blood glucose, modulated insulin secretion and improved insulin resistance, as demonstrated by the HOMA-IR index in Wistar STZ-induced diabetic rats fed with an HFD. The comparison between the two extracts indicated the polar one as the most active on fasting blood glucose and the lipophilic on plasma insulin levels [124]. Similar results were achieved in HFD-fed mice supplemented with an anthocyanin-rich black elderberry extract displaying a reduction in fasting insulin, HOMA-IR index and fasting serum triacylglycerol. Systemic inflammation was attenuated as serum MCP-1 and TNFα were reduced by 30% [123,224].

Some observations suggest a potential role for black elderberry also in diabetes complications. Acute and sub-chronic in vivo effects of elderberry were evaluated in an alloxan-induced diabetic mouse model of neuropathic pain. The study confirms elderberry's hypoglycemic effect: the extract, containing the most active fraction of quercetin, kaempferol and isorhamnetin, reduced the blood glucose level both acutely (after 6 h) and chronically (on the 8th day of treatment). Moreover, a significant increase in body weight (suggesting an improvement of the catabolic deterioration associated with the alloxan-diabetes model mimicking the T1DM) was registered. Interestingly the elderberry extract at 50–200 mg/kg acutely improved the tactile allodynia and chronically ameliorated thermal hyperalgesia in a dose-dependent manner [225]. In the same year, the effects on immune system imbalance of a dried and powdered fruit extract obtained using acidulated methanol were studied in Wistar white male rats with diabetes induced by the injection of a single dose of STZ [226]. Elderberry extract reduced TNF-α and interferon (IFN)-γ levels in diabetic animals. It limited the production of fibrinogen, thus evidencing an anti-inflammatory effect helpful in reducing the development of microvascular diabetic complications. Consistently, in activated RAW 264.7 macrophages, the extract of elderberry down-regulated the expression of proinflammatory genes such as TNF-α, IL-6, COX-2 and iNOS [127]. Altogether, these findings suggested a potential role of elderberry in reducing insulin resistance, lipo-toxicity and inflammation, with a possibly beneficial effect on both prevention of diabetes and its complications.

4.8. Rosehip

Used as a folk remedy in treating diabetes in Turkey [227], rosehip has been extensively studied for its anti-inflammatory, anti-obesity and anti-diabetic activity [57,228,229]. Two primary mechanisms are proposed for the anti-diabetic properties: the inhibition of α-amylase enzymes and the stimulation of insulin secretion by protecting pancreas damage. The first mechanism was pointed out evaluating the anti-oxidant activity and α-amylase inhibition of a methanolic extract (1:10; total phenolic content expressed in GAE = 21.918 mg/g and complete flavonoids as rutin equivalents (RE) = 2.65 mg/g) of pseudo-fruit and flowers from rosehip [130]. Results of the preliminary agar diffusion of amylase inhibition assays indicate that rosehip extracts exhibit a complete inhibition of α-amylase enzyme; that was confirmed by the α-amylase inhibition quantitative assay, with a 100% inhibition for extract concentration of 5.5 mg/mL. This effect was ascribed to vitamin C, one of the active ingredients of rosehip alongside tocopherols, carotenoids, phenolic, organic acids, sugars and essential fatty acids [57]. Consistently, a significant free radical scavenging activity was observed [130].

Both in vitro and in vivo studies demonstrated the protective effect of rosehip on β-cells. An aqueous extract of ripe fruits at 0.001 mg/mL significantly enhanced cell number and morphological alterations in pancreatic βTC6 cells, acting as a growth factor for the pancreatic β-cell line [129]. This could contribute to the beneficial effect of rosehip in diabetes, although no significant protective effect was observed after STZ cell exposure. However, STZ cytotoxicity mimics alterations in T1DM, but the proliferative effect observed on βTC6 cells suggests that this plant could be effective when β-cells are still at least partially preserved, as in at least the early phases of T2DM. Moreover, the amount of active ingredients, especially the oligosaccharide fraction, could account for that. Indeed, in vivo demonstrations again favor a beneficial effect on the pancreas in models of STZ-induced diabetes. The daily oral administration of a hydroalcoholic extract (70% ethanol) of rosehip (5.9% ± 0.4% of gallic acid equivalent/100 g fraction) at 250 mg/kg or 500 mg/kg for 6 weeks was effective as glibenclamide (600 μg/kg) in reducing fasting blood glucose levels, improving in the pancreas the necrotic islets and increasing the number of islets [128]. In parallel, a reduction in serum triglycerides has been observed. This data is consistent with the findings that treating the hydroalcoholic extract of rosehip improves liver histology and liver enzyme activity compared to untreated diabetic rats [230]. Interestingly, rosehip extract showed positive effects if administered to animals with pre-diabetic status. In particular, it was demonstrated that spontaneously diabetic Torii rats supplemented with rosehip extract at 100 mg/kg body weight for 12 weeks preserved pancreatic β-cell function, promoted insulin secretion, improved glucose tolerance and suppressed advanced plasma glycation end-products formation [131]. These effects are probably due to the oligosaccharide fraction in rosehip extract. Indeed, the authors purified and characterized this fraction and demonstrated that the supplementation with isolate oligosaccharide (8–40 mg/kg bid) for 21 days to STZ-diabetic rats increased β-cell mass, suggesting the regenerative potential of the oligosaccharide fraction from rosehip [133]. An analysis of the expression of genes involved in the function and proliferation of β-cells showed that the extract also increases *Ngn3*, *Nkx6.1* and *Ins* expression. Moreover, this supplementation decreased blood glucose levels, gluconeogenesis and α-glucosidase activity and improved the OGTT test. Similarly, the purified oligosaccharide (2 mg/kg) was even more effective than the crude extract of rosehip (40%) in decreasing fasting blood glucose in diabetic rats. Oligosaccharides also increased blood insulin levels and the number and size of pancreatic islets, indicating an increase in the number of β-cells [132].

Moreover, oligosaccharide was able to increase insulin secretion similarly to glibenclamide. Molecular analysis revealed that oligosaccharides increased the expression of *Ins1* and pancreatic and duodenal homeobox 1 (*Pdx1*) insulin-producing genes, which led to an increase in the expression of insulin-dependent genes such as *Gck* and *Ptp1b*. On the other hand, the expression of the solute carrier family 2 member 2 (*Slc2a2*) gene is related to the GLUT-2 transporter [231], was significantly reduced by increased insulin concentrations. Finally, the same authors deepened the molecular mechanism of the anti-diabetic effect of the oligosaccharide fraction from rosehip, focusing on its effect on DNA methylation, whose alteration has been reported in diabetic humans and animal models [128]. Diabetic rats receiving isolated oligosaccharides (10, 20, or 30 mg/kg) twice daily for 4 and 8 weeks showed a significant increase in body weight and a decrease in blood sugar level compared to their untreated counterpart. In parallel, treated groups displayed reduced DNA methylation, as demonstrated by the decreased expression of DNA methyltransferases (DNMT) such as *Dnmt1* and *Dnmt3α* in both pancreas and the blood. On the other side, Dnmt3β expression was reduced in the pancreas and increased in the blood, suggesting that oligosaccharides may interact with some upstream signaling factors of DNMT. Furthermore, the increased expression levels of the *Pdx1*, *PTP1B2*, *Ins1* and *Gck* genes observed in the oligosaccharide-treated group may be related to the hypomethylation effect of this fraction in diabetic rats [232]. Interestingly, the evidence against a possible role of rosehip on the metabolism of glucose by hepatic cells was provided. Indeed, the extract did not cause a significant glucose-lowering effect in the human hepatocellular carcinoma cell line exposed to hyperglycemia (11.1 mmol/L) [129]. On the contrary, an additional mechanism of action for rosehip oligosaccharides could be related to the modulation of the autophagic pathway. The activation of autophagy has been associated with the downregulation of the microRNAs (short non-coding RNAs involved in silencing RNA targets to regulate gene expression and cellular function) miR-21 and miR-22, in particular [233,234]. These two microRNAs have also been implicated in diabetes mellitus [235]. Using Rin-5F, a secondary clone of the rat islet tumor cell line RIN-m used as a model of pancreatic β-cells, the oligosaccharide fraction from rosehip was demonstrated to increase the expression of autophagy markers, such as miR-21, miR-22, AKT (Figure 2), autophagy-related 5 (ATG5), Beclin1, LC3A and LC3B, suggesting that rosehip anti-diabetic effects are mediated at least in part by the modulation of the autophagy pathway [236].

The anti-diabetic effects of cornelian cherry fruits seem to be related to their significant amounts of bioflavonoids (anthocyanins), ursolic acid and vitamin C, in particular [152,237–240]. Anthocyanins have been recognized to exert anti-diabetic effects increasing insulin sensitivity, protecting hepatic function and exercising a general anti-oxidant effect.

4.9. Cornelian Cherry

The cornelian cherry ability to reduce hyperglycemia is well documented and consistent between different models such as the alloxan model [241], the STZ model [135] and the T2DM model of ZDF [134]. However, the increase in insulin sensitivity by cornelian cherry is under debate. The effects of red and yellow fruit extracts [20 mg/kg/day (mean of total polyphenols content of about 350 GAE/g and mean anti-oxidant activity of about 2.70 Trolox equivalent/g), by oral gavage for 14 days] was evaluated in STZ-induced diabetes rats. However, the same authors did not confirm these positive results when administered a pure loganic acid extract from cornelian cherry being this the main iridoid glycoside in the cornelian cherry fruits [135]. The authors specified in the discussion that it is possible that the 20 mg/kg/day for 14 days dosage of pure loganic acid extract could be too low to be effective. Similar effect on hyperglycemia, but not on HbAc1 or insulin sensitivity were reported by Capcarova et al. [134]. However, insulin sensitivity was measured in terms of HOMA-IR instead of OGTT. Moreover, it is essential to underline that the study from Dzydzan et al. [135] was focused on the T1DM model, while that from Capcarova et al. [134] used the T2DM model of ZDF rat. Also, time and dosages were different: 20 mg/kg/day for 14 days [135] versus 1000 mg/kg/day for 10 weeks [134], respectively. Another investigated mechanism underlying the anti-hyperglycemic effect of cornelian cherry extract is the ability to inhibit the α-glucosidase and α-amylase enzymes. Loganic acid, cornuside, cyanidin 3-galactoside and pelargonidin 3-glucoside were identified as the most dominant compounds, presenting $\geq 90\%$ of all detected iridoid and phenolic constituents in the extracts and pointed out pelargonidin 3-robinobiosby analyzing ide as the most potent inhibitor of α-glucosidase [137]. A 2 g/day homogenous powder of cornelian cherry for 4 weeks was also ascribed to liver beneficial effect, reducing hepatic inflammation in diabetic Wistar rats [241]. This anti-inflammatory effect could be related to the anti-oxidant property of the plant extracts. They induced an increase of GSH and thio-barbituric acid reactive substances (TBARS), a decrease in protein carbonyl groups and advanced oxidation protein product levels, all markers of oxidative stress and subsequent damage to proteins [135]. The loganic acid extract, in particular, increases the GSH content, GPx and glutathione reductase in leukocytes, thus reducing ROS production [136].

5. The Anti-Diabetic Effects of Plants from the Occitan Valleys of the Piedmont Alps: Clinical Evidence of Efficacy

The clinical studies demonstrating the anti-diabetic efficacy of Alpine plants are scant and sparse. As expected, the Alpine extracts were tested in T2DM patients as supplements. Consistently, the tested common hypothesis is that these plants could be effective as an add-on-therapy of T2DM in a combinatorial approach with traditional anti-diabetic drugs to improve their efficacy.

As observed for experimental studies, the number of trials published between 2011 and 2021 is highly variable, with no analysis, to the best of our knowledge, on the use of elderberry in diabetic patients. Similarly, in 2011 the lack of human trials on yarrow was pointed out [65]. Nowadays, to the best of our knowledge, on the clinicaltrials.gov register, only one study aims to evaluate the use of this plant in diabetic patients. The study, completed in February 2019 (ClinicalTrials.gov Identifier: NCT04054284), seeks to assess the safety and efficacy of an herbal tea mixture in T2DM, including yarrow among the components of the mix without anti-diabetic properties (control treatment) in comparison with a combination with anti-diabetic properties and including stinging nettle and bilberry. The data has not been published yet, but the use of yarrow as a plant without anti-diabetic properties should be carefully interpreted, considering the type of preparation (herbal tea), the exact dosage and the active ingredient titration.

5.1. Stinging Nettle

The number of clinical trials on stinging nettle underlines the interest in the therapeutic potential of this plant. As reported by a recent meta-analysis, the effect on fasting glycaemia in T2DM is apparent [242]. It is reasonable to expect similar results from the ongoing study NCT00422357, investigating the hypoglycemic effects of stinging nettle tea bags in T2DM patients. On the contrary, no significant reduction was observed in insulin levels, HOMA-IR index, or HbA1c percentage [242]. Conflicting results have been reported from the different trials included. As written by the authors, the discrepancy is almost on biochemical outcomes, which are favorable for some [243–248] and unaffected for others [249,250]. For instance, it was observed that adding to the standard anti-diabetic therapy (mainly metformin or glibenclamide) hydroalcoholic extract of stinging nettle (45% ethanol, 55% water and 2.7 g of dry plant in 1 L extract) 100 mg/kg in 3 portions a day (after each main meals) for 8 weeks, was not superior to placebo in affecting insulin concentration and insulin sensitivity in patients of both genders (29.2% man and 49.3% women) with T2DM (mean duration of diabetes at baseline more than 8 years, HbAc1 $\leq$ 10%) [251]. In parallel, no differences were registered in circulating TNF-α, but the extract significantly reduced serum IL-6 and high-sensitivity C-reactive protein (hs-CRP) [251]. However, the clinical results confirm the anti-oxidant activity of the extract that increased the total anti-oxidant capacity and SOD, although no significant differences were registered for MDA and GPx [252]. Similar results on SOD levels were also reported in a third randomized double-blinded clinical trial involving just women over 50 years old affected by T2DM (mean duration of diabetes 13 years, HbA1c $\leq$ 10%) who received, as add-on therapy, placebo or the hydro-alcoholic extract of stinging nettle 100 mg/kg in 3 portions a day for 8 weeks [253]. Nitroxide levels were also increased in the treated group, suggesting that the extract could prevent the development of cardiovascular complications [254], as indicated by in vivo data cited above [187].

5.2. Wild Strawberry

Another plant widely studied in clinics is the wild strawberry. Notably, in the period included in the present review (2011–2021), very few experimental and preclinical papers were found in the PubMed database, while a plethora of studies demonstrated that consuming strawberries attenuates unfavorable postprandial responses or reduces the cardiovascular risk in subjects with metabolic syndrome clinical trials [255–258]. Moreover, a search on ClinicalTrial.gov in June 2022 retrieved five studies for wild strawberry as active treatment and diabetes: the NCT05362968, just approved and not yet recruiting; the NCT02610179, recruiting; and the completed studies NCT01199848 [259,260], NCT02607007 and NCT01766570 (no results available for both). This could reflect a higher stage of development of a strawberry-based approach as an add-on therapy for diabetes compared to other Alpine plants. For instance, a randomized, single-blind, placebo-controlled, cross-over trial demonstrated in overweight adults that the aqueous acetone strawberry extract (70%) containing 10 g of freeze-dried powder 81.6 mg/of anthocyanins (54% pelargonidin-3-O-glucoside) and 94.7 mg of phenols was effective in reducing the postprandial insulin response, potentially affecting the postprandial inflammatory response as suggested by the levels of hs-CRP and IL-6 [261]. In obese individuals with insulin resistance, the beneficial effects of a strawberry-based drink at 40 g at breakfast improved insulin sensitivity [260]. Concordant results were also observed in another study, where authors demonstrated that consuming strawberries (12 g) before 2 h of one meal reduced glucose concentrations and attenuated IL-6 responses [261]. The critical point approached by the papers published in the last decade dealing with dose selection is demonstrated by several papers [262–264]. In particular, the most recent paper is a multicenter, randomized, double-blind controlled, crossover trial on 33 obese adults (median age 53 $\pm$ 13 y; BMI: 33 ± 3.0 kg/m^2) examining the effect of two achievable dietary doses of strawberries (13 g (one serving a day) or 32 g/day (two-and-a-half servings a day) of freeze-dried strawberry powder) on glycemic control and lipid profiles in obese adults with elevated serum LDL cholesterol. This study demonstrated that the strawberry-based beverage has a dose-dependent effect on cardiometabolic risk. The highest dose reduces both circulating insulin and insulin sensitivity, as shown by the decrease in HOMA-IR. Among adipokines, the treatment at 32 g/day significantly decreased serum plasminogen activator inhibitor-1 (PAI-1) and increased serum glucagon. However, only a borderline significant effect on serum LDL cholesterol was registered and no effects were reported for serum glucose, HbAc1 (5.5% for all treatment groups), total and HDL cholesterol, triglycerides, body weight, BMI and waist circumference [263].

5.3. Cornelian Cherry

Contrary to wild strawberries, for which in the last decade clinical trials have been more abundant than experimental investigations, the clinical evidence on cornelian cherry for glycemic control is lacking. However, the effect of the supplementation with fruit extract capsules (150mg of total anthocyanin for each tablet, four capsules/day for 6 weeks) was evaluated in a randomized, double-blind, placebo-controlled clinical trial involving 60 patients (median age 49 years) with T2DM (HbA1c between 7% and10% and duration of diabetes from at least 2 years). Compared to placebo, the extract significantly decreased HbAc1 and increased insulin levels; in parallel, a decrease in triglycerides was observed, while only a trend toward a reduction of BMI, fasting plasma glucose and 2-h postprandial glucose was registered [265]. Although these findings favor the potential use of cornelian cherry as a nutritional supplement for adult patients with T2DM, the data cannot be conclusive due to the paucity of evidence. A recruiting study aimed to determine the efficacy of 3 months of supplementation with lyophilized dried cornelian cherry on women with insulin resistance (ClinicalTrials.gov Identifier: NCT05292300) could significantly contribute to this issue.

5.4. Raspberry

Coming back to Rosaceae and focusing on the *Rubus* genus, red raspberry is extensively studied in the clinics. A search on the ClinicalTrials.gov database retrieved 65 results, of which 8 were related to pre-diabetes (NCT03049631 [266]), T2D (NCT03403582), blood sugar response (NCT02763020), insulin resistance (NCT02479035, NCT02479035 [267], NCT04306406), or metabolic syndrome (NCT03620617 [268], NCT01414647), aimed to test the effect of a diet rich in berries, including strawberry, raspberry and bilberry, on glucose and lipid metabolism and inflammatory marker [269–271].

In particular, to evaluate the effect of red raspberries consumption at breakfast in pre-diabetic women with insulin resistance (fasting glucose between 5.5 mmol/L and 7.0 mmol/L and insulin resistance value $\geq$ 2.5) it was performed a randomized, single-blind, three-arm, 24-h clinical [268]. The study demonstrated a reduction of the peak of both glucose and insulin and the 2-h glucose area under the curve (AUC) following 250 mg of red raspberry consumption. Similar trends of reduced insulin concentrations in response to raspberry were also observed in metabolically healthy subjects displaying obesity [267]. Similar results were obtained in a study including 25 adults with T2DM for at least 5 years, on oral hypoglycemic agents and elevated waist circumference (>89 cm for women and >102 cm for men), that demonstrated that postprandial blood glucose and the mean AUC of serum glucose significantly decreased in patients with postprandial raspberries supplementation versus control conditions, following a high-fat breakfast [272]. However, these effects were observed only in the acute phase of the study and were not confirmed after a 4-week supplementation. The trial was structured into two phases: the first postprandial phase with acute raspberry supplementation, followed by a 1-week washout and the second phase of a 10-week diet supplement with and without raspberry supplementation periods of 4 weeks each, separated by a 2-week washout phase. Despite the similarities in the acute effects of red raspberry on glycaemia and oxidized LDL, other studies report conflicting results on inflammation and oxidative stress markers [268,272]. Raspberry intake did not significantly modify postprandial inflammation and oxidative stress markers, IL-6 and IL-10. On the contrary, red raspberry reduced serum IL-6 and TNF-α in acute and chronic supplementation [273].

Interestingly, the titration of the frozen red raspberries revealed that among the active ingredients, 59.0% were anthocyanins (cyanidin sophoroside and cyanidin-3-glucoside in particular), followed by ellagic acid and ellagitannins (28.5%, with sanguiin H-6 the major ellagitannin) [272]. Insulin concentration was positively correlated with methylcyanidin sophoroside and negatively correlated with 4′-hydroxycinnamic acid and 4′-hydroxy-3′-methoxycinnamic acid and its isomer; no correlation was observed between circulating polyphenolic metabolites and glucose concentrations. Red raspberry at 250 mg increased the AUC_{0-24h} for methyl-cyanidin 3-O-glucoside in the pre-diabetic group [272]. Notably, it was demonstrated that ellagitannins bioavailability is dependent on the composition of gut microbiota, which could affect the clinical benefits of a supplementation base on ellagitannin-containing berries [269].

5.5. Blackberry

In 2018 Solverson et al. published the results obtained from a clinical study in which the effects of blackberry Lochness variety intake on energy substrate uses and glucoregulation in 27 volunteers of overweight or obese men (BMI > 25 kg/m^2) consuming an HFD [274]. Subjects were fed with a controlled HFD which contained either 600 g/day blackberries (1500 mg/day flavonoids, 360 mg/day anthocyanins) or a calorie and carbohydrate matched amount of gelatin (flavonoid-free control) for seven days before a meal-based glucose tolerance test in combination with a 24 h stay in a room-size indirect calorimeter. Blackberry supplementation significantly reduced in average 24 h respiratory quotient, indicating increased fat oxidation. The respiratory quotient during the glucose tolerance test was significantly lower in patients supplemented with blackberry and a similar result was obtained in a 4 h isolation during dinner. Blackberry supplementation did not modify the glucose AUC, but the insulin AUC was significantly lower and HOMA-IR improved, with more prominent results in lower age subjects [274]. This study was conducted on a healthy subject, trials registered on ClinicalTrial.gov investigate the effects of blackberry on metabolic syndrome (NCT01944579) or T2D and gestational diabetes (NCT01474525), but their results are not yet available.

5.6. Bilberry

As briefly described above, the study NCT01414647 [269–271] compared, in subjects with metabolic syndrome (plasma glucose 5.6–6.9 mmol/L, BMI 26–39 kg/m^2), the effects of an 8 week of dietary supplement of 300 g of strawberry, raspberry and cloudberry against a dietary supplement of 400 g of bilberry and a standard diet (daily consumption of berries restricted to maximum 80 g). The bilberry-rich diet reduced the low-grade inflammation associated with metabolic syndrome reducing serum hs-CRP, IL-6, IL-12 and LPS and possibly triggering an immunomodulatory response as demonstrated by the regulation of the genes related to Toll-like receptors (TLR) and B-cell receptor signaling together with the expression of genes associated to monocyte and macrophage function [270]. Consistently, the blackberry diet reduced vascular cell adhesion molecule (VCAM), an immunoglobulin involved in leukocyte trafficking [275], TNF-α and adiponectin levels [276], as also confirmed by the more recent NCT02689765 study [277]. However, the same concordance was not registered in glucose metabolism: not affected in the study by de Mello et al. [271] or in the study by Chan et al. [278], in which just a slight but not significant reduction in HbA1c was observed, but in the NCT03185676 trial bilberry supplements increased both HbA1c and fasting plasma insulin [276]. However, in this latter study, the berry diet was equivalent to 100g fresh berries [276] instead of 400 g [270], a dose that, when administered for a more extended period (1-year supplementation vs. 8 weeks), improved the glucose and insulin metabolism possibly by increasing the fasting serum hippuric acid levels [271]. An apparent effect on fasting glucose was observed already after 12 weeks of 320 mg anthocyanins extracted from natural bilberry and blackcurrant administrated to subjects with newly diagnosed diabetes in the NCT02689765 study [277,279–281], in which a parallel increase in serum insulin-like growth factor binding protein-4 (IGFBP-4) fragment was reported. The same treatment moderately reduced HbA1c [279] and had beneficial effects on the adipocyte dysfunction associated with T2DM increasing serum adipsin and reducing visfatin [281], an adipokine whose deficiency has been associated with β-cell failure [282] firstly, and a pleiotropic protein, also known as pre-B cell colony-enhancing factor (PBEF), whose circulating levels are increased in T2DM patients [283] secondly. Similarly, the anthocyanin extract supplement at 320 mg for 24 weeks prevented insulin resistance and reduced low-grade inflammation in T2DM patients (NCT02317211 [284]).

Collectively, these results should be interpreted in terms of both supplement dosage and duration, as underlined by Chan et al. [278], in which 1.4 g extract (containing $\geq$ 25% anthocyanidins)/day were administered but for just 4 weeks. The recruiting study NCT04637945 was designed to evaluate the effect of 8 weeks of supplementation with a mixture of anthocyanin from bilberry, black currant extract and black rice on glucose homeostasis and inflammation should contribute to the debate. However, the study by Hoggard et al. [285], registered as NCT01180712 trial, demonstrated in subjects with T2DM controlled by diet and lifestyle alone or with impaired glucose tolerance that the supplementation with a standardized bilberry extract (Mirtoselect®, 36% anthocyanins and equivalent to 50 g of fresh bilberries) at 0.47 g acutely lower the incremental plasma glucose AUC by 18% compared with the placebo. The extract did not affect plasma glucose-dependent insulinotropic polypeptide (gastric inhibitory polypeptide, GIP), glucagon-like peptide 1 (GLP-1), glucagon, or amylin, as well as the inflammatory adipokine monocyte chemoattractant protein 1 (MCP-1). These results led the authors to hypothesize that these acute effects could deal with the polyphenols content other than anthocyanins (about 18%) which could reduce carbohydrate digestion and absorption (NCT01245270 [285]). The recruiting studies NCT04004182 could probably contribute to defining bilberry's contribution to inhibiting intestinal glucose transporters. Nevertheless, the high phenolic content of not only bilberry but also rosehip was associated with the reduction in the very early phase (0–30 min) of postprandial insulin response (NCT03159065 [286]). These promising results should be extended in the long-term study NCT03185676, completed in 2017, but whose results have not yet been widespread. Finally, the beneficial effect of bilberry on diabetic retinopathy is also under investigation. Apart from the NCT02984813 study, oral supplementation with Mirtoselect® was tested in T2DM patients with severe pre-proliferative retinopathy with a preventive effect on the worsening of retinopathy symptoms [287], confirming the potential of this supplement in the retinopathies [288], diabetic included.

5.7. Rosehip

Apart from the NCT03159065 trial, investigating the comparative effect of different supplements, rosehip included, no other trials dialing with anti-diabetic properties of this plant are reported on the ClinicalTrial.gov database. However, the combinatory effect of a standardized extract from seven plants, *Rosa canina* L. (225mg), *Capparis spinosa* L. (170 mg), *Securidaca securigera* L. (170 mg), *Silybum marianum* L. (65 mg), *Urtica dioica* L. (170 mg),

Trigonella foenum-graecum L. (115 mg) and *Vaccinium arctostaphylos* L. (85 mg), was recently investigated in T2DM patients (men and woman 40 to 60 years old with fasting plasma glucose level from 130 to 160 mg/dL and HbA1c 7.5% to 8.5% and under the therapy of oral anti-hyperglycemic drugs maximum 10 mg glyburide and 1000 mg metformin daily) [289]. The study was designed as a three-arm (herbal composition, metformin and placebo), double-blind, randomized placebo-controlled clinical trial. The interventions herbal combination (1000 mg), metformin (250 mg), or placebo were administered once a day before the meal. The authors showed a significant decrease in fasting plasma glucose and HbAc1 compared to the placebo, but no differences were observed in lipid outcomes, renal, hepatic and hematological serum parameters. These results are only partially in accord with the data from previous studies [230]. A randomized, double-blind placebo-controlled trial involving 60 patients with T2DM (aged 35–60 years with fasting blood glucose levels between 130 to 200 mg/dL and HbA1c between 7–9% despite using conventional oral hypoglycemic drugs) compared the effect of 750 mg rosehip fruit extract to placebo in a 3-month treatment [230]. The long-term treatment was able to significantly reduce both fasting blood glucose and cholesterol/HDL-C ratio. However, no significant differences were measured for postprandial blood glucose, HbA1c, total cholesterol, triglyceride, LDL-C/HDL-C ratio, blood urea nitrogen (BUN), creatinine, serum glutamic-oxaloacetic transaminase (SGOT) and serum glutamic pyruvic transaminase (SGPT). The effects on glucose and lipid profile observed in the present study in part may be due to phenolic compounds present in rosehip [290]. Another randomized, double-blind, cross-over study involving 30 obese subjects with normal or impaired glucose tolerance, showed that a daily intake of a rosehip powder drink over 6 weeks, compared with a control drink, did not induce significative changes in glucose tolerance, plasma levels of HDL cholesterol, triglycerides, incretins and markers of inflammation. However, the authors observed a significant systolic blood pressure and LDL/HDL ratio reduction in the treatment group. In particular, the rosehip drink assumption significantly reduced the Reynolds risk assessment score for cardiovascular disease, thus suggesting that the rosehip extract could reduce the risk markers of T2DM and cardiovascular disease [291].

6. Discussion

Although the use of herbal remedies in developed countries is growing [292], many challenges must still be faced before validating them as both functional food and eventually effective drugs. First, the traditional knowledge of local plants should be reinterpreted based on modern science [293]. This more and more general approach can also be applied to the linguistic islet minority, where traditions are usually well settled but linked to oral transmission and customary use with the evident risk of getting lost. The process of traditional knowledge reinterpretation according to modern science could have evident positive effects, which include the economic development of the regions to which the specific herbal tradition belongs, the preservation of the cultural identity of linguistic islets minority and, more importantly, the promotion of a culture of traditional herbal remedies based on evidence. This latter aspect is in line with the World Health Organization's (WHO) strategy 2014–2023, which aimed to strengthen the role of traditional medicine and promote the utilization of medicinal plants in the different health systems of its member countries [278].

In our analysis, we evaluated the last 10 years of evidence-based data on the potential anti-diabetic properties of nine plants used in the Occitan valley of the Piedmont Alps. Notably, the local uses of the plants described are not related to diabetes and only a few of them have been traditionally used with this indication in other countries, such as yarrow in Peru [66], rosehip in Turkey [227], or bilberry in Russia [294]. The literature points out experimental and clinical evidence suggesting their anti-diabetic properties. However, before a consensus could be reached, a rigorous and quantitative analysis of their consumption should be performed. This data should be cross-linked with the prevalence and diffusion of T2DM in the selected geographical area to which the traditional heritage belongs. Unfortunately, this data is unavailable, thus affecting any conclusion on the preventive role of the consumption of the plants herein reported.

6.1. Herbal Remedies for Diabetes: The Bench-to-Bedside Challenge

Herbal remedies could be focused on different therapeutic anti-diabetic approaches: (1) the prevention of T2DM by consuming plants, for which the traditional uses are primarily as food; (2) supplementation strategy as an add-on treatment to potentiate the effect of the current anti-diabetics; (3) use the plant extracts as anti-diabetics.

The first two strategies converge on the concept of functional foods. They are the most reliable, at least nowadays, as demonstrated by the current clinical studies in this review evaluating the use of herbal remedies as add-on therapy for T2DM or metabolic syndrome. Notably, the mechanism(s) of glycemic control shown by the plants considered in our study converges on the ability to reduce intestinal glucose absorption (Table 2 and Figure 1). Therefore, they could be an exciting option in pre-diabetic patients' borderline for T2DM development, for which no specific drugs are currently approved. Under this task, evaluating the effect of herbal remedies on risk factors for T2DM development, inflammation, obesity and lipid profile, including [295], is desirable.

On the contrary, the use of the plant extracts as anti-diabetics is still far from being proven, even if some experimental studies herein reported demonstrated that several extracts had a similar effect to acarbose [110,210], glibenclamide [103,105,108,128,232,251], or even metformin [192,251,289]. However, appropriate non-inferior or, even better, superiority studies and comparative clinical trials are mandatory to assess the anti-diabetic effect and the therapeutic positioning.

The reported data indicates that both preclinical and clinical studies are ongoing for almost all the selected plants. However, the majority are preclinical investigations. In vivo studies rely primarily on an experimental design that mimics T1DM, using the STZ-induced diabetes model. This is the most reproducible and consistent model of diabetes. However, it mimics T1DM, as STZ destroys the β cells leading to a dramatic deprivation of insulin [296,297]. Some other authors used alloxan-induced diabetes, again a model of insulin-dependent diabetes [298]. Due to these models based on pancreatic damage could help evaluate the protective effect of the natural extract on β cell integrity, as reported for yarrow [77,103] or red raspberry [110], but do not produce results completely transferable on T2DM.

T2DM accounts for 90% of diabetes cases and has a multifactorial etiology, where lifestyle and diet have a pivotal role. T2DM models could validate specific herbal remedies as a preventive or add-on therapy. Therefore, implementing studies using HDF or STZ+nicotinamide models, both reproducing T2DM [299,300], should be encouraged. Moreover, a stricter indication of the supplement initiation should also be included in all the experimental studies, thus clarifying the prophylactic or therapeutic design and improving the transfer from the bench to the bedside of the results obtained.

Overall, we must recognize that the studies published between 2011–2021 present several limitations that necessary affect our review. Basically, their quality is sub-optimal, for instance, a proper qualitative and quantitative characterization of the extract use is often missing, as well as the part of the plant utilized sometimes not reported. Moreover, the experimental reports mostly rely on endpoints that do not allow efficacy extrapolation and many studies only replicate and confirm previous data, eventually pointing out plants of different variety, genotype, storage, processing, climate, soil and harvest time or remedy preparations (i.e., aqueous or ethanol extract), but again the papers usually do not specify the cultivar differences and are not comparative.

Looking at the clinical studies, evaluating any efficacy is still far from conclusive. The number of studies is limited and the quality is affected by the number of individuals included and the absence of a stratification considering the co-variables that could occur (such as sex, age, duration of diabetes, anti-diabetic and non-anti-diabetic drugs, BMI at baseline, co-morbidities).

6.2. Herbal Remedies for Diabetes: The Standardization of Titration Challenge

Apart from the experimental design, the challenge of understanding the actual contribution to diabetes management of the plants in this review (but for all plants in general) deals with many issues, starting from the several different preparations that could be obtained and including extraction, fractionation, purification, concentration, fermentation and distillation [99]. All of these could have different active ingredient titrations that make difficult any attempt at standardization. This could be affected by the production process reported for raspberry. Their anthocyanins content could decrease by 42% in frozen berries compared to their raw counterparts [230]. Differences in the raw material account for the extract titration variability. This point is even more critical when considering the differences between cultivars' conditions. Variety, genotype, storage, processing, climate, soil and harvest time affect the phenolic composition of plants [82]. A comparison of nutritional and nutraceutical properties in wild strawberry fruits cultivated in Italy evidenced that in the same region, at the same altitude and under the same agricultural practices, the genotype exerts a significant effect on the synthesis of polyphenols: higher in *Sara* and *Alpine* than in *Regina delle Valli* and *Valitutto* varieties [301]. Thus, underling how the specific variety could affect the results in active ingredients should be declared in the experimental design to allow comparison and obtain conclusive data. Interestingly, higher amounts of sugar and organic acid were found in wild strawberries grown at the highest altitude despite the altitude did not affect the secondary metabolites and radical scavenging activity [302]. According to the organ or the plant development stage, the sugar composition of some polysaccharides or the proportion of substituents linked to sugar could be affected. These differences could significantly affect the active ingredients' biological activity and the whole plant. It has been reported that the yield of oligosaccharides differs based on the plant's growth level, which affects the anti-coagulant activity [174]. Moreover, looking at oligosaccharides, their standardization is challenged by isolating individual oligosaccharides with characterized structures [303]. The isolation and chemical characterization of unique active ingredients are one of the most dramatic issues. Indeed, the literature generically reports phenols, polyphenols, anthocyanins and tannins without grouping them correctly. Proper identification and quantification of bioactive constituents should be necessary to predict the biological properties of extracts of plants [29].

Even when extracts are enriched in one of the active ingredients, such as anthocyanins, it is unlikely this should be the only active ingredient present. The role of the single active ingredient is based on the relative amount compared to the others present. Therefore, it is possible to argue that anti-diabetic effects are the results of the synergism of all the components of the extract. We identified as major actors in the described anti-diabetic effects, several substances summarized in polyphenols and oligosaccharides. However, also vitamins, carbohydrates, minerals and dietary fibers are present and their contribution to the whole extract effect should not be ruled out. For instance, dietary fibers, affecting the digestive processes could indirectly reduce the glycemic index [304]. The complex mechanism includes the selective promotion of gut bacteria [305]. These issues inevitably bring out the bias generated using different extracts without rigorous standardization, which makes the comparison between studies just elusive and not conclusive. Nevertheless, other studies not included in this review used more than one plant as an active treatment, thus making it even more challenging to dissect each plant's relative contribution and the functional role of each active ingredient. Even if titration standardization is solved, bioavailability should still be considered.

6.3. Herbal Remedies for Diabetes: The Bioavailability Challenge

The chemical structure of polyphenols can influence their bioavailability, as their binding to sugars (as glycoside) can increase solubility. Additionally, polyphenols must be released from the food matrix to be absorbable after oral administration. Intestinal absorption can be influenced by intestinal enzymes and colonic microflora [262], which hydrolyze polyphenols that the native form cannot absorb (polymers, esters, glycosides).

Due to the low permeation through the intestinal barrier, the amount of bioactive compound in the bloodstream is often very low. They also undergo complex metabolism in the gut and liver [306]. In this regard, the effect of gut microbiota is not negligible. For instance, it was demonstrated that ellagitannin bioavailability is dependent on the composition of gut microbiota, which could affect the clinical benefits of a supplementation base on ellagitannin-containing berries [270]. Indeed, ellagitannins are not absorbed in the small intestine but in the colon, where microbiota converts them to urolithins [307].

Notably, a bidirectional interaction exists: the ellagitannin-rich berry consumption stimulated a predominance of Ruminococcaceae and Lachnospiraceae [269]. Consistently, gallic acid had been reported to enhance the number of *Atopobium* spp. Bacteria and reduced that of Clostridium histolyticum and anthocyanins increase *Bifidobacterium* spp. and *Lactobacillus-Enterococcus* spp. [308,309]. Similarly, cornelian cherry promotes the development of prebiotic microorganisms with a similar detrimental effect on the growth of potentially pathogenic microorganisms [310,311]. This reciprocal influence further supports the hypothesis of a prominent gastrointestinal activity of herbal plants that could inhibit the α-glucosidase enzyme before systemic absorption and, in parallel, improve the microbiota system contributing to a "healthy" gut microbiota [312]. On the other side, this data points out the possible inter-individual variability in bioavailability and consequently in response to dietary interventions [313].

The problem of sub-optimal bioavailability deals with the high dosages requested to obtain the desired effects. It is not unusual that the most active polyphenol is not the most bioavailable [314]. For instance, the number of bilberries eaten for a clinical effect corresponded to 400 g of fresh berries daily [271]. As reported by the authors, the average intake of berries in the Finnish population on which the study was conducted is 50–70 g/day. Increasing dosages could lead to the onset of toxicity. Overconsumption of polyphenols could have hypoglycemic effects [5]. Therefore, their overconsumption could dramatically cause a pharmacodynamic interaction with hypoglycemic agents. Pharmacodynamic interaction can also occur beyond the hypoglycemic effect: for instance, *Vaccinium Mirtillus* L. could increase bleeding in patients with anti-coagulants treatment [315]. In parallel, pharmacokinetics interaction could be overcome. Therefore, monitoring the acute and long-term side effects of natural compounds and their interaction with food and drugs is strictly needed. The 400 g/day of bilberries could be realistically obtained only using freeze/dried berries [232], which could be lower concentrated in active ingredients.

To increase polyphenol bioavailability and protect them from metabolism, several strategies could be considered, including condensed forms. The new frontiers focus on natural products encapsulation in nanocarriers, which also allow to control and target their release [316]. An example is the use of gold nanoparticles functionalized with elderberry extract tested in a model of STZ-induced diabetes [317]. This formulation was more active than elderberry extract in reducing the HbA1c. However, new formulations based on nanoparticles deserve deep and long-term studies to evaluate the safety profile.

7. Conclusions

This review reveals scantly sparse literature, with plants growing in different areas and under other conditions, without rigorous standardization. Nevertheless, specific variety from the linguistic 'isles' of Piedmont has never been tested for their anti-diabetic properties. However, the lack of homogeneity in the provenience of the plants and the consistency among the results obtained favor a generalization of the evidence collected. Therefore, we can conclude that using the selected Alpine plants belonging to the "linguistic isles" of Piedmont as a functional food for T2DM is reasonable. The most vital data are experimental, but the clinical assessment is ongoing. However, studies on single varieties grown in the geographical area, with strict standardization and titration of all the active ingredients, are warranted before applying the WHO strategy 2014–2023.

Author Contributions: Conceptualization, A.C.R., E.B., V.B. (Valentina Boscaro) and M.R.; ethnobotanical analysis, M.R., B.S. and A.P.; literature search, V.B. (Valentina Bordano), F.D. and A.P.;

writing—original draft preparation, V.B. (Valentina Boscaro) and M.G.; writing—review & editing A.C.R. and E.B.; supervision A.C.R. and E.B. All authors have read and agreed to the published version of the manuscript.

Funding: This research was funded by the University of Turin (ex 60%).

Institutional Review Board Statement: Not applicable.

Informed Consent Statement: Not applicable.

Conflicts of Interest: The authors declare no conflict of interest.

References

1. Zheng, Y.; Ley, S.H.; Hu, F.B. Global aetiology and epidemiology of type 2 diabetes mellitus and its complications. *Nat. Rev. Endocrinol.* **2018**, *14*, 88–98. [CrossRef] [PubMed]
2. IDF. IDF Diabetes Atlas. Available online: https://www.diabetesatlas.org (accessed on 15 January 2022).
3. Lankatillake, C.; Huynh, T.; Dias, D.A. Understanding glycaemic control and current approaches for screening antidiabetic natural products from evidence-based medicinal plants. *Plant Methods* **2019**, *15*, 105. [CrossRef] [PubMed]
4. American Diabetes Association. Introduction: Standards of Medical Care in Diabetes-2020. *Diabetes Care* **2020**, *43*, S1–S2. [CrossRef] [PubMed]
5. Blahova, J.; Martiniakova, M.; Babikova, M.; Kovacova, V.; Mondockova, V.; Omelka, R. Pharmaceutical Drugs and Natural Therapeutic Products for the Treatment of Type 2 Diabetes Mellitus. *Pharmaceuticals* **2021**, *14*, 806. [CrossRef] [PubMed]
6. Kim, Y.; Keogh, J.B.; Clifton, P.M. Polyphenols and Glycemic Control. *Nutrients* **2016**, *8*, 17. [CrossRef]
7. Malviya, N.; Jain, S.; Malviya, S. Antidiabetic potential of medicinal plants. *Acta Pol. Pharm.* **2010**, *67*, 113–118.
8. Patel, D.K.; Prasad, S.K.; Kumar, R.; Hemalatha, S. An overview on antaidiabetic medicinal plants having insulin mimetic property. *Asian Pac. J. Trop. Biomed.* **2012**, *2*, 320–330. [CrossRef]
9. American Diabetes Association. 2. Classification and Diagnosis of Diabetes: Standards of Medical Care in Diabetes-2022. *Diabetes Care* **2022**, *45*, S17–S38. [CrossRef]
10. Silveira, D.; Prieto-Garcia, J.M.; Boylan, F.; Estrada, O.; Fonseca-Bazzo, Y.M.; Jamal, C.M.; Magalhães, P.O.; Pereira, E.O.; Tomczyk, M.; Heinrich, M. COVID-19: Is There Evidence for the Use of Herbal Medicines as Adjuvant Symptomatic Therapy? *Front. Pharmacol.* **2020**, *11*. [CrossRef]
11. Eggers, M.; Jungke, P.; Wolkinger, V.; Bauer, R.; Kessler, U.; Frank, B. Antiviral activity of plant juices and green tea against SARS-CoV-2 and influenza virus. *Phytother. Res.* **2022**, *36*, 2109–2115. [CrossRef]
12. Sabzian-Molaei, F.; Nasiri Khalili, M.A.; Sabzian-Molaei, M.; Shahsavarani, H.; Fattah Pour, A.; Molaei Rad, A.; Hadi, A. Urtica dioica Agglutinin: A plant protein candidate for inhibition of SARS-COV-2 receptor-binding domain for control of Covid19 Infection. *PLoS ONE* **2022**, *17*, e0268156. [CrossRef]
13. Vilhelmova-Ilieva, N.; Petrova, Z.; Georgieva, A.; Tzvetanova, E.; Trepechova, M.; Mileva, M. Anti-Coronavirus Efficiency and Redox-Modulating Capacity of Polyphenol-Rich Extracts from Traditional Bulgarian Medicinal Plants. *Life* **2022**, *12*, 1088. [CrossRef]
14. Xu, J.; Gao, L.; Liang, H.; Chen, S.-d. In silico screening of potential anti–COVID-19 bioactive natural constituents from food sources by molecular docking. *Nutrition* **2021**, *82*, 111049. [CrossRef]
15. Tilwani, K.; Patel, A.; Parikh, H.; Thakker, D.J.; Dave, G. Investigation on anti-Corona viral potential of Yarrow tea. *J. Biomol. Struct. Dyn.* **2022**, 1–13. [CrossRef]
16. Alqathama, A.; Alluhiabi, G.; Baghdadi, H.; Aljahani, L.; Khan, O.; Jabal, S.; Makkawi, S.; Alhomoud, F. Herbal medicine from the perspective of type II diabetic patients and physicians: What is the relationship? *BMC Complement. Med. Ther.* **2020**, *20*, 65. [CrossRef]
17. Medagama, A.B.; Bandara, R. The use of complementary and alternative medicines (CAMs) in the treatment of diabetes mellitus: Is continued use safe and effective? *Nutr. J.* **2014**, *13*, 102. [CrossRef]
18. Gruppo di lavoro Diabetologia e Futuro. Focus Diabete alcuni esempi regionali. *Il Sole 24 Ore*, 24 July 2018.
19. Gnavi, R.; Picariello, R.; Pilutti, S.; Di Monaco, R.; Oleandri, S.; Costa, G. Epidemiology in support of intervention priorities: The case of diabetes in Turin (Piedmont Region, Northern Italy). *Epidemiol. Prev.* **2020**, *44*, 172–178. [CrossRef]
20. Camoletto, S.; Bellandi, M. Forms of «hybrid» local development: A case study on the province of Cuneo. *Stato Mercato* **2021**, *3*, 389–420.
21. Acharya, K.P.; Prajapati, D.; Sherpa, K.; Khanal, A.; Poudel, P.; Hirachan, A.; Bogati, A.; Adhikari, C. Prevalence and determinants of diabetes mellitus in high altitude: A cross sectional study in mountainous region of Nepal. *Asian J. Med. Sci.* **2020**, *11*, 44–48. [CrossRef]
22. Regis, R. On this side of the Alps: A sociolinguistic overview of Francoprovençal in northwestern Italy. *Int. J. Sociol. Lang.* **2018**, *249*, 119–133. [CrossRef]
23. Pieroni, A.; Giusti, M.E. Alpine ethnobotany in Italy: Traditional knowledge of gastronomic and medicinal plants among the Occitans of the upper Varaita valley, Piedmont. *J. Ethnobiol. Ethnomed.* **2009**, *5*, 32. [CrossRef] [PubMed]

24. Mattalia, G.; Quave, C.; Pieroni, A. Traditional uses of wild food and medicinal plants among Brigasc, Kyé, and Provençal communities on the Western Italian Alps. *Genet. Resour. Crop Evol.* **2012**, *60*, 587–603. [CrossRef]
25. Hasler, C.M. Functional foods: Benefits, concerns and challenges-a position paper from the american council on science and health. *J. Nutr.* **2002**, *132*, 3772–3781. [CrossRef] [PubMed]
26. Ali, N.; Shaoib, M.; Shah, S.W.; Shah, I.; Shuaib, M. Pharmacological profile of the aerial parts of Rubus ulmifolius Schott. *BMC Complement. Altern. Med.* **2017**, *17*, 59. [CrossRef] [PubMed]
27. Hummer, K.E. Rubus Pharmacology: Antiquity to the Present. *HortScience* **2010**, *45*, 1587–1591. [CrossRef]
28. Schulz, M.; Seraglio, S.K.T.; Della Betta, F.; Nehring, P.; Valese, A.C.; Daguer, H.; Gonzaga, L.V.; Costa, A.C.O.; Fett, R. Blackberry (Rubus ulmifolius Schott): Chemical composition, phenolic compounds and antioxidant capacity in two edible stages. *Food Res. Int.* **2019**, *122*, 627–634. [CrossRef]
29. Zia-Ul-Haq, M.; Riaz, M.; De Feo, V.; Jaafar, H.Z.; Moga, M. *Rubus fruticosus* L.: Constituents, biological activities and health related uses. *Molecules* **2014**, *19*, 10998–11029. [CrossRef]
30. Jaberg, K.; Jud, J.; Scheuermeier, P. *Prach-und Sachatlas Italiens und der Südschweiz (AIS) (Linguistic and Ethnographic Atlas of Italy and Southern Switzerland) Electronic Version NavigAIS from Tisato, G.*; Verlagsanstalt Ringier & Co.: Zofingen, Switzerland, 1928–1940; pp. 606, 609, 610, 611, 622.
31. Canobbio, S.; Telmon, T. *Atlante Linguistico ed Etnografico del Piemonte Occidentale: Il Mondo Vegetale. Alberi e Arbusti. (ALEPO)*; Priuli & Verlucca: Pavone Canavese-Scarmagno, Italy, 2009; Volume 1, pp. 65, 69, 89, 165, 169.
32. Cusan, F. *Parola alle piante. Saggio di Fitotoponomastica di una Valle Alpina*; Edizioni dell'Orso: Alessandria, Italy, 2020; p. 208.
33. Acta Plantarum-Flora delle Regioni Italiane. Available online: https://www.actaplantarum.org/ (accessed on 1 June 2022).
34. European Directorate for the Quality of Medicines & HealthCare. *Council of Europe*, 10th ed.; European Pharmacopoeia: Strasbourg Cedex, France; Available online: https://pheur.edqm.eu/home (accessed on 1 June 2022).
35. Committee on Herbal Medicinal Products. European Medicines Agency (EMA). Available online: https://www.ema.europa.eu/en (accessed on 1 June 2022).
36. Calvano, A.; Izuora, K.; Oh, E.C.; Ebersole, J.L.; Lyons, T.J.; Basu, A. Dietary berries, insulin resistance and type 2 diabetes: An overview of human feeding trials. *Food Funct.* **2019**, *10*, 6227–6243. [CrossRef]
37. Bellia, G.; Martina, P.A.; Pons, A.; Richard, F. Available online: http://www.coltivareparole.it/ (accessed on 1 May 2022).
38. Pons, A. Coltivare parole. Un Piccolo Atlante Fitonimico del Pinerolese e delle Valli Valdesi. *Boll. Atlante Linguist. Ital.* **2016**, *III*, 153–162.
39. Burton-Freeman, B.M.; Sandhu, A.K.; Edirisinghe, I. Red Raspberries and Their Bioactive Polyphenols: Cardiometabolic and Neuronal Health Links. *Adv. Nutr.* **2016**, *7*, 44–65. [CrossRef]
40. Krauze-Baranowska, M.; Glod, D.; Kula, M.; Majdan, M.; Halasa, R.; Matkowski, A.; Kozlowska, W.; Kawiak, A. Chemical composition and biological activity of Rubus idaeus shoots—A traditional herbal remedy of Eastern Europe. *BMC Complement. Altern. Med.* **2014**, *14*, 480. [CrossRef]
41. Gesek, J.; Jakimiuk, K.; Atanasov, A.G.; Tomczyk, M. Sanguiins-Promising Molecules with Broad Biological Potential. *Int. J. Mol. Sci.* **2021**, *22*, 12972. [CrossRef]
42. CABI. Invasive Species Compendium. Available online: www.cabi.org/isc. (accessed on 15 January 2022).
43. de Souza, V.R.; Pereira, P.A.; da Silva, T.L.; de Oliveira Lima, L.C.; Pio, R.; Queiroz, F. Determination of the bioactive compounds, antioxidant activity and chemical composition of Brazilian blackberry, red raspberry, strawberry, blueberry and sweet cherry fruits. *Food Chem.* **2014**, *156*, 362–368. [CrossRef]
44. Nile, S.H.; Park, S.W. Edible berries: Bioactive components and their effect on human health. *Nutrition* **2014**, *30*, 134–144. [CrossRef]
45. Uncini Manganelli, R.E.; Tomei, P.E. Ethnopharmacobotanical studies of the Tuscan Archipelago. *J. Ethnopharmacol.* **1999**, *65*, 181–202. [CrossRef]
46. Gowd, V.; Bao, T.; Wang, L.; Huang, Y.; Chen, S.; Zheng, X.; Cui, S.; Chen, W. Antioxidant and antidiabetic activity of blackberry after gastrointestinal digestion and human gut microbiota fermentation. *Food Chem.* **2018**, *269*, 618–627. [CrossRef]
47. Riaz, M.; Mansoor, A.; Najmur, R. Antimicrobial screening of fruit, leaves, root and stem of Rubus fruticosus. *J. Med. Plants Res.* **2011**, *5*, 5920–5924. [CrossRef]
48. Verma, R.; Gangrade, T.; Punasiya, R.; Ghulaxe, C. Rubus fruticosus (blackberry) use as an herbal medicine. *Pharm. Rev.* **2014**, *8*, 101–104. [CrossRef]
49. Maugini, E.; Maleci Bini, L.; Mariotti Lippi, M. *Botanica Farmaceutica*, IX ed.; Piccin-Nuova Libraria: Padua, Italy, 2014.
50. Canobbio, S.; Telmon, T. *Atlante Linguistico ed Etnografico del Piemonte Occidentale: Il Mondo Vegetale. Erbaceei. (ALEPO)*; Priuli & Verlucca: Pavone Canavese-Scarmagno, Italy, 2008; Volume 1, pp. 220, 259.
51. Liberal, J.; Francisco, V.; Costa, G.; Figueirinha, A.; Amaral, M.T.; Marques, C.; Girao, H.; Lopes, M.C.; Cruz, M.T.; Batista, M.T. Bioactivity of *Fragaria vesca* leaves through inflammation, proteasome and autophagy modulation. *J. Ethnopharmacol.* **2014**, *158 Pt A*, 113–122. [CrossRef]
52. Mudnic, I.; Modun, D.; Brizic, I.; Vukovic, J.; Generalic, I.; Katalinic, V.; Bilusic, T.; Ljubenkov, I.; Boban, M. Cardiovascular effects in vitro of aqueous extract of wild strawberry (*Fragaria vesca*, L.) leaves. *Phytomedicine* **2009**, *16*, 462–469. [CrossRef]

53. Couto, J.; Figueirinha, A.; Batista, M.T.; Paranhos, A.; Nunes, C.; Goncalves, L.M.; Marto, J.; Fitas, M.; Pinto, P.; Ribeiro, H.M.; et al. *Fragaria vesca* L. Extract: A Promising Cosmetic Ingredient with Antioxidant Properties. *Antioxidants* **2020**, *9*, 154. [CrossRef] [PubMed]

54. A.A.V.V. *Piante e Fiori delle Nostre Montagne: Plantes et Fleurs de Nos Montagnes*; Valchisone, C.S., Ed.; Standard Perosa Argentina: Perosa Argentina, Italy, 2019; p. 224.

55. Bernard, G. *Lou Saber*; Vivo, O., Ed.; dizionario enciclopedico dell'occitano di Blins: Venasca, Italy, 1996; pp. 24, 62, 368.

56. Winther, K.; Campbell-Tofte, J.; Hansen, A. Bioactive ingredients of rose hips (*Rosa canina* L.) with special reference to antioxidative and anti-inflammatory properties: In vitro studies. *Bot. Targets Ther.* **2016**, *6*, 11–23. [CrossRef]

57. Barros, L.; Carvalho, A.M.; Ferreira, I.C.F.R. Exotic fruits as a source of important phytochemicals: Improving the traditional use of rosa canina fruits in portugal. *Food Res. Int.* **2011**, *44*, 2233–2236. [CrossRef]

58. Marmol, I.; Sanchez-de-Diego, C.; Jimenez-Moreno, N.; Ancin-Azpilicueta, C.; Rodriguez-Yoldi, M.J. Therapeutic Applications of Rose Hips from Different Rosa Species. *Int. J. Mol. Sci.* **2017**, *18*, 1137. [CrossRef] [PubMed]

59. Pekacar, S.; Bulut, S.; Özüpek, B.; Orhan, D.D. Anti-Inflammatory and Analgesic Effects of Rosehip in Inflammatory Musculoskeletal Disorders and Its Active Molecules. *Curr. Mol. Pharm.* **2021**, *14*, 731–745. [CrossRef]

60. Chu, W.K.; Cheung, S.C.; Lau, R.A.; Benzie, I.F. Bilberry (*Vaccinium myrtillus* L.). In *Herbal Medicine: Biomolecular and Clinical Aspects*; Benzie, I.F.F., Wachtel-Galor, S., Eds.; CRC Press: Boca Raton, FL, USA, 2012.

61. Mlynarczyk, K.; Walkowiak-Tomczak, D.; Lysiak, G.P. Bioactive properties of *Sambucus nigra* L. as a functional ingredient for food and pharmaceutical industry. *J. Funct. Foods* **2018**, *40*, 377–390. [CrossRef]

62. Baser, K.H.; Demirci, B.; Demirci, F.; Kocak, S.; Akinci, C.; Malyer, H.; Guleryuz, G. Composition and antimicrobial activity of the essential oil of Achillea multifida. *Planta Med.* **2002**, *68*, 941–943. [CrossRef]

63. Benedek, B.; Rothwangl-Wiltschnigg, K.; Rozema, E.; Gjoncaj, N.; Reznicek, G.; Jurenitsch, J.; Kopp, B.; Glasl, S. Yarrow (*Achillea millefolium* L. s.l.): Pharmaceutical quality of commercial samples. *Die Pharm.* **2008**, *63*, 23–26. [CrossRef]

64. Guarrera, P.M. *Usi e Tradizioni della Flora Italiana. Medicina Popolare ed Etnobotanica*; Aracne: Rome, Italy, 2007; p. 436.

65. Applequist, W.L.; Moerman, D.E. Yarrow (*Achillea millefolium* L.): A Neglected Panacea? A Review of Ethnobotany, Bioactivity, and Biomedical Research1. *Econ. Bot.* **2011**, *65*, 209–225. [CrossRef]

66. Bussmann, R.W.; Sharon, D.; Vandebroek, I.; Jones, A.; Revene, Z. Health for sale: The medicinal plant markets in Trujillo and Chiclayo, Northern Peru. *J. Ethnobiol. Ethnomed.* **2007**, *3*, 37. [CrossRef]

67. Ali, S.I.; Gopalakrishnan, B.; Venkatesalu, V. Pharmacognosy, Phytochemistry and Pharmacological Properties of *Achillea millefolium* L.: A Review. *Phytother. Res.* **2017**, *31*, 1140–1161. [CrossRef]

68. Karamenderes, C.; Apaydin, S. Antispasmodic effect of *Achillea nobilis* L. subsp. sipylea (O. Schwarz) Bassler on the rat isolated duodenum. *J. Ethnopharmacol.* **2003**, *84*, 175–179. [CrossRef]

69. Stojanovic, G.; Radulovic, N.; Hashimoto, T.; Palic, R. In vitro antimicrobial activity of extracts of four Achillea species: The composition of *Achillea clavennae* L. (Asteraceae) extract. *J. Ethnopharmacol.* **2005**, *101*, 185–190. [CrossRef]

70. Cavalcanti, A.M.; Baggio, C.H.; Freitas, C.S.; Rieck, L.; de Sousa, R.S.; Da Silva-Santos, J.E.; Mesia-Vela, S.; Marques, M.C. Safety and antiulcer efficacy studies of *Achillea millefolium* L. after chronic treatment in Wistar rats. *J. Ethnopharmacol.* **2006**, *107*, 277–284. [CrossRef]

71. Si, X.T.; Zhang, M.L.; Shi, Q.W.; Kiyota, H. Chemical constituents of the plants in the genus Achillea. *Chem. Biodivers* **2006**, *3*, 1163–1180. [CrossRef]

72. Lazarevic, J.; Radulovic, N.; Zlatkovic, B.; Palic, R. Composition of Achillea distans Willd. subsp. distans root essential oil. *Nat. Prod. Res.* **2010**, *24*, 718–731. [CrossRef]

73. Falconieri, D.; Piras, A.; Porcedda, S.; Marongiu, B.; Goncalves, M.J.; Cabral, C.; Cavaleiro, C.; Salgueiro, L. Chemical composition and biological activity of the volatile extracts of Achillea millefolium. *Nat. Prod. Commun.* **2011**, *6*, 1527–1530. [CrossRef]

74. Akram, M. Minireview on Achillea millefolium Linn. *J. Membr. Biol.* **2013**, *246*, 661–663. [CrossRef]

75. Dias, M.I.; Barros, L.; Dueñas, M.; Pereira, E.; Carvalho, A.M.; Alves, R.C.; Oliveira, M.B.; Santos-Buelga, C.; Ferreira, I.C. Chemical composition of wild and commercial *Achillea millefolium* L. and bioactivity of the methanolic extract, infusion and decoction. *Food Chem.* **2013**, *141*, 4152–4160. [CrossRef]

76. Benedek, B.; Geisz, N.; Jäger, W.; Thalhammer, T.; Kopp, B. Choleretic effects of yarrow (*Achillea millefolium* s.l.) in the isolated perfused rat liver. *Phytomedicine* **2006**, *13*, 702–706. [CrossRef]

77. Zolghadri, Y.; Fazeli, M.; Kooshki, M.; Shomali, T.; Karimaghayee, N.; Dehghani, M. *Achillea Millefolium*, L. Hydro-Alcoholic Extract Protects Pancreatic Cells by Down Regulating IL-1β and iNOS Gene Expression in Diabetic Rats. *Int. J. Mol. Cell. Med.* **2014**, *3*, 255–262. [PubMed]

78. Gadgoli, C.; Mishra, S. Antihepatotoxic activity of 5-hydroxy 3, 40, 6, 7-tetramethoxy flavone from Achillea millefolium. *Pharmacology* **2007**, *1*, 391–399.

79. Yaeesh, S.; Jamal, Q.; Khan, A.-u.; Gilani, A.H. Studies on hepatoprotective, antispasmodic and calcium antagonist activities of the aqueous-methanol extract of Achillea millefolium. *Phytother. Res.* **2006**, *20*, 546–551. [CrossRef] [PubMed]

80. Adhikari, B.M.; Bajracharya, A.; Shrestha, A.K. Comparison of nutritional properties of Stinging nettle (*Urtica dioica*) flour with wheat and barley flours. *Food Sci. Nutr.* **2016**, *4*, 119–124. [CrossRef] [PubMed]

81. Rutto, L.K.; Xu, Y.; Ramirez, E.; Brandt, M. Mineral Properties and Dietary Value of Raw and Processed Stinging Nettle (*Urtica dioica* L.). *Int. J. Food Sci.* **2013**, *2013*, 857120. [CrossRef]

82. Kregiel, D.; Pawlikowska, E.; Antolak, H. *Urtica* spp.: Ordinary Plants with Extraordinary Properties. *Molecules* **2018**, *23*, 1664. [CrossRef]
83. Gul, S.; Demirci, B.; Baser, K.H.; Akpulat, H.A.; Aksu, P. Chemical composition and in vitro cytotoxic, genotoxic effects of essential oil from *Urtica dioica* L. *Bull. Environ. Contam. Toxicol.* **2012**, *88*, 666–671. [CrossRef]
84. Orcic, D.; Franciskovic, M.; Bekvalac, K.; Svircev, E.; Beara, I.; Lesjak, M.; Mimica-Dukic, N. Quantitative determination of plant phenolics in Urtica dioica extracts by high-performance liquid chromatography coupled with tandem mass spectrometric detection. *Food Chem.* **2014**, *143*, 48–53. [CrossRef]
85. Otles, S.; Yalcin, B. Phenolic compounds analysis of root, stalk, and leaves of nettle. *Sci. World J.* **2012**, *2012*, 564367. [CrossRef]
86. Pinelli, P.; Ieri, F.; Vignolini, P.; Bacci, L.; Baronti, S.; Romani, A. Extraction and HPLC analysis of phenolic compounds in leaves, stalks, and textile fibers of *Urtica dioica* L. *J. Agric. Food Chem.* **2008**, *56*, 9127–9132. [CrossRef]
87. Said, A.A.H.; Otmani, I.S.E.; Derfoufi, S.; Benmoussa, A. Highlights on nutritional and therapeutic value of stinging nettle (*Urtica dioica*). *Int. J. Pharm. Pharm. Sci.* **2015**, *7*, 8–14.
88. Upton, R. Stinging nettles leaf (*Urtica dioica* L.): Extraordinary vegetable medicine. *J. Herb. Med.* **2013**, *3*, 9–36. [CrossRef]
89. El Haouari, M.; Rosado, J.A. Phytochemical, Anti-diabetic and Cardiovascular Properties of *Urtica dioica* L. (Urticaceae): A Review. *Mini Rev. Med. Chem.* **2019**, *19*, 63–71. [CrossRef]
90. De Biaggi, M.; Donno, D.; Mellano, M.G.; Riondato, I.; Rakotoniaina, E.N.; Beccaro, G.L. *Cornus mas* (L.) Fruit as a Potential Source of Natural Health-Promoting Compounds: Physico-Chemical Characterisation of Bioactive Components. *Plant Foods Hum. Nutr.* **2018**, *73*, 89–94. [CrossRef]
91. Ahangarpour, A.; Mohammadian, M.; Dianat, M. Antidiabetic effect of hydroalcholic urticadioica leaf extract in male rats with fructose-induced insulin resistance. *Iran. J. Med. Sci.* **2012**, *37*, 181–186.
92. Dar, S.A.; Ganai, F.A.; Yousuf, A.R.; Balkhi, M.U.; Bhat, T.M.; Sharma, P. Pharmacological and toxicological evaluation of Urtica dioica. *Pharm. Biol.* **2013**, *51*, 170–180. [CrossRef]
93. Kadan, S.; Saad, B.; Sasson, Y.; Zaid, H. In Vitro Evaluations of Cytotoxicity of Eight Antidiabetic Medicinal Plants and Their Effect on GLUT4 Translocation. *Evid.-Based Complement. Altern. Med.* **2013**, *2013*, 549345. [CrossRef]
94. Qujeq, D.; Tatar, M.; Feizi, F.; Parsian, H.; Sohan Faraji, A.; Halalkhor, S. Effect of Urtica dioica Leaf Alcoholic and Aqueous Extracts on the Number and the Diameter of the Islets in Diabetic Rats. *Int. J. Mol. Cell. Med.* **2013**, *2*, 21–26.
95. Rahimzadeh, M.; Jahanshahi, S.; Moein, S.; Moein, M.R. Evaluation of alpha- amylase inhibition by Urtica dioica and Juglans regia extracts. *Iran. J. Basic Med. Sci.* **2014**, *17*, 465–469.
96. Ranjbari, A.; Azarbayjani, M.A.; Yusof, A.; Halim Mokhtar, A.; Akbarzadeh, S.; Ibrahim, M.Y.; Tarverdizadeh, B.; Farzadinia, P.; Hajiaghaee, R.; Dehghan, F. In vivo and in vitro evaluation of the effects of Urtica dioica and swimming activity on diabetic factors and pancreatic beta cells. *BMC Complement. Altern. Med.* **2016**, *16*, 101. [CrossRef]
97. Obanda, D.N.; Zhao, P.; Richard, A.J.; Ribnicky, D.; Cefalu, W.T.; Stephens, J.M. Stinging Nettle (*Urtica dioica* L.) Attenuates FFA Induced Ceramide Accumulation in 3T3-L1 Adipocytes in an Adiponectin Dependent Manner. *PLoS ONE* **2016**, *11*, e0150252. [CrossRef] [PubMed]
98. Obanda, D.N.; Ribnicky, D.; Yu, Y.; Stephens, J.; Cefalu, W.T. An extract of *Urtica dioica* L. mitigates obesity induced insulin resistance in mice skeletal muscle via protein phosphatase 2A (PP2A). *Sci. Rep.* **2016**, *6*, 22222. [CrossRef] [PubMed]
99. Gohari, A.; Noorafshan, A.; Akmali, M.; Zamani-Garmsiri, F.; Seghatoleslam, A. *Urtica dioica* Distillate Regenerates Pancreatic Beta Cells in Streptozotocin-Induced Diabetic Rats. *Iran. J. Med. Sci.* **2018**, *43*, 174–183. [PubMed]
100. Abedinzade, M.; Rostampour, M.; Mirzajani, E.; Khalesi, Z.B.; Pourmirzaee, T.; Khanaki, K. *Urtica dioica* and Lamium Album Decrease Glycogen Synthase Kinase-3 beta and Increase K-Ras in Diabetic Rats. *J. Pharmacopunct.* **2019**, *22*, 248–252. [CrossRef] [PubMed]
101. Fan, S.; Raychaudhuri, S.; Kraus, O.; Shahinozzaman, M.; Lofti, L.; Obanda, D.N. Urtica dioica Whole Vegetable as a Functional Food Targeting Fat Accumulation and Insulin Resistance—A Preliminary Study in a Mouse Pre-Diabetic Model. *Nutrients* **2020**, *12*, 1059. [CrossRef]
102. Salim, B.; Said, G.; Kambouche, N.; Kress, S. Identification of Phenolic Compounds from Nettle as New Candidate Inhibitors of Main Enzymes Responsible on Type-II Diabetes. *Curr. Drug Discov. Technol.* **2020**, *17*, 197–202. [CrossRef]
103. Mustafa, K.G.; Ganai, B.A.; Akbar, S.; Dar, M.Y.; Masood, A. ß-Cell protective efficacy, hypoglycemic and hypolipidemic effects of extracts of Achillea millifolium in diabetic rats. *Chin. J. Nat. Med.* **2012**, *10*, 0185–0189. [CrossRef]
104. Ramirez, G.; Zavala, M.; Perez, J.; Zamilpa, A. In vitro screening of medicinal plants used in Mexico as antidiabetics with glucosidase and lipase inhibitory activities. *Evid.-Based Complement. Altern. Med.* **2012**, *2012*, 701261. [CrossRef]
105. Chavez-Silva, F.; Ceron-Romero, L.; Arias-Duran, L.; Navarrete-Vazquez, G.; Almanza-Perez, J.; Roman-Ramos, R.; Ramirez-Avila, G.; Perea-Arango, I.; Villalobos-Molina, R.; Estrada-Soto, S. Antidiabetic effect of Achillea millefollium through multitarget interactions: Alpha-glucosidases inhibition, insulin sensitization and insulin secretagogue activities. *J. Ethnopharmacol.* **2018**, *212*, 1–7. [CrossRef]
106. Brader, L.; Overgaard, A.; Christensen, L.P.; Jeppesen, P.B.; Hermansen, K. Polyphenol-rich bilberry ameliorates total cholesterol and LDL-cholesterol when implemented in the diet of Zucker diabetic fatty rats. *Rev. Diabet. Stud. RDS* **2013**, *10*, 270–282. [CrossRef]
107. Kim, J.; Kim, C.S.; Lee, Y.M.; Sohn, E.; Jo, K.; Kim, J.S. *Vaccinium myrtillus* extract prevents or delays the onset of diabetes—Induced blood-retinal barrier breakdown. *Int. J. Food Sci. Nutr.* **2015**, *66*, 236–242. [CrossRef]

108. Asgary, S.; RafieianKopaei, M.; Sahebkar, A.; Shamsi, F.; Goli-malekabadi, N. Anti-hyperglycemic and anti-hyperlipidemic effects of *Vaccinium myrtillus* fruit in experimentally induced diabetes (antidiabetic effect of *Vaccinium myrtillus* fruit). *J. Sci. Food Agric.* **2016**, *96*, 764–768. [CrossRef]

109. Buchholz, T.; Melzig, M.F. Medicinal Plants Traditionally Used for Treatment of Obesity and Diabetes Mellitus—Screening for Pancreatic Lipase and alpha-Amylase Inhibition. *Phytother. Res.* **2016**, *30*, 260–266. [CrossRef]

110. Bljajic, K.; Petlevski, R.; Vujic, L.; Cacic, A.; Sostaric, N.; Jablan, J.; Saraiva de Carvalho, I.; Zovko Koncic, M. Chemical Composition, Antioxidant and alpha-Glucosidase-Inhibiting Activities of the Aqueous and Hydroethanolic Extracts of *Vaccinium myrtillus* Leaves. *Molecules* **2017**, *22*, 703. [CrossRef]

111. Karcheva-Bahchevanska, D.P.; Lukova, P.K.; Nikolova, M.M.; Mladenov, R.D.; Iliev, I.N. Effect of Extracts of Bilberries (*Vaccinium myrtillus* L.) on Amyloglucosidase and alpha-Glucosidase Activity. *Folia Med.* **2017**, *59*, 197–202. [CrossRef]

112. Xiao, T.; Guo, Z.; Sun, B.; Zhao, Y. Identification of Anthocyanins from Four Kinds of Berries and Their Inhibition Activity to alpha-Glycosidase and Protein Tyrosine Phosphatase 1B by HPLC-FT-ICR MS/MS. *J. Agric. Food Chem.* **2017**, *65*, 6211–6221. [CrossRef]

113. Schreck, K.; Melzig, M.F. Traditionally Used Plants in the Treatment of Diabetes Mellitus: Screening for Uptake Inhibition of Glucose and Fructose in the Caco2-Cell Model. *Front. Pharm.* **2021**, *12*, 692566. [CrossRef]

114. Zhu, M.J.; Kang, Y.; Xue, Y.; Liang, X.; Garcia, M.P.G.; Rodgers, D.; Kagel, D.R.; Du, M. Red raspberries suppress NLRP3 inflammasome and attenuate metabolic abnormalities in diet-induced obese mice. *J. Nutr. Biochem.* **2018**, *53*, 96–103. [CrossRef]

115. Xiong, S.-L.; Yue, L.-M.; Lim, G.T.; Yang, J.-M.; Lee, J.; Park, Y.-D. Inhibitory effect of raspberry ketone on α-glucosidase: Docking simulation integrating inhibition kinetics. *Int. J. Biol. Macromol.* **2018**, *113*, 212–218. [CrossRef]

116. Zhao, L.; Zou, T.; Gomez, N.A.; Wang, B.; Zhu, M.J.; Du, M. Raspberry alleviates obesity-induced inflammation and insulin resistance in skeletal muscle through activation of AMP-activated protein kinase (AMPK) alpha1. *Nutr. Diabetes* **2018**, *8*, 39. [CrossRef]

117. Xing, T.; Kang, Y.; Xu, X.; Wang, B.; Du, M.; Zhu, M.J. Raspberry Supplementation Improves Insulin Signaling and Promotes Brown-Like Adipocyte Development in White Adipose Tissue of Obese Mice. *Mol. Nutr. Food Res.* **2018**, *62*, 1701035. [CrossRef] [PubMed]

118. Gutierrez-Albanchez, E.; Kirakosyan, A.; Bolling, S.F.; Garcia-Villaraco, A.; Gutierrez-Manero, J.; Ramos-Solano, B. Biotic elicitation as a tool to improve strawberry and raspberry extract potential on metabolic syndrome-related enzymes in vitro. *J. Sci. Food Agric.* **2019**, *99*, 2939–2946. [CrossRef] [PubMed]

119. Bispo, K.; Amusquivar, E.; Garcia-Seco, D.; Ramos-Solano, B.; Gutierrez-Manero, J.; Herrera, E. Supplementing diet with blackberry extract causes a catabolic response with increments in insulin sensitivity in rats. *Plant Foods Hum. Nutr.* **2015**, *70*, 170–175. [CrossRef] [PubMed]

120. Takács, I.; Szekeres, A.; Takáco, Á.; Rakk, D., Mézes, M.; Polyák, Á.; Lakatos, L.; Gyémánt, G.; Csupor, D.; Kovács, K.J. Wild strawberry, blackberry, and blueberry leaf extracts alleviate starch-induced hyperglycemia in prediabetic and diabetic mice. *Planta Med.* **2020**, *86*, 790–799. [CrossRef] [PubMed]

121. Schrader, E.; Wein, S.; Kristiansen, K.; Christensen, L.P.; Rimbach, G.; Wolffram, S. Plant extracts of winter savory, purple coneflower, buckwheat and black elder activate PPAR-γ in COS-1 cells but do not lower blood glucose in db/db mice in vivo. *Plant Foods Hum. Nutr.* **2012**, *67*, 377–383. [CrossRef]

122. Bhattacharya, S.; Christensen, K.B.; Olsen, L.C.; Christensen, L.P.; Grevsen, K.; Faergeman, N.J.; Kristiansen, K.; Young, J.F.; Oksbjerg, N. Bioactive components from flowers of *Sambucus nigra* L. increase glucose uptake in primary porcine myotube cultures and reduce fat accumulation in Caenorhabditis elegans. *J. Agric. Food Chem.* **2013**, *61*, 11033–11040. [CrossRef]

123. Farrell, N.J.; Norris, G.H.; Ryan, J.; Porter, C.M.; Jiang, C.; Blesso, C.N. Black elderberry extract attenuates inflammation and metabolic dysfunction in diet-induced obese mice. *Br. J. Nutr.* **2015**, *114*, 1123–1131. [CrossRef]

124. Salvador, A.C.; Krol, E.; Lemos, V.C.; Santos, S.A.; Bento, F.P.; Costa, C.P.; Almeida, A.; Szczepankiewicz, D.; Kulczynski, B.; Krejpcio, Z.; et al. Effect of Elderberry (*Sambucus nigra* L.) Extract Supplementation in STZ-Induced Diabetic Rats Fed with a High-Fat Diet. *Int. J. Mol. Sci.* **2016**, *18*, 13. [CrossRef]

125. Ho, G.T.; Kase, E.T.; Wangensteen, H.; Barsett, H. Phenolic Elderberry Extracts, Anthocyanins, Procyanidins, and Metabolites Influence Glucose and Fatty Acid Uptake in Human Skeletal Muscle Cells. *J. Agric. Food Chem.* **2017**, *65*, 2677–2685. [CrossRef]

126. Ho, G.T.; Kase, E.T.; Wangensteen, H.; Barsett, H. Effect of Phenolic Compounds from Elderflowers on Glucose- and Fatty Acid Uptake in Human Myotubes and HepG2-Cells. *Molecules* **2017**, *22*, 90. [CrossRef]

127. Zielinska-Wasielica, J.; Olejnik, A.; Kowalska, K.; Olkowicz, M.; Dembczynski, R. Elderberry (*Sambucus nigra* L.) Fruit Extract Alleviates Oxidative Stress, Insulin Resistance, and Inflammation in Hypertrophied 3T3-L1 Adipocytes and Activated RAW 264.7 Macrophages. *Foods* **2019**, *8*, 326. [CrossRef]

128. Taghizadeh, M.; Rashidi, A.A.; Taherian, A.A.; Vakili, Z.; Sajad Sajadian, M.; Ghardashi, M. Antidiabetic and Antihyperlipidemic Effects of Ethanol Extract of *Rosa canina* L. fruit on Diabetic Rats: An Experimental Study With Histopathological Evaluations. *J. Evid.-Based Complement. Altern. Med.* **2016**, *21*, NP25–NP30. [CrossRef]

129. Fattahi, A.; Niyazi, F.; Shahbazi, B.; Farzaei, M.H.; Bahrami, G. Antidiabetic Mechanisms of *Rosa canina* Fruits: An In Vitro Evaluation. *J. Evid.-Based Complement. Altern. Med.* **2017**, *22*, 127–133. [CrossRef]

130. Jemaa, H.B.; Jemia, A.B.; Khlifi, S.; Ahmed, H.B.; Slama, F.B.; Benzarti, A.; Elati, J.; Aouidet, A. Antioxidant Activity and a-Amylase Inhibitory Potential of *Rosa Canina*, L. *Afr. J. Tradit. Complement. Altern. Med. AJTCAM* **2017**, *14*, 1–8. [CrossRef]

131. Chen, S.J.; Aikawa, C.; Yoshida, R.; Kawaguchi, T.; Matsui, T. Anti-prediabetic effect of rose hip (*Rosa canina*) extract in spontaneously diabetic Torii rats. *J. Sci. Food Agric.* **2017**, *97*, 3923–3928. [CrossRef]

132. Bahrami, G.; Miraghaee, S.S.; Mohammadi, B.; Bahrami, M.T.; Taheripak, G.; Keshavarzi, S.; Babaei, A.; Sajadimajd, S.; Hatami, R. Molecular mechanism of the anti-diabetic activity of an identified oligosaccharide from Rosa canina. *Res. Pharm. Sci.* **2020**, *15*, 36–47. [CrossRef]

133. Rahimi, M.; Sajadimajd, S.; Mahdian, Z.; Hemmati, M.; Malekkhatabi, P.; Bahrami, G.; Mohammadi, B.; Miraghaee, S.; Hatami, R.; Mansouri, K.; et al. Characterization and anti-diabetic effects of the oligosaccharide fraction isolated from *Rosa canina* in STZ-Induced diabetic rats. *Carbohydr. Res.* **2020**, *489*, 107927. [CrossRef]

134. Capcarova, M.; Kalafova, A.; Schwarzova, M.; Schneidgenova, M.; Svik, K.; Prnova, M.S.; Slovak, L.; Kovacik, A.; Lory, V.; Zorad, S.; et al. Cornelian cherry fruit improves glycaemia and manifestations of diabetes in obese Zucker diabetic fatty rats. *Res. Vet. Sci.* **2019**, *126*, 118–123. [CrossRef]

135. Dzydzan, O.; Bila, I.; Kucharska, A.Z.; Brodyak, I.; Sybirna, N. Antidiabetic effects of extracts of red and yellow fruits of cornelian cherries (*Cornus mas* L.) on rats with streptozotocin-induced diabetes mellitus. *Food Funct.* **2019**, *10*, 6459–6472. [CrossRef]

136. Dzydzan, O.; Brodyak, I.; Sokol-Letowska, A.; Kucharska, A.Z.; Sybirna, N. Loganic Acid, an Iridoid Glycoside Extracted from *Cornus mas* L. Fruits, Reduces of Carbonyl/Oxidative Stress Biomarkers in Plasma and Restores Antioxidant Balance in Leukocytes of Rats with Streptozotocin-Induced Diabetes Mellitus. *Life* **2020**, *10*, 349. [CrossRef]

137. Blagojevic, B.; Agic, D.; Serra, A.T.; Matic, S.; Matovina, M.; Bijelic, S.; Popovic, B.M. An in vitro and in silico evaluation of bioactive potential of cornelian cherry (*Cornus mas* L.) extracts rich in polyphenols and iridoids. *Food Chem.* **2021**, *335*, 127619. [CrossRef] [PubMed]

138. Tsao, R. Chemistry and biochemistry of dietary polyphenols. *Nutrients* **2010**, *2*, 1231–1246. [CrossRef] [PubMed]

139. Shahwan, M.; Alhumaydhi, F.; Ashraf, G.M.; Hasan, P.M.Z.; Shamsi, A. Role of polyphenols in combating Type 2 Diabetes and insulin resistance. *Int. J. Biol. Macromol.* **2022**, *206*, 567–579. [CrossRef] [PubMed]

140. Pandey, K.B.; Rizvi, S.I. Plant polyphenols as dietary antioxidants in human health and disease. *Oxid. Med. Cell. Longev.* **2009**, *2*, 270–278. [CrossRef] [PubMed]

141. Cheynier, V. Polyphenols in foods are more complex than often thought. *Am. J. Clin. Nutr.* **2005**, *81*, 223S–229S. [CrossRef]

142. Tadera, K.; Minami, Y.; Takamatsu, K.; Matsuoka, T. Inhibition of alpha-glucosidase and alpha-amylase by flavonoids. *J. Nutr. Sci. Vitam.* **2006**, *52*, 149–153. [CrossRef]

143. Iwai, K. Antidiabetic and antioxidant effects of polyphenols in brown alga Ecklonia stolonifera in genetically diabetic KK-A(y) mice. *Plant Foods Hum. Nutr.* **2008**, *63*, 163–169. [CrossRef]

144. Dao, T.M.; Waget, A.; Klopp, P.; Serino, M.; Vachoux, C.; Pechere, L.; Drucker, D.J.; Champion, S.; Barthelemy, S.; Barra, Y.; et al. Resveratrol increases glucose induced GLP-1 secretion in mice: A mechanism which contributes to the glycemic control. *PLoS ONE* **2011**, *6*, e20700. [CrossRef]

145. Johnston, K.L.; Clifford, M.N.; Morgan, L.M. Coffee acutely modifies gastrointestinal hormone secretion and glucose tolerance in humans: Glycemic effects of chlorogenic acid and caffeine. *Am. J. Clin. Nutr.* **2003**, *78*, 728–733. [CrossRef]

146. Jung, U.J.; Lee, M.K.; Jeong, K.S.; Choi, M.S. The hypoglycemic effects of hesperidin and naringin are partly mediated by hepatic glucose-regulating enzymes in C57BL/KsJ-db/db mice. *J. Nutr.* **2004**, *134*, 2499–2503. [CrossRef]

147. Wolfram, S.; Raederstorff, D.; Preller, M.; Wang, Y.; Teixeira, S.R.; Riegger, C.; Weber, P. Epigallocatechin gallate supplementation alleviates diabetes in rodents. *J. Nutr.* **2006**, *136*, 2512–2518. [CrossRef]

148. Zhang, B.; Kang, M.; Xie, Q.; Xu, B.; Sun, C.; Chen, K.; Wu, Y. Anthocyanins from Chinese bayberry extract protect beta cells from oxidative stress-mediated injury via HO-1 upregulation. *J. Agric. Food Chem.* **2011**, *59*, 537–545. [CrossRef]

149. Vetterli. Resveratrol potentiates glucose-stimulated insulin secretion in INS-1E β-cells and human islets through a SIRT1-dependent mechanism. *J. Biol. Chem.* **2011**, *286*, 6049–6060. [CrossRef]

150. Rowley, T.J.t.; Bitner, B.F.; Ray, J.D.; Lathen, D.R.; Smithson, A.T.; Dallon, B.W.; Plowman, C.J.; Bikman, B.T.; Hansen, J.M.; Dorenkott, M.R.; et al. Monomeric cocoa catechins enhance beta-cell function by increasing mitochondrial respiration. *J. Nutr. Biochem.* **2017**, *49*, 30–41. [CrossRef]

151. Clifford, M.N. Anthocyanins—Nature, occurrence and dietary burden. *J. Sci. Food Agric.* **2000**, *80*, 1063–1072. [CrossRef]

152. Ghosh, D.; Konishi, T. Anthocyanins and anthocyanin-rich extracts: Role in diabetes and eye function. *Asia Pac. J. Clin. Nutr.* **2007**, *16*, 200–208.

153. Yoshida, K.; Mori, M.; Kondo, T. Blue flower color development by anthocyanins: From chemical structure to cell physiology. *Nat. Prod. Rep.* **2009**, *26*, 884–915. [CrossRef]

154. Upton, R. Bilberry Fruit *Vaccinium myrtillus* L. Standards of Analysis, Quality Control, and Therapeutics. In *American Herbal Pharmacopoeia and Therapeutic Compendium*; Santa Cruz: Santa Cruz, CA, USA, 2001.

155. Kong, J.M.; Chia, L.S.; Goh, N.K.; Chia, T.F.; Brouillard, R. Analysis and biological activities of anthocyanins. *Phytochemistry* **2003**, *64*, 923–933. [CrossRef]

156. Prior, R.L.; Wu, X. Anthocyanins: Structural characteristics that result in unique metabolic patterns and biological activities. *Free Radic. Res.* **2006**, *40*, 1014–1028. [CrossRef]

157. Zoratti, L.; Jaakola, L.; Haggman, H.; Giongo, L. Anthocyanin Profile in Berries of Wild and Cultivated *Vaccinium* spp. along Altitudinal Gradients in the Alps. *J. Agric. Food Chem.* **2015**, *63*, 8641–8650. [CrossRef]

158. Wilkes, K.; Howard, L.R.; Brownmiller, C.; Prior, R.L. Changes in chokeberry (*Aronia melanocarpa* L.) polyphenols during juice processing and storage. *J. Agric. Food Chem.* **2014**, *62*, 4018–4025. [CrossRef] [PubMed]
159. Wojdylo, A.; Figiel, A.; Oszmianski, J. Effect of drying methods with the application of vacuum microwaves on the bioactive compounds, color, and antioxidant activity of strawberry fruits. *J. Agric. Food Chem.* **2009**, *57*, 1337–1343. [CrossRef] [PubMed]
160. Rozanska, D.; Regulska-Ilow, B. The significance of anthocyanins in the prevention and treatment of type 2 diabetes. *Adv. Clin. Exp. Med. Off. Organ Wroc. Med. Univ.* **2018**, *27*, 135–142. [CrossRef] [PubMed]
161. Cassidy, A.; Mukamal, K.J.; Liu, L.; Franz, M.; Eliassen, A.H.; Rimm, E.B. High anthocyanin intake is associated with a reduced risk of myocardial infarction in young and middle-aged women. *Circulation* **2013**, *127*, 188–196. [CrossRef] [PubMed]
162. Hollman, P.C.; Geelen, A.; Kromhout, D. Dietary flavonol intake may lower stroke risk in men and women. *J. Nutr.* **2010**, *140*, 600–604. [CrossRef]
163. Jennings, A.; Welch, A.A.; Fairweather-Tait, S.J.; Kay, C.; Minihane, A.M.; Chowienczyk, P.; Jiang, B.; Cecelja, M.; Spector, T.; Macgregor, A.; et al. Higher anthocyanin intake is associated with lower arterial stiffness and central blood pressure in women. *Am. J. Clin. Nutr.* **2012**, *96*, 781–788. [CrossRef]
164. Inaguma, T.; Han, J.; Isoda, H. Improvement of insulin resistance by Cyanidin 3-glucoside, anthocyanin from black beans through the up-regulation of GLUT4 gene expression. *BMC Proc.* **2011**, *5* (Suppl. 8), 21. [CrossRef]
165. Takikawa, M.; Inoue, S.; Horio, F.; Tsuda, T. Dietary anthocyanin-rich bilberry extract ameliorates hyperglycemia and insulin sensitivity via activation of AMP-activated protein kinase in diabetic mice. *J. Nutr.* **2010**, *140*, 527–533. [CrossRef]
166. Kurimoto, Y.; Shibayama, Y.; Inoue, S.; Soga, M.; Takikawa, M.; Ito, C.; Nanba, F.; Yoshida, T.; Yamashita, Y.; Ashida, H.; et al. Black soybean seed coat extract ameliorates hyperglycemia and insulin sensitivity via the activation of AMP-activated protein kinase in diabetic mice. *J. Agric. Food Chem.* **2013**, *61*, 5558–5564. [CrossRef]
167. Rios, J.L.; Francini, F.; Schinella, G.R. Natural Products for the Treatment of Type 2 Diabetes Mellitus. *Planta Med.* **2015**, *81*, 975–994. [CrossRef]
168. Bedekar, A.; Shah, K.; Koffas, M. Natural products for type II diabetes treatment. *Adv. Appl. Microbiol.* **2010**, *71*, 21–73. [CrossRef]
169. McDougall, G.J.; Stewart, D. The inhibitory effects of berry polyphenols on digestive enzymes. *BioFactors* **2005**, *23*, 189–195. [CrossRef]
170. Mussatto, S.; Mancilha, I. Non-digestible oligosaccharides: A review. *Carbohydr. Polym.* **2007**, *68*, 587–597. [CrossRef]
171. Zhu, D.; Yan, Q.; Liu, J.; Wu, X.; Jiang, Z. Can functional oligosaccharides reduce the risk of diabetes mellitus? *FASEB J. Off. Publ. Fed. Am. Soc. Exp. Biol.* **2019**, *33*, 11655–11667. [CrossRef]
172. Cheng, W.; Lu, J.; Li, B.; Lin, W.; Zhang, Z.; Wei, X.; Sun, C.; Chi, M.; Bi, W.; Yang, B.; et al. Effect of Functional Oligosaccharides and Ordinary Dietary Fiber on Intestinal Microbiota Diversity. *Front. Microbiol.* **2017**, *8*, 1750. [CrossRef]
173. Zhou, W.; Chen, G.; Dan, C.; Ye, H. The antidiabetic effect and potential mechanisms of natural polysaccharides based on the regulation of gut microbiota. *J. Funct. Foods* **2020**, *75*, 104222. [CrossRef]
174. Courtois, J. Oligosaccharides from land plants and algae: Production and applications in therapeutics and biotechnology. *Curr. Opin. Microbiol.* **2009**, *12*, 261–273. [CrossRef]
175. Dhouibi, R.; Affes, H.; Ben Salem, M.; Hammami, S.; Sahnoun, Z.; Zeghal, K.M.; Ksouda, K. Screening of pharmacological uses of Urtica dioica and others benefits. *Prog. Biophys. Mol. Biol.* **2020**, *150*, 67–77. [CrossRef]
176. Pourmorad, F.; Hosseinimehr, S.J.; Shahabimajd, N. Antioxidant activity of phenol and flavonoid contents of some Iranian medicinal plants. *Afr. J. Biotechnol.* **2006**, *5*, 1142–1145.
177. Bnouham, M.; Merhfour, F.Z.; Ziyyat, A.; Mekhfi, H.; Aziz, M.; Legssyer, A. Antihyperglycemic activity of the aqueous extract of Urtica dioica. *Fitoterapia* **2003**, *74*, 677–681. [CrossRef]
178. Onal, S.; Timur, S.; Okutucu, B.; Zihnioglu, F. Inhibition of alpha-glucosidase by aqueous extracts of some potent antidiabetic medicinal herbs. *Prep. Biochem. Biotechnol.* **2005**, *35*, 29–36. [CrossRef] [PubMed]
179. Farzami, B.; Ahmadvand, D.; Vardasbi, S.; Majin, F.J.; Khaghani, S. Induction of insulin secretion by a component of Urtica dioica leave extract in perifused Islets of Langerhans and its in vivo effects in normal and streptozotocin diabetic rats. *J. Ethnopharmacol.* **2003**, *89*, 47–53. [CrossRef]
180. Obanda, D.N.; Hernandez, A.; Ribnicky, D.; Yu, Y.; Zhang, X.H.; Wang, Z.Q.; Cefalu, W.T. Bioactives of *Artemisia dracunculus* L. mitigate the role of ceramides in attenuating insulin signaling in rat skeletal muscle cells. *Diabetes* **2012**, *61*, 597–605. [CrossRef] [PubMed]
181. Kraegen, E.W.; Cooney, G.J.; Ye, J.M.; Thompson, A.L.; Furler, S.M. The role of lipids in the pathogenesis of muscle insulin resistance and beta cell failure in type II diabetes and obesity. *Exp. Clin. Endocrinol. Diabetes Off. J. Ger. Soc. Endocrinol. Ger. Diabetes Assoc.* **2001**, *109* (Suppl. 2), S189–S201. [CrossRef]
182. Blouin, C.M.; Prado, C.; Takane, K.K.; Lasnier, F.; Garcia-Ocana, A.; Ferre, P.; Dugail, I.; Hajduch, E. Plasma membrane subdomain compartmentalization contributes to distinct mechanisms of ceramide action on insulin signaling. *Diabetes* **2010**, *59*, 600–610. [CrossRef] [PubMed]
183. Jope, R.S.; Yuskaitis, C.J.; Beurel, E. Glycogen synthase kinase-3 (GSK3): Inflammation, diseases, and therapeutics. *Neurochem. Res.* **2007**, *32*, 577–595. [CrossRef]
184. Kusminski, C.M.; Scherer, P.E. Lowering ceramides to overcome diabetes. *Science* **2019**, *365*, 319–320. [CrossRef]
185. Patel, S.S.; Udayabanu, M. Effect of Urtica dioica on memory dysfunction and hypoalgesia in an experimental model of diabetic neuropathy. *Neurosci. Lett.* **2013**, *552*, 114–119. [CrossRef]

186. Golalipour, M.J.; Kabiri Balajadeh, B.; Ghafari, S.; Azarhosh, R.; Khori, V. Protective Effect of *Urtica dioica* L. (Urticaceae) on Morphometric and Morphologic Alterations of Seminiferous Tubules in STZ Diabetic Rats. *Iran. J. Basic Med. Sci.* **2011**, *14*, 472–477.
187. Shokrzadeh, M.; Sadat-Hosseini, S.; Fallah, M.; Shaki, F. Synergism effects of pioglitazone and Urtica dioica extract in streptozotocin-induced nephropathy via attenuation of oxidative stress. *Iran. J. Basic Med. Sci.* **2017**, *20*, 497–502. [CrossRef]
188. Tesch, G.H.; Allen, T.J. Rodent models of streptozotocin-induced diabetic nephropathy. *Nephrology* **2007**, *12*, 261–266. [CrossRef]
189. Seyydi, S.M.; Tofighi, A.; Rahmati, M.; Tolouei Azar, J. Exercise and *Urtica dioica* extract ameliorate mitochondrial function and the expression of cardiac muscle Nuclear Respiratory Factor 2 and Peroxisome proliferator-activated receptor Gamma Coactivator 1-alpha in STZ-induced diabetic rats. *Gene* **2022**, *822*, 146351. [CrossRef]
190. Rabinovitch, A.; Suarez-Pinzon, W.L. Cytokines and their roles in pancreatic islet beta-cell destruction and insulin-dependent diabetes mellitus. *Biochem. Pharmacol.* **1998**, *55*, 1139–1149. [CrossRef]
191. Clark, R.B. The role of PPARs in inflammation and immunity. *J. Leukoc. Biol.* **2002**, *71*, 388–400. [CrossRef]
192. Rezaei, S.; Ashkar, F.; Koohpeyma, F.; Mahmoodi, M.; Gholamalizadeh, M.; Mazloom, Z.; Doaei, S. Hydroalcoholic extract of Achillea millefolium improved blood glucose, liver enzymes and lipid profile compared to metformin in streptozotocin-induced diabetic rats. *Lipids Health Dis.* **2020**, *19*, 81. [CrossRef]
193. Furman, B.L. Streptozotocin-Induced Diabetic Models in Mice and Rats. *Curr. Protoc.* **2021**, *1*, e78. [CrossRef]
194. Pini, A.; Grange, C.; Veglia, E.; Argenziano, M.; Cavalli, R.; Guasti, D.; Calosi, L.; Ghe, C.; Solarino, R.; Thurmond, R.L.; et al. Histamine H4 receptor antagonism prevents the progression of diabetic nephropathy in male DBA2/J mice. *Pharm. Res.* **2018**, *128*, 18–28. [CrossRef] [PubMed]
195. Verta, R.; Grange, C.; Gurrieri, M.; Borga, S.; Nardini, P.; Argenziano, M.; Ghe, C.; Cavalli, R.; Benetti, E.; Miglio, G.; et al. Effect of Bilastine on Diabetic Nephropathy in DBA2/J Mice. *Int. J. Mol. Sci.* **2019**, *20*, 2554. [CrossRef] [PubMed]
196. Mouhid, L.; Gomez de Cedron, M.; Quijada-Freire, A.; Fernandez-Marcos, P.J.; Reglero, G.; Fornari, T.; Ramirez de Molina, A. Yarrow Supercritical Extract Ameliorates the Metabolic Stress in a Model of Obesity Induced by High-Fat Diet. *Nutrients* **2019**, *12*, 72. [CrossRef]
197. Gharibi, S.; Tabatabaei, B.E.; Saeidi, G.; Goli, S.A. Effect of Drought Stress on Total Phenolic, Lipid Peroxidation, and Antioxidant Activity of Achillea Species. *Appl. Biochem. Biotechnol.* **2016**, *178*, 796–809. [CrossRef]
198. Nematy, M.; Mazidi, M.; Jafari, A.; Baghban, S.; Rakhshandeh, H.; Norouzy, A.; Esmaily, H.; Etemad, L.; Patterson, M.; Mohammadpour, A.H. The effect of hydro-alcoholic extract of Achillea millefolium on appetite hormone in rats. *Avicenna J. Phytomed.* **2017**, *7*, 10–15. [PubMed]
199. Sidorova, Y.; Shipelin, V.; Mazo, V.; Zorin, S.; Petrov, N.; Kochetkova, A. Hypoglycemic and hypolipidemic effect of *Vaccinium myrtillus* L. leaf and *Phaseolus vulgaris* L. seed coat extracts in diabetic rats. *Nutrition* **2017**, *41*, 107–112. [CrossRef] [PubMed]
200. Chen, K.; Wei, X.; Zhang, J.; Pariyani, R.; Jokioja, J.; Kortesniemi, M.; Linderborg, K.M.; Heinonen, J.; Sainio, T.; Zhang, Y.; et al. Effects of Anthocyanin Extracts from Bilberry (*Vaccinium myrtillus* L.) and Purple Potato (*Solanum tuberosum* L. var. 'Synkea Sakari') on the Plasma Metabolomic Profile of Zucker Diabetic Fatty Rats. *J. Agric. Food Chem.* **2020**, *68*, 9436–9450. [CrossRef] [PubMed]
201. Shiota, M.; Printz, R.L. Diabetes in Zucker diabetic fatty rat. *Methods Mol. Biol.* **2012**, *933*, 103–123. [CrossRef] [PubMed]
202. Mykkanen, O.T.; Kalesnykas, G.; Adriaens, M.; Evelo, C.T.; Torronen, R.; Kaarniranta, K. Bilberries potentially alleviate stress-related retinal gene expression induced by a high-fat diet in mice. *Mol. Vis.* **2012**, *18*, 2338–2351.
203. Matysek, M.; Borowiec, K.; Szwajgier, D.; Szalak, R.; Arciszewski, M.B. Insulin receptors in the CA1 field of hippocampus and selected blood parameters in diabetic rats fed with bilberry fruit. *Ann. Agric. Environ. Med.* **2021**, *28*, 430–436. [CrossRef]
204. Velickov, A.; Mitrovic, O.; Djordjevic, B.; Sokolovic, D.; Zivkovic, V.; Velickov, A.; Pantovic, V.; Urlih, N.P.; Radenkovic, G. The effect of bilberries on diabetes-related alterations of interstitial cells of Cajal in the lower oesophageal sphincter in rats. *Histol. Histopathol.* **2017**, *32*, 639–647. [CrossRef]
205. Cho, H. Protein tyrosine phosphatase 1B (PTP1B) and obesity. *Vitam. Horm.* **2013**, *91*, 405–424. [CrossRef]
206. Nieto-Vazquez, I.; Fernandez-Veledo, S.; de Alvaro, C.; Rondinone, C.M.; Valverde, A.M.; Lorenzo, M. Protein-tyrosine phosphatase 1B-deficient myocytes show increased insulin sensitivity and protection against tumor necrosis factor-alpha-induced insulin resistance. *Diabetes* **2007**, *56*, 404–413. [CrossRef]
207. Delibegovic, M.; Bence, K.K.; Mody, N.; Hong, E.G.; Ko, H.J.; Kim, J.K.; Kahn, B.B.; Neel, B.G. Improved glucose homeostasis in mice with muscle-specific deletion of protein-tyrosine phosphatase 1B. *Mol. Cell. Biol.* **2007**, *27*, 7727–7734. [CrossRef]
208. Silva, F.M.; Kramer, C.K.; de Almeida, J.C.; Steemburgo, T.; Gross, J.L.; Azevedo, M.J. Fiber intake and glycemic control in patients with type 2 diabetes mellitus: A systematic review with meta-analysis of randomized controlled trials. *Nutr. Rev.* **2013**, *71*, 790–801. [CrossRef]
209. Derrick, S.A.; Kristo, A.S.; Reaves, S.K.; Sikalidis, A.K. Effects of Dietary Red Raspberry Consumption on Pre-Diabetes and Type 2 Diabetes Mellitus Parameters. *Int. J. Environ. Res. Public Health* **2021**, *18*, 9364. [CrossRef]
210. Spinola, V.; Pinto, J.; Llorent-Martinez, E.J.; Tomas, H.; Castilho, P.C. Evaluation of *Rubus grandifolius* L. (wild blackberries) activities targeting management of type-2 diabetes and obesity using in vitro models. *Food Chem. Toxicol. Int. J. Publ. Br. Ind. Biol. Res. Assoc.* **2019**, *123*, 443–452. [CrossRef]

211. Wedick, N.M.; Pan, A.; Cassidy, A.; Rimm, E.B.; Sampson, L.; Rosner, B.; Willett, W.; Hu, F.B.; Sun, Q.; van Dam, R.M. Dietary flavonoid intakes and risk of type 2 diabetes in US men and women. *Am. J. Clin. Nutr.* **2012**, *95*, 925–933. [CrossRef]
212. Luo, T.; Miranda-Garcia, O.; Sasaki, G.; Shay, N.F. Consumption of a single serving of red raspberries per day reduces metabolic syndrome parameters in high-fat fed mice. *Food Funct.* **2017**, *8*, 4081–4088. [CrossRef]
213. Fan, R.; You, M.; Toney, A.M.; Kim, J.; Giraud, D.; Xian, Y.; Ye, F.; Gu, L.; Ramer-Tait, A.E.; Chung, S. Red Raspberry Polyphenols Attenuate High-Fat Diet-Driven Activation of NLRP3 Inflammasome and its Paracrine Suppression of Adipogenesis via Histone Modifications. *Mol. Nutr. Food Res.* **2020**, *64*, e1900995. [CrossRef]
214. Stefanut, M.N.; Cata, A.; Pop, R.; Tanasie, C.; Boc, D.; Ienascu, I.; Ordodi, V. Anti-hyperglycemic effect of bilberry, blackberry and mulberry ultrasonic extracts on diabetic rats. *Plant Foods Hum. Nutr.* **2013**, *68*, 378–384. [CrossRef]
215. Jouad, H.; Maghrani, M.; Eddouks, M. Hypoglycaemic effect of *Rubus fructicosis* L. and *Globularia alypum* L. in normal and streptozotocin-induced diabetic rats. *J. Ethnopharmacol.* **2002**, *81*, 351–356. [CrossRef]
216. Lalanza, J.F.; Snoeren, E.M. The cafeteria diet: A standardized protocol and its effects on behavior. *Neurosci. Biobehav. Rev.* **2021**, *122*, 92–119. [CrossRef]
217. Fierascu, R.C.; Temocico, G.; Fierascu, I.; Ortan, A.; Babeanu, N.E. Fragaria Genus: Chemical Composition and Biological Activities. *Molecules* **2020**, *25*, 498. [CrossRef]
218. Roy, S.; Wu, B.; Liu, W.; Archbold, D.D. Comparative analyses of polyphenolic composition of *Fragaria* spp. color mutants. *Plant Physiol. Biochem. PPB* **2018**, *125*, 255–261. [CrossRef] [PubMed]
219. Abdulazeez, S.S. Effects of freeze-dried Fragaria x ananassa powder on alloxan-induced diabetic complications in Wistar rats. *J. Taibah Univ. Med. Sci.* **2014**, *9*, 268–273. [CrossRef]
220. Yella, V.; Dass, A. Effect of Ethanolic Extract of Fragaria Vesca on serum glucose levels and body weight in diet induced obese rats. *Int. J. Pharmacol. Res.* **2015**, *5*, 236–238. [CrossRef]
221. Bailey, C.J.; Flatt, P.R. Development of antidiabetic drugs. In *Drugs, Diet and Disease*; Ioannides, C., Flatt, P.R., Eds.; Mechanistic Approaches to Diabetes; Ellis Horwood Ltd.: Chichester, UK, 1995; Volume 2, pp. 279–326.
222. Gray, A.M.; Abdel-Wahab, Y.H.; Flatt, P.R. The traditional plant treatment, *Sambucus nigra* (elder), exhibits insulin-like and insulin-releasing actions in vitro. *J. Nutr.* **2000**, *130*, 15–20. [CrossRef] [PubMed]
223. Gray, A.M.; Flatt, P.R. Pancreatic and extra-pancreatic effects of the traditional anti-diabetic plant, Medicago sativa (lucerne). *Br. J. Nutr.* **1997**, *78*, 325–334. [CrossRef] [PubMed]
224. Farrell, N.; Norris, G.; Lee, S.G.; Chun, O.K.; Blesso, C.N. Anthocyanin-rich black elderberry extract improves markers of HDL function and reduces aortic cholesterol in hyperlipidemic mice. *Food Funct.* **2015**, *6*, 1278–1287. [CrossRef]
225. Raafat, K.; El-Lakany, A. Acute and subchronic in-vivo effects of *Ferula hermonis* L. and *Sambucus nigra* L. and their potential active isolates in a diabetic mouse model of neuropathic pain. *BMC Complement. Altern. Med.* **2015**, *15*, 257. [CrossRef]
226. Badescu, M.; Badulcocu, O.; Badescu, L.; Ciocolu, M. Effects of *Sambucus nigra* and Aronia melanocarpa extracts on immune system disorders within diabetes mellitus. *Pharm. Biol.* **2015**, *53*, 533–539. [CrossRef]
227. Orhan, N.; Aslan, M.; Hosbas, C.S.; Didem, D.D. Antidiabetic Effect and Antioxidant Potential of *Rosa canina* Fruits. *Pharmacogn. Mag.* **2009**, *20*, 309–315. [CrossRef]
228. Orhan, D.; Hartevioglu, A.; Kupeli, E.; Yesilada, E. In vivo anti-inflammatory and antinociceptive activity of the crude extract and fractions from *Rosa canina* L. fruits. *J. Ethnopharmacol.* **2007**, *112*, 394–400. [CrossRef]
229. Wenzig, E.M.; Widowitz, U.; Kunert, O.; Chrubasik, S.; Bucar, F.; Knauder, E.; Bauer, R. Phytochemical composition and in vitro pharmacological activity of two rose hip (*Rosa canina* L.) preparations. *Phytomedicine* **2008**, *15*, 826–835. [CrossRef] [PubMed]
230. Hashem Dabaghian, F.; Abdollahifard, M.; Khalighi, S.F.; Taghavi, S.M.; Shojaee, A.; Sabet, Z.; Fallah, H.H. Effects of *Rosa canina* L. Fruit on Glycemia and Lipid Profile in Type 2 Diabetic Patients: A Randomized, Double-Blind, Placebo-Controlled Clinical Trial. *J. Med. Plants* **2015**, *14*, 95–104.
231. Michau, A.; Guillemain, G.; Grosfeld, A.; Vuillaumier-Barrot, S.; Grand, T.; Keck, M.; L'Hoste, S.; Chateau, D.; Serradas, P.; Teulon, J.; et al. Mutations in SLC2A2 gene reveal hGLUT2 function in pancreatic beta cell development. *J. Biol. Chem.* **2013**, *288*, 31080–31092. [CrossRef] [PubMed]
232. Bahrami, G.; Sajadimajd, S.; Mohammadi, B.; Hatami, R.; Miraghaee, S.; Keshavarzi, S.; Khazaei, M.; Madani, S.H. Anti-diabetic effect of a novel oligosaccharide isolated from *Rosa canina* via modulation of DNA methylation in Streptozotocin-diabetic rats. *DARU J. Fac. Pharm. Tehran Univ. Med. Sci.* **2020**, *28*, 581–590. [CrossRef] [PubMed]
233. Guo, Y.B.; Ji, T.F.; Zhou, H.W.; Yu, J.L. Effects of microRNA-21 on Nerve Cell Regeneration and Neural Function Recovery in Diabetes Mellitus Combined with Cerebral Infarction Rats by Targeting PDCD4. *Mol. Neurobiol.* **2018**, *55*, 2494–2505. [CrossRef]
234. Zhang, Y.; Zhao, S.; Wu, D.; Liu, X.; Shi, M.; Wang, Y.; Zhang, F.; Ding, J.; Xiao, Y.; Guo, B. MicroRNA-22 Promotes Renal Tubulointerstitial Fibrosis by Targeting PTEN and Suppressing Autophagy in Diabetic Nephropathy. *J. Diabetes Res.* **2018**, *2018*, 4728645. [CrossRef]
235. Qadir, M.M.F.; Klein, D.; Alvarez-Cubela, S.; Dominguez-Bendala, J.; Pastori, R.L. The Role of MicroRNAs in Diabetes-Related Oxidative Stress. *Int. J. Mol. Sci.* **2019**, *20*, 5423. [CrossRef]
236. Sajadimajd, S.; Bahrami, G.; Mohammadi, B.; Nouri, Z.; Farzaei, M.H.; Chen, J.T. Protective effect of the isolated oligosaccharide from *Rosa canina* in STZ-treated cells through modulation of the autophagy pathway. *J. Food Biochem.* **2020**, *44*, e13404. [CrossRef]
237. Seeram, N.P.; Schutzki, R.; Chandra, A.; Nair, M.G. Characterization, quantification, and bioactivities of anthocyanins in Cornus species. *J. Agric. Food Chem.* **2002**, *50*, 2519–2523. [CrossRef]

238. Jayaprakasam, B.; Olson, L.K.; Schutzki, R.E.; Tai, M.H.; Nair, M.G. Amelioration of obesity and glucose intolerance in high-fat-fed C57BL/6 mice by anthocyanins and ursolic acid in Cornelian cherry (*Cornus mas*). *J. Agric. Food Chem.* **2006**, *54*, 243–248. [CrossRef]

239. Duthie, G.G.; Duthie, S.J.; Kyle, J.A. Plant polyphenols in cancer and heart disease: Implications as nutritional antioxidants. *Nutr. Res. Rev.* **2000**, *13*, 79–106. [CrossRef]

240. Tural, S.; Koca, I. Physico-chemical and antioxidant properties of cornelian cherry fruits (*Cornus mas* L.) grown in Turkey. *Sci. Hortic.* **2008**, *116*, 362–366. [CrossRef]

241. Asgary, S.; Rafieian-Kopaei, M.; Shamsi, F.; Najafi, S.; Sahebkar, A. Biochemical and histopathological study of the anti-hyperglycemic and anti-hyperlipidemic effects of cornelian cherry (*Cornus mas* L.) in alloxan-induced diabetic rats. *J. Complement. Integr. Med.* **2014**, *11*, 63–69. [CrossRef]

242. Ziaei, R.; Foshati, S.; Hadi, A.; Kermani, M.A.H.; Ghavami, A.; Clark, C.C.T.; Tarrahi, M.J. The effect of nettle (*Urtica dioica*) supplementation on the glycemic control of patients with type 2 diabetes mellitus: A systematic review and meta-analysis. *Phytother. Res.* **2020**, *34*, 282–294. [CrossRef]

243. Korani, B.; Mirzapour, A.; Moghadamnia, A.; Khafri, S.; Neamati, N.; Parsian, H. The Effect of Urtica dioica Hydro-Alcoholic Extract on Glycemic Index and AMP-Activated Protein Kinase Levels in Diabetic Patients: A Randomized Single-Blind Clinical Trial. *Iran. Red Crescent Med. J.* **2017**, *19*, e40572. [CrossRef]

244. Khajeh-Mehrizi, R.; Mozaffari-khosravi, H.; Ghadiri-anari, A.; Dehghani, A. The Effect of *Urtica dioica* Extract on Glycemic Control and Insulin Resistance Indices in Patients with Type2 Diabetes: A Randomized, Double-Blind Clinical Trial. *Iran. J. Diabetes Obes.* **2014**, *6*, 149–155.

245. Dabagh, S.; Nikbakht, M. Glycemic Control by Exercise and *Urtica dioica* Supplements in Men with Type 2 Diabetes. *Jundishapur J. Chronic Dis. Care* **2016**, *5*, e31745. [CrossRef]

246. Ghalavand, A.; Motamedi, P.; Delaramnasab, M.; Khodadoust, M. The Effect of Interval Training and Nettle Supplement on Glycemic Control and Blood Pressure in Men WithType 2 Diabetes. *Int. J. Basic Sci. Med.* **2017**, *2*, 33–40. [CrossRef]

247. Kianbakht, S.; Khalighi-Sigaroodi, F.; Dabaghian, F.H. Improved glycemic control in patients with advanced type 2 diabetes mellitus taking Urtica dioica leaf extract: A randomized double-blind placebo-controlled clinical trial. *Clin. Lab.* **2013**, *59*, 1071–1076. [CrossRef]

248. Tarighat, E.A.; Namazi, N.; Bahrami, A.M.E. Effect of Hydroalcoholic Extract Of Nettle (Urtica Dioica) On Glycemic Index And Insulin Resistance Index In Type 2 Diabetic Patients. *Iran. J. Endocrinol. Metab. (IJEM)* **2012**, *13*, 561–568.

249. Hassani, A.; Maryam, E.; Reza, R. Survey on the Effect of Eight Weeks of Regular Aerobic Eexercise with Consumption of Nettle Extract on Blood Glucose and Insulin Resistance Index among women with Type II Diabetes. *J. Knowl. Health* **2016**, *10*, 57–64.

250. Dadvar, N.; Ghalavand, K.; Zakerkish, M.; Hojat, S.; Alijani, E.; Mahmoodkhani, K.R. The Effect Of Aerobic Training And *Urtica dioica* On Lipid Profile And Fasting Blood Glucose In Middle Age Female With Type Ii Diabetes. *Jundishapur Sci. Med. J. (JSMJ)* **2016**, *15*, 707–716.

251. Namazi, N.; Esfanjani, A.T.; Heshmati, J.; Bahrami, A. The effect of hydro alcoholic Nettle (*Urtica dioica*) extracts on insulin sensitivity and some inflammatory indicators in patients with type 2 diabetes: A randomized double-blind control trial. *Pak. J. Biol. Sci. PJBS* **2011**, *14*, 775–779. [CrossRef] [PubMed]

252. Namazi, N.; Tarighat, A.; Bahrami, A. The effect of hydro alcoholic nettle (*Urtica dioica*) extract on oxidative stress in patients with type 2 diabetes: A randomized double-blind clinical trial. *Pak. J. Biol. Sci. PJBS* **2012**, *15*, 98–102. [CrossRef] [PubMed]

253. Amiri Behzadi, A.; Kalalian-Moghaddam, H.; Ahmadi, A.H. Effects of Urtica dioica supplementation on blood lipids, hepatic enzymes and nitric oxide levels in type 2 diabetic patients: A double blind, randomized clinical trial. *Avicenna J. Phytomed.* **2016**, *6*, 686–695.

254. Cohen, R.A. Role of nitric oxide in diabetic complications. *Am J. Ther.* **2005**, *12*, 499–502. [CrossRef]

255. Zhu, Y.; Miao, Y.; Meng, Z.; Zhong, Y. Effects of Vaccinium Berries on Serum Lipids: A Meta-Analysis of Randomized Controlled Trials. *Evid.-Based Complement. Altern. Med.* **2015**, *2015*, 790329. [CrossRef]

256. Basu, A.; Lyons, T.J. Strawberries, blueberries, and cranberries in the metabolic syndrome: Clinical perspectives. *J. Agric. Food Chem.* **2012**, *60*, 5687–5692. [CrossRef]

257. Grundy, S.M.; Stone, N.J.; Bailey, A.L.; Beam, C.; Birtcher, K.K.; Blumenthal, R.S.; Braun, L.T.; de Ferranti, S.; Faiella-Tommasino, J.; Forman, D.E.; et al. 2018 AHA/ACC/AACVPR/AAPA/ABC/ACPM/ADA/AGS/APhA/ASPC/NLA/PCNA Guideline on the Management of Blood Cholesterol: A Report of the American College of Cardiology/American Heart Association Task Force on Clinical Practice Guidelines. *J. Am. Coll. Cardiol.* **2019**, *73*, e285–e350. [CrossRef]

258. Mattiuzzi, C.; Sanchis-Gomar, F.; Lippi, G. Worldwide burden of LDL cholesterol: Implications in cardiovascular disease. *Nutr. Metab. Cardiovasc. Dis.* **2020**, *30*, 241–244. [CrossRef]

259. Burton-Freeman, B.; Linares, A.; Hyson, D.; Kappagoda, T. Strawberry modulates LDL oxidation and postprandial lipemia in response to high-fat meal in overweight hyperlipidemic men and women. *J. Am. Coll. Nutr.* **2010**, *29*, 46–54. [CrossRef]

260. Park, E.; Edirisinghe, I.; Wei, H.; Vijayakumar, L.P.; Banaszewski, K.; Cappozzo, J.C.; Burton-Freeman, B. A dose-response evaluation of freeze-dried strawberries independent of fiber content on metabolic indices in abdominally obese individuals with insulin resistance in a randomized, single-blinded, diet-controlled crossover trial. *Mol. Nutr. Food Res.* **2016**, *60*, 1099–1109. [CrossRef]

261. Huang, Y.; Park, E.; Edirisinghe, I.; Burton-Freeman, B.M. Maximizing the health effects of strawberry anthocyanins: Understanding the influence of the consumption timing variable. *Food Funct.* **2016**, *7*, 4745–4752. [CrossRef]
262. Basu, A.; Betts, N.M.; Nguyen, A.; Newman, E.D.; Fu, D.; Lyons, T.J. Freeze-dried strawberries lower serum cholesterol and lipid peroxidation in adults with abdominal adiposity and elevated serum lipids. *J. Nutr.* **2014**, *144*, 830–837. [CrossRef]
263. Basu, A.; Izuora, K.; Betts, N.M.; Kinney, J.W.; Salazar, A.M.; Ebersole, J.L.; Scofield, R.H. Dietary Strawberries Improve Cardiometabolic Risks in Adults with Obesity and Elevated Serum LDL Cholesterol in a Randomized Controlled Crossover Trial. *Nutrients* **2021**, *13*, 1421. [CrossRef]
264. Schell, J.; Scofield, R.H.; Barrett, J.R.; Kurien, B.T.; Betts, N.; Lyons, T.J.; Zhao, Y.D.; Basu, A. Strawberries Improve Pain and Inflammation in Obese Adults with Radiographic Evidence of Knee Osteoarthritis. *Nutrients* **2017**, *9*, 949. [CrossRef]
265. Soltani, R.; Gorji, A.; Asgary, S.; Sarrafzadegan, N.; Siavash, M. Evaluation of the Effects of *Cornus mas* L. Fruit Extract on Glycemic Control and Insulin Level in Type 2 Diabetic Adult Patients: A Randomized Double-Blind Placebo-Controlled Clinical Trial. *Evid.-Based Complement. Altern. Med.* **2015**, *2015*, 740954. [CrossRef]
266. Zhang, X.; Zhao, A.; Sandhu, A.K.; Edirisinghe, I.; Burton-Freeman, B.M. Red Raspberry and Fructo-Oligosaccharide Supplementation, Metabolic Biomarkers, and the Gut Microbiota in Adults with Prediabetes: A Randomized Crossover Clinical Trial. *J. Nutr.* **2022**, *152*, 1438–1449. [CrossRef]
267. Xiao, D.; Zhu, L.; Edirisinghe, I.; Fareed, J.; Brailovsky, Y.; Burton-Freeman, B. Attenuation of Postmeal Metabolic Indices with Red Raspberries in Individuals at Risk for Diabetes: A Randomized Controlled Trial. *Obesity* **2019**, *27*, 542–550. [CrossRef]
268. Franck, M.; de Toro-Martin, J.; Varin, T.V.; Garneau, V.; Pilon, G.; Roy, D.; Couture, P.; Couillard, C.; Marette, A.; Vohl, M.C. Raspberry consumption: Identification of distinct immune-metabolic response profiles by whole blood transcriptome profiling. *J. Nutr. Biochem.* **2022**, *101*, 108946. [CrossRef]
269. Puupponen-Pimia, R.; Seppanen-Laakso, T.; Kankainen, M.; Maukonen, J.; Torronen, R.; Kolehmainen, M.; Leppanen, T.; Moilanen, E.; Nohynek, L.; Aura, A.M.; et al. Effects of ellagitannin-rich berries on blood lipids, gut microbiota, and urolithin production in human subjects with symptoms of metabolic syndrome. *Mol. Nutr. Food Res.* **2013**, *57*, 2258–2263. [CrossRef] [PubMed]
270. Kolehmainen, M.; Mykkanen, O.; Kirjavainen, P.V.; Leppanen, T.; Moilanen, E.; Adriaens, M.; Laaksonen, D.E.; Hallikainen, M.; Puupponen-Pimia, R.; Pulkkinen, L.; et al. Bilberries reduce low-grade inflammation in individuals with features of metabolic syndrome. *Mol. Nutr. Food Res.* **2012**, *56*, 1501–1510. [CrossRef] [PubMed]
271. de Mello, V.D.; Lankinen, M.A.; Lindstrom, J.; Puupponen-Pimia, R.; Laaksonen, D.E.; Pihlajamaki, J.; Lehtonen, M.; Uusitupa, M.; Tuomilehto, J.; Kolehmainen, M.; et al. Fasting serum hippuric acid is elevated after bilberry (*Vaccinium myrtillus*) consumption and associates with improvement of fasting glucose levels and insulin secretion in persons at high risk of developing type 2 diabetes. *Mol. Nutr. Food Res.* **2017**, *61*, 1700019. [CrossRef] [PubMed]
272. Zhang, X.; Fan, J.; Xiao, D.; Edirisinghe, I.; Burton-Freeman, B.M.; Sandhu, A.K. Pharmacokinetic Evaluation of Red Raspberry (Poly)phenols from Two Doses and Association with Metabolic Indices in Adults with Prediabetes and Insulin Resistance. *J. Agric. Food Chem.* **2021**, *69*, 9238–9248. [CrossRef]
273. Schell, J.; Betts, N.M.; Lyons, T.J.; Basu, A. Raspberries Improve Postprandial Glucose and Acute and Chronic Inflammation in Adults with Type 2 Diabetes. *Ann. Nutr. Metab.* **2019**, *74*, 165–174. [CrossRef]
274. Solverson, P.M.; Rumpler, W.V.; Leger, J.L.; Redan, B.W.; Ferruzzi, M.G.; Baer, D.J.; Castonguay, T.W.; Novotny, J.A. Blackberry Feeding Increases Fat Oxidation and Improves Insulin Sensitivity in Overweight and Obese Males. *Nutrients* **2018**, *10*, 1048. [CrossRef]
275. Gallicchio, M.; Rosa, A.C.; Dianzani, C.; Brucato, L.; Benetti, E.; Collino, M.; Fantozzi, R. Celecoxib decreases expression of the adhesion molecules ICAM-1 and VCAM-1 in a colon cancer cell line (HT29). *Br. J. Pharm.* **2008**, *153*, 870–878. [CrossRef]
276. Lehtonen, H.M.; Suomela, J.P.; Tahvonen, R.; Yang, B.; Venojarvi, M.; Viikari, J.; Kallio, H. Different berries and berry fractions have various but slightly positive effects on the associated variables of metabolic diseases on overweight and obese women. *Eur. J. Clin. Nutr.* **2011**, *65*, 394–401. [CrossRef]
277. Yang, L.; Ling, W.; Qiu, Y.; Liu, Y.; Wang, L.; Yang, J.; Wang, C.; Ma, J. Anthocyanins increase serum adiponectin in newly diagnosed diabetes but not in prediabetes: A randomized controlled trial. *Nutr. Metab.* **2020**, *17*, 78. [CrossRef]
278. Chan, S.W.; Chu, T.T.; Choi, S.W.; Benzie, I.F.; Tomlinson, B. Impact of short-term bilberry supplementation on glycemic control, cardiovascular disease risk factors, and antioxidant status in Chinese patients with type 2 diabetes. *Phytother. Res.* **2021**, *35*, 3236–3245. [CrossRef]
279. Yang, L.; Ling, W.; Yang, Y.; Chen, Y.; Tian, Z.; Du, Z.; Chen, J.; Xie, Y.; Liu, Z.; Yang, L. Role of Purified Anthocyanins in Improving Cardiometabolic Risk Factors in Chinese Men and Women with Prediabetes or Early Untreated Diabetes—A Randomized Controlled Trial. *Nutrients* **2017**, *9*, 1104. [CrossRef]
280. Yang, L.; Liu, Z.; Ling, W.; Wang, L.; Wang, C.; Ma, J.; Peng, X.; Chen, J. Effect of Anthocyanins Supplementation on Serum IGFBP-4 Fragments and Glycemic Control in Patients with Fasting Hyperglycemia: A Randomized Controlled Trial. *Diabetes Metab. Syndr Obes* **2020**, *13*, 3395–3404. [CrossRef]
281. Yang, L.; Qiu, Y.; Ling, W.; Liu, Z.; Yang, L.; Wang, C.; Peng, X.; Wang, L.; Chen, J. Anthocyanins regulate serum adipsin and visfatin in patients with prediabetes or newly diagnosed diabetes: A randomized controlled trial. *Eur. J. Nutr.* **2021**, *60*, 1935–1944. [CrossRef]

282. Lo, J.C.; Ljubicic, S.; Leibiger, B.; Kern, M.; Leibiger, I.B.; Moede, T.; Kelly, M.E.; Chatterjee Bhowmick, D.; Murano, I.; Cohen, P.; et al. Adipsin is an adipokine that improves beta cell function in diabetes. *Cell* **2014**, *158*, 41–53. [CrossRef]
283. Chen, M.P.; Chung, F.M.; Chang, D.M.; Tsai, J.C.; Huang, H.F.; Shin, S.J.; Lee, Y.J. Elevated plasma level of visfatin/pre-B cell colony-enhancing factor in patients with type 2 diabetes mellitus. *J. Clin. Endocrinol. Metab.* **2006**, *91*, 295–299. [CrossRef]
284. Li, D.; Zhang, Y.; Liu, Y.; Sun, R.; Xia, M. Purified anthocyanin supplementation reduces dyslipidemia, enhances antioxidant capacity, and prevents insulin resistance in diabetic patients. *J. Nutr.* **2015**, *145*, 742–748. [CrossRef]
285. Hoggard, N.; Cruickshank, M.; Moar, K.M.; Bestwick, C.; Holst, J.J.; Russell, W.; Horgan, G. A single supplement of a standardised bilberry (*Vaccinium myrtillus* L.) extract (36% wet weight anthocyanins) modifies glycaemic response in individuals with type 2 diabetes controlled by diet and lifestyle. *J. Nutr. Sci.* **2013**, *2*, e22. [CrossRef]
286. Xu, J.; Jonsson, T.; Plaza, M.; Hakansson, A.; Antonsson, M.; Ahren, I.L.; Turner, C.; Spegel, P.; Granfeldt, Y. Probiotic fruit beverages with different polyphenol profiles attenuated early insulin response. *Nutr. J.* **2018**, *17*, 34. [CrossRef]
287. Mazzolani, F.; Togni, S.; Franceschi, F.; Eggenhoffner, R.; Giacomelli, L. The effect of oral supplementation with standardized bilberry extract (Mirtoselect®) on retino-cortical bioelectrical activity in severe diabetic retinopathy. *Minerva Oftalmol* **2017**, *59*, 38–41. [CrossRef]
288. Gizzi, C.; Belcaro, G.; Gizzi, G.; Feragalli, B.; Dugall, M.; Luzzi, R.; Cornelli, U. Bilberry extracts are not created equal: The role of non anthocyanin fraction. Discovering the "dark side of the force" in a preliminary study. *Eur. Rev. Med. Pharmacol. Sci.* **2016**, *20*, 2418–2424.
289. Mehrzadi, S.; Mirzaei, R.; Heydari, M.; Sasani, M.; Yaqoobvand, B.; Huseini, H.F. Efficacy and Safety of a Traditional Herbal Combination in Patients with Type II Diabetes Mellitus: A Randomized Controlled Trial. *J. Diet. Suppl.* **2021**, *18*, 31–43. [CrossRef]
290. Roman, I.; Stanila, A.; Stanila, S. Bioactive compounds and antioxidant activity of *Rosa canina* L. biotypes from spontaneous flora of Transylvania. *Chem. Cent. J.* **2013**, *7*, 73. [CrossRef]
291. Andersson, U.; Berger, K.; Hogberg, A.; Landin-Olsson, M.; Holm, C. Effects of rose hip intake on risk markers of type 2 diabetes and cardiovascular disease: A randomized, double-blind, cross-over investigation in obese persons. *Eur. J. Clin. Nutr.* **2012**, *66*, 585–590. [CrossRef]
292. Sanchez, M.; Gonzalez-Burgos, E.; Iglesias, I.; Lozano, R.; Gomez-Serranillos, M.P. Current uses and knowledge of medicinal plants in the Autonomous Community of Madrid (Spain): A descriptive cross-sectional study. *BMC Complement. Med. Ther.* **2020**, *20*, 306. [CrossRef] [PubMed]
293. Li, F.S.; Weng, J.K. Demystifying traditional herbal medicine with modern approach. *Nat. Plants* **2017**, *3*, 17109. [CrossRef] [PubMed]
294. Shikov, A.N.; Narkevich, I.A.; Akamova, A.V.; Nemyatykh, O.D.; Flisyuk, E.V.; Luzhanin, V.G.; Povydysh, M.N.; Mikhailova, I.V.; Pozharitskaya, O.N. Medical Species Used in Russia for the Management of Diabetes and Related Disorders. *Front. Pharm.* **2021**, *12*, 697411. [CrossRef] [PubMed]
295. Zunino, S.J.; Parelman, M.A.; Freytag, T.L.; Stephensen, C.B.; Kelley, D.S.; Mackey, B.E.; Woodhouse, L.R.; Bonnel, E.L. Effects of dietary strawberry powder on blood lipids and inflammatory markers in obese human subjects. *Br. J. Nutr.* **2012**, *108*, 900–909. [CrossRef]
296. Wu, J.; Yan, L.J. Streptozotocin-induced type 1 diabetes in rodents as a model for studying mitochondrial mechanisms of diabetic beta cell glucotoxicity. *Diabetes Metab. Syndr. Obes.* **2015**, *8*, 181–188. [CrossRef]
297. Grange, C.; Gurrieri, M.; Verta, R.; Fantozzi, R.; Pini, A.; Rosa, A.C. Histamine in the kidneys: What is its role in renal pathophysiology? *Br. J. Pharm.* **2020**, *177*, 503–515. [CrossRef]
298. Ighodaro, O.M.; Adeosun, A.M.; Akinloye, O.A. Alloxan-induced diabetes, a common model for evaluating the glycemic-control potential of therapeutic compounds and plants extracts in experimental studies. *Medicina* **2017**, *53*, 365–374. [CrossRef]
299. Heydemann, A. An Overview of Murine High Fat Diet as a Model for Type 2 Diabetes Mellitus. *J. Diabetes Res.* **2016**, *2016*, 2902351. [CrossRef]
300. Benetti, E.; Mastrocola, R.; Vitarelli, G.; Cutrin, J.C.; Nigro, D.; Chiazza, F.; Mayoux, E.; Collino, M.; Fantozzi, R. Empagliflozin Protects against Diet-Induced NLRP-3 Inflammasome Activation and Lipid Accumulation. *J. Pharm. Exp. Ther.* **2016**, *359*, 45–53. [CrossRef]
301. Doumett, S.; Fibbi, D.; Cincinelli, A.; Giordani, E.; Nin, S.; Del Bubba, M. Comparison of nutritional and nutraceutical properties in cultivated fruits of *Fragaria vesca* L. produced in Italy. *Food Res. Intern.* **2011**, *44*, 1209–1216. [CrossRef]
302. Peñarrieta, J.M.; Alvarado, J.A.; Bergenståhl, B.; Åkesson, B. Total antioxidant capacity and content of phenolic compounds in wild strawberries (*Fragaria vesca*) collected in Bolivia. *Int. J. Fruit Sci.* **2009**, *9*, 344–359. [CrossRef]
303. Larskaya, I.A.; Gorshkova, T.A. Plant oligosaccharides—Outsiders among elicitors? *Biochemistry* **2015**, *80*, 881–900. [CrossRef]
304. De Natale, C.; Annuzzi, G.; Bozzetto, L.; Mazzarella, R.; Costabile, G.; Ciano, O.; Riccardi, G.; Rivellese, A.A. Effects of a plant-based high-carbohydrate/high-fiber diet versus high-monounsaturated fat/low-carbohydrate diet on postprandial lipids in type 2 diabetic patients. *Diabetes Care* **2009**, *32*, 2168–2173. [CrossRef]
305. Zhao, L.; Zhang, F.; Ding, X.; Wu, G.; Lam, Y.Y.; Wang, X.; Fu, H.; Xue, X.; Lu, C.; Ma, J.; et al. Gut bacteria selectively promoted by dietary fibers alleviate type 2 diabetes. *Science* **2018**, *359*, 1151–1156. [CrossRef]
306. Annunziata, G.; Jimenez-Garcia, M.; Capo, X.; Moranta, D.; Arnone, A.; Tenore, G.C.; Sureda, A.; Tejada, S. Microencapsulation as a tool to counteract the typical low bioavailability of polyphenols in the management of diabetes. *Food Chem. Toxicol. Int. J. Publ. Br. Ind. Biol. Res. Assoc.* **2020**, *139*, 111248. [CrossRef]

307. Cerda, B.; Periago, P.; Espin, J.C.; Tomas-Barberan, F.A. Identification of urolithin a as a metabolite produced by human colon microflora from ellagic acid and related compounds. *J. Agric. Food Chem.* **2005**, *53*, 5571–5576. [CrossRef]
308. Cerda, B.; Tomas-Barberan, F.A.; Espin, J.C. Metabolism of antioxidant and chemopreventive ellagitannins from strawberries, raspberries, walnuts, and oak-aged wine in humans: Identification of biomarkers and individual variability. *J. Agric. Food Chem.* **2005**, *53*, 227–235. [CrossRef]
309. Hidalgo, M.; Oruna-Concha, M.J.; Kolida, S.; Walton, G.E.; Kallithraka, S.; Spencer, J.P.; de Pascual-Teresa, S. Metabolism of anthocyanins by human gut microflora and their influence on gut bacterial growth. *J. Agric. Food Chem.* **2012**, *60*, 3882–3890. [CrossRef]
310. Sip, S.; Szymanowska, D.; Chanaj-Kaczmarek, J.; Skalicka-Wozniak, K.; Budzynska, B.; Wronikowska-Denysiuk, O.; Slowik, T.; Szulc, P.; Cielecka-Piontek, J. Potential for Prebiotic Stabilized *Cornus mas* L. Lyophilized Extract in the Prophylaxis of Diabetes Mellitus in Streptozotocin Diabetic Rats. *Antioxidants* **2022**, *11*, 380. [CrossRef]
311. Oledzka, A.; Cichocka, K.; Wolinski, K.; Melzig, M.F.; Czerwinska, M.E. Potentially Bio-Accessible Metabolites from an Extract of *Cornus mas* Fruit after Gastrointestinal Digestion In Vitro and Gut Microbiota Ex Vivo Treatment. *Nutrients* **2022**, *14*, 2287. [CrossRef]
312. Hou, K.; Wu, Z.X.; Chen, X.Y.; Wang, J.Q.; Zhang, D.; Xiao, C.; Zhu, D.; Koya, J.B.; Wei, L.; Li, J.; et al. Microbiota in health and diseases. *Signal. Transduct. Target Ther.* **2022**, *7*, 135. [CrossRef]
313. Healey, G.R.; Murphy, R.; Brough, L.; Butts, C.A.; Coad, J. Interindividual variability in gut microbiota and host response to dietary interventions. *Nutr. Rev.* **2017**, *75*, 1059–1080. [CrossRef]
314. Bohn, T. Dietary factors affecting polyphenol bioavailability. *Nutr. Rev.* **2014**, *72*, 429–452. [CrossRef]
315. Paoletti, A.; Gallo, E.; Benemei, S.; Vietri, M.; Lapi, F.; Volpi, R.; Menniti-Ippolito, F.; Gori, L.; Mugelli, A.; Firenzuoli, F.; et al. Interactions between Natural Health Products and Oral Anticoagulants: Spontaneous Reports in the Italian Surveillance System of Natural Health Products. *Evid.-Based Complement. Altern. Med.* **2011**, *2011*, 612150. [CrossRef]
316. Yang, B.; Dong, Y.; Wang, F.; Zhang, Y. Nanoformulations to enhance the bioavailability and physiological functions of polyphenols. *Molecules* **2020**, *25*, 4613. [CrossRef]
317. Opris, R.; Tatomir, C.; Olteanu, D.; Moldovan, R.; Moldovan, B.; David, L.; Nagy, A.; Decea, N.; Kiss, M.L.; Filip, G.A. The effect of *Sambucus nigra* L. extract and phytosinthesized gold nanoparticles on diabetic rats. *Colloids Surf. B Biointerfaces* **2017**, *150*, 192–200. [CrossRef]

Review

Medicinal Components in Edible Mushrooms on Diabetes Mellitus Treatment

Arpita Das [1], Chiao-Ming Chen [2], Shu-Chi Mu [3,4], Shu-Hui Yang [5], Yu-Ming Ju [6] and Sing-Chung Li [1,*]

[1] School of Nutrition and Health Sciences, College of Nutrition, Taipei Medical University, Taipei 11031, Taiwan; da07110008@tmu.edu.tw

[2] Department of Food Science, Nutrition and Nutraceutical Biotechnology, Shih Chien University, Taipei 10462, Taiwan; charming@g2.usc.edu.tw

[3] Department of Pediatrics, Shin-Kong Wu Ho-Su Memorial Hospital, Taipei 11101, Taiwan; musc1004@gmail.com

[4] School of Medicine, College of Medicine, Fu-Jen Catholic University, New Taipei City 24205, Taiwan

[5] Fengshan Tropical Horticultural Experiment Branch, Taiwan Agricultural Research Institute, Kaohsiung City 83052, Taiwan; debbie@tari.gov.tw

[6] Institute of Plant and Microbial Biology, Academia Sinica, Taipei 11529, Taiwan; yumingju@gate.sinica.edu.tw

* Correspondence: sinchung@tmu.edu.tw; Tel.: +886-2-27361661 (ext. 6560)

Abstract: Mushrooms belong to the family "Fungi" and became famous for their medicinal properties and easy accessibility all over the world. Because of its pharmaceutical properties, including anti-diabetic, anti-inflammatory, anti-cancer, and antioxidant properties, it became a hot topic among scientists. However, depending on species and varieties, most of the medicinal properties became indistinct. With this interest, an attempt has been made to scrutinize the role of edible mushrooms (EM) in diabetes mellitus treatment. A systematic contemporary literature review has been carried out from all records such as Science Direct, PubMed, Embase, and Google Scholar with an aim to represents the work has performed on mushrooms focuses on diabetes, insulin resistance (IR), and preventive mechanism of IR, using different kinds of mushroom extracts. The final review represents that EM plays an important role in anticipation of insulin resistance with the help of active compounds, i.e., polysaccharide, vitamin D, and signifies α-glucosidase or α-amylase preventive activities. Although most of the mechanism is not clear yet, many varieties of mushrooms' medicinal properties have not been studied properly. So, in the future, further investigation is needed on edible medicinal mushrooms to overcome the research gap to use its clinical potential to prevent non-communicable diseases.

Keywords: edible mushroom; insulin resistance; diabetes; polysaccharide; vitamin D; α-glucosidase; α-amylase

Citation: Das, A.; Chen, C.-M.; Mu, S.-C.; Yang, S.-H.; Ju, Y.-M.; Li, S.-C. Medicinal Components in Edible Mushrooms on Diabetes Mellitus Treatment. *Pharmaceutics* **2022**, *14*, 436. https://doi.org/10.3390/pharmaceutics14020436

Academic Editor: Anna Rita Bilia

Received: 20 January 2022
Accepted: 14 February 2022
Published: 17 February 2022

Publisher's Note: MDPI stays neutral with regard to jurisdictional claims in published maps and institutional affiliations.

1. Introduction

The word "mushroom" is derived from the Latin and Greek words "fungus" and "mykes" and is considered edible if consumption does not cause health disorders. Based on the edibility, it can be divided into three groups-edible (*Lepiota procera*), inedible or poisonous (*Lepiota margani*), and non-poisonous [1]. For thousands of years, mushrooms have been used as food and medicine (38%), and many Asian and south-Asian countries use traditional wild edible mushrooms as appealing and nutritious foods, including China, Japan, India, and Taiwan. As a result, mushrooms received a high demand in the global market with 34 billion kg production and per capita consumption of 4.7 kg in 2013. Some of the significant ordinarily available edible mushrooms are chanterelles (*Cantharellaceae*), puffballs (*Lycoperdon* spp. and *Calvatia* spp.), shaggy mane (*Coprinus comatus*), two oyster mushrooms (*Pleurotus ostreatus* and *Pleurotus cystidiosus*), boletes (*Boletaceae*), sulfer shelf (*Laetiporus sulphurous*), hen of the woods (*Grifora frondosa*), button mushroom (*Agricus bisporus*), golden oyster mushroom

(*Pleurotus citrinopileatus*) [2–4], morels (*Morachella esculenta*), bearded tooth (*Hericium erinaceus*), straw mushroom (*Volvariella volvacea*) [5,6], Eenoki (*Flammulina velutipes*) [7], shiitake (*Lentinula edodes*) [8], beech mushroom (*Hypsizygus marmoreus*) [9], french horn mushroom (*Pleurotus eryngii*) [10], dancing mushroom (*Grifola frondosa*) [11], and black poplar mushroom (*Agrocybe aegerita*) [12]. Wild edible mushrooms are not only superior for the chemical and nutritional characteristics but also the protein and vitamin contents, B vitamins, vitamin D, vitamin K, and rarely vitamin A and C [13–15]. Moreover, mushrooms have low-fat content and are high in dietary fiber, nutraceuticals, and polysaccharides, which show positive health benefits on several diseases through their immunomodulatory and anti-inflammation properties [16,17]. The core structural polysaccharides from mushrooms include homoglucans (β-1, 3 glucan), heteroglycans, heterogalactans, and heteromannans. These polysaccharides and terpenoids (secondary metabolites) play a pivotal role in glucose homeostasis by inhibiting α-glucosidase, assisting glucose transporter 4 actions, and reducing inflammatory factors to improve insulin resistance and lipid metabolism [18,19].

The term "diabetes mellitus" or "DM" is derived from the Greek word "diabetes", which means siphon, to pass through, and the Latin word "mellitus", which means sweet. Diabetes mellitus is a group of non-communicable metabolic diseases characterized by prolonged hyperglycemia resulting from defects in insulin secretion, insulin action, or both [20]. According to American Diabetes Association, 1997 (ADA), diabetes mellitus is classified as type 1 diabetes mellitus (T1DM) (insulin-dependent or juvenile-onset; accounting for 3–10% cases), type 2 diabetes mellitus (T2DM) (non-insulin-dependent or adult-onset; accounting 85–90% cases), and gestational diabetes mellitus (hyperglycemia occurs during the second or third trimester of pregnancy and generally resolves after delivery; accounting 2–5% cases) [21,22]. According to the International Diabetes Federation Data in 2015, 415 million people (80% from middle- and low-income family) was suffering from diabetes, and if it continues to grow will reach 642 million people by 2040 [23]. T1DM is generally accompanied by an autoimmune disorder that triggers the destruction of pancreatic beta cells, alterations in lipid metabolism, enhanced hyperglycemia-mediated oxidative stress, endothelial cell dysfunction, and apoptosis [24,25]. Whereas T2DM causes glucotoxicity, lipotoxicity, endoplasmic reticulum-induced stress, and apoptosis, which finally leads to progressive loss of beta cells [26]. Specific symptom includes polydipsia, polyphagia, polyuria, and nocturia, whereas complication comprises microvascular (retinopathy, nephropathy, and neuropathy) and macrovascular (ischemic heart disease, peripheral vascular disease, and cerebrovascular disease) abnormalities. Several studies have proven that insulin resistance (IR) is the main factor to be concerned about in complications of DM [26–28]. Besides IR, some of the common risk factors for insulin resistance are oxidative stress, hydrolytic enzymatic inhibition, inflammation, genetic habitual, environmental, dyslipidemia, obesity, and epigenetic modulations [29,30]. Thus, many pathological factors use to contribute to insulin resistance, although the exact mechanism is not clear yet.

By considering the following factors, the main aim of this review is to elucidate the role of edible mushroom (EM) in diabetes mellitus (DM) treatment by monitoring bioactive compounds of mushrooms, pathophysiology of insulin resistance (IR), and the preventive mechanism of IR using EMs. The data were retrieved by searching scientific publications (research and/or review papers) from databases including Science Direct, PubMed, Embase, and Google Scholar with keywords such as "diabetes", "insulin resistance", "mushroom extracts", "in vivo and in vitro studies", etc. A total of 100 publications were collected, including in vivo and in vitro studies (Figure 1), in which 23 common edible mushroom varieties have been identified (Table 1), and further discussion has been carried out based on scientific literature availability on DM.

We further investigated different species of edible mushrooms and presented the data based on hypoglycemic compounds. After considering the potential bioactive compounds, in vivo and in vitro research analysis was carried out, and the information is represented in Table 2.

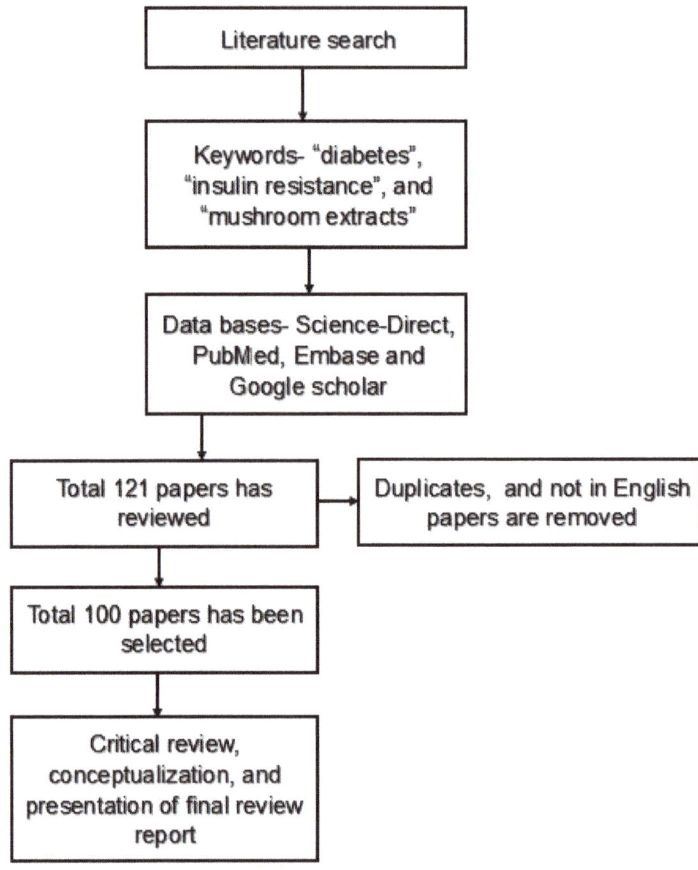

Figure 1. Review research methodology.

Table 1. Types of edible mushrooms.

S. No.	Scientific Name	Vernacular Name	Photos	Reference
1	*Craterellus aureus*	Cantharellus, chanterelles		[31]
2	*Calvatia rugosa*	Puffballs		[32]

Table 1. *Cont.*

S. No.	Scientific Name	Vernacular Name	Photos	Reference
3	*Coprinus comatus*	Shaggy mane		[33]
4	*Pleurotus ostreatus*	Oyster mushroom		[34]
5	*Boletaceae Boletales*	Boletes		[35]
6	*Laetiporus sulphurous*	Sulfer shelf		[36]
7	*Grifora frondosa*	Hen of the woods		[37]
8	*Agaricus bisporus*	Button mushroom		[38]

Table 1. *Cont.*

S. No.	Scientific Name	Vernacular Name	Photos	Reference
9	*Ramariopsis subarctica*	Coral fungi		[39]
10	*Morachella esculenta*	Morels		[40]
11	*Hericium erinaceus*	Bearded tooth		[41]
12	*Volvariella volvacea*	Straw mushroom		[42]
13	*Ganoderma lucidium*	Lingzhi mushroom		[43]
14	*Tremella fuciformis*	Snow fungus		[44]

Table 1. *Cont.*

S. No.	Scientific Name	Vernacular Name	Photos	Reference
15	*Lentinus concentricus*	-		[45]
16	*Calocybe indica*	Milky white mushroom		[46]
17	*Lenzites betulina*	Wood-rooting fungi		[47]
18	*Pholiota microspora*	Slippery mushroom		[48]
19	*Flammulina velutipes*	Enoki mushroom		[49]
20	*Lentinula edodes*	Shiitake mushroom		[50]

Table 1. *Cont.*

S. No.	Scientific Name	Vernacular Name	Photos	Reference
21	*Hypsizygus tessellatus*	Buna shimeji		[51]
22	*Agrocybe aegerita*	Poplar mushroom, Chestnut mushroom, Velvet pioppini		[52]
23	*Termitomyces robustus*	Termitomyces mushrooms		[53]

Table 2. Bioactive compounds of edible mushrooms and their effect on in vivo and in vitro study model.

S. No.	Scientific Name	Compounds	Functions	Models	Mushroom Doses	Mechanism/Action	Reference
1	*Calvatia gigantea*	2-Pyrrolidinone, 1-Dodecene, ergosterol, hexadecane, benzeneacetic acid.	Anti-diabetic, antioxidant, anti-inflammatory.	Alloxan-induced diabetic rat.	100, 200, and 400 mg/kg BW/day.	-Alpha-amylase inhibitory activity.	[54]
2	*Coprinus comatus*	Mycelium, polysaccharides.	Immunomodulatory, anti-diabetic, antioxidant, anti-cancer.	High-fat diet and STZ-induced mice.	400 mg/kg BW/day.	-Reduce BG level, relieves oxidative stress, ameliorate DN via PI3K/Akt and Wnt-1/β-catenin pathways.	[55]
3	*Pleurotus ostreatus, Pleurotus pulmonarius, and Pleurotus fossulatus*	Terpenoids, heterocyclic amines, phenols, glucan, proteoglycan.	Anti-cholesterol, anti-cancer effect, anti-inflammatory, anti-diabetic.	STZ and metformin-induced rat.	5–10% powder, 50, 80, 200, 250, 400, and 500 mg/kg BW/day extract.	-Decreases serum glucose level, alpha-amylase activity; -Increases P-AMPK and GLUT4 in muscle and adipose tissue; -Improves liver functions and maintains AST, ALT, and ALP levels.	[56,57]

Table 2. *Cont.*

S. No.	Scientific Name	Compounds	Functions	Models	Mushroom Doses	Mechanism/Action	Reference
4	*Boletus*	Tocopherol, quinic acid, hydroxy benzoic acid.	Antioxidant, anti-inflammatory, hypoglycemic.	STZ-induced rat.	400 mg extract/kg BW/day.	-Decreases TC, TG, TNF-alpha, and NF-kB level; -Maintains MDA level; -Improves antioxidant (CAT, SOD, and GSH) and CYP7A1 levels.	[58]
5	*Grifora frondosa*	Grifolan polysaccharide, D-fraction, MD-fraction, polysaccharide, galactomannan, heteroglycan.	Hypoglycemic, anti-inflammatory, anti-modulatory, anti-tumor.	STZ, alloxan-induced rat and palmitate-induced C2C12 cells.	0.5–20 μM (introduced to C2C12 cell), 112.5, 200, and 675 mg/kg BW/day extract (introduced to STZ- and alloxan-induced rat).	-Inhibits serum levels of IL-2, IL-6; -Modulates serum level of oxidant factors such as superoxide dismutase, glutathione peroxidase, catalase, malondialdehyde, and reactive oxygen species; -Increases glucose uptake and decreases ROS formation and up-regulates IRS-1, p-IRS-1, PI3K, Akt, pAkt and GLUT4 protein, and down-regulates p-JNK and p-p38 expression; -Improves insulin resistance and gut microflora content.	[59,60]
6	*Agricus bisporus*	Pyrogallol, hydroxybenzoic acid derivatives glavonoid.	Anti-inflammatory, anti-diabetic.	Alloxan-induced rat.	15–30 g/day, 250, 500 and 750 mg/kg BW/day.	-Improves antioxidant (SOD) level; -Improves ALP, AST, ALT level; -Reduces hyperlipidemia.	[38,61]
7	*Morachella esculenta*	Polysaccharides (mannose, galactose, and glucose), phenolic compounds.	Antioxidant, anti-inflammation, immunoregulation, hypoglycemic.	-	-	-	[62]
8	*Hericium erinaceus*	4-chloro-3,5-dimethoxybenzoic acid-O-arabitol ester, 2-hydroxymethyl-5-α-hydroxyethyl-γ-pyranone, 6-methyl-2,5-dihydroxymethyl-γ-pyranone, 4-chloro-3,5-dihydroxybenzaldehyde, 4-chloro-3,5-dihydroxybenzyl alcohol.	Immunomodulatory, hypoglycemic, antimicrobial.	STZ-treated rat.	400–600 mg/kg BW/day.	-Reduces blood glucose, BUN, and CRT level; -Maintains ALP, ALT, and AST levels; -Improves antioxidant level (SOD, glutathione).	[63,64]

Table 2. *Cont.*

S. No.	Scientific Name	Compounds	Functions	Models	Mushroom Doses	Mechanism/Action	Reference
9	*Ganoderma lucidium*	Ganoderic acid, danoderiol, danderenic acid, lucidenic acid, ganoderma leucidum polysaccharide.	Anti-diabetic, anti-inflammatory.	Metformin-, STZ-, and high-fat-treated rat.	1–3% freeze-dried mushroom, 25, 50, 100, 250, 500 and 1000 mg/kg BW/day extract.	-Decreases HBA1c and improves AST, ALT level.	[60,65]
10	*Lenzites betulina*	α-glucan, β-glucan, β-glucan protein, galacturonic acid.	Antioxidant, anti-hyperglycaemic, anti-inflammatory, anti-proliferative, antibacterial.	-	-	-	[66]
11	*Flammulina velutipes*	Flammulinolide, enokipodin, proflamin, other polysaccharide.	Anti-tumor, anti-hypertension, anti-hypercholesterolemia, hypoglycemic.	STZ-induced mice.	400 mg/kg, 600 mg/kg, and 800 mg/kg BW/day.	-Improves PI3K/AKT pathway.	[67,68]
12	*Lentinula edodes*	Lentinan, eritadenina.	Anti-carcinogenic, antioxidant, hypocholesterolemic action.	STZ-treated rat (gestational diabetes).	100 mg/kg BW/day.	-Improves maternal insulin level; -Reduces aminotransferase, aspartate aminotransferase, triglyceride, and total cholesterol level.	[69,70]
13	*Termitomyces robustus*	γ glutamyl-β-phenylethylamine, tryptophan 1,4-hydroxyphenylacetic acid, hydroxyphenyl propionic acid and phenyllactic acid.	Hypoglycemic effect.	In vitro assay, Wister rat.	Crude extract 78.05 and 86.10 µg/mL. 500, 1000, and 1500 mg/kg BW/day. In vitro assay, rate acute toxicity test (10 g/kg extract) and subacute toxicity test (500, 1000, and 1500 mg/kg) BW/day.	-α-glucosidase and α-amylase inhibitory activity.	[15,71]

DN—diabetic nephropathy, Wnt-1—wingless-related integration site, P-AMPK—activated protein kinase, GLUT4—glucose transporter type 4, TC—total cholesterol, STZ—streptozotocin, FBG—fasting blood glucose, BGL—blood glucose level, TG—triglyceride, CRT—creatinine, BUN—blood urea nitrogen, AST—aspartate aminotransferase, ALT—alanine aminotransferase, ALP—alkaline phosphatase, TNF—alpha-tumor necrosis factor alpha, NF-kB—nuclear factor kappa B, MDA—malondialdehyde, CAT—chloramphenicol acetyltransferase, SOD—superoxide dismutase, GSH—glutathione, CYP7A1—cholesterol 7-alpha-monooxygenase gene, IL-6—interleukin 6, IL-2—interleukin 2, IRS-1—insulin receptor substrate 1, PI3K—phosphoinositide-3-kinase, Akt—protein kinase B, p-JNK-c—Jun N-terminal kinase, HBA1c—glycosylated hemoglobin, PGL—plasma glucose level.

2. Diabetes Mellitus and Insulin Resistance

Nowadays, the global incidence of diabetes is escalating in both developed and developing countries. Generally, obesity is directly involved in the pathogenesis of T2DM along with insulin resistance (IR), and the incidence of IR in T1DM is also increasing frequently. Therefore, it is crucial to understand the mechanism of glucose homeostasis and insulin resistance.

Pancreatic cell plays an important role in glucose homeostasis. T2DM reduces insulin secretion by 50% and also lessens the sensitivity of peripheral tissue to insulin up to 70%. So, the study on IR has a great clinical significance in medical interventions. Basically, insulin exerts its effect through phosphorylation of phosphoinositide 3-kinase (PI3K) and protein kinase B (PkB, Akt) [72,73]. Next, through PI3k phosphorylation, it activates glucose transporter 4 (GLUT4). Other factors involved in insulin action and carbohydrate metabolism are serine/threonine kinase Akt phosphorylates GSk3 *ß*, FOXOs, IRS (insulin-regulated sequence), and SREBP-1c (sterol-regulated element-binding protein transcription factors) (Figure 2) [74].

Insulin sensitivity can be defined as a pathological condition in which normal plasma insulin level in targeted tissues fails to maintain normal blood glucose levels via enhancement of endogenous glucose production, lipogenesis, and glycolysis [75]. Thereby, IR refers to a state that exhibits the reduced biological effect of a given insulin concentration. So, IR increases insulin secretion and thus enhances fasting plasma insulin level, which is finally considered IR [76]. Moreover, IR is not only associated with DM but also several pathological conditions such as cardiovascular disease, non-alcoholic fatty liver disease, and cancer as well [77].

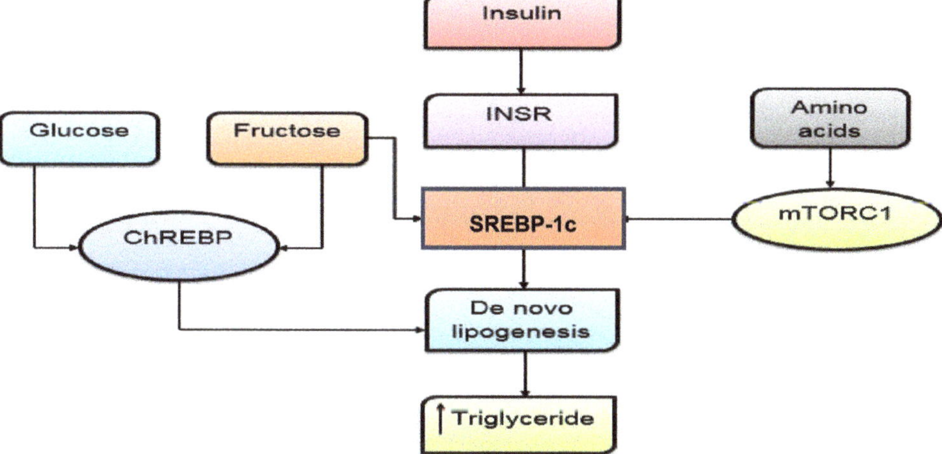

Figure 2. Relationship between liver insulin resistance and macronutrients. ChREBP—carbohydrate response element-binding protein, INSR—insulin receptor, SREBP-1c—sterol-regulated element-binding protein, mTORC1—mammalian target of rapamycin complex 1 or mechanistic target of rapamycin complex 1.

3. Diabetes Mellitus and Insulin Resistance Preventive Mechanism by Edible Mushroom

Mushrooms possess medicinal components due to the presence of different types of secondary metabolites such as polysaccharides, lectins, lactones, terpenoids, alkaloids, antibiotics, and metal-chelating agents [78]. Mechanism of insulin resistance using mushroom is as follows.

3.1. Blood Glucose-Lowering Effect of Polysaccharide

Polysaccharides are ubiquitous biopolymers made up of simple sugar or monosaccharides linked together by glycosidic linkage. Based on structure, polysaccharides are divided into two groups-homopolysaccharides (linear or highly branched, composed of the same monosaccharide molecules) and heteropolysaccharides (made up of different monosaccharide units) [79]. Earlier studies show that mushrooms are rich in *ß*-D-glucans, *ß*-glucan, a type of dietary fiber that shows promising health benefits against T2DM [80]. Mushroom extracts from *Pleurotus species* [56], *Boletu* [58], *Grifola frondosa* [81], *Agaricus bisporus* [38], *Hericium erinaceus* [63], and *Ganoderma lucidum* [82] regulate the synthesis of glycogen and lowers blood glucose levels by regulating gene expression of glycogen synthase kinase (GSK-3 *ß*), glycogen synthase (GS) and glucose transporter 4 (GLUT4) in liver and muscle. Therefore GSK-3 *ß* could be identified as an insulin-mediated GS-regulated negative regulator [83]. Other mechanisms by which polysaccharides prevent insulin resistance are by reducing α-amylase activity, α-glucosidase activities, and finally facilitating PI3K/AKT pathways, which are directly involved in glucose homeostasis [78].

Other mechanisms by which polysaccharides lowers blood glucose level is as follows-

3.1.1. Inhibition of Glucose Absorption

Due to the presence of water-soluble dietary fiber, mushrooms delay glucose absorption and slow down digestion rates, thereby postprandial glucose upsurge [84,85]. Several studies have proven that mushrooms, especially *Pleurotus* spp. [56], *Grifola frondosa* [81], *Agaricus bisporus* [38], *Hericium erinaceus* [63], and *Ganoderma lucidum* [65] have significant blood glucose-lowering effect, as they delay the absorption of glucose and therefore improves hyperglycaemic condition [78].

3.1.2. Maintains Pancreatic *ß* Cells Activity

Mushrooms polysaccharides (*ß*-D-glucan) act as a potent immune modulator and prevent activations of pro-inflammatory cytokines by reducing the activity of NF-kB, and it also outlawed oxidative damage. Bioactive compounds from mushrooms, especially polysaccharides, prevent pancreatic *ß*-cell apoptosis and hinder glucotoxicity [80]. Studies also showed that mushroom extracts from *Pleurotus* spp., *Boletus, Agaricus bisporus,* and *Hericium erinaceus* have a significant effect on *ß*-cell functionality and thus maintain *ß*-cell growth [63].

3.2. Blood Glucose-Lowering Effect of Terpenoids

Enzymes like α-glucosidase and α-amylase hydrolyze oligosaccharides to monosaccharides and increase blood glucose levels [86]. Mushrooms terpenoids (monoterpenes, diterpenes, sesquiterpenes, and triterpenes) (Figure 3) [19] from *Pleurotus* spp. [56], *Laetiporus sulphurous* [87], *Tremella fuciformis* [88], *Ganoderma lucidum* [60], and *Pholiota microspore* [89] are believed to have an α-glucosidase inhibitory activity that prevents the formation of monosaccharide molecules and facilitates glycogen formation in the liver and muscle.

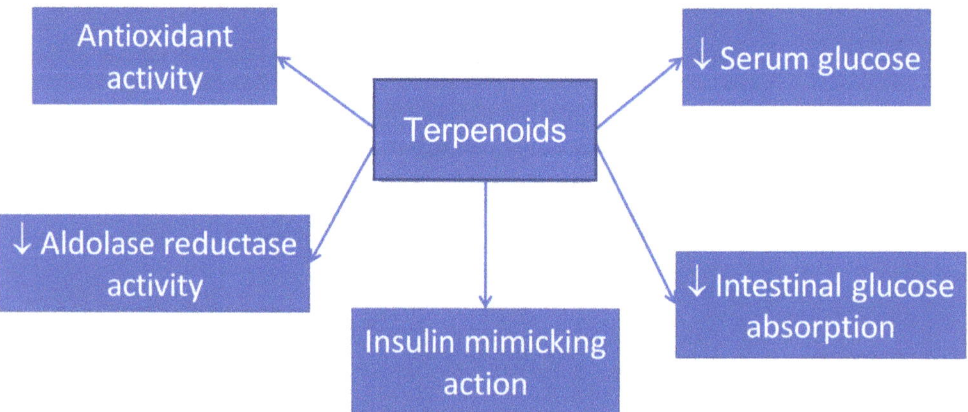

Figure 3. Beneficial effects of mushroom terpenoids (monoterpenes, diterpenes, sesquiterpenes, and triterpenes).

3.3. Role of Vitamin D in Blood Glucose Regulations

Mushroom resides from the fungal kingdom, and unlike a plant, it has a high concentration of ergosterol in the cell wall. In the presence of sunlight, ergosterol in the mushroom cell wall is transformed to pre-vitamin D2 and thermally isomerized to ergocalciferol, commonly known as vitamin D2 [90]. 1, 5 (OH) 2D or 1,25-dihydroxy vitamin D plays an important role in glucose homeostasis. It also protects *ß*-cells from harmful immune attacks by its direct action on *ß*-cells, and indirect action on different immune cells, including inflammatory macrophages, dendritic cells, and a variety of T cells [91]. Molecular mechanisms by which vitamin D maintains insulin secretion is by regulating intracellular calcium concentration. Vitamin D, with the help of calbindin, facilitates Ca+ absorption, PKA activation, and PLC synthesis, which facilitates calcium secretion that, in consequence, leads to insulin secretion [92] (Figure 4).

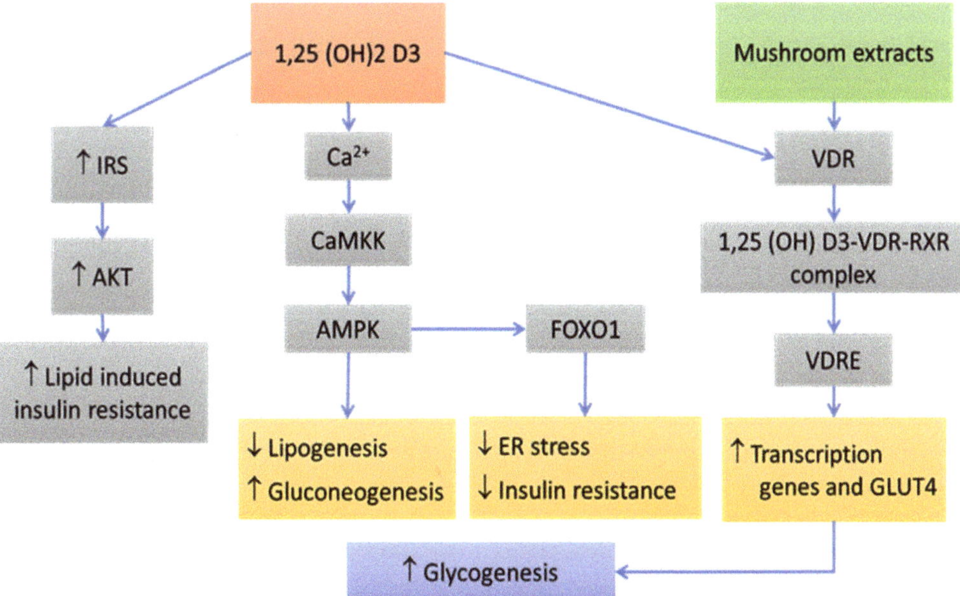

Figure 4. Relationship between vitamin D, mushroom extracts, and insulin resistance. IRS—insulin receptor substrate, AKT—protein kinase B, CaMKK—calcium/calmodulin-dependent protein kinase, AMPK—activated protein kinase, FOXO1—forkhead box transcription factor 1, VDR-, VDRE—vitamin D-responsive elements.

Subsequently, the genomic mechanism of vitamin D action is mediated by the vitamin D receptor (VDR). The active form of vitamin D 1,25 (OH)2 D3 binds to VDR and forms a heterodimer with the retinoid receptors (RXR). The complex of 1,25 (OH)2 D3-VDR-RXR is translocated to the nucleus and binds to vitamin D-responsive elements (VDRE), thereby facilitating epigenetic modulations and preventing insulin resistance [93].

In short, vitamin D shows its function in different ways, such as inherited gene polymorphism, immune-regulatory functions, proliferation preventive actions, anti-inflammatory activities (decreases the functions of pro-inflammatory cytokines, TNF-α, IL-8b, and IL-6), and finally regulates the production of adipokines and thereby prevents insulin resistance (via IRS, AKT, PPARγ, and VDRE gene regulations) [94] (Table 3).

Table 3. Functions of vitamin D on different organs [94].

Organ Name	Functions
Pancreas	Increases insulin secretion and enhances the transformation of pro-insulin to insulin
Skeletal muscle	Through VDR expression maintains glucose homeostasis
Skin	Improves skin micro-circulations and fasten wound healing
Nervous system	Improves nerve conduction and shows the analgesic effect
Kidney	Controls urinary albuminuria
Retina	Defend against oxidative stress

The bioavailability of vitamin D in diabetic treatment is still controversial, and based on the available data, it is not yet transparent among scientists [95]. However, recent findings from randomized placebo trial data by Urbain et al., 2011 shows that button mushroom treated with UV-B can improve vitamin D2 bioavailability among human subjects, and the significant value does not differ with vitamin D2 supplements [96]. In addition, studies by Wenclewska et al., 2019; Mutt et al., 2020; Hajj et al., 2020; and Cojic et al., 2021 demonstrated the positive impact of vitamin D supplementation (2000IU and 30,000IU) on T2DM, which requires further trials [97–100]. Therefore, the question remains the exact mechanism and dose-dependent action of vitamin D on diabetes mellitus treatment.

4. Conclusions

In the present review, a total of 23 edible mushrooms have been scrutinized to review the medicinal components and diabetes mellitus preventive activities. Among all the varieties, 13 varieties have anti-diabetic properties. Mushrooms' anti-diabetic activity is generally dependent on their polysaccharide (ß-D-glucan) and vitamin D contents. Therefore, in vivo and in vitro studies show that among 13 varieties, only 11 verities demonstrated anti-diabetic activities. Additionally, the most widely studied variations are *Pleurotus*, *Grifola*, and *Ganoderma* species. Thus, based on the available data, it can be concluded that mushrooms are beneficial fungi that have a great potential to treat non-communicable diseases, especially diabetes. However, future research work is necessary for the clinical field such as animal study (in vivo and in vitro), enzyme inhibition assay (α-amylase, α-glucosidase, pancreatic lipase, and DPP4-dipeptidyl peptidase 4), human trial, pilot study, as well as prospective and retrospective studies to use the possible therapeutic applications of mushrooms. Special focus must be given to the link between vitamin D and insulin resistance along with enzymatic assay by considering its potential effect. Furthermore, without adequate investigation, the conclusion is relatively challenging. So, further revelation is recommended in the clinical probe.

Author Contributions: Conceptualization, S.-C.L. and A.D.; validation, C.-M.C. and S.-C.M.; drafting of the manuscript, S.-C.L. and A.D.; critical revision of the manuscript, S.-H.Y. and Y.-M.J., and final approval of the published version, S.-C.L. All authors have read and agreed to the published version of the manuscript.

Funding: This study was supported by the Council of Agriculture in Taiwan. The funding number is 111AS-11.1.2-CI-C1.

Institutional Review Board Statement: Not applicable.

Informed Consent Statement: Not applicable.

Data Availability Statement: The data that support the findings of this study are available from the corresponding author upon reasonable request.

Acknowledgments: We thank Seng-Kai Vong and Hsin-Yi Huang for their technical assistance.

Conflicts of Interest: The authors declare no conflict of interest.

Abbreviations

EM—edible mushroom, IR—insulin resistance, DM—diabetes mellitus, T1DM—type 1 diabetes mellitus, T2DM—type 2 diabetes mellitus, PI3k—phosphoinositide 3-kinase, PkB—protein kinase B, GSK-3 β—glycogen synthase kinase-3 β, GS—glycogen synthase, FOXO—forkhead box transcription factor, SREBP-1c—sterol-regulated element-binding protein, PkA—protein kinase, PLC—phospholipase C, RXR—retenoid receptor, VDRE—vitamin D-responsive elements, IRS—insulin receptor substrate, PPARy—peroxisome proliferator-activated receptor gamma, TNF-α—tumor necrosis factor alpha, IL-8b—interleukin 8 receptor, IL-6—interleukin 6 receptor, CaMKK—calcium/calmodulin-dependent protein kinase, AMPK—activated protein kinase, 1,25 (OH)2 D3—calcitriol/1,25 dihydroxycholecalciferol, ChREBP—carbohydrate response element-binding protein, mTORC1—mammalian target of rapamycin complex 1 or mechanistic target of rapamycin complex 1.

References

1. Ukwuru, M.; Muritala, A.; Eze, L. Edible and non-edible wild mushrooms: Nutrition, toxicity and strategies for recognition. *J. Clin. Nutr. Metab.* **2018**, *2*, 9.
2. Ślusarczyk, J.; Adamska, E.; Czerwik-Marcinkowska, J. Fungi and algae as sources of medicinal and other biologically active compounds: A review. *Nutrients* **2021**, *13*, 3178. [CrossRef] [PubMed]
3. Hoa, H.T.; Wang, C.L.; Wang, C.H. The effects of different substrates on the growth, yield, and nutritional composition of two oyster mushrooms (*Pleurotus ostreatus* and *Pleurotus cystidiosus*). *Mycobiology* **2015**, *43*, 423–434. [CrossRef]
4. Koutrotsios, G.; Tagkouli, D.; Bekiaris, G.; Kaliora, A.; Tsiaka, T.; Tsiantas, K.; Chatzipavlidis, I.; Zoumpoulakis, P.; Kalogeropoulos, N.; Zervakis, G.I. Enhancing the nutritional and functional properties of *Pleurotus citrinopileatus* mushrooms through the exploitation of winery and olive mill wastes. *Food Chem.* **2022**, *370*, 131022. [CrossRef]
5. Yadav, D.R. Edible Mushrooms. 2010. Available online: https://www.researchgate.net/publication/322210506_EDIBLE_MUSHROOMS?channel=doi&linkId=5a4bbf790f7e9b8284c2ded5&showFulltext=true (accessed on 19 January 2022).
6. Das, A.K.; Nanda, P.K.; Dandapat, P.; Bandyopadhyay, S.; Gullón, P.; Sivaraman, G.K.; McClements, D.J.; Gullón, B.; Lorenzo, J.M. Edible mushrooms as functional ingredients for development of healthier and more sustainable muscle foods: A flexitarian approach. *Molecules* **2021**, *26*, 2463. [CrossRef] [PubMed]
7. Sharma, V.P.; Barh, A.; Bairwa, R.; Annepu, S.K.; Thakur, B.; Kamal, S. Enoki mushroom (*Flammulina velutipes* (Curtis) Singer) breeding. In *Advances in Plant Breeding Strategies: Vegetable Crops*; Springer: Cham, Switzerland, 2021; pp. 423–441.
8. Tiane, C.; Finimundy, T.; José, A.; Dillon, P.; Antônio, J.; Henriques, J.A.; Ely, M. A review on general nutritional compounds and pharmacological properties of the *Lentinula edodes* mushroom. *Food Nutr. Sci.* **2014**, *5*, 1095–1105. [CrossRef]
9. Song, Q.; Teng, A.G.; Zhu, Z. Chemical structure and inhibition on α-glucosidase of a novel polysaccharide from *Hypsizygus marmoreus*. *J. Mol. Struct.* **2020**, *1211*, 128110. [CrossRef]
10. Kleftaki, S.A.; Simati, S.; Amerikanou, C.; Gioxari, A.; Tzavara, C.; Zervakis, G.I.; Kalogeropoulos, N.; Kokkinos, A.; Kaliora, A.C. *Pleurotus eryngii* improves postprandial glycaemia, hunger and fullness perception, and enhances ghrelin suppression in people with metabolically unhealthy obesity. *Pharmacol. Res.* **2022**, *175*, 105979. [CrossRef]
11. He, X.; Wang, X.; Fang, J.; Chang, Y.; Ning, N.; Guo, H.; Huang, L.; Huang, X.; Zhao, Z. Polysaccharides in *Grifola frondosa* mushroom and their health promoting properties: A review. *Int. J. Biol. Macromol.* **2017**, *101*, 910–921. [CrossRef]
12. Diyabalanage, T.; Mulabagal, V.; Mills, G.; Dewitt, D.; Nair, M. Health-beneficial qualities of the edible mushroom, *Agrocybe aegerita*. *Food Chem.* **2008**, *108*, 97–102. [CrossRef]
13. Rahman, M.; Akter, R. Diabetes ameliorating effect of mushrooms. *J. Nov. Physiother.* **2021**, *2*, 9–13.
14. Thu, Z.M.; Myo, K.K.; Aung, H.T.; Clericuzio, M.; Armijos, C.; Vidari, G. Bioactive phytochemical constituents of wild edible mushrooms from southeast Asia. *Molecules* **2020**, *25*, 1972. [CrossRef] [PubMed]
15. Ogbole, O.O.; Noleto-Dias, C.; Kamdem, R.S.T.; Akinleye, T.E.; Nkumah, A.; Ward, J.L.; Beale, M.H. γ-Glutamyl-β-phenylethylamine, a novel α-glucosidase and α-amylase inhibitory compound from *Termitomyces robustus*, an edible Nigerian mushroom. *Nat. Prod. Res.* **2021**, 1–11. [CrossRef]
16. Ma, G.; Yang, W.; Zhao, L.; Pei, F.; Fang, D.; Hu, Q. A critical review on the health promoting effects of mushrooms nutraceuticals. *Food Sci. Hum. Wellness* **2018**, *7*, 125–133. [CrossRef]
17. Tung, Y.T.; Pan, C.H.; Chien, Y.W.; Huang, H.Y. Edible mushrooms: Novel medicinal agents to combat metabolic syndrome and associated diseases. *Curr. Pharm. Des.* **2020**, *26*, 4970–4981. [CrossRef] [PubMed]
18. Yang, S.; Yan, J.; Yang, L.; Meng, Y.; Wang, N.; He, C.; Fan, Y.; Zhou, Y. Alkali-soluble polysaccharides from mushroom fruiting bodies improve insulin resistance. *Int. J. Biol. Macromol.* **2019**, *126*, 466–474. [CrossRef]
19. Dasgupta, A.; Acharya, K. Mushrooms: An emerging resource for therapeutic terpenoids. *3 Biotech* **2019**, *9*, 369. [CrossRef]
20. Kharroubi, A.T.; Darwish, H.M. Diabetes mellitus: The epidemic of the century. *World J. Diabetes* **2015**, *6*, 850–867. [CrossRef]
21. Gabir, M.M.; Hanson, R.L.; Dabelea, D.; Imperatore, G.; Roumain, J.; Bennett, P.H.; Knowler, W.C. The 1997 American Diabetes Association and 1999 World Health Organization criteria for hyperglycemia in the diagnosis and prediction of diabetes. *Diabetes Care* **2000**, *23*, 1108–1112. [CrossRef]

22. Alam, F.; Islam, M.A.; Mohamed, M.; Ahmad, I.; Kamal, M.A.; Donnelly, R.; Idris, I.; Gan, S.H. Efficacy and safety of pioglitazone monotherapy in type 2 diabetes mellitus: A systematic review and meta-analysis of randomised controlled trials. *Sci. Rep.* **2019**, *9*, 5389. [CrossRef]

23. Vitak, T.; Yurkiv, B.; Wasser, S.; Nevo, E.; Sybirna, N. Effect of medicinal mushrooms on blood cells under conditions of diabetes mellitus. *World J. Diabetes* **2017**, *8*, 187–201. [CrossRef] [PubMed]

24. Misra, A.; Ramchandran, A.; Jayawardena, R.; Shrivastava, U.; Snehalatha, C. Diabetes in South Asians. *Diabet. Med.* **2014**, *31*, 1153–1162. [CrossRef]

25. Kosiborod, M.; Gomes, M.B.; Nicolucci, A.; Pocock, S.; Rathmann, W.; Shestakova, M.V.; Watada, H.; Shimomura, I.; Chen, H.; Cid-Ruzafa, J.; et al. Vascular complications in patients with type 2 diabetes: Prevalence and associated factors in 38 countries (the DISCOVER study program). *Cardiovasc. Diabetol.* **2018**, *17*, 150. [CrossRef] [PubMed]

26. Ndisang, J.F.; Vannacci, A.; Rastogi, S. Insulin resistance, type 1 and type 2 diabetes, and related complications 2017. *J. Diabetes Res.* **2017**, *2017*, 1478294. [CrossRef] [PubMed]

27. Ndisang, J.F.; Rastogi, S.; Vannacci, A. Insulin resistance, type 1 and type 2 diabetes, and related complications 2015. *J. Diabetes Res.* **2015**, *2015*, 234135. [CrossRef]

28. Taylor, R. Insulin resistance and type 2 diabetes. *Diabetes* **2012**, *61*, 778–779. [CrossRef] [PubMed]

29. Duvnjak, L.; Duvnjak, M. The metabolic syndrome—an ongoing story. *J. Physiol. Pharmacol. Off. J. Pol. Physiol. Soc.* **2009**, *60* (Suppl. 7), 19–24.

30. Ndisang, J.F.; Rastogi, S.; Vannacci, A. Insulin resistance, type 1 and type 2 diabetes, and related complications: Current status and future perspective. *J. Diabetes Res.* **2014**, *2014*, 276475. [CrossRef]

31. Montoya, L.; Herrera, M.; Bandala, V.M.; Ramos, A. Two new species and a new record of yellow *Cantharellus* from tropical *Quercus* forests in eastern Mexico with the proposal of a new name for the replacement of *Craterellus confluens*. *MycoKeys* **2021**, *80*, 91. [CrossRef]

32. Hermawan, R.; Putra, I. *Calvatia rugosa*: Epigeous *Puffball* mushroom reported from West Java. *Sci. Educ. Appl. J.* **2021**, *3*, 1. [CrossRef]

33. Nowakowski, P.; Naliwajko, S.K.; Markiewicz-Żukowska, R.; Borawska, M.H.; Socha, K. The two faces of *Coprinus comatus*-functional properties and potential hazards. *Phytother. Res.* **2020**, *34*, 2932–2944. [CrossRef] [PubMed]

34. Starzycki, M.; Paukszta, D.; Starzycka, E. Vitro growth of oyster mushroom (pleurotus ostreatus) mycelium on composites filled with rapeseed. *Phytopathologia* **2014**, *65*, 33–37.

35. Frank, J.; Siegel, N.; Schwarz, C.; Araki, B.; Vellinga, E. Xerocomellus (Boletaceae) in western North America. *Fungal Syst. Evol.* **2020**, *6*, 265–288. [CrossRef] [PubMed]

36. Elkhateeb, W.; Ezzat el ghwas, D.; Gundoju, N.; Tiruveedhula, S.; Akram, M.; Daba, G. Chicken of the woods *Laetiporus Sulphureus* and *Schizophyllum Commune* treasure of medicinal mushrooms. *J. Microbiol. Biotechnol.* **2021**, *6*, 1–7. [CrossRef]

37. Dissanayake, A.A.; Zhang, C.R.; Mills, G.L.; Nair, M.G. Cultivated maitake mushroom demonstrated functional food quality as determined by in vitro bioassays. *J. Funct. Foods* **2018**, *44*, 79–85. [CrossRef]

38. Ekowati, N.; Yuniati, N.I.; Hernayanti, H.; Ratnaningtyas, N.I. Antidiabetic potentials of button mushroom (*Agaricus bisporus*) on alloxan-induced diabetic rats. *Biosaintifika J. Biol. Biol. Educ.* **2018**, *10*, 655–662. [CrossRef]

39. Halama, M.; Pech, P.; Shiryaev, A.G. Contribution to the knowledge of *Ramariopsis subarctica* (*Clavariaceae, Basidiomycota*). *Pol. Bot. J.* **2017**, *62*, 123–133. [CrossRef]

40. Masaphy, S.; Zabari, L.; Goldberg, D.; Jander-Shagug, G. The complexity of *Morchella* systematics: A case of the yellow morel from Israel. *Fungi* **2010**, *3*, 14–18.

41. Wu, D.; Yang, S.; Tang, C.; Liu, Y.; Li, Q.; Zhang, H.; Cui, F.; Yang, Y. Structural properties and macrophage activation of cell wall polysaccharides from the fruiting bodies of *Hericium erinaceus*. *Polymers* **2018**, *10*, 850. [CrossRef]

42. Bao, D.; Gong, M.; Zheng, H.; Chen, M.; Zhang, L.; Wang, H.; Jiang, J.; Wu, L.; Zhu, Y.; Zhu, G.; et al. Sequencing and comparative analysis of the straw mushroom (*Volvariella volvacea*) genome. *PLoS ONE* **2013**, *8*, e58294. [CrossRef]

43. Liang, C.; Tian, D.; Liu, Y.; Li, H.; Zhu, J.; Li, M.; Xin, M.; Xia, J. Review of the molecular mechanisms of *Ganoderma lucidum* triterpenoids: Ganoderic acids A, C2, D, F, DM, X and Y. *Eur. J. Med. Chem.* **2019**, *174*, 130–141. [CrossRef] [PubMed]

44. Lin, C.P.; Tsai, S.-Y. Differences in the moisture capacity and thermal stability of *Tremella fuciformis* polysaccharides obtained by various drying processes. *Molecules* **2019**, *24*, 2856. [CrossRef] [PubMed]

45. Senthilarasu, G. The lentinoid fungi (Lentinus and Panus) from Western Ghats, India. *IMA Fungus* **2015**, *6*, 119–128. [CrossRef] [PubMed]

46. Chakraborty, B.; Chakraborty, U.; Barman, S.; Roy, S. Effect of different substrates and casing materials on growth and yield of *Calocybe indica* (P&C) in North Bengal, India. *J. Appl. Nat. Sci.* **2016**, *8*, 683–690.

47. Karun, N.; Sharma, B.; Sridhar, K. Biodiversity of macrofungi in Yenepoya Campus, Southwest India. *Microb. Biosyst.* **2018**, *3*, 1–11. [CrossRef]

48. Adhikari, M.K.; Watanabe, K.; Parajuli, G. A new variety of *Pholiota microspora* (Agaricales) from Nepal. *Biodiversitas J. Biol. Divers.* **2014**, *15*, 101–103. [CrossRef]

49. Hu, Y.-N.; Sung, T.-J.; Chou, C.-H.; Liu, K.-L.; Hsieh, L.-P.; Hsieh, C.-W. Characterization and antioxidant activities of yellow Strain *Flammulina velutipes* (Jinhua Mushroom) polysaccharides and their effects on ROS content in L929 cell. *Antioxidants* **2019**, *8*, 298. [CrossRef]

50. Chicatto, J.; Castamann, V.; Helm, C.; Tavares, L. Optimization of the production process of enzymatic activity of *Lentinula edodes* (Berk.) pegler in holocelulases. *Nat. Resour.* **2014**, *5*, 241–255. [CrossRef]
51. Shah, S.R.; Ukaegbu, C.I.; Hamid, H.A.; Alara, O.R. Evaluation of antioxidant and antibacterial activities of the stems of *Flammulina velutipes* and *Hypsizygus tessellatus* (white and brown var.) extracted with different solvents. *J. Food Meas. Charact.* **2018**, *12*, 1947–1961. [CrossRef]
52. Nagalakshmi, M.; Krishnakumari, S.; Kathiravan, S. Comparitive study of various substrate supplements in the growth and yield of *Agrocybe aegerita*, black poplar mushroom. *World J. Pharm. Res.* **2014**, *3*, 487–496.
53. Rahmad, N.; Al-Obaidi, J.; Mohd Nor Rashid, N.; Zean, N.; Yuswan, M.H.; Shaharuddin, N.; Mohd Jamil, N.A.; Mohd Saleh, N. Comparative proteomic analysis of different developmental stages of the edible mushroom *Termitomyces heimii*. *Biol. Res.* **2014**, *47*, 30. [CrossRef] [PubMed]
54. Ogbole, O.O.; Nkumah, A.O.; Linus, A.U.; Falade, M.O. Molecular identification, in vivo and in vitro activities of *Calvatia gigantea* (macro-fungus) as an antidiabetic agent. *Mycology* **2019**, *10*, 166–173. [CrossRef] [PubMed]
55. Gao, Z.; Kong, D.; Cai, W.; Zhang, J.; Jia, L. Characterization and anti-diabetic nephropathic ability of mycelium polysaccharides from *Coprinus comatus*. *Carbohydr. Polym.* **2021**, *251*, 117081. [CrossRef] [PubMed]
56. Asrafuzzaman, M.; Rahman, M.; Mandal, M.; Marjuque, M.; Bhowmik, D.; Begum, R.; Hassan, M.Z.; Faruque, M. Oyster mushroom functions as an anti-hyperglycaemic through phosphorylation of AMPK and increased expression of GLUT4 in type 2 diabetic model rats. *J. Taibah Univ. Med. Sci.* **2018**, *13*, 465–471. [CrossRef]
57. Ali Sangi, S.; Bawadekji, A.; Al Ali, M. Comparative effects of metformin, *Pleurotus ostreatus*, *Nigella sativa*, and *Zingiber officinale* on the streptozotocin-induced diabetes mellitus in rats. *Pharmacogn. Mag.* **2018**, *14*, 268–273. [CrossRef]
58. Xiao, Y.; Chen, L.; Fan, Y.; Yan, P.; Li, S.; Zhou, X. The effect of boletus polysaccharides on diabetic hepatopathy in rats. *Chem.-Biol. Interact.* **2019**, *308*, 61–69. [CrossRef]
59. Xiao, C.; Jiao, C.; Xie, Y.; Ye, L.; Li, Q.; Wu, Q. *Grifola frondosa* GF5000 improves insulin resistance by modulation the composition of gut microbiota in diabetic rats. *J. Funct. Foods* **2021**, *77*, 104313. [CrossRef]
60. Patel, D.K.; Dutta, S.D.; Ganguly, K.; Cho, S.J.; Lim, K.T. Mushroom-derived bioactive molecules as immunotherapeutic agents: A review. *Molecules* **2021**, *26*, 1359. [CrossRef]
61. Blumfield, M.; Abbott, K.; Duve, E.; Cassettari, T.; Marshall, S.; Fayet-Moore, F. Examining the health effects and bioactive components in *Agaricus bisporus* mushrooms: A scoping review. *J. Nutr. Biochem.* **2020**, *84*, 108453. [CrossRef]
62. Wu, H.; Chen, J.; Li, J.; Liu, Y.; Park, H.; Yang, L. Recent advances on bioactive ingredients of *Morchella esculenta*. *Appl. Biochem. Biotechnol.* **2021**, *193*, 4197–4213. [CrossRef]
63. Zhang, C.; Li, J.; Hu, C.; Wang, J.; Zhang, J.; Ren, Z.; Song, X.; Jia, L. Antihyperglycaemic and organic protective effects on pancreas, liver and kidney by polysaccharides from *Hericium erinaceus* SG-02 in streptozotocin-induced diabetic mice. *Sci. Rep.* **2017**, *7*, 10847. [CrossRef] [PubMed]
64. Thongbai, B.; Rapior, S.; Hyde, K.D.; Wittstein, K.; Stadler, M. *Hericium erinaceus*, an amazing medicinal mushroom. *Mycol. Prog.* **2015**, *14*, 91. [CrossRef]
65. Huang, C.-H.; Lin, W.-K.; Chang, S.H.; Tsai, G.-J. Evaluation of the hypoglycaemic and antioxidant effects of submerged *Ganoderma lucidum* cultures in type 2 diabetic rats. *Mycology* **2021**, *12*, 82–93. [CrossRef] [PubMed]
66. Zeb, M.; Lee, C.H. Medicinal properties and bioactive compounds from wild mushrooms native to North America. *Molecules* **2021**, *26*, 251. [CrossRef]
67. Wang, Y.; Zhang, H. Advances in the extraction, purification, structural-property relationships and bioactive molecular mechanism of *Flammulina velutipes* polysaccharides: A review. *Int. J. Biol. Macromol.* **2021**, *167*, 528–538. [CrossRef]
68. Song, X.; Fu, H.; Chen, W. Effects of *Flammulina velutipes* polysaccharides on quality improvement of fermented milk and antihyperlipidemic on streptozotocin-induced mice. *J. Funct. Foods* **2021**, *87*, 104834. [CrossRef]
69. Rivera, O.A.; Albarracín, W.; Lares, M. Bioactive components of shiitake (*Lentinula edodes* Berk. Pegler) and its impact on health. *Arch. Venez. De Farmacol. Y Ter.* **2017**, *36*, 67–71.
70. Laurino, L.F.; Viroel, F.J.M.; Caetano, E.; Spim, S.; Pickler, T.B.; Rosa-Castro, R.M.; Vasconcelos, E.A.; Jozala, A.F.; Hataka, A.; Grotto, D.; et al. *Lentinus edodes* exposure before and after fetus implantation: Materno-fetal development in rats with gestational diabetes mellitus. *Nutrients* **2019**, *11*, 2720. [CrossRef]
71. Ugbogu, E.A.; Akubugwo, E.; Ude, V.; Okezie, E.; Nduka, O.; Ibeh, C.; Onyero, O. Safety evaluation of aqueous extract of *Termitomyces robustus* (Agaricomycetes) in Wistar Rats. *Int. J. Med. Mushrooms* **2018**, *21*, 193–203. [CrossRef]
72. Wondmkun, Y.T. Obesity, insulin resistance, and type 2 diabetes: Associations and therapeutic implications. *Diabetes Metab. Syndr. Obes.* **2020**, *13*, 3611–3616. [CrossRef]
73. Shyur, L.-F.; Varga, V.; Chen, C.-M.; Mu, S.-C.; Chang, Y.-C.; Li, S.-C. Extract of white sweet potato tuber against TNF-α-induced insulin resistance by activating the PI3K/Akt pathway in C2C12 myotubes. *Bot. Stud.* **2021**, *62*, 7. [CrossRef] [PubMed]
74. Nolan, C.J.; Prentki, M. Insulin resistance and insulin hypersecretion in the metabolic syndrome and type 2 diabetes: Time for a conceptual framework shift. *Diabetes Vasc. Dis. Res.* **2019**, *16*, 118–127. [CrossRef] [PubMed]
75. Junttila, M.J.; Kiviniemi, A.M.; Lepojärvi, E.S.; Tulppo, M.; Piira, O.-P.; Kenttä, T.; Perkiömäki, J.S.; Ukkola, O.H.; Myerburg, R.J.; Huikuri, H.V. Type 2 diabetes and coronary artery disease: Preserved ejection fraction and sudden cardiac death. *Heart Rhythm.* **2018**, *15*, 1450–1456. [CrossRef]
76. Czech, M.P. Insulin action and resistance in obesity and type 2 diabetes. *Nat. Med.* **2017**, *23*, 804–814. [CrossRef] [PubMed]

77. Shulman, G.I. Ectopic fat in insulin resistance, dyslipidemia, and cardiometabolic disease. *N. Engl. J. Med.* **2014**, *371*, 1131–1141. [CrossRef] [PubMed]
78. Aramabašić Jovanović, J.; Mihailović, M.; Usković, A.; Grdović, N.; Dinić, S.; Vidaković, M. The effects of major mushroom bioactive compounds on mechanisms that control blood glucose level. *J. Fungi* **2021**, *7*, 58. [CrossRef] [PubMed]
79. Grondin, J.M.; Tamura, K.; Déjean, G.; Abbott, D.W.; Brumer, H. Polysaccharide utilization loci: Fueling microbial communities. *J. Bacteriol.* **2017**, *199*, e00860-16. [CrossRef]
80. Dubey, S.K.; Chaturvedi, V.K.; Mishra, D.; Bajpeyee, A.; Tiwari, A.; Singh, M.P. Role of edible mushroom as a potent therapeutics for the diabetes and obesity. *3 Biotech* **2019**, *9*, 450. [CrossRef]
81. Kou, L.; Du, M.; Liu, P.; Zhang, B.; Zhang, Y.; Yang, P.; Shang, M.; Wang, X. Anti-diabetic and anti-nephritic activities of *Grifola frondosa* mycelium polysaccharides in diet-streptozotocin-induced diabetic rats via modulation on oxidative stress. *Appl. Biochem. Biotechnol.* **2019**, *187*, 310–322. [CrossRef]
82. Ratnaningtyas, N.I.; Hernayanti, H.; Andarwanti, S.; Ekowati, N.; Purwanti, E.S.; Sukmawati, D. Effects of *Ganoderma lucidum* extract on diabetic rats. *Biosaintifika J. Biol. Biol. Educ.* **2018**, *10*, 6. [CrossRef]
83. Lin, X.; Brennan-Speranza, T.C.; Levinger, I.; Yeap, B.B. Undercarboxylated osteocalcin: Experimental and human evidence for a role in glucose homeostasis and muscle regulation of insulin sensitivity. *Nutrients* **2018**, *10*, 847. [CrossRef] [PubMed]
84. Zhang, L.; Hu, Y.; Duan, X.; Tang, T.; Shen, Y.; Hu, B.; Liu, A.; Chen, H.; Li, C.; Liu, Y. Characterization and antioxidant activities of polysaccharides from thirteen boletus mushrooms. *Int. J. Biol. Macromol.* **2018**, *113*, 1–7. [CrossRef] [PubMed]
85. Khursheed, R.; Singh, S.K.; Wadhwa, S.; Gulati, M.; Awasthi, A. Therapeutic potential of mushrooms in diabetes mellitus: Role of polysaccharides. *Int. J. Biol. Macromol.* **2020**, *164*, 1194–1205. [CrossRef]
86. Panigrahy, S.; Bhatt, R.; Kumar, A. Targeting type II diabetes with plant terpenes: The new and promising antidiabetic therapeutics. *Biologia* **2020**, *76*, 241–254. [CrossRef]
87. Kolundzic, M.; Grozdanić, N.; Stanojkovic, T.; Milenkovic, M.; Dinic, M.; Golić, N.; Kojic, M.; Kundaković, T. Antimicrobial and cytotoxic activities of the sulphur shelf medicinal mushroom *laetiporus sulphureus* (Agaricomycetes) from Serbia. *Int. J. Med. Mushrooms* **2016**, *18*, 469–476. [CrossRef]
88. Ma, X.; Yang, M.; He, Y.; Zhai, C.; Li, C. A review on the production, structure, bioactivities and applications of *Tremella* polysaccharides. *Int. J. Immunopathol. Pharmacol.* **2021**, *35*, 20587384211000541. [CrossRef]
89. Zhu, G.; Hayashi, M.; Shimomura, N.; Yamaguchi, T.; Aimi, T. Differential expression of three α-amylase genes from the basidiomycetous fungus *Pholiota microspora. Mycoscience* **2017**, *58*, 188–191. [CrossRef]
90. Cardwell, G.; Bornman, J.F.; James, A.P.; Black, L.J. A review of mushrooms as a potential source of dietary vitamin D. *Nutrients* **2018**, *10*, 1498. [CrossRef]
91. Sung, C.-C.; Liao, M.-T.; Lu, K.-C.; Wu, C.-C. Role of vitamin D in insulin resistance. *J. Biomed. Biotechnol.* **2012**, *2012*, 634195. [CrossRef]
92. Szymczak-Pajor, I.; Drzewoski, J.; Śliwińska, A. The molecular mechanisms by which vitamin D prevents insulin resistance and associated disorders. *Int. J. Mol. Sci.* **2020**, *21*, 6644. [CrossRef]
93. Talaei, A.; Mohamadi, M.; Adgi, Z. The effect of vitamin D on insulin resistance in patients with type 2 diabetes. *Diabetol. Metab. Syndr.* **2013**, *5*, 8. [CrossRef]
94. Zakhary, C.M.; Rushdi, H.; Hamdan, J.A.; Youssef, K.N.; Khan, A.; Abdalla, M.A.; Khan, S. Protective role of vitamin D therapy in diabetes mellitus type II. *Cureus* **2021**, *13*, e17317. [CrossRef]
95. Al-Shoumer, K.A.; Al-Essa, T.M. Is there a relationship between vitamin D with insulin resistance and diabetes mellitus? *World J. Diabetes* **2015**, *6*, 1057–1064. [CrossRef] [PubMed]
96. Urbain, P.; Singler, F.; Ihorst, G.; Biesalski, H.K.; Bertz, H. Bioavailability of vitamin D$_2$ from UV-B-irradiated button mushrooms in healthy adults deficient in serum 25-hydroxyvitamin D: A randomized controlled trial. *Eur. J. Clin. Nutr.* **2011**, *65*, 965–971. [CrossRef] [PubMed]
97. Wenclewska, S.; Szymczak-Pajor, I.; Drzewoski, J.; Bunk, M.; Śliwińska, A. Vitamin D supplementation reduces both oxidative DNA damage and insulin resistance in the elderly with metabolic disorders. *Int. J. Mol. Sci.* **2019**, *20*, 2891. [CrossRef] [PubMed]
98. Mutt, S.J.; Raza, G.S.; Mäkinen, M.J.; Keinänen-Kiukaanniemi, S.; Järvelin, M.R.; Herzig, K.H. Vitamin D deficiency induces insulin resistance and re-supplementation attenuates hepatic glucose output via the PI3K-AKT-FOXO1 mediated pathway. *Mol. Nutr. Food Res.* **2020**, *64*, e1900728. [CrossRef]
99. El Hajj, C.; Walrand, S.; Helou, M.; Yammine, K. Effect of vitamin D supplementation on inflammatory markers in non-obese Lebanese patients with type 2 diabetes: A randomized controlled trial. *Nutrients* **2020**, *12*, 2033. [CrossRef]
100. Cojic, M.; Kocic, R.; Klisic, A.; Kocic, G. The effects of vitamin D supplementation on metabolic and oxidative stress markers in patients with type 2 diabetes: A 6-month follow up randomized controlled study. *Front. Endocrinol.* **2021**, *12*, 610893. [CrossRef]

 *pharmaceutics*

Review

The Pivotal Role of Oleuropein in the Anti-Diabetic Action of the Mediterranean Diet: A Concise Review

Andrea Da Porto *, Gabriele Brosolo, Viviana Casarsa, Luca Bulfone, Laura Scandolin, Cristiana Catena and Leonardo A. Sechi

Clinica Medica, Department of Medicine, University of Udine, 33100 Udine, Italy;
gabriele.brosolo@uniud.it (G.B.); viviana.casarsa@gmail.com (V.C.); luca.bulfone1@gmail.com (L.B.);
scandolin.laura@spes.it (L.S.); cristiana.catena@uniud.it (C.C.); leonardo.sechi@uniud.it (L.A.S.)
* Correspondence: andrea.daporto@uniud.it; Tel.: +39-432-559801

Abstract: Type 2 diabetes currently accounts for more than 90% of all diabetic patients. Lifestyle interventions and notably dietary modifications are one of the mainstays for the prevention and treatment of type 2 diabetes. In this context, the Mediterranean diet with its elevated content of phytonutrients has been demonstrated to effectively improve glucose homeostasis. Oleuropein is the most abundant polyphenolic compound contained in extra-virgin olive oil and might account for some of the anti-diabetic actions of the Mediterranean diet. With the aim to provide an overview of the possible contributions of oleuropein to glucose metabolism, we conducted a PubMed/Medline search in order to provide an update to the available evidence regarding this interesting compound. This narrative review summarizes the data that was obtained in in vitro and animal studies and the results of clinical investigations. Preclinical studies indicate that oleuropein improves glucose transport, increases insulin sensitivity, and facilitates insulin secretion by pancreatic β-cells, thereby supporting the hypothesis of the possible benefits of the control of hyperglycemia. However, on the clinical side, the available evidence is still preliminary and requires more extensive investigations. Thus, many questions remain unanswered in regards to the potential benefits of oleuropein in diabetes prevention and treatment. These questions should be addressed in appropriately designed studies in the future.

Keywords: oleuropein; Mediterranean diet; diabetes; metabolic syndrome

Citation: Da Porto, A.; Brosolo, G.;
Casarsa, V.; Bulfone, L.; Scandolin, L.;
Catena, C.; Sechi, L.A. The Pivotal
Role of Oleuropein in the
Anti-Diabetic Action of the
Mediterranean Diet: A Concise
Review. *Pharmaceutics* **2022**, *14*, 40.
https://doi.org/10.3390/
pharmaceutics14010040

Academic Editor: Maria
Camilla Bergonzi

Received: 16 November 2021
Accepted: 21 December 2021
Published: 25 December 2021

Publisher's Note: MDPI stays neutral
with regard to jurisdictional claims in
published maps and institutional affiliations.

1. Introduction

It is now more than 50 years since the benefits of the Mediterranean diet on the prevention of cardiovascular and neurodegenerative diseases, cancers, and diet-related metabolic conditions, like diabetes, have been convincingly demonstrated [1]. Diabetes is a major health problem in economically developed countries and is one of the leading causes of morbidity and mortality worldwide. Type 2 diabetes is the most common form of disease and can be prevented, mainly through lifestyle interventions, notably dietary modifications. In fact, the reduced incidence of type 2 diabetes has been reported in prospective studies, which have demonstrated the important benefits of the Mediterranean diet [2,3]. Dietary interventions are also one of the mainstays of the treatment of type 2 diabetes and large studies have consistently demonstrated that these interventions, including the Mediterranean diet, have significant impacts on the clinical outcomes of this condition [4]. Evidence of the benefits of these dietary interventions has therefore triggered a wide interest in the role of functional foods and the bioactive components of the Mediterranean diet for diabetes prevention and treatment.

Substantial consumption of extra-virgin olive oil (EVOO), vegetables, legumes, fruits, and whole grain cereals is the distinctive feature of the Mediterranean diet that replicates some of the traditional eating habits of the countries bordering the Mediterranean Sea. Since the time of Hippocrates of Kos who named it "The Great Healer", EVOO has shown medical properties that have been held in high regard by countless generations of physicians [5]. Nowadays, the biological effects of EVOO have received broad recognition in the clinical setting and many dietary intervention studies have consistently reported substantial benefits on human health [6]. Along with the Mediterranean diet, which has continuously proven to be a valuable contributor to the dietary treatment of diabetic patients [7], benefits of regular consumption of EVOO have been reported in the management of type 2 diabetes [8]. The effects on the glucose metabolism of specific components of EVOO, however, are still debated.

The medical proprieties of EVOO can be ascribed to some of its chemical components, including fatty acids (triacylglycerols, free fatty acids, mono and diacylglycerols) and other lipids (hydrocarbons, sterols, aliphatic alcohols, and tocopherols), and volatile compounds, including polyphenols, which develop in response to insect injuries in olive trees. The main sources of olive polyphenols are olive leaves that possess the highest antioxidant and scavenging power among the different parts of the olive tree. There are five different chemical groups of phenolic compounds in olive leaves: oleuropeoides, flavones, flavonols, flavan-3-ols, and substituted phenols. Polyphenolic compounds are particularly abundant in EVOO, and multiple beneficial biological effects of these compounds have been reported, including antioxidant, anti-inflammatory, antiproliferative, and finally, antiobesity and antidiabetic properties [9–11].

In recent years, oleuropein has developed a growing interest in regard to the phenolic compounds contained in EVOO because of its broad metabolic effects. These effects might contribute to the benefits that are commonly seen in subjects who eat a Mediterranean diet. This narrative article has been written with the purpose of highlighting the possible metabolic benefits of oleuropein, with a specific focus on its effects on glucose metabolism and diabetes. Information was obtained by a PubMed/MEDLINE systematic search of articles that included oleuropein and diabetes as key words. Experimental data obtained in vitro and in animal studies and the results of clinical investigations are included and overviewed. This article will provide the readership with an update on the currently available evidence on this interesting compound and hopefully trigger further clinical research in order to develop better knowledge and to set the stage for its routine use.

2. Oleuropein and Diabetes: Experimental Studies

Oleuropein is an iridoid molecule of oleuropeoide and is esterified with a dihydro-caffeoyl alcohol residue that provides the phenolic characteristic to this molecule. It is present in EVOO, both in a glycated and aglycone (iridoid esterified with the phenylpropanoid alcohol) (Figure 1) form, and is the most abundant among the polyphenolic compounds of olive leaves. Like other phenolic compounds contained in EVOO, oleuropein exhibits antioxidant [12], anti-inflammatory [13], and antiproliferative [14] properties and exerts protective actions at the cardiovascular [15], metabolic [12,15], neurological [16], and hepatic [17] level (Table 1).

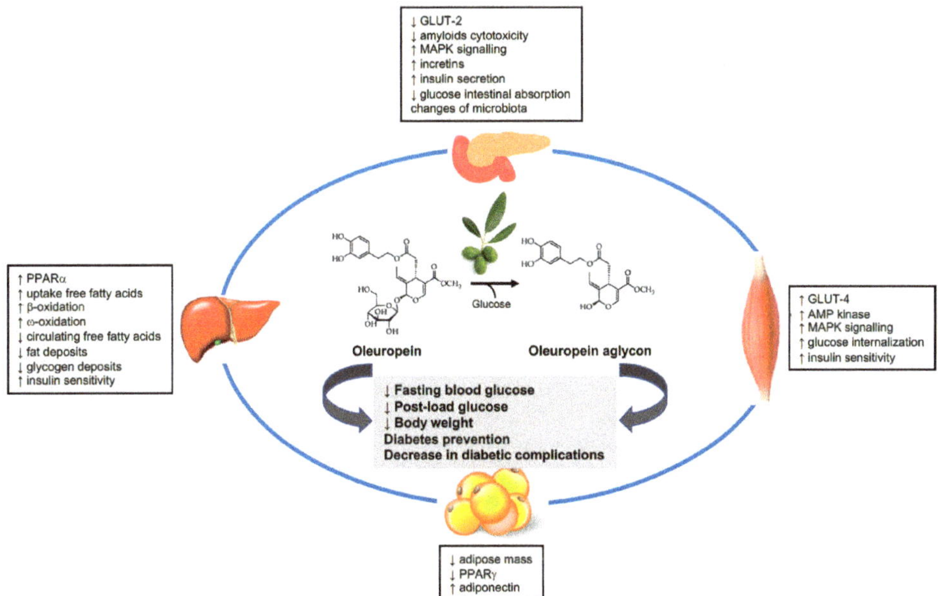

Figure 1. Effects of oleuropein on glucose homeostasis and the related mechanisms.

Table 1. Biological properties and effects of oleuropein.

Properties	Possible Mechanisms
Anti-Oxidation [12]	reactive oxygen species scavenging
	improved free radical stability
	increased catalase, superoxide dismutase, glutathione peroxidase, thioredoxin reductase activity
	decreased malondialdehyde, advanced glycation endproducts
Anti-Inflammation [13]	decreased C-reactive protein, neutral factor-κB, interleukin-1β, interleukin-6, adipocytokines, tumor necrosis factor
	lipooxygenase inhibition
Anti-Cancer [14]	inhibition of cell proliferation, angiogenesis, cell migration
	induction of apoptosis
	reactive oxygen species scavenging
	inhibition of human epidermal growth factor receptor, Bcl-2A pathways, protein kinases, neutral factor-κB, cyclinD1
	activation of Bax, Janus kinase
Cardiovascular Protection [15]	reduced oxidative stress
	increased nitric oxide formation
	lipid lowering and reduced lipid peroxidation
	reduced blood pressure
Metabolic Protection [12,15]	decreased obesity
	reduced blood glucose
	diabetes prevention
Neuroprotection [16]	reduction of oxidative stress
	stabilization of amyloid fibers
Hepatoprotection [17]	reduction of oxidative stress
	reduction of fat deposition

Moreover, oleuropein has already demonstrated interesting blood glucose lowering properties [18]. The results of experimental studies suggest that the hypoglycemic effects of oleuropein could be mediated by modulation of multiple intracellular signaling mechanisms that are directly involved in the regulation of blood glucose concentration [12]. Exposure of cultured mouse myoblasts to physiological concentrations of oleuropein promoted translocation of the glucose transporter GLUT-4 (glucose transporter-4) to the plasma membrane. This effect was mediated by the activation of adenosine monophosphate-activated protein kinase (MAPK) and was associated with the increased cellular internalization of glucose [19]. The same effect was replicated in vivo in oleuropein-fed mice in whom insulin sensitivity was significantly improved [19]. In vitro studies indicate that the activity of the high-capacity, low-affinity, glucose transporter GLUT-2 (glucose transporter-2) is decreased in cultured cells exposed to oleuropein [20]. GLUT-2 is highly expressed in the gut and is part of the glucose sensor of pancreatic β-cells, thereby playing a central role in the regulation of intestinal glucose absorption and glucose-stimulated insulin secretion. Furthermore, experiments conducted in cultured hepatocytes and in mice exposed to oleuropein (100 mg/kg/day p.o.) for six weeks demonstrate that oleuropein acts as a ligand of the peroxisome proliferator activated-receptor alpha (PPARα), a nuclear receptor that activates a family of genes that are critically involved in many metabolic pathways [21]. PPARα activates gene encoding enzymes that are involved in fatty acid cell uptake, mitochondrial β-oxidation, microsomal ω-oxidation, and genes that encode many apolipoproteins [21]. Therefore, PPARα participates in the control of circulating fatty acids with all their related effects, including those on insulin sensitivity and glucose metabolism. Finally, it was reported that oral administration of oleuropein (50 mg/kg/day p.o.) to cholesterol-fed rats for eight weeks decreased body weight and adipose tissue mass and attenuated liver fat deposits. These effects on lipid uptake and adipogenesis were associated with the significant inhibition of peroxisome proliferator activated-receptor gamma (PPARγ) and increase in serum adiponectin levels [22] with important implications for the development of obesity.

In addition to the effects on metabolic pathways and insulin sensitivity, oleuropein facilitates glucose-stimulated insulin secretion by pancreatic β-cells. This experimental evidence was initially reported by Wu et al. who demonstrated a dose-dependent effect of oleuropein in INS-1 (rat insulinoma cell line) β-cells that was linked to activation of the MAPK pathway [23]. Moreover, it was demonstrated that oleuropein prevents cytotoxicity of β-cells that is induced by amyloids, which is a distinctive feature of type 2 diabetes. More recently, it has been demonstrated that oleuropein aglycone inhibits pancreatic β-cell cytotoxicity by human islet amyloid polypeptide aggregates and protects the plasma membrane from permeabilization, thereby preventing cell death [24].

Benefits of oleuropein on blood glucose control have been reported in many studies conducted using animal models of diabetes. Initial evidence was obtained in experiments conducted on rabbits with alloxan-induced diabetes who were treated with oleuropein (25 mg/kg/day p.o.) for 16 weeks. In these animals, oleuropein significantly reduced blood glucose and malondialdehyde, which is a marker of oxidative stress that is markedly increased in this experimental model and might contribute to diabetic complications [25]. Similarly, the administration of oleuropein (8–16 mg/kg/day p.o.) for four weeks reduced blood glucose and improved glucose tolerance in rats with alloxan-induced diabetes [26] and other rodent models of insulin-deficient (streptozotocin-diabetic rats) and insulin-resistant (leptin-receptor deficient ob/ob, Tsumura-Suzuky, fat diet obese mice) diabetes [18]. The effects of oleuropein have also been tested in gestational diabetes using a mice model [27]. In these animals, oleuropein (5–10 mg/day intraperitoneally) attenuated the body weight increase and efficiently decreased blood glucose, plasma insulin, and hepatic glycogen levels, with an overall improvement in the gestational outcome after 20 gestational days. Very recently, Zheng et al. tested the effects of oleuropein (200 mg/kg/day p.o.) administration for 15 weeks in diabetic *db/db* mice, reporting a significant decrease in fasting blood glucose levels, an improvement to glucose tolerance, a lowering of homeostasis model assessment insulin-resistance index, and a promotion of protein kinase B activation [28]. The *db/db* mice model is characterized by pancreatic islet damage, lipid deposition in hepatocytes, and disordered myocardial fibers, which are all histopathological changes that were restored by oleuropein administration. [28]. Additionally, and most importantly, this study examined the composition of the gut microbiota, which shows remarkable changes in the intestinal bacteria in diabetic mice fed with oleuropein. At the phylum level, the relative abundance of *Verrucomicrobia* and *Deferribacteres* was increased by oleuropein, whereas the abundance of *Bacteroidetes* decreased; at the genus level, the relative abundance of *Akkermansia* increased with oleuropein, whereas *Prevotella*, *Odoribacter*, *Ruminococcus*, and *Parabacteroides* decreased [28]. In agreement with the above findings obtained with oleuropein, robust experimental evidence suggests that polyphenols contained in EVOO have important beneficial effects on intestinal bacteria [29]. These findings support the intriguing hypothesis that oleuropein may improve glucose metabolism in diabetes through the modulation of the composition and function of the gut microbiota and open an interesting area of investigation for future studies.

It is well known that gut microbiota has a central role in the maintenance of physiological homeostasis and that its dysregulation (dysbiosis) can modify intestinal permeability, thereby affecting several pathophysiological mechanisms that can have an impact on glucose metabolism. Diverse families of intestinal bacteria mediate their beneficial effects through the fermentation of dietary fibers with the production of short-chain fatty acids. These fatty acids are endogenous signals that play important roles in the regulation of metabolic pathways, including those involved in glucose homeostasis.

Important differences in the doses of oleuropein, used both in acute and chronic experiments, different durations of exposure, and differences in the route of administration to experimental animals do not permit the definition of the optimal doses of this compound, which would be needed to obtain the best metabolic responses.

In summary, data obtained in the animal models of diabetes indicate that oleuropein interferes beneficially with glucose metabolism. These benefits could result from changes in the glucose transport and intracellular metabolism, increased insulin sensitivity, and more efficient glucose-stimulated insulin secretion by pancreatic β-cells (Figure 1).

Oleuropein could regulate the glucose metabolism that acts at many levels and in different organs. On the one hand, oleuropein facilitates insulin secretion by pancreatic β-cells and reduces intestinal glucose absorption. On the other hand, oleuropein improves insulin sensitivity in skeletal muscles and the liver, thereby increasing glucose internalization and decreasing the circulating levels of free fatty acids, thus leading to reduced fat accumulation. The final effects demonstrated in animal studies and suggested by the results of early clinical trials are weight loss and the reduction of fasting and post-oral load glucose. These actions of oleuropein may result in the effective prevention of new-onset diabetes and renal and cardiovascular renal complications. GLUT, glucose transporter; PPAR, peroxisome proliferator activated-receptor; MAPK, adenosine monophosphate-activated protein kinase; AMP, adenosine mono phosphate.

3. Oleuropein and Diabetes: Clinical Evidence

As previously stated, the benefits of the regular consumption of EVOO in the management of diabetes have been repeatedly reported. Eleven obese patients with type 2 diabetes, who were treated with oral antidiabetic agents, ate refined oil polyphenol-free for four weeks. This was then replaced by polyphenol-rich EVOO for a further four weeks [30]. Polyphenol-rich EVOO significantly reduced body weight, fasting plasma glucose, and glycated hemoglobin and these changes were associated with a significant decrease in serum visfatin, a pro-inflammatory adipocytokine. Schwingshackl et al. conducted a meta-analysis in order to examine the association between EVOO consumption and the risk of type 2 diabetes, in addition to the effects of EVOO on its management [8]. The analysis included 15,784 subjects who were included in four cohort studies and 29 intervention trials. The results of this meta-analysis demonstrated that the highest EVOO intake category had a 16% lower risk of developing type 2 diabetes, as compared to the lowest. Additionally, in type 2 diabetics, EVOO supplementation resulted in a significantly greater reduction of fasting plasma glucose and glycated hemoglobin than in the control subjects. These studies provided solid evidence that the intake of EVOO could be beneficial for the prevention and management of type 2 diabetes, but did not provide any insight into the contribution of single components of EVOO.

With specific regard to oleuropein, the effects on glucose metabolism and diabetes have been examined in clinical studies and recent data provide initial evidence of the potential benefits (Table 2). In a randomized, double-blind, crossover trial conducted in New Zealand, 46 middle-aged, overweight men received oleuropein containing capsules (51 mg/day) or a placebo for 12 weeks [31]. As compared to the placebo, oleuropein supplementation was associated with a significant improvement in insulin sensitivity and pancreatic β-cells secretory capacity. Kerimi et al. conducted seven separate randomized, crossover, double-blind, placebo controlled, intervention trials on healthy volunteers to examine the effect of oleuropein on post-prandial blood glucose after the consumption of bread, glucose, or sucrose [20]. Oleuropein (35 to 200 mg/day) in solution attenuated the post-prandial blood glucose response after the consumption of sucrose, but did not affect post-prandial glucose after the ingestion of bread or glucose. Examination of the effects of oleuropein on enzymes involved in carbohydrate digestion showed the inhibition of sucrase and GLUT-2-mediated transport, but no significant effect on α-amylase, thus explaining the findings regarding the post-prandial blood glucose changes. Finally, in an open study of hypertensive patients, many of whom had obesity and/or diabetes, oral supplementation of oleuropein (100 mg/day) was administered for two months. In these patients, oleuropein decreased fasting blood glucose, together with other markers of a metabolic syndrome, such as waist circumference and serum triglycerides [32].

Table 2. Clinical trials that examined the effects of oleuropein on glucose homeostasis.

Reference	Study Design	Source, Main Content and Time of Exposure to Oleuropein	Effects on Glucose Metabolism
Kerimi et al. [17]	RCT, double-blind, crossover 24 healthy volunteers	Supplement vs. Placebo 35–200 mg—Single dose	Reduction of Post-prandial glucose Inhibition of GLUT2 and maltase
De Bock et al. [28]	RCT, double-blind, crossover 46 overweight volunteers	Olive Leaf vs. Placebo 51.1 mg vs. Placebo 12 weeks	Improvement in insulin sensitivity Improvement in pancreatic β-cell responsiveness
Hermans et al. [29]	Prospective, open observational 663 Hypertensive patients	Supplement 100 mg 8 weeks	Reduction of fasting glucose
Violi et al. [30]	RCT, double-blind, crossover 25 healthy volunteers	EVOO vs. Corn Oil 20 mg Single dose	Reduction of post-prandial glucose Inhibition of DPP-4 Improvement in GLP-1 mediated insulin secretion
Carnevale et al. [31]	RCT, double-blind 20 healthy volunteers	EVOO vs. Corn Oil 20 mg Single dose	Reduction of post-prandial glucose Inhibition of DPP-4 Improvement in GLP-1 mediated insulin secretion
Carnevale et al. [32]	RCT, double-blind 30 patients with IGT	EVOO vs. Corn Oil 20 mg Single dose	Reduction of post-prandial glucose Inhibition of DPP-4 activity Improvement in GLP-1 mediated insulin secretion
Del Ben et al. [33]	RCT, single-blind 25 patients with Type 2 Diabetes 20 healthy volunteers	EVOO Enriched vs. Standard Chocolate 40 mg Single dose	Reduction of post-prandial glucose Inhibition of DPP-4 activity Improvement in GLP-1 mediated insulin secretion

EVOO, Extra Virgin Olive Oil; IGT, Impaired Fasting Glucose; DPP-4, Dipeptidlpepdisase-4; GLP-1, Glucagon Like Peptide-1.

Clinical investigations have also tried to clarify some of the mechanisms that might mediate the effects of oleuropein on glucose metabolism. These investigations have brought the possibility that the incretin axis could be modulated by oleuropein to the forefront. The post-prandial glycemic profile was investigated in a crossover study of 25 healthy subjects that were randomly allocated to a Mediterranean diet with or without supplementation of oleuropein (20 mg). Two hours after the meal, subjects who ate oleuropein supplements had a significantly lower blood glucose and dipeptidyl-peptidase 4 (DPP-4) protein concentration and activity, and higher serum insulin and glucacon-like peptide-1 (GLP1) levels [33]. As an extension of this study, the same research group reported that the effects of oleuropein on blood glucose and DPP-4 were associated with a significant reduction in markers of oxidative stress, such as soluble NADPH oxidase-derived peptide activity and 8-iso-prostaglandin-2α [34]. The same protocol with a Mediterranean diet meal with or without oleuropein supplementation was later applied to 30 subjects with impaired fasting glucose [35]. In agreement with the findings obtained in healthy subject, the meal containing oleuropein was associated with a reduction of blood glucose and DPP-4 activity and an increase in insulin and GLP-1, as compared to the meal without oleuropein. Finally, similar experiments were conducted with a crossover design in 25 type 2 diabetic patients who were randomized to receive 40 g of oleuropein in the form of oleuropein-enriched chocolate or control chocolate [36]. In diabetic patients who received oleuropein-enriched (40 mg) chocolate, the increase in blood glucose following ingestion was smaller than in diabetic patients who received plain chocolate, and even in that case, the effect of oleuropein was associated with decreased DPP-4 activity.

In summary, initial clinical observations suggest that there is a potential for oleuropein use in the control of hyperglycemia. The findings suggest that oleuropein might decrease post-prandial blood glucose with a mechanism that counteracts oxidative stress-mediated incretin down-regulation. While more comprehensive evidence is required, the effects of oleuropein might provide both preventive and therapeutic benefits to patients with type 2 diabetes. Due to the significant variability in doses of refined oleuropein that were used in acute and chronic studies, and in the amount of oleuropein contained in EVOO, the current evidence does not allow us to establish suitable doses of this compound for clinical use.

4. Oleuropein and Renal Complications: Experimental Evidence

It is important to note that the potential of oleuropein to be beneficial in the context of diabetes is not limited to the control of metabolic balance with the reduction of hyperglycemia, but this compound might also have an impact on diabetic complications. For instance, some studies suggest the possibility that oleuropein may exert protective actions on the kidney, which is one of the main targets of diabetes. These protective actions of oleuropein were recently demonstrated in experimental animal models of renal damage. First, oleuropein (50 mg/kg of body weight) was administered to rats in whom acute kidney injury was induced by a glycerol injection [37]. This model of renal damage is closely linked to the activation of proinflammatory mechanisms and oxidative stress demonstrated by elevated cytokines and malondialdehyde and decreased glutathione content. Furthermore, renal damage is associated with rhabdomyolysis and multiple molecular, biochemical, and histological alterations. All these alterations were reversed by oleuropein, thus suggesting that this compound might have the potential to be a treatment for this condition. Second, oleuropein was tested in rats with a kidney ischemia-reperfusion injury, and even in this case, renal damage was reduced with a significant improvement in creatinine, urea, uric acid, and lactate dehydrogenase levels [38]. In this model, oleuropein administration (10, 50, and 100 mg/kg) decreased proinflammatory (C-reactive protein and expression of cyclooxygenase 2) and pro-apoptotic (caspase-3) markers, and up-regulated antioxidant capacities that were identified by activated catalase, superoxide dismutase, and glutathione peroxidase activity. Third, kidney injury was induced in rats by three-day unilateral ureteral obstruction and oleuropein (50, 100, and 200 mg/kg) was administered in increasing doses, which improved renal response in a dose-dependent way [39]. Even in this model, renal improvement was associated with activated superoxide dismutase and glutathione peroxidase and down-regulation of pro-inflammatory (tumor necrosis factor-α) and pro-apoptotic (caspase-3, Bax protein) molecules. Thus, all these studies linked the observed benefits of oleuropein on renal function to the evidence of important antioxidant, anti-inflammatory, and anti-apoptotic effects. These effects of oleuropein might also come into play in other more chronic renal conditions, including diabetic nephropathy.

In summary, although oleuropein has been beneficial in different experimental models of renal damage, no data is currently available on the use of oleuropein in the context of diabetic nephropathy. This could be an issue worth testing in future animal and human studies.

5. Oleuropein and Cardiovascular Complications: Experimental Evidence

Another major issue is related to the possible role of oleuropein in the protection from cardiovascular events that are the principal cause of morbidity, disability, and mortality in patients with type 2 diabetes. Oleuropein (3% as oral supplements) has been shown to reduce body weight gain and abdominal fat in high fat-fed rats with a mechanism that appears to be linked to the repression of mitochondrial activity during the differentiation of adipocytes [40]. Inhibition of clonal expansion with the prevention of differentiation and intracellular triglyceride accumulation was also demonstrated in cultured 3T3-L1 preadipocytes because of the reduced expression of adipogenesis-related genes [41]. Animal studies suggest that oleuropein might favorably affect some risk factors that, together with diabetes, contribute to the global cardiovascular risk [15]. In alloxan-diabetic rats, oral supplementation with oleuropein (8–16 mg/kg/day for 4 weeks) has been shown to reduce total cholesterol, serum LDL, and triglycerides and to increase serum HDL [26]. Oleuropein increases LDL-receptors of hepatocytes and regulates the expression of genes that are involved in metabolism and the disposal of triglycerides [19]. Most of these effects of oleuropein on lipid metabolism appear to be related to increased activation of PPARα, with many of its target genes having a mechanism like that of fibrates [21]. In a rat model of simultaneous diabetes and hypertension, oleuropein was administered at increasing doses (20, 40, and 60 mg/kg/day) for four weeks and was compared to the vehicle [42]. Oleuropein significantly reduced blood pressure, blood glucose, total and LDL cholesterol, and triglycerides, and improved glucose tolerance. These effects were apparently linked to antioxidant and simpatho-inhibitory mechanisms [43], and released nitric oxide, which, at physiological levels, activate the soluble guanylate cyclase signal transduction pathway and thereby leads to various beneficial effects [44]. Moreover, in the same experimental diabetic model (20, 40, and 60 mg/kg/day for 4 weeks), oleuropein reduced infarct size and coronary effluent creatine kinase after cardiac ischemia-reperfusion injury, in comparison to controls, thus supporting the hypothesis of a cardioprotective effect of this compound [45]. Even in this case, the benefit of oleuropein was associated with activation of superoxide dismutase, thus suggesting the involvement of antioxidant mechanisms. Regarding hypertension, in a double-blind, randomized, crossover dietary intervention study, 24 women with high-normal blood pressure or stage 1 hypertension received either polyphenol-rich oil (30 mg/day) or polyphenol-free oil for two months. Only the polyphenol-rich oil led to a significant fall in systolic and diastolic blood pressure, which was associated with significantly decreased serum asymmetric dimethylarginine, thus suggesting improved endothelial functioning [46].

In summary, there is evidence of the possible benefits of oleuropein on overweight/obese, dyslipidemia and hypertension patients that might result in better protection from cardiovascular events in high-risk conditions, such as type 2 diabetes.

6. Perspectives

The increase in life expectancy observed in the Western World is paying a price to the progressively greater incidence of many chronic diseases associated with ageing and lifestyle. For this reason, medical research is progressively shifting its focus from a cure to a prevention, with greater relevance attributed to the "lifestyle" concept that primarily involves diet and thus food. Epidemiological evidence has provided support to the idea that "healthy diets", such as the Mediterranean diet, are associated with a lower incidence of chronic conditions. Therefore, there is now rapidly growing interest in the effects of the bioactive components of food.

An important feature of the Mediterranean diet is the elevated content of phytonutrients that affect multiple biological functions, including glucose homeostasis. Type 2 diabetes is the most common form of diabetes, currently accounting for more than 90% of all diabetic patients. Lifestyle interventions and notably dietary modifications are one of the mainstays of type 2 diabetes prevention and treatment. For this reason, dietary components have been extensively investigated in order to find which of them could sensibly affect glucose metabolism. For instance, the dietary content of advanced glycation end products (AGE) has been investigated and a recent meta-analysis has reported that a poor diet of AGE has beneficial effects on fasting insulin and insulin resistance [47]. However, several bioactive compounds have been reported to favorably affect glucose homeostasis in type 2 diabetes [48]. Polyphenols are highly represented in EVOO, and in turn, it is one of the main components of the Mediterranean diet. Among the polyphenols contained in EVOO, oleuropein has gathered growing interest because of its antioxidant, anti-inflammatory, and antiproliferative properties, which have been demonstrated in experimental in vitro and in vivo studies. With specific regard to glucose metabolism, initial clinical observations suggest that there is a potential for oleuropein use in the prevention and treatment of diabetes.

Oleuropein is only one of a myriad of bioactive compounds of food that deserve careful evaluation by both basic and clinical researchers. Expansion of knowledge on the effects of oleuropein and other food components on human health, together with the refinement of manufacturing for market use, will hopefully permit future, effective preventive and therapeutic interventions to reduce the burden of chronic disease.

7. Conclusions

A Mediterranean dietary pattern is effective in managing nutrition-related metabolic disorders, including diabetes. Polyphenols are under the spotlight as they are compounds with greater biological interest among the nutraceuticals contained in the Mediterranean diet. Therefore, the potential of oleuropein, as a major component of EVOO in the Mediterranean diet for modulation of glucose metabolism, is rapidly gaining interest within the scientific community. Preclinical studies indicate that oleuropein improves glucose transport and intracellular metabolism, increases insulin sensitivity, and facilitates insulin secretion by pancreatic β-cells, thus supporting the intriguing hypothesis that this phenolic compound might be beneficial for the prevention of diabetes and control of hyperglycemia. However, the available clinical evidence is too preliminary and must be strengthened in larger and appropriately designed studies. Strengthening the evidence in the clinical field would require some essential steps. First, sources of oleuropein should be carefully defined by standardization of olive leaf extraction and purification procedures in order to have control of the administered amount that, up to now, has been extremely variable. Second, dose finding studies should be planned to define the threshold for oleuropein effects that may differ depending on the outcome of various interests. Third, appropriate assessment of the exposure time that is necessary to obtain biologically relevant effects will be needed. Finally, like any other dietary intervention, it would be useful to have laboratory tests that permit the definition of adherence to oleuropein use in chronic studies.

While many commercial preparations of oleuropein obtained from olive leaves extraction are currently invading the market, more still needs to be learnt regarding the best sources, the appropriate intake, and the exposure time that is necessary to obtain clinical benefits in people with type 2 diabetes or that are at risk of diabetes. Future randomized clinical trials will hopefully assess the evidence needed for the widespread use of oleuropein in daily clinical practice.

Author Contributions: Conceptualization, A.D.P., L.A.S.; methodology A.D.P., G.B., V.C., L.B., L.S., C.C., L.A.S.; validation, G.B., V.C., L.B., L.S., C.C.; writing—original draft preparation, A.D.P., L.A.S.; writing—review and editing, G.B., V.C., L.B., L.S., C.C. All authors have read and agreed to the published version of the manuscript.

Funding: This research received no external funding.

Institutional Review Board Statement: Not applicable.

Informed Consent Statement: Not applicable.

Data Availability Statement: Not applicable.

Conflicts of Interest: The authors declare no conflict of interest.

References

1. Davis, C.; Bryan, J.; Hodgson, J.; Murphy, K. Definition of the mediterranean diet: A literature review. *Nutrients* **2015**, *7*, 9139–9153. [CrossRef]
2. Georgoulis, M.; Kontogianni, M.D.; Yiannakouris, N. Mediterranean diet and diabetes: Prevention and treatment. *Nutrients* **2014**, *6*, 1406–1423. [CrossRef]
3. Estruch, R.; Ros, E.; Salas-Salvado, J.; Covas, M.I.; Corella, D.; Aros, F.; Gomez-Garcia, E.; Ruiz-Gutierrez, V.; Fiol, M.; Lapetra, J.; et al. PREDIMED Study investigators. Primary prevention of cardiovascular disease with a Mediterranean diet supplemented with extra-virgin olive oil or nuts. *N. Engl. J. Med.* **2013**, *368*, 1279–1290. [CrossRef]
4. Salas-Salvadò, J.; Bullò, M.; Estruch, R.; Ros, E.; Covas, M.I.; Ibarrola-Jurado, N.; Corella, D.; Aros, F.; Gomez-Garcia, E.; Ruiz-Gutierrez, V.; et al. Prevention of diabetes with Mediterranean diets: A subgroup analysis of a randomized trial. *Ann. Intern. Med.* **2014**, *160*, 1–10. [CrossRef]
5. Clodoveo, M.L.; Camposeo, S.; De Gennaro, B.; Pascuzzi, S.; Roselli, L. In the ancient world, virgin olive oil was called "liquid gold" by Homer and "the great healer" by Hippocrates. Why has this mythic image been forgotten? *Food Res. Int.* **2014**, *62*, 1062–1068. [CrossRef]
6. Martínez-González, M.A.; Gea, A.; Ruiz-Canela, M. The Mediterranean Diet and Cardiovascular Health. *Circ. Res.* **2019**, *124*, 779–798. [CrossRef] [PubMed]
7. Milenkovic, T.; Bozhinovska, N.; Macut, D.; Bjekic-Makut, J.; Rahelic, D.; Asimi, Z.V.; Burekovic, A. Mediterrenean diet and type 2 diabetes mellitus: A perpetual inspiration for the scientific world. A review. *Nutrients* **2021**, *13*, 1307. [CrossRef] [PubMed]
8. Schwingshackl, L.; Lampousi, A.M.; Portillo, M.P.; Romaguera, D.; Hoffmann, G.; Boeing, H. Olive oil in the prevention and management of type 2 diabetes mellitus: A systematic review and meta-analysis of cohort studies and intervention trials. *Nutr. Diabetes* **2017**, *7*, e262. [CrossRef]
9. Omar, S.H. Oleuropein in oil and its pharmacological effects. *Sci. Pharm.* **2010**, *78*, 133–154. [CrossRef]
10. Jin, T.; Song, Z.; Weng, J.; Fantus, I.G. Curcumin and other dietary polyphenols: Potential mechanisms of metabolic actions and therapy for diabetes and obesity. *Am. J. Physiol. Endocrinol. Metab.* **2018**, *314*, E201–E205. [CrossRef]
11. Da Porto, A.; Cavarape, A.; Colussi, G.; Casarsa, V.; Catena, C.; Sechi, L.A. Polyphenols Rich Diets and Risk of Type 2 Diabetes. *Nutrients* **2021**, *13*, 1445. [CrossRef]
12. Ahamad, J.; Toufeeq, I.; Khan, M.A.; Ameen, M.S.M.; Anwer, E.T.; Uthirapathy, S.; Mir, S.R.; Ahmad, J. Oleuropein: A natural antioxidant molecule in the treatment of metabolic syndrome. *Phytother. Res.* **2019**, *33*, 3112–3128. [CrossRef]
13. Castejon, M.L.; Rosillo, M.A.; Montoya, T.; Gonzalez-Benjumea, A.; Fernandez-Bolanos, J.G.; Alarcon-de-la-Lastra, C. Oleuropein down-regulated IL-1β-induced inflammation and oxidative stress in human synovial fibroblast cell line SW982. *Food. Funct.* **2017**, *8*, 1890–1898. [CrossRef]
14. Imran, M.; Nadeem, M.; Gilani, S.A.; Khan, S.; Sajid, M.W.; Amir, R.M. Antitumor perspectives of oleuropeinand its metabolite hydroxytyrosol: Recent updates. *J. Food Sci.* **2018**, *83*, 1781–1791. [CrossRef] [PubMed]
15. Bulotta, S.; Celano, M.; Lepore, S.M.; Montalcini, T.; Pujia, A.; Russo, D. Beneficial effects of the olive oil phenlic components oleuropein and hydroxytyrosol: Focus on protection against cardiovascular and metabolic diseases. *J. Tsansl. Med.* **2014**, *12*, 219. [CrossRef] [PubMed]
16. Zhang, W.; Liu, X.; Li, Q. Protective effects of oleuropein against cerebral ischemia/reperfusion by inhibiting neuronal apoptosis. *Med. Sci. Monit.* **2018**, *24*, 6587–6598. [CrossRef] [PubMed]
17. Fki, I.; Sayadi, S.; Mahmoudi, A.; Daoued, I.; Marrekchi, R.; Ghorbel, H. Comparative study on beneficial effects of hydroxytyrosol and oleuropein rich olive leaf extracts on high-fat diet induced lipid metabolism disturbance and liver injuries in rats. *Biomed. Res. Int.* **2020**, *2020*, 1315202. [CrossRef]
18. Zheng, S.; Huang, K.; Tong, T. Efficacy and mechanisms of oleuropein in mitigating diabetes and diabetes complications. *J. Agric. Food Chem.* **2021**, *69*, 6145–6155. [CrossRef] [PubMed]

19. Fujiwara, Y.; Tsukahara, C.; Ikeda, N.; Sone, Y.; Ishikawa, T.; Ichi, I.; Koike, T.; Aoki, Y. Oleuropein improves insulin resistance in skeletal muscle by promoting the translocation of GLUT4. *J. Clin. Biochem. Nutr.* **2017**, *61*, 196–202. [CrossRef]
20. Kerimi, A.; Nyambe-Silavwe, H.; Pyner, A.; Oladele, E.; Gauer, J.S.; Stevens, Y.; Williamson, G. Nutritional implications of olives and sugar: Attenuation of post-prandial glucose spikes in healthy volunteers by inhibition of sucrose hydrolysis and glucose transport by oleuropein. *Eur. J. Nutr.* **2019**, *58*, 1315–1330. [CrossRef]
21. Malliou, F.; Andreadou, I.; Gonzalez, F.J.; Lazou, A.; Xepapadaki, E.; Vallianou, I.; Lambrinidis, G.; Mikros, E.; Marselos, M.; Skaltsounmis, A.L.; et al. The olive constituent oleuropein, as PPARα agonist, markedly reduces serum triglycerides. *J. Nutr. Biochem.* **2018**, *59*, 17–28. [CrossRef]
22. Hadrich, F.; Mahmoudi, A.; Bouallagui, Z.; Feki, I.; Isoda, H.; Feve, B.; Sayadi, S. Evaluation of hypocholesterolemic effect of oleuropein in cholesterol-fed rats. *Chem. Biol. Interact.* **2016**, *252*, 54–60. [CrossRef] [PubMed]
23. Wu, L.; Velander, P.; Liu, D.; Xu, B. Olive component oleuropein promotes beta-cell insulin secretion and protects beta-cells from amylin amyloid-induced cytotoxicity. *Biochemistry* **2017**, *56*, 5035–5039. [CrossRef] [PubMed]
24. Chaari, A. Inhibition of human islet amyloid polypeptide aggregation an cellular toxicity by oleuropein and derivatives from olive oil. *Int. J. Biol. Macromol.* **2020**, *162*, 284–300. [CrossRef] [PubMed]
25. Al-Azzawie, H.F.; Alhamdani, M.-S.S. Hypoglicemic and oxidant effect of oleuropein in alloxan-diabetic rabbits. *Life Sci.* **2006**, *78*, 1371–1377. [CrossRef]
26. Jemai, H.; El Feki, A.; Sayadi, S. Antidiabetic and antioxidant effects of hydroxytyrosol and oleuropein from olive leaves in alloxan-diabetic rats. *J. Agric. Food Chem.* **2009**, *57*, 8798–8804. [CrossRef]
27. Zhang, Z.; Zhao, H.; Wang, A. Oleuropein alleviated gestational diabetes mellitus by activating AMPK signalling. *Endocr. Connect.* **2021**, *10*, 45–53. [CrossRef] [PubMed]
28. Zheng, S.; Wang, Y.; Fang, J.; Geng, R.; Li, M.; Zhao, Y.; Kang, S.G.; Huang, K.; Tong, T. Oleuropein ameliorates advanced stage of type 2 diabetes in *db/db* mice by regulating gut microbiota. *Nutrients* **2021**, *13*, 2131. [CrossRef]
29. Marcelino, G.; Hiane, P.A.; de Cassia Freitas, K.; Figueiredo Santana, L.; Pott, A.; Rodrigues Donadon, J.; de Cassia Avellaneda Guimares, R. Effects of olive oil and its minor components on cardiovascular diseases, inflammation, and gut microbiota. *Nutrients* **2019**, *11*, 1826. [CrossRef]
30. Santangelo, C.; Filesi, C.; Vari, R.; Scazzocchio, B.; Filardi, T.; Fogliano, V.; D'Archivio, M.; Giovannini, C.; Lenzi, A.; Morano, S.; et al. Consumption of extra-virgin oil rich in phenolic compounds improves metabolic control in patients with type 2 diabetes mellitus: A possible involvement of reduced levels of circulating visfatin. *J. Endocrinol. Invest.* **2016**, *39*, 1295–1301. [CrossRef]
31. de Bock, M.; Derraik, J.G.; Brennan, C.M.; Biggs, J.B.; Morgan, P.E.; Hodgkinson, S.C.; Hofman, P.L.; Cutfield, W.S. Olive (*Olea europaea* L.) leaf polyphenols improve insulin sensitivity in middle-aged overweight men: A randomized, placebo-controlled, crossover trial. *PLoS ONE* **2013**, *8*, e57622. [CrossRef] [PubMed]
32. Hermans, M.P.; Lempereur, P.; Salembier, J.P.; Maes, N.; Albert, A.; Jansen, O.; Pincemail, J. Supplementation effect of a combination of olive (*Olea europea* L.) leaf and fruit extracts in the clinical management of hypertension and metabolic syndrome. *Antioxidants* **2020**, *9*, 872. [CrossRef]
33. Violi, F.; Loffredo, L.; Pignatelli, P.; Angelico, F.; Bartimoccia, S.; Nocella, C.; Cangemi, R.; Petruccioli, A.; Monticolo, R.; Pastori, D.; et al. Extra virgin olive oil use is associated with improved post-prandial blood glucose and LDL cholesterol in healthy subjects. *Nutr. Diabetes.* **2015**, *5*, e172. [CrossRef]
34. Carnevale, R.; Silvestri, R.; Loffredo, L.; Novo, M.; Cammisotto, V.; Castellani, V.; Bartimoccia, S.; Nocella, C.; Violi, F. Oleuropein, a component of extra virgin olive oil, lowers postprandial glycaemia in healthy subjects. *Br. J. Clin. Pharmacol.* **2018**, *84*, 1566–1574. [CrossRef]
35. Carnevale, R.; Loffredo, L.; Del Ben, M.; Angelico, F.; Nocella, C.; Petruccioli, A.; Bartimoccia, S.; Monticolo, R.; Cava, E.; Violi, F. Extra virgin olive oil improves post-prandial glycemic and lipid profile in patients with impaired fasting glucose. *Clin. Nutr.* **2017**, *36*, 782–787. [CrossRef]
36. Del Ben, M.; Nocella, C.; Loffredo, L.; Bartimoccia, S.; Cammisotto, V.; Mancinella, M.; Angelico, F.; Valenti, V.; Cavarretta, E.; Carnevale, R.; et al. Oleuropein-enriched chocolate by extra virgin olive oil blunts hyperglycaemia in diabetic patients: Results from a one-time 2-hour post-prandial cross over study. *Clin. Nutr.* **2020**, *39*, 2187–2191. [CrossRef] [PubMed]
37. Yin, M.; Jang, N.; Guo, L.; Ni, Z.; Al-Brakati, A.Y.; Othman, M.S.; Moneim, A.E.A.; Kassab, R.B. Oleuropein suppresses oxidative, inflammatory, and apoptoric responses following glycerol-induced acute kidney injury in rats. *Life Sci.* **2019**, *232*, 116634. [CrossRef]
38. Nasrallah, H.; Aissa, I.; Slim, C.; Boujbiha, M.A.; Zaouali, M.A.; Bejaoui, M.; Wilke, V.; Jannet, H.B.; Mosbah, H.; Abdennebi, H.B. Effect of oleuropein on oxidative stress, inflammation and apoptosis induced by ischemia-reperfusion injury in rat kidney. *Life Sci.* **2020**, *255*, 117833. [CrossRef] [PubMed]
39. Kaeidi, A.; Sahamsizadeh, A.; Allahtavakoli, M.; Fatemi, I.; Rahmani, M.; Hakimizadeh, E.; Hassanshahi, J. The effects of oleuropein on unilateral ureteral obstruction induced-kidney injury in rats: The role of oxidative stress, inflammation and apoptosis. *Mol. Biol. Rep.* **2020**, *47*, 1371–1379. [CrossRef]
40. Poudyal, H.; Campbell, F.; Brown, L. Olive leaf extract attenuates cardiac, hepatic, and metabolic changes in high carbohydrate, high fat-fed rats. *J. Nutr.* **2010**, *140*, 946–953. [CrossRef] [PubMed]

41. Drira, R.; Chen, S.; Sakamoto, K. Oleuropein and hydroxytyrosol inhibit adipocyte differentiation in 3 T3-L1 cells. *Life Sci.* **2011**, *89*, 708–716. [CrossRef]
42. Khalili, A.; Nekooeian, A.A.; Khosravi, M.B. Oleuropein improves glucose tolerance and lipid profile in rats with simultaneous renovascular hypertension and type 2 diabetes. *J. Asian Nat. Prod. Res.* **2017**, *19*, 1011–1021. [CrossRef]
43. Nekooeian, A.A.; Khalili, A.; Khosravi, M.B. Effects of oleuropein in rats with simultaneous type 2 diabetes and renal hypertension: A study of antihypertensive mechanisms. *J. Asian Nat. Prod. Res.* **2014**, *16*, 953–962. [CrossRef]
44. Pacher, P.; Beckman, J.S.; Liaudet, L. Nitric oxide and peroxynitrite in health and disease. *Physiol. Rev.* **2007**, *87*, 315–424. [CrossRef]
45. Nekooeian, A.A.; Khalili, A.; Khosravi, M.B. Oleuropein offers cardioprotection in rats with simultaneous type 2 diabetes and renal hypertension. *Indian J. Pharmacol.* **2014**, *46*, 398–403. [CrossRef] [PubMed]
46. Moreno-Luna, R.; Munoz-Hernandez, R.; Miranda, M.L.; Costa, A.F.; Jimenez-Jimenez, L.; Vallejo-Vaz, A.J.; Muriana, F.J.G.; Villar, J.; Stiefel, P. Olive oil polyphenols decrease blood pressure and improve endothelial function in young women with mild hypertension. *Am. J. Hypertens.* **2012**, *25*, 1299–1304. [CrossRef] [PubMed]
47. Sohouli, M.H.; Fatahi, S.; Sharifi-Zahabi, E.; Santos, H.O.; Tripathi, N.; Lari, A.; Pourrajab, B.; Kord-Varkaneh, H.; Gaman, M.-A.; Shidfar, F. The impact of low advanced glycation end products diet on metabolic risk factors: A systematic review and meta-analysis of randomized controlled trials. *Adv. Nutr.* **2021**, *12*, 766–776. [CrossRef] [PubMed]
48. Egbuna, C.; Awuchi, C.G.; Kushwaha, G.; Rudrapal, M.; Patrick-Iwuanyanwu, K.C.; Singh, O.; Odoh, U.E.; Khan, J.; Jeevanandam, J.; Kumarasamy, S.; et al. Bioactive compounds effective against type 2 diabetes mellitus: A systematic review. *Curr. Top. Med. Chem.* **2021**, *21*, 1067–1095. [CrossRef]

Review

N-3 PUFA and Pregnancy Preserve C-Peptide in Women with Type 1 Diabetes Mellitus

Josip Delmis [1,*], Marina Ivanisevic [1] and Marina Horvaticek [2]

[1] Clinical Department of Obstetrics and Gynecology, University Hospital Centre Zagreb, School of Medicine, University of Zagreb, 10000 Zagreb, Croatia; marina.ivanisevic@pronatal.hr

[2] Institute Rudjer Boskovic, 10000 Zagreb, Croatia; marina.horvaticek@gmail.com

* Correspondence: josip.djelmis@zg.t-com.hr; Tel.: +385-98460485

Abstract: Type 1 diabetes (T1DM) is an autoimmune disease characterized by the gradual loss of β-cell function and insulin secretion. In pregnant women with T1DM, endogenous insulin production is absent or minimal, and exogenous insulin is required to control glycemia and prevent ketoacidosis. During pregnancy, there is a partial decrease in the activity of the immune system, and there is a suppression of autoimmune diseases. These changes in pregnant women with T1DM are reflected by Langerhans islet enlargement and improved function compared to pre-pregnancy conditions. N-3 polyunsaturated fatty acids (n-3 PUFA) have a protective effect, affect β-cell preservation, and increase endogenous insulin production. Increased endogenous insulin production results in reduced daily insulin doses, better metabolic control, and adverse effects of insulin therapy, primarily hypoglycemia. Hypoglycemia affects most pregnant women with T1DM and is several times more common than that outside of pregnancy. Strict glycemic control improves the outcome of pregnancy but increases the risk of hypoglycemia and causes maternal complications, including coma and convulsions. The suppression of the immune system during pregnancy increases the concentration of C-peptide in women with T1DM, and n-3 PUFA supplements serve as the additional support for a rise in C-peptide levels through its anti-inflammatory action.

Keywords: pregnancy; type 1 diabetes mellitus; beta-cell; insulin; C-peptide; hypoglycemia; n-3 PUFA; placenta; fetus

Citation: Delmis, J.; Ivanisevic, M.; Horvaticek, M. N-3 PUFA and Pregnancy Preserve C-Peptide in Women with Type 1 Diabetes Mellitus. *Pharmaceutics* **2021**, *13*, 2082. https://doi.org/10.3390/pharmaceutics13122082

Academic Editors: Fátima Regina Mena Barreto Silva and Diana Marcela Aragon Novoa

Received: 31 October 2021
Accepted: 2 December 2021
Published: 4 December 2021

1. Introduction

Type 1 diabetes is a condition caused by autoimmune damage to the insulin-producing β-cells of the pancreatic islets, usually leading to severe endogenous insulin deficiency [1]. The natural course of type 1 diabetes includes four very different stages of the disease:

- preclinical autoimmunity towards beta cells with progressive decline in insulin production;
- the onset of clinical diabetes;
- transient remission; and
- clinically pronounced diabetes with acute and chronic complications (Figure 1).

Early introduction of insulin into the treatment of type 1 diabetes, in addition to the primary effect of good metabolic control, appears to have an additional effect on delaying the destruction of β-cells. There are considerations according to which the use of exogenous insulin in the early phase of the disease puts β-cells in a state of relative metabolic rest and a lower load on insulin production, which makes them resistant to destruction [2]. Intensive insulin therapy effectively delays the onset and slows the progression of diabetic retinopathy, nephropathy, and neuropathy in patients with T1DM [3].

The goal of treating women with type 1 diabetes who plan a pregnancy is to achieve optimal glycemic control before and during pregnancy in order to reduce the incidence of miscarriage, congenital malformations, macrosomia, and neonatal complications. Poor metabolic control in pregnant women with type 1 diabetes mellitus is associated with an increased risk of miscarriage, preeclampsia, congenital malformations, asphyxia, macrosomia, and perinatal morbidity and mortality [4–6]. A successful perinatal outcome requires intensive clinical care in order to achieve normoglycemia before conception and during pregnancy. Proper diet and intensive insulin therapy are effective in achieving the desired glycemic targets. Good metabolic control is achieved when the venous plasma glucose values are between 4.0 and 5.0 mmol/L before a meal and 5.0 and 7.8 mmol/L after a meal and HbA1c values are lower than 6.0% (<42 mmol/mol) [7]. Significantly improved metabolic control of pregnant women with T1DM, intensive antenatal and neonatal care, more frequent antenatal ultrasound examinations, and cardiotocographic monitoring with a more liberal attitude towards cesarean delivery are mirrored by perinatal mortality rate below ten per mill; it shows a steady downward trend [5–7].

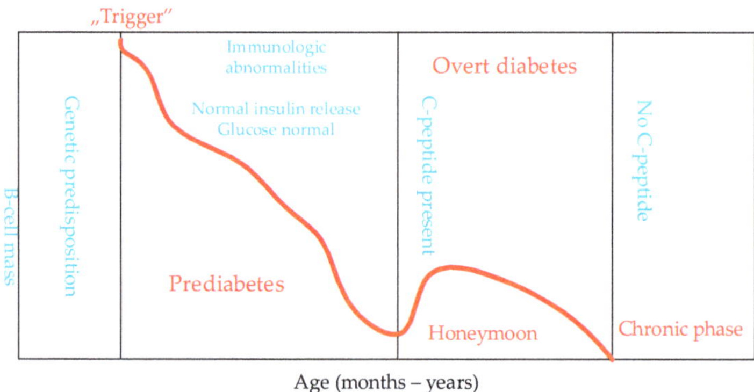

Figure 1. Stages in the development of type 1 diabetes mellitus. Phase 1 is the preclinical phase with the development of antibodies (ICA, IAA, GAD, ZnT8). Phase 2 is the preclinical phase which can last for months and years. Phase 3 is the onset of clinical diabetes with transient remission ("Honeymoon"). Phase 4 is clinically advanced diabetes with acute and chronic complications.

Intensive glycemic control improves pregnancy outcomes; however, it increases the risk of hypoglycemia [8,9] and causes maternal complications, including coma, convulsions, and death [10]. Hypoglycemia affects many pregnant women with T1DM and is several times more common than that outside of pregnancy [11,12].

2. Materials and Methods

The literature was reviewed using relevant databases (Pubmed/Medline, Scopus, Science Direct). The following inclusion criteria were considered: studies that investigated hypoglycemia (hypoglycemia in pregnant women with type 1 diabetes), n-6 PUFA and n-3 PUFA (in pregnant women; in pregnant women with type 1 diabetes), supplementation of EPA and DHA (in pregnant women with type 1 diabetes) C-peptide (in pregnant women with type 1 diabetes; in nonpregnant people), the immune system in pregnancy, vitamin D T1DM and relevant observational, randomized control trial and review articles were selected to provide information and data for the text (published from 2000 to 2021). Only papers written in English were selected.

Exclusion Criteria

We excluded papers with gestational and type 2 diabetes. Articles on immunosuppressive drugs (corticosteroids, azathioprine, methotrexate, cyclophosphamide, and cyclosporine) associated with C-peptide preservation were also excluded.

3. Hypoglycemia in Pregnant Women with T1DM

Known risk factors for hypoglycemia in pregnancy are the duration of type 1 diabetes mellitus, history of previous severe hypoglycemia, hypoglycemia unawareness, changes in insulin treatment, and HbA1c < 6.5% (48 mmol/mol). Reducing the risk of hypoglycemia is a major challenge for physicians taking care of pregnant women with T1DM.

The International Hypoglycaemic Research Group (IHSG) unequivocally considers a glucose concentration of <3.0 mmol/L as a hypoglycemic value detected by plasma self-monitoring, continuous glucose monitoring (at least 20 min), or laboratory plasma glucose measurement [13]. IHSG considers a glucose concentration of <3.0 mmol/L low enough to indicate severe, clinically significant hypoglycemia [13]. The same group suggested that a glucose value of 3.9 mmol/L or less should indicate possible hypoglycemia. Severe hypoglycemia is considered a hypoglycemic episode that requires assistance from another person to treat.

3.1. Insulin and C-Peptide

Insulin is a low-molecular-weight protein (5808 daltons) secreted as a primary response to elevated blood glucose levels [14]. Insulin is the primary hormone that controls glucose metabolism and facilitates the entry of glucose into cells [14]. Some nerve stimuli and elevated concentrations of other energy sources, such as amino acids and fatty acids, stimulate insulin secretion [15]. The central role of insulin is to control the entry of glucose from the blood into the cells and its utilization in peripheral tissues. Insulin also participates in lowering blood glucose by stimulating glycolysis, inhibiting gluconeogenesis and glycogenolysis in the liver. Insulin's action is counter-regulated by hormones that increase glucose concentration, such as glucagon, adrenaline, growth hormone, and cortisol. Healthy pancreatic beta cells produce and secrete as much insulin as is needed to control glucose homeostasis [15].

Insulin is synthesized from a pre-proinsulin precursor molecule. C-peptide is a peptide consisting of 31 amino acids that, in the proinsulin molecule, connects the amino end of the alpha chain and the carboxyl end of the beta chain of insulin, hence the name (connecting peptide (C-peptide)) [16]. Insulin and C-peptide are excreted together in equimolar amounts into the portal circulation. About 50% of insulin is metabolized (absorbed) in the liver during the first pass; this is called the "first-pass effect." Residual insulin plays an important regulatory role in peripheral tissues. Unlike insulin, C-peptide is not absorbed by the liver and has a longer half-life (about 30 min) than insulin (about 4 min) and is suitable for measuring the concentration of endogenous insulin in the peripheral circulation [16]. The close association of C-peptide in the systemic circulation and endogenous insulin in the portal system has been well established [16,17]. C-peptide concentration under standardized conditions in the peripheral circulation is the most appropriate measure of endogenous insulin secretion and β-cell function in pregnant women with T1DM.

From the time when a clinical diagnosis of T1DM is made, patients typically retain a limited ability to produce endogenous insulin for several months or years [18–20] (Figure 1). Excretion is gradually lost as β-cell function is selectively destroyed. Autoimmune destruction of β-cells is more pronounced/aggressive when T1DM develops in childhood and youth, while this process is longer, more variable, and beta cells more slowly deteriorate entirely in patients with T1DM later in life [21–23].

3.2. The Immune System in Pregnancy

Pregnancy is a unique event in which the genetically and immunologically different fetus survives until full term without being rejected from its mother's immune system [24]. It is associated with partial suppression of the immune and pro-inflammatory system, which is why autoimmune diseases such as diabetes often go into remission during pregnancy [25].

The maternal immune system undergoes significant changes, including developing specific pathways to protect the fetus from the mother's cytotoxic attack. One mechanism decreases the expression of classical HLA class I molecules, while other mechanisms are associated with an altered balance of Th1 and Th2 [25]. During pregnancy, cellular immune function and pro-inflammatory Th1 cytokines (e.g., IL-2, TNF-α, and INF-γ) are suppressed, while humoral immunity and the production of anti-inflammatory Th2 cytokines (e.g., IL-4 and IL-10) are enhanced [25]. Such a pattern of immune function is reversed in the postpartum period [25]. A partial reduction in the activity of the inflammatory immune system leads to an improvement in various autoimmune diseases during pregnancy, including diabetes. In type 1 diabetes, these changes are expressed by the growth and functional modification of the Langerhans pancreatic islets. The most significant change that these islets undergo during pregnancy is the increased secretion of insulin. Numerous animal models have shown that β-cell mass increases three to four times during pregnancy with significant hypertrophy and β-cell proliferation [26–28]. In a pilot study, Amouyal C et al. found that higher levels of regulatory T cells and IL-2 improved endogenous insulin secretion in pregnant women with T1DM [29]. The authors [30] showed that C-peptide concentrations increase gradually during pregnancy in women with type 1 diabetes.

Meeck CL et al. [31] have measured maternal serum C-peptide concentrations at 12, 24 and 34 weeks of gestation in 127 pregnant women with type 1 diabetes and cord blood C-peptide concentrations. In 74 (58%) pregnant women, C-peptide was not detected; in 22 (17%), it was confirmed at the beginning of pregnancy, and in 31 (24%), it was detected in the 34th week of pregnancy. Neonates born to the mothers in whom C-peptide was detected at 34 weeks of gestation had elevated cord blood C-peptide and more frequent neonatal complications than others. Based on the results, the authors suggest a transfer of C-peptide from fetal to maternal serum without the regeneration of pregnancy-related beta cells. Pregnant women with detected C-peptide had better-regulated glycemia, fewer hypoglycemia events according to CGM, reduced total insulin dose, and lower incidence of macrosomia [31]. According to other authors' research, it is plausible that pregnancy yields immunological tolerance and stimulates endogenous insulin production in women with type 1 diabetes mellitus [30]. Due to the small number of participants, the authors [32] did not prove an increase in the concentration of C-peptide during pregnancy, which, nonetheless, does not contradict the findings of other authors.

4. Fatty Acids

Fatty acids (FA) are found in most lipids and have several physiologically significant roles in mammalian tissues, such as structural roles, storage, and energy production, and as precursors of biologically active substances. Fatty acids are constituent units of phospholipids and glycolipids and are, therefore, a fundamental component of cell membranes that give the cell its structural integrity and delimit the organelles [32]. They are rarely found in free form in the body due to their detergent and cytotoxic effects but are generally esterified into larger molecules such as phospholipids (PL) and triacylglycerols (TAG). They are divided based on the degree of saturation:

- saturated fatty acids (SFA), which do not contain double (unsaturated) bonds;
- monounsaturated fatty acids (MUFAs) in which there is one double bond between carbon atoms; and
- polyunsaturated fatty acids (PUFA) in which there are several double bonds.

The human body can produce saturated and monounsaturated fatty acids with a double bond on the ninth carbon atom counting from the end of the chain (n-9 unsaturated fatty acids). However, due to the lack of the Δ-12 and Δ-15 desaturase enzymes, humans cannot de-novo synthesize the n-6 and n-3 fatty acid families [32]. Linoleic acid (LA, C18: 2n-6), from which an n-6 series of unsaturated fatty acids are further synthesized in the body, and α-linolenic acid (ALA, C18: 3n-3), from which n-3 series of unsaturated fatty acids are synthesized acids, must, therefore, be taken into the body through food and are called essential fatty acids (EFA). The biosynthesis of long-chain polyunsaturated fatty acids (LCPUFA) from precursors involves a series of alternating activities of the enzymes elongase and desaturase, with competition between n-6 and n-3 acids for desaturation enzymes [33]. However, the effectiveness of endogenous conversion of ALA from food to EPA, DPA, and DHA in the adult human body is not significant [33]. Due to the low n-3 PUFA intake, supplementation of these fatty acids is recommended to gain and preserve cardiovascular health and establish the normal neurological development of fetuses and newborns [34].

Inflammatory Processes and n-3 Polyunsaturated Fatty Acids

Long-chain PUFAs such as AA, EPA, and DHA in the body can originate directly from the food consumed or are synthesized endogenously from their food-derived precursors [35]. The main representative of n-3 PUFA is ALA, found in green leafy vegetables and some seeds, nuts, and legumes. It can be metabolized in a limited capacity to EPA and DHA, two n-3 PUFAs whose primary source is algae, fish, and seafood [35]. The most abundant n-6 PUFA in the diet is LA, which is primarily found in nuts, seeds, and vegetable oils. It is a precursor to AA synthesis. Since ALA and LA compete for crucial enzymes involved in fatty acid metabolism and conversion to pro-inflammatory or anti-inflammatory eicosanoids, it is important to investigate the effect of co-intake of n-3 and n-6 PUFAs [36]. Prostanoids are cyclooxygenase-formed eicosanoids and include prostaglandins (PG), prostacyclins (PGI), and thromboxanes (TX). Leukotrienes (LT) and lipoxins (LX) are eicosanoids formed by lipoxygenase. Precursors, especially AA and EPA, can be derived directly from food or are metabolically derived from LA and ALA. Fatty acids for eicosanoid synthesis are usually released from the sn-2 position of phospholipids from the cell membrane (the site where PUFAs are most commonly esterified) by the action of the enzyme phospholipase A2 (PLA2) [36]. Therefore, the activity of PLA2 is a limiting factor in the rate of eicosanoid formation from AA and EPA. Since PLA2 has the same affinity for AA and EPA, the fatty acid composition of phospholipids depends on the ratio of released n-3 and n-6 PUFAs used as starting compounds for eicosanoid synthesis [36]. With the Western diet, significant amounts of AA are introduced into the body, which is consequently the most abundant PUFA in membrane phospholipids, so the most common forms of PG series 2 in the human body are eicosanoids synthesized from fatty acids n-6 (AA) and n-3 (EPA) order they interact antagonistically. AA metabolism produces pro-inflammatory eicosanoids, prostaglandins, and thromboxanes of series 2 (PG2 and TX2) and leukotrienes of series 4 (LT4).

In contrast, EPA metabolism produces anti-inflammatory eicosanoids, prostaglandins, and thromboxanes of series 3 (PG3 and TX3) and leukotrienes of series 5 (LT5) [36]. Therefore, an elevated AA/EPA or AA/DHA ratio in the body may indicate inflammatory processes. Eicosanoids are considered to be a link between n-3 PUFA, inflammation, and immunity [37]. In addition to anti-inflammatory effects by suppressing LT4 synthesis, n-3 PUFAs suppress the excessive ability of monocytes to synthesize interleukin-1 (IL-1) and tumor necrosis factor (TNF). The anti-inflammatory properties of n-3 PUFA, especially EPA and DHA, result from the competition with AA as a substrate for cyclooxygenase (COX) and 5-lipoxygenase. Hence, EPA and DHA increase the synthesis of LT5 (which has a weaker inflammatory effect than LT4) and decrease the synthesis of LT4 derived from AA [35]. PG3 is a potent vasodilator and inhibitor of platelet aggregation like PG2, while TX3 is a significantly weaker platelet aggregator with minimal vasoconstrictive effect, unlike TX2. These changes in PG, TX, and LT production alter the balance of their activities, such as vasodilation and inhibition of platelet aggregation, resulting in decreased inflammatory activity. DHA is a precursor of resolvins, protectins, and maresins, characterized by anti-inflammatory effects [36]. Several recent studies show that changes in the amount and type of fatty acids in the diet can affect the formation of various eicosanoids.

Replacement of food containing oils with predominant n-6 fatty acids with oils with prevailing n-3 fatty acids results in increased synthesis of less potent eicosanoids than those derived from n-6 fatty acids.

5. n-3 PUFA and C-Peptide Preservation

5.1. n-3 PUFA and C-Peptide Preservation in Pregnant Women with T1DM

A prospective randomized placebo-controlled clinical trial [37] included 90 pregnant women with T1DM. Forty-seven of them took n-3 supplementation (60 mg of EPA and 308 mg of DHA) from the 9th week of pregnancy, while 43 took a placebo and were on a regular diabetic diet. In this clinical trial, the duration of T1DM was between 5 and 30 years, and pregnant women were divided into two subgroups. The first subgroup included 36 pregnant women with a duration of T1DM from 5 to 10 years, where 21 were on n-3 supplementation, and nine were on placebo. The second subgroup included 54 pregnant women with a duration of T1DM from 11 to 30 years, while 26 were on n-3 supplementation and 28 were on placebo. In both fasting C-peptide tended to increase during pregnancy. The increase in FC-peptide during pregnancy was statistically significant in the intervention group, while in the control group, there was no statistically significant increase in FC-peptide in the third trimester compared to the FC-peptides values from the first and second trimester [37] (Table 1).

In both subgroups of pregnant women, based on the duration of T1DM (5–10 and 11–30 years), a statistically significant difference was found in the intervention group between FC-peptide values in the first and third trimester, while this difference was not found in the control group. A significant reduction in the dose of long-acting insulin was found in both study subgroups [37].

The results unequivocally demonstrate the importance of supplementation with EPA and DHA in pregnant women with T1DM. Measurement of C-peptide concentration provides a validated way to quantify endogenous insulin secretion. As the C-peptide does not cross the placenta in either direction, the C-peptide values detected during pregnancy are derived from maternal beta cells and not from the fetus. The increase in C-peptide concentration in both groups of pregnant women is due to suppression of the pro-inflammatory system during pregnancy. Therefore, the maternal chances of having a genetically and immunologically diverse fetus are significantly increased, and EPA and DHA have an additional synergistic impact as they equally produce anti-inflammatory effects. After childbirth, there is a rapid decrease in C-peptide concentration, which indicates the active role of the placenta in increasing the concentration of C-peptide during pregnancy [30].

5.2. n-3 PUFA and C-Peptide Preservation in Nonpregnant Patient with T1DM

Mayer-Davis EJ. et al. included surveillance of a large number (1316) of young people with eating habits in a prospective two-year study called SEARCH for Diabetes in Youth [38]. Participants were measured for DHA and EPA concentrations in venous plasma and venous serum FC-peptide. An association was found between the concentration of n-3 PUFA and FC-peptide. Higher concentrations of EPA and DHA were associated with higher concentrations of FC-peptide. According to FFQ (Food Frequency Questionnaires) data, an association was found between increased leucine intake and higher FC-peptide concentrations [38].

The result of this study strongly confirms the effect of n-3 PUFA on C-peptide conservation in young individuals with T1DM (Table 1).

6. Vitamin D Has a Protective Effect in Preventing T1DM

Studies show that vitamin D can have a protective effect in preventing the formation of T1DM [39,40].

The active form of vitamin D (D3, 1.25 dihydrocolecalciferol) has an immunomodulatory role in preventing type 1 diabetes through the activation of β-cells. At the level of pancreatic islets, vitamin D3 reduces proinflammatory cytokines (IL-1, IL6), making β-cells less chemoattractive and more resistant to inflammatory mediators, which ultimately results in a decrease in total T-lymphocytes and an increase in the number of regulatory (helper/suppressor) mononuclear cells, which prevent the autoimmune process. Therefore, dietary vitamin D intake during pregnancy and early childhood is considered to reduce the risk of developing autoimmune diabetes. Hyppönen E et al. [40] proved, in a long-term prospective study, that vitamin D supplementation, started in the first year of life in children, reduces the incidence (prevalence) of T1DM compared to those without it (RR 0.12, 95% CI 0.03–0.05) [40].

7. C-Peptide, Insulin Doses, and Glycemic Control

The most reliable indicator of maintaining beta-cell function is the concentration of C-peptides in the blood [16]. Measurement of the C-peptide concentration provides a validated method for quantifying secreted endogenous insulin. The close association between C-peptide in the systemic circulation and endogenous insulin in the portal system is well established [16]. The C-peptide concentration gradually increases during pregnancy, independent of blood glucose concentration, in pregnant women suffering from type 1 diabetes mellitus [29,38]. A significant increase in fasting C-peptide occurred in the third trimester of pregnancy compared to the first trimester [37].

Preservation of β-cell function is associated with a lower risk of hypoglycemic events, as well as lower HbA1c and less frequent microvascular complications [17]. Nutritional factors previously identified as potentially protective against the development of type 1 diabetes [38], or those associated with insulin production may also be valuable for the preservation of the β-cell function.

Ilić et al. [41] found an association between higher C-peptide concentrations and lower total insulin doses in the first trimester of pregnancy. A lower prevalence of severe hypoglycemia and a decline in total insulin units during pregnancy were associated with increased endogenous insulin secretion [41]. Pregnancy induces an increase in fasting C-peptide concentration in most women with diabetes mellitus but is more pronounced in those with a shorter duration of diabetes [38]. A lower prevalence of severe hypoglycemia and a decline in required total insulin units during pregnancy were associated with increased endogenous insulin secretion. C-peptide serves as the mediator in the correlation between the duration of type 1 diabetes and the required total insulin dose [42,43] (Table 1).

Table 1. Reviews of studies that evaluate C-peptide levels in pregnant women and young population with type 1 diabetes.

Reference	Study Population	No of Participants	Evaluated Parameters	Results/Conclusions
Ilic, S. [41]	Pregnant women T1DM At 10 w of gestation	10	Fasting C-peptide	C-peptide non-detectable before pregnancy to detectable at 10 weeks of gestation.
Nielsen L. [30]	Pregnant women T1DM Early and late pregnancy, postpartum	108; two groups based on serum glucose, >5.0 or <5.0 mmol/L	C-peptide, glucose, placental GH, IGF-I.	C-peptide in women with long-term T1DM C-peptide in early pregnancy: raised up to 97% by 33 weeks gestation.
Murphy, HR [32]	Pregnant women T1DM Early and late pregnancy	10	Fasting or meal-stimulated C-peptide concentration.	No change in fasting or meal-stimulated plasma C-peptide from early to late pregnancy.
Mayer-Davis, EJ. [38]	Young people (up to 20 years)/T1DM	1316	EPA and DHA, vitamins D E. Intake of leucine	Intake of n-3 PUFA sustained β-cell function, higher DHA and EPA were associated with higher fasting C-peptide.
Horvaticek, M. [37]	Pregnant women T1DM Three trimesters of pregnancy	90; two groups, n-3 supplementation or placebo	Fasting C-peptide, FPG EPA and DHA in maternal and cord blood serum.	n-3 PUFAs and pregnancy stimulates the production of endogenous insulin in women with T1DM Rise in FC-peptide during pregnancy was significant in exposed group

Table 1. *Cont.*

Reference	Study Population	No of Participants	Evaluated Parameters	Results/Conclusions
Amouyal, C. [29]	Pregnant women T1DM Three trimesters of pregnancy and post-partum	13	Clinical, immunological and diabetes parameters.	One group ($n = 7$) rise in C-peptide between the 2nd and 3rd trimesters, while second group had no β-cell function improvement during pregnancy.
Meek, L. [31]	Pregnant women T1DM Three trimesters of pregnancy	127; three groups based on detection of maternal C-peptide	Serum C-peptide concentration in a maternal and cord blood samples	Most women had undetectable C-peptide throughout pregnancy. Second group with detectable C-peptide characterized by lower BMI, later onset, shorter duration of T1DM improved glycemic control. A third group had the first appearance of C-peptide in maternal serum at 34 weeks gestation.

Hypoglycemia in pregnant women with type 1 diabetes most often occurs due to insulin overdose or untimely insulin delivery, missed or reduced (insufficient) meals, emesis or hyperemesis in the first trimester of pregnancy, gastroparesis, or increased glucose consumption during and immediately after exercise [8]. Reducing the risk of hypoglycemia is a major challenge for doctors caring for pregnant women with T1DM. Technological advances in insulin production (e.g., analogues), insulin delivery devices (e.g., insulin pumps) and glucose monitoring devices have contributed to overall metabolic improvement, but in clinical practice, severe hypoglycemia rates, especially in pregnant women, remain high. Continuous glucose monitoring (CGM) is a method of continuous monitoring of glucose levels in interstitial fluid which was coined for improving metabolic control and expected result of glycemic monitoring is a reduction in hyperglycemia and a reduction in low glucose levels including symptomatic hypoglycemia [7,44].

8. Conclusions and Recommendation

Intake of n-3 long-chain polyunsaturated fatty acids affects the health of the mother, fetus, and child.

N-3 PUFA increases endogenous insulin production, resulting in reduced daily insulin doses, better metabolic control, reduced risk of long-term diabetic complications, and adverse effects of intensified insulin therapy, primarily hypoglycemia. There is evidence that n-3 PUFA prevents preterm birth before 34 gestational weeks [45].

In addition to the beneficial effects for the mother, n-3 PUFA affects the health of the child. Docosahexaenoic acid is incorporated into the fetal brain and retinal phospholipids, resulting in a high degree of fluidity and flexibility of neural and endothelial membranes. It is involved in neurotransmission, regulates ion channel activity and gene expression, and can be metabolized to neuroprotective metabolites [46]. Offspring from the mothers taking docosahexaenoic acid supplementation had higher IQs than those in the control group [47].

Due to the many beneficial effects for both mother and fetus, n-3 PUFA supplementation is recommended in women who plan pregnancy and pregnant women with T1DM. The European Food Safety Authority (EFSA) allows reaching the Dietary Reference Value for LC-PUFA (250 mg EPA plus DHA, plus an additional 100–200 mg of DHA) by food and supplements [48].

Author Contributions: Writing—review and editing, M.I., M.H. and J.D. All authors have read and agreed to the published version of the manuscript.

Funding: This research was funded by the CROATIAN SCIENCE FOUNDATION, grant number project PRE-HYPO No. IP-2018-01-1284. https://mef.unizg.hr/znanost/istrazivanje/web-stranice-projekata/projekt-hrzz-pre-hypo/ accessed on 1 February 2019.

Institutional Review Board Statement: Not applicable.

Informed Consent Statement: Not applicable.

Conflicts of Interest: The authors declare no conflict of interest.

References

1. Holt, R.I.G.; DeVries, J.H.; Hess-Fischl, A.; Hirsch, I.B.; Kirkman, M.S.; Klupa, T.; Ludwig, B.; Nørgaard, K.; Pettus, J.; Renard, E.; et al. The Management of Type 1 Diabetes in Adults. A Consensus Report by the American Diabetes Association (ADA) and the European Association for the Study of Diabetes (EASD). *Diabetes Care* **2021**, *44*, 2589–2625. [CrossRef]
2. Brown, R.J.; Rother, K.I. Effects of beta-cell rest on beta-cell function: A review of clinical and preclinical data. *Pediatric Diabetes* **2008**, *9 Pt 2*, 14–22. [CrossRef] [PubMed]
3. Murphy, H.R. Glycemic treatment during type 1 diabetes pregnancy: A story of (mostly) sweet success! *Diabetes Care* **2018**, *41*, 1563–1571. [CrossRef]
4. Persson, M.; Norman, M.; Hanson, U. Obstetric and perinatal outcomes in type 1 diabetic pregnancies: A large, population-based study. *Diabetes Care* **2009**, *32*, 2005–2009. [CrossRef]
5. Djelmis, J. Clinical management of pregnancy complicated with type1/type2 diabetes mellitus. In *Diabetology of Pregnancy*; Djelmis, J., Desoye, G., Ivanisevic, M., Eds.; Karger: Basel, Switzerland, 2005; Volume 17, pp. 161–173.
6. Alexopoulos, A.S.; Blair, R.; Peters, A.L. Management of Preexisting Diabetes in Pregnancy: A Review. *JAMA* **2019**, *321*, 1811–1819. [CrossRef] [PubMed]
7. Feig, D.S.; Donovan, L.E.; Corcoy, R.; Murphy, K.E.; Amiel, S.A.; Hunt, K.F.; Asztalos, E.; Barrett, J.F.R.; Sanchez, J.J.; de Leiva, A.; et al. Continuous glucose monitoring in pregnant women with type 1 diabetes (CONCEPTT): A multicentre international randomised controlled trial. *Lancet* **2017**, *390*, 2347–2359. [CrossRef]
8. Ringholm, L.; Pedersen-Bjergaard, U.; Thorsteinsson, B.; Damm, P.; Mathiesen, E.R. Hypoglycaemia during pregnancy in women with type 1 diabetes. *Diabet. Med.* **2012**, *29*, 558–566. [CrossRef]
9. Kinsley, B. Achieving better outcomes in pregnancies complicated by type 1 and type 2 diabetes mellitus. *Clin. Ther.* **2007**, *29* (Suppl. SD), S153–S160. [CrossRef]
10. Leinonen, P.J.; Hiilesmaa, V.K.; Kaaja, R.J.; Teramo, K.A. Maternal Mortality in Type 1 Diabetes. *Diabetes Care* **2001**, *24*, 1501–1502. [CrossRef]
11. Rosenn, B.M.; Miodovnik, M. Glycemic control in the diabetic pregnancy: Is tighter always better? *J. Matern. Fetal Med.* **2000**, *9*, 29–34.
12. Ng, D.; Noor, N.M.; Yong, S.L. Prevalence of Hypoglycaemia among Insulin-Treated Pregnant Women with Diabetes Who Achieved Tight Glycaemic Control. *J. ASEAN Fed. Endocr. Soc.* **2019**, *34*, 29–35. [CrossRef]
13. International Hypoglycaemia Study Group. Glucose Concentrations of Less Than 3.0 mmol/L (54 mg/dL) Should Be Reported in Clinical Trials: A Joint Position Statement of the American Diabetes Association and the European Association for the Study of Diabetes. *Diabetes Care* **2017**, *40*, 155–157. [CrossRef]
14. Mayer, J.P.; Zhang, F.; DiMarchi, R.D. Insulin structure and function. *Biopolymers* **2007**, *88*, 687–713. [CrossRef] [PubMed]
15. Rahman, S.; Hossain, K.S.; Das, S.; Kundu, S.; Adegoke, E.O.; Rahman, A.; Hannan, A.; Uddin, J.J.; Pang, M.-D. Role of Insulin in Health and Disease: An Update. *Int. J. Mol. Sci.* **2021**, *22*, 6403. [CrossRef]
16. Wahren, J.; Ekberg, K.; Johansson, J.; Henriksson, M.; Pramanik, A.; Johansson, B.-L.; Rigler, R.; Rnvall, H.J. Role of C-peptide in human physiology. *Am. J. Physiol. Endocrinol. Metab.* **2000**, *278*, E759–E768. [CrossRef]
17. Palmer, J.P.; Fleming, G.A.; Greenbaum, C.J.; Herold, K.C.; Jansa, L.D.; Kolb, H.; Lachin, J.M.; Polonsky, K.S.; Pozzilli, P.; Skyler, J.S. ADA Workshop Report: C-Peptide. Is the Appropriate Outcome Measure for Type 1 Diabetes Clinical Trials to Preserve β-Cell Function. *Diabetes* **2004**, *53*, 250–264. [CrossRef]
18. Steffes, M.W.; Sibley, S.; Jackson, M.; Thomas, W. Beta-cell function and the development of diabetes-related complications in the diabetes control and complications trial. *Diabetes Care* **2003**, *26*, 832–836. [CrossRef] [PubMed]
19. Greenbaum, C.J.; Anderson, A.M.; Dolan, L.M.; Mayer-Davis, E.J.; Dabelea, D.; Imperatore, G.; Marcovina, S.; Pihoker, C.; SEARCH Study Group. Preservation of beta-cell Function in auto antibody-positive youth with diabetes. *Diabetes Care* **2009**, *32*, 1839–1844. [CrossRef]
20. Evans-Molina, C.; Sims, E.K.; DiMeglio, L.A.; Ismail, H.M.; Steck, A.K.; Palmer, J.P.; Krischer, J.P.; Geyer, S.; Xu, P.; Sosenko, J.M. Type 1 Diabetes TrialNet Study Group. β Cell dysfunction exists more than 5 years before type 1 diabetes diagnosis. *JCI Insight* **2018**, *3*, e120877. [CrossRef] [PubMed]
21. Barbetti, F.; Simeon, I.; Taylor, S.I. Insulin: Still a miracle after all these years. *J. Clin. Investig.* **2019**, *129*, 3045–3047. [CrossRef]

22. Poudel, A.; Savari, O.; Striegel, D.A.; Periwal, V.; Taxy, J.; Millis, J.M.; Witkowski, P.; Atkinson, M.A.; Hara, M. Beta-cell destruction and preservation in childhood and adult-onset type 1 diabetes. *Endocrine* **2015**, *49*, 693–702. [CrossRef]

23. Lam, C.J.; Jacobson, D.R.; Rankin, M.M.; Cox, A.R.; Kushner, J.A. β Cells Persist in T1D Pancreata without Evidence of Ongoing β-Cell Turnover or Neogenesis. *J. Clin. Endocrinol. Metab.* **2017**, *102*, 2647–2659. [CrossRef]

24. Ziegler, A.-G.; Pflueger, M.; Winkler, C.; Achenbach, P.; Akolkar, B.; Krischer, J.P.; Bonifacio, E. Accelerated progression from islet autoimmunity to diabetes is causing the escalating incidence of type 1 diabetes in young children. *J. Autoimmun.* **2011**, *37*, 3–7. [CrossRef]

25. Poole, J.A.; Claman, H.N. Immunology of pregnancy: Implications for the mother. *Clin. Rev. Allergy Immunol.* **2004**, *26*, 161–170. [CrossRef]

26. Nalla, A.; Ringholm, L.; Sørensen, S.N.; Damm, P.; Mathiesen, E.R.; Nielsen, J.H. Possible mechanisms involved in improved beta cell function in pregnant women with type 1 diabetes. *Heliyon* **2020**, *19*, e04569. [CrossRef] [PubMed]

27. Baeyens, L.; Hindi, S.; Sorenson, R.L.; German, M.S. β-Cell adaptation in pregnancy. *Diabetes Obes. Metab.* **2016**, *18* (Suppl. S1), 63–70. [CrossRef] [PubMed]

28. Rieck, S.; Kaestner, K.H. Expansion of beta-cell mass in response to pregnancy. *Trends Endocrinol. Metab.* **2010**, *21*, 151–158. [CrossRef]

29. Amouyal, C.; Klatzmann, D.; Tibi, E.; Salem, J.-E.; Halbron, M.; Popelier, M.; Jacqueminet, S.C.; Ciangura, C.O.; Bourron, O.; Andreelli, F.A.; et al. Pregnant type 1 diabetes women with rises in C-peptide display higher levels of regulatory T cells: A pilot study. *Diabetes Metab.* **2021**, *47*, 101188. [CrossRef]

30. Nielsen, L.R.; Rehfeld, J.F.; Pedersen-Bjergaard, U.; Damm, P.; Mathiesen, E.R. Pregnancy-induced rise in serum C-peptide concentrations in women with type 1 diabetes. *Diabetes Care* **2009**, *32*, 1052–1057. [CrossRef] [PubMed]

31. Meek, C.L.; Oram, R.A.; McDonald, T.J.; Feig, D.S.; Hattersley, A.T.; Murphy, H.R.; CONCEPTT Collaborative Group. Reappearance of C-Peptide During the Third Trimester of Pregnancy in Type 1 Diabetes: Pancreatic Regeneration or Fetal Hyperinsulinism? *Diabetes Care* **2021**, *44*, 1826–1834. [CrossRef] [PubMed]

32. Murphy, H.R.; Elleri, D.; Allen, J.M.; Simmons, D.; Nodale, M.; Hovorka, R. Plasma C-peptide concentration in women with Type 1 diabetes during early and late pregnancy. *Diabet. Med.* **2012**, *29*, e361–e364. [CrossRef]

33. Burdge, G.C.; Wootton, S.A. Conversion of alpha-linolenic acid to eicosapentaenoic, docosapentaenoic and docosahexaenoic acids in young women. *Br. J. Nutr.* **2002**, *88*, 411–420. [CrossRef] [PubMed]

34. Williams, C.M.; Burdge, G. Long-chain n-3 PUFA: Plant v. marine sources. *Proc. Nutr. Soc.* **2006**, *65*, 42–50. [CrossRef]

35. Simopoulos, A.P. Omega-3 fatty acids in inflammation and autoimmune diseases. *J. Am. Coll. Nutr.* **2002**, *21*, 495–505. [CrossRef] [PubMed]

36. Serhan, C.N.; Chiang, N.; Dalli, J.; Levy, B.D. Lipid mediators in the resolution of inflammation. *Cold Spring Harb. Perspect. Biol.* **2014**, *7*, a016311. [CrossRef] [PubMed]

37. Horvaticek, M.; Djelmis, J.; Ivanisevic, M.; Oreskovic, S.; Herman, M. Effect of eicosapentaenoic acid and docosahexaenoic acid supplementation on C-peptide preservation in pregnant women with type-1 diabetes: Randomized placebo controlled clinical trial. *Eur. J. Clin. Nutr.* **2017**, *71*, 968–972. [CrossRef] [PubMed]

38. Mayer-Davis, E.J.; Dabelea, D.; Crandell, J.L.; Crume, T.; D'Agostino, R.B., Jr.; Dolan, L.; King, I.B.; Lawrence, J.M.; Norris, J.M.; Pihoker, C.; et al. Nutritional factors and preservation of C-peptide in youth with recently diagnosed type 1 diabetes: SEARCH Nutrition Ancillary Study. *Diabetes Care* **2013**, *36*, 1842–1850. [CrossRef] [PubMed]

39. Chiu, K.C.; Chu, A.; Go, V.L.; Saad, M.F. Hypovitaminosis D is associated with insulin resistance and beta cell dysfunction. *Am. J. Clin. Nutr.* **2004**, *79*, 820–825. [CrossRef]

40. Hyppönen, E.; Läärä, E.; Reunanen, A.; Järvelin, M.R.; Virtanen, S.M. Intake of vitamin D and risk of type 1 diabetes: A birth-cohort study. *Lancet* **2001**, *358*, 1500–1503. [CrossRef]

41. Ilic, S.; Jovanovic, L.; Wolitzer, A.O. Is the paradoxical first trimester drop in insulin requirement due to an increase in C-peptide concentration in pregnant Type 1 diabetic women? *Diabetologia* **2000**, *43*, 1329–1336.

42. Marren, S.M.; Hammersley, S.; McDonald, B.M.; Shields, T.J.; xKnight, B.A.; Hill, A.; Bolt, R.; Tree, T.I.; Roep, B.O.; Hattersley, A.T.; et al. Persistent C-peptide is associated with reduced hypoglycaemia but not HbA$_{1c}$ in adults with longstanding Type 1 diabetes: Evidence for lack of intensive treatment in UK clinical practice? *Diabet. Med.* **2019**, *36*, 1092–1099. [CrossRef] [PubMed]

43. Wellens, M.J.; Vollenbrock, C.; Dekker, P.; Boesten, L.S.M.; Geelhoed-Duijvestijn, P.H.; Martine, M.C.; de Vries-Velraeds, M.M.C.; Nefs, G.; Wolffenbuttel, B.H.R.; Aanstoot, H.-J.; et al. Residual C-peptide secretion and hypoglycemia awareness in people with type 1 diabetes. *BMJ Open Diabetes Res. Care* **2021**, *9*, e002288. [CrossRef]

44. Feig, D.S.; Murphy, H.R. Continuous glucose monitoring in pregnant women with Type 1 diabetes: Benefits for mothers, using pumps or pens, and their babies. *Diabet. Med.* **2018**, *35*, 430–435. [CrossRef]

45. Szajewska, H.; Horvath, A.; Koletzko, B. Effect of n-3 long-chain polyunsaturated fatty acid supplementation of women with low-risk pregnancies on pregnancy outcomes and growth measures at birth: A meta-analysis of randomized controlled trials. *Am. J. Clin. Nutr.* **2006**, *83*, 1337–1344. [CrossRef] [PubMed]

46. Innis, S.M. Dietary (n-3) fatty acids and brain development. *J. Nutr.* **2007**, *137*, 855–859. [CrossRef] [PubMed]

47. Helland, I.B.; Smith, L.; Saarem, K.; Saugstad, O.D.; Drevon, C.A. Maternal supplementation with very-long-chain n-3 fatty acids during pregnancy and lactation augments children's IQ at 4 years of age. *Pediatrics* **2003**, *111*, e39–e44. [CrossRef]
48. EFSA Panel on Dietetic Products, Nutrition, and Allergies (NDA). Scientific Opinion on Dietary Reference Values for fats, including saturated fatty acids, polyunsaturated fatty acids, monounsaturated fatty acids, trans fatty acids, and cholesterol. *EFSA J.* **2010**, *8*, 1461.

 *pharmaceutics*

Review

The Role of Natural Products on Diabetes Mellitus Treatment: A Systematic Review of Randomized Controlled Trials

Lucía Vivó-Barrachina [1], María José Rojas-Chacón [1], Rocío Navarro-Salazar [1], Victoria Belda-Sanchis [1], Javier Pérez-Murillo [1], Alicia Peiró-Puig [1], Mariana Herran-González [1] and Marcelino Pérez-Bermejo [2,*]

[1] School of Medicine and Health Sciences, Department of Nutrition, Catholic University of Valencia San Vicente Mártir, C/Quevedo nº 2, 46001 Valencia, Spain; luviba@mail.ucv.es (L.V.-B.); mariarojaas96@mail.ucv.es (M.J.R.-C.); rocioinmaculada.navarro@mail.ucv.es (R.N.-S.); victoria.belda@mail.ucv.es (V.B.-S.); javierperezmurillo@mail.ucv.es (J.P.-M.); alicia.peiro@mail.ucv.es (A.P.-P.); mariana.herran@mail.ucv.es (M.H.-G.)

[2] SONEV Research Group, School of Medicine and Health Sciences, Catholic University of Valencia San Vicente Mártir, C/Quevedo nº 2, 46001 Valencia, Spain

* Correspondence: marcelino.perez@ucv.es; Tel.: +34-620-984-639

Abstract: The present study was carried out to relate the role of natural products in the metabolism of an increasingly prevalent disease, type 2 diabetes mellitus. At present, in addition to the pharmacological resources, an attempt is being made to treat diabetes mellitus with natural products. We carried out a systematic review of studies focusing on the role of natural products on diabetes mellitus treatment. The bibliographic search was done through Medline (Pubmed) and Web of Science. From 193 records, the title and summary of each were examined according to the criteria and whether they met the selection criteria. A total of 15 articles were included; after reviewing the literature, it is apparent that the concept of natural products is ambiguous as no clear boundary has been established between what is natural and what is synthetic, therefore we feel that a more explicit definition of the concept of "natural product" is needed. Gut microbiota is a promising therapeutic target in the treatment of diabetes. Therefore, it would be necessary to work on the relationship between the microbiome and the benefits in the treatment of diabetes mellitus. Treatment based solely on these natural products is not currently recommended as more studies are needed.

Keywords: diabetes mellitus; type 2; biological products; plants; antioxidants; therapeutics; natural products; phenols

Citation: Vivó-Barrachina, L.; Rojas-Chacón, M.J.; Navarro-Salazar, R.; Belda-Sanchis, V.; Pérez-Murillo, J.; Peiró-Puig, A.; Herran-González, M.; Pérez-Bermejo, M. The Role of Natural Products on Diabetes Mellitus Treatment: A Systematic Review of Randomized Controlled Trials. *Pharmaceutics* **2022**, *14*, 101. https://doi.org/10.3390/pharmaceutics14010101

Academic Editor: Joseph Kost

Received: 13 December 2021
Accepted: 29 December 2021
Published: 2 January 2022

Publisher's Note: MDPI stays neutral with regard to jurisdictional claims in published maps and institutional affiliations.

1. Introduction

The rapid development of society over the 21st century has brought about a complete change in lifestyle of the population in both a positive and negative way; new risk factors have emerged that have conditioned an increase in the prevalence of chronic diseases worldwide, such as type 2 diabetes mellitus (T2D), which in turn increases the morbidity and mortality of the world population [1].

Type 2 diabetes is a multifactorial metabolic pathology. The WHO puts its prevalence at over 422 million subjects, and a total of 1.6 million die per year. It is estimated that eight out of every 1000 inhabitants suffer from it and its prevalence increases in the elderly; however, it is becoming increasingly common in children and adolescents [2]. The constant increase in the population diagnosed with T2D, together with its associated complications, makes it a first-order healthcare and economic problem whose impact is reflected in its treatment and complications that cause poor quality of life. In fact, annual healthcare costs are calculated to be between 121.97–141.6 million euros, which is why it is estimated that people with T2D generate twice the healthcare costs of persons who do not suffer from this pathology. Type 2 diabetes is currently deemed one of the pandemics of the 21st century [1,2].

The development of T2D is frequently associated with a combination of a failure in the functioning of the β cells of the pancreas and insulin resistance in various target tissues, such as liver, muscle and adipocytes [3]. Healthy β cells compensate for insulin resistance with an increase in insulin secretion, but a failure in this compensatory mechanism leads to glucose intolerance. Once hyperglycemia occurs, β-cell function deteriorates and insulin resistance worsens, a process known as glucose toxicity [4]. Prior to the onset of diabetes, there is a stage known as pre-diabetes; as it progresses, alterations occur in the cells of the pancreas that make up the islets of Langerhans [1,2].

A sedentary lifestyle, being overweight and malnutrition generate an increase in the production of different reactive oxygen species which produce a chronic state of oxidative stress. This alters the secretion of insulin by the pancreas and the action of hormones in target cells, generating a greater risk of macro- and microvascular complications [5]. There is scientific evidence showing that β cells have very low levels of antioxidant enzymes compared to other tissues and there lies their high vulnerability to oxidative stress. As mentioned, T2D is a chronic disease that combines different metabolic disorders that coexist in a positive feedback loop guided by inflammation. Therefore, reducing and fighting inflammation, as well as oxidative stress, is one of the main therapeutic objectives [6].

There are numerous chemical compounds included in drugs that help control blood glucose levels, such as oral antidiabetics and preloaded insulins. At present, in addition to these pharmacological resources, an attempt is being made to treat this pathology with natural products [7]. Natural products started being used to control blood glucose levels ever since they took center stage in experimental investigations; some of these plants are *Bauhinia forficata*, *Cecropia obtusifolia* (Bertol), *Equisetum myriochaetum* and *Cucurbita ficifolia* bouche, among others [6–8]. Treatment using them has a notable local connotation, since natural products generally vary depending on the country and its culture. An example of this can be seen in Latin America, where *Bauhinia forficata* is more often used, while in Sri Lanka *Senna auriculata* (L.) is used more routinely [9]. Other products, such as alfalfa, *Ginkgo biloba*, ginseng and turmeric are found more frequently. Familiarity with the properties of plants and scientific evidence are the starting point for using this type of product in the treatment of chronic diseases such as diabetes [10].

Current drugs for treatment of diabetes employ antidiabetic mechanisms of action that include the inhibition of alpha-glucosidase and alpha-amylase in the digestive tract, decreasing the uptake of glucose through its transporters by stimulating the release of insulin [11]. However, there are also many others, especially the most recent ones, which try to imitate the mechanism of action of natural products by incorporating several of their active principles. At present, the most common drug on the market for glucose control is metformin, which comes from *Galega officinalis*, a guanidine-rich plant used in European folk medicine [12,13]. In an attempt to find a similar drug, the study of *Ginkgo biloba*, ginseng and turmeric, among many others, has been implemented. This shows the true importance of the investigation of plants that can be sources of new compounds with clinical activities for the treatment of chronic diseases, such as diabetes [10,13].

Although reviews on the use of natural products in T2D have previously been published [14–16], these are works that analyze in depth the therapeutic effects and biochemical mechanisms of different natural products against T2D, reporting that most analyzed studies provide preliminary or inadequately documented results and that is necessary to continue working on their research and development. This review, focused exclusively on randomized clinical trials, tries to delve deeper into the work that has been carried out in recent years on the application of natural products to treat the disease.

2. Search Methodology

This systematic review was conducted in accordance with the criteria set out in the Preferred Reporting Items for Systematic Reviews and Meta-Analysis (PRISMA) [17]. The literature search was carried out in PubMed and Web of Science. The search strategy was carried out by combining the terms "Diabetes Mellitus", "Type 2", "Biological Products",

"Antioxidants", "Plants", "Therapeutics" and "Phenols", combined with each other using Boolean operators. Randomized controlled trials from within the last five years were selected, giving a total of 15 to review. The flowchart in Figure 1 details the search and selection process.

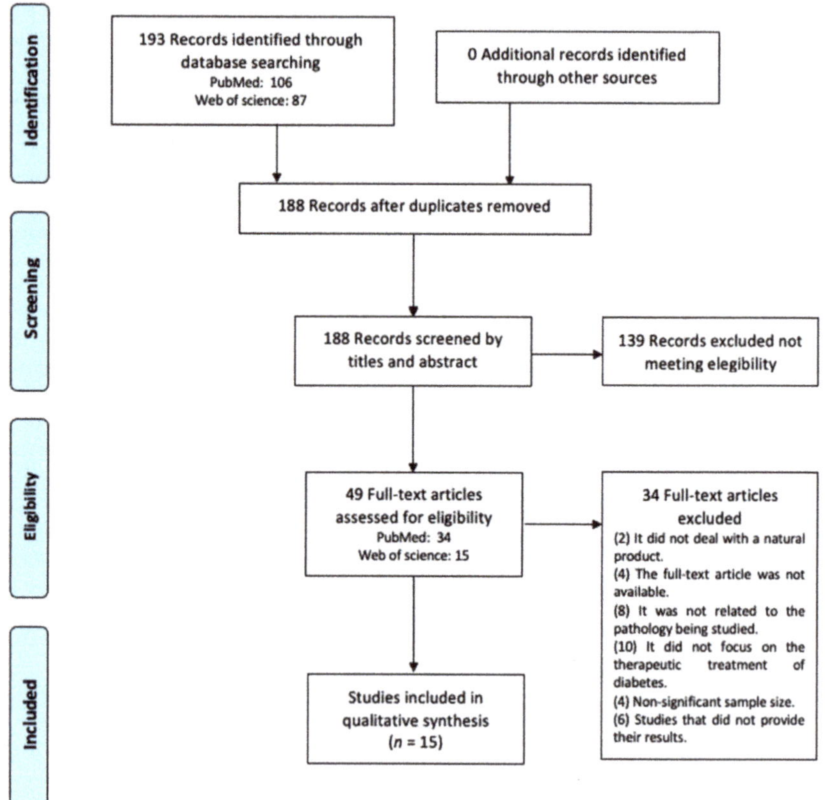

Figure 1. PRISMA flowchart of study selection process.

The Joanna Briggs Institute (JBI) [18] checklist for randomized controlled trials was used to assess study design and quality. One point was ascribed to each criterion achieved on the checklist. The quality of the studies was rated as a percentage of the total available points on the checklist.

3. Results

As shown in Figure 1, the literature search identified 193 records. After removing duplicates and screening titles and abstracts, 49 articles were selected for full-text review, of which 15 studies met the inclusion criteria [19–33]. Table 1 shows the main characteristics of each study. Table 2 shows the quality assessment of the studies.

Table 1. Main characteristics of each study analyzed.

Ref.	Year	Type of Study	Sample	Results	Conclusions
[19]	2020	Randomized controlled trial	580	HbA1c improved similarly in both treatment groups. Changes in fasting plasma glucose and plasma glucose profile and insulin doses were similar in both groups.	This study concludes that SAR-Asp was well tolerated and demonstrated effective glycemic control with a safety and immunogenicity profile similar to that of commercially available insulin aspart formulations in people with diabetes treated for 26 weeks.
[20]	2020	Randomized controlled trial	17	BS with metformin treatment improved glucose tolerance and the expression of inflammatory markers in patients with T2D. Furthermore, a considerable number of bacterial taxa were correlated with clinical markers, especially with blood glucose during OGTT. Furthermore, metagenomic data indicated that the metabolism of seleno compounds increased after treatment with BS.	The results obtained suggest that treatment with BS and metformin can improve glucose metabolism through modulation of the intestinal microbiota in patients with T2D.
[21]	2018	Randomized controlled trial	39	Blood pressure decreased in patients treated with extracts of J. regia. It also showed a significant decrease in body mass. Thus, hydroalcoholic extracts of J. regia do not produce beneficial effects in patients with T2D.	We cannot conclude that the extracts obtained from walnut leaves may be a future treatment for T2D.
[22]	2018	Randomized controlled trial	450	Only AMC improved insulin resistance and triglyceride levels. Both metformin and AMC treatments improved blood glucose, glycosylated hemoglobin, cellular function of beta cells; they both increased gut microbiota diversity, thereby decreasing the number of subjects in the groups related to the pathogenesis of the disease. AMC showed a greater modulating effect on the microbiota as well as an increase in the concentration of beneficial microorganisms for the treatment of T2D.	This study suggests that the beneficial effects of metformin and AMC are due to the modification in the intestinal microbiota. It was possible to correlate the increase in *Blautia spp.* and *Faecalibacterium spp.* with the improvement in both lipid and glucose homeostasis. The results show that the gut microbiota is a promising therapeutic target for diabetes. It will be interesting for future studies on the diabetes treatment capacity of mangiferin and berberine.
[23]	2018	Randomized controlled trial	47	In the placebo group, no significant decrease in the concentration of glycated hemoglobin was observed. On the other hand, it was in the group treated with GKB and the same for fasting blood glucose. Regarding the decrease in insulin levels in the treatment group, no significant differences were observed. Body mass index, visceral adiposity and hip circumference decreased in GKB patients in contrast to the group treated only with metformin or placebo.	The results therefore show that *Ginkgo biloba* extract is a promising adjuvant to metformin. However, more studies are needed to analyze its long-term effects as well as a study with a larger number of participants.

Table 1. *Cont.*

Ref.	Year	Type of Study	Sample	Results	Conclusions
[24]	2018	Randomized controlled trial	118	Six weeks of intake of pinitol-enriched beverages resulted in a significant increase in two proteins involved in the insulin secretion pathway, the acid labile subunit of insulin-like growth factor and complement C4A in glucose intolerant subjects, but not in healthy volunteers.	This study concluded that substituting a common sugar source (such as sucrose) for a natural drink enriched with pinitol in subjects with glucose intolerance could benefit glucose metabolism by reducing and stimulating insulin secretion. This mechanism may play an important role in the prevention of insulin resistance and the progression of diabetes.
[25]	2017	Randomized controlled trial	60	The mean values of fasting blood glucose, glycated hemoglobin and triglycerides in the group of herbal medicines, were significantly lower than the values of the placebo group.	The study shows a potential antihyperglycemic and triglyceride-reducing effect through the use of a mix of silymarin, nettle and olibanum extracts. It caused a significant reduction in glycated hemoglobin, triglycerides, and fasting blood glucose. It did not reduce cholesterol or blood pressure. It stated that other hypoglycemic agents should be used to confirm these effects.
[26]	2019	Randomized controlled trial	81	The HbA1C and the fasting blood glucose remained the same. Insulin sensitivity was significantly improved in the group with the curcumin supplement compared to the double placebo group. The triglyceride level increased in the double placebo group and the group with the curcumin and LCn-3 PUFA supplementation decreased triglyceride levels.	This study does not provide evidence of a curcumin and LCn-3 PUFA-based supplementation. This can be due to various factors such as the study population, balance in the groups or interactions between bioactives. A curcumin-based supplementation does have a positive effect on improving insulin sensitivity. It is believed that these results may indicate that a better strategy can be carried out to reduce risk factors in the progression of T2D.
[27]	2019	Randomized controlled trial	94	The effect of propolis on glucose metabolism shows that after the intervention, the mean HbA1C, 2hpp FBS and insulin decreased significantly compared to the placebo group. It also significantly decreases HOMA-IR and HOMA-β. It is suggested that glycemic control is related to the intestinal reduction of carbohydrates and increases the level of glycolysis and the use of glucose in the liver through increasing the absorption of glucose by peripheral tissue by activating the glucose-sensitive transporter.	Iranian propolis has effects on post-pandial blood glucose, serum insulin, and insulin resistance. The study also mentions the decrease in inflammatory cytokines, associated with oxidative stress and chronic inflammation. This pathogenesis is related to T2D.

Table 1. *Cont.*

Ref.	Year	Type of Study	Sample	Results	Conclusions
[28]	2017	Randomized controlled trial	39	After the 8-week intervention supplemented with the three grams of cinnamon, there were no changes or findings on the effects that cinnamon may have.	This study was designed to evaluate whether cinnamon intake influenced glycemic markers, glycation end products, and inflammatory indicators in patients with T2D.
[29]	2018	Randomized controlled trial	74	The study analyzed the cytokines, glutathione, peroxidase, HbA1c and the fecal microbiome composition. Probiotics being killed by heat have shown to positively affect immunodeficient patients. HbA1c levels were significantly reduced in patients taking L. reuteri ADR-1in the tests taken at the visits 2, 3 and 4. In this sense, the patients were able to maintain a stable HbA1c level for 3 months after the intervention. A decrease in blood lipids was observed: cholesterol, free fatty acids and LDL in groups with ADR-1, although only cholesterol was significant. In the ADR-3 group, a significant reduction in arterial pressure and loss of body weight was observed. The inflammatory cytokines IL-1Beta showed a significant reduction in the group with ADR-3. Regarding the microbiome, L. reuteri increased significantly in the group with ADR-1. A significant increase in bifidobacterium was also observed in the ADR-1 group.	The conclusion of the results suggest that the reduced levels of HbA1c are positively affected by L. reuteri after the regulated consumption of ADR-1 and ADR-3 and that changes in the microbiome are the result of its ingestion. These changes should have a study conducted examining regulating blood sugar levels and further TD2 complications.
[30]	2019	Randomized controlled trial	52	The study showed that HbA1c levels did not vary significantly between the two groups: patients taking the supplement decreased significantly in terms of TG compared to the placebo group after 3–6 months of taking the supplement. In the group taking the supplement, BMI increased and adiponectin decreased, probably since curcumin tends to increase appetite, although the mechanism is not clear. Leptin also decreased, probably due to the improvement in the leptin resistance.	The study had a limitation which was the small sample size and the short duration of the intervention. We would have to make a further study to reach a conclusion regarding the beneficial effects of curcumin on adiponectin and antioxidative LDL in patients with T2D. However, the study showed that curcumin inhibits the increase in oxidative LDL in patients with T2D, so it can be used to prevent cardiovascular diseases and common diabetes complications.

Table 1. *Cont.*

Ref.	Year	Type of Study	Sample	Results	Conclusions
[31]	2019	Randomized controlled trial	36	The results of the study show that when in treatment, patients were asked to eat a high glucose meal in which blood samples were taken. Insulin levels were high and serum unesterified fatty acids were low. The berries showed an improvement in serum and insulin value close to the statistical significance. There was a significant reduction in the insulin response of the patients who consumed berries compared to those consuming gelatin.	The effect of each kind of berry in terms of glucoregulatory powers cannot be determined possibly due to the variability between their inclusion in the samples given to the patients during the study. Some berries have a more potent effect and this effect was diluted because of their mixture. Evidence has shown that berries have both short- and long-term effects, but further investigation might be needed to conclude these results.
[32]	2016	Randomized controlled trial	14	Some previous studies have shown an increase of the release of GLP-1 in diabetic mice. However, in this study, a dose of 500 mg of resveratrol administered twice a day for 5 weeks showed no effect on GLP-1 secretion, glycemic control, gastric emptying and body weight and did not suppress energy intake on type 2 diabetic humans	In total, 14 subjects (10 men and 4 women) with type 2 diabetes, managed only by diet, showed no effects after 5 weeks of resveratrol administration compared to the placebo group. It is worth bearing in mind the short duration of the study, the lack of obese subjects and the small sample size.
[33]	2017	Randomized controlled trial	86	An herbal combination capsule of 600 mg was composed of Terminalia chebula fruit extract (200 mg), Commiphora mukul (200 mg) and Commiphora myrrha oleo-gum-resin (200 mg). The participants were randomly assigned to either herbal combination or placebo group. The capsule was taken 3 times a day and showed an improvement of glycemic control, total cholesterol and low-density lipoprotein cholesterol. Moreover, it increased high-density lipoprotein cholesterol levels.	Eighty-six hyperlipidemic type 2 diabetic women between 40 and 60 years, fasting serum glucose levels between 150 and 180 mg/dL; blood glycosylated hemoglobin levels between 7.5% and 8.5%; low-density lipoprotein cholesterol > 100 mg/dL; and daily oral intake of not more than 10 mg glyburide and 1000 mg metformin at maximum were included in the study.The herbal combination was well tolerated and did not cause any hepatic, renal or other adverse effects. However, it is worth bearing in mind the short duration of the study and the different durations of the study between participants. The actual mechanism of drug action remains unknown at the moment. More studies are necessary to ensure the safety and effectiveness of the compounds.

Abbreviations: HbA1c: Glycated Hemoglobin. SAR-Asp: Insulin Aspart Biosimilar/follow-on biologic product SAR341402; BS: Scutellaria Baicalensis. T2D: Type 2 Diabetes. OGTT: Oral Glucose Tolerance Test. AMC: Traditional Chinese Herbal Formula (Extract of the dry corolla of A. manihot). GKB: *Ginkgo biloba* Extract. C4A: Complement C4A. LCn-3 PUFA: Long-Chain n-3 Polyunsaturated Fatty Acid. 2hpp: 2-h post prandia. FBS: Fasting Blood Sugar. HOMA-IR: Homeostasis Model Assessment-Insulin Resistance. HOMA-β: Homeostasis Model Assessment of β-cell function. LDL: Low-Density Lipoprotein. TG: Triglycerides. BMI: Body Mass Index. GLP-1: Glucagon-Like Peptide 1.

Table 2. Studies appraised using the Joanna Briggs Institute critical appraisal checklist for randomized controlled trials.

Study	Was True Randomization Used for Assignment of Participants to Treatment Groups?	Was Allocation to Treatment Groups Concealed?	Were Treatment Groups Similar at the Baseline?	Were Participants Blind to Treatment Assignment?	Were Those Delivering Treatment Blind to Treatment Assignment?	Were Outcomes Assessors Blind to Treatment Assignment?	Were Treatment Groups Treated Identically Other Than the Intervention of Interest?	Was Follow up Complete and If Not, Were Differences between Groups in Terms of Their Follow up Adequately Described and Analyzed?	Were Participants Analyzed in the Groups to Which They Were Randomized?	Were Outcomes Measured in the Same Way for Treatment Groups?	Were Outcomes Measured in a Reliable Way?	Was Appropriate Statistical Analysis Used?	Was the Trial Design Appropriate, and Any Deviations from the Standard RCT Design (Individual Randomization, Parallel Groups) Accounted for in the Conduct and Analysis of the Trial?	Score out of 13 (100%)
Garg et al., 2020 [19]	Y	Y	Y	N	U	Y	Y	U	Y	Y	Y	Y	Y	76.90%
Shin et al., 2020 [20]	Y	Y	Y	Y	Y	Y	Y	Y	Y	Y	Y	Y	Y	100%
Rabiie et al., 2018 [21]	Y	Y	Y	Y	U	Y	Y	Y	Y	Y	Y	Y	Y	92.30%
Tong et al., 2018 [22]	Y	Y	Y	U	U	U	Y	Y	N	Y	Y	Y	Y	69.20%
Aziz et al., 2018 [23]	Y	Y	Y	Y	Y	U	Y	Y	Y	Y	Y	Y	Y	92.30%
Lambert et al., 2018 [24]	Y	Y	Y	Y	Y	U	Y	Y	Y	Y	Y	Y	Y	92.30%
Khalili et al., 2017 [25]	Y	Y	Y	Y	Y	U	Y	Y	Y	Y	Y	Y	Y	92.30%
Thota RN et al., 2019 [26]	Y	Y	Y	Y	Y	U	Y	Y	Y	Y	Y	Y	Y	92.30%
Zakerkihs et al., 2019 [27]	Y	Y	Y	Y	Y	U	Y	Y	Y	Y	Y	Y	Y	92.30%
Talaei et al., 2017 [28]	Y	Y	Y	Y	Y	U	Y	Y	Y	Y	Y	Y	Y	92.30%
Hsieh et al., 2018 [29]	Y	Y	Y	Y	Y	Y	Y	Y	Y	Y	Y	Y	Y	100%
Funamoto et al., 2019 [30]	Y	Y	Y	Y	Y	Y	Y	Y	Y	Y	Y	Y	Y	100%
Solverson et al., 2019 [31]	Y	Y	Y	Y	Y	Y	Y	Y	Y	Y	Y	Y	Y	100%
Thazhath et al., 2016 [32]	Y	Y	Y	Y	Y	U	Y	Y	U	Y	Y	Y	Y	84.60%
Shokoohi et al., 2017 [33]	Y	Y	Y	Y	Y	U	Y	Y	Y	Y	Y	Y	Y	92.30%

Y = Yes. N = No. U = Unclear.

4. Discussion

The present study was carried out to relate the role of natural products in the metabolism of an increasingly prevalent disease, type 2 diabetes mellitus (T2D) [1,2]. After reviewing the literature, it is apparent that the concept of natural products is ambiguous as no clear boundary has been established between what is natural and what is synthetic, since a product can be synthesized from a natural extract. Metformin, for example, is obtained from a derivative of the guanidines of the French lilac, *Galega officinalis* [34]. Insulin can also be synthesized from bacterium *E. coli*, building or transforming an analogue of human insulin [35]. This has made it difficult to search bibliographies as it has not been possible to delimit some keywords to make an exact search.

The studies analyzed do not show enough scientific evidence to use the methods investigated in a population. More research is required to be able to observe possible side effects caused, in the short and long term, by the use of these treatments with natural products at the individual and collective level [21,23,26–28,32,33]. Some studies [32,33,36,37] indicate that more research is required to determine evidence of beneficial effects through the use of natural products.

Studies show that natural products have more than one beneficial effect in addition to being insulin-sensitizing or hypoglycemic as they can be anti-inflammatory, antioxidant and cholesterol-lowering [22–25,27,32]. Zakerkish et al. [27] used Iranian propolis collected from beehives and its use was found to provide various beneficial effects in patients with T2D. These results indicate that a promising treatment can be achieved with long-term studies since the use of natural products has good potential.

Insulin (SAR-Asp) has been shown to have an effective glycemic control effect very similar to the different insulins found on the market [19]. It is important to evaluate the significant effect of this insulin to implement a longer study time. In another study, the insulin-sensitizing effect of *Scutellaria baicalensis* (SB) was proposed as an adjunct to metformin in the treatment of type 2 diabetic patients and also indicates that SB improves metabolism and the number of microbial taxa, which suggests that this treatment can improve glucose metabolism through modulation of the gut microbiota in patients with T2D [20].

The studies analyzed in this work presented some difficulties when deciding whether or not they should be part of the bibliographic review. As mentioned above, this was due to the indistinct boundary between the concepts of natural and synthetic products as well as the border between treatment and patient improvement. Another limitation is that the analysis of the action of natural products is restricted to an adjuvant action together with drugs in current use [20,23], which makes it difficult to define the benefits of natural compounds. However, it can be stated that as an adjunct to metformin, Gingko biloba extract [23] as well as treatment with *Scutellaria baicalensis* extract [20] present benefits compared to metformin therapy alone [20,23]. Among the outstanding natural products in the treatment for T2D, we can highlight *Ginkgo biloba* extract as an adjuvant to Metformin [23], pinitol [24], propolis [27], live probiotic L. Reuteri ADR-1 and heat-killed probiotic L. reuteri ADR-3 [29], mixed berries [30], a Chinese plant extract [22] and resveratrol [32].

A limitation of this review is that the results of the studies were obtained over a short period of time, from three to six months, so these results may be affected if an increase in the temporality of the treatments is protocolized [23–28,34]. Within the application of natural products as a treatment for patients with T2D, there are single products that can have effects at different levels, that is, a natural product can generate a modulating effect on glucose and lipid metabolism hypoglycemia, hypolipidemia, modulation of the microbiota and anti-inflammatory and antioxidant agents [22–24,27,29,30,32]. Their multiple applicability and their effect in the short and long term must be evaluated and taken into account; consequently, studies are necessary in which the greatest number of these aspects are measured so that more significant results can be obtained.

The studies analyzed show the close relationship between the diversity of microorganisms in the intestinal microbiota and the functioning and metabolism of the host, as well as linking certain microorganisms with specific effects [36,38,39]. The studies by Tong et al. [22], Na Rae Shin et al. [20] and Ming-Chia et al. [29] show promising benefits for the treatment of T2D by increasing diversity and certain species in the intestinal microbiota, making it a promising therapeutic target for the disease in question. Specifically, it would be interesting to study in depth the benefits of the increase in microbiota of butyrate-producing species as well as *Blautia* spp. and *Faecalibacterium* spp. since they show an improvement in carbohydrate and lipid homeostasis [22]. By contrast, other studies reported no benefit for the treatment of T2D [21,26,28,32].

In general, it can be concluded that the most notable benefits of treatments and protocols using natural products in our review studies are the improvement in insulin resistance of the subjects [19,22,27,29,31], improvement in biochemical parameters of glycosylated hemoglobin [23,25,27,29], improvement in the lipid profile in general [22,23,27,29,30] and a reduction in glucose levels in pre-prandial blood [20,23].

In this context, it is also a source of discussion that the presence of environmental factors and factors specific to individuals can interfere with the results obtained, since, for example, when carrying out a therapy with natural products accompanied by adjuvant drugs, it is difficult to determine if the effect obtained is due to the use of the natural treatment or the drug. In addition, the presence of toxic habits such as an unbalanced diet, among other aspects, can be the cause of the possible negative effects that are associated with a natural product. The individuality of the patient makes the response to the treatments totally different, since individual factors, such as lifestyle, eating habits, genetic factors, underlying pathologies and treatments, can lead to totally different results. That is why when considering this type of study, it is necessary to evaluate the personal factor that characterizes each individual. This fact, together with the small sample size, can interfere with how results can be extrapolated to a larger population.

5. Conclusions

The purpose of this study was to analyze the effect of natural products as a therapy in T2D. After conducting this systematic review, we feel that a more explicit definition of both the concept of "natural product" and that of "treatment" is needed. This would make it easier to select and filter different studies based on whether they focus on therapeutic treatment, nutritional treatment or quality of life improvement.

Gut microbiota is a promising therapeutic target in the treatment of diabetes. We also observe that there is a correlation between the presence of specific species and therapeutic properties. Therefore, it would be necessary to work on the relationship between the microbiome and the benefits in the treatment of T2D.

Natural products that improve comorbidities related to metabolic syndrome and T2D, such as oxidative stress and chronic systemic inflammation, have beneficial effects on patients. However, treatment based solely on these natural products is not currently recommended as more studies are needed. Nonetheless, natural products favoring T2D treatment may be a promising adjunct to current therapies. However, the present review shows some limitations in the protocols of the different studies that make drawing conclusions difficult, which is why studies with more specific protocols are needed, such as long-term vision and larger samples to confirm the benefits and to standardize the results obtained.

Author Contributions: Conceptualization, M.J.R.-C., M.H.-G., L.V.-B., V.B.-S., A.P.-P., R.N.-S., J.P.-M. and M.P.-B.; methodology, M.J.R.-C., L.V.-B., M.H.-G., V.B.-S., A.P.-P., R.N.-S., J.P.-M. and M.P.-B.; formal analysis, M.J.R.-C., L.V.-B., V.B.-S., R.N.-S., J.P.-M. and M.P.-B.; writing—original draft preparation, M.J.R.-C., M.H.-G., L.V.-B., V.B.-S., A.P.-P., R.N.-S., J.P.-M. and M.P.-B.; writing—review and editing, M.P.-B.; supervision, M.P.-B. All authors have read and agreed to the published version of the manuscript.

Funding: This research received no external funding.

Institutional Review Board Statement: Not applicable.

Informed Consent Statement: Not applicable.

Acknowledgments: The authors thank the Catholic University of Valencia San Vicente Mártir for their contribution and help in the payment of the Open Access publication fee.

Conflicts of Interest: The authors declare no conflict of interest.

References

1. Mendoza Romo, M.A.; Padrón Salas, A.; Cossío Torres, P.E.; Soria Orozco, M. Global prevalence of type 2 diabetes mellitus and its relationship with the Human Development Index. *Rev. Panam. Salud Pública* **2017**, *41*, e103. [CrossRef] [PubMed]
2. Kaneto, H. Pathophysiology of type 2 diabetes mellitus. *Nihon Rinsho Jpn. J. Clin. Med.* **2015**, *73*, 2003–2007.
3. Quinn, L. Type 2 diabetes: Epidemiology, pathophysiology, and diagnosis. *Nurs. Clin. N. Am.* **2001**, *36*, 175–192.
4. Khalid, M.; Alkaabi, J.; Khan, M.A.B.; Adem, A. Insulin Signal Transduction Perturbations in Insulin Resistance. *Int. J. Mol. Sci.* **2021**, *22*, 8590. [CrossRef]
5. Poblete-Aro, C.; Russell-Guzmán, J.; Parra, P.; Soto-Muñoz, M.; Villegas-González, B.; Cofré-BolaDos, C.; Herrera-Valenzuela, T. Exercise and oxidative stress in type 2 diabetes mellitus. *Rev. Med. Chile* **2018**, *146*, 362–372. [CrossRef] [PubMed]
6. Yeung, A.W.K.; Tzvetkov, N.T.; Durazzo, A.; Lucarini, M.; Souto, E.B.; Santini, A.; Gan, R.-Y.; Jozwik, A.; Grzybek, W.; Horbańczuk, J.O.; et al. Natural products in diabetes research: Quantitative literature analysis. *Nat. Prod. Res.* **2021**, *35*, 5813–5827. [CrossRef] [PubMed]
7. Gori, M.; Campbell, R.K. Natural Products and Diabetes Treatment. *Diabetes Educ.* **1998**, *24*, 201–208. [CrossRef]
8. Pepato, M.T.; Keller, E.H.; Baviera, A.M.; Kettelhut, I.C.; Vendramini, R.C.; Brunetti, I.L. Anti-diabetic activity of Bauhinia forficata decoction in streptozotocin-diabetic rats. *J. Ethnopharmacol.* **2002**, *81*, 191–197. [CrossRef]
9. Patle, D.; Vyas, M.; Khatik, G.L. A Review on Natural Products and Herbs Used in the Management of Diabetes. *Curr. Diabetes Rev.* **2021**, *17*, 186–197. [CrossRef]
10. Abu-Odeh, A.M.; Talib, W.H. Middle East Medicinal Plants in the Treatment of Diabetes: A Review. *Molecules* **2021**, *26*, 742. [CrossRef]
11. Salehi, B.; Ata, A.; Kumar, N.V.A.; Sharopov, F.; Ramírez-Alarcón, K.; Ruiz-Ortega, A.; Ayatollahi, S.A.; Fokou, P.V.T.; Kobarfard, F.; Zakaria, Z.A.; et al. Antidiabetic Potential of Medicinal Plants and Their Active Components. *Biomolecules* **2019**, *9*, 551. [CrossRef]
12. Pereira, A.S.P.; Banegas-Luna, A.J.; Peña-García, J.; Pérez-Sánchez, H.; Apostolides, Z. Evaluation of the Anti-Diabetic Activity of Some Common Herbs and Spices: Providing New Insights with Inverse Virtual Screening. *Molecules* **2019**, *24*, 4030. [CrossRef] [PubMed]
13. Odeyemi, S.; Bradley, G. Medicinal Plants Used for the Traditional Management of Diabetes in the Eastern Cape, South Africa: Pharmacology and Toxicology. *Molecules* **2018**, *23*, 2759. [CrossRef] [PubMed]
14. Ríos, J.L.; Francini, F.; Schinella, G.R. Natural Products for the Treatment of Type 2 Diabetes Mellitus. *Planta Med.* **2015**, *81*, 975–994. [CrossRef] [PubMed]
15. Xu, L.; Li, Y.; Dai, Y.; Peng, J. Natural products for the treatment of type 2 diabetes mellitus: Pharmacology and mechanisms. *Pharmacol. Res.* **2018**, *130*, 451–465. [CrossRef] [PubMed]
16. Bedekar, A.; Shah, K.; Koffas, M. Natural Products for Type II Diabetes Treatment. *Adv. Appl. Microbiol.* **2010**, *71*, 21–73. [CrossRef] [PubMed]
17. Moher, D.; Liberati, A.; Tetzlaff, J.; Altman, D.G.; The PRISMA Group. Preferred reporting items for systematic reviews and meta-analyses: The PRISMA Statement. *PLoS Med.* **2009**, *6*, e1000097. [CrossRef] [PubMed]
18. Joanna Briggs Institute. Critical Appraisal Tools. 2017. Available online: https://jbi.global/critical-appraisal-tools (accessed on 9 December 2021).
19. Garg, S.K.; Wernicke-Panten, K.; Wardecki, M.; Kramer, D.; Delalande, F.; Franek, E.; Sadeharju, K.; Monchamp, T.; Mukherjee, B.; Shah, V.N. Efficacy and Safety of Insulin Aspart Biosimilar SAR341402 Versus Originator Insulin Aspart in People with Diabetes Treated for 26 Weeks with Multiple Daily Injections in Combination with Insulin Glargine: A Randomized Open-Label Trial (GEMELLI 1). *Diabetes Technol. Ther.* **2020**, *22*, 85–95. [CrossRef]
20. Shin, N.R.; Gu, N.; Choi, H.-S.; Kim, H. Combined effects of Scutellaria baicalensis with metformin on glucose tolerance of patients with type 2 diabetes via gut microbiota modulation. *Am. J. Physiol. Endocrinol. Metab.* **2020**, *318*, E52–E61. [CrossRef]
21. Rabiei, K.; Ebrahimzadeh, M.A.; Saeedi, M.; Bahar, A.; Akha, O.; Kashi, Z. Effects of a hydroalcoholic extract of Juglans regia (walnut) leaves on blood glucose and major cardiovascular risk factors in type 2 diabetic patients: A double-blind, placebo-controlled clinical trial. *BMC Complement. Altern. Med.* **2018**, *18*, 206. [CrossRef]
22. Tong, X.; Xu, J.; Lian, F.; Yu, X.; Zhao, Y.; Xu, L.; Zhang, M.; Zhao, X.; Shen, J.; Wu, S.; et al. Structural Alteration of Gut Microbiota during the Amelioration of Human Type 2 Diabetes with Hyperlipidemia by Metformin and a Traditional Chinese Herbal Formula: A Multicenter, Randomized, Open Label Clinical Trial. *mBio* **2018**, *9*, e02392-17. [CrossRef]
23. Aziz, T.A.; Hussain, S.A.; Mahwi, T.O.; Ahmed, Z.A.; Rahman, H.S.; Rasedee, A. The efficacy and safety of *Ginkgo biloba* extract as an adjuvant in type 2 diabetes mellitus patients ineffectively managed with metformin: A double-blind, randomized, placebo-controlled trial. *Drug Des. Dev. Ther.* **2018**, *12*, 735–742. [CrossRef]

24. Lambert, C.; Cubedo, J.; Padró, T.; Vilahur, G.; López-Bernal, S.; Rocha, M.; Hernández-Mijares, A.; Badimon, L. Effects of a Carob-Pod-Derived Sweetener on Glucose Metabolism. *Nutrients* **2018**, *10*, 271. [CrossRef]
25. Khalili, N.; Fereydoonzadeh, R.; Mohtashami, R.; Mehrzadi, S.; Heydari, M.; Huseini, H.F. Silymarin, Olibanum, and Nettle, A Mixed Herbal Formulation in the Treatment of Type II Diabetes: A Randomized, Double-Blind, Placebo-Controlled, Clinical Trial. *J. Evid.-Based Complementary Altern. Med.* **2017**, *22*, 603–608. [CrossRef]
26. Thota, R.N.; Acharya, S.H.; Garg, M.L. Curcumin and/or omega-3 polyunsaturated fatty acids supplementation reduces insulin resistance and blood lipids in individuals with high risk of type 2 diabetes: A randomised controlled trial. *Lipids Health Dis.* **2019**, *18*, 31. [CrossRef]
27. Zakerkish, M.; Jenabi, M.; Zaeemzadeh, N.; Hemmati, A.A.; Neisi, N. The Effect of Iranian Propolis on Glucose Metabolism, Lipid Profile, Insulin Resistance, Renal Function and Inflammatory Biomarkers in Patients with Type 2 Diabetes Mellitus: A Randomized Double-Blind Clinical Trial. *Sci. Rep.* **2019**, *9*, 7289. [CrossRef]
28. Talaei, B.; Amouzegar, A.; Sahranavard, S.; Hedayati, M.; Mirmiran, P.; Azizi, F. Effects of Cinnamon Consumption on Glycemic Indicators, Advanced Glycation End Products, and Antioxidant Status in Type 2 Diabetic Patients. *Nutrients* **2017**, *9*, 991. [CrossRef] [PubMed]
29. Hsieh, M.-C.; Tsai, W.-H.; Jheng, Y.-P.; Su, S.-L.; Wang, S.-Y.; Lin, C.-C.; Chen, Y.-H.; Chang, W.-W. The beneficial effects of Lactobacillus reuteri ADR-1 or ADR-3 consumption on type 2 diabetes mellitus: A randomized, double-blinded, placebo-controlled trial. *Sci. Rep.* **2018**, *8*, 16791. [CrossRef] [PubMed]
30. Funamoto, M.; Shimizu, K.; Sunagawa, Y.; Katanasaka, Y.; Miyazaki, Y.; Kakeya, H.; Yamakage, H.; Satoh-Asahara, N.; Wada, H.; Hasegawa, K.; et al. Effects of Highly Absorbable Curcumin in Patients with Impaired Glucose Tolerance and Non-Insulin-Dependent Diabetes Mellitus. *J. Diabetes Res.* **2019**, *2019*, 8208237. [CrossRef] [PubMed]
31. Solverson, P.M.; Henderson, T.R.; Debelo, H.; Ferruzzi, M.G.; Baer, D.J.; Novotny, J.A. An Anthocyanin-Rich Mixed-Berry Intervention May Improve Insulin Sensitivity in a Randomized Trial of Overweight and Obese Adults. *Nutrients* **2019**, *11*, 2876. [CrossRef]
32. Thazhath, S.S.; Wu, T.; Bound, M.J.; Checklin, H.L.; Standfield, S.; Jones, K.L.; Horowitz, M.; Rayner, C.K. Administration of resveratrol for 5 wk has no effect on glucagon-like peptide 1 secretion, gastric emptying, or glycemic control in type 2 diabetes: A randomized controlled trial. *Am. J. Clin. Nutr.* **2016**, *103*, 66–70. [CrossRef]
33. Shokoohi, R.; Kianbakht, S.; Faramarzi, M.; Rahmanian, M.; Nabati, F.; Mehrzadi, S.; Huseini, H.F. Effects of an Herbal Combination on Glycemic Control and Lipid Profile in Diabetic Women: A Randomized, Double-Blind, Placebo-Controlled Clinical Trial. *J. Evid.-Based Complementary Altern Med.* **2017**, *22*, 798–804. [CrossRef]
34. Foretz, M.; Guigas, B.; Bertrand, L.; Pollak, M.; Viollet, B. Metformin: From Mechanisms of Action to Therapies. *Cell Metab.* **2014**, *20*, 953–966. [CrossRef]
35. Hwang, H.-G.; Kim, K.-J.; Lee, S.-H.; Kim, C.-K.; Min, C.-K.; Yun, J.-M.; Lee, S.U.; Son, Y.-J. Recombinant Glargine Insulin Production Process Using *Escherichia coli*. *J. Microbiol. Biotechnol.* **2016**, *26*, 1781–1789. [CrossRef] [PubMed]
36. Moszak, M.; Szulińska, M.; Bogdański, P. You Are What You Eat—The Relationship between Diet, Microbiota, and Metabolic Disorders—A Review. *Nutrients* **2020**, *12*, 1096. [CrossRef] [PubMed]
37. Rashad, H.; Metwally, F.M.; Ezzat, S.M.; Salama, M.M.; Hasheesh, A.; Motaal, A.A. Randomized double-blinded pilot clinical study of the antidiabetic activity of Balanites aegyptiaca and UPLC-ESI-MS/MS identification of its metabolites. *Pharm. Biol.* **2017**, *55*, 1954–1961. [CrossRef] [PubMed]
38. Qin, Q.; Chen, Y.; Li, Y.; Wei, J.; Zhou, X.; Le, F.; Hu, H.; Chen, T. Intestinal Microbiota Play an Important Role in the Treatment of Type I Diabetes in Mice with BefA Protein. *Front. Cell. Infect. Microbiol.* **2021**, *11*, 719542. [CrossRef] [PubMed]
39. Merkevičius, K.; Kundelis, R.; Maleckas, A.; Veličkienė, D. Microbiome Changes after Type 2 Diabetes Treatment: A Systematic Review. *Medicina* **2021**, *57*, 1084. [CrossRef] [PubMed]

 *pharmaceutics*

Article

Extract of Calyces from *Physalis peruviana* Reduces Insulin Resistance and Oxidative Stress in Streptozotocin-Induced Diabetic Mice

Ivonne Helena Valderrama [1], Sandra Milena Echeverry [1], Diana Patricia Rey [1], Ingrid Andrea Rodríguez [1], Fátima Regina Mena Barreto Silva [2], Geison M. Costa [3], Luis Fernando Ospina-Giraldo [1] and Diana Marcela Aragón [1,*]

[1] Departamento de Farmacia, Universidad Nacional de Colombia, Av. Carrera 30 # 45-03 Edif. 450, Bogotá 111321, Colombia
[2] Departamento de Bioquímica, Centro de Ciências Biológicas, Universidade Federal de Santa Catarina (UFSC), Campus Universitário, Bairro Trindade, Cx. Postal 5069, Florianópolis 88040, SC, Brazil
[3] Departamento de Química, Pontificia Universidad Javeriana, Av. Carrera 7 # 40–62, Bogotá 110231, Colombia
* Correspondence: dmaragonn@unal.edu.co; Tel.: +57-601-3165000 (ext. 14630)

Citation: Valderrama, I.H.; Echeverry, S.M.; Rey, D.P.; Rodríguez, I.A.; Silva, F.R.M.B.; Costa, G.M.; Ospina-Giraldo, L.F.; Aragón, D.M. Extract of Calyces from *Physalis peruviana* Reduces Insulin Resistance and Oxidative Stress in Streptozotocin-Induced Diabetic Mice. *Pharmaceutics* **2022**, *14*, 2758. https://doi.org/10.3390/pharmaceutics14122758

Academic Editor: Maria Camilla Bergonzi

Received: 10 October 2022
Accepted: 2 December 2022
Published: 9 December 2022

Publisher's Note: MDPI stays neutral with regard to jurisdictional claims in published maps and institutional affiliations.

Abstract: Diabetes mellitus is a metabolic disorder mainly characterized by obesity, hyperglycemia, altered lipid profile, oxidative stress, and vascular compromise. *Physalis peruviana* is a plant used in traditional Colombian medicine for its known activities of glucose regulation. This study aimed to evaluate the anti-diabetic activity of the butanol fraction from an extract of *Physalis peruviana* calyces in two doses (50 mg/kg and 100 mg/kg) in induced type 2 diabetic mice. Blood glucose levels were evaluated once a week, demonstrating that a dose of 100 mg/kg resulted in greater regulation of blood glucose levels in mice throughout the experiment. The same overall result was found for the oral glucose tolerance test (OGTT) and the homeostatic model assessment for insulin resistance (HOMA- IR). The lipid profile exhibited improvement compared to the non-treated group, a dose of 100 mg/kg having greater protection against oxidative stress (catalase, superoxide dismutase, and malondialdehyde levels). Histopathological findings in several tissues showed structure preservation in most of the animals treated. The butanol fraction from *Physalis peruviana* at 100 mg/kg showed beneficial results in improving hyperglycemia, lipidemia, and oxidative stress status, and can therefore be considered a beneficial coadjuvant in the therapy of diabetes mellitus.

Keywords: flavonoid; rutin; antioxidant; lipidosis; hyperglycemia; diabetes

1. Introduction

Physalis peruviana (Solanaceae) grows at high altitudes in countries of the South American Andes; in Colombia, it grows at 2200 m above sea level (m.a.s.l) [1]. Colombia produces a significant amount of fruits annually; around 15,000 tons/hectare of fruits were collected in 2019 [2]. *P. peruviana* is known as Gold Cape Gooseberry, Poha berry, and Peruvian groundcherry. In Latin counties it is called *uvilla, aguaymanto, topo, uchuva* and *capulí* [3]. Calyces cover the fruits and protect them from different exogen factors, such as birds, insects, inclement changes in weather, and pathogens [4]. In traditional medicine and ethnomedicine around the world, *Physalis peruviana* is used to treat different disorders: the leaves and stems to treat cancer, fungal infections, and malaria; the fruits to treat diabetes, cataracts and conjunctivitis, and typhoid fever, and the aerial parts to treat diabetes, cholic in children, malaria, pneumonia, and postpartum pain [5].

Several reports have confirmed the antihyperglycemic and antioxidant effects of several parts of the plant. Authors have previously shown the anti-diabetic activity from fruits, regulating blood glucose levels and attenuating oxidative stress [6]. Studies have also shown the inhibition of some intestinal carbohydrases using an extract from *P. peruviana* [7].

In 2006, different extracts obtained from the leaves of this plant were studied; scavenging activity on superoxide anion and xanthine oxidase inhibition was studied, as well as its anti-inflammatory activities. All these activities were attributable to the flavonoid content [8]. The calyces have demonstrated antioxidant and anti-inflammatory activities of this harvest by-product [6,9,10] and it has been clearly demonstrated that in extract of *P. peruviana* calyces, rutin (Quercetin 3-*O*-rutinoside) is the main flavonoid [11]. Beneficial effects, such as antioxidant effects and blood glucose level regulation, have been attributed to this flavonoid [12]. This study aimed to determine the effect of the butanol fraction from an extract from calyces of *P. peruviana*, to determine the content of the most active extract obtained in the order of related in vitro and in vivo experiments, and to determine its anti-diabetic potential. A diabetic mice model, induced by a high-fat diet and administration of streptozotocin (STZ) in repeated low doses, was chosen because it is the model most commonly used in the physiopathological study of diabetes mellitus, and because it enables outcomes to be determined in a dose-dependent manner [13].

2. Materials and Methods

2.1. Materials

Ethanol 99.5% (PanReac AppliChem, Darmstadt, DEU) was used to prepare the *P. peruviana* extract. Formic acid reagent grade (Merck, Rahway, NJ, USA), acetonitrile chromatography grade (Merck), and ultrapure water (Milli-Q system Millipore®, Rahway, NJ, USA) were used for the liquid chromatographic analysis.

Standar LabDiet® (5001, Richmond, IN, USA), DIO Rodent Purified Diet with 45% Energy from Fat-Red (58V8), Tween® 80, carboxymethyl cellulose-CMC (419273), glucose (G7021), streptozotocin (572201), a catalase assay kit (CAT100), a lipid peroxidation (MDA) assay kit (MAK085), and a SOD assay kit (19160) were purchased from Sigma-Aldrich® (St. Louis, MI, USA). An ELISA monobind insulin kit (2425-300A) was used to determine serum insulin levels. A triglyceride MR kit, cholesterol MR kit (1118005), LDL-cholesterol kit (1133105), and HDL-cholesterol kit (1133010) were used to measure lipid markers. These were purchased from Linear Chemicals SLU (Montgat, ESP). Pentobarbital sodium was purchased from Invet (Bogotá, COL).

2.2. Plant Material

The calyces of *Physalis peruviana* were collected in Granada-Cundinamarca (Colombia), dried at 40 °C, ground, and sifted [14]. A voucher specimen was stored in the Herbarium of the National University of Colombia (COL 512200). The plant material was identified by taxonomist Parra C.

Preparation of the Extract and Fractions

Hydroethanolic extract of *P. peruviana* (HEPP) was prepared according to the standardized methodology of the same research group [14]. For every 10 g of vegetal material, 150 mL of solvent was added (drug:solvent ratio 1:15) using 70% ethanol. The extraction was performed by percolation after 72 h, collecting the percolated product every 24 h. Finally, the ethanol of the extracts was evaporated under reduced pressure, and the remaining water was eliminated by freeze-drying for 24 h (yield: 9.56%). HEPP was submitted to fractionation using, successively, dichloromethane, ethyl acetate, butanol, and water [15]. The dichloromethane fraction was not used due to its low yield (0.9%). The aqueous fraction (WFPP) was frozen and then lyophilized (yield: 10.84%). Ethyl acetate (EFPP) (yield: 39.27%) and BFPP (yield: 12.34%) were evaporated to dryness in a vacuum centrifuge. The fractions were stored at 4 ± 2 °C until further analysis. Preliminary screening studies were conducted to select the most promising fraction.

2.3. Dereplication Analysis

From the butanol fraction obtained from the hydroethanolic extract of calyces of *P. peruviana*, separation was performed using an RP-18 column (Phenomenex Luna Omega

1.6 μ, 2.1 × 150 mm, 100 A) and, as mobile phases, 0.2% formic acid aqueous solution (A) and acetonitrile: methanol (80:20) acidified with 0.2% formic acid (B). Gradient elution was performed as follows: 0–5 min 5% B isocratic conditions; 5–28 min 5–70% B; 28–33 min 70–98% B; 33–43 min 98% B; 43–43.1 min 98-5% B; and 43.1–45 min 5% isocratic conditions. The samples were dissolved in methanol. The flow rate of the mobile phase was 0.25 mL/min at 40 °C. A Thermo Scientific Ultimate 3000 HPLC coupled with A maXis HD QTOF mass spectrometer (Bruker, Germany) was used. Chromatography parameters were determined using positive and negative ionization modes (100–1500 Da). A capillary voltage from +2700 V to −2500 V), a nebulizer pressure of 0.4 Bar, a dry gas flow rate of 4 L/min and a dry gas temperature of 200 °C were used as the chromatographic conditions. The results were analyzed using the software program Bruker Compass Data Analysis 4.2.

Complementary analysis was performed to corroborate the presence of molecules previously reported in the literature. This was done using a Nexera UHPLC coupled to a mass spectrometer with ESI ionization and triple quadrupole (TQ) (LCMS 9030 Shimadzu Scientific Instruments®, Columbia, MD, USA). Spectrophotometric conditions were the same as those mentioned above. The mass spectrometer was operated in positive and negative ionization modes, and data were obtained at between 100 and 2000 atomic mass. Capillarity conditions were a voltage +3 kV, a sample cone voltage of 33 kV, a dry temperature of 250 °C, nitrogen nebulizer gas at 350 L/h, argon collision gas at 50 L/h, and a collision energy of 2.5 eV. The data were processed using the software program Lab Solutions.

2.4. Animals

For the screening OGTT assay, normoglycemic female Wistar rats 7–10 weeks old (180–200 g) were used to evaluate the hypoglycemic activity of the HEPP and its fractions. During acclimatization, water and food were available ad libitum [16].

The diabetic model was performed using CD-1 female mice at 16 weeks old, weighing about 20–25 g. From the beginning until the end of the experiment, animals received a high-fat diet (Test Diet® DIO Rodent Purified Diet 45% Energy from Fat-Red 58V8) and water ad libitum. Body weight and blood glucose levels were checked once a week.

All animals were obtained from the Pharmacy Department of the National University of Colombia. The mice were housed in optimal humidity and temperature conditions (22 °C ± 2 °C) with 12-h light/dark cycles. The study was approved by the local Research Ethics Committee (Act 08 of 16 April 2020, Faculty of Science).

2.5. Diabetic Chronic Model Induction and Treatments

In the first part of the experiment, all animals received a daily high-fat diet for eight weeks (protein 20.8%; fat 23.6%; fiber 5.8%; carbohydrates 41.2%; metabolizable energy (kcal/g)^2 4.6). At the end of this period, streptozotocin was administered via intraperitoneal (i.p.) injection (STZ-40 mg/kg) (Sigma Chemical Company, St. Louis, MO, USA) twice, with an interval of five days between each injection [17]. STZ was prepared in citrate buffer (pH 4.5) and applied once, attempered and stabilized [18]. Immediately before and during the first 24 h, animals were provided with a 5% glucose solution overnight to avoid drug-induced hypoglycemia. Three days after the last administration of STZ, blood glucose levels (BGL) were measured using an Accu-Chek Performa® device; blood samples were obtained from a small incision in the caudal vein. For the classification of diabetic animals, mice with BGL above 150 mg/dL were considered suitable for the next phase of the experiment. The diabetic mice were randomly distributed into four groups (n = 6). All treatments were administered for 21 days through an orogastric tube: vehicle CMC 0.5%-Tween 80 0.5%; metformin 250 mg/kg; BFPP 50 mg/kg, and BFPP 100 mg/kg. Additionally, a normoglycemic group was fed on a regular diet (protein 24.1%; fat 6.4%; fiber 5.3%; carbohydrates 25.2%;metabolizable energy (kcal/g)^2: 3.35). This group was assigned as additional control, and received saline solution as treatment.

At the end of the experiment, all the animals were sacrificed by cervical decapitation after i.p. injection of pentobarbital sodium (60 mg/kg) [19]. Blood samples were obtained immediately and centrifuged (3500 rpm for 10 min) to collect the serum supernatant. The organs were removed, washed with saline buffer, conserved at −80 °C for subsequent assays, and then placed in formalin 10% for histopathological analysis.

2.5.1. Oral Glucose Tolerance Test (OGTT)

After 4 h of fasting, animals received each treatment 30 min before an oral glucose overload (2000 mg/kg). BGL was measured before and after glucose administration and every 30 min (30, 60, and 90 min) [20].

2.5.2. HOMA-IR

An ELISA kit was used to measure insulin levels. Insulin resistance (IR) was determined according to the homeostasis model assessment index, which was calculated based on fasting serum insulin and blood glucose levels at the end of the experiment. HOMA-IR was determined by the following formula [21]:

$$\text{HOMA} - \text{IR} = \frac{(\text{plasma insulin level } [\mu U/mL] \times \text{ fasting plasma glucose } [mg/dL])}{405} \tag{1}$$

2.6. Biochemical Parameters

2.6.1. Oxidative Stress Parameters

Commercial kits were used to evaluate different parameters of oxidative status in the tissues of the euthanized animals. Malondialdehyde (MDA), catalase (CAT), and superoxide dismutase (SOD) in liver, pancreas, and kidney homogenates were determined.

2.6.2. Serum Lipid Profile

Total cholesterol (TC), triglycerides (TG), low-density lipoprotein (LDL), and high-density lipoprotein (HDL) levels were measured in the diagnostic kit described previously.

2.7. Histological Analysis

Postmortem, the pancreas was removed, weighed, and placed in a 10% phosphate-buffered formaldehyde solution at a ratio of 1:20. The samples were placed and cut into thin slices of 5 μm. Finally, the sections were stained with hematoxylin & eosin.

3. Statistical Analysis

The software program GraphPad Prism® (version 6, San Diego, CA, USA) was used for statistical analysis. The results are expressed as mean ± SEM. ANOVA analysis of variance (ANOVA) and Bonferroni or Dunnett's test for multiple comparisons. $p \leq 0.05$ was considered statistically significant.

4. Results

4.1. Chemical Characterization

The chemical composition of BFPP confirmed the content of the flavonoid quercetin 3-O-rutinoside (#10). Figure 1 shows the compounds detected, followed by a list of the possible chemical identification using ACD/Labs C NMR Predictor software.

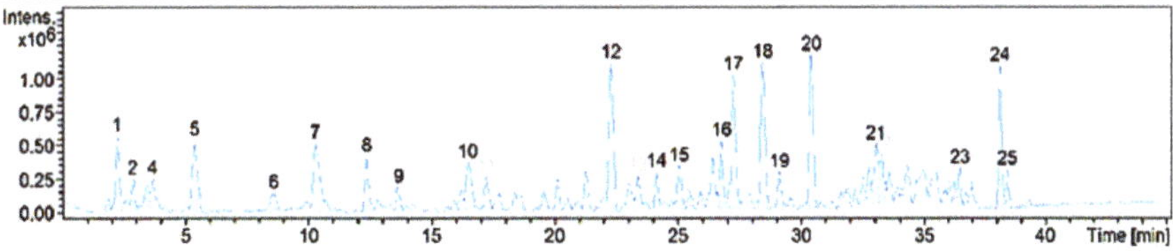

Figure 1. Total ion chromatogram of BFPP in the positive ion mode.

As can be seen in Tables 1–3, several compounds constitute the matrix of the fraction. Several of these compounds have been studied for their protective activities against cellular damage, as well as their regulatory activity on some diabetes disease markers. Phenolic acid derivates of cinnamic acid (caffeic acid and coumaric acid) obtained from red wine were shown to provide a protective effect on human serum and low-density oxidation [22]. The presence of derivates from hydroxycinnamic acids, such as chlorogenic acid, feruloylquinic acid, and neochlorogenic acid, reduced proinflammatory markers and improved the anti-inflammatory response and parameters over blood oxidative stress in the presence of fractions from *Herniaria polygama* [23,24].

Table 1. Compounds identified in BFPP by the dereplication method.

#	Max. *m/z*	RT.	Area	Possible Compound	Possible ID (CAS.)
1	116.0708	2.2	6680138	Amino Acid: L-Proline	147-85-3
2	142.1230	2.9	2810985	Tropane Alkaloids:	
				(1) Tropine	120-29-6
				(2) Physoperuvine	60723-27-5
				(3) Cycloheptanone, 3-(methylamino)	73744-99-7
3	132.1024	3.4	2759332	Amino Acid: L-Isoleucine	61-90-5
4	132.1024	3.7	3675771	Amino Acid: S-Leucine	73-32-5
5	166.0868	5.4	8761530	Amino Acid: Phenylalanine	63-91-2
6	251.1399	8.6	2877906	Cinnamic Acid Derivatives: Cinnamamide, N-(4-aminobutyl)-3,4-dihydroxy-	8CI; 29554-26-5
7	205.0979	10.3	9139461	Amino Acid: L-Tryptophan	73-22-3
8	265.1555	12.4	45393229	Cinnamic Acid Derivates: Cinnamamide, N-(4-aminobutyl)-4-hydroxy-3-methoxy-	7CI,8CI; 501-13-3
9	474.2610	13.7	2022964	N/I	N/I
10	611.1622	16.5	5663280	Flavanoid: Flavone, 2′,3,4′,5,7-pentahydroxy-	8CI; 480-16-0
11	1037.5117	17.2	2652622	Withanolides: Withaperuvin E	92125-38-7
12	1005.5214	22.3	14971186	Withanolides: Withaperuvin M	1353093-20-5
				Phyperunolide A	1198400-48-4
13	233.1023	23.4	3245775	N/I	N/I
14	331.0817	24.1	2913771	Terpene: Dihydroactinidiolide	17092-92-1
15	247.1182	25.1	4087021	N/I	N/I
16	977.5637	26.8	4565672	Withanolides:	
				(1) Withaferin A	5119-48-2
				(2) Withanone	7CI; 27570-38-3
				(3) Withanolide D	30655-48-2
				(4) 27-Hydroxywithanolide B	60124-17-6
17	203.1797	27.2	12844741	Monoterpene: Calamenene	73209-42-4
18	345.0974	28.4	14689013	Retinol	68-26-8
19	316.2853	29.1	3375421	Monoterpene: β-Vetivenene	27840-40-0
20	345.2429	30.4	13726374	N/I	N/I
21	289.2535	33.0	5550203	Biotin	58-85-5
22	289.2533	33.2	3910919	Fatty Acid: Linolenic acid, ethyl ester	1783-84-2
23	273.2581	36.5	2264958	Flavanoid: Flavanone, 4′,5,7-trihydroxy-	8CI; 480-41-1
24	621.3086	38.1	10006030	N/I	N/I
25	621.3088	38.4	2128388	N/I	N/I

N/I: Not identified.

Table 2. Additional flavonoids and phytoprostanes present in BFPP detected by LC-MS analysis in negative mode, and corroborated by previous reports in the literature.

Compounds	Rt	[M-H]⁻ *m/z*	MSⁿ	Hexose (-162)	Aglicone	Author
Quercetin 7-*O*-glucoside 3-*O*-rutinoside	4.17	771 (T)		609	301	[25,26]
Kaempferol 7-*O*-glucoside-3-rutinoside	4.8	755		593		[25,26]
Quercetin 3-*O*-rutinoside	6.8	609 *			301	[10,25–27]
Quercetin-3-*O*-glucoside	6.8	463			301	[25,26]
Kaempferol 3-*O*-rutinoside	7.7	593 *			285	[10,25–27]
9-Flt-PhytoP	5	327.2 (T)	171.1			[25]

T: Traces. * The greater amount with respect to the other compounds.

Table 3. Additional phenolic compounds present in BFPP, detected by LC-MS analysis in negative mode, and corroborated by previous reports in the literature.

Compounds	Rt	[M-H]⁻ m/z	MSⁿ	[AF-H]⁻	[AQ-H]⁻	[AC-H]⁻	[AF-H-18]⁻	[AQ-H-18]⁻	[p.CoA-H]⁻	Author
3-*O*-Caffeoylquinic acid	1.9	179 (T)								[26]
3-*p*-Cumaroylquinic acid	2.6	337						173 (T)	166 *	[26]
3-*O*-Feluroylquinic acid	3.2	367		193 *				173 (T)		[26]
5-*O*-Caffeoylquinic acid	3.3	353			191 *	179 (T)				[25,26]
Ferulic acid hexoside	3.5	355	193 *				175			[26]
Ferulic acid hexoside	3.8	355	193 *				175			[26]
Ferulic acid hexoside	4.1	431	385 205 153 *							[25]

T: Traces. * The greater amount with respect to the other compounds.

4.2. Diabetic Model

4.2.1. Effect of Butanol Fraction from *P. peruviana* on Blood Glucose Levels

The hypoglycemic activity from BFPP was evaluated following BGL over 3 weeks. As shown in Figure 2, the diabetic group not treated (vehicle) reached BGL of up to 400 mg/dL, remaining above that range over time. All the diabetic animals started the experiment with BGL up to 200 mg/dL (fasting BGL from 150 mg/dL were considered diabetic [18]); however, as the graph shows, the group treated with metformin maintained the lowest levels of BGL during treatment, with statistical differences between the vehicle group. Similar behavior from BFPP 100 mg/kg was also observed. BFPP of 50 mg/kg also demonstrated significant differences from the vehicle, reaching BLG of up to 300 mg/kg. Although BGL tended to increase over time when compared to the vehicle group, an anti-hyperglycemic effect was evidenced; this outcome is attributed to content of the flavonoid rutin as the major flavonoid [20]. Previous reports assessed the activity of withanolide, 4-OH-withanolide, and perulactone as anti-diabetic agents present in the ethanolic extract of *P. peruviana*, showing mild action on the insulin tyrosine kinase receptor. Manzano et al., attributed the inhibitory activity on SGLT 1 channels to quercetin 3 O glucoside, quercetin-4-*O*-glucoside. The inhibitory effect of quercetin on GLUT 2 in Caco-2 cells has also been demonstrated [28]. These mechanisms prevent hyperglycemia and support the anti-diabetic treatment target.

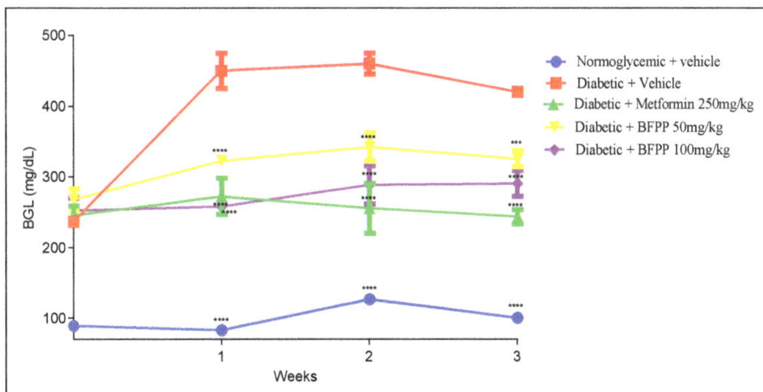

Figure 2. Blood glucose levels (BGL) of animals treated for 21 days. Normoglycemic (blue), vehicle (red), metformin 250 mg/kg (green), BFPP 50 mg/Kg (yellow), BFPP 100 mg/Kg (purple). The data are expressed as mean ± SEM, n = 6 animals per group. Two-way ANOVA post-test Bonferroni; *** $p < 0.001$ and **** $p < 0.0001$ compared with the vehicle group.

Once ingested, most of the rutin flavonoid undergoes a deglycosylation in intestinal tissue mediated by the catalytic effect of α-rhamnosidase and β-glucosidase from intestinal microbiota. The flavonoid quercetin remains after of conjugation with sulfate or glucuronic acid in intestinal villi. It was shown that quercetin-3-O-sulfate (Q3OS) and quercetin-3-O-glucuronide (Q3OG) are the principal molecules found in plasma when rutin is administrated orally. Quercetin molecules can flow in the blood circulation through P-glycoprotein membrane efflux [29]. Previous studies of our research group demonstrated that the matrix from the extract of calyces from *P. peruviana*, improved rutin clearance and volume of distribution. Additionally, the absorption and oral bioavailability of the conjugates was evidenced [30].

As can be seen in Figure 3, weight loss became apparent from the start of treatments until after the last administration of STZ. However, the animals remained overweight when compared with the control group. This finding is consistent with those of other studies, without significant statistical differences between treatments [17]. Only the NO-treated group (vehicle) continued to lose weight, which is consistent with the progress of the disease and with the injury caused by STZ, including damage to the pancreatic islets, massive β-cell damage, sustained inflammation, damage to other organs, and insufficient insulin production resulting in hyperglycemia [18]. Regarding Metformin and BFPP, although they caused weight loss, this weight loss was not progressive, and the mice did not regain weight but stayed at the same weight. Catabolic processes were improved.

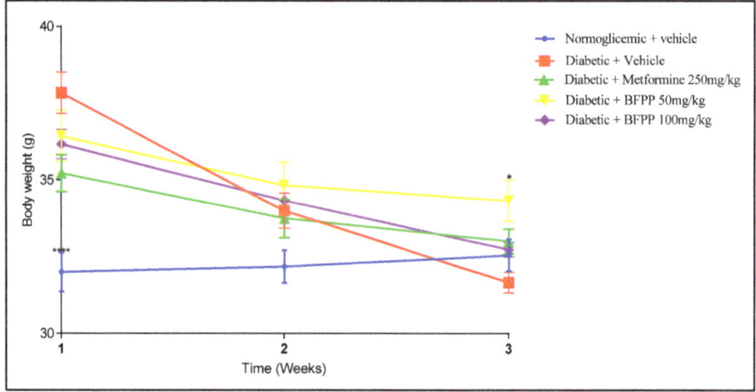

Figure 3. Effects of BFPP on body weight of mice with induced diabetes. Normoglycemic (blue), vehicle (red), metformin 250 mg/kg (green), BFPP 50 mg/kg (yellow), BFPP 100 mg/kg (purple). The data are expressed as mean ± SEM. Two-way ANOVA post-test Bonferroni. * $p \leq 0.05$ and **** $p \leq 0.0001$ compared with the vehicle group.

4.2.2. Oral Glucose Tolerance Test (OGTT)

At the end of the treatments, an OGTT was carried out to determine the capability of the animals to regulate postprandial glucose levels. Figure 4 shows similar levels of BGL between the metformin group and BFPP 100 mg/kg at 60 min, with statistical differences when compared with the vehicle group. A BFPP of 100 mg/kg represented the lowest BGL compared to the other treatment at 60 min.

Several studies support these results in relation to several means of regulation. One such means is the inhibitory activity on different intestinal carbohydrases from the *P. peruviana* extract, as has been reported in previous studies [7,31]. Additionally, effects on α- amylase have been attributed to the presence of O-glycosyl flavonoids, as identified in BPFF [32]. Other mechanisms include muscular glucose uptake and insulin secretagogue, which are attributed to the flavonoid rutin content [33]. Additional investigations cite polyphenols as being responsible for inducing GLP-1 secretion by activation of GPCRs (G-protein coupled receptors) to regulate intracellular signaling and modulate its production [34]. The n-BuOH and aqueous residual fractions of *M. × paradisiaca* leaves were shown to improve glucose homeostasis OGTT in transitory hyperglycemic rats; the authors identified the presence of rutin as its major compound [33].

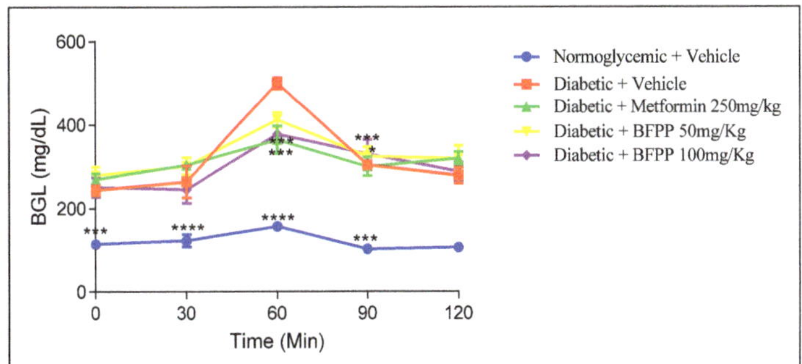

Figure 4. Hypoglycemic activity of BFPP on diabetic mice. Normoglycemic (blue), vehicle (red), metformin 250 mg/kg (green), BFPP 50 mg/Kg (yellow), BFPP 100 mg/Kg (purple). The data are expressed as mean $\pm$ SEM, n = 6 animals per group. Two-way ANOVA post-test Bonferroni; * $p \leq 0.05$; *** $p < 0.001$ and **** $p \leq 0.0001$ compared with the vehicle group.

4.2.3. Homa-IR

Homeostasis model assessment-estimated insulin resistance (HOMA-IR) is a mathematical model that includes interactions between fasting plasma insulin and fasting plasma glucose concentrations. The model has been widely used to estimate insulin resistance in preclinical research, such as epidemiological studies [35,36].

According to the results shown in Table 4, after 21 days of treatment, fasting glucose, insulin, and the HOMA index decreased in mice that received metformin, as was expected, and in agreement with what was previously reported [37]. Furthermore, 50 mg/kg of BFPP decreased fasting glucose by 27.7%, fasting insulin by 57.9%, and HOMA-IR by 70%. Meanwhile, treatment with 100 mg/kg of BFPP reduced fasting glucose by 35.4%, fasting insulin by 58.8%, and HOMA-IR by 73.2%. Notably, the higher dose of the butanol fraction showed more improvement in the three parameters measured than the dose of 50 mg/kg, which may be explained by the higher proportion of flavonoids and phenolic acids present. Other extracts for which the presence of rutin has been reported, and which have shown to improve the HOMA-IR index in a high-fat diet with low-dose streptozotocin-induced type 2 diabetic rats are *Capparis spinosa* fruit extract, also reported as containing kaempferol-3-rutinoside [38], and *Morus alba* ethanol leaf extract, reported as containing chlorogenic acid [39]. It is remarkable that rutin has previously been reported to diminish glycemia by enhancing insulin secretion and stimulating calcium uptake in rat pancreatic islets [36]. The decrease in insulin resistance can be attributed to this flavonoid. Finally, it should be noted that 25 mg/kg or 50 mg/kg of rutin has been shown to decrease the HOMA-IR index in twenty-month-old rats [40]. We can conclude that the decrease in insulin resistance caused by BFPP can be attributed to the presence of rutin.

Table 4. Effect of butanol fraction of *P. peruviana* (BFPP) on fasting glucose, fasting insulin, and HOMA Index.

Treatment	Fasting Glucose mg/dL	Fasting Insulin (µUI/mL)	HOMA-IR
Normoglycemic	113 ± 5 ****	3.2 ± 0.3 ****	1 ± 0.1 ****
Vehicle	463 ± 18	27 ± 3.5	31 ± 4.1
Metformin 250 mg/kg	339 ± 29 ****	18 ± 3.9 ***	15 ± 2.5 ****
BFPP 50 mg/kg	335 ± 37 ****	11 ± 1.3 ****	9 ± 0.4 ****
BFPP 100 mg/kg	299 ± 18 ****	11 ± 1.4 ****	8 ± 1 ****

The data are expressed as mean $\pm$ SD. n = 6 animals per group. One-way ANOVA post-test Dunnet; **** $p \leq 0.0001$ or *** $p \leq 0.001$ compared with the vehicle group.

4.2.4. Effect of BFPP on Lipid Profile

Lipid profile markers showed a protective effect of BFPP at both doses (Figure 5); however, the group which received 100 mg/kg showed a more marked attenuating effect on TG, reaching a lower level than even the group treated with metformin. LDL levels were similar among the same groups. This attenuation in the other groups suggests the beneficial effect of BFPP on lipid profile. HDL showed no differences compared to the vehicle group.

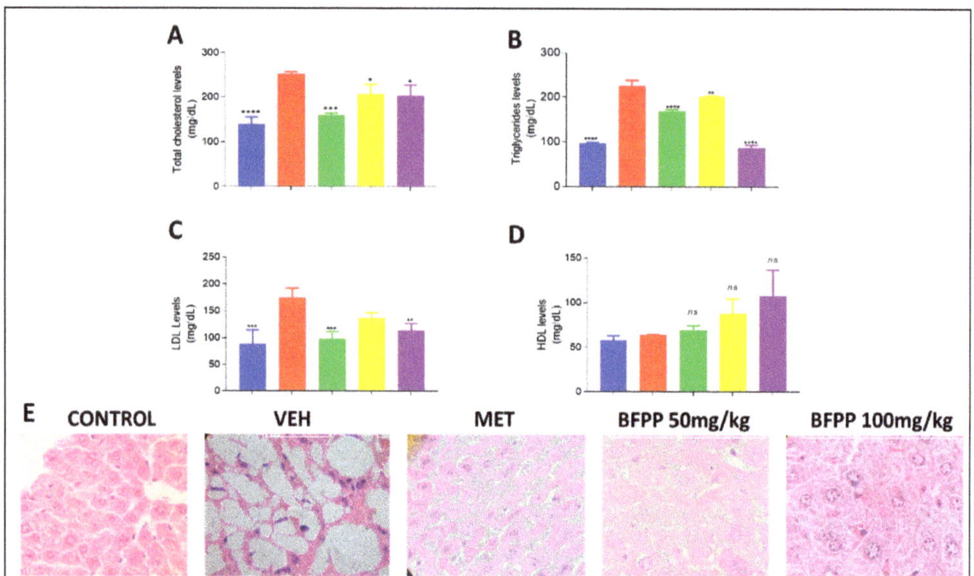

Figure 5. Effect of BFPP on lipid profile parameters. Normoglycemic (blue), vehicle (red), metformin 250 mg/kg (green), BPFF 50 mg/kg (yellow), BFPP 100 mg/kg (purple). The data are expressed as mean ± SEM n = 6 animals per group. One-way ANOVA post-test Dunnet; *ns*: not significant; * $p \leq 0.05$; ** $p \leq 0.01$, *** $p < 0.001$ and **** $p \leq 0.0001$ with respect to the vehicle group. (**A**) Total cholesterol levels; (**B**) triglyceride levels; (**C**) LDL levels; (**D**) HDL levels; (**E**) Liver histology (CONTROL: Normoglycemic group; VEH: diabetic group which received only vehicle; MET: Metformin 250 mg/kg; BFPP 50 mg/kg and BFPP 100 mg/kg.

Lipidemic disorders have been considered responsible for human mortality in the last decade, and studies on the effects of polyphenol compounds have demonstrated beneficial improvements on hyperlipidemic status [41]. For example, previous studies have demonstrated the effects of ferulic acid in decreasing TG and cholesterol levels [42]. Similar findings were reported when studying the antilipidemic effects of zedoary (*Curcuma zedoaria Roscoe*) herbal tea in humans. These effects were related to the phenols contained among its constituents [43]. It also improved the lipid profile in young rats with induced insulin resistance and then supplemented with peel flour of *Passiflora edulis* Var. Flavicarpa. These effects were also attributable to a high amount of flavonoids (quercetin-3-*O*-glucoside) and caffeic acid [44]. A derivate of rutin was demonstrated to attenuate cholesterol levels in rats induced to diabetes by alloxan application [45]. *P. peruviana* pomace also decreased the total cholesterol between 23% and 35% in rats and increased HDL cholesterol levels [46]. Similar results were found when STZ diabetic rats were treated with pomace and juice from *P. peruviana*, corroborating the beneficial effect on lipid profile parameters. These effects were related to the presence of flavonoids and vitamin C [47]. Figure 5E shows the difference between the livers of animals that received the different treatments. The normal group shows normal liver architecture, with clearly-defined hepatic cords, while the vehicle group presents hypertrophic adipocytes, a finding compatible with marked steatosis, which would be expected based on the triglyceride levels achieved. The metformin group shows a small peripheral white area of cytoplasm corresponding to lipid deposits, making this more similar to the BFPP 50 mg/kg group. In fact, BFPP 100 mg/kg showed the greatest preservation of liver structure than the vehicle group, with no evidence of steatosis and a better response with respect to triglyceride levels. In general, BFPP treatments showed a limited presence of binucleated cells as a typical sign of regeneration, demonstrating the protective response in liver tissue. These results are in agreement with previous reports that attribute dyslipidemia attenuation by the flavonoid content in an extract from *Sophora alopecuroides* L. in HFD and STZ-induced diabetic mice [17]. Another study demonstrated the effect of the flavonoid rutin on lipid accumulation in older rats, decreasing liver injury and related complications, such as insulin resistance and chronic systemic inflammation [40].

4.2.5. Effect of BFPP on Oxidative Stress Markers

ROS production is essential for the development of tissue damage in diabetes mellitus [48]. This study evaluated some parameters of oxidative stress in diabetic mice in association with other findings observed during the treatments and post-mortem. ROS are a critical cause of beta cell deterioration and cellular damage produced after STZ treatment, as determined by DNA alkylation and DNA fragmentation. The author hypothesizes as to the diabetogenic capability of streptozotocin such as nitric oxide (NO) speeding up pancreatic cell damage [49].

As can be observed in Figure 6A,B, enzyme activities were maintained in the pancreas, with CAT being active in the group treated with metformin and showing very similar behavior to BFPP 100 mg/kg. These results were statistically significant compared to the vehicle group. SOD activity was greater in the animals that received BFPP 100 mg/kg, and even greater than in the metformin group, suggesting the protective activity of the fraction to this dose. An important protective activity against lipid peroxidation (MDA) is shown in Figure 6C, where a protective response can be observed concerning the negative control group with significant differences compare with untreated group. BFPP significantly inhibited MDA production in a dose-dependent manner. These findings are consistent with previous research that reported the protective effect against lipid peroxidation and increased SOD and CAT levels in the pancreas of STZ-diabetic rats treated with *P. peruviana* juice and pomace [6,47].

As shown in Figure 6D, the livers from all groups had high CAT activity compared with the negative control, with similar levels of enzymatic activity. Figure 6E shows SOD activity in groups treated with BFPP in a dose-dependent manner. The BFPP 100 mg/kg group showed similar behavior to the group treated with metformin. MDA levels were statistically different in all groups compared with the vehicle group. Both doses of BFPP showed protective enzyme activity from the butanol fraction of *P. peruviana* on this organ (Figure 6F). Unlike other organs, the liver is much better equipped to prevent the formation of hydrogen peroxide by CAT endogen levels, compared to the pancreas, for example [49]. Previous findings corroborate the hepatoprotective effect of *P. peruviana* calyces by neutralization of ROS by CAT and SOD enzymes in a liver inflammation model induced by CCl_4 [50]. Marked differences between the residual activity of SOD in the pancreas and kidney, as compared to the liver, can be explained by the lower intrinsic capability of these organs to express this antioxidant enzyme gene, with the pancreas expressing it to a lesser extent [51].

As shown in Figure 6G, CAT in the kidney was active in all the treated groups, but in BFPP, the best outcome was observable at a dose of 100 mg/kg. Figure 6H shows the activity of SOD but without statistical differences in relation to the vehicle. MDA levels in Figure 6G show the protective activity of all treatments compared to the untreated group. The importance of the antioxidant effect on the kidney in a diabetic individual is the risk of gradual structural and functional loss, which was countered in diabetic rats induced by streptozotocin and treated with *P. peruviana* juice [6,47]. Other studies have reported that these results can be associated, in part, with the presence of rutin in BFPP, which corroborates the finding that its flavonoid improves oxidative stress [40].

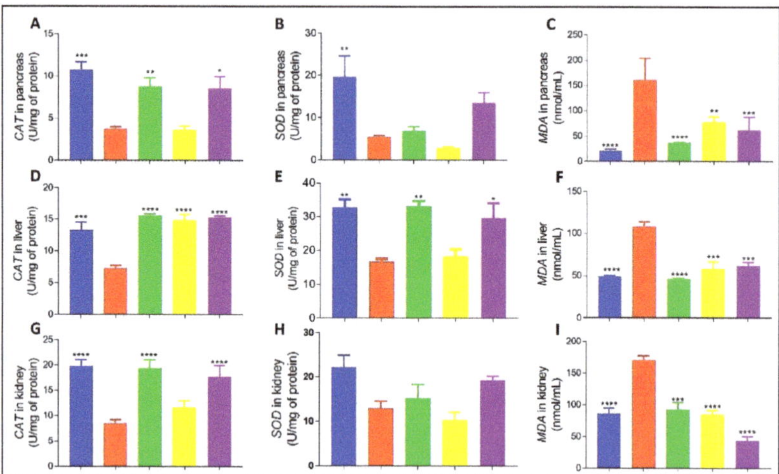

Figure 6. Effect of BFPP on oxidative stress markers. Normoglycemic (blue), vehicle (red), metformin 250 mg/kg (green), BFFF 50 mg/kg (yellow), BFPP 100 mg/kg (purple); (**A**) pancreas CAT; (**B**) pancreas SOD; (**C**) pancreas MDA levels; (**D**) liver CAT; (**E**) liver SOD; (**F**) liver MDA levels; (**G**) kidney CAT; (**H**) kidney SOD; (**I**) kidney MDA levels. The data are expressed as mean $\pm$ SEM, n = 6 animals per group. One-way ANOVA post-test Dunnet; * $p \leq 0.05$; ** $p \leq 0.01$, *** $p < 0.001$ and **** $p \leq 0.0001$ compared with the vehicle group.

4.2.6. Histopathological Analysis

As shown in Figure 7, animals treated with BFPP maintained the presence of granules, especially FBPP 100 mg/kg (Figure 7E). The architecture of the islets was preserved without hypertrophy. At the dosage of 50 mg/kg (Figure 7D), the glandular tissue was protected, but not enough to preserve the entire islet compartment to avoid diffusion of the islets. Comparing the diabetic untreated group (Figure 7A) with the other groups, we saw an evident absence of glandular tissue, which is usually a consequence of the initial tissue inflammation and hypertrophy. The metformin group (Figure 7C) demonstrated the presence of Langerhans islets, without enlargement, and a preserved architecture similar to that of the control group (Figure 7B), which is consistent with the respective responses on BGL. However, BFPP 100 mg/kg showed the best parenchymal and granular protection and BGL attenuation.

As seen in Figure 8A, the control group showed a normal tissue architecture with the presence of glomeruli, capillary loops and Bowman space and capsule. Sections from the vehicle group showed a total absence of the glomeruli and tubular disruptions (Figure 8B). The groups treated with BFPP fraction showed the presence of glomeruli, and at a higher dose (Figure 8E), greater preservation of the architecture. Some hemorrhagic foci were evidenced at the lower dose of BFPP (Figure 8D). The metformin group showed a normal appearance with the presence of glomeruli and preserved tissue architecture. A previous study, which tested ethanolic extract from fruits of *P. peruviana* and its ethyl acetate subfraction, showed the beneficial effects on nephropathy in rats with streptozotocin-induced diabetes. Among the identified compounds, several of them are also present in BFPP, such as kaempferol, quercetin and *O*-caffeoylquinic acid derivates. This mode of action has been associated with apoptosis inhibition and the promotion of autophagy. The authors suggest that *P. peruviana* offers a reno-protective effect on diabetic disorders [52]. To further investigate the effect of BFPP on kidney tissue, further studies are recommended to analyze the markers of kidney disease. Previous reports have mentioned the beneficial and protective effect of the polyphenols contained in an extract of *Hibiscus sab dariffa* calyx, as a coadjuvant in the treatment of type 2 diabetic rats. Caffeoylquinic acid derivates, caffeic acid, feruloyl derivative, quercetin derivative and, kaempferol-3-glucoside were some of the molecules identified [53].

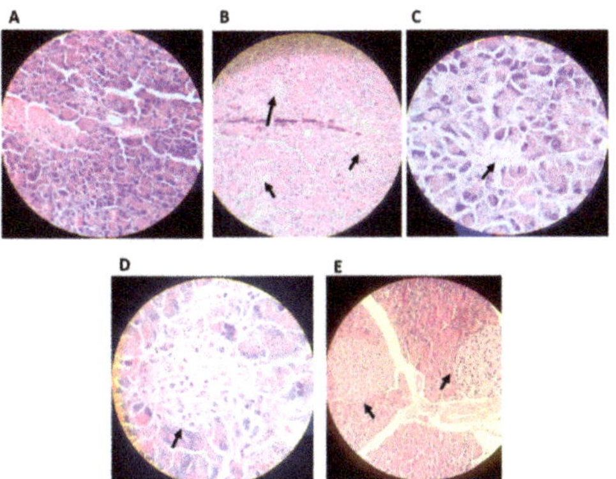

Figure 7. Effect of the butanol fraction from calyces of *P. peruviana* on pancreatic islets at 50 mg/kg and 100 mg/kg. (**A**) Untreated diabetic group (vehicle); (**B**) normal control group; (**C**) diabetic group treated with metformin; (**D**) diabetic group treated with BFPP 50 mg/kg; (**E**) diabetic group treated with BFPP 100 mg/kg. Micrographs were stained with H&E, with a magnification of 40× and 100×. Arrows shows the presence of Langerhans islets.

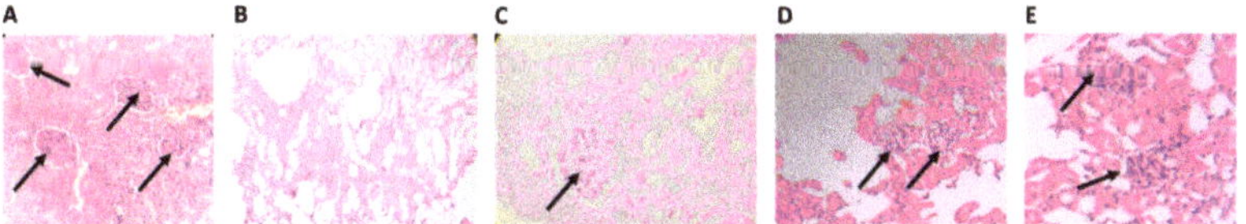

Figure 8. Effect of the butanol fraction from calyces of *P. peruviana* on kidney tissue at 50 mg/kg and 100 mg/kg. (**A**) Untreated diabetic group (vehicle); (**B**) normal control group; (**C**) diabetic group treated with metformin 250 mg/kg; (**D**) diabetic group treated with BFPP 50 mg/kg; (**E**) diabetic group treated with BFPP 100 mg/kg. Micrographs were stained with H&E, with a magnification of 40× and 100×. Arrows show the presence of renal glomeruli.

5. Conclusions

Natural products with antioxidant activities are used as adjuvants to treat several diseases. Several parts of *P. peruviana* have been studied for their antioxidant activity and anti-diabetic effects [54], and they have been used in traditional medicine to treat several diseases related to oxidative stress. In this study, the butanol fraction from *P. peruviana* showed remarkable antioxidant activity in several tissues, as well as an antihyperlipemic effect and glucose regulation. As observed, in addition to the content of the rutin flavonoid (major compound), BFPP is rich in several molecules that can act synergistically, with potentially beneficial effects. Caffeic acid derivates present in BFPP have demonstrated antioxidant activity in other studies, and withanolide derivates (withanona) have been related to the antioxidant potential of *Withania somnifera* [22,55,56]. Some amino acids present in BFPP (alanine, leucine, and isoleucine) have been previously reported for their activity on insulin secretion stimulation. This is preliminary information that may open the way for more in-depth study of the anti-diabetic modes action of the butanol fraction extracted from the calyx. Molecules present in BFPP can explain the protective activities of the previously studied oxidative process.

The results of this study suggest that the butanol fraction from *P. peruviana* is a natural source of anti-diabetic compounds. *P. peruviana* could be considered for use as an adjuvant for the treatment of diabetes mellitus. At the end of the treatment, different lipid and oxidative stress markers and BGL were attenuated in a dose-dependent manner.

Author Contributions: Conceptualization, D.M.A., L.F.O.-G., G.M.C., I.H.V., D.P.R. and S.M.E.; methodology, I.H.V., D.P.R. and S.M.E.; performed experiments, formal analysis, and data curation I.H.V., D.P.R., S.M.E. and I.A.R.; writing—original, project D.M.A. and L.F.O.-G.; funding resources D.M.A., draft preparation I.H.V., S.M.E., D.P.R., F.R.M.B.S. and I.A.R.; supervision and review and editing, D.M.A., L.F.O.-G., F.R.M.B.S. and G.M.C. All authors have read and agreed to the published version of the manuscript.

Funding: This project was funded by Universidad Nacional de Colombia, Grant 41660.

Institutional Review Board Statement: The bioassays were carried out following Guide for the Care and Use of Laboratory Animals (1996, published by National Academy Press, 2101 Constitution Ave. NW, Washington, DC 20055, USA) and approved by the Ethics Committee of Universidad Nacional de Colombia (Act 08 of 16 April 2020, Faculty of Science).

Informed Consent Statement: Not applicable.

Acknowledgments: The authors I.H.V., D.P.R. and S.M.E. would like to thank COLCIENCIAS-Convocatoria 647/2014 and MADS Access contract 249 of 2018.

Conflicts of Interest: The authors declare no conflict of interest.

References

1. Fischer, G.; Melgarejo, L.U.Z.M. The ecophysiology of cape gooseberry (*Physalis peruviana* L.)—An Andean fruit crop. A review. La ecofisiología de uchuva (*Physalis peruviana* L.)—Un frutal andino. Una revisión. *Rev. Colomb. Cienc. Hortíc.* **2020**, *14*, 76–89.
2. Granados Pérez, W.; Muñoz Vanegas, C.A.; Aguillón Mayorga, D.M. Cadena de la uchuva. Dir Cadenas Agrícolas y For [Internet]. 2019, pp. 1–18. Available online: https://bit.ly/3Ik0gE2 (accessed on 1 December 2022).
3. Saltveit, M.E. *Postharvest Biology and Technology of Tropical and Subtropical Fruits*; Woodhead Publishing: Sawston, UK, 2011. [CrossRef]
4. Fischer, G.; Almanza-Merchán, P.J.; Miranda, D. Importancia y cultivo de la uchuva (*Physalis peruviana* L.). *Rev. Bras. Frutic.* **2014**, *36*, 1–15. [CrossRef]
5. Kasali, F.M.; Tusiimire, J.; Kadima, J.N.; Tolo, C.U.; Weisheit, A.; Agaba, A.G. Ethnotherapeutic Uses and Phytochemical Composition of *Physalis peruviana* L.: An Overview. *Sci. World J.* **2021**, *2021*, 5212348. [CrossRef] [PubMed]
6. Mora, Á.C.; Aragón, D.M.; Ospina, L.F. Effects of *Physalis peruviana* fruit extract on stress oxidative parameters in streptozotocin-diabetic rats. *Lat. Am. J. Pharm.* **2010**, *29*, 1132–1136.
7. Rey, D.P.; Ospina, L.F.; Aragón, D.M. Inhibitory effects of an extract of fruits of *Physalis peruviana* on some intestinal carbohydrases. *Rev. Colomb. Cienc. Quim.-Farm.* **2015**, *44*, 72–89. [CrossRef]
8. Wu, S.J.; Tsai, J.Y.; Chang, S.P.; Lin, D.L.; Wang, S.S.; Huang, S.N.; Ng, L.T. Supercritical carbon dioxide extract exhibits enhanced antioxidant and anti-inflammatory activities of *Physalis peruviana*. *J. Ethnopharmacol.* **2006**, *108*, 407–413. [CrossRef]
9. Franco, L.A.; Ocampo, Y.C.; Gómez, H.A.; de la Puerta, R.; Espartero, J.L.; Ospina, L.F. Sucrose esters from *Physalis peruviana* calyces with anti-inflammatory activity. *Planta Med.* **2014**, *80*, 1605–1614. [CrossRef]
10. Toro, R.M.; Aragón, D.M.; Ospina, L.F.; Ramos, F.A.; Castellanos, L. Phytochemical analysis, antioxidant and anti-inflammatory activity of calyces from *Physalis peruviana*. *Nat. Prod. Commun.* **2014**, *9*, 1573–1575. [CrossRef] [PubMed]
11. Toro, R.; Arangon, M. Propuesta de un Marcador Analítico Como Herramienta en la Microencapsulación de un Extracto con Actividad Antioxidante de Cálices de *Physalis peruviana*. Master's Thesis, Universidad Nacional de Colombia, Bogota, Colombia, 2014.
12. Gutierres, J.M.; Lencina, C.L.; Carvalho, F.B.; Soares, M.S.P.; Bona, N.P.; Stefanello, F.M.; Vieira, A.; Chaves, V.C.; Spanevello, R.M.; Reginatto, F.H.; et al. Southern Brazilian native fruit shows neurochemical, metabolic and behavioral benefits in an animal model of metabolic syndrome. *Metab. Brain Dis.* **2018**, *33*, 1551–1562. [CrossRef]
13. Rendon, L.M.; Zuluaga, A.F.; Rodríguez, C.A.; Agudelo, M.; Vesga, O.; Ospina, L.F. Obtaining a Murine Model of Streptozotocin-Induced Diabetes Useful in the Pharmacodynamic Evaluation of Regular Insulin. *Biomedica* **2017**, *2*, 11–18. [CrossRef]
14. Cardona, M.I.; Toro, R.M.; Costa, G.M.; Ospina, L.F.; Castellanos, L.; Ramos, F.A.; Aragón, D.M. Influence of extraction process on antioxidant activity and rutin content in *Physalis peruviana* calyces extract. *J. Appl. Pharm. Sci.* **2017**, *7*, 164–168. [CrossRef]
15. Costa, G.M. Vitexin Derivatives as Chemical Markers in the Differentiation of the Closely Related Species *Passiflora quadrangularis* Linn. *J. Liq. Chromatogr.* **2013**, *36*, 37–41.
16. Miranda, C.A.; Schönholzer, T.E.; Klöppel, E.; Sinzato, Y.K.; Volpato, G.T.; Damasceno, D.C.; Campos, E.K. Repercussions of low fructose-drinking water in male rats. *An. Acad. Bras. Cienc.* **2019**, *91*, e20170705. [CrossRef] [PubMed]

17. Lv, Y.; Zhao, P.; Pang, K.; Ma, Y.; Huang, H.; Zhou, T.; Yang, X. Antidiabetic effect of a flavonoid-rich extract from Sophora alopecuroides L. in HFD- and STZ-induced diabetic mice through PKC/GLUT4 pathway and regulating PPARα and PPARγ expression. *J. Ethnopharmacol.* **2021**, *268*, 113654. [CrossRef]
18. Furman, B.L. Streptozotocin-Induced Diabetic Models in Mice and Rats. *Curr. Protoc. Pharmacol.* **2015**, *70*, 1–20. [CrossRef]
19. Janssen, B.J.A.; de Celle, T.; Debets, J.J.M.; Brouns, A.E.; Callahan, M.F.; Smith, T.L. Effects of anesthetics on systemic hemodynamics in mice. *Am. J. Physiol.—Heart Circ. Physiol.* **2004**, *287*, 1618–1625. [CrossRef]
20. Aragón, D.M.; Echeverry, S.M.; Valderrama, I.H.; Costa, G.M.; Ospina, L.F. Development and optimization of microparticles containing a hypoglycemic fraction of calyces from *Physalis peruviana*. *J. Appl. Pharm. Sci.* **2018**, *8*, 10–18. [CrossRef]
21. Abdelhameed, R.F.A.; Ibrahim, A.K.; Elfaky, M.A.; Habib, E.S.; Mahamed, M.I.; Mehanna, E.T.; Darwish, K.M.; Khodeer, D.M.; Ahmed, S.A.; Elhady, S.S. Antioxidant and anti-inflammatory activity of *Cynanchum acutum* L. isolated flavonoids using experimentally induced type 2 diabetes mellitus: Biological and in silico investigation for nf-κb pathway/mir-146a expression modulation. *Antioxidants* **2021**, *10*, 1713. [CrossRef]
22. Abu-Amsha, R.; Croft, K.D.; Puddey, I.B.; Proudfoot, J.M.; Beilin, L.J. Phenolic content of various beverages determines the extent of inhibition of human serum and low-density lipoprotein oxidation in vitro: Identification and mechanism of action of some cinnamic acid derivatives from red wine. *Clin. Sci.* **1996**, *91*, 449–458. [CrossRef]
23. Ou, L.; Kong, L.Y.; Zhang, X.M.; Niwa, M. Oxidation of ferulic acid by *Momordica charantia* peroxidase and related anti-inflammation activity changes. *Biol. Pharm. Bull.* **2003**, *26*, 1511–1516. [CrossRef]
24. Kolodziejczyk-Czepas, J.; Kozachok, S.; Pecio, Ł.; Marchyshyn, S.; Oleszek, W. Determination of phenolic profiles of *Herniaria polygama* and *Herniaria incana* fractions and their in vitro antioxidant and anti-inflammatory effects. *Phytochemistry* **2021**, *190*, 112861. [CrossRef]
25. Gironés-Vilaplana, A.; Baenas, N.; Villaño, D.; Speisky, H.; García-Viguera, C.; Moreno, D.A. Evaluation of Latin-American fruits rich in phytochemicals with biological effects. *J. Funct. Foods* **2014**, *7*, 599–608. [CrossRef]
26. Medina, S.; Collado-gonzález, J.; Londoño-londoño, J.; Gil-izquierdo, A. Potential of *Physalis peruviana* calyces as a low-cost valuable resource of phytoprostanes and phenolic compounds. *J. Sci. Food Agric.* **2019**, *99*, 2194–2204. [CrossRef]
27. Franco, L.A.O.L.; Matiz, G.E.; Calle, J.; Pinzón, R. Actividad antinflamatoria de extractos y fracciones obtenidas de cálices de *Physalis peruviana* L. Luis. *Biomedica* **2007**, *27*, 110–115. [CrossRef] [PubMed]
28. Manzano, S.; Williamson, G. Polyphenols and phenolic acids from strawberry and apple decrease glucose uptake and transport by human intestinal Caco-2 cells. *Mol. Nutr. Food Res.* **2010**, *54*, 1773–1780. [CrossRef]
29. Liu, Z.; Hu, M. Natural polyphenol disposition via coupled metabolic pathways. *Expert Opin. Drug Metab. Toxicol.* **2007**, *3*, 389–406. [CrossRef]
30. Moré, G.P.D.; Cardona, M.I.; Sepúlveda, P.M.; Echeverry, S.M.; Simões, C.M.O.; Aragón, D.M. Matrix effects of the hydroethanolic extract of calyces of *Physalis peruviana* L. On rutin pharmacokinetics in wistar rats using population modeling. *Pharmaceutics* **2021**, *13*, 535. [CrossRef] [PubMed]
31. Monzón Daza, G.; Meneses Macías, C.; Forero, A.M.; Rodríguez, J.; Aragón, M.; Jiménez, C.; Ramos, F.A.; Castellanos, L. Identification of α-Amylase and α-Glucosidase Inhibitors and Ligularoside A, a New Triterpenoid Saponin from *Passiflora ligularis* Juss (Sweet Granadilla) Leaves, by a Nuclear Magnetic Resonance-Based Metabolomic Study. *J. Agric. Food Chem.* **2021**, *69*, 2919–2931. [CrossRef]
32. Novoa, D.M.A.; Giraldo, L.F.O.; Rodríguez, F.A.R.; Hernández, L.C.; Modesti, G.C.; Silva, F.R.M.B. *Passiflora ligularis Juss. (Granadilla): Farmacológicos de Una Estudios Químicos y Planta con Potencial Terapéutico*; Coordinaci, Universidad Nacional de Colombia: Bogotá, Colombia, 2021. [CrossRef]
33. Kappel, V.D.; Cazarolli, L.H.; Pereira, D.F.; Postal, B.G.; Madoglio, F.A.; Buss, Z.d.; Reginatto, F.H.; Silva, F.R.M.B. Beneficial effects of banana leaves (Musa x paradisiaca) on glucose homeostasis: Multiple sites of action. *Braz. J. Pharmacogn.* **2013**, *23*, 706–715. [CrossRef]
34. Wang, Y.; Alkhalidy, H.; Liu, D. The emerging role of polyphenols in the management of type 2 diabetes. *Molecules* **2021**, *26*, 703. [CrossRef] [PubMed]
35. Qu, H.-Q.; Li, Q.; Rentfro, A.R.; Fisher-Hoch, S.P.; Mccormick, J.B. The Definition of Insulin Resistance Using HOMA-IR for Americans of Mexican Descent Using Machine Learning. *PLoS ONE* **2011**, *6*, e21041. [CrossRef]
36. Antunes, L.C.; Elkfury, J.L.; Jornada, M.N.; Foletto, K.C.; Bertoluci, M.C. Validation of HOMA-IR in a model of insulin-resistance induced by a high-fat diet in Wistar rats. *Arch. Endocrinol. Metab.* **2016**, *60*, 138–142. [CrossRef]
37. Mather, K.J.; Verma, S.; Anderson, T.J. Improved endothelial function with metformin in type 2 diabetes mellitus. *J. Am. Coll. Cardiol.* **2001**, *37*, 1344–1350. [CrossRef]
38. Assadi, S.; Shafiee, S.M.; Erfani, M.; Akmali, M. Antioxidative and antidiabetic effects of *Capparis spinosa* fruit extract on high-fat diet and low-dose streptozotocin-induced type 2 diabetic rats. *Biomed. Pharmacother.* **2021**, *138*, 111391. [CrossRef]
39. Ji, S.; Zhu, C.; Gao, S.; Shao, X.; Chen, X.; Zhang, H.; Tang, D. *Morus alba* leaves ethanol extract protects pancreatic islet cells against dysfunction and death by inducing autophagy in type 2 diabetes. *Phytomedicine* **2021**, *83*, 153478. [CrossRef]
40. Li, T.; Chen, S.; Feng, T.; Dong, J.; Li, Y.; Li, H. Rutin protects against aging-related metabolic dysfunction. *Food Funct.* **2016**, *7*, 1147–1154. [CrossRef]
41. Imran, A.; Butt, M.S.; Arshad, M.S.; Arshad, M.U.; Saeed, F.; Sohaib, M.; Munir, R. Exploring the potential of black tea based flavonoids against hyperlipidemia related disorders. *Lipids Health Dis.* **2018**, *17*, 57. [CrossRef] [PubMed]

42. De Paiva, L.B.; Goldbeck, R.; Santos, W.D.d.; Squina, F.M. Ferulic acid and derivatives: Molecules with potential application in the pharmaceutical field. *Braz. J. Pharm. Sci.* **2013**, *49*, 395–411. [CrossRef]
43. Tariq, S.; Imran, M.; Mushtaq, Z.; Asghar, N. Phytopreventive antihypercholesterolmic and antilipidemic perspectives of zedoary (*Curcuma zedoaria* Roscoe.) herbal tea. *Lipids Health Dis.* **2016**, *15*, 39. [CrossRef]
44. Goss, M.J.; Nunes, M.L.O.; Machado, I.D.; Merlin, L.; Macedo, N.B.; Silva, A.M.O.; Bresolin, T.M.B.; Santin, J.R. Peel flour of *Passiflora edulis* Var. Flavicarpa supplementation prevents the insulin resistance and hepatic steatosis induced by low-fructose-diet in young rats. *Biomed. Pharmacother.* **2018**, *102*, 848–854. [CrossRef]
45. Albu, E.; Lupaşcu, D.; Filip, C.; Jaba, I.M.; Zamosteanu, N. The influence of a new rutin derivative on homocysteine, cholesterol and total antioxidative status in experimental diabetes in rat. *Farmacia* **2013**, *61*, 1167–1177. Available online: https://pdfs.semanticscholar.org/2e5c/6e7f3fde85fcc7591dc13e4f1ab8604e731a.pdf (accessed on 12 March 2018).
46. Ramadan, M.F. *Physalis peruviana* pomace suppresses high-cholesterol diet-induced hypercholesterolemia in rats. *Grasas Aceites* **2012**, *63*, 411–422. [CrossRef]
47. Darwish, A.G.; Mahmoud, H.I.; Refaat, I. Antioxidative and Antidiabetic Effect of Goldenberries juice and pomace on Experimental Rats Induced with streptozotocin In Vitro. *J. Food Dairy Sci.* **2020**, *11*, 277–283. [CrossRef]
48. Mansuroğlu, B.; Derman, S.; Yaba, A.; Kizilbey, K. Protective effect of chemically modified SOD on lipid peroxidation and antioxidant status in diabetic rats. *Int. J. Biol. Macromol.* **2015**, *72*, 79–87. [CrossRef]
49. Lenzen, S. The mechanisms of alloxan- and streptozotocin-induced diabetes. *Diabetologia* **2008**, *51*, 216–226. [CrossRef]
50. Toro, A.R.M.; Aragón, N.D.M.; Ospina, G.L.F. Hepatoprotective effect of calyces extract of *Physalis peruviana* on hepatotoxicity induced by CCl4 in wistar rats. *Vitae* **2013**, *20*, 125–132.
51. Lenzen, S.; Drinkgern, J.; Tiedge, M. Low antioxidant enzyme gene expression in pancreatic islets compared with various other mouse tissues. *Free Radic. Biol. Med.* **1996**, *20*, 463–466. [CrossRef]
52. Ezzat, S.M.; Abdallah, H.M.I.; Yassen, N.N.; Radwan, R.A.; Mostafa, E.S.; Salama, M.M.; Salem, M.A. Phenolics from *Physalis peruviana* fruits ameliorate streptozotocin-induced diabetes and diabetic nephropathy in rats via induction of autophagy and apoptosis regression. *Biomed. Pharmacother.* **2021**, *142*, 111948. [CrossRef]
53. Yang, Y.S.; Huang, C.N.; Wang, C.J.; Lee, Y.J.; Chen, M.L.; Peng, C.H. Polyphenols of *Hibiscus sabdariffa* improved diabetic nephropathy via regulating the pathogenic markers and kidney functions of type 2 diabetic rats. *J. Funct. Foods* **2013**, *5*, 810–819. [CrossRef]
54. Lan, Y.H.; Chang, F.R.; Pan, M.J.; Wu, C.C.; Wu, S.J.; Chen, S.L.; Wang, S.S.; Wu, M.J.; Wu, Y.C. New cytotoxic withanolides from *Physalis peruviana*. *Food Chem.* **2009**, *116*, 462–469. [CrossRef]
55. Devkar, S.T.; Muthal, A.P.; Patil, P.V.; Mukherjee-Kandhare, A.A.; Kandhare, A.D.; Jagtap, S.D.; Bodhankar, S.L.; Hegde, M.V. Evaluation of the physicochemical stability and biological activity of withanolide rich fraction from withania somnifera root by hptlc and cyclic voltammetry: A simple, reliable, and cost-effective approach. *Lat. Am. J. Pharm.* **2021**, *40*, 946–956.
56. Bharti, S.K.; Krishnan, S.; Kumar, A.; Kumar, A. Antidiabetic phytoconstituents and their mode of action on metabolic pathways. *Ther. Adv. Endocrinol. Metab.* **2018**, *9*, 81–100. [CrossRef]

 *pharmaceutics*

Article

In Vivo and In Vitro Antidiabetic Efficacy of Aqueous and Methanolic Extracts of Orthosiphon Stamineus Benth

Najlaa Bassalat [1,2], Sleman Kadan [3], Sarit Melamed [4], Tamar Yaron [5], Zipora Tietel [4], Dina Karam [1], Asmaa Kmail [1], Mahmud Masalha [3] and Hilal Zaid [2,3,*]

1   Faculty of Sciences, Arab American University, Jenin P.O. Box 240, Palestine
2   Faculty of Medicine, Arab American University, Jenin P.O. Box 240, Palestine
3   Qasemi Research Center, Al-Qasemi Academic College, P.O. Box 124, Baqa El-Gharbia 3010000, Israel
4   Department of Food Science, Gilat Research Center, Agricultural Research Organization, M.P. Negev, Gilat 8531100, Israel
5   Faculty of Science, Beit Berl College, Kfar Saba 4490500, Israel
*   Correspondence: hilalz@qsm.ac.il

Abstract: *Orthosiphon stamineus* is a popular folk herb used to treat diabetes and some other disorders. Previous studies have shown that *O. stamineus* extracts were able to balance blood glucose levels in diabetic rat animal models. However, the antidiabetic mechanism of *O. stamineus* is not fully known. This study was carried out to test the chemical composition, cytotoxicity, and antidiabetic activity of *O. stamineus* (aerial) methanol and water extracts. GC/MS phytochemical analysis of *O. stamineus* methanol and water extracts revealed 52 and 41 compounds, respectively. Ten active compounds are strong antidiabetic candidates. Oral treatment of diabetic mice with *O. stamineus* extracts for 3 weeks resulted significant reductions in blood glucose levels from 359 ± 7 mg/dL in diabetic non-treated mice to 164 ± 2 mg/dL and 174 ± 3 mg/dL in water- and methanol-based-extract-treated mice, respectively. The efficacy of *O. stamineus* extracts in augmenting glucose transporter-4 (GLUT4) translocation to the plasma membrane (PM) was tested in a rat muscle cell line stably expressing myc-tagged GLUT4 (L6-GLUT4myc) using enzyme-linked immunosorbent assay. The methanol extract was more efficient in enhancing GLUT4 translocation to the PM. It increased GLUT4 translocation at 250 µg/mL to $279 \pm 15\%$ and $351 \pm 20\%$ in the absence and presence of insulin, respectively. The same concentration of water extract enhanced GLUT4 translocation to $142 \pm 2.5\%$ and $165 \pm 5\%$ in the absence and presence of insulin, respectively. The methanol and water extracts were safe up to 250 µg/mL as measured with a Methylthiazol Tetrazolium (MTT) cytotoxic assay. The extracts exhibited antioxidant activity as measured by 2,2-diphenyl-1-picrylhydrazyl (DPPH) assay. *O. stamineus* methanol extract reached the maximal inhibition of $77 \pm 10\%$ at 500 µg/mL, and *O. stamineus* water extract led to $59 \pm 3\%$ inhibition at the same concentration. These findings indicate that *O. stamineus* possesses antidiabetic activity in part by scavenging the oxidants and enhancing GLUT4 translocation to the PM in skeletal muscle.

Keywords: *Orthosiphon stamineus*; GLUT4; anti-oxidant; diabetes mellitus; phytochemicals

Citation: Bassalat, N.; Kadan, S.; Melamed, S.; Yaron, T.; Tietel, Z.; Karam, D.; Kmail, A.; Masalha, M.; Zaid, H. In Vivo and In Vitro Antidiabetic Efficacy of Aqueous and Methanolic Extracts of Orthosiphon Stamineus Benth. *Pharmaceutics* **2023**, *15*, 945. https://doi.org/10.3390/pharmaceutics15030945

Academic Editors: Diana Marcela Aragon Novoa and Fátima Regina Mena Barreto Silva

Received: 29 November 2022
Revised: 14 February 2023
Accepted: 7 March 2023
Published: 14 March 2023

1. Introduction

Type 2 diabetes (T2DM) is considered a global health issue threatening the life of 537 million people worldwide in 2021. Diabetic cases are estimated to rise up to 784 million by 2045 [1]. T2DM is responsible for a 5% increase in premature mortality due to its complications, which start with hyperglycemia and proceed to a combination of resistance to insulin action, insufficient insulin secretion, and excessive glucagon breakdown and secretion [2]. Diabetes is also considered a major cause of blindness, kidney failure, heart attacks, stroke, and lower limb amputation. T2DM is a multifactorial disorder that can be triggered by aging and genetic factors, environmental factors, and factors related to the patient's lifestyle, such as diet, physical activity, and obesity [3].

Increased glucose levels in bloodstream stimulate secretion of insulin from the islets of Langerhans in the pancreas. Insulin is the responsible hormone for regulation of circulation glucose levels by increasing glucose transport into adipose and muscle tissues and by suppressing the hepatic glucose production. Binding of the insulin to its receptors on cell surfaces induces the translocation of glucose transporter-4 (GLUT4) from intracellular vessels to plasma membranes, which results in diffusion of glucose in muscle, hepatocytes and adipocytes [4]. Insulin binds to the β-subunit of the insulin receptor (IR), leading to autophosphorylation and then recruitment of the insulin receptor substrate-1 (IRS-1), which in turn activates phosphatidylinositol 3-kinase (PI3K). PI3-K phosphorylates phosphatidylinositol-4, 5-bisphosphate (PIP2) to yield phosphatidylinositol-3, 4, 5-triphosphate (PIP3). PIP3 activates Akt (protein kinase B), which then phosphorylates many substrates, including Akt substrate of 160 kDa (AS160), leading to its inhibition and thus augmenting GLUT4-containing vesicle translocation and fusion with the plasma membrane (PM) [5]. GLUT4 translocation to the PM is also triggered by AMP-activated protein kinase (AMPK) under certain conditions such as muscle contraction, increased cellular [AMP]/[ATP] ratio and deprivation of glucose or oxygen. AMPK activation in skeletal muscle promotes GLUT4 trafficking to the PM and enhanced glucose uptake in the insulin independent pathway [4,5].

Medicinal plants are traditionally used in folk medicine as natural healing remedies with therapeutic effects for many diseases, including diabetes. Antidiabetic herbs balance blood glucose and delay the progression of diabetic complications. More than 800 plant species worldwide are reported as potentially antidiabetic herbs [6].

Orthosiphon stamineus Benth. (Lamiaceae) is a perennial herb found in tropical and subtropical regions [7]. *O. stamineus* is used in folk medicine as an antidiabetic and diuretic and for treating abdominal pain, kidney and bladder inflammation, edema, and gout [8]. Pharmacological studies have shown antimicrobial, antioxidant, hepatoprotective, antigenotoxic, antiplasmodial, cytotoxic, cardioactive, anti-inflammatory and antidiabetic activities [8,9]. Reports have shown that *O. stamineus* contains a variety of groups of phytochemicals such as flavones and other polyphenols, bioactive proteins, glycosides, volatile oils, as well as large quantities of potassium [9].

One animal study has shown that the water extract of *O. stamineus* has hypoglycemic effects on diabetic rats [10]. In a recent study, 80% ethanol extract (about 0.4 g/kg) reduced blood glucose levels in an oral glucose tolerance test in normal C57BL/6J mice and high-fat-diet (HFD) C57BL/6 mice after 1.5 [11] and 8 weeks [12] administration of the extract, respectively. Water extract of *O. stamineus* was also administrated orally (0.5 g/kg) to normal and diabetic rats loaded with glucose. In normal rats, the aqueous extract reduced plasma glucose after 1 h of glucose loading by 15% and 21% in diabetic rats, respectively [13].

Yet the antidiabetic mechanism of action of these plant extracts is not known. According to a recent systematic review [7], no study was published on the effect of *O. stamineus* extracts on GLUT4 translocation and activity. This study thus aims to examine the effect of *O. stamineus* extracts on GLUT4 translocation to the PM in L6 skeletal muscle cell line, its antioxidant scavenging activity, and its chemical composition

2. Materials and Methods

2.1. Materials

α-MEM (modified Eagle's medium), fetal bovine serum, and all other tissue culture reagents were purchased from biological industries (Beit Haemek, Israel). Horseradish-peroxidase (HRP-) -conjugated goat anti-rabbit antibodies were obtained from Promega (Madison, WI, USA). Polyclonal anti-myc (A-14) and other standard chemicals were purchased from Sigma-Aldrich (Saint Louis, MO, USA).

2.2. Plant Extract Preparation

The aerial parts of *Orthosiphon stamineus* Benth were purchased from a medicinal plants trader in Nablus, Palestine. The plant was identified by Prof. Nidal Jaradat (An-Najah National University, Nablus, Palestine). Two extracts were prepared: in water and in methanol. Forty grams of air-dried aerial parts of *O. Stamineus* was powdered and packed in an Erlenmeyer flask with 200 mL solvent. The flasks were then sonicated for 2 h at 60 °C (Elmasonic, Singen, Germany) and left in dark glass bottles for 24 h for complete extraction. The yield of the extract in methanol and water was 10.5% and 12.7%, respectively. The stock extracts were kept at -20 °C in air-tight glass containers.

2.3. Silylation Derivatization

One mL of the water and methanol extract was transferred to a glass vial, and the solvents were evaporated under a gentle stream of nitrogen at ambient temperature, and 150 µL of N,O-Bis (trimethylsilyl) trifluoroacetamide (BSTFA) containing 1% trimethylchlorosilane reagent used for GC silylation derivatization (>99%, Sigma-Aldrich) was added to each dry *O. Stamineus* crude extract followed by heating up to 70 °C for 20 min [14].

2.4. Gas Chromatography—Mass Spectrometry Analysis and Compounds Identification

One µL of each silylated sample was injected into the gas chromatograph (GC) coupled with mass spectrometer detector (MS) as previously described by our group [15]. Component relative percentages of the samples were calculated from the GC peak areas. NIST GC/MS Library and mass spectra from the literature were used to annotate the compounds.

2.5. Cell Growth and Treatment

Cells from the rat L6 muscle cell line stably expressing myc-tagged GLUT4 (L6-GLUT4myc) were purchased from (Kerafast, Boston, MA, USA). The cells were grown at 37 °C, 95% air, and 5% CO_2 in a-MEM supplemented with 10% fetal calf serum (FCS), 100 U/mL penicillin, and 100 µg/mL streptomycin.

2.6. MTT Cytotoxic Assay

The Methylthiazol Tetrazolium (MTT) assay was used to detect the viability of the cells as described in [16]. This assay relies on the colorimetric change from yellow to purple, indicating that the cells are active. On the first day, cells with a density of 2×10^4 were seeded in 96 well-plates, each well containing 100 µL of medium. The cells were incubated for 24 h in an incubator (37 °C and 5% CO_2). On the next day, 100 µL of the *O. Stamineus* extracts (water and methanol) were added to each well at increasing concentrations up to 1 mg/mL and incubated for 20 h. The old medium was removed and a fresh medium containing 0.5 mg/mL MTT was added into each well and incubated for 4 h. Medium containing MTT was removed and 100 µL of isopropanol/HCl (1 mM HCl in 100% isopropanol) were added to each well. Absorbance at 570 nm was measured using a plate reader. Experiments were repeated three times, with six replicates each time.

The following formula was used to detect the viability of the cells:

Percent viability = (A570 nm of plant extract treated cells/A570 of untreated cells) $\times$ 100%

2.7. GLUT4 Translocation

Surface myc-tagged GLUT4 was measured in intact cells as described previously [14]. Briefly, L6-GLUT4 myc cells were seeded in 24 well-plates and incubated for 24 h. *O. Stamineus* extracts (water and methanol) were added to the cells for 20 h and followed by serum starvation for 3 h and treated with or without 1 µM insulin for 20 min. The cells were washed twice with ice-cold PBS, fixed with 3% paraformaldehyde for 15 min, then blocked with 3% (*v/v*) goat serum for 10 min, incubated with polyclonal anti-myc antibody (1:200) for 1 h at 4 °C, washed ten times with PBS and incubated with goat-anti-rabbit secondary antibody conjugated with horseradish peroxidase (1:1000) for 1 h at 4 °C, then washed ten times with PBS at room temperature. One milliliter of o-phenylenediamine dihydrochloride reagent was added to each well and incubated in the dark at room temperature for 20–30 min and the reaction was stopped by adding 0.5 mL of 3 M HCl. The absorbance was measured by using a spectrophotometer at 492 nm. Background absorbance obtained from 3 wells in each 24-well plate untreated with anti-myc antibody was subtracted from all values.

2.8. DPPH Scavenging Activity

2,2-diphenyl-1-picrylhydrazyl (DPPH) free radical scavenging ability of the extracts was tested as described by Blois [17] with slight modifications. The hydrogen atom donating ability of the plant extractives was determined by the decolorization of methanol solution of DPPH which produces violet/purple color in methanol solution and fades to yellow in the presence of antioxidants.

DPPH (0.002% w/v) dissolved in methanol was mixed with *O. stamineus* extract or Trolox as a positive control at a 1:1:1 ratio. A negative control solution was prepared by mixing the mentioned DPPH solution with methanol in a 1:1 ratio. The mixtures were incubated at room temperature in the darkness for 30 min. The color density was determined by a Spectrophotometer at 517 nm (Jenway 72000, Cole-Palmer, Vernon Hills, IL, USA). The antioxidant activity of Trolox and *O. stamineus* extract was calculated using the following formula:

$$\% \text{ Inhibition of DPPH activity} = (\text{Blank absorbance} - \text{extract absorbance}/\text{Blank absorbance}) \times 100\%$$

2.9. Animals and Induction of Type-2 Diabetes

The experimental protocols, ethical procedures, and policies were authorized according to the Animal Care and Use Committee of Arab American university. In the current study, male C57BL/6 mice (14 weeks of age and weighing approximately 30 g) obtained from the Arab American university experimental animals care facility were used. The animal facility maintained an environment temperature of 25 ± 2 °C, a humidity of 55 ± 5%, with a 12 h controlled light/dark cycle. Free access to tap water and standard laboratory food was given to the mice and separated to groups 1 week before starting the experimental part.

Induction of diabetes: Mice were treated with streptozotocin (STZ) (40 mg/kg) in 50 mM sodium citrate buffer (pH, 4.5) for 5 consecutive days via the intraperitoneal route. To prevent fatal hypoglycemia, mice freely received 10% sucrose water for 5 days after STZ treatment. On experimental day 14 (9 days after the last STZ injection), the blood sugar levels were measured with a diabetes test strips in a glucometer apparatus (Abbott, Abbott Park, IL, USA). Mice with a blood glucose levels more than 200 mg/dL were considered diabetic and taken for further experimentation procedures [18].

Grouping of animals: All the experimental mice were given an essential diet during the experimental period and separated into six groups (n = 4): Control non-diabetic groups without and with WOS and MOS and STZ-induced diabetic mice without and with WOS and MOS. The extracts were diluted in water daily and administered by gavage at a concentration of 100 mg/kg.

2.10. Statistical Analysis

The data were normally distributed and variables were expressed as mean ± SEM. T test was used to investigate statistically significant differences. A level of $p < 0.05$ was accepted as statistically significant. Statistical analyses were conducted using SPSS version 23.0 for Windows.

3. Results and Discussion

This study focused on testing the chemical composition, cytotoxicity, and antidiabetic activity of two *O. stamineus* extracts: water (WOS) and methanol (MOS).

3.1. Toxicity of O. stamineus Extracts

MTT assay was performed to define the non-toxic concentrations of the two *O. stamineus* extracts: water (WOS) and methanol (MOS). Extract concentrations that led to more than 90% of cell viability were considered safe, non-toxic concentrations. It was shown that concentrations up to 250 μg/mL were safe for both WOS and MOS extracts (Figure 1). Therefore, the efficacy experiments of WOS and MOS were performed up to 250 μg/mL.

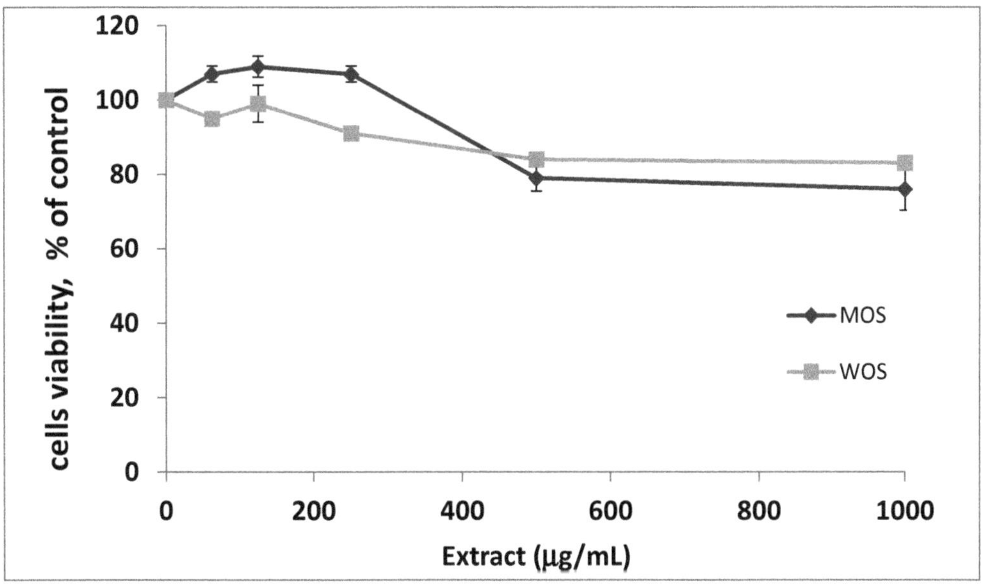

Figure 1. Effect of *O. stamineus* methanol (MOS) and water (WOS) extract on cell viability examined by MTT assay. L6-GLUT4myc cells (20,000 cell/well) were seeded in 96-well plate and were exposed to the extract for 24 h. Values given represent means ± SEM (% of untreated control cells) of three independent experiments carried out in triplicates.

3.2. Effects of O. stamineus Extracts on GLUT4 Translocation

In type 2 diabetes, GLUT4 translocation to the plasma membrane is impaired. Some antidiabetic medicinal plants can increase the translocation of GLUT4 [19]. Therefore, the effect of the two extracts (WOS and MOS) on the translocation of GLUT4 to the plasma membrane was measured by the GLUT4 translocation assay as described in the methods in the presence and absence of insulin. WOS extract enhanced GLUT4 translocation slightly; at 250 μg/mL, the translocation increased from 100% to 142 ± 2.5% and 165 ± 5% with and without insulin, respectively (Figure 2A). MOS extract exhibited a more potent effect on GLUT4 translocation. GLUT4 translocation reached 209 ± 8% and 306 ± 9% in cells exposed to 125 μg/mL MOS in the presence and absence of insulin, respectively. GLUT4 translocation increased more when exposed to 250 μg/mL and reached 279 ± 15% and 351 ± 20% in the absence and presence of insulin, respectively (Figure 2B).

The extent of increase in insulin-stimulated GLUT4 translocation, as depicted in Figure 2, was additive to that of basal GLUT4 translocation in *O. stamineus*-exposed cells, especially MOS. This result suggests a possible additive efficacy between *O. stamineus* active ingredients and insulin. Consecutively, *O. stamineus* active phytochemicals might enhance GLUT4 translocation in a non-insulin dependent pathway, such as the AMP-activated protein kinase (AMPK) pathway. It is possible that *O. stamineus* active compounds might possess "insulin-sensitizing" activity.

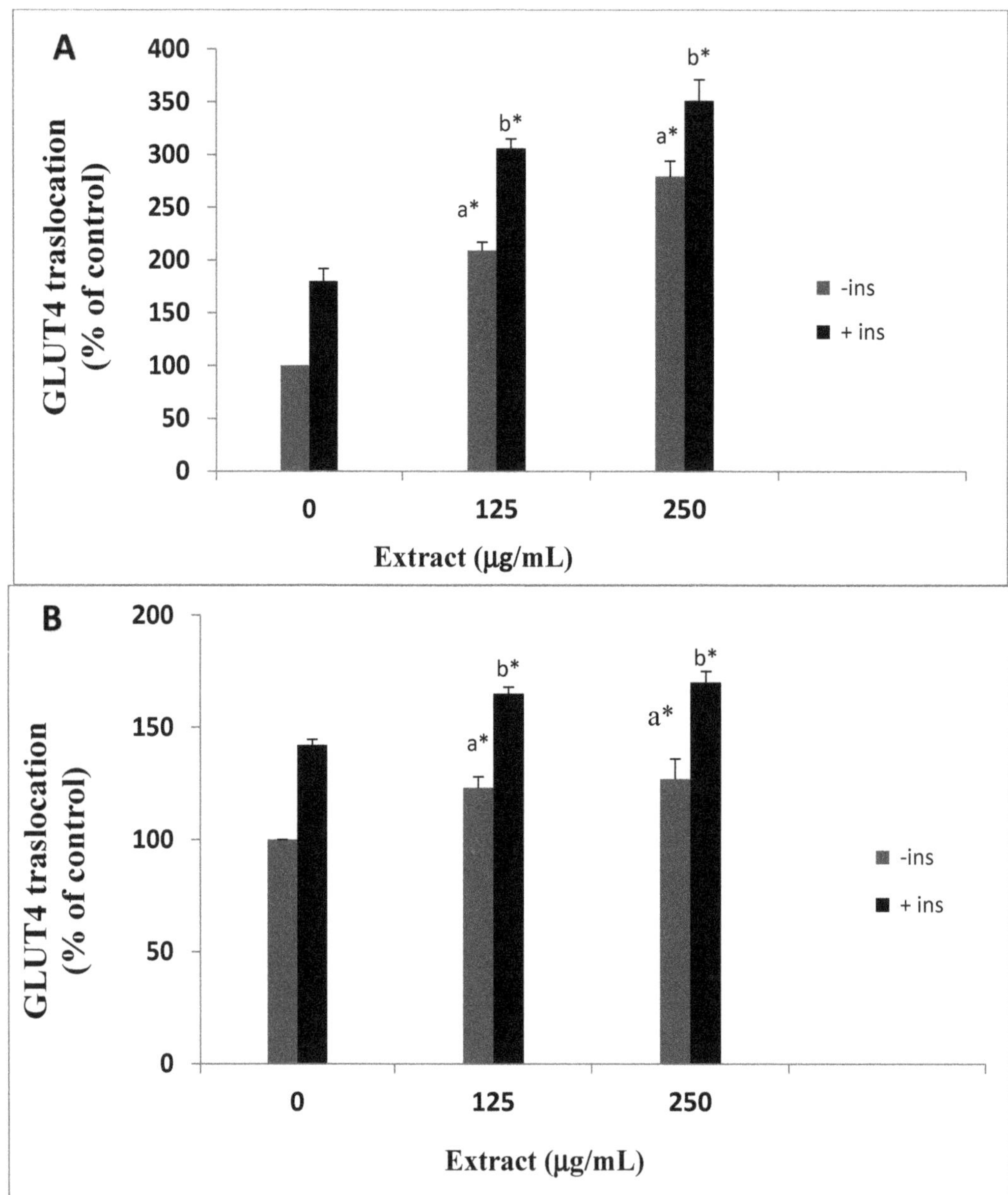

Figure 2. GLUT4 translocation to the plasma membrane. L6-GLUT4myc cells (150,000 cell/well) were seeded in a 24-well plate and were exposed to WOS (**A**) and MOS (**B**) without (−) or with (+) insulin as described in the methods. Surface *myc*-tagged GLUT4 density was quantified using the antibody coupled colorimetric assay. Values given represent means ± SEM (relative to untreated control cells) of three independent experiments carried out in triplicates. Statistical significance: (a) compared with (-ins) control group, (b) compared with (+ins) control group.

3.3. DPPH Scavenging Activity

The effect of antioxidants on DPPH is believed to be due to their hydrogen-donating ability [17]. The ability of the WOS and MOS extracts to act as antioxidants was tested by the DPPH scavenging activity assay. WOS and MOS scavenging activity was tested up to 100 µg/mL and reached 53.3 ± 4.8% and 79 ± 0.6%, respectively (Figure 3). Trolox was used as a positive control and led to maximum scavenging activity at around 20 µg/mL. MOS reached the maximal inhibition of 77 ± 10% at 500 µg/mL, while WOS led to 59 ± 3% at the same concentration (Figure 3). None of the extracts led to maximal scavenging like Trolox, yet MOS was more efficient in DPPH scavenging, indicating its higher content of antioxidant compounds (Table 1).

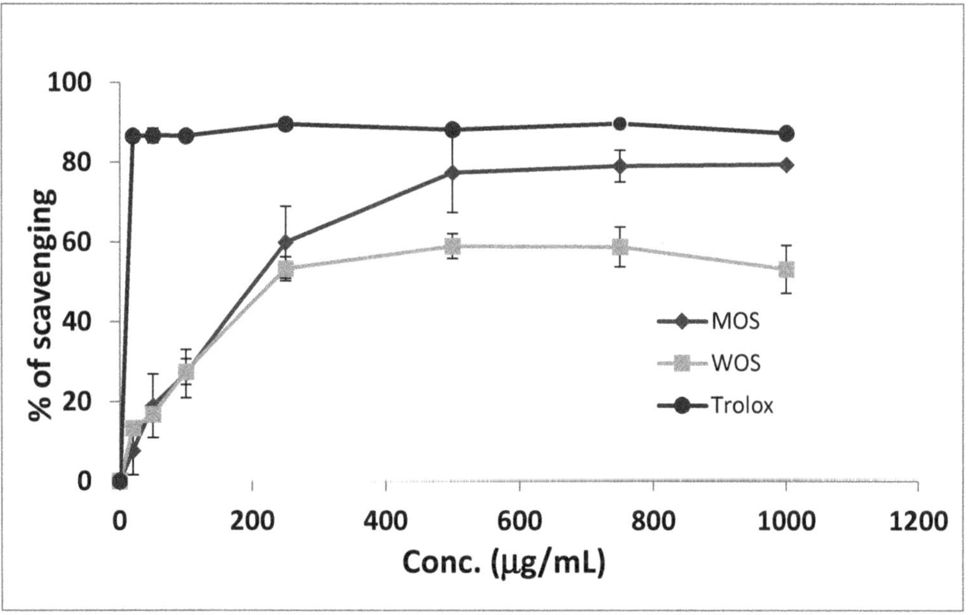

Figure 3. Determination of DPPH radical scavenging activity of MOS and WOS. Trolox was used as a positive control. All experiments were performed in triplicates. Data are expressed as mean ± SEM.

Table 1. Phytochemicals of *O. stamineus* from methanol extract verified by GC/MS.

Peak	Name	R_t	% Area	Match Factor
1	Propane-1,2-diol	12.46	0.081	67.8
2	Lactic Acid	13.42	0.043	90.5
3	1-Heptanol	13.60	0.008	66.9
4	Phosphonic acid	19.99	0.557	86.8
5	Glycerol	20.11	13.714	98.3
6	Butanedioic acid (Succinic acid)	20.93	0.384	96.9
7	Glyceric acid	21.59	0.407	97
8	Erythritol	26.11	0.034	93.3
9	Threitol	26.29	0.177	97.4
10	4-Hydroxybenzeneacetic acid	28.87	0.207	87.9
11	Neophytadiene	32.75	0.153	83.6
12	Pinitol	33.23	2.331	82.7
13	Saccharide -unknown	33.52	2.613	67.4
14	Fructose	34.07	29.749	95.7
15	Galactose	34.49	1.049	80

Table 1. *Cont.*

Peak	Name	R_t	% Area	Match Factor
16	Mannitol	34.71	18.749	92.3
17	Sorbitol	34.97	11.994	95.9
18	saccharide -unknown	35.07	0.890	86.3
19	Gluconic acid	35.82	0.061	88.9
20	Palmitic Acid	35.61	2.584	98.4
21	Myo-Inositol	36.48	2.672	98.1
22	Phytol	37.01	0.173	90.5
23	9,12-Octadecadienoic acid (Linoleic acid)	37.36	0.632	73
24	alpha-Linolenic acid	37.44	6.102	98.4
25	Stearic acid	37.62	0.161	92.3
26	Glyceryl-glycoside	38.69	0.262	91.7
27	1,3-Dihydroxyanthraquinone	39.27	0.022	67.5
28	Uridine	39.66	0.064	68.9
29	Sucrose	41.75	0.812	82.1
30	Trehalose	43.10	0.536	77.5
31	1-Octacosanol	48.91	0.037	81
32	alpha-Tocopherol	49.16	0.010	77.1
33	Campesterol	51.80	0.065	80.8
34	Stigmasterol	52.33	0.338	91.6
35	beta-Sitosterol	53.22	0.547	97.7
36	Sterol-unknown	53.31	0.260	89.3
37	Sterol-unknown	53.41	0.182	65.8
38	Sterol-unknown	53.78	0.208	84.5
39	Sterol-unknown	53.89	0.275	88.1
40	Sterol-unknown	54.85	0.410	76.8
41	Sterol-unknown	54.97	0.448	89.5

3.4. Effect of O. stamineus Extracts on Diabetic Mice Blood Glucose Levels and Mass

The effects of WOS and MOS on blood glucose levels in the STZ-injected mice are shown in Figure 4A. The treatment with *O. stamineus* extracts for 3 weeks resulted significant reductions in blood glucose levels compared to diabetic control group ($p < 0.05$). The effect was significantly appreciated after 1 week of administration. At day 36, the blood glucose level in the diabetic control group mice was 359 ± 7 mg/dL compared to 164 ± 2 mg/dL and 174 ± 3 mg/dL in WOS and MOS treated mice, respectively (Figure 4A). Blood glucose levels in non-diabetic mice were 104 to 109 mg/dL with and without WOS and MOS during the experiment period. Moreover, WOS and MOS did not affect the non-diabetic mice mass, which was 31 to 34 gr during the experiment period. As expected, diabetic mice mass was reduced from 32 ± 0.4 gr at the first day of the experiment to 28 ± 0.5 gr at day 36. MOS but not WOS rescued the mice mass as mice mass treated with MOS decreased only from 31 ± 0.5 gr to 29 ± 0.2 gr. WOS-treated mice mas decreased from 31 ± 0.3 to 26 ± 0.3 gr (Figure 4B).

STZ is widely used as a diabetogenic agent in mice [18]. It is a potent alkylating agent that enters the pancreatic β cells via GLUT2 and enhances DNA methylation and enhances hydrogen peroxide generation, causing DNA fragmentation and apoptosis and necrosis induction. STZ thus leads to insulin depletion leading to hyperglycemia. WOS and MOS significantly lowered blood glucose levels by 54% and 57%, respectively, compared with diabetic non-treated mice. MOS also maintained the diabetic mice mass. It decreased only by 6% at day 36 compared to 15% and 16% in the diabetic non-treated group and WOS-treated group, respectively. Those results are in line with the in vitro GLUT4 translocation results, where MOS was more effective in augmenting GLUT4 to the PM surface (Figure 2). GC/MS tests also showed that MOS is more rich in antidiabetic phytochemicals compared with WOS.

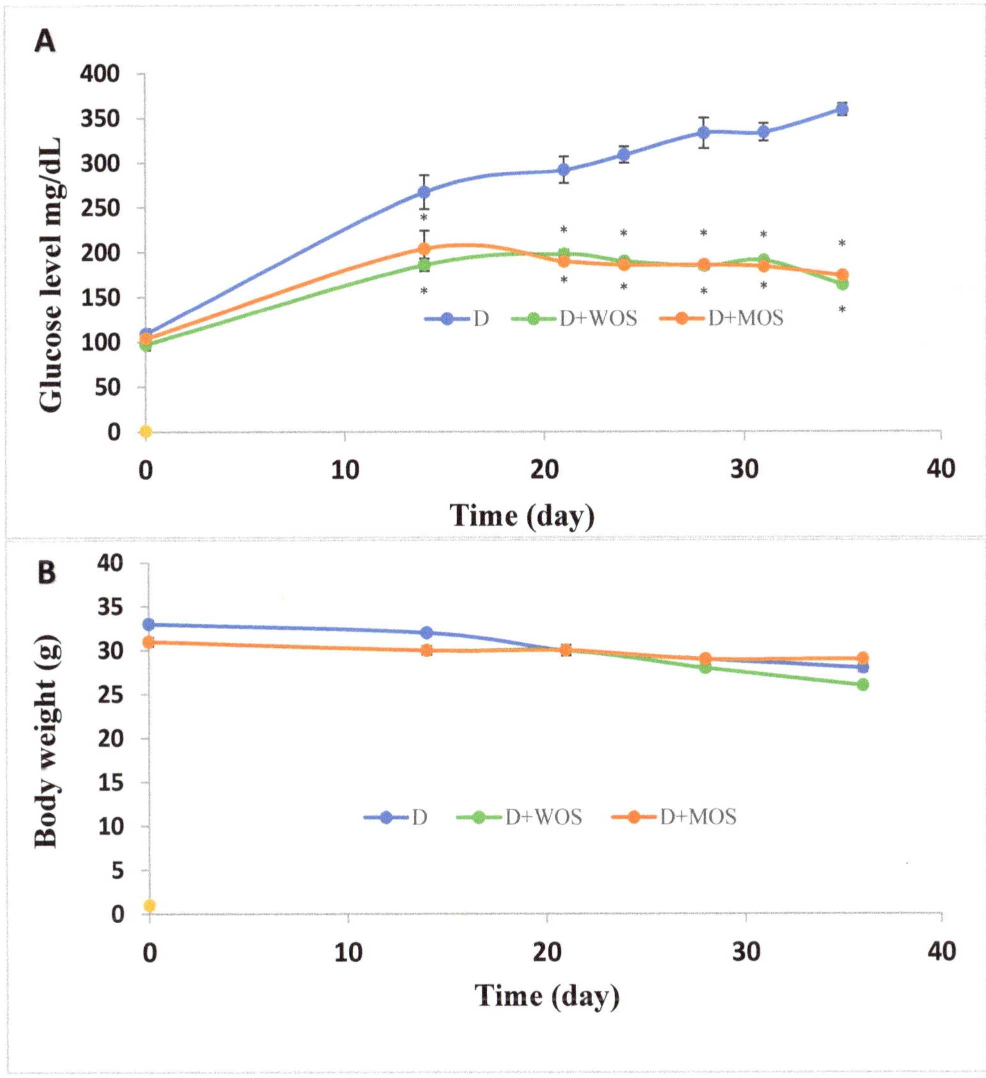

Figure 4. The effect of oral administration of 100 mg/kg WOS and MOS on blood glucose levels (**A**) and body weight (**B**) of STZ-induced diabetic mice. Values are expressed as mean ± SEM. * $p < 0.05$, significant as compared with diabetic group.

3.5. Chemical Analysis of O. stamineus Extracts

Phytochemical screening using GC/MS revealed 52 compounds in WOS (Table 1) and 41 in MOS (Table 2), including sterols, esters, phenolic compounds, saturated and unsaturated fatty acids, and aromatic compounds. Fourteen components, namely, phosphonic acid, glycerol, butanedioic acid, glyceric acid, pinitol, fructose, galactose, mannitol, gluconic acid, myo-inositol, glyceryl-glycoside, uridine, sucrose, and trehalose, were conjoined in the two extracts.

Table 2. Phytochemicals of *O. stamineus* from water extract verified by GC/MS.

Peak	Name	R_t	% Area	Match Factor
1	Glycin	12.24	0.02	87.7
2	Alanine	14.75	0.09	96
3	Leucine	16.23	0.03	86
4	Proline	16.70	0.39	91.4
5	Isoleucine	16.84	0.03	84.1
6	Malonic acid	17.87	0.07	95.3
7	Valine	18.21	0.12	96.2
8	Serine	19.34	0.07	86.4
9	Phosphonic acid	19.99	2.81	90.1
10	Glycerol	20.09	1.94	98.5
11	Butanedioic acid (Succinic acid)	20.91	0.28	98.3
12	Glyceric acid	21.64	4.05	97.5
13	2-Butenedioic acid (Fumaric acid)	21.79	0.04	95.4
14	Threonine	23.07	0.19	95.9
15	Aspartic acid	23.77	0.05	94.7
16	Malic acid	25.76	19.19	97.8
17	5-Oxoproline	26.25	0.19	96.8
18	4-Aminobutanoic acid (GABA)	26.46	0.35	97.1
19	Phenylalanine	26.64	0.02	79
20	Threonic acid	27.14	0.07	99.1
21	Erythronic acid	27.57	0.32	85.4
22	Glutaric acid	28.65	0.27	88
23	Tartaric acid	29.37	2.38	98.3
24	Asparagine	29.81	0.18	96.1
25	2-amino-Adipic acid	30.78	0.02	85.2
26	Ribose	31.48	0.02	82.1
27	Glutamine	31.92	0.17	96.1
28	Citric acid	32.94	7.54	87.1
29	Pinitol	33.22	0.11	96.7
30	Adenine	33.34	0.11	97.1
31	Quininic acid	33.67	5.71	89.6
32	Fructose	34.17	1.25	96.4
33	Galactose	34.28	21.88	77.2
34	Saccharide -unknown5	34.49	4.11	96.2
35	Mannitol	34.65	0.36	97.8
36	saccharide -unknown5	34.98	11.62	85.8
37	Gluconic acid	35.55	3.34	93.8
38	Ferulic acid	36.19	0.01	85.1
39	Myo-Inositol	36.50	8.37	96.6
40	Guanine	36.65	0.04	90.7
41	Caffeic acid	36.71	0.29	95.7
42	Tryptophan	37.60	0.04	83.1
43	Glyceryl-glycoside	38.69	0.58	94.5
44	Uridine	39.69	0.18	94.7
45	Sucrose	41.69	0.11	88.7
46	Cytidine	41.89	0.10	78.8
47	Trehalose	42.95	0.37	48.1
48	Chlorogenic acid	49.79	0.20	86.5
49	Cellobiose	50.97	0.09	83.2
50	Quercetin	51.17	0.06	88.3
51	Trisaccharide -unknown1	55.08	0.13	91.1
52	Trisaccharide -unknown2	55.86	0.02	68.8

The recognized antidiabetic compounds, especially those previously reported to enhance glucose uptake and increase GLUT4 activity, are highlighted in bold in Tables 1 and 2, and their chemical structure is drawn in the GC/MS chromatogram (Figure 5). In MOS, palmitic acid, phytol, alpha-linolenic acid, stearic acid, 1,3-dihydroxyanthraquinone, and stigmasterol (Table 1 and Figure 5B) are reported to enhance glucose disposal. Palmitic acid augmented GLUT4 translocation in muscle cells [14], and phytol was reported to increase AS160 and GLUT4 gene expression and activate the PI3K/Akt signaling pathway in mouse white adipose tissue [20]. Alpha-linolenic acid lowered blood glucose levels in diabetic mice as it increased GLUT4 amount at the muscle membrane [21]. Stearic acid enhanced basal glucose uptake in myotubes [22]. 1,3 Dihydroxyanthraquinone enhanced glucose uptake in C2C12 muscle cells in an AMPK-signaling-dependent pathway [23]. Stigmasterol augmented GLUT4 translocation and expression in L6 muscle cells [24]. Among the detected compounds in MOS was erythritol. It is 60–70% as sweet as sucrose; however, it provides only 6% of the calories in an equal amount of sugar and does not affect blood sugar levels [25].

In WOS, four compounds were found to be antidiabetic, namely, caffeic acid, chlorogenic acid, 4-aminobutanoic acid, and quercetin (Table 2 and Figure 5D). Only the last two compounds were reported to enhance GLUT4 translocation. 4-aminobutanoic, also known as gamma-aminobutyric acid (GABA), improved insulin resistance in diabetic patients by increasing the expression of GLUT4 [26]. Quercetin increased expression of GLUT4 [5]. Caffeic acid and chlorogenic acid were reported to possess antidiabetic activity, yet their activity was not associated directly with GLUT4 activity. Recently, caffeic acid was shown to decrease blood glucose levels and improve glucose tolerance in diabetic rats in an unknown mechanism [27]. Chlorogenic acid reduced insulin resistance and modulated glucose uptake in HepG2 cell line [28].

To our best of knowledge, this is the first reported study that compares water and methanol-based extracts from *O. stamineus* arial part (leaves and barks) in terms of detected antidiabetic active ingredients, antioxidant activity, and antidiabetic activity in vitro and in vivo at low doses (100 mg/kg). Others have treated diabetic mice and rats with up to 1000 mg/kg. For instance, in STZ-induced diabetic rats, only the group treated with 1000 mg/kg of the 50% ethanolic extract of *O. stamineus* [29] and water extract at 500 mg/kg [13] showed significantly lower plasma glucose levels. Others reported 200 and 400 mg/kg of ethanol extract of *O. stamineus* reduced fasting blood glucose levels in high-fat-diet (HFD) C57BL/6 mice after 8-week administration of the extract [12]. In the present study, the WOS and MOS reduced the diabetic mice blood glucose levels at low doses, and its antidiabetic activity was appreciated the first week of treatment.

The present results showed that *O. stamineus* methanol extract (and, to a lesser extent, the water extract) significantly enhanced GLUT4 translocation to the PM at non-toxic concentrations, reduced blood sugar levels of diabetic mice, and possessed antioxidant activity. Thus, *O. stamineus* extract may be beneficial for diabetes treatment. Ten reported effective antidiabetic constituents were detected in the extracts, especially in the MOS extract, which is in line with our current findings that MOS is more effective in augmenting GLUT4 translocation. It is crucial to separate *O. stamineus* detected phytochemicals in order to identify their cellular/molecular target and point out their specific antidiabetic mechanism and cellular pathways.

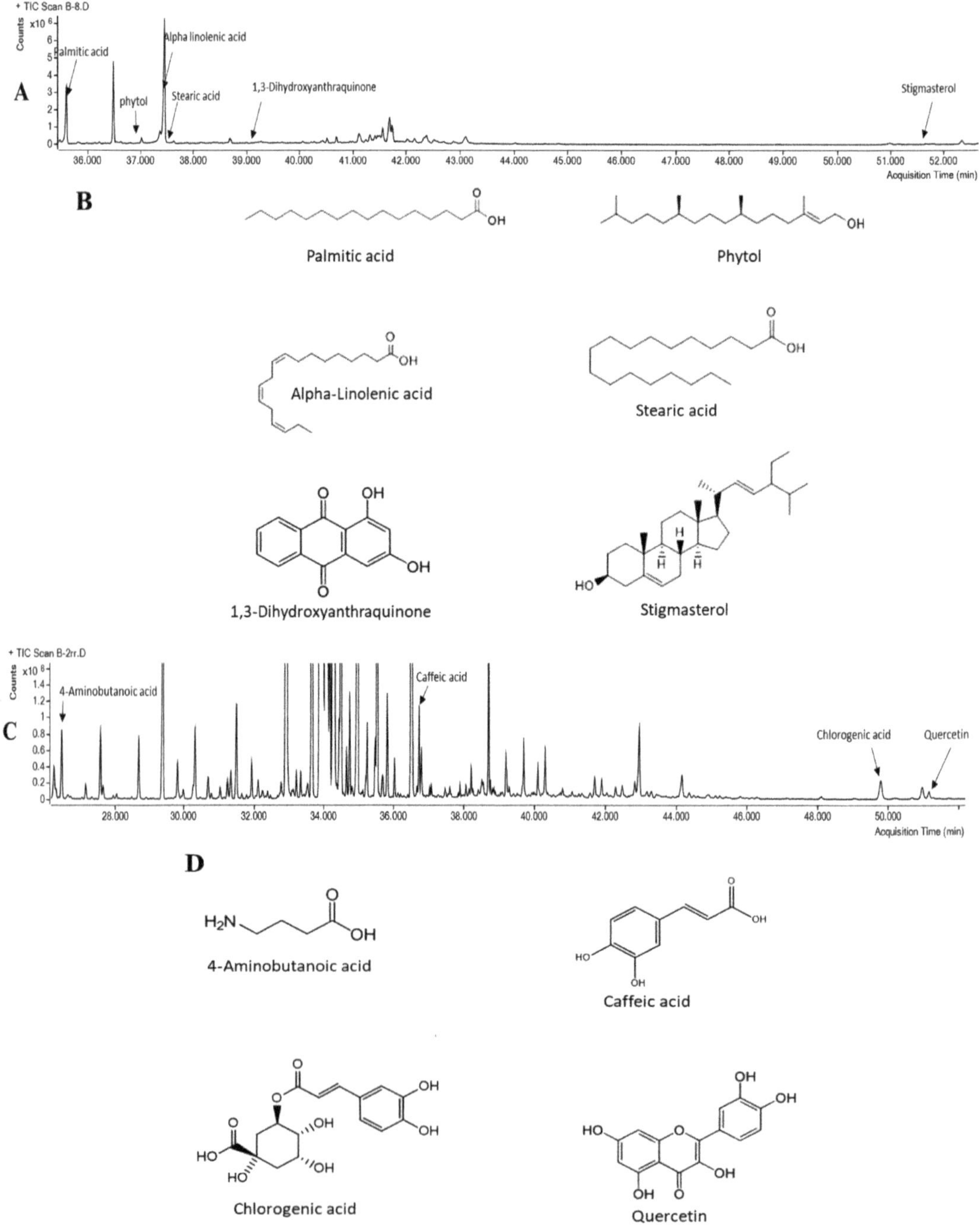

Figure 5. Total ion chromatogram (TIC) of *O. stamineus* methanol (**A**) and water (**C**) extract. Chemical structure of the identified antidiabetic components in *O. stamineus* methanol and water extract is depicted in (**B**) and (**D**), respectively.

4. Conclusions

O. stamineus water and especially methanol extracts significantly reduced the plasma glucose concentrations of STZ-induced diabetic mice. Concomitantly, methanol extract was more efficient in augmenting GLUT4 translocation to the PM of L6 myocytes. Among the compounds detected, 10 are reported to enhance either GLUT4 transport or translocation to the PM. Moreover, *O. stamineus* extracts exhibited antioxidant activity that might be associated with its antidiabetic activity and glucose disposal. *O. stamineus* might be considered an antidiabetic agent once its activity is tested and proven on diabetic subjects.

Author Contributions: Conceptualization, H.Z.; methodology, N.B., S.K., S.M., T.Y., M.M., D.K. and A.K.; validation, Z.T., S.K. and H.Z.; formal analysis, N.B., S.K., Z.T. and H.Z.; writing—original draft preparation, N.B., S.K., T.Y., A.K., Z.T. and H.Z.; writing—review and editing, M.M., S.K. and H.Z. All authors have read and agreed to the published version of the manuscript.

Funding: German Federal Ministry of Education and Research (PALGER), MOFET, AAUP and Al-Qasemi research foundations.

Institutional Review Board Statement: Not applicable.

Informed Consent Statement: Not applicable.

Data Availability Statement: All data contained within the article.

Acknowledgments: We would like to acknowledge MOFET institute and the Al-Qasemi and AAUP Research Foundations for providing financial support.

Conflicts of Interest: The authors declare no conflict of interest. The funding received did not lead to any conflicts of interest regarding the publication of this manuscript.

References

1. Singer, M.E.; Dorrance, K.A.; Oxenreiter, M.M.; Yan, K.R.; Close, K.L. The type 2 diabetes 'modern preventable pandemic' and replicable lessons from the COVID-19 crisis. *Prev. Med. Rep.* **2022**, *25*, 101636. [CrossRef] [PubMed]
2. Kadan, S.; Saad, B.; Sasson, Y.; Zaid, H. In vitro evaluation of anti-diabetic activity and cytotoxicity of chemically analysed *Ocimum basilicum* extracts. *Food Chem.* **2016**, *196*, 1066–1074. [CrossRef] [PubMed]
3. Zaid, H.; Saad, B. State of the art of diabetes treatment in Greco-Arab and islamic medicine. In *Bioactive Food as Dietary Interventions for Diabetes*, 1st ed.; Ronald Ross Watson, V.R.P., Ed.; Elsevier: Amsterdam, The Netherlands, 2013; pp. 327–335.
4. Zaid, H.; Antonescu, C.N.; Randhawa, V.K.; Klip, A. Insulin action on glucose transporters through molecular switches, tracks and tethers. *Biochem. J.* **2008**, *413*, 201–215. [CrossRef]
5. Shanak, S.; Bassalat, N.; Barghash, A.; Kadan, S.; Ardah, M.; Zaid, H. Drug Discovery of Plausible Lead Natural Compounds That Target the Insulin Signaling Pathway: Bioinformatics Approaches. *Evid. Based Complement. Altern. Med.* **2022**, *2022*, 2832889. [CrossRef] [PubMed]
6. Saad, B.; Zaid, H.; Shanak, S.; Kadan, S. Introduction to medicinal plant safety and efficacy. In *Antidiabetes and Anti-Obesity Medicinal Plants and Phytochemicals*, 1st ed.; Springer International Publishing: Berlin/Heidelberg, Germany, 2017; pp. 21–55.
7. Wang, Q.; Wang, J.; Li, N.; Liu, J.; Zhou, J.; Zhuang, P.; Chen, H. A Systematic Review of *Orthosiphon stamineus* Benth. in the Treatment of Diabetes and Its Complications. *Molecules* **2022**, *27*, 444. [CrossRef]
8. Ashraf, K.; Sultan, S.; Adam, A. *Orthosiphon stamineus* Benth. is an Outstanding Food Medicine: Review of Phytochemical and Pharmacological Activities. *J. Pharm. Bioallied Sci.* **2018**, *10*, 109–118. [CrossRef]
9. Li, Z.; Qu, B.; Zhou, L.; Chen, H.; Wang, J.; Zhang, W.; Chen, C. A New Strategy to Investigate the Efficacy Markers Underlying the Medicinal Potentials of *Orthosiphon stamineus* Benth. *Front. Pharmacol.* **2021**, *12*, 748684. [CrossRef]
10. Mohamed, E.A.; Mohamed, A.J.; Asmawi, M.Z.; Sadikun, A.; Ebrika, O.S.; Yam, M.F. Antihyperglycemic effect of *Orthosiphon stamineus* benth leaves extract and its bioassay-guided fractions. *Molecules* **2011**, *16*, 3787–3801. [CrossRef]
11. Luo, Y.; Liu, Y.; Wen, Q.; Feng, Y.; Tan, T. Comprehensive chemical and metabolic profiling of anti-hyperglycemic active fraction from Clerodendranthi Spicati Herba. *J. Sep. Sci.* **2021**, *44*, 1805–1814. [CrossRef]
12. Seyedan, A.; Alshawsh, M.A.; Alshagga, M.A.; Mohamed, Z. Antiobesity and Lipid Lowering Effects of *Orthosiphon stamineus* in High-Fat Diet-Induced Obese Mice. *Planta Med.* **2017**, *83*, 684–692. [CrossRef]
13. Sriplang, K.; Adisakwattana, S.; Rungsipipat, A.; Yibchok-Anun, S. Effects of *Orthosiphon stamineus* aqueous extract on plasma glucose concentration and lipid profile in normal and streptozotocin-induced diabetic rats. *J. Ethnopharmacol.* **2007**, *109*, 510–514. [CrossRef] [PubMed]
14. Kadan, S.; Melamed, S.; Benvalid, S.; Tietel, Z.; Sasson, Y.; Zaid, H. *Gundelia tournefortii*: Fractionation, Chemical Composition and GLUT4 Translocation Enhancement in Muscle Cell Line. *Molecules* **2021**, *26*, 3785. [CrossRef]

15. Kadan, S.; Sasson, Y.; Saad, B.; Zaid, H. *Gundelia tournefortii* Antidiabetic Efficacy: Chemical Composition and GLUT4 Translocation. *Evid. Based Complement. Alternat. Med.* **2018**, *2018*, 8294320. [CrossRef] [PubMed]

16. Shahinaz Mahajna, M.A.; Zaid, H.; Farich, B.A.; Al Battah, F.; Mashner, S.; Saad, B. In vitro Evaluations of Cytotoxicity and Anti-inflammatory Effects of *Peganum harmala* Seed Extracts in THP-1-derived Macrophages. *Eur. J. Med. Plants* **2015**, *5*, 165–172. [CrossRef]

17. Blois, M.S. Antioxidant Determinations by the Use of a Stable Free Radical. *Nature* **1958**, *181*, 2. [CrossRef]

18. Furman, B.L. Streptozotocin-Induced Diabetic Models in Mice and Rats. *Curr. Protoc.* **2021**, *1*, e78. [CrossRef]

19. Sleman Kadan, Y.S.; Abu-Reziq, R.; Saad, B.; Benvalid, S.; Linn, T.; Cohen, G.; Zaid, H. Teucrium polium extracts stimulate GLUT4 translocation to the plasma membrane in L6 muscle cells. *Adv. Med. Plant Res.* **2018**, *6*, 8.

20. Wang, J.; Hu, X.; Ai, W.; Zhang, F.; Yang, K.; Wang, L.; Zhu, X.; Gao, P.; Shu, G.; Jiang, Q.; et al. Phytol increases adipocyte number and glucose tolerance through activation of PI3K/Akt signaling pathway in mice fed high-fat and high-fructose diet. *Biochem. Biophys. Res. Commun.* **2017**, *489*, 432–438. [CrossRef]

21. Motoshi Kato, T.M.; Nakao, M.; Iwamoto, N.; Ishida, T.; Tanigawa, K. Effect of Alpha-Linolenic Acid on Blood Glucose, Insulin and GLUT4 Protein Content of Type 2 Diabetic Mice. *J. Health Sci.* **2000**, *46*, 4.

22. Yore, M.M.; Syed, I.; Moraes-Vieira, P.M.; Zhang, T.; Herman, M.A.; Homan, E.A.; Patel, R.T.; Lee, J.; Chen, S.; Peroni, O.D.; et al. Discovery of a class of endogenous mammalian lipids with anti-diabetic and anti-inflammatory effects. *Cell* **2014**, *159*, 318–332. [CrossRef]

23. Zhou, R.; Wang, L.; Xu, X.; Chen, J.; Hu, L.H.; Chen, L.L.; Shen, X. Danthron activates AMP-activated protein kinase and regulates lipid and glucose metabolism in vitro. *Acta Pharmacol. Sin.* **2013**, *34*, 1061–1069. [CrossRef] [PubMed]

24. Wang, J.; Huang, M.; Yang, J.; Ma, X.; Zheng, S.; Deng, S.; Huang, Y.; Yang, X.; Zhao, P. Anti-diabetic activity of stigmasterol from soybean oil by targeting the GLUT4 glucose transporter. *Food Nutr. Res.* **2017**, *61*, 1364117. [CrossRef] [PubMed]

25. Regnat, K.; Mach, R.L.; Mach-Aigner, A.R. Erythritol as sweetener-wherefrom and whereto? *Appl. Microbiol. Biotechnol.* **2018**, *102*, 587–595. [CrossRef] [PubMed]

26. Rezazadeh, H.; Sharifi, M.R.; Sharifi, M.; Soltani, N. Gamma-aminobutyric acid attenuates insulin resistance in type 2 diabetic patients and reduces the risk of insulin resistance in their offspring. *Biomed. Pharmacother.* **2021**, *138*, 111440. [CrossRef] [PubMed]

27. Salau, V.F.; Erukainure, O.L.; Ijomone, O.M.; Islam, M.S. Caffeic acid regulates glucose homeostasis and inhibits purinergic and cholinergic activities while abating oxidative stress and dyslipidaemia in fructose-streptozotocin-induced diabetic rats. *J. Pharm. Pharmacol.* **2022**, *74*, 973–984. [CrossRef] [PubMed]

28. Chen, L.; Teng, H.; Cao, H. Chlorogenic acid and caffeic acid from Sonchus oleraceus Linn synergistically attenuate insulin resistance and modulate glucose uptake in HepG2 cells. *Food Chem. Toxicol.* **2019**, *127*, 182–187. [CrossRef] [PubMed]

29. Mohamed, E.A.; Ahmad, M.; Ang, L.F.; Asmawi, M.Z.; Yam, M.F. Evaluation of alpha-Glucosidase Inhibitory Effect of 50% Ethanolic Standardized Extract of *Orthosiphon stamineus* Benth in Normal and Streptozotocin-Induced Diabetic Rats. *Evid. Based Complement. Alternat. Med.* **2015**, *2015*, 754931. [CrossRef] [PubMed]

 *pharmaceutics*

Article

Antihyperglycemic Effect of *Lavandula pedunculata*: In Vivo, In Vitro and Ex Vivo Approaches

Salima Boutahiri [1,2,3], Mohamed Bouhrim [4,*], Chayma Abidi [5], Hamza Mechchate [6,*], Ali S. Alqahtani [7], Omar M. Noman [7], Ferdinand Kouoh Elombo [1,8], Bernard Gressier [1], Sevser Sahpaz [3], Mohamed Bnouham [4], Jehan-François Desjeux [9], Touriya Zair [2] and Bruno Eto [1]

[1] Laboratoires TBC, Laboratory of Pharmacology, Pharmacokinetics and Clinical Pharmacy, Faculty of Pharmacy, University of Lille, F-59000 Lille, France; boutahirisalima@gmail.com (S.B.); ferdinand.kouohelombo@univ-lille.fr (F.K.E.); gressier.bernard@univ-lille.fr (B.G.); eto.bruno@univ-lille.fr (B.E.)

[2] Research Team of Chemistry of Bioactive Molecules and the Environment, Laboratory of Innovative Materials and Biotechnology of Natural Resources, Faculty of Sciences, Moulay Ismaïl University, B.P. 11201 Zitoune, Meknes 50070, Morocco; zair.touriya@umi.ac.ma

[3] Univ. Lille, University of Liège, University of Picardie Jules Verne, JUNIA, UMRT 1158 BioEcoAgro, Specialized Metabolites of Plant Origin, F-59000 Lille, France; sahpaz.sevser@univ-lille.fr

[4] Laboratory of Bioresources, Biotechnology, Ethnopharmacology and Health, Faculty of Sciences, Mohammed First University, B.P. 717, Oujda 60000, Morocco; bnouham.mohamed@ump.ac.ma

[5] Laboratory of Functional Physiology and Valorization of Bio-Resources, Higher Institute of Biotechnology of Beja, University of Jendouba, B.P. 382, Beja 9000, Tunisia; abidichayma07@gmail.com

[6] Laboratory of Biotechnology, Environment, Agri-Food, and Health, Faculty of Sciences Dhar El Mahraz, University Sidi Mohamed Ben Abdellah, P.O. Box 1796, Fez 30000, Morocco

[7] Department of Pharmacognosy, College of Pharmacy, King Saud University, Riyadh 11451, Saudi Arabia; alalqahtani@ksu.edu.sa (A.S.A.); onoman@ksu.edu.sa (O.M.N.)

[8] Laboratory de Pharmacology and Toxicology (LPT), Unit of Aromatic and Medicinal Plants Valorization, Department of Biochemistry, Faculty of Sciences, University of Yaoundé 1, Yaoundé BP 812, Cameroon

[9] National Academy of Medicine, 16, rue Bonaparte, 75006 Paris, France; jehanfrancois.desjeux@gmail.com

* Correspondence: mohamed.bourhim@gmail.com (M.B.); hamza.mechchate@usmba.ac.ma (H.M.)

Citation: Boutahiri, S.; Bouhrim, M.; Abidi, C.; Mechchate, H.; Alqahtani, A.S.; Noman, O.M.; Elombo, F.K.; Gressier, B.; Sahpaz, S.; Bnouham, M.; et al. Antihyperglycemic Effect of *Lavandula pedunculata*: In Vivo, In Vitro and Ex Vivo Approaches. *Pharmaceutics* **2021**, *13*, 2019. https://doi.org/10.3390/pharmaceutics13122019

Academic Editors: Fátima Regina Mena Barreto Silva and Diana Marcela Aragon Novoa

Received: 19 September 2021
Accepted: 19 November 2021
Published: 26 November 2021

Publisher's Note: MDPI stays neutral with regard to jurisdictional claims in published maps and institutional affiliations.

Abstract: *Lavandula pedunculata* (Mill.) Cav. (LP) is one of lavender species traditionally used in Morocco to prevent or cure diabetes, alone or in the form of polyherbal preparations (PHP). Therefore, the primary objective of this study was to test the antihyperglycemic effect of the aqueous extract of LP, alone and in combination with *Punica granatum* L. (PG) and *Trigonella foenum-graecum* L. (FGK). The secondary objective was to explore some mechanisms of action on the digestive functions. The antihyperglycemic effect of the aqueous extract of LP, alone and in combination with PG and FGK, was studied in vivo using an oral glucose tolerance test (OGTT). In addition, LP extract was tested on the activities of some digestive enzymes (pancreatic α-amylase and intestinal α-glucosidase) in vitro and on the intestinal absorption of glucose ex vivo using a short-circuit current (I_{sc}) technique. Acute and chronic oral administration of LP aqueous extract reduced the peak of the glucose concentration (30 min, $p < 0.01$) and the area under the curve (AUC, $p < 0.01$). The effect of LP + PG was at the same amplitude to that of the positive control Metformin (MET). LP aqueous extract inhibited the pancreatic α-amylase with an IC_{50} almost identical to acarbose (0.44 ± 0.05 mg/mL and 0.36 ± 0.02 mg/mL, respectively), as well as the intestinal α-glucosidase, ($IC_{50} = 131 \pm 20$ μg/mL) and the intestinal glucose absorption ($IC_{50} = 81.28 \pm 4.01$ μg/mL) in concentration-dependent manners. LP aqueous extract exhibited potent actions on hyperglycemia, with an inhibition on digestive enzymes and glucose absorption. In addition, the combination with PG and FGK enhanced oral glucose tolerance in rats. These findings back up the traditional use of LP in type 2 diabetes treatment and the effectiveness of the alternative and combinative poly-phytotherapy (ACPP).

Keywords: *Lavandula pedunculata* (Mill.) Cav.; *Punica granatum* L.; *Trigonella foenum-graecum* L.; diabetes; glucose intestinal absorption; oral glucose tolerance test

1. Introduction

Diabetes is a group of metabolic diseases of which blood hyperglycaemia is the main characteristic. This hyperglycaemia results from a lack of insulin secretion and/or a failure in insulin action. Type 1 diabetes is mostly induced by autoimmune diseases causing the destruction of the pancreatic β-cells responsible of the insulin production. In the more common type 2 diabetes, the main mechanism is insulin resistance. The latter is related to abnormalities in the insulin action on the target tissues (decrease in tissue responses and/or in insulin secretion) [1,2]. The long-term exposure to hyperglycaemia causes serious damages in the eyes (retinopathy), kidneys (nephropathy), and nerves (peripheral and autonomic neuropathy), as well as in the heart and blood vessels (cardiovascular and cerebrovascular diseases) [1].

In 2019, more than one million of the population under the age of 20 (children and teenagers) were living with type 1 diabetes. Moreover, 463 million people aged from 20 to 79 years, and representing 9.3% of adults, were living with diabetes. By this rate, the number of adults with diabetes is estimated to reach 578 million by 2030, and 700 million by 2045. The increasing prevalence of diabetes is due to an upsurge in type 2 diabetes. Changing lifestyle habits, like unhealthy diets, obesity, and sedentary lifestyles, are the main contributory factors leading to this increase. However, levels of childhood-onset type 1 diabetes are also on the rise [2].

Diabetes management requires the following of specific diets and physical exercises, as well as the use of medicines that regulate the blood sugar level. Individuals with type 1 diabetes (5–10% of the diabetic population) are dependent on insulin intake for survival. Type 2 diabetes patients (90–95% of the diabetic population) are treated with a variety of drugs with different modes of action. Among these drugs, biguanides (MET and phenformin) and thiazolidinediones (rosiglitazone and pioglitazone) decrease insulin resistance, and the α-glucosidase inhibitors (miglitol and acarbose) decrease glucose absorption from intestine. There are also other drugs like meglitinides which are insulin secretagogues, sulfonylureas that are blockers of the ATP sensitive potassium channels and glycosurics that inhibit reabsorption of glucose in the kidney. As for peptide analogs, gastric inhibitory peptide analogs glucagon like peptide-1 (GLP-1), injectable incretin mimetics, and injectable Amylin analogues, they increase incretin levels, which increases insulin secretion and inhibits glucagon release [3–5]. These conventional drugs may cause moderate to severe side effects, that is why research for alternative treatments is needed [6].

Medicinal plants are used across the world for their antidiabetic activity [7,8]. In Morocco, lavender is largely cited in different ethnobotanical and ethnopharmacological studies for its antidiabetic activity [9–11]. Moreover, it is often used by the population in the treatment of various diseases. Furthermore, researchers have been investigating the different activities of lavender and they were able to prove that it has antidiabetic, antimicrobial, anti-inflammatory, antirheumatic, antioxidant, antispasmodic, and antidepressant properties [12–18]. However, lavender is usually studied for the properties of its essential oils, regardless the large use of its aqueous extracts in traditional medicine [19,20].

According to Ghourri et al. (2013), medicinal plants are used essentially in the form of PHP to treat diabetes in the Moroccan Sahara (Tan-Tan). *Trigonella foenum-graecum*, *Nigella sativa*, *Lavandula stoechas*, *Origanum species*, *Rosmarinus officinalis*, and *Ammodaucus leucotrichus* are among the most used plants [21]. It was also reported that the antidiabetic potentials of PHPs are more important when compared to individual plant extracts [15]. This enhancing effect might be due to a synergetic effect between all the plant extracts when mixed together [15].

The previously mentioned information was the basis on which we relied to carry out this study that aims to investigate the antihyperglycemic activity of LP, one of lavender species that has never been studied for its antidiabetic activity. Moreover, its combination with PG and FGK was also studied. A special focus was given to some of its potential mechanisms of action on hyperglycemia, i.e., inhibition of activities of the digestive enzymes involved in the carbohydrate digestion (pancreatic α-amylase and intestinal α-glucosidase) and inhibition of D-glucose intestinal absorption.

2. Materials and Methods

2.1. Chemicals and Reagents

The used standards are protocatechic acid (Koch-Light Laboratories LTD, Bucks, UK), cinnamic acid (Rhône-Poulenc, Paris, France), chlorogenic acid (Sigma-Aldrich, St. Louis, MO, USA), vanillic acid (Merck, Darmstadt, Germany), gallic acid (Prolabo, Paris, France), rosmarinic acid (Extrasynthèse, Genay, France), Coumarin (Behringer, Willich, Germany), apigenin (obtained from Carl Roth, Karlsruhe, Germany), ferulic acid, and caffeic acid (purchased from Sigma, USA) as well as herniarine, luteolin, and myricetin (purchased from Sarsyntex, Merignac, France). Methanol (Carlo Erba Reagents, Val-de-Reuil, France) and formic acid (Carlo Erba Reagents^{TM}, Cornaredo, Italy) were of HPLC grades. Acarbose, α-amylase and α-glucosidase were purchased from Sigma Aldrich (St. Louis, MO, USA). D(+)-glucose anhydrous, sucrose, starch and phloridzin were purchased from Sigma Aldrich (Riedel-de Haen, Seelze, Germany). Glucose oxidase-peroxidase (GOD-POD) kit was purchased from Biosystems (Barcelone, Spain). All the chemicals used were of analytical grades.

2.2. Plant Material

The fresh flowering tops of *Lavandula pedunculata* (Mill.) Cav. (Voucher number: RAB111854) and *Punica granatum* L. pericarp (Voucher number: RAB65559) were collected respectively from Azrou (2019) and Boulemane (2018) regions (Middle Atlas, Morocco). Identification of the plants was carried out at the Scientific Institute of Rabat (Rabat, Morocco) by Dr. Hamid Khamar where a specimen was deposited. The collected plants were then dried for thirteen days in the open air and in the shade. Aqueous dry extract powder of *Galega officinalis* L. (GO) (PR 3199) was purchased from Plantex (Saint-Michel-sur-Orge, France) and *Trigonella foenum-graecum* L. (FENU60001) from Natural functional ingredients (Saint-Sylvain-d'Anjou, France).

2.2.1. Preparation of the Aqueous Extracts

Air dried flowering tops of LP or PG pericarps were mixed with distilled water (1:20; w/v) and heated for one hour at 70 to 80 °C. The mixtures were then filtered and the obtained filtrates were dried in the oven at 70 °C until obtaining the dry extract powder. The extracts were put in closed flasks away from light and humidity until further use.

2.2.2. UHPLC Analysis of Aqueous Extracts

Chromatographic analysis of the aqueous extracts was carried out on an AQUITY UPLC H-Class System (Waters Corporation, Manchester M23 9LZ, UK) equipped with two independent pumps, an automatic injector, a controller, a diode array UV detector (DAD), a mass spectrometer with ESI ionization source and a quadrupole as an analyzer. The stationary phase was a reverse phase Waters® Acquity BEH C18 column (2.1 × 50 mm, 1.7 μm) connected to a 0.2 μm in-line filter. The mobile phase was composed of two solvents: (A) ultrapure water (Milli-Q® Integral 5, Merck^{TM}, Allemagne, Germany) + 0.1% formic acid; (B) methanol + 0.1% formic acid. The elution gradient established was 0–5% B (1 min), 5–20% B (0.5 min), 20% B (3.5 min), 20–100% B (4 min), rinsing of the column 100% B (2 min) and re-equilibration 100-0% B (0.5 min), 0% B (2.5 min). Methanol 70% was required for washing the system.

The aqueous extracts were solubilized in a methanol / water mixture (1:1, v/v), so as to obtain a concentration of 1 mg/mL, then they were filtered through 0.2 μm PTFE filter. For each analysis, 4 μL of the extract were injected. The temperature was set at 30 °C and flow rate was set at 0.3 mL/min.

This analysis was carried out on few standards chosen according to bibliographic data, and which were injected under the same conditions as those of the extracts. These standards are cinnamic acid, luteolin, apigenin, myricetin, ferulic acid, protocatechuic acid, vanillic acid, chlorogenic acid, caffeic acid, rosmarinic acid, gallic acid, herniarine, and coumarin.

Compounds were identified by matching their retention time, UV spectrum and molecular weight to those of the used standards.

2.3. Animals

For OGTT and toxicological studies, healthy adult *Wistar* rats (150–250 g body weight, 2–3 months of age) and Swiss *albino* mice (20–30 g body weight) were provided by the animal house of the Faculty of Science, Mohammed First University, Oujda-Morocco.

For the intestinal glucose absorption studies, 7 week old male mice (Black mice C57BL/6JRj, 20–25 g) were purchased from Janvier SASA (Route des chênes, Le Genest-st-Isle, St Berthevin, France) and were acclimated for a week in the animal house conditions of the Faculty of Pharmacy, Lille University, France (approval N° D5935010). The animals were grouped in polycarbonate cages with free access to food and water, in an environmentally controlled room (22–26 °C, ventilation, 12/12 h light/dark cycle). Experiments on animals were carried out in accordance with the internationally approved "Guide for the care and use of laboratory animals" prepared by the National Academy of Sciences and published by the National Institutes of Health [22]. All efforts were made to minimize animal suffering and the number of animals used. For OGTT studies, ethical approval was obtained from Mohammed First University, Oujda-Morocco (31/2019/LBBEH-05 and 21/011/2019).

2.4. Pharmacological Studies

2.4.1. Inhibition Assay of Pancreatic α-Amylase Enzyme Activity

The inhibitory effect of LP aqueous extract on the enzymatic activity of pancreatic α-amylase was studied in vitro according to the method described by Bouhrim et al. (2021) with some modifications [23]. Then, 200 μL of the α-amylase enzyme solution (13 IU) and 200 μL of phosphate buffer (0.02 M; pH = 6.9) were placed in different tubes. On one hand, 200 μL of the LP extract were added at different concentrations: 2.27, 1.82, 1.36, 0.91, 0.45, and 0.23 mg/mL. On the other hand, 200 μL of acarbose were used as a positive control at several concentrations: 2.27, 0.91, 0.45, 0.23, 0.11, and 0.06 mg/mL. All the reagents were dissolved in phosphate buffer and 200 μL of this solution were used as a control. The mixtures were pre-incubated for 10 min at 37 °C before adding 200 μL of starch (1%) to each tube. The different solutions were then incubated for 20 min at 37 °C. Then, 600 μL of dinitrosalicylic acid (DNSA) color reagent (2.5%) was added to stop the enzymatic reaction. The tubes were then incubated for 8 min at 100 °C, and they were put in an ice-water bath for few minutes. At the end, the mixtures were diluted by the addition of 1 mL of distilled water and the absorbance of each tube was measured at 540 nm. The inhibition percentage was then calculated using the formula below:

$$\mathrm{IP} = \frac{A_{control} - A_{test}}{A_{control}} \times 100 \tag{1}$$

where IP: inhibition percentage (%); $A_{control}$: absorbance of the final mixture without inhibitor; A_{test}: absorbance of the final mixture in the presence of LP extract or acarbose.

The concentration of the tested substance (LP extract/acarbose) inhibiting 50% (IC$_{50}$) of the enzymatic activity of α-amylase was determined graphically.

2.4.2. Inhibition Assay of Intestinal α-Glucosidase Enzyme Activity

The inhibitory effect of LP aqueous extract on the enzymatic activity of intestinal α-glucosidase was evaluated in vitro according to the method described by Bouhrim et al. (2021) with some modifications [23]. The method consists of the quantification of D-glucose released from sucrose degradation. 10 μL of different concentrations of acarbose solutions (41, 82, 165, 328 and 656 μg/mL) or LP extract solutions (80, 170, 250, 330 and 650 μg/mL) were added to a mixture containing 100 μL of sucrose (50 mM), 100 μL of α-glucosidase enzyme solution (10 IU), and 1000 μL of phosphate buffer (50 mM; pH = 7.5). All the reagents were dissolved in phosphate buffer and 10 µL of this solution was used as a control. The mixtures were then incubated in a water bath at 37 °C for 25 min. Immediately after incubation, the different solutions were heated for 5 min at 100 °C to stop the enzymatic reaction. The D-glucose oxidase method using GOD/POD kit was used to determine the released D-glucose. The absorbance was measured at 500 nm and the inhibition percentage was calculated using the formula below:

$$IP = \frac{A_{control} - A_{test}}{A_{control}} \times 100 \tag{2}$$

where IP: inhibition percentage (%); $A_{control}$: absorbance of the final mixture without inhibitor; A_{test}: absorbance of the final mixture in the presence of LP extract or acarbose.

The concentration of the tested substance (LP extract/acarbose) inhibiting 50% (IC_{50}) of the enzymatic activity of α-glucosidase was determined graphically.

2.4.3. Intestinal Glucose Absorption

D-glucose absorption through the intestinal tissue was studied ex vivo using Ussing chambers. The Black mice were left on empty stomachs with free access to drinking water for 18 h before the experiments. After killing the animals by CO_2 inhalation, their intestines were removed and emptied of their content using a cold Ringer's solution. Then the jejunum was cut into small fragments (1–1.5 cm) and opened along the mesenteric border. Every fragment was placed as a flat sheet between the two sides of the plexiglas Ussing chambers (exposed area: 0.3 cm^2), allowing a mucosal compartment (luminal side) and a serosal compartment (serosal side) to be determined (blood side). Every half-chamber was filled in with 3 mL of Ringer's solution (pH = 7.4) thermostatically controlled at 37 °C. The oxygen saturation of this solution was constantly maintained by a permanent bubbling with carbogen (95% O_2/5% CO_2) circulation the chambers [24].

The normal isotonic Ringer's solution that was used contains NaCl (115 mM), $NaHCO_3$ (25 mM), $MgCl_2$ (1.2 mM), $CaCl_2$ (1.2 mM), K_2HPO_4 (2.4 mM), and KH_2PO_4 (0.4 mM).

The spontaneous transmural potential difference (PD) was measured using agar bridges containing KCl solution (3 M) in agar (4 percent, *w/v*) to represent asymmetry of electrical charges between the mucosal and serosal sides of the gut. These bridges were placed on both sides of the tissue and connected to calomel half-cells, connected to a high-impedance voltmeter. The PD was short-circuited and maintained at 0 mV by a short-circuit current (I_{sc}) throughout the experiment via two stainless steel 316L working electrodes directly placed in each compartment [25], and linked to a voltage-clamp system (JFD-1V, Laboratoires TBC, France). The delivered I_{sc} (in µA/cm^2), that was recorded continuously using Biodaqsoft software and corrected for fluid resistance. It represents the sum of the net ion fluxes transported across the epithelium in the absence of an electrochemical gradient (mainly Cl^-, Na^+ and $HCO_3{}^-$).

LP extract solutions were prepared extemporaneously when added to the mucosal compartment (prewarmed at 37 °C) in a volume of 250 µL, 5 min before adding the glucose (10 mM) to obtain final concentrations of 0.01 to 1000 µg/mL (glucose was replaced with mannitol at 10 mM in the serosal compartment).

Results are expressed as the difference between the I_{sc} plateau measured after 20 min of the glucose addition (without extract), and the I_{sc} plateau measured after the glucose addition preceded by the addition of LP extract. The principle is to stimulate the intestinal absorption of glucose (increase of I_{sc}) in the presence and absence of LP extract at different concentrations to detect any blocking of this absorption.

2.4.4. Antihyperglycemic Study in Healthy Rats
Acute Oral Glucose Tolerance Test

Acute OGTT was carried out using normal *Wistar* rats, left on empty stomachs for 14 h with free access to water, and which were grouped in six groups of four rats each (two males and two females). The control group (CT) was orally administered distilled water (10 mL/kg b.w.). The five test groups (LP, PG, FGK, GO, MET) were administered the aqueous extracts at 1 g/kg b.w. The rats were force-fed with the different products immediately after the measurement of their glycaemia; 30 min later, another measurement of the glycaemia was carried out, and then the rats were overloaded with D-glucose at a dose of 2 g/kg b.w. Subsequently, the variation in blood sugar was monitored every 30 min for two hours [23].

Chronic Oral Glucose Tolerance Test

To assess the effect of the repeated extracts administration, a chronic OGTT was performed using the same groups of rats used above (Acute oral glucose tolerance test). The testing products were orally administered to rats once a day for four weeks (chronic treatment). After the experimental period, the rats were left on empty stomachs for 14 h, and the glycaemia measurement was carried out according to the same procedure described above (Acute oral glucose tolerance test).

Acute Oral Glucose Tolerance Test for Plant Mixtures

The principle of this test is similar to the one used for individual plants which is described above (Acute oral glucose tolerance test). Four groups were studied; starting with the control group which was administered distilled water orally (10 mL/kg b.w.). Test group (LP + PG) was administered a mixture containing 50% of LP extract and 50% of PG extract at a final dose of 1 g/kg b.w. Test group (LP + FGK) was administered a mixture containing 50% of LP extract and 50% of FGK extract at a final dose of 1 g/kg b.w. MET group was administered MET (1 g/kg b.w.).

2.5. Toxicological Studies

2.5.1. Acute Toxicity

Swiss albino mice were used to test the acute toxicity of lavender aqueous extract (LP). Mice were left on empty stomachs for 16 h before the test, and then they were randomly divided into five groups of six mice each (three males and three females). The control group was force-fed with distilled water. Test groups 1, 2, 3, 4, and 5 were administered different concentrations of LP extract (1, 3, 5, 7, and 10 g/kg b.w., respectively). The mice were given the extract only once, and then they were put under observation for 15 days. Any clinical signs and/or behavioral changes would have been considered signs of acute toxicity.

2.5.2. Subchronic Toxicity

Subchronic toxicity test was performed using the same groups of rats used for the chronic OGTT (Chronic oral glucose tolerance test). The body weights of the rats were measured weekly as they were force-fed with the plant extracts. After the experimental period, the rats were left on empty stomachs for 14 h before being sacrificed to collect their blood and organs. The biochemical parameters of the collected blood were then measured in plasma: Alanine aminotransferase (ALT), aspartate aminotransferase (AST), direct bilirubin, total bilirubin, albumin, total cholesterol, triglycerides, high-density lipoproteins (HDL-c), low-density lipoproteins (LDL-c), plasma glucose, lipase, total protein, creatinine, urea, uric acid, calcium, and phosphate. All tests were performed with the COBAS INTEGRA® 400-Plus analyzer using standard clinical diagnostic kits. Also, the relative weights of the livers and the kidneys were measured.

2.5.3. Histology

For histological examination, different tissues (stomach, small intestine, liver, spleen, and kidneys of rats) were fixed in 10% neutral formalin, embedded in paraffin, sectioned to a thickness of approximately 5 μm, stained with hematoxylin and eosin, and examined for histopathological changes under the microscope (Olympus, Tokyo, Japan).

2.6. Statistical Analysis

Pharmacological responses for separate experiments using n tissues are presented as means $\pm$ SEM (standard error of the mean). Graphs of the concentration-response curves were determined using nonlinear regression and were fitted to the Hill equation by an iterative least-squares method (GraphPad Prism version 8.0 for Windows, GraphPad Software, San Diego, CA, USA). One-way analysis of variance (ANOVA) was performed for the comparison of the different effects with the control (Dunnett) and for the multiple-group comparisons (Newman-Keuls). GraphPad Prism computed the AUC using the trapezoid rule. Statistical significance was set as $p < 0.05$.

3. Results

3.1. UHPLC Analysis

The chemical composition of LP aqueous extract was determined using UHPLC-MS. The results (Table 1) show that the most abundant compound in this extract is rosmarinic acid. Figure 1 shows the UHPLC-MS chromatograms of rosmarinic acid detected in LP aqueous extract. Moreover, coumarin, protocatechuic acid, herniarin, caffeic acid, apigenin, luteolin, myricetin, and chlorogenic acid are present in this extract but with a lower abundance, as well as gallic acid and cinnamic acid that are present as traces. We can notice that vanillic acid and ferulic acid are absent in LP aqueous extract.

Table 1. Results of UHPLC-MS analysis regarding the presence or the absence of some chemical compounds in LP aqueous extract.

Rt (min)	λ_{max} (nm)	$[M - H]^-$ (m/z)	$[M + H]^+$ (m/z)	Compounds	LP
6.869	278.1	WD	147	Coumarin	+
8.259	255.5, 297.2	147	149	Cinnamic acid	T
2.697	259.1, 293.6	153	155	Protocatechuic acid	+
3.923	260.3, 292.4	167	169	Vanillic acid	−
1.518	371.0	169	171	Gallic acid	T
7.761	307.9	WD	177	Herniarin	+
3.976	324.7	179	181	Caffeic acid	+
6.536	322.3	193	195	Ferulic acid	−
8.606	338.8	269	271	Apigenin	+
8.368	254.3, 350.4	285	287	Luteolin	+
7.792	253.2, 372.1	317	319	Myricetin	+
3.680	240.1, 325.8	353	355	Chlorogenic acid	+
7.831	329.4	359	WD	Rosmarinic acid	++

(WD) weak detection; (T) traces; (−) Absence; (+) presence; (++) high presence; (LP) *L. pedunculata*.

3.2. Antihyperglycemic Study in Healthy Rats

3.2.1. Oral Glucose Tolerance Test

To evaluate the effect of LP aqueous extract on the systemic glucose homeostasis, an OGTT was performed in conscious fasted rats after acute and chronic oral administration of this extract (LP), and we compared its effect with other plant aqueous extracts such as PG, FGK, and GO.

During the OGTT, acute oral dose of LP decreased the peak glucose concentration (30 min, $p < 0.01$, Figure 2A) and the AUC ($p < 0.01$, Figure 2B). We noticed that this effect was lower when compared to that of GO and the positive control MET ($p < 0.001$), but remained close to that of PG ($p < 0.01$).

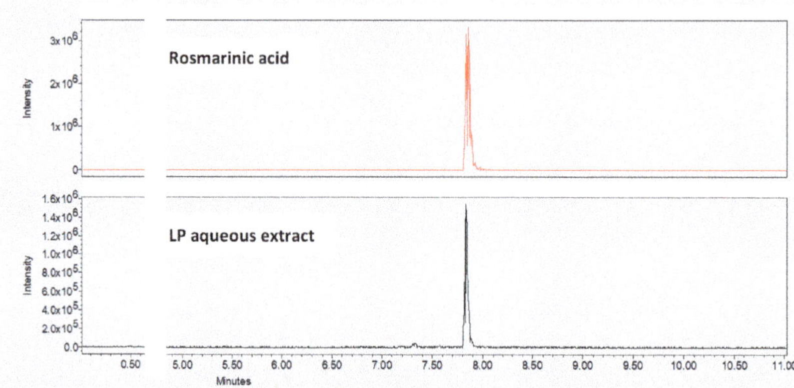

Figure 1. UHPLC-MS chromatograms showing the dominance of rosmarinic acid peak in LP aqueous extract.

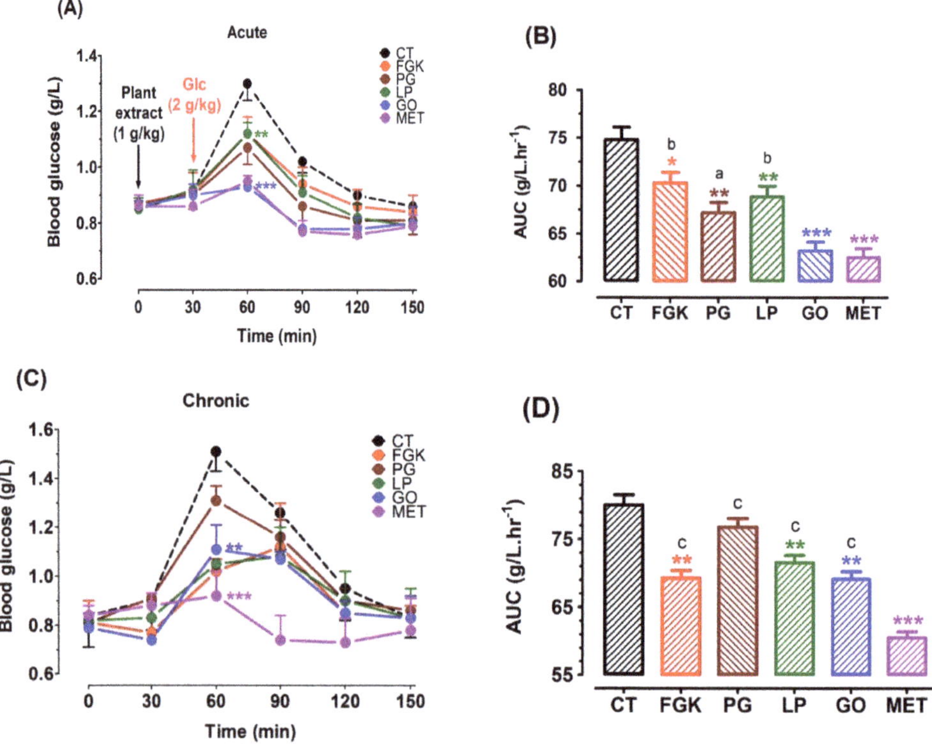

Figure 2. Acute and chronic OGTT after oral treatment of rats with plant extracts. Rats were force-fed intragastrically with distilled water as a negative control (CT, ■), 1 g/kg of the different aqueous plants extracts (LP, ■; PG, ■; FGK, ■; GO, ■), or 1 g/kg of the reference antidiabetic drug metformin (MET, ■) for 4 weeks. OGTT was performed on fasted rats every 30 min during 2 h after the first glucose (2 g/kg) overload (acute, panel (**A**)) and at the end of the 4-week treatment (chronic, panel (**C**)). Panels (**B**) and (**D**) show the AUCs (CT, ■; LP, ■; PG, ■; FGK, ■; GO, ■; MET, ■) for panels (**A**) and (**C**) respectively and were calculated using GraphPad Prism Version 8.0 software. * $p < 0.05$; ** $p < 0.01$; *** $p < 0.001$ in comparison with the control (Dunnett's test). (a) $p < 0.05$; (b) $p < 0.01$; (c) $p < 0.001$ in comparison with MET (Newman-Keuls test).

The same findings were observed after the chronic OGTT with LP when compared to the control ($p < 0.01$, Figure 2C,D). Finally, no statistical differences between male and female animals were observed (data not illustrated).

3.2.2. Oral Glucose Tolerance Test with Combinations of Plants

To evaluate the effect of plants aqueous extracts combined with LP aqueous extract on the systemic glucose homeostasis. After acute treatment of the combination of LP + PG and LP + FGK, an OGTT was conducted in awoke fasting rats. We compared the obtained effect with the positive control (MET). Acute oral administration of LP + PG and LP + FGK reduced the peak of the glucose concentration and the AUC ($p < 0.001$ with LP + PG and $p < 0.01$ with LP + FGK, Figure 3A,B). The effect of LP + PG was similar to that of the positive control (MET).

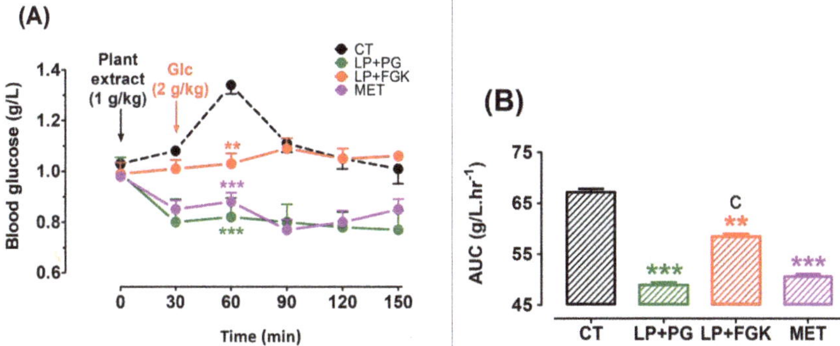

Figure 3. Acute OGTT after oral treatment of rats with combinations of plant extracts. Rats were force-fed intragastrically with distilled water as a negative control (CT, ■), 1 g/kg of the different combinations of the plants aqueous extracts (LP + PG, ■; LP + FGK, ■), or 1 g/kg of the reference antidiabetic drug metformin (MET, ■). OGTT was performed on fasted rats every 30 min during 2 h after the glucose (2 g/kg) overload (panel (**A**)). Panels (**B**) show the AUCs that were calculated using GraphPad Prism Version 8.0 software. ** $p < 0.01$; *** $p < 0.001$ in comparison with the control (Dunnett's test). (c) $p < 0.001$ in comparison with MET (Newman-Keuls test).

3.3. Inhibition of Pancreatic α-Amylase Activity

The enzymatic activity of pancreatic α-amylase, incubated with the substrate (starch) in the presence of LP extract, was inhibited in concentration-dependent manner (Figure 4) and this inhibition was close to that observed with acarbose with IC_{50} respectively of 0.44 ± 0.05 mg/mL for LP and 0.36 ± 0.02 mg/mL for acarbose.

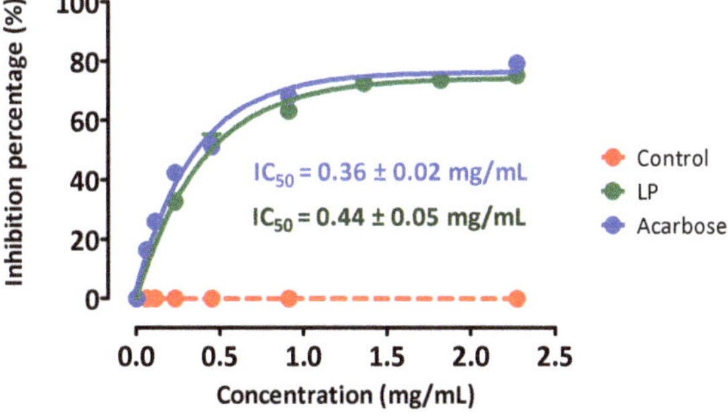

Figure 4. Inhibiting effect of the LP aqueous extract and acarbose on the pancreatic α-amylase enzyme activity. α-amylase was mixed and incubated with starch in the presence or absence of LP extract. Arcabose (medicine) was used as positive control.

3.4. Inhibition of Intestinal α-Glucosidase Activity

The enzymatic activity of intestinal α-glucosidase, incubated with the substrate (sucrose) in the presence of LP extract, was inhibited in a concentration-dependent manner (Figure 5). The effect of Acarbose was more pronounced than that of LP extract. The IC_{50} was respectively of 131 ± 20 µg/mL for LP, whereas that of Acarbose was 52 ± 2 µg/mL.

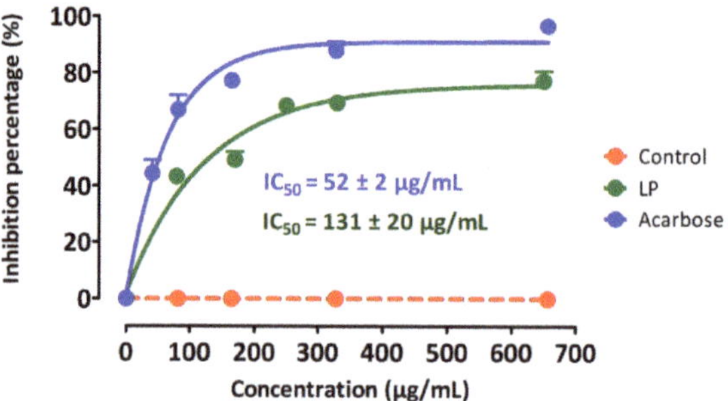

Figure 5. Inhibiting effect of the LP aqueous extract and acarbose on the intestinal α-glucosidase enzyme activity. Arcabose (medicine) was used as positive control.

3.5. Lavandula Pedunculata Inhibits Absorption of Sodium-Dependent D-Glucose

The impact of LP on the intestinal transit of glucose was studied using a short-circuit current method in Ussing chambers.

After a stabilization time (minimum 40 min), 10 mM D-glucose were added to the mucosal side of the mouse jejunum and the activity of the sodium-dependent glucose transporter-1 (SGLT-1) was followed as the rise in the I_{sc} sodium-dependent (Figure 6A).

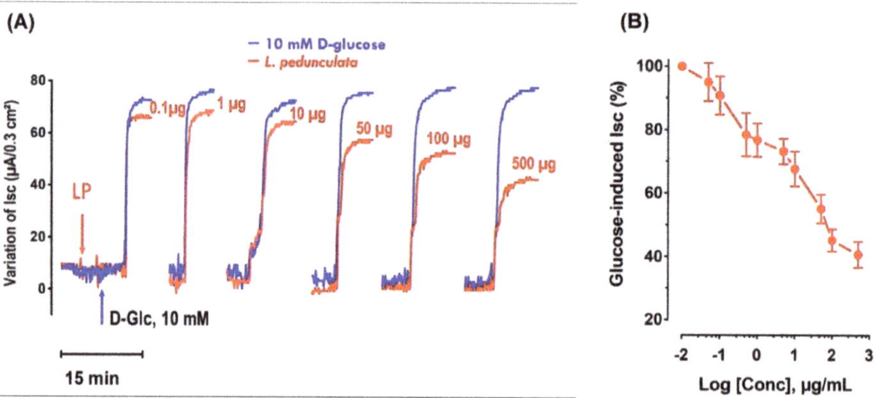

Figure 6. Effect of LP aqueous extract on the glucose-induced I_{sc}. (**A**) Typical recording of the I_{sc} ($\mu A / 0.3$ cm^2) inhibition across mouse jejunum fragment in Ussing chambers by the addition of LP on the mucosal side 5 min before adding the D-glucose (10 mM) into the same side. This figure showed that LP induced a concentration-response inhibition of I_{sc}. Increase in I_{sc} reflects a sodium-dependent glucose absorption through the sodium-glucose cotransporter-1 (SGLT1). The plateau represents the maximum increase in I_{sc}. (**B**) Represents the concentration–response curve related to the use of LP extract. n = 6–7 tissues studied.

The introduction of LP aqueous extract into the mucosal compartment, 5 min before adding the D-glucose (Figure 5), induced a significant and concentration-dependent inhibition of the I_{sc} (IC_{50} = 81.28 ± 4.01 µg/mL) as shown in Figure 6B. The maximum inhibition of the I_{sc} reached almost 60% of the inhibition induced by phloridzin (0.5 mM) and was obtained with a concentration of 500 µg/mL.

The inhibitory impact of LP aqueous extract on mucosal intestinal glucose absorption was not shown to be glucose concentration dependent.

3.6. Toxicological Studies

3.6.1. Acute Toxicity

Oral administration of the LP aqueous extract to mice at different doses (1, 3, 5, 7, and 10 g/kg b.w.) did not cause any mortality, and no signs of toxicity were noticed (diarrhea, vomiting, abnormal mobility, etc.) after a follow-up of 15 days.

3.6.2. Subchronic Toxicity

Effect on Body Weights and Relative Weights of Livers and Kidneys

The safety of the LP, FGK, PG, and GO aqueous extracts was evaluated by monitoring the variation in the body weight of normal rats during four weeks of the extract consumption at a dose of 1 g/kg b.w. (Figure 7). Also, this safety was assessed by measuring the relative weights of the livers and the kidneys at the end of the experimental period (Table 2). The obtained results showed that the variation in body weights and relative weights of organs of the treated rats is not significant in comparison with the control.

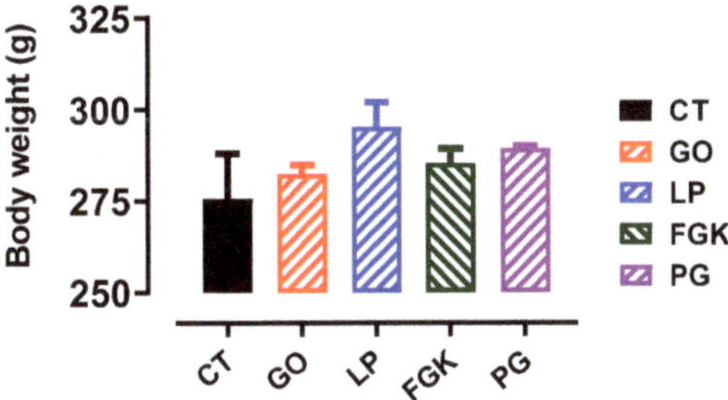

Figure 7. Effect of the oral administration of the plant aqueous extracts on the body weight variation in normal rats after four weeks.

Table 2. Effect of the oral administration of the plant aqueous extracts on relative weights of livers and kidneys in normal rats.

Extracts	Liver	Left Kidney	Right Kidney
Control	5.08 ± 0.40	0.65 ± 0.06	0.69 ± 0.07
LP	5.83 ± 0.62	0.54 ± 0.11	0.61 ± 0.11
FGK	5.98 ± 0.52	0.59 ± 0.07	0.66 ± 0.12
PG	6.75 ± 0.72	0.57 ± 0.07	0.61 ± 0.05
GO	4.92 ± 0.61	0.53 ± 0.04	0.53 ± 0.05

Effect on the Blood Biochemical Parameters

Biochemical parameters of the studied rats that were administered the LP, FGK, PG, and GO aqueous extracts for four weeks is shown in Table 3. The daily administration of these extracts at a dose of 1 g/kg b.w. did not cause any significant variation in these parameters in comparison with the control group.

Table 3. Analysis of the biochemical parameters of the rat's blood.

Biochemistry	Control	LP	FGK	PG	GO
ALT (UI/L)	18.33 ± 5.86	17.33 ± 0.59	25.33 ± 4.73	20.33 ± 3.79	11.00 ± 2.65
AST (UI/L)	59.33 ± 28.38	55.67 ± 24.83	63.67 ± 16.92	59.33 ± 19.35	50.00 ± 14.53
Direct bilirubin (mg/mL)	1.00 ± 0.00	1.00 ± 0.00	1.00 ± 0.00	1.00 ± 0.00	1.00 ± 0.00

Table 3. *Cont.*

Biochemistry	Control	LP	FGK	PG	GO
Total bilirubin (mg/mL)	1.43 ± 0.49	1.45 ± 0.49	1.27 ± 0.06	1.45 ± 0.07	1.30 ± 0.42
Albumin (g/L)	28.00 ± 2.00	25.00 ± 9.64	22.33 ± 2.08	24.67 ± 6.11	19.30 ± 6.51
Cholesterol (g/L)	0.79 ± 0.07	0.66 ± 0.25	0.86 ± 0.04	0.91 ± 0.26	0.71 ± 0.18
Triglyceride (g/L)	0.45 ± 0.12	0.40 ± 0.25	0.38 ± 0.17	0.40 ± 0.10	0.26 ± 0.04
HDL (g/L)	0.21 ± 0.04	0.23 ± 0.09	0.26 ± 0.07	0.24 ± 0.05	0.20 ± 0.03
LDL (g/L)	0.11 ± 0.02	0.10 ± 0.04	0.13 ± 0.05	0.14 ± 0.04	0.10 ± 0.06
Glucose (g/L)	1.58 ± 0.30	1.93 ± 0.80	1.85 ± 0.09	1.63 ± 0.33	1.41 ± 0.62
Lipase (UI/L)	9.67 ± 3.06	7.33 ± 2.89	11.50 ± 2.12	11.00 ± 6.24	5.67 ± 2.08
Protein (g/L)	55.00 ± 16.55	44.85 ± 25.39	58.83 ± 7.26	58.23 ± 16.39	31.50 ± 7.78
Creatinine (mg/L)	6.37 ± 1.13	5.45 ± 1.95	6.19 ± 0.68	5.98 ± 1.29	4.53 ± 1.68
Urea (g/L)	0.23 ± 0.06	0.16 ± 0.06	0.23 ± 0.05	0.18 ± 0.03	0.17 ± 0.03
Uric acid (mg/L)	17.70 ± 9.72	16.33 ± 8.20	14.83 ± 2.81	23.03 ± 11.57	16.77 ± 11.72
Calcium (mg/L)	88.23 ± 15.03	84.87 ± 26.58	88.20 ± 7.62	86.80 ± 20.78	50.50 ± 5.23
Phosphate (mg/L)	52.50 ± 8.92	70.15 ± 7.71	62.65 ± 12.52	53.10 ± 18.10	59.20 ± 27.72

3.6.3. Histology

Histological analysis of rat stomach sections (Figure 8A,B) and small intestinal sections (Figure 8C,D) did not show any modifications after the chronic administration of LP (B,D) while compared to the control (A,C). Moreover, histological analysis of rat liver (Figure 9a,b), spleen (Figure 9c,d) and kidney (Figure 9e,f) slices did not show any cellular modifications after the chronic administration of LP (b,d,f) while compared to the control (a,c,e).

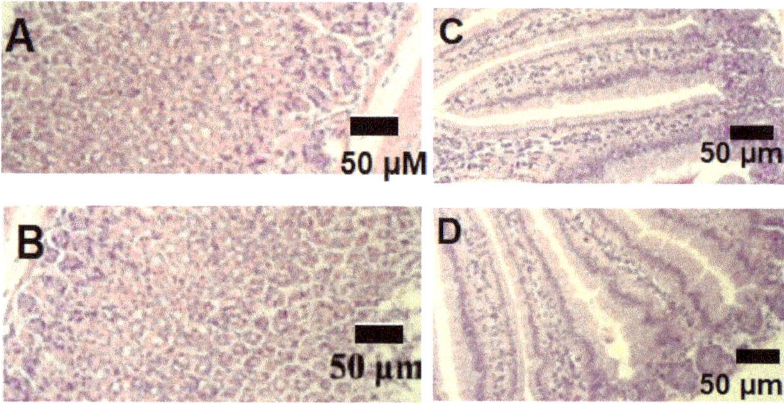

Figure 8. Histological analysis of rat stomach sections (**A,B**) and small intestine sections (**C,D**). The control groups (**A,C**) were given water and the treated groups (**B,D**) were given *Lavandula pedunculata* (LP) aqueous extract every day for four weeks. The figure shows the absence of any cellular modifications after the chronic administration of LP while compared to the control.

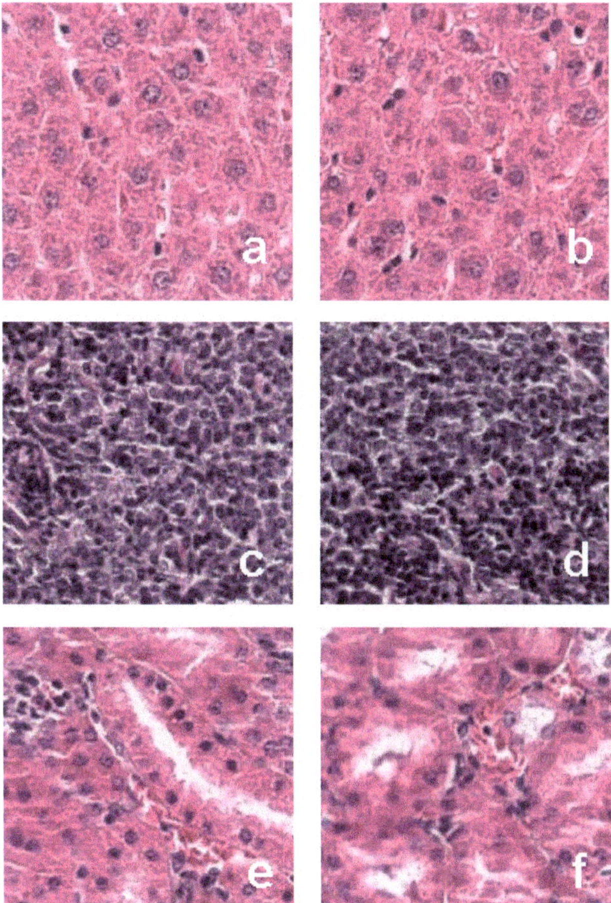

Figure 9. Histological analysis of tissues from control or *Lavandula pedunculata* (LP) treated rats. Rats were given LP or water every day for 4 weeks by gavage. Liver, spleen, and kidney slices were analyzed from control (**a,c,e**, respectively) or LP-treated rats (**b,d,f**, respectively). No cellular modifications were observed in the rats receiving LP in comparison with the control.

4. Discussion

UHPLC analysis of LP aqueous extract studied in this work demonstrated its high content in rosmarinic acid. Moreover, coumarin, protocatechuic acid, herniarin, caffeic acid, apigenin, luteolin, myricetin, and chlorogenic acid were also identified in this plant extract. Lopes and his collaborators identified some phenolic compounds of LP aqueous extract from different populations in Portugal using HPLC analysis. The results showed that phenolic acids are the most abundant, especially salvianolic acid B and rosmarinic acid present in large concentrations, and caffeic acid present in smaller ones. As for flavonoids, luteolin-7-O-glucuronide was the major compound in the studied extracts [26]. Another study was also conducted on different extracts of LP from Portugal that were analyzed using HPLC. The results showed high concentrations of rosmarinic acid and smaller ones of luteolin. Moreover, apigenin was present in the ethanolic and hydroethanolic extracts and absent in the aqueous ones [27].

Different studies demonstrated the inhibitory effects exerted by rosmarinic acid on the enzymatic activity of α-amylase and α-glucosidase. Other studies showed that this compound is able to improve insulin sensitivity and glucose uptake in animal models [28,29]. Furthermore, the antidiabetic activity of luteolin was assessed in mice and the results showed its potential in improving diabetes [30].

To our knowledge, this is the first study demonstrating that LP directly inhibits the pancreatic α-amylase and intestinal α-glucosidase enzyme activities in vitro, as well as the electrogenic intestinal glucose absorption in vitro, and that it improves the glucose tolerance in rats after acute and chronic oral administration in vivo. Although this inhibition of the intestinal glucose absorption remains low compared to that of *Arbutus unedo* root crude aqueous extract [31] and *Boscia senegalensis* seeds [32], the maximum effect is close to 60%. This work shows that LP could join the list of natural α-amylase, α-glucosidase and SGLT1 inhibitors. Acarbose, Phlorizin (a glucoside of phloretin) and other natural α-amylase, α-glucosidase, and SGLT1 inhibitors have been utilized as pharmacological agents or in the treatment of type 2 diabetes [33,34]. Moreover, LP could join the list of natural α-glucosidase inhibitors.

LP aqueous extract exhibited potent actions on hyperglycemia. It ameliorates the oral glucose tolerance in rats, and it exhibits an inhibiting effect on digestive enzymes and glucose absorption. Although there are no available studies on the antidiabetic activity of LP, there are few limited studies concerning other lavender species like *L. stoechas* and *L. officinalis*. Indeed, Bint Mustafa and her collaborators have shown in a previous study that the treatment of alloxan-induced diabetic mice with a hydroalcoholic (7:3) maceration of *L. stoechas* roots showed significant ($p < 0.05$) effects on fasting mice's blood glucose levels in a dose-dependent manner. The results were similar to the ones obtained by the reference drug (pioglitazone; 1 mg/kg b.w. intraperitoneal). The prepared plant extract was used at different concentrations (50, 100, and 150 mg/kg b.w.). It was used for intraperitoneal injection into the experimental mice at a dose of 0.1 mL/kg b.w. [15]. The same lavender species (*L. stoechas*) was also studied by Sebai et al. (2013). These researchers used its essential oil that they had obtained by hydrodistillation of the aerial parts. After 15 consecutive days of treatment with the essential oil (50 mg/kg b.w.), the obtained chronic effect showed that *L. stoechas* essential oil treatment significantly protected against the increase of blood glucose induced by alloxan treatment of *Wistar* rats [35]. In another research, the effect of *L. officinalis* aerial parts macerated in ethanol (96%) was studied in alloxan-induced diabetic *Wistar* rats. Two doses of the extract (100 and 200 mg/kg b.w.) were used during 21 days of treatment. The results showed that the two doses of the *L. officinalis* ethanolic extract significantly ($p < 0.01$) diminished the blood glucose level in comparison with the diabetic non-treated group [36].

In addition, the combination of LP with PG and FGK improved the oral glucose tolerance in rats after acute oral administration in vivo and it exhibited a higher antihyperglycemic activity in comparison with the individual plant extracts. Moreover, the activity obtained after combining LP with PG was judged to be similar to the reference drug MET. This amplifier effect obtained after combining the plant extracts could be either the result of the accumulation of their common active compounds, or a synergic reaction between their different compounds. The obtained results confirm and support the traditional use of LP in ACPP in type 2 diabetes therapy. In fact, LP is one of lavender species widely used by traditional healers in Morocco to prevent or cure diabetes in combination with other antidiabetic plants. This traditional method consists of combining plant extracts in the same homemade beverage to ameliorate or reduce the effect of healer potion. In Africa, it is one of the methods used to personalize treatment of patients, according to their own physical condition (child, elderly, pregnant woman) [21].

It was found in another study that the polyherbal preparation of *L. stoechas*, *Curcuma longa*, *Aegle marmelos*, and *Glycyrrhiza glabra* hydroalcoholic extracts at a dose of 150 mg/kg b.w. is significantly ($p < 0.05$) more antihyperglycemic than individual extracts [15]. Likewise, it was cited in an ethnobotanical study realized in Morocco that the population use a mixture of *L. stoechas*, *Artemisia herba alba*, and *Mentha rotundifolia* inflorescences in a decoction to treat diabetes [10].

ACPP is the possibility to use, in the same formulation, several plant extracts. It is also the possibility to combine/replace or alternate one or several plant extracts in the formula depending on the goal. ACPP offers phytopharmacologists a multitude of possibilities to formulate several combinations of plant extracts [32]. Polyphytotherapy offers other benefits in the fight against diabetes. It makes it possible to make a learned association of plants whose extracts act on several targets. This is perhaps the main reason why traditional healers combine and/or alternate different plants in their preparations against diabetes, because they can combine in the same preparation the extracts of plants which (1) stimulate the pancreas to release more insulin (secretagogues), (2) help the body use insulin (sensitizers), and (3) block the breakdown of starches and sugars (α-amylase and α-glucosidase inhibitors) such as LP.

For the time being, the present results are not supposed to alter the treatment of type 1 diabetes, which is based on life-saving insulinotherapy. However, it may be of interest to explore the possibility that LP, in combinative poly-phytotherapy, could decrease the amount of daily insulin requirement, as a consequence of improved oral glucose tolerance.

This study showed that LP aqueous crude extract inhibits glucose absorption, as well as the pancreatic α-amylase and intestinal α-glucosidase enzyme activities. These two cumulative effects could justify its use as an antidiabetic by the traditional healers. In a study on the traditional use of *Arbutus unedo* as an antidiabetic in Morocco [31], various pharmacological actions attributed to natural or synthetic SGLT inhibitors were listed, which makes them a potential new class of drugs that could be used to manage type 2 diabetes, including:

1. Reducing the glycemic index: in addition to their improving effect on the glycemic control, some SGLT inhibitors, like *Boscia senegalensis* (Boscisucrophage) and *Nigella sativa* (Biodiabétine), have been shown to prevent diabetic nephropathy by lowering glycosylated hemoglobin (HbAlC) levels in experimental animals [32].
2. Intestinal SGLT1 inhibition: The impact of LP aqueous extract on glucose absorption in the intestine is concentration dependent and comparable to that of other antidiabetics [31,37].
3. Limited action on renal SGLT1: plasma glucose levels are maintained near normal thanks to SGLT1 that has the ability to reabsorb glucose in the proximal tubule, preventing thereby the fasting hypoglycemia [38].
4. Furthermore, the aim of SGLT inhibitor therapy is to lowering postprandial blood glucose levels [39].

The findings of the present study indicate that systemic glucose homeostasis is improved in OGTT in conscious fasted rats. This improvement is better when LP is used in combination with PG (Figure 1) than when LP and PG are used alone. In addition, this combination is as good as the clinical chemical drug Metformin. These results confirm the effectiveness of combining plant extracts to reduce systemic blood sugar.

5. Conclusions

Lavandula pedunculata (Mill.) Cav. (LP) aqueous extract inhibited enzymatic activities of pancreatic α-amylase and intestinal α-glucosidase in vitro. In addition, this extract directly inhibited the electrogenic intestinal glucose absorption ex vivo. Moreover, LP improved the antihyperglycemic effect of PG and FGK in vivo in rats after acute oral administration (Figure 10). These findings back up the traditional use of LP in type 2 diabetes treatment and the effectiveness of the alternative and combinative poly-phytotherapy (ACPP). Still, more studies are needed to better understand all pharmacological parameters of this plant extract to be used as alternative medicine to treat hyperglycemia.

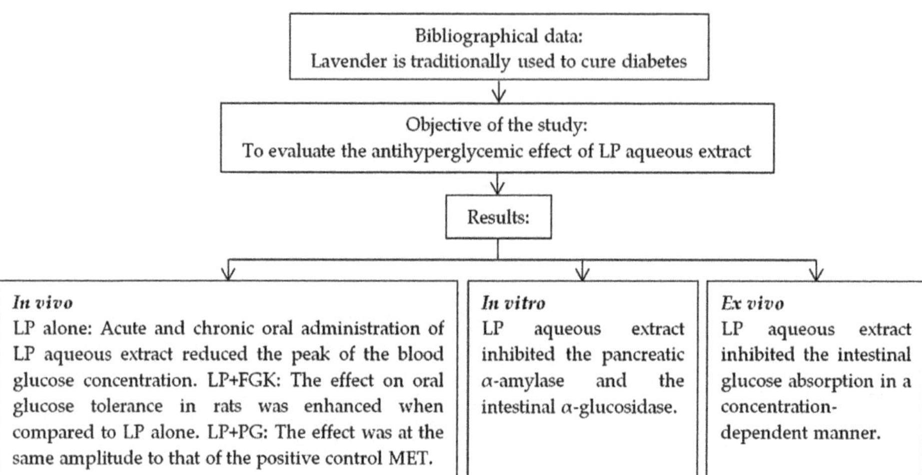

Figure 10. A schematic diagram of the study.

Author Contributions: Conceptualization, S.B. and B.E.; methodology, M.B. (Mohamed Bouhrim), M.B. (Mohamed Bnouham) and C.A.; data curation, H.M.; writing—original draft preparation, S.B., M.B. (Mohamed Bouhrim) and H.M.; writing—review and editing, T.Z. and B.E.; supervision, F.K.E., B.G., S.S. and J.-F.D. Validation A.S.A. and O.M.N. All authors have read and agreed to the published version of the manuscript.

Funding: This research was funded by researchers supporting project-under the group number (RSP-2021/132), King Saud University, Riyadh, Saudi Arabia and also researchers Supporting Project AP-ANPMA and UMI of September 2018, Morocco.

Institutional Review Board Statement: The study was conducted according to the guidelines of the Declaration of Helsinki and approved by the Institutional Review Board at the Faculty of Sciences, Oujda, Morocco (31/2019/LBBEH-05 and 21/011/2019).

Informed Consent Statement: Not applicable.

Data Availability Statement: Data are available upon request.

Acknowledgments: The authors acknowledge the generous support from researchers supporting project-under the group number (RSP-2021/132), King Saud University, Riyadh, Saudi Arabia.

Conflicts of Interest: The authors declare no conflict of interest.

References

1. American Diabetes Association. Diagnosis and Classification of Diabetes Mellitus. *Diabetes Care* **2014**, *37*, S81–S90. [CrossRef] [PubMed]
2. International Diabetes Federation Worldwide Toll of Diabetes. Available online: https://www.diabetesatlas.org/en/sections/worldwide-toll-of-diabetes.html (accessed on 2 May 2021).
3. Verma, S.; Gupta, M.; Popli, H.; Aggarwal, G. Diabetes Mellitus Treatment Using Herbal Drugs. *Int. J. Phytomedicine* **2018**, *10*, 1–10. [CrossRef]
4. American Diabetes Association. Classification and Diagnosis of Diabetes: Standards of Medical Care in Diabetes. *Diabetes Care* **2021**, *44*, S15–S33. [CrossRef]
5. Padhi, S.; Nayak, A.K.; Behera, A. Type II Diabetes Mellitus: A Review on Recent Drug Based Therapeutics. *Biomed. Pharmacother.* **2020**, *131*, 110708. [CrossRef] [PubMed]
6. Es-safi, I.; Mechchate, H.; Amaghnouje, A.; Elbouzidi, A.; Bouhrim, M.; Bencheikh, N.; Hano, C.; Bousta, D. Assessment of Antidepressant-Like, Anxiolytic Effects and Impact on Memory of *Pimpinella anisum* L. Total Extract on Swiss Albino Mice. *Plants* **2021**, *10*, 1573. [CrossRef]
7. Mechchate, H.; Es-safi, I.; Mohamed Al kamaly, O.; Bousta, D. Insight into Gentisic Acid Antidiabetic Potential Using In Vitro and In Silico Approaches. *Molecules* **2021**, *26*, 1932. [CrossRef]

8.   Bouhrim, M.; Boutahiri, S.; Kharchoufa, L.; Mechchate, H.; Mohamed Al Kamaly, O.; Berraaouan, A.; Eto, B.; Ziyyat, A.; Mekhfi, H.; Legssyer, A.; et al. Acute and Subacute Toxicity and Cytotoxicity of Opuntia Dillenii (Ker-Gawl) Haw. Seed Oil and Its Impact on the Isolated Rat Diaphragm Glucose Absorption. *Molecules* **2021**, *26*, 2172. [CrossRef]
9.   Barkaoui, M.; Katiri, A.; Boubaker, H.; Msanda, F. Ethnobotanical Survey of Medicinal Plants Used in the Traditional Treatment of Diabetes in Chtouka Ait Baha and Tiznit (Western Anti-Atlas), Morocco. *J. Ethnopharmacol.* **2017**, *198*, 338–350. [CrossRef]
10.   Benkhnigue, O.; Ben Akka, F.; Salhi, S.; Fadli, M.; Douira, A.; Zidane, L. Catalogue des plantes médicinales utilisées dans le traitement du diabète dans la région d'Al Haouz-Rhamna (Maroc). *J. Anim. Plant Sci.* **2014**, *23*, 3539–3568.
11.   Orch, H.; Douira, A.; Zidane, L. Étude Ethnobotanique Des Plantes Médicinales Utilisées Dans Le Traitement Du Diabète, et Des Maladies Cardiaques Dans La Région d'Izarène (Nord Du Maroc). *J. Appl. Biosci.* **2015**, *86*, 7940–7956. [CrossRef]
12.   Almohawes, Z.N.; Alruhaimi, H.S. Effect of *Lavandula dentata* Extract on Ovalbumin-Induced Asthma in Male Guinea Pigs. *Braz. J. Biol.* **2019**, *80*, 87–96. [CrossRef]
13.   Baptista, R.; Madureira, A.M.; Jorge, R.; Adão, R.; Duarte, A.; Duarte, N.; Lopes, M.M.; Teixeira, G. Antioxidant and Antimycotic Activities of Two Native *Lavandula* Species from Portugal. *Evid. Based Complement. Alternat. Med.* **2015**, *2015*, 1–10. [CrossRef] [PubMed]
14.   Kulabas, S.S.; Ipek, H.; Tufekci, A.R.; Arslan, S.; Demirtas, I.; Ekren, R.; Sezerman, U.; Tumer, T.B. Ameliorative Potential of *Lavandula Stoechas* in Metabolic Syndrome via Multitarget Interactions. *J. Ethnopharmacol.* **2018**, *223*, 88–98. [CrossRef] [PubMed]
15.   Bint Mustafa, S.; Akram, M.; Muhammad Asif, H.; Qayyum, I.; Mehmood Hashmi, A.; Munir, N.; Said Khan, F.; Riaz, M.; Ahmad, S. Antihyperglycemic Activity of Hydroalcoholic Extracts of Selective Medicinal Plants *Curcuma longa, Lavandula stoechas, Aegle marmelos*, and *Glycyrrhiza glabra* and Their Polyherbal Preparation in Alloxan-Induced Diabetic Mice. *Dose-Response Int. J.* **2019**, *17*, 1–6. [CrossRef]
16.   Rahmati, B.; Kiasalari, Z.; Roghani, M.; Khalili, M.; Ansari, F. Antidepressant and Anxiolytic Activity of *Lavandula officinalis* Aerial Parts Hydroalcoholic Extract in Scopolamine-Treated Rats. *Pharm. Biol.* **2017**, *55*, 958–965. [CrossRef] [PubMed]
17.   Sadraei, H.; Asghari, G.; Rahmati, M. Study of Antispasmodic Action of *Lavandula angustifolia* Mill Hydroalcoholic Extract on Rat Ileum. *J. Herbmed Pharmacol.* **2019**, *8*, 56–63. [CrossRef]
18.   Shahdadi, H.; Bahador, R.S.; Eteghadi, A.; Boraiinejad, S. Lavender a Plant for Medical Uses: A Literature Review. *Indian J. Public Health Res. Dev.* **2017**, *8*, 328–332. [CrossRef]
19.   Chaachouay, N.; Benkhnigue, O.; Fadli, M.; El Ibaoui, H.; El Ayadi, R.; Zidane, L. Ethnobotanical and Ethnopharmacological Study of Medicinal and Aromatic Plants Used in the Treatment of Respiratory System Disorders in the Moroccan Rif. *Ethnobot. Res. Appl.* **2019**, *18*, 1–17. [CrossRef]
20.   Teixidor-Toneu, I.; Martin, G.J.; Ouhammou, A.; Puri, R.K.; Hawkins, J.A. An Ethnomedicinal Survey of a Tashelhit-Speaking Community in the High Atlas, Morocco. *J. Ethnopharmacol.* **2016**, *188*, 96–110. [CrossRef]
21.   Ghourri, M.; Zidane, L.; Douira, A. Usage des plantes médicinales dans le traitement du Diabète Au Sahara marocain (Tan-Tan). *J. Anim. Plant Sci.* **2013**, *17*, 2388–2411.
22.   National Research Council. *Guide for the Care and Use of Laboratory Animals*, 8th ed.; The National Academies Press: Washington, DC, USA, 2011.
23.   Bouhrim, M.; Ouassou, H.; Boutahiri, S.; Daoudi, N.E.; Mechchate, H.; Gressier, B.; Eto, B.; Imtara, H.; Alotaibi, A.A.; Al-zharani, M.; et al. *Opuntia dillenii* (Ker Gawl.) Haw., Seeds Oil Antidiabetic Potential Using In Vivo, In Vitro, In Situ, and Ex Vivo Approaches to Reveal Its Underlying Mechanism of Action. *Molecules* **2021**, *26*, 1677. [CrossRef]
24.   Eto, B.; Boisset, M.; Griesmar, B.; Desjeux, J.-F. Effect of Sorbin on Electrolyte Transport in Rat and Human Intestine. *Am. J. Physiol.* **1999**, *276*, G107–G114. [CrossRef]
25.   Mathieu, J.; Mammar, S.; Eto, B. Automated Measurement of Intestinal Mucosa Electrical Parameters Using a New Digital Clamp. *Methods Find. Exp. Clin. Pharmacol.* **2008**, *30*, 591–598. [CrossRef]
26.   Lopes, C.L.; Pereira, E.; Soković, M.; Carvalho, A.M.; Barata, A.M.; Lopes, V.; Rocha, F.; Calhelha, R.C.; Barros, L.; Ferreira, I.C.F.R. Phenolic Composition and Bioactivity of *Lavandula pedunculata* (Mill.) Cav. Samples from Different Geographical Origin. *Molecules* **2018**, *23*, 1037. [CrossRef]
27.   Costa, P.; Gonçalves, S.; Valentão, P.; Andrade, P.B.; Almeida, C.; Nogueira, J.M.F.; Romano, A. Metabolic Profile and Biological Activities of *Lavandula pedunculata* Subsp. *Lusitanica* (Chaytor) Franco: Studies on the Essential Oil and Polar Extracts. *Food Chem.* **2013**, *141*, 2501–2506. [CrossRef]
28.   Ngo, Y.L.; Lau, C.H.; Chua, L.S. Review on Rosmarinic Acid Extraction, Fractionation and Its Anti-Diabetic Potential. *Food Chem. Toxicol.* **2018**, *121*, 687–700. [CrossRef]
29.   Tolmie, M.; Bester, M.J.; Apostolides, Z. Inhibition of α-Glucosidase and α-Amylase by Herbal Compounds for the Treatment of Type 2 Diabetes: A Validation of in Silico Reverse Docking with in Vitro Enzyme Assays. *J. Diabetes* **2021**, *13*, 779–791. [CrossRef]
30.   Zang, Y.; Igarashi, K.; Li, Y. Anti-Diabetic Effects of Luteolin and Luteolin-7-O-Glucoside on KK-A(y) Mice. *Biosci. Biotechnol. Biochem.* **2016**, *80*, 1580–1586. [CrossRef] [PubMed]
31.   Naceiri Mrabti, H.; Faouzi, M.E.A.; Massako Mayuk, F.; Makrane, H.; Limas-Nzouzi, N.; Dibong, S.D.; Cherrah, Y.; Kouoh Elombo, F.; Gressier, B.; Desjeux, J.-F.; et al. *Arbutus unedo* L., (Ericaceae) Inhibits Intestinal Glucose Absorption and Improves Glucose Tolerance in Rodents. *J. Ethnopharmacol.* **2019**, *235*, 385–391. [CrossRef] [PubMed]
32.   Eto, B. *Clinical Phytopharmacology Improving Health Care in Developing Countries*; LAP LAMBERT Academic Publishing: Paris, France, 2019.

33. Crane, R.K. Intestinal Absorption of Sugars. *Physiol. Rev.* **1960**, *40*, 789–825. [CrossRef] [PubMed]
34. Ríos, J.L.; Francini, F.; Schinella, G.R. Natural Products for the Treatment of Type 2 Diabetes Mellitus. *Planta Med.* **2015**, *81*, 975–994. [CrossRef] [PubMed]
35. Sebai, H.; Selmi, S.; Rtibi, K.; Souli, A.; Gharbi, N.; Sakly, M. Lavender (*Lavandula stoechas* L.) Essential Oils Attenuate Hyperglycemia and Protect against Oxidative Stress in Alloxan-Induced Diabetic Rats. *Lipids Health Dis.* **2013**, *12*. [CrossRef] [PubMed]
36. Azarmi, M.; Saei, H.; Zehtab Salmasi, S.; Dehghan, G. Investigating the Treatable Effect of *Lavandula officinalis* L. Ethanolic Extract in Alloxan-Induced Male Diabetic Rats. *Pharmacol. Line* **2016**, *1*, 53–63.
37. Meddah, B.; Ducroc, R.; Faouzi, M.E.A.; Eto, B.; Mahraoui, L.; Benhaddou-Andaloussi, A.; Martineau, L.C.; Cherrah, Y.; Haddad, P.S. Nigella Sativa Inhibits Intestinal Glucose Absorption and Improves Glucose Tolerance in Rats. *J. Ethnopharmacol.* **2009**, *121*, 419–424. [CrossRef]
38. Katsuno, K.; Fujimori, Y.; Takemura, Y.; Hiratochi, M.; Itoh, F.; Komatsu, Y.; Fujikura, H.; Isaji, M. Sergliflozin, a Novel Selective Inhibitor of Low-Affinity Sodium Glucose Cotransporter (SGLT2), Validates the Critical Role of SGLT2 in Renal Glucose Reabsorption and Modulates Plasma Glucose Level. *J. Pharmacol. Exp. Ther.* **2007**, *320*, 323–330. [CrossRef]
39. Du, F.; Hinke, S.A.; Cavanaugh, C.; Polidori, D.; Wallace, N.; Kirchner, T.; Jennis, M.; Lang, W.; Kuo, G.-H.; Gaul, M.D.; et al. Potent Sodium/Glucose Cotransporter SGLT1/2 Dual Inhibition Improves Glycemic Control Without Marked Gastrointestinal Adaptation or Colonic Microbiota Changes in Rodents. *J. Pharmacol. Exp. Ther.* **2018**, *365*, 676–687. [CrossRef]

pharmaceutics

Article

Protective Effect of Natural Antioxidant, Curcumin Nanoparticles, and Zinc Oxide Nanoparticles against Type 2 Diabetes-Promoted Hippocampal Neurotoxicity in Rats

Shaymaa Abdulmalek [1,2], Mayada Nasef [1], Doaa Awad [1] and Mahmoud Balbaa [1,*]

[1] Department of Biochemistry, Faculty of Science, Alexandria University, Alexandria 21511, Egypt; shimaa_salamy@yahoo.com (S.A.); mayadanasef90@gmail.com (M.N.); doaaelsayed363@hotmail.com (D.A.)
[2] Center of Excellency for Preclinical Study (CE-PCS), Pharmaceutical and Fermentation Industries Development Centre, City of Scientific Research and Technological Applications (SRTA-City), New Borg El-Arab City 21934, Egypt
* Correspondence: mahmoud.balbaa@alexu.edu.eg; Fax: +20-39-1179-4320

Abstract: Numerous epidemiological findings have repeatedly established associations between Type 2 Diabetes Mellitus (T2DM) and Alzheimer's disease. Targeting different pathways in the brain with T2DM-therapy offers a novel and appealing strategy to treat diabetes-related neuronal alterations. Therefore, here we investigated the capability of a natural compound, curcumin nanoparticle (CurNP), and a biomedical metal, zinc oxide nanoparticle (ZnONP), to alleviate hippocampal modifications in T2DM-induced rats. The diabetes model was induced in male Wistar rats by feeding a high-fat diet (HFD) for eight weeks followed by intraperitoneal injection of streptozotocin (STZ). Then model groups were treated orally with curcumin, zinc sulfate, two doses of CurNP and ZnONP, as well as metformin, for six weeks. HFD/STZ-induced rats exhibited numerous biochemical and molecular changes besides behavioral impairment. Compared with model rats, CurNP and ZnONP boosted learning and memory function, improved redox and inflammation status, lowered Bax, and upregulated Bcl2 expressions in the hippocampus. In addition, the phosphorylation level of the MAPK/ERK pathway was downregulated significantly. The expression of amyloidogenic-related genes and amyloid-beta accumulation, along with tau hyperphosphorylation, were lessened considerably. In addition, both nanoparticles significantly improved histological lesions in the hippocampus. Based on our findings, CurNP and ZnONP appear to be potential neuroprotective agents to mitigate diabetic complications-associated hippocampal toxicity.

Keywords: diabetes mellitus; neurotoxicity; curcumin nanoparticle; zinc oxide nanoparticle; metformin; inflammation; oxidative stress; apoptosis

Citation: Abdulmalek, S.; Nasef, M.; Awad, D.; Balbaa, M. Protective Effect of Natural Antioxidant, Curcumin Nanoparticles, and Zinc Oxide Nanoparticles against Type 2 Diabetes-Promoted Hippocampal Neurotoxicity in Rats. *Pharmaceutics* **2021**, *13*, 1937. https://doi.org/10.3390/pharmaceutics13111937

Academic Editors: Marta González-Álvarez

Received: 13 October 2021
Accepted: 8 November 2021
Published: 16 November 2021

Publisher's Note: MDPI stays neutral with regard to jurisdictional claims in published maps and institutional affiliations.

1. Introduction

Type 2 diabetes mellitus (T2DM) is associated with an increased risk of neurocognitive dysfunction and Alzheimer's disease (AD). As a progressive and irreversible disease, AD is characterized by neuronal cell death, increased neurofibrillary tangles deposition inside cells, and amyloid plaque production in the gaps between neurons [1]. There is currently no viable treatment for this complex condition [2]. Additionally, in AD, tau undergoes various post-translational changes in addition to phosphorylation, which are thought to be involved in its pathological assembly, causing the breakdown of neuronal cells [3].

While the etiology of AD remains unknown, numerous studies increasingly focus on the risk factors that may shed light on the pathophysiology of the disease, such as overnutrition, aging, and T2DM [4]. Amongst various risk factors, obesity, a public health problem in Western nations, is thought to be the prominent link between T2DM and neurocognitive impairment. Furthermore, obese animal models have shown impaired insulin signaling before developing T2DM [5]. It has also been documented those changes in carbohydrate metabolism play an essential role in the pathophysiology of T2DM and act as an important contributor to the pathogenic mechanisms related to reduced cognitive performance [6,7].

Moreover, it is worth noting that chronic inflammation is a common risk factor for developing T2DM and AD. Many dementia-related neurodegenerative disorders are linked to the start and progression of neuroinflammation [8]. It is characterized by enhanced activation of microglia and is frequently accompanied by an increase in inflammatory cytokines, such as interleukin-1 (IL-1) and tumor necrosis factor (TNF) [9]. Various investigations have found that neuroinflammatory alterations develop with age, even in healthy brains, and include increased levels of cytokines in serum and CSF [10,11]. These accumulated alterations can be cytotoxic and affect critical neural functions, resulting in the development of neurodegenerative diseases [12].

Among several potential molecular pathways, the MAPK/ERK pathway is an evolutionarily well-preserved signaling mode that regulates fundamental cellular processes, such as proliferation, cellular survival, and differentiation. In addition, it provides an excellent example of such kinase cascades and as a therapeutic target for neurodegeneration [13,14]. There are three protein families: extracellular signal kinase (ERK), p38 kinase, and c-Jun N-terminal kinase (JNK) families. ERK1/2 and p38 MAPK are triggered under different cellular processes, including stress and inflammatory conditions, and regulate the expression of pro-inflammatory cytokines [15]. They also modulate the apoptotic pathway [16]. Moreover, Amyloid-beta (Aβ) can modify the ERK/MAPK pathway and trigger memory impairment [17].

Considering the biochemical link between AD and T2DM [18], a similar therapeutic possibility may exist for both diseases. Due to their effectiveness, relatively few side effects, and simple accessibility, natural compounds may provide an alternative medication for diabetes and the promise of a new therapy for AD [19]. Historically, natural medicinal plants have played and will continue to play an important role in drug discovery as they possess valuable therapeutic properties. The isolated bioactive compounds derived from herbal plants can be used for the discovery of new drugs. Curcumin is one of the natural phytochemicals in turmeric and is a potent anti-amyloidogenic agent which has been found to effectively disaggregate Aβ species directly, which in turn hinders the formation of fibrils and oligomers and improves cognitive performance [20,21]. In addition, curcumin attenuated the oxidative stress in a streptozotocin (STZ)-induced model by enhancing reduced glutathione (GSH) content and reducing the acetylcholinesterase activity in the hippocampus and cerebral cortex [22,23]. Previous studies showed that curcumin increased lifespan and decreased amyloid neurotoxicity in an invertebrate model [24,25].

Unfortunately, some studies revealed that pure curcumin has numerous toxic effects on living tissues [26,27]. In addition to this, the insufficient absorption and low bioavailability of pure curcumin are the major problems when it comes to oral administration [28]. A possible solution to these problems would be to develop curcumin nanoparticles (CurNP) to enhance their stability, bioavailability, and reduce the undesirable effects [29].

Recently, biomedical nanomaterials have received more attention because of their salient biological features and biomedical applications. With the development of nanomaterials, metal oxide nanoparticles exhibit promising and far-ranging prospects in biomedicine [30]. Zinc oxide nanoparticles (ZnONP)—one of the most important classes of metal oxide nanoparticles—are commonly utilized in various fields due to their physical and chemical properties [31]. ZnONP, a new type of low-toxicity and low-cost nanomaterial, has attracted great interest in different biomedical fields, including anticancer, antioxidant, anti-inflammatory, and antidiabetic activities, as well as for bioimaging applications and drug delivery [32,33].

Although studies are implicating the adverse effects of ZnONP on the animal; and indicating its harmful effect on oxidative stress and inflammation, not enough studies are being carried out to investigate the neuroprotective actions of ZnONP in obese-T2DM animal models. However, the capability of ZnONP in mitigating diabetes-induced neurotoxicity in the hippocampus and improving diabetes-associated cognitive changes compared to its classical form, zinc sulfate, is still incomplete and partially contradictory.

Through these biological effects of CurNP and ZnONP, the study reported here focused on the therapeutic efficacy of prepared, naturally occurring CurNP and a biomedical metal oxide, ZnONP, compared with their classical forms, curcumin and zinc sulfate, as well as an antidiabetic drug, metformin, on learning and memory functions, the main amyloidogenic cascade, neuroinflammation, oxidative stress, and neuro-apoptosis caused by a high-fat diet (HFD) and STZ administration in the hippocampus of rats. We also examined the role of prepared nanoparticles on the MAPK/ERK pathway in the hippocampus of T2DM-induced rats to figure out the most effective and safe compound(s) among the current therapies and find the effective dose, and to support the rationale for using naturally occurring nanoparticles and metal oxide-based nanomaterials in clinical trials for preventing or treating T2DM and its complications, particularly AD.

2. Materials and Methods

2.1. Preparation and Characterization of CurNP and ZnONP Nanoparticles

Curcumin nanoparticles were prepared using a dropwise method. Briefly, pure curcumin powder (Sigma-Aldrich, St. Louis, MO, USA) was dissolved in Dichloromethane (Sigma-Aldrich, MO, USA) to prepare a stock solution, 5 mg/mL. Then 1 mL from curcumin stock solution was added to 50 mL boiling water under ultrasonication for around 30 min, then stirred for 20 min until the orange-colored precipitate was obtained. Zinc oxide nanoparticles were made by adding 2 M sodium hydroxide (Sigma-Aldrich, MO, USA) to an aqueous solution of 1M zinc sulfate heptahydrate (Sigma-Aldrich, MO, USA), followed by 2 mL 0.01% polyvinyl alcohol and vigorously stirred for about 18 h. Curcumin and ZnO nanoparticles were fixed for TEM investigation by placing a drop of the suspension on carbon-coated copper grids. The samples were dried under an infrared lamp and then the images were recorded using TEM (JEOL JEM-1400Flash instrument, Tokyo, Japan). The average particle size and polydispersity index (PDI) were evaluated by Zetasizer (Malvern Instruments, Malvern, WR14 1XZ, United Kingdom).

2.2. Experimental Animals

The animal study was performed in the laboratory animal house of the Medical Research Institute, Alexandria University, and compliance with the policy of animal care. Eighty male Wistar rats (Medical Research Institute, Alexandria University, Alexandria, Egypt), approximately 8–10 months of age, and weighing 120–150 g, were used in this study and were kept (five rats/cage) in polycarbonate cages with stainless-steel wire led with ad libitum feeding, a constant ambient temperature of 22 ± 2 °C, a humidity of 55 ± 5%, and a light-dark cycle of 12 h. Animals were humanely cared for during all experiments, and all efforts were made to minimize animal suffering.

2.3. Type 2 Diabetes and Neurodegeneration Rat Model

After two weeks of acclimatization, the rats were further divided into nine groups as follows: control group, received saline; HFD/STZ-induced group, induced to establish a simultaneous T2DM and neurodegeneration disease model; HFD/STZ-Cur, model rats treated with pure curcumin at a dose of 50 mg/kg body weight (b.w.); HFD/STZ-CurNP-10, model rats treated with CurNP at a dose of 10 mg/kg b.w.; HFD/STZ-CurNP-50, model rats treated with CurNP at a dose of 50 mg/kg b.w.; HFD/STZ-Zinc sulfate, model rats treated with zinc sulfate at a dose of 50 mg/kg b.w.; HFD/STZ-ZnONP-10, model rats treated with ZnONP at a dose of 10 mg/kg b.w; HFD/STZ-ZnONP-50, model rats treated with ZnONP at a dose of 50 mg/kg b.w.; and HFD/STZ-Metformin, model rats treated with metformin at a dose of 100 mg/kg b.w. All current treatment options were orally administered to rats daily for six weeks. The doses were administered daily at the same time and prepared by dissolving suitable doses of each therapy in saline according to the body weight of each rat and then ingested by oral gavage. The induction model of type

2 diabetes and neurodegeneration was established by feeding rats HFD for eight weeks followed by intraperitoneal injection of 35 mg/kg streptozotocin dissolved in 0.1 M citrate buffer, pH 4.5. Blood samples were taken three days after injection from overnight fasting rats to measure blood glucose levels. Type 2 diabetes was defined as a fasting blood glucose level of more than 350 mg/dL.

2.4. Morris Water Maze Test (MWM)

The spatial learning and memory function of rats was assessed with an MWM test. The MWM contained a circular pool of 200 cm diameter and 50 cm deep, which filled with water $22 \pm 2\,^{\circ}$C, plus a 10 cm diameter escape platform submerged below the surface of the water by 1 cm. Firstly, rats were trained to seek the platform that was submerged in water and exit the pool. The test was carried out for five consecutive days and the rats went through three training trials each day. Each rat was given a trial for only 180 s to successfully find the hidden platform. The time required for each rat to detect the platform, escape latency, the distance swum, and the escape distance were measured. The shorter the escape latency, the stronger spatial learning ability. On the sixth day, for the second phase of testing, a probe test to detect memory maintenance or deficit after removal of the platform, each trial was 60 s in duration. The swimming path length, speed, and time spent in the target quadrant were measured.

2.5. Sample Preparation and Biochemical Evaluations

After six weeks of treatment, rats were fasted overnight, then anesthetized using sodium pentobarbital (100 mg/kg, i.p.) and blood was drawn from the abdominal aorta with a syringe puncture, and serum was separated. Some brains were immediately removed and stored at $-80\,^{\circ}$C for RNA isolation and western blot testing and others were handled by isolating the hippocampus on ice. The separated hippocampus tissue was homogenized in lysis buffer with protease inhibitor (150 mM NaCl, 1% Triton X-100, 10 mM Tris, pH 7.4) and centrifuged at $10{,}000\times g$ at $4\,^{\circ}$C for 15 min. The total protein contents were measured with the Lowry method [34]. The remaining part of the hippocampus was stored in 10% formalin for histopathological study. The blood glucose levels were monitored with an ACCU-CHEK active blood glucose meter. Serum insulin was performed according to the ELISA technique using a commercial kit specific for rats (MyBioSource, MBS724709) according to the manufacturer's instructions provided with the kit. Insulin resistance estimation was performed using the homeostasis model assessment method; HOMA-IR was calculated by the following formula: plasma glucose (mg/dL) $\times$ fasting plasma insulin (IU mg/L in the fasting state divided by 405) [35]. In addition, AGEs levels were determined in serum by a specific ELISA kit (MyBioSource, MBS261131), and the assays were performed following the manufacturer's manual.

2.6. Measurement of Hippocampal Oxidant and Antioxidant Biomarkers

Malondialdehyde levels were assessed in the form of thiobarbituric acid-reactive substances (TBARS) using a spectrophotometric colorimetric assay at 532 nm; the concentration of TBARS in the hippocampus was expressed as nmol/mg protein [36]. Nitric oxide (NO) was assayed according to the standard procedure of the Griess reaction, and the absorbance was read at 540 nm against the blank, then expressed as μM/mg protein [37]. GSH was examined and the generated yellow color was detected at 412 nm immediately, and the GSH concentration was reported as mg/mg sample protein [38]. Glutathione-S-Transferase (GST) activity was measured in the hippocampus at 310 nm and expressed as μmol/min/mg protein [39]. In the assay of superoxide dismutase (SOD), the absorbance was measured at 420 nm and the sample enzyme activity was reported in U/mg protein [40]. The activity of catalase (CAT) was measured by measuring the breakdown of its substrate H_2O_2 at 240 nm and calculating the change in absorbance per minute [41]. The activity of glutathione peroxidase (GPx) was measured and estimated using the method described previously and expressed as mol/min/mg protein [42].

2.7. Quantification of Serum Adipokines Levels and Hippocampal IL-6 and TNF-α

Leptin (MyBioSource, MBS012834) and adiponectin (MyBioSource, MBS068220) levels were detected in serum, while IL-6 (MyBioSource, MBS355410) and TNF-α (MyBioSource, MBS2507393) were detected in the hippocampus using rat specific ELISA kits following the manufacture's instruction.

2.8. Quantification of Acetylcholine Esterase Activity (AChE), IDE, and Aβ-42 Concentrations

As previously mentioned, the activity of AChE was evaluated using Ellman's reagent colorimetric assay [43]. AChE activity was measured at 412 nm using 0.75 mM acetylthiocholine and 0.5 mM 5,5-dithiobis (2-nitrobenzoic acid) (DTNB) in 5 mM HEPES buffer (pH 7.5). Additionally, Aβ-42 (MyBioSource, MBS726579) and insulin-degrading enzyme (IDE) (MyBioSource, MBS9139258) levels were detected in the hippocampus using rat specific ELISA kits following the manufacture's protocol.

2.9. Real-Time Polymerase Chain Reaction (RT-PCR)

Total RNA was isolated from the frozen hippocampus and cortex of rats using QIAzol Lysis Reagent (Qiagen, 79306) according to the manufacturer's instructions. Then, 1 µg of total RNA was subjected to reverse transcription in triplicate for cDNA preparation, and the PCR reaction was carried out using SYBER green master mix (Qiagen, 204143); the mRNA expression levels were calculated relative to β-actin gene's mRNA levels using $2^{-\Delta\Delta Ct}$ method [44]. PCR conditions were set with an initial incubation at 95 °C for 10 min, and 40 cycles at 95 °C for 15 s, 60 °C for 1 min, and 72 °C for 40 s. Primer sequence (Sigma-Aldrich): APP (NM_019288.2), F-5′-AGAGGTCTACCCTGAACTGC-3′, R-5′-ATCGCTTACAAACTCACCAAC-3′; BACE1 (NM_019204.2), F-5′-CGGGAGTGGTATTAT GAAGTG-3′, R-5′-AGGATGGTGATGCGGAAG-3′; BDNF (NM_012513.4), F-5′-ATGGGAC TCTGGAGAGCGTGAA-3′, R-5′-CGC CAGCCA ATTCTC TTT TTGC-3′; ADAM10 (NM_01 9254.1), F-5′-GTTAATTCTGCTCCTCTCCTGG-3′, R-5′-TGGATATCTGGGCAATCACAGC-3′; Bcl-2 (NM_016993.2), sense: 5′-GCAGCTTCTTTCCCCGGAAGGA-3′, antisense: 5′-AGGTGCAGCTGACTGGACATCT-3′; Bax (NM_017059.2), sense: 5′-AACTTCAACTGGG GCCGCGTGGTT-3′, antisense: 5′-CATCTTCTTCCAGATGGTGAGCGAG-3′; β-actin (NM_ 031144.3), sense: 5′-TGAGAGGGAAATCGTGCGT-3′, anti-sense 5′-TCATGGATGCCACA GGATTCC-3′.

2.10. Western Blotting

The frozen hippocampus tissue was homogenized in lysis buffer (100 mM NaCl, 100 mM EDTA, 0.5% Nonidet p-40, 0.5% Na-deoxycholate, 10 mM Tris, pH 7.5, containing protease inhibitors). After centrifuging the sample for 10 min at $2000 \times g$ and 4 °C, the supernatant was collected, and the protein content was determined. The protocol for Western blot testing was followed as stated previously [45]. Briefly, 50 µg of protein samples were mixed with 2X loading buffer and boiled at 100 °C for 10 min. The proteins were then separated using SDS-PAGE (12%) and transferred to nitrocellulose membranes. Then the membranes were blocked by soaking for 1 h at room temperature in a solution of 5% nonfat milk in TBST buffer (10 mM Tris–HCl, pH 8.0, 150 mM NaCl, and 0.2 percent Tween-20). The membranes were then treated with primary antibodies overnight at 4 °C (β-actin (NB600-501), p-p38 MAPK (9215), p38 MAPK (9211), p-Erk1/2 (4377), ERK1/2 (9102), MEK1/2 (9122), p-MEK1/2 (Ser 217/221; 9154)), then washed with TBST after primary incubation and then incubated with secondary antibodies for 1 h at room temperature. The signal was developed using a TMB Western Blotting Detection kit. Image Lab 6.1 software (BioRad) was then used to quantify the produced bands.

2.11. Histological Analysis

The fixed brain tissues in 10% formalin were processed for dehydration in ascending grades of alcohol, followed by impregnation. The specimens were then embedded in paraffin and allowed to solidify at room temperature. Using a rotatory microtome, serial sections of 5 μm thick were cut. After that, sections were stained with Haematoxylin and Eosin (H&E) and observed for histopathological changes.

2.12. Statistical Analysis

Results are shown as mean $\pm$ SEM (standard error of the mean) for eight rats in each group. Significant differences between groups were compared by one-way analysis of variance (ANOVA), followed by Tukey's post hoc test for multiple comparisons; with a p-value < 0.01, the difference between the groups was considered statistically significant.

3. Results

3.1. Characterization of ZnONP and CurNP

TEM analysis displayed that ZnONP had a diameter between 11 and 22 nm while CurNP had a diameter between 44 and 52 nm and both nanoparticles had a spherical shape (Figure 1A). Additionally, the particle size of ZnONPs and CurNPs (Figure 1B) was 21.00 nm and 58.80 nm, respectively, confirming the results obtained from TEM analysis. Where the PDI was (0.201 ± 0.01) for ZnONP and (0.141 ± 0.01) for CurNP. CurNP and curcumin FTIR spectra were measured in the region of 4000–400 cm^{-1}. The stretching vibration of hydrogen-bonded OH found in curcumin correlates to the intense band at wavenumber 3508 cm^{-1}. The frequency of aromatic CH was determined to be 1628 cm^{-1}. The aromatic stretching vibrations of the benzene ring are represented by 1427 cm^{-1}. The intense characteristic band centered at 1507 cm^{-1} represents the stretching vibration of the conjugated carbonyl (C=O). The stretching vibration of hydrogen-bonded OH observed in nano curcumin corresponds to the broad, strong band at wavenumbers 3268 and 3268 cm^{-1}. The stretching vibrations of the benzene ring are shown by bands at 1455 and 1430 cm^{-1}. The characteristic band centered at 1513 cm^{-1} represents the stretching vibration of the conjugated carbonyl (C=O). Stretching vibrations of C=C bonds are shown at 1598 (Figure 2C). The chain absorption peaks in the FTIR spectra of ZnONP were 400 and 4000 cm^{-1}. The C=C stretch of alkenes or the C=O stretch of amides is represented by the absorption peaks at 1631, 1505, and 1381 cm^{-1}. Stretching vibrations of the C–O bond are responsible for the bands at 1076 cm^{-1}. Peaks at 692 and 899 cm^{-1} may also be related to C–N stretching amine groups. Furthermore, in the absorption peaks of 3433 and 3396 cm^{-1}, hydroxyl group stretching can be noticed. The primary absorption bands 422, 434, and 470 were assigned to the Zn–O bond and are distinctive of it (Figure 2D).

3.2. Effects of CurNP and ZnONP on Memory Deficits in T2DM-Induced Rat Models

The MWM test was used to determine the time it took rats to identify and mount a water maze platform (escape latency) and the total swimming distance they covered before escaping (escape distance). There was no significant difference in escape latency (Figure 2A, $p < 0.01$) or distance (Figure 2B, $p < 0.01$) for either group for the first three days. On the fourth day, however, the model group had significantly longer escape latency and distance than the controls ($p < 0.01$). Furthermore, compared to model rats, the rats receiving current medications had considerably lower escape latency and distance ($p < 0.01$). However, rats given 50 mg/kg ZnONP or 10 mg/kg CurNP had significantly longer escape latency and distance than rats given other treatments ($p < 0.01$). Probe testing showed that the model group had a considerably shorter swim time (Figure 2C) and distance swum (Figure 2D) in the target quadrant (represented as a percentage of total time and distance swum) than the control group ($p < 0.01$). Treatment groups had considerably higher percentages of time and distance swum compared to model rats ($p < 0.01$). However, rats given 50 mg/kg ZnONP or 10 mg/kg CurNP spent significantly less time and swam significantly lesser distances than rats given other treatments ($p < 0.01$).

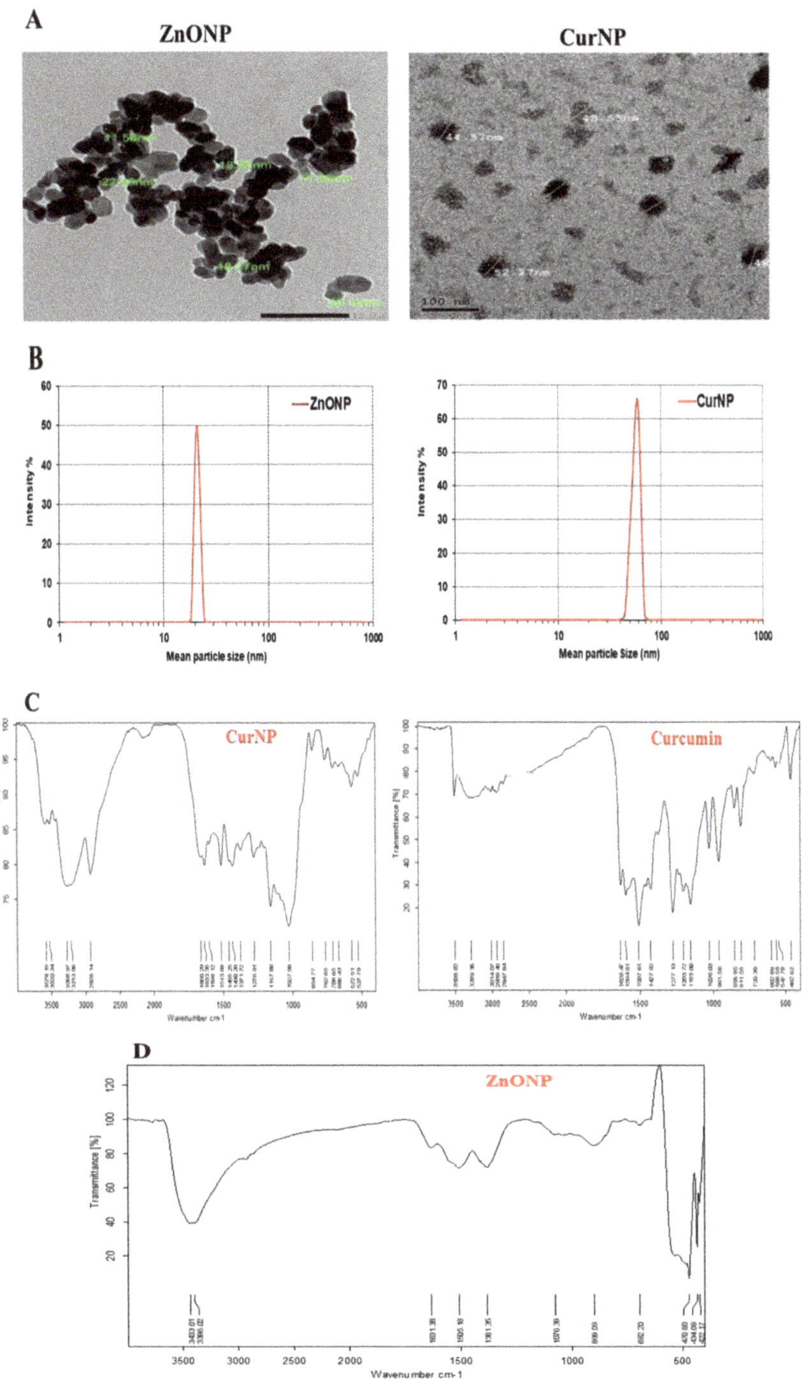

Figure 1. Characterization of ZnONP and CurNP. (**A**) TEM analysis of nanoparticles. (**B**) The particle size of nanoparticles. (**C**) FTIR analysis of curcumin and CurNP. (**D**) FTIR analysis of ZnONP.

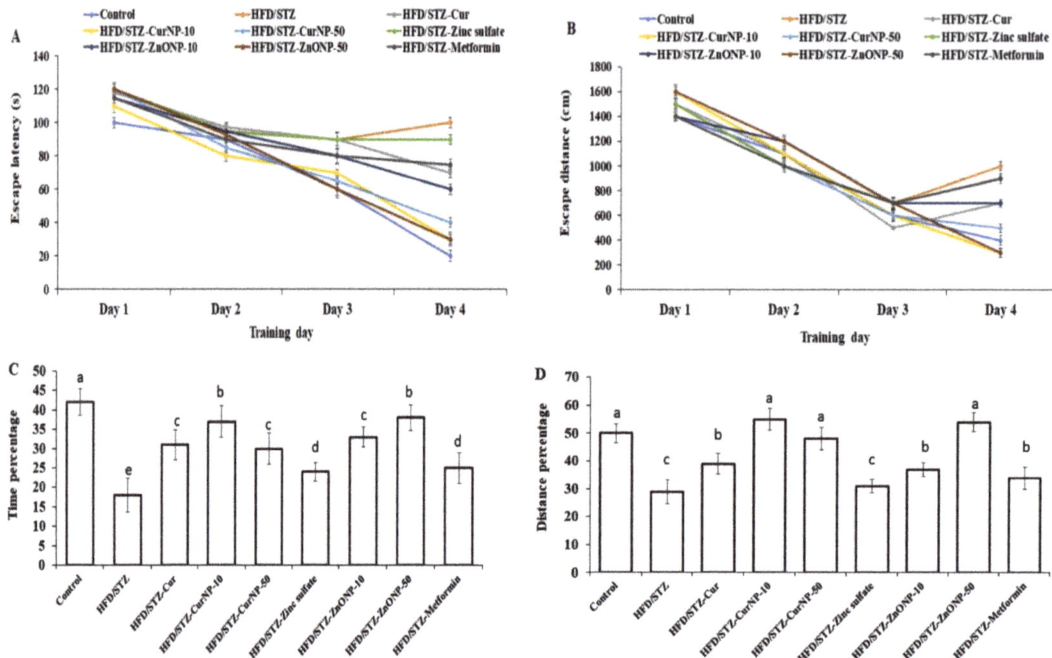

Figure 2. Effects of ZnONP and CurNP on memory impairment in rats. The Morris water maze (MWM) was used to test rats' memories by measuring (**A**) escape latency and (**B**) escape distances. A probe test was used to analyze the maintenance of memory in the MWM by measuring the percentage of (**C**) time spent and (**D**) distance swum in the target quadrant. Three independent experiments were performed. Data are expressed as mean $\pm$ SEM ($n = 8$); means with different letters (a–e) in each bar are significantly different ($p < 0.01$), in Figure 2C the largest data value take the letter (a) and the smallest data value take the letter (e) and in Figure 2D the largest data value take the letter (a) and the smallest data value take the letter (c).

3.3. Blood Glucose, Insulin, and AGEs Levels

The fasting glucose and insulin levels in type 2 diabetic rats after HFD/STZ induction displayed a significant increase versus the control group ($p < 0.01$). In contrast, after the treatment period with all current treatment options, the fasting glucose and insulin levels were reduced significantly relative to the untreated group ($p < 0.01$). Changes examined in levels of glucose and insulin are also represented in the HOMA-IR index; marked differences between model and healthy groups were found in HOMA-IR, and the treatment of model rats for six constitutive weeks was associated with a significant decrease in HOMA-IR index. In addition to this, T2DM may promote AD pathology through chronic hyperglycemia, which is associated with elevation of AGEs levels in serum of T2DM-induced rats compared to healthy control ($p < 0.01$). All treatments of T2DM may affect AD indirectly via effects on circulating levels of glucose and insulin. In our study, the administration of curcumin, zinc sulfate, two doses of CurNP and ZnONP, as well as metformin, resulted in a marked decrease in AGEs levels relative to untreated rats; the best results were shown in groups treated with 10 mg/kg CurNP and 50 mg/kg ZnONP, which reached levels comparable to those of the control group (Table 1, $p < 0.01$).

Table 1. Effects of CurNP and ZnONP on the serum fasting blood glucose, insulin, HOMA-IR, and AGEs levels.

Groups	Fasting Blood Glucose (mg/dL)	Insulin (μU/mL)	HOMA-IR	Serum AGEs (ng/mg Protein)
Control	106.80 ± 0.66 [c]	14.72 ± 0.22 [d]	3.88 ± 0.06 [e]	12.34 ± 0.03 [d]
HFD/STZ	489.20 ± 1.80 [a]	169.60 ± 1.40 [a]	204.85 ± 1.70 [a]	33.21 ± 0.01 [a]
HFD/STZ-Cur	88.80 ± 0.58 [d]	57.24 ± 0.30 [c]	12.55 ± 0.12 [c,d]	22.41 ± 0.01 [b]
HFD/STZ-CurNP-10	89.20 ± 0.37 [d]	71.24 ± 0.56 [b]	15.68 ± 0.12 [c]	14.31 ± 0.06 [c,d]
HFD/STZ-CurNP-50	92.20 ± 0.86 [d]	68.76 ± 0.40 [b]	15.65 ± 0.13 [c]	17.99 ± 0.02 [c]
HFD/STZ-Zinc sulfate	119.00 ± 1.20 [b]	70.58 ± 0.50 [b]	20.74 ± 0.33 [b]	25.41 ± 0.02 [b]
HFD/STZ-ZnONP-10	75.60 ± 0.50 [e]	60.58 ± 0.45 [c]	11.30 ± 0.12 [d]	20.32 ± 0.04 [b]
HFD/STZ-ZnONP-50	89.80 ± 0.37 [d]	69.34 ± 0.60 [b]	15.37 ± 0.14 [c]	15.67 ± 0.03 [c]
HFD/STZ-Metformin	70.60 ± 0.93 [e]	61.22 ± 0.26 [c]	10.67 ± 0.12 [d]	21.34 ± 0.05 [b]

Results are tabulated as mean ± SEM (n = 8). Statistical analyses were performed using one-way ANOVA; means for the Control same parameter with different letters (a–e) in each column are significantly different ($p < 0.01$), the largest data value take the letter (a) and the smallest data value take the letter (e).

3.4. Hippocampal Concentrations of Oxidant and Antioxidant Biomarkers

As shown in Table 2, there was a marked elevation in hippocampal levels of NO and TBARS relative to controls ($p < 0.01$). Additionally, simultaneous marked decreases in GST, CAT, GPx, and SOD activities, as well as GSH content, were noticed in rats from the type 2 diabetes model relative to controls ($p < 0.01$). In contrast, a marked decrease in hippocampal TBARS and NO levels and concurrent increases in GST, CAT, GPx, and SOD activities, as well as GSH content, were observed in all treated rats, especially 10 mg/kg CurNP and 50 mg/kg ZnONP, relative to model groups ($p < 0.01$). These results suggest a better aptitude of CurNP at a dose of 10 mg/kg to fight oxidative stress in the brain of rats.

Table 2. Effect of CurNP and ZnONP on hippocampus oxidant and antioxidant biomarkers levels in HFD/STZ-induced and treated rats.

Groups	TBARS (μmoL/mg Protein)	NO (μmoL/mg Protein)	CAT (U/mg Protein)	SOD	GPx (μmoL/min/mg Protien)	GST	GSH (μmoL/mg Protein)
Control	174.29 ± 5.11 [d]	0.80 ± 0.04 [c]	0.04 ± 0.00 [d]	105.63 ± 5.40 [g]	44.28 ± 0.08 [a]	40.10 ± 0.10 [c]	24.31 ± 1.32 [e]
HFD/STZ	381.07 ± 24.50 [a]	5.41 ± 0.01 [a]	0.01 ± 0.00 [f]	18.89 ± 2.00 [i]	19.42 ± 0.14 [c]	10.44 ± 0.30 [f]	8.06 ± 0.90 [f]
HFD/STZ-Cur	188.95 ± 5.56 [d]	0.64 ± 0.02 [c]	0.04 ± 0.00 [d]	121.30 ± 11.70 [f]	38.3 ± 0.09 [b]	36.22 ± 0.50 [d]	33.23 ± 0.90 [d]
HFD/STZ-CurNP-10	148.19 ± 5.94 [f]	0.52 ± 0.01 [d]	0.10 ± 0.00 [a]	733.31 ± 10.15 [a]	47.58 ± 0.12 [a]	77.30 ± 0.40 [a]	71.78 ± 2.35 [a]
HFD/STZ-CurNP-50	165.54 ± 15.41 [e]	0.71 ± 0.03 [c]	0.06 ± 0.00 [c]	380.48 ± 7.50 [c]	44.14 ± 0.07 [a]	65.33 ± 0.30 [b]	60.18 ± 2.36 [b]
HFD/STZ-Zinc sulfate	226.20 ± 7.76 [c]	0.61 ± 0.02 [c,d]	0.04 ± 0.00 [d]	84.82 ± 7.88 [h]	33.26 ± 0.10 [b]	25.43 ± 0.20 [e]	36.45 ± 1.70 [d]
HFD/STZ-ZnONP-10	177.88 ± 5.35 [d]	0.77 ± 0.04 [c]	0.06 ± 0.00 [c]	227.82 ± 7.50 [d]	40.12 ± 0.05 [a]	50.43 ± 0.10 [c]	55.97 ± 1.80 [b]
HFD/STZ-ZnONP-50	158.05 ± 6.90 [e]	0.66 ± 0.03 [c]	0.08 ± 0.00 [b]	529.68 ± 24.40 [b]	44.78 ± 0.65 [a]	60.33 ± 0.50 [b]	65.00 ± 2.90 [a]
HFD/STZ-Metformin	250.51 ± 4.00 [b]	1.27 ± 0.02 [b]	0.02 ± 0.00 [e]	160.35 ± 16.32 [e]	35.68 ± 0.15 [b]	34.55 ± 0.50 [d]	47.86 ± 0.90 [c]

Results are tabulated as mean ± SEM (n = 8). Statistical analyses were performed using one-way ANOVA; means for the same parameter with different letters (a–i) in each column are significantly different ($p < 0.01$), the largest data value takes the letter (a) and the smallest data value takes the letter (i).

3.5. Alteration of Concentrations of Serum Adipokines and Hippocampal Inflammatory Mediators

These results indicate a significant decrease in the serum adiponectin levels of type 2 diabetes-induced rats versus the control group ($p < 0.01$), while serum leptin levels were found to be increased in type 2 diabetes-induced rats as compared to control animals ($p < 0.01$). Treatment of diabetic rats with ZnONP and CurNP improved serum adiponectin levels significantly in type 2 diabetes-induced rats versus untreated rats ($p < 0.01$). In contrast, administration of all current treatment options to type 2 diabetes-induced rats produced a significant reduction in serum leptin levels compared to the untreated group ($p < 0.01$), and 10 mg/kg CurNP and 50 mg/kg ZnONP showed a more efficient decrease in serum leptin levels than other doses (Table 3, $p < 0.01$). Further, to investigate the effect of CurNP and ZnONP on proinflammatory cytokines, IL-6 and TNF-α protein levels were detected in all groups. Our results show an increase in TNF-α and IL-6 protein levels in the hippocampus of model rats compared with the healthy control rats, TNF-α and IL-6 levels in type 2 diabetes-induced rats increased 10.2-fold and 3.9-fold in the hippocampus, respectively ($p < 0.01$), while among all current treatment options, 10 mg/kg CurNP and

50 mg/kg ZnONP treatment significantly attenuated the increased TNF-α and IL-1β levels in type 2 diabetes-induced rats relative to the untreated model (Table 3; $p < 0.01$).

Table 3. Effect of CurNP and ZnONP on serum adipokines levels and hippocampus inflammatory biomarkers.

Groups	Adiponectin (ng/mL)	Leptin (ng/mL)	IL-6 (ng/mg Protein)	TNF-α (ng/mg Protein)
Control	7.46 ± 0.17 [b]	15.56 ± 0.08 [c]	17.45 ± 0.023 [e]	4.63 ± 0.02 [e]
HFD/STZ	3.40 ± 0.23 [d]	29.70 ± 0.15 [a]	68.45 ± 0.03 [a]	47.36 ± 0.04 [a]
HFD/STZ-Cur	5.63 ± 0.22 [c]	23.56 ± 0.12 [b]	26.41 ± 0.02 [d]	15.34 ± 0.02 [c]
HFD/STZ-CurNP-10	6.63 ± 0.15 [b]	21.23 ± 0.16 [b]	22.60 ± 0.02 [e]	9.43 ± 0.01 [d]
HFD/STZ-CurNP-50	7.93 ± 0.26 [b]	16.63 ± 0.25 [c]	25.71 ± 0.02 [d]	14.03 ± 0.01 [c]
HFD/STZ-Zinc sulfate	5.46 ± 0.06 [c]	25.63 ± 0.17 [b]	55.34 ± 0.02 [b]	22.85 ± 0.04 [b]
HFD/STZ-ZnONP-10	7.10 ± 0.11 [b]	17.63 ± 0.18 [c]	31.24 ± 0.03 [d]	17.53 ± 0.01 [b]
HFD/STZ-ZnONP-50	9.03 ± 0.19 [a]	19.00 ± 0.12 [c]	24.30 ± 0.01 [d]	10.95 ± 0.04 [d]
HFD/STZ-Metformin	6.27 ± 0.03 [b]	22.30 ± 0.17 [b]	43.11 ± 0.01 [c]	19.83 ± 0.02 [b]

Results are tabulated as mean ± SEM ($n = 8$). Statistical analyses were performed using one-way ANOVA; means for the same parameter with different letters (a–e) in each column are significantly different ($p < 0.01$), the largest data value take the letter (a) and the smallest data value take the letter (e).

3.6. Regulation of AChE Activity, IDE Level, and Aβ-42 Clearance in the Hippocampus of Rats

As presented in Table 4, AChE activity in the hippocampus was significantly increased in the type 2 diabetes model group versus healthy controls ($p < 0.01$). Alternatively, a significant decrease in AChE activity was found in the hippocampus of rats obtaining CurNP and ZnONP treatment versus model rats ($p < 0.01$). Consistently, IDE levels tended to be lower in the hippocampus of rats with type 2 diabetes-induced neurodegeneration compared to control rats ($p < 0.01$). The treatment of the induced groups with ZnONP and CurNP resulted in an increase in hippocampal IDE levels that was similar to that for the control group compared to untreated rats (Table 4, $p < 0.01$). To date, Aβ-42 production and aggregation are considered the center of attention of neurodegeneration models like AD. In our study, a significant elevation of Aβ-42 levels in the hippocampus of rats with type 2 diabetes-induced neurodegeneration was detected compared to healthy rats ($p > 0.01$). The treatment of the induced groups with ZnONP and CurNP, particularly those with 10 mg/kg CurNP and 50 mg/Kg ZnONP, exhibited a marked decrease in hippocampal Aβ-42 levels with values near to the control group.

Table 4. Regulation of Aβ-42 clearance, levels of IDE, and activity of AChE in the hippocampus of HFD/STZ-induced rats after treatment with CurNP and ZnONP.

Groups	AChE (mmoL/min/mg Protein)	IDE (ng/mg Protein)	Hippocampal Aβ-42 (ng/mg Protein)
Control	15.58 ± 0.30 [b]	30.10 ± 0.05 [a]	23.43 ± 0.01 [f]
HFD/STZ	84.01 ± 4.10 [a]	7.12 ± 0.01 [c]	140.87 ± 0.34 [a]
HFD/STZ-Cur	17.37 ± 0.57 [b]	23.00 ± 0.01 [b]	52.00 ± 0.00 [d]
HFD/STZ-CurNP-10	12.16 ± 0.14 [c]	24.24 ± 0.12 [a,b]	31.53 ± 0.02 [e]
HFD/STZ-CurNP-50	14.60 ± 0.47 [b]	29.16 ± 0.08 [a]	37.42 ± 0.01 [e]
HFD/STZ-Zinc sulfate	16.63 ± 0.46 [b]	17.09 ± 0.00 [b]	101.33 ± 0.02 [b]
HFD/STZ-ZnONP-10	17.02 ± 0.49 [b]	27.17 ± 0.09 [a]	72.55 ± 0.01 [c]
HFD/STZ-ZnONP-50	11.30 ± 0.80 [c]	25.08 ± 0.04 [a]	33.06 ± 0.03 [e]
HFD/STZ-Metformin	11.12 ± 0.54 [c]	22.04 ± 0.02 [b]	60.03 ± 0.02 [c]

Results are tabulated as mean ± SEM ($n = 8$). Statistical analyses were performed using one-way ANOVA; means for the same parameter with different letters (a–f) in each column are significantly different ($p < 0.01$), the largest data value takes the letter (a) and the smallest data value takes the letter (f).

3.7. Gene Expression Profile of APP, BACE-1, BDNF, and ADAM-10 in the Hippocampus of Rats

Type 2 diabetes-induced neurodegeneration in rats revealed a significant upregulation of gene expression of APP and BACE-1 when compared with control rats ($p < 0.01$). Interestingly, these elevations were markedly suppressed following the administration of curcumin, zinc sulfate, two doses of CurNP, ZnONP, and metformin to model rats versus healthy rats ($p < 0.01$). Conversely, gene expression of BDNF and ADAM-10 were significantly downregulated in the hippocampus of model rats when compared with the control group ($p < 0.01$). Further, administration of curcumin, zinc sulfate, two doses of CurNP, ZnONP, and metformin displayed a marked upregulation of gene expression of BDNF and ADAM-10 compared to model rats (Figure 3, $p < 0.01$). Of all treatment options, 10 mg/kg CurNP and 50 mg/kg ZnONP significantly ameliorated the gene expression of the evaluated genes (Figure 3).

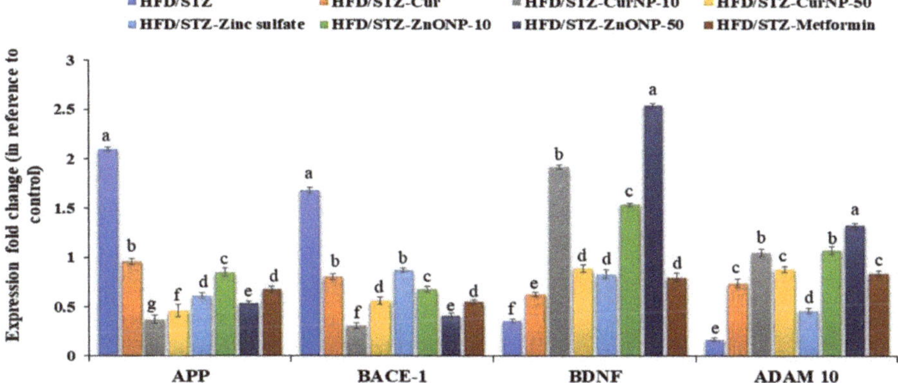

Figure 3. Gene expression profile of APP, BACE-1, BDNF, and ADAM-10 in the hippocampus of rats. Data are expressed as mean $\pm$ SEM ($n = 3$); means for the same parameter with different letters (a–g) in each bar are significantly different ($p < 0.01$), the largest data value takes the letter (a) and the smallest data value takes the letter (g).

3.8. Effect of CurNP and ZnONP on an Apoptotic Pathway in the Hippocampus and Cortex of Rats

Under stress conditions, Bcl-2 expression levels in the hippocampi and cortices of rats with type 2 diabetes-induced neurodegeneration were significantly low ($p < 0.01$), while Bax expression was significantly high compared with the control group ($p < 0.01$). At the end of the CurNP and ZnONP treatments of model rats, the proapoptotic member, Bax expression was significantly suppressed along with increased Bcl-2 expression in both hippocampi and cortices (Figure 4A,B, $p < 0.01$).

3.9. p38-MAPK Signaling Cascade and Tau Protein in the Hippocampus of Rats

It has been reported that the MAPK pathway, p38, ERK, and MEK activate many signaling cascades including inflammation and apoptosis. Consequently, we here examined the protein phosphorylation levels of MAPK signaling in the hippocampus of rats by western blotting. As shown in Figure 5, the levels of phosphorylated-p38MAPK, phosphorylated-ERK 1,2 (p-ERK), and phosphorylated-MEK (p-MEK) in rats with type 2 diabetes-induced neurodegeneration were increased significantly compared with the control group ($p < 0.01$). After CurNP and ZnONP treatment, the phosphorylation levels of p-p38MAPK, p-ERK, and p-MEK were markedly decreased compared with those of the untreated group ($p < 0.01$). Our data indicated that 10 mg/kg CurNP or 50 mg/kg ZnONP might be associated with the regulation of neuroinflammation and apoptosis in the hippocampus of type 2 diabetes-induced rats at least partly, by adjusting the activation of the MAPK signaling pathway. Additionally, type 2 diabetes-induced rats revealed significant hyperphosphorylation of tau proteins when compared with the control rat group ($p < 0.01$). Interestingly, this elevation was markedly suppressed following administration of all current treatment options to model rats, with marked inhibition of tau phosphorylation observed in groups treated with 10 mg/kg CurNP or 50 mg/kg ZnONP (Figure 5A–E).

3.10. Histopathological Study

The histopathological study of the hippocampi was seen in Figure 6. Paraffin section photomicrographs of control brains show the hippocampus area with arranged pyramidal cells with large pyramidal nuclei and minimum vacuolated cytoplasm, few pyknotic neuropil cells, and homogenous brain tissue (Figure 6A). Conversely, the H&E-stained sections of type 2 diabetes-induced group display a marked proliferating pyramidal cell forming gland feature and have dark nuclei with minimum vacuolated cytoplasm and many pyknotic neuropil cells, marked congested, dilated blood vessels and capillaries, beside edema and reduction of the myelinated sheath, indicating AD induction (Figure 6B). Moreover, paraffin section photomicrographs of rat brains treated with curcumin-50 show the hippocampus area with some pyramidal cells, hyperchromatic nuclei, and marked vacuolated cytoplasm with many pyknotic pyramidal cell and neuropil cells, as well as reduction of the myelinated sheath, and a mildly dilated and hemorrhaged blood capillary was seen (Figure 6C). Besides, the brain photomicrographs of the induced group treated with CurNP-10 shows the hippocampus region with dark pyramidal nuclei and low vacuolated cytoplasm (normal architecture) and few pyknotic pyramidal cells; few neuroglial cells have rounded nuclei; prominent nucleolus, marked hemorrhage dilated blood vessels, and capillaries with the homogenous field of myelinated sheath and edema were seen (Figure 6D). In addition, the brain photomicrographs of the induced group treated with CurNP-50 show hippocampi with few neuroglial cells and proliferating pyramidal cells, having dark pyramidal nuclei and mild vacuolated cytoplasm. Furthermore, rearrangement of a pyramidal cell for the organized hippocampus region and a few necrotic ones were noticed. A mildly dilated and hemorrhaged blood vessel with homogenous brain tissue was seen (Figure 6E). In addition, rat brains treated with zinc sulfate-50 showed a hippocampus area with some pyramidal cells, hyperchromatic nuclei with homogenous cytoplasm, few necrotic and many pyknotic pyramidals, as well as neuroglial cells. A mildly dilated and hemorrhaged blood capillary with a mildly degenerative myelinated sheath was also seen (Figure 6F). Furthermore, the brain photomicrographs of the ZnONP-10-treated group showed a hippocampus with regenerative pyramidal cells and hyperchromatic nuclei, with mild vacuolated cytoplasm and few necrotic neuroglial cells. A mildly dilated blood capillary with an area of homogenous brain tissue was also seen (Figure 6G). In addition to this, the brain photomicrographs for the ZnONP-50-treated group showed a hippocampus with the normal features of many neuropil cells and few pyknotic ones, and mild proliferating pyramidal cells with hyperchromatic nuclei and mildly vacuolated cytoplasm. A mildly dilated and hemorrhaged blood vessel with homogenous brain tissue was also observed (Figure 6H). The brain photomicrographs of metformin-treated rats showed the hippocampus region with many neuroglial cells with rounded nuclei and a prominent nucleolus, as well as necrotic neuroglial cells. Many necrotic pyramidal cells and others with small nuclei (atrophied pyramidal cells) and pyknotic pyramidal cells were seen. Markedly dilated and hemorrhaged blood vessels and capillaries were also noticed, as was the reduction of a myelinated sheath in the necrotic area (Figure 6I).

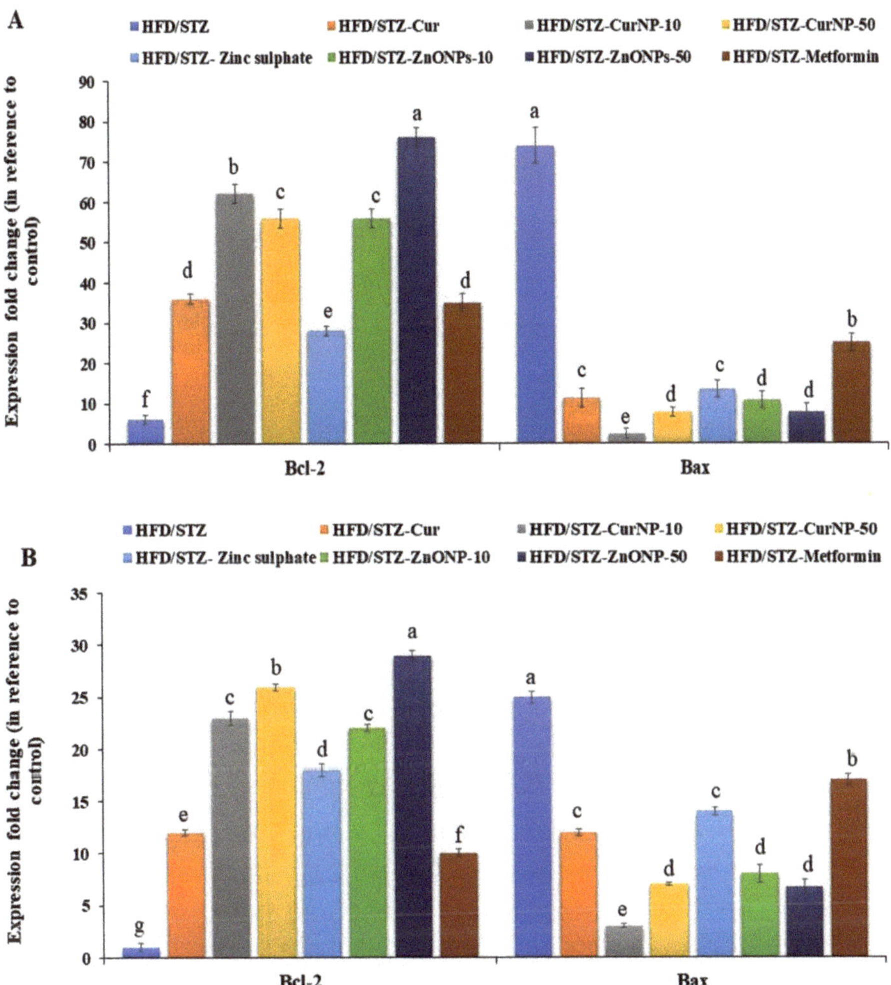

Figure 4. Gene expression profile of Bcl-2 and Bax in (**A**) hippocampus and (**B**) cortex of rats. Data are expressed as mean ± SEM (*n* = 3); means for the same parameter with different letters (a–g) in each bar are significantly different ($p < 0.01$), the largest data value takes the letter (a) and the smallest data value takes the letter (g).

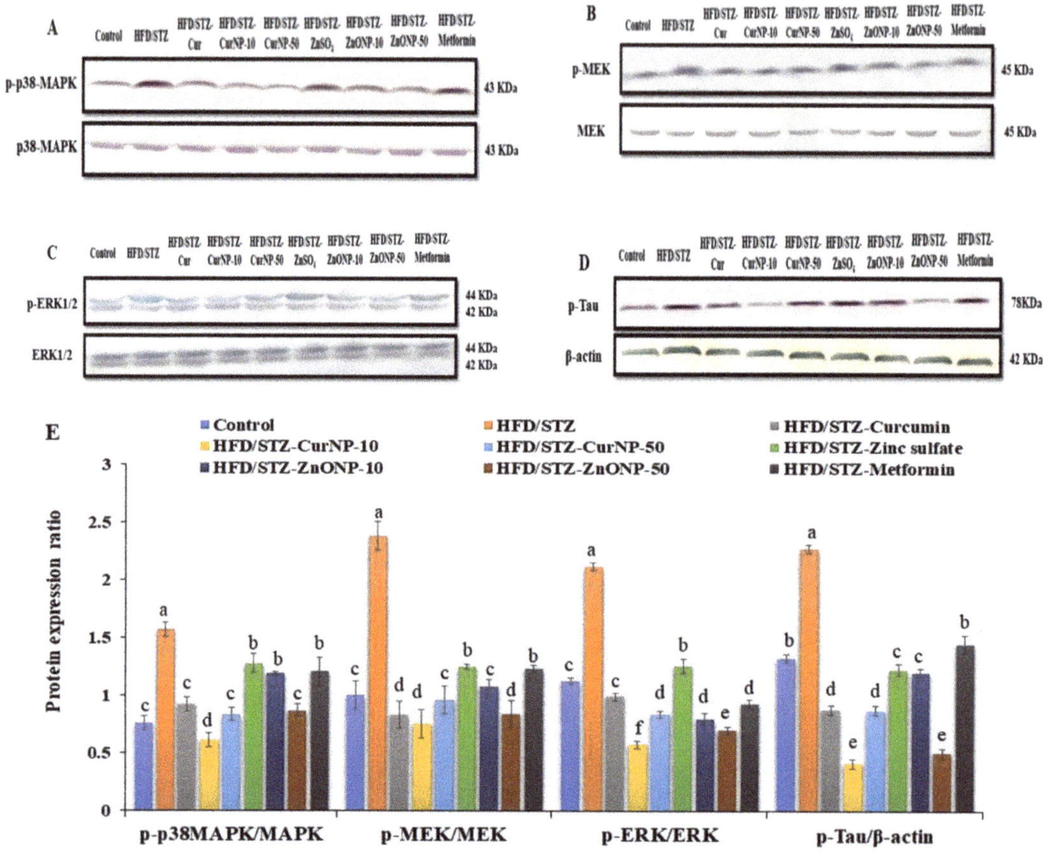

Figure 5. Protein expression ratio profile. (**A**) p-p38MAPK/p38MAPK, (**B**) p-ERK/ERK, (**C**) p-MEK/MEK, (**D**) p-Tau/β-actin in the hippocampus of rats. (**E**) Quantitative analysis of p-p38MAPK/p38MAPK, p-ERK/ERK, p-MEK/MEK, and p-Tau/β-actin in the hippocampus of rats. Data are expressed as mean ± SEM (n = 3); means for the same parameter with different letters (a–f) in each bar are significantly different ($p < 0.01$), the largest data value takes the letter (a) and the smallest data value takes the letter (f).

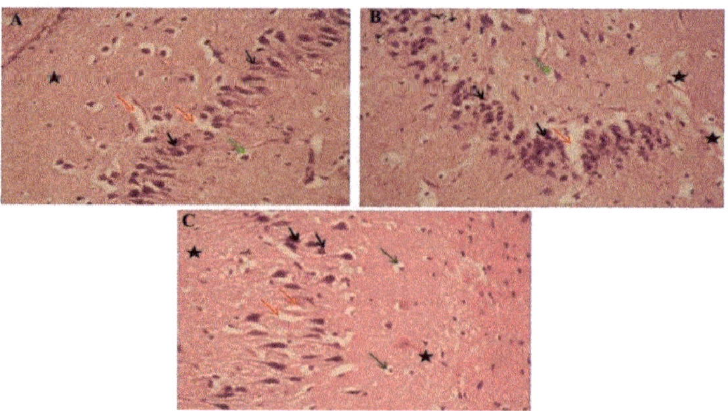

Figure 6. *Cont.*

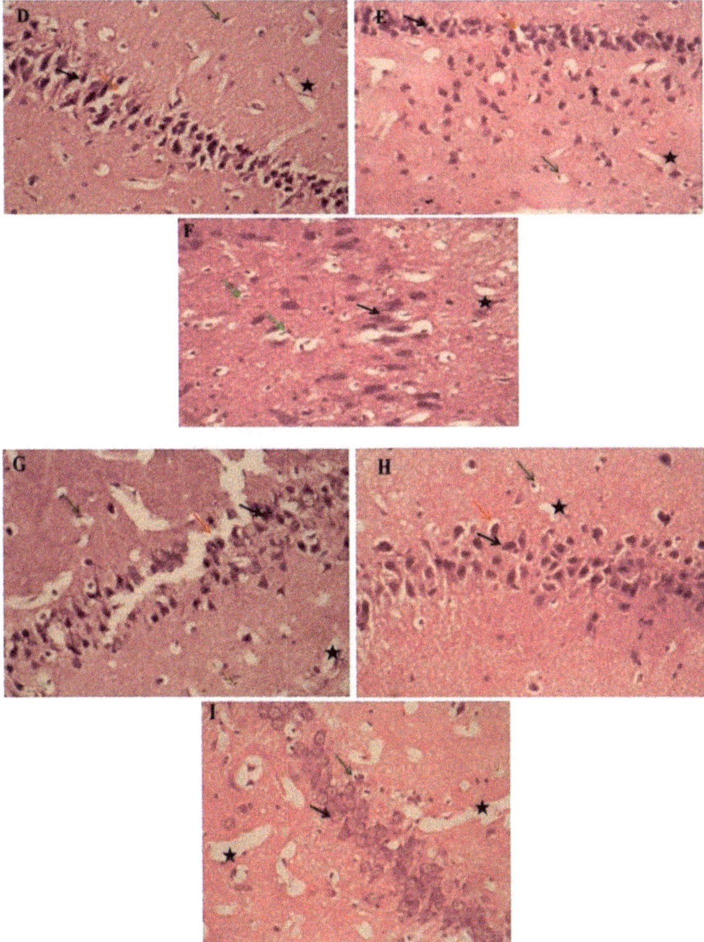

Figure 6. The effects of ZnONP and CurNP on the histology of hippocampi in rat groups. Hippocampi of rats were evaluated via hematoxylin-eosin (H&E) staining at 400 X magnification. (**A**) The control group: a photomicrograph of the hippocampus showing arranged pyramidal cells (black arrows) with minimum vacuolated cytoplasm (Red arrows), a few pyknotic neuropil cells (Green arrows), and homogenous brain tissue (black asterisk). (**B**) Untreated HFD/STZ-induced group: a photomicrograph of the hippocampus showing marked proliferating pyramidal cells with dark nuclei (black arrows) with minimum vacuolated cytoplasm (red arrows) and many pyknotic neuropile cells (green arrows), and reduction of the myelinated sheath (black asterisk). (**C**) HFD/STZ-induced rats treated with curcumin: a photomicrograph of the hippocampus showing some pyramidal cells with hyperchromatic nuclei (black arrows) and marked vacuolated cytoplasm (red arrows) and many neuropile cells (green arrows), with mildly dilated and hemorrhaged blood capillaries (black asterisk). (**D**) HFD/STZ-induced rats treated with CurNP-10: a photomicrograph of the hippocampus showing normal features of many neuropil cells and few pyknotic ones (green arrows), mildly proliferating pyramidal cells with hyperchromatic nuclei (black arrows) and mildly vacuolated cytoplasm (red arrows) with homogenous brain tissue (black asterisk). (**E**) HFD/STZ-induced rats treated with CurNP-50: a photomicrograph of the hippocampus showing pyramidal cells having dark pyramidal nuclei (black arrows) and mildly vacuolated cytoplasm (red arrows) with few necrotic ones (green arrows) and homogenous brain tissue (black asterisk). (**F**) HFD/STZ-induced

rats treated with zinc sulfate: a photomicrograph of the hippocampus showing some pyramidal cells with hyperchromatic nuclei and homogenous cytoplasm (black arrows), many neuroglial cells (green arrows), with mildly degenerative myelinated sheaths (black asterisk). (**G**) HFD/STZ-induced rats treated with ZnONP-10: a photomicrograph of the hippocampus showing pyramidal cells, dark pyramidal nuclei (black arrows) and low vacuolated cytoplasm (normal architecture) (red arrows), prominent nucleolus (green arrows), and markedly hemorrhaged dilated blood vessels and edema (black asterisk). (**H**) HFD/STZ-induced rats treated with ZnONP-50: a photomicrograph of the hippocampus showing hyperchromatic nuclei (black arrows) with mild vacuolated cytoplasm (red arrows), and pyknotic pyramidal (green arrows), with the area of homogenous brain tissue (black asterisk). (**I**) HFD/STZ-induced rats treated with metformin: a photomicrograph of the hippocampus showing many prominent nucleoli (green arrows), pyknotic pyramidal cells (black arrows), and reduction of myelinated sheaths in the necrotic area (black asterisk).

4. Discussion

Accumulating data have established that T2DM-induced cognitive deficits increase the risk for AD. Further, T2DM-associated complications, such as oxidative stress and inflammation, are thought to play a critical role in cognitive decline. Therefore, the current study was aimed to evaluate novel therapeutic effects and the mechanisms of nanoparticles from a natural source like CurNP, as well as metal oxide-base nanomaterials like ZnONP, on the hippocampus of T2DM-induced rats.

T2DM develops in case of insulin resistance and dysfunction of insulin secretion [46]. We and others have shown that the HFD/STZ model appears to produce a degree of neurodegeneration and mimics pre-AD symptoms in the hippocampus, including memory loss, insulin resistance, Aβ aggregation, and tau hyperphosphorylation. It is also considered a useful experimental model to study the activity of hypoglycemic agents [47,48].

Constant hyperglycemia, hyperinsulinemia, and a significant elevation in HOMA-IR were observed in model rats and are consistent with our previous results [47]. The long-term administration of either CurNP and ZnONP is effective in reducing T2DM-associated alterations through decreasing blood glucose and insulin levels and enhancing HOMA-IR. These results are not surprising considering curcumin's hypoglycemic effect on pancreatic cells and insulin sensitivity, which have been extensively studied [49]. Furthermore, ZnONP can improve serum insulin levels, glucose utilization, and metabolism by influencing hepatic glycogenesis and possibly acting on the insulin signaling pathway [50–52]. Nevertheless, a preceding study presented an adverse effect of short-term administration of ZnONP (100 mg/kg) that increased blood glucose levels in diabetic and healthy rats depending on the dose and route of administration [53].

Our results revealed that hyperglycemia leads to tissue damage by several mechanisms, including the AGEs formation. Furthermore, accumulating evidence suggested that AGEs play a key role in the pathogenesis of neurodegenerative disorders, such as AD and diabetic neuropathy, a diabetes-related complication [54]. Until now, it has been well recognized that blood proteins, including IgG, are glycated in AD patients by a variety of reducing sugars and metabolites, manifesting in AGEs formation leading to protein structure and function distortion [55]. The current findings show an increase in AGEs levels in serum of model rats which confirm our previous findings [47]. All current therapies, especially CurNP and ZnONP, according to our findings, reduced the production of AGEs. Therefore, both nanoparticles may have therapeutic potential in delaying disease development and/or reversing disease pathogenesis in AGEs-related neurodegenerative disorders.

Interestingly, the hippocampus is an important structure in rats and humans, and it plays a crucial integrative role in learning and memory [56]. As a result, hyperglycemia-induced alterations in the structure of the hippocampus may represent the pathological basis of cognitive deterioration in diabetic rats [57]. Our behavioral analyses demonstrate that HFD/STZ induction prolonged escape latency and distance in MWM training, indicating spatial learning deficiencies. CurNP and ZnONP treatments reduced the escape latency and distance of model rats, protecting them against HFD/STZ-induced spatial learning impairments. In the probe trial, shorter times spent in the target quadrant and fewer platform zone crossings were seen in model rats, indicating impairments in spatial memory. CurNP and ZnONP treatments protected the spatial memory of model rats; they spent more time in the target quadrant and crossed the platform zone twice more. Our study suggests that CurNP and ZnONP may protect the spatial and memory functions in the T2DM rat model.

Certainly, the combination of the pressure from oxidative stress with lowered antioxidant defense creates a harmful effect that contributes to disrupting the functions of cells and leads to cell death. In the current work, the antioxidant capacity parameters (SOD, GPx, CAT, and GST enzyme activities, as well as GSH level) and typical oxidative stress biomarkers (TBARS and NO levels) were used to investigate the redox profile in the hippocampus of the rats model. The results demonstrated a remarkable elevation in oxidative stress biomarkers, with a significant reduction of enzymatic and non-enzymatic antioxidants in the hippocampus of model rats. These findings are consistent with prior research that linked amyloid plaque formation to oxidative stress, including lipid peroxidation markers as TBARS [58]. Additionally, hyperglycemia activates multiple signaling pathways, which lead to increased ROS production and induce insulin resistance [59]. Indeed, oxidative stress is one of the initial signs of AD disease and is the first consequence of Aβ accumulation in the brain [60]. Treatment of model rats with 10 mg/kg CurNP exhibited the most potent effect in terms of amelioration of hippocampal redox profiles, followed by 50 mg/kg ZnONP, compared with other doses of nanoparticles and classical curcumin and zinc sulfate. Curcumin and CurNP are considered antioxidant agents and could enhance the expression of antioxidant enzymes and thus increase total antioxidant capacity [61]. Moreover, our data for diabetic rats demonstrated that treatment with ZnONP efficiently restored GPx, SOD, and CAT activities by increasing the biosynthesis of GSH or reducing oxidative stress [62–64]. Conversely, as mentioned before, the intraperitoneal injection of 25 mg/kg ZnONP had no significant effect on the levels of CAT, SOD, and GSH in the brain, while treatment with ZnONP at a high dose decreased brain GSH and SOD and increased brain MDA levels [65].

Oxidative stress excites the translocation of the transcription factor nuclear factor kappa B (NF-κB) to the nucleus, which adjusts pro-inflammatory gene expressions, such as TNF-α and IL-6 [66]. By further analyzing hippocampal inflammatory cytokines, we were able to pin down the anti-inflammatory effect of nanoparticles. We showed that the administration of our current treatments produced a significant reduction in TNF-α and IL-6 levels, with the maximum inhibition level in both 10 mg/kg CurNP and 50 mg/kg ZnONP treated groups. Curcumin and CurNP confirmed a reduced level of inflammatory cytokines: TNF-α and IL-6 [67]. Curcumin prohibited the elevation of the cytokines since it can decrease the inflammatory reactions by interfering with NF-κB activation [68]. Likewise, it has been found that zinc and ZnONP were attributed to the anti-inflammatory effect of this ion [62,69], which inhibits the gene expression of inflammatory cytokines, including TNF-α and IL-6, which are known to produce ROS [70,71]. In contrast to our study, it was demonstrated that daily administration of 100 mg/kg ZnONP significantly increased the concentrations of TNF-α and IL-1β [72].

Our findings showed high leptin levels and significant adiponectin reduction in model rats, and after six weeks of supplementation of CurNP, ZnONP, a significant amelioration of serum adipokines was observed. Hence, the protective curcumin effects associated with adiponectin expression upregulation were suggested [61], these inhibiting lipogenic expression and adipose tissue inflammation [73]. Inflammation due to visceral adiposity is improved and this effect comes with an increase in adiponectin and an improvement in sensitivity to insulin [74]. The ZnONP-enhanced mechanism is remarkable, and this huge effect is achieved through a direct entry into the intracellular region, which leads to the adjustment of many enzymes, adipogenesis, and transcription factors.

Further, the anti-apoptotic Bcl-2 and pro-apoptotic Bax are two critical molecules involved in cell death [75]. The enhanced expression of Bax and decreased expression of Bcl-2 following the HFD/STZ induction observed in the current study suggest that neuro-apoptosis is elevated by T2DM induction in the hippocampi and cortices of induced rats. The equilibrium of the antiapoptotic (represented by Bcl-2) and pro-apoptotic (represented by Bax) molecules were significantly changed post-treatment with CurNP and ZnONP, which effectively inhibited apoptosis in the hippocampus and cortex and increased neuronal survival. Further studies of the apoptotic pathways like caspase(s) and cytochrome c are needed to determine their involvement in our pathway, like the mitochondrial-dependent intrinsic apoptotic pathway, as reported in other studies [76].

The regulation of cholinergic function is a critical aspect of the management of AD development. We supported this by showing a partial inhibition in AChE activity in the model rat after administration of both nanoparticles [77,78]. It is noteworthy that the molecular pathogenesis of AD is underlain by accelerated Aβ production, aggregation, and compromised Aβ clearance [79]. Current results revealed an elevation in the Aβ-generating enzyme BACE1, which is responsible for Aβ production [80], and depletion in Aβ-degrading enzyme (IDE), which is responsible for Aβ clearance [81]. As stated previously, the level of Aβ-42 in the hippocampus of HFD/STZ-induced rats showed a significant elevation compared to control rats, proving the inability of the brain to clear the Aβ deposits in T2DM. This may be due to hyperinsulinemia perceived in diabetic rats and a struggle between degradation of Aβ-42 and insulin for the same substrate, IDE [81].

As it happens, the administration of CurNP and ZnONP significantly downregulated the expression of APP and BACE-1. In addition to this, they decrease the Aβ-42 levels and significantly elevate the IDE concentration in the hippocampus of rats, with the maximum effective results, compared with curcumin, zinc sulfate, and metformin. Numerous earlier studies have shown that natural antioxidants, such as CurNP, prevent neuronal death occurring in AD, and lead to a reduction in Aβ plaque deposition [24]. These results were linked with a marked upregulation of gene expression of BDNF and ADAM-10, providing a new strategy for attenuating neurodegeneration developed in T2DM [82].

The present study demonstrates that the HFD/STZ rat model is a well-established model for evaluating the efficacies of medication intended for the treatment of AD, which causes cognitive deficits, microglial activation, and phosphorylation of hippocampal MAPK/ERK. The expression of MAPK family proteins (MEK, ERK1/2, and p38-MAPK) was studied to further understand the molecular pathways related to neuroprotection by naturally occurring nanoparticles, CurNP, and the biomedical metal oxide nanoparticle, ZnONP. The levels of phosphorylated p38-MAPK, ERK1/2, and MEK in the hippocampus of rats following HFD/STZ induction were considerably increased [83]. MAPK signaling is important for synaptic plasticity and hippocampus-dependent memory [84]; also, p38-MAPK is an important kinase in tau phosphorylation [85]. In the hippocampus and microglia of rats under pre-AD, hyperphosphorylation of the MAPK pathway, besides tau-hyperphosphorylation, has been seen [86]. Our results in the present study showed that our current treatments inhibit microglial activation by decreasing the phosphorylation of the MAPK family (p38-MAPK, ERK1/2, MEK) and phosphorylation of tau protein in the hippocampus of HFD/STZ induced rats [87]. Consistent with our results, as reported before, CurNP plays a role in inhibiting the phosphorylation of MAPKs-regulated IL-6 and TNF-α production and suppresses the phosphorylation of tau protein [88]. In addition to this, the results showed a significant effect of ZnONP at a dose of 50 mg/kg on the MAPK/ERK pathway via the marked inhibition of its phosphorylation, and this is the first time the effect of ZnONP on the MAPK pathway in the hippocampus of T2DM-induced rats has been studied. Overall, our results indicated that the MAPK signaling pathway appears to be implicated in CurNP and ZnONP-mediated neuroprotection.

The damage in any cell in the hippocampus can cause gross effects on the learning process of the individual. After histopathological examination of normal and diabetic hippocampal sections, results revealed that diabetes had a significant effect in the form of cell death in various areas, as well as disruption of normal layer organization. This was associated with clumping of neuronal processes, a sign of neuronal injury [89]. Fortunately, these changes were significantly improved by the administration of CurNP and ZnONP, showing well-preserved pyramidal cells in both groups, which confirms the neuroprotective effect of both nanoparticles on brain cells through the improvement of neurogenesis.

5. Conclusions

In conclusion, the findings of the present study demonstrate that naturally occurring CurNP and the biomedical metal-oxide ZnONP inhibit HFD/STZ-induced neuro-apoptosis, neuroinflammation, cognitive dysfunction, amyloidogenesis, and tau hyperphosphorylation in the hippocampi of rats. These exhibited modifications likely occur by modifications to the critical pathways involved in these complications, such as p38-MAPK/ERK. In general, new information has been discovered that adds to the puzzle of nanoparticle use in biomedicine. As a result of our findings, taking 10 mg/kg naturally CurNP or 50 mg/kg metal oxide ZnONP may potentially reduce diabetic complications-induced neurotoxicity and aid in proper care, particularly in the case of T2DM. However, more research is needed to confirm this, particularly regarding the use of ZnONP.

Author Contributions: Conceptualization, M.B. and S.A.; methodology, M.N.; software, S.A.; validation, S.A., D.A. and M.B.; formal analysis, M.N.; investigation, M.N.; resources, S.A.; data curation, D.A.; writing—original draft preparation, S.A.; writing—review and editing, M.B.; visualization, S.A.; supervision, M.B. All authors have read and agreed to the published version of the manuscript.

Funding: This research received no external funding.

Institutional Review Board Statement: The study was conducted according to the guidelines of the Declaration of Alexandria University and approved by the Ethics Committee of Faculty of Science, Alexandria University, Egypt (AU 04190824102).

Informed Consent Statement: Not applicable.

Data Availability Statement: All data are available in the manuscript.

Acknowledgments: The authors thank the late Elsayed El-Ashry, Department of Chemistry, Alexandria University for the gift of pure curcumin. The authors express thanks to Mona Yehia, Department of Pathology, Medical Research Institute for her support in the histopathological analysis.

Conflicts of Interest: The authors declare no conflict of interest.

References

1. Verbeek, M.M.; Eikelenboom, P.; de Waal, R.M. Differences between the pathogenesis of senile plaques and congophilic angiopathy in Alzheimer disease. *J. Neuropathol. Exp. Neurol.* **1997**, *56*, 751–761. [CrossRef]
2. Rao, C.V.; Asch, A.S.; Carr, D.J.; Yamada, H.Y. Amyloid-beta accumulation cycle as a prevention and/or therapy target for Alzheimer's disease. *Aging Cell* **2020**, *19*, e13109. [CrossRef]
3. Marcus, J.N.; Schachter, J. Targeting post-translational modifications on tau as a therapeutic strategy for Alzheimer's disease. *J. Neurogenet.* **2011**, *25*, 127–133. [CrossRef] [PubMed]
4. Rönnemaa, E.; Zethelius, B.; Sundelöf, J.; Sundström, J.; Degerman-Gunnarsson, M.; Berne, C.; Lannfelt, L.; Kilander, L. Impaired insulin secretion increases the risk of Alzheimer disease. *Neurology* **2008**, *71*, 1065–1071. [CrossRef]
5. Carvalheira, J.B.; Ribeiro, E.B.; Araújo, E.P.; Guimarães, R.B.; Telles, M.M.; Torsoni, M.; Gontijo, J.A.R.; Velloso, L.A.; Saad, M.J.A. Selective impairment of insulin signalling in the hypothalamus of obese Zucker rats. *Diabetologia* **2003**, *46*, 1629–1640. [CrossRef]
6. Soumya, D.; Srilatha, B. Late-stage complications of diabetes and insulin resistance. *J. Diabetes Metab.* **2011**, *2*, 1672.
7. Weinstein, G.; Maillard, P.; Himali, J.J.; Beiser, A.S.; Au, R.; Wolf, P.A.; Seshadri, S.; DeCarli, C. Glucose indices are associated with cognitive and structural brain measures in young adults. *Neurology* **2015**, *84*, 2329–2337. [CrossRef]
8. Lu, M.J.; Zhu, Y.; Sun, J.; Yang, X.R. Microglia mediates inflammation injury in mouse models of Parkinson's disease. *Zhongguo Zuzhi Gongcheng Yanjiu* **2011**, *15*, 1945–1948.
9. Hein, A.M.; O'Banion, M.K. Neuroinflammation and cognitive dysfunction in chronic disease and aging. *J. Neuroimmune Pharmacol.* **2012**, *7*, 3–6. [CrossRef] [PubMed]
10. Park, J.C.; Han, S.H.; Mook-Jung, I. Peripheral inflammatory biomarkers in Alzheimer's disease: A brief review. *BMB Rep.* **2020**, *53*, 10. [CrossRef] [PubMed]
11. e Silva, N.M.L.; Gonçalves, R.A.; Pascoal, T.A.; Lima-Filho, R.A.; Resende, E.D.P.F.; Vieira, E.L.; Teixeira, A.L.; de Souza, L.C.; Peny, J.A.; Fortuna, J.T.; et al. Pro-inflammatory interleukin-6 signaling links cognitive impairments and peripheral metabolic alterations in Alzheimer's disease. *Transl. Psychiatry* **2021**, *11*, 1–15.

12. Pizza, V.; Agresta, A.; D'Acunto, C.W.; Festa, M.; Capasso, A. Neuroinflammation and ageing: Current theories and an overview of the data. *Rev. Recent Clin. Trials* **2011**, *6*, 189–203. [CrossRef]
13. Arthur, J.S.; Ley, S.C. Mitogen-activated protein kinases in innate immunity. *Nat. Rev. Immunol.* **2013**, *13*, 679–692. [CrossRef]
14. Albert-Gascó, H.; Ros-Bernal, F.; Castillo-Gómez, E.; Olucha-Bordonau, F.E. MAP/ERK signaling in developing cognitive and emotional function and its effect on pathological and neurodegenerative processes. *Int. J. Mol. Sci.* **2020**, *21*, 4471. [CrossRef] [PubMed]
15. Jayaraj, R.L.; Azimullah, S.; Beiram, R.; Jalal, F.Y.; Rosenberg, G.A. Neuroinflammation: Friend and foe for ischemic stroke. *J. Neuroinflamm.* **2019**, *16*, 142. [CrossRef] [PubMed]
16. Xing, L.; Larsen, R.S.; Bjorklund, G.R.; Li, X.; Wu, Y.; Philpot, B.D.; Snider, W.D.; Newbern, J.M. Layer specific and general requirements for ERK/MAPK signaling in the developing neocortex. *eLife* **2016**, *5*, e11123. [CrossRef]
17. Faucher, P.; Mons, N.; Micheau, J.; Louis, C.; Beracochea, D.J. Hippocampal injections of oligomeric amyloid β-peptide (1-42) induce selective working memory deficits and longlasting alterations of ERK signaling pathway. *Front. Aging Neurosci.* **2015**, *7*, 245. [PubMed]
18. Lenzen, S. The mechanisms of alloxan-and streptozotocin-induced diabetes. *Diabetologia* **2008**, *51*, 216–226. [CrossRef]
19. Szkudelski, T. The mechanism of alloxan and streptozotocin action in B cells of the rat pancreas. *Physiol. Res.* **2001**, *50*, 537–546.
20. Yang, F.; Lim, G.P.; Begum, A.N.; Ubeda, O.J.; Simmons, M.R.; Ambegaokar, S.S.; Chen, P.P.; Kayed, R.; Glabe, C.G.; Frautschy, S.A.; et al. Curcumin inhibits formation of amyloid β oligomers and fibrils, binds plaques, and reduces amyloid in vivo. *J. Biol. Chem.* **2005**, *280*, 5892–5901. [CrossRef]
21. Khan, A.; Jahan, S.; Imtiyaz, Z.; Alshahrani, S.; Antar Makeen, H.; Mohammed Alshehri, B.; Kumar, A.; Arafah, A.; Rehman, M.U. Neuroprotection: Targeting Multiple Pathways by Naturally Occurring Phytochemicals. *Biomedicines* **2020**, *8*, 284. [CrossRef]
22. Agrawal, R.; Mishra, B.; Tyagi, E.; Nath, C.; Shukla, R. Effect of curcumin on brain insulin receptors and memory functions in STZ (ICV) induced dementia model of rat. *Pharm. Res.* **2010**, *61*, 247–252. [CrossRef] [PubMed]
23. ELBini-Dhouib, I.; Doghri, R.; Ellefi, A.; Degrach, I.; Srairi-Abid, N.; Gati, A. Curcumin Attenuated Neurotoxicity in Sporadic Animal Model of Alzheimer's Disease. *Molecules* **2021**, *26*, 3011. [CrossRef] [PubMed]
24. Caesar, I.; Jonson, M.; Nilsson, K.P.R.; Thor, S.; Hammarström, P. Curcumin promotes A-beta fibrillation and reduces neurotoxicity in transgenic Drosophila. *PLoS ONE* **2012**, *7*, e31424. [CrossRef]
25. Ege, D. Action Mechanisms of Curcumin in Alzheimer's Disease and Its Brain Targeted Delivery. *Materials* **2021**, *14*, 3332. [CrossRef] [PubMed]
26. Burgos-Morón, E.; Calderón-Montaño, J.M.; Salvador, J.; Robles, A.; López-Lázaro, M. The dark side of curcumin Estefanía. *Int. J. Cancer* **2010**, *126*, 1771–1775.
27. Basnet, P.; Skalko-Basnet, N. Curcumin: An anti-inflammatory molecule from a curry spice on the path to cancer treatment. *Molecules* **2011**, *16*, 4567–4598. [CrossRef] [PubMed]
28. Modasiya, M.; Patel, V. Studies on solubility of curcumin. *Int. J. Pharm. Life Sci.* **2012**, *3*, 1490–1497.
29. Mohanty, C.; Sahoo, S.K. The in vitro stability and in vivo pharmacokinetics of curcumin prepared as an aqueous nanoparticulate formulation. *Biomaterials* **2010**, *31*, 6597–6611. [CrossRef]
30. Mishra, P.K.; Mishra, H.; Ekielski, A.; Talegaonkar, S.; Vaidya, B. Zinc oxide nanoparticles: A promising nanomaterial for biomedical applications. *Drug Discov.* **2017**, *22*, 1825–1834. [CrossRef]
31. Smijs, T.G.; Pavel, S. Titanium dioxide and zinc oxide nanoparticles in sunscreens: Focus on their safety and effectiveness. *Nanotechnol. Sci. Appl.* **2011**, *4*, 95. [CrossRef] [PubMed]
32. Ruszkiewicz, J.A.; Pinkas, A.; Ferrer, B.; Peres, T.V.; Tsatsakis, A.; Aschner, A. Neurotoxic effect of active ingredients in sunscreen products, a contemporary review. *Toxicol. Rep.* **2017**, *4*, 245–259. [CrossRef]
33. Xiong, H.M. ZnO nanoparticles applied to bioimaging and drug delivery. *Adv. Mater.* **2013**, *25*, 5329–5335. [CrossRef] [PubMed]
34. Lowry, O.H.; Rosebrough, N.J.; Farr, A.L.; Randall, R.J. Protein measurement with the Folin phenol reagent. *J. Biol. Chem.* **1951**, *193*, 265–275. [CrossRef]
35. Matthews, D.R.; Hosker, J.P.; Rudenski, A.S.; Naylor, B.A.; Treacher, D.F.; Turner, R.C. Homeostasis model assessment: Insulin resistance and beta-cell function from fasting plasma glucose and insulin concentrations in man. *Diabetologia* **1981**, *28*, 412–419. [CrossRef]
36. Tappel, A.; Zalkin, H. Lipid peroxidation in isolated mitochondria. *Arch. Biochem. Biophys.* **1959**, *80*, 326–332. [CrossRef]
37. Montgomery, H.; Dymock, J. Colorimetric determination of nitric oxide. *Analyst* **1961**, *86*, 414–417.
38. Jollow, D.; Mitchell, J.; Zampaglione, N.; Gillette, J.R. Bromobenzene-induced liver necrosis. Protective role of glutathione and evidence for 3, 4-bromobenzene oxide as the hepatotoxic metabolite. *Pharmacology* **1974**, *11*, 151–169. [CrossRef] [PubMed]
39. Habig, W.; Pabst, M.; Jakoby, W. Glutathione S-transferases the first enzymatic step in mercapturic acid formation. *J. Biol. Chem.* **1974**, *249*, 7130–7139. [CrossRef]
40. Marklund, S.; Marklund, G. Involvement of the superoxide anion radical in the autoxidation of pyrogallol and a convenient assay for superoxide dismutase. *Eur. J. Biochem.* **1974**, *47*, 469–474. [CrossRef] [PubMed]
41. Aebi, H. Catalase in vitro. *Methods Enzymol.* **1984**, *105*, 121–126.
42. Paglia, D.; Valentine, W. Studies on the quantitative and qualitative characterization of erythrocyte glutathione peroxidase. *J. Lab. Clin. Med.* **1967**, *70*, 158–169.

43. Ellman, G.L.; Courtney, K.D.; Andres, V.; Featherstone, R.M. A new and rapid colorimetric determination of acetylcholinesterase activity. *Biochem. Pharmacol.* **1961**, *7*, 88–95. [CrossRef]
44. Livak, K.; Schmittgen, T. Analysis of relative gene expression data using real-time quantitative PCR and the 2(-Delta Delta C(T)) method. *Methods* **2001**, *25*, 402–408. [CrossRef] [PubMed]
45. Burnette, W. Western blotting: Electrophoretic transfer of proteins from sodium dodecyl sulfate-polyacrylamide gels to unmodified nitrocellulose and radiographic detection with antibody and radioiodinated protein A. *Anal. Biochem.* **1981**, *112*, 195–203. [CrossRef]
46. Ramachandran, S.; Asokkumar, K.; Uma Maheswari, M.; Ravi, T.; Sivashanmugam, A.; Saravanan, S.; Rajasekaran, A.; Dharman, J. Investigation of antidiabetic, antihyperlipidemic, and in vivo antioxidant properties of Sphaeranthus indicus Linn. in type 1 diabetic rats: An identification of possible biomarkers. *Evid. Based Complement. Altern. Med.* **2011**, *2011*, 571721. [CrossRef]
47. Balbaa, M.; Abdulmalek, S.A.; Khalil, S. Oxidative stress and expression of insulin signaling proteins in the brain of diabetic rats: Role of Nigella sativa oil and antidiabetic drugs. *PLoS ONE* **2017**, *12*, e0172429. [CrossRef]
48. Salkovic-Petrisic, M.; Hoyer, S. Central insulin resistance as a trigger for sporadic Alzheimer-like pathology: An experimental approach. *Neuropsychiatr. Disord. Integr. Approach* **2007**, *72*, 217–233.
49. Shehzad, A.; Ha, T.; Subhan, F.; Lee, Y.S. New mechanisms and the anti-inflammatory role of curcumin in obesity and obesity-related metabolic diseases. *Eur. J. Nutr.* **2011**, *50*, 151–161. [CrossRef]
50. Nazarizadeh, A.; Asri-Rezaie, S. Comparative study of antidiabetic activity and oxidative stress induced by zinc oxide nanoparticles and zinc sulfate in diabetic rats. *AAPS PharmSciTech* **2016**, *17*, 834–843. [CrossRef]
51. Umrani, R.D.; Paknikar, K.M. Zinc oxide nanoparticles show antidiabetic activity in streptozotocin-induced type 1 and 2 diabetic rats. *Nanomedicine* **2014**, *9*, 89–104. [CrossRef]
52. Alkaladi, A.; Abdelazim, A.M.; Afifi, M. Antidiabetic activity of zinc oxide and silver nanoparticles on streptozotocin-induced diabetic rats. *Int. J. Mol. Sci.* **2014**, *15*, 2015–2023. [CrossRef]
53. Virgen-Ortiz, A.; Apolinar-Iribe, A.; Díaz-Reval, I.; Parra-Delgado, H.; Limón-Miranda, S.; Sánchez-Pastor, E.A.; Castro-Sánchez, L.; Castillo, S.J.; Dagnino-Acosta, A.; Bonales-Alatorre, E.; et al. Zinc Oxide Nanoparticles Induce an Adverse Effect on Blood Glucose Levels Depending on the Dose and Route of Administration in Healthy and Diabetic Rats. *Nanomaterials* **2020**, *10*, 2005. [CrossRef]
54. Thornalley, P.J. Glycation in diabetic neuropathy: Characteristics, consequences, causes, and therapeutic options. *Int. Rev. Neurobiol.* **2002**, *50*, 37–57.
55. Loeffler, D.A. Should development of Alzheimer's disease-specific intravenous immunoglobulin be considered? *J. Neuroinflamm.* **2014**, *11*, 1–10. [CrossRef]
56. Reaven, G.M.; Thompson, L.W.; Nahum, D.; Haskins, E. Relationship between hyperglycemia and cognitive function in older NIDDM patients. *Diabetes Care* **1990**, *13*, 16–21. [CrossRef] [PubMed]
57. Bottino, C.M.; Castro, C.C.; Gomes, R.L.; Buchpiguel, C.A.; Marchetti, R.L.; Neto, M.R. Volumetric MRI measurements can differentiate Alzheimer's disease, mild cognitive impairment, and normal aging. *Int. Psychogeriatr.* **2002**, *14*, 59–72. [CrossRef]
58. Padurariu, M.; Ciobica, A.; Hritcu, L.; Stoica, B.; Bild, W.; Stefanescu, C. Changes of some oxidative stress markers in the serum of patients with mild cognitive impairment and Alzheimer's disease. *Neurosci. Lett.* **2010**, *469*, 6–10. [CrossRef] [PubMed]
59. Yu, T.; Robotham, J.L.; Yoon, Y. Increased production of reactive oxygen species in hyperglycemic conditions requires dynamic change of mitochondrial morphology. *Proc. Natl. Acad. Sci. USA* **2006**, *103*, 2653–2658. [CrossRef]
60. Shen, C.; Chen, Y.; Liu, H.; Zhang, K.; Zhang, T.; Lin, A.; Jing, N. Hydrogen peroxide promotes Aβ production through JNK-dependent activation of γ-secretase. *J. Biol. Chem.* **2008**, *283*, 17721–17730. [CrossRef] [PubMed]
61. Hodaei, H.; Adibian, M.; Nikpayam, O.; Hedayati, M.; Sohrab, G. The effect of curcumin supplementation on anthropometric indices, insulin resistance and oxidative stress in patients with type 2 diabetes: A randomized, double-blind clinical trial. *Diabetol. Metab. Syndr.* **2019**, *11*, 1–8. [CrossRef] [PubMed]
62. Ukperoro, J.U.; Offiah, N.; Idris, T.; Awogoke, D. Antioxidant effect of zinc, selenium and their combination on the liver and kidney of alloxan-induced diabetes in rats. *Mediterr. J. Nutr. Metab.* **2010**, *3*, 25–30. [CrossRef]
63. El-Bahr, S.M.; Shousha, S.; Albokhadaim, I.; Shehab, A.; Khattab, W.; Ahmed-Farid, O.; El-Garhy, O.; Abdelgawad, A.; El-Naggar, M.; Moustafa, M.; et al. Impact of dietary zinc oxide nanoparticles on selected serum biomarkers, lipid peroxidation and tissue gene expression of antioxidant enzymes and cytokines in Japanese quail. *BMC Vet. Res.* **2020**, *16*, 1–12. [CrossRef] [PubMed]
64. Rahdar, A.; Hajinezhad, M.R.; Bilal, M.; Askari, F.; Kyzas, G.Z. Behavioral effects of zinc oxide nanoparticles on the brain of rats. *Inorg. Chem. Commun.* **2020**, *119*, 108131. [CrossRef]
65. Dkhil, M.A.; Diab, M.S.; Aljawdah, H.M.; Murshed, M.; Hafiz, T.A.; Al-Quraishy, S.; Bauomy, A.A. Neuro-biochemical changes induced by zinc oxide nanoparticles. *Saudi J. Biol. Sci.* **2020**, *27*, 2863–2867. [CrossRef]
66. Castranova, V. Signaling pathways controlling the production of inflammatory mediators in response to crystalline silica exposure: Role of reactive oxygen/nitrogen species. *Free Radic. Biol. Med.* **2004**, *37*, 916–925. [CrossRef]
67. Boarescu, P.-M.; Boarescu, I.; Bocșan, I.C.; Gheban, D.; Bulboacă, A.E.; Nicula, C.; Pop, R.M.; Râjnoveanu, R.-M.; Bolboacă, S.D. Antioxidant and anti-inflammatory effects of Curcumin nanoparticles on drug-induced acute myocardial infarction in diabetic rats. *Antioxidants* **2019**, *8*, 504. [CrossRef] [PubMed]
68. Kim, Y.S.; Park, H.J.; Joo, S.Y.; Hong, M.H.; Kim, K.H.; Hong, Y.J.; Kim, J.H.; Park, H.W.; Jeong, M.H.; Cho, J.G.; et al. The protective effect of curcumin on myocardial ischemia-reperfusion injury. *Korean Circ. J.* **2008**, *38*, 353–359. [CrossRef]

69. Gammoh, N.Z.; Rink, L. Zinc in infection and inflammation. *Nutrients* **2017**, *9*, 624. [CrossRef]
70. Hassan, R.M.; Elsayed, M.; Kholief, T.E.; Hassanen, N.H.; Gafer, J.A.; Attia, Y.A. Mitigating effect of single or combined administration of nanoparticles of zinc oxide, chromium oxide, and selenium on genotoxicity and metabolic insult in fructose/streptozotocin diabetic rat model. *ESPR* **2021**, *28*, 1–18.
71. Bettger, W.J.; O'Dell, B.L. A critical physiological role of zinc in the structure and function of biomembranes. *Life Sci.* **1981**, *28*, 1425–1438. [CrossRef]
72. Attia, H.; Nounou, H.; Shalaby, M. Zinc oxide nanoparticles induced oxidative DNA damage, inflammation and apoptosis in rat's brain after oral exposure. *Toxics* **2018**, *6*, 29. [CrossRef]
73. Bai, Y.N.; Zheng, J.; Song Zf, D.Y.; Wen, Y. Effect of curcumin on expression of adiponectin in mice with insulin resistance. *J. Shanghai Jiaotong Univ. Sci.* **2013**, *33*, 136.
74. Shao, W.; Yu, Z.; Chiang, Y.; Yang, Y.; Chai, T.; Foltz, W.; Lu, H.; Fantus, I.G.; Jin, T. Curcumin prevents high fat diet induced insulin resistance and obesity via attenuating lipogenesis in liver and inflammatory pathway in adipocytes. *PLoS ONE* **2012**, *7*, e28784. [CrossRef]
75. Zhao, H.; Yenari, M.A.; Cheng, D.; Sapolsky, R.M.; Steinberg, G.K. Bcl-2 overexpression protects against neuron loss within the ischemic margin following experimental stroke and inhibits cytochrome c translocation and caspase-3 activity. *J. Neurochem.* **2008**, *85*, 1026–1036. [CrossRef]
76. Zhao, Y.; Zhao, H.; Zhai, X.; Dai, J.; Jiang, X.; Wang, G.; Li, W.; Cai, L. Effects of Zn deficiency, antioxidants, and low-dose radiation on diabetic oxidative damage and cell death in the testis. *Toxicol. Mech. Methods* **2013**, *23*, 42–47. [CrossRef]
77. Sharma, D.; Sethi, P.; Hussain, E.; Singh, R. Curcumin counteracts the aluminium-induced ageing-related alterations in oxidative stress, Na+, K+ ATPase and protein kinase C in adult and old rat brain regions. *Biogerontology* **2009**, *10*, 489–502. [CrossRef]
78. Hamza, R.Z.; Al-Salmi, F.A.; El-Shenawy, N.S. Evaluation of the effects of the green nanoparticles zinc oxide on monosodium glutamate-induced toxicity in the brain of rats. *PeerJ* **2019**, *7*, e7460. [CrossRef]
79. Saido, T.; Leissring, M.A. Proteolytic degradation of amyloid β-protein. *Cold Spring Harb. Perspect. Med.* **2012**, *2*, a006379. [CrossRef] [PubMed]
80. Devi, L.; Ohno, M. PERK mediates eIF2α phosphorylation responsible for BACE1 elevation, CREB dysfunction and neurodegeneration in a mouse model of Alzheimer's disease. *Neurobiol. Aging* **2014**, *35*, 2272–2281. [CrossRef]
81. Carrasquillo, M.M.; Belbin, O.; Zou, F.; Allen, M.; Ertekin-Taner, N.; Ansari, M.; Wilcox, S.L.; Kashino, M.R.; Ma, L.; Younkin, L.H.; et al. Concordant association of insulin degrading enzyme gene (IDE) variants with IDE mRNA, Aß, and Alzheimer's disease. *PLoS ONE* **2010**, *5*, e8764. [CrossRef]
82. Hurley, L.L.; Akinfiresoye, L.; Nwulia, E.; Kamiya, A.; Kulkarni, A.A.; Tizabi, Y. Antidepressant-like effects of curcumin in WKY rat model of depression is associated with an increase in hippocampal BDNF. *Behav. Brain Res.* **2013**, *239*, 27–30. [CrossRef] [PubMed]
83. Wang, L.Y.; Tang, Z.J.; Han, Y.Z. Neuroprotective effects of caffeic acid phenethyl ester against sevoflurane-induced neuronal degeneration in the hippocampus of neonatal rats involve MAPK and PI3K/Akt signaling pathways. *Mol. Med. Rep.* **2016**, *14*, 3403–3412. [CrossRef] [PubMed]
84. Sweatt, J.D. The neuronal MAP kinase cascade: A biochemical signal integration system subserving synaptic plasticity and memory. *J. Neurochem.* **2001**, *76*, 1–10. [CrossRef]
85. Avila, J. Tau kinases and phosphatases. *J. Cell Mol. Med.* **2008**, *12*, 258–259. [CrossRef]
86. Lee, K.M.; Bang, J.; Kim, B.Y.; Lee, I.S.; Han, J.S.; Hwang, B.Y.; Jeon, W.K. *Fructus mume* alleviates chronic cerebral hypoperfusion-induced white matter and hippocampal damage via inhibition of inflammation and downregulation of TLR4 and p38 MAPK signaling. *BMC Complement. Altern. Med.* **2015**, *15*, 1–10. [CrossRef]
87. Wang, B.; Li, W.; Jin, H.; Nie, X.; Shen, H.; Li, E.; Wang, W. Curcumin attenuates chronic intermittent hypoxia-induced brain injuries by inhibiting AQP4 and p38 MAPK pathway. *Respir. Physiol. Neurobiol.* **2018**, *255*, 50–57. [CrossRef] [PubMed]
88. Jung, K.K.; Lee, H.S.; Cho, J.Y.; Shin, W.C.; Rhee, M.H.; Kim, T.G.; Kang, J.H.; Kim, S.H.; Hong, S.; Kang, S.Y. Inhibitory effect of curcumin on nitric oxide production from lipopolysaccharide-activated primary microglia. *Life Sci.* **2006**, *79*, 2022–2031. [CrossRef]
89. Amin, S.N.; Younan, S.M.; Youssef, M.F.; Rashed, L.A.; Mohamady, I. A histological and functional study on hippocampal formation of normal and diabetic rats. *F1000Research* **2013**, *2*, 151. [CrossRef]

 *pharmaceutics*

Article

A Combination Approach in Inhibiting Type 2 Diabetes-Related Enzymes Using *Ecklonia radiata* Fucoidan and Acarbose

Blessing Mabate [1], Chantal Désirée Daub [1], Samkelo Malgas [2], Adrienne Lesley Edkins [3] and Brett Ivan Pletschke [1,*]

[1] Enzyme Science Programme (ESP), Department of Biochemistry and Microbiology, Faculty of Science, Rhodes University, Makhanda 6140, South Africa; g18M0025@campus.ru.ac.za (B.M.); g15D2439@campus.ru.ac.za (C.D.D.)

[2] Department of Biochemistry, Genetics and Microbiology, University of Pretoria, Pretoria 0028, South Africa; samkelo.malgas@up.ac.za

[3] Biomedical Biotechnology Research Unit, Department of Biochemistry and Microbiology, Faculty of Science, Rhodes University, Makhanda 6140, South Africa; a.edkins@ru.ac.za

* Correspondence: b.pletschke@ru.ac.za; Tel.: +27-46-603-8081; Fax: +27-46-603-7576

Citation: Mabate, B.; Daub, C.D.; Malgas, S.; Edkins, A.L.; Pletschke, B.I. A Combination Approach in Inhibiting Type 2 Diabetes-Related Enzymes Using *Ecklonia radiata* Fucoidan and Acarbose. *Pharmaceutics* **2021**, *13*, 1979. https://doi.org/10.3390/pharmaceutics13111979

Academic Editors: Diana Marcela Aragon Novoa and Fátima Regina Mena Barreto Silva

Received: 11 September 2021
Accepted: 12 November 2021
Published: 22 November 2021

Abstract: Although there are chemotherapeutic efforts in place for Type 2 diabetes mellitus (T2DM), there is a need for novel strategies (including natural products) to manage T2DM. Fucoidan, a sulphated polysaccharide was extracted from *Ecklonia radiata*. The integrity of the fucoidan was confirmed by structural analysis techniques such as FT-IR, NMR and TGA. In addition, the fucoidan was chemically characterised and tested for cell toxicity. The fucoidan was investigated with regards to its potential to inhibit α-amylase and α-glucosidase. The fucoidan was not cytotoxic and inhibited α-glucosidase (IC_{50} 19 µg/mL) more strongly than the standard commercial drug acarbose (IC_{50} 332 µg/mL). However, the fucoidan lacked potency against α-amylase. On the other hand, acarbose was a more potent inhibitor of α-amylase (IC_{50} of 109 µg/mL) than α-glucosidase. Due to side effects associated with the use of acarbose, a combination approach using acarbose and fucoidan was investigated. The combination showed synergistic inhibition (>70%) of α-glucosidase compared to when the drugs were used alone. The medicinal implication of this synergism is that a regimen with a reduced acarbose dose may be used, thus minimising side effects to the patient, while achieving the desired therapeutic effect for managing T2DM.

Keywords: acarbose; combination approach; *Ecklonia radiata*; fucoidan; type 2 diabetes management

1. Introduction

Type 2 diabetes mellitus (T2DM) remains one of the prominent emerging chronic diseases in humans, with about 400 million people living with diabetes worldwide [1]. T2DM is a burden to the world in mortality, morbidity and disability-adjusted life years (DALYs), which significantly steers economic pressure in medium or low-income countries [1]. Diabetes is projected to be the seventh-highest global cause of death by 2030 [2]. T2DM is characterised by hyperglycemia, leading to considerable damage to organs, nervous and cardiovascular systems [3]. The main source of blood glucose levels is carbohydrate (starch) hydrolysis by catabolic enzymes in the alimentary canal, namely α-amylase and α-glucosidase [4]. Carbohydrate digestion directly impacts the amount of glucose and sugars absorbed into the bloodstream, causing hyperglycaemia, especially in insulin-resistant individuals [5]. Therefore, slowing down carbohydrate digestion or inhibiting glucose absorption is a promising approach for treating diabetes and its complications [6].

Several strategies to manage T2DM exist, including insulin secretagogues, insulin sensitisers and acarbose, a semi-synthetic compound that inhibits carbohydrate digesting enzymes [4,7]. However, as with most chemotherapeutics, the use of acarbose is associated with various side effects, including flatulence, abdominal discomfort and diarrhoea [8]. Despite the promising success of chemotherapeutic drugs in clinical trials, side effects associated with cytotoxicity often surface. Therefore, there is a need for developing better-tolerated treatments with improved therapeutic properties. Natural bio-compounds, including marine products, are known to have therapeutic advantages over chemotherapeutics [9]. Fucoidans are among the major marine polysaccharides that have gained interest as α-amylase and α-glucosidase inhibitors [10,11]. Besides their inhibition potency on the digestive enzymes mentioned, fucoidans have demonstrated anti-diabetic potential by inhibiting dipeptidyl peptidase IV (DPPIV) [12]. Also, fucoidans have shown anti-diabetic potential in vivo where they have reduced blood sugar levels in induced hyperglycemic rats [13]. The anti-diabetic relevance of fucoidans is profound and must be further elucidated.

Fucoidans are complex anionic sulphated polysaccharides composed mainly of fucose and varying monosaccharides in their structure. Fucose residues are linked by α-(1-3), α-(1-3)-α-(1-4) or α-(1-3)-α-(1-2) linkages and sulphate groups that are attached to the O-2 and/or O-4 position of the sugar residues composing the polysaccharide backbone [4]. The biological activity of fucoidan has gained remarkable attention in biomedical studies, including T2DM alleviation efforts [10,11]. However, there is limited literature on fucoidan extracted from South African seaweeds despite the fact the country has one of the largest coastlines globally, with vast algal biodiversity [14].

Combining multiple drugs is a common approach in the biomedical and pharmaceutical sciences [15]. The rationale behind drug combinations with non-overlapping side effects is to obtain regimens with lowered adverse effects, avoiding resistance to single compounds and possibly obtaining a direct empirical synergy of the compounds against a targeted biological process [16]. Although there are so many conflicting definitions of synergy —in this context, it is defined as a combined experimental effect greater than the additive effects of individual compounds. Drug synergism has been quantified in vitro and in vivo by two widely accepted reference models: the Loewe Additivity and the Bliss Independence [17]. The Loewe model describes the additive effect of two compounds with the same mode of action, while the Bliss model describes the additive effect of two ingredients with different modes of action [17]. Among the reference models for measuring synergy, the Chou-Talalay method has become the most popular and used model [17]. The median effect equation, characteristic of the Chou-Talalay method, was derived from a unified theory of Michaelis-Menten, Hill, Henderson-Hasselbalch and Scatchard equations [18]. Despite the limitations of this reference model, it is best suited for drug combination studies involving enzymes [18].

It has been established that acarbose inhibits α-amylase significantly more than it does α-glucosidase; the inverse has been observed for most characterised fucoidans [10,19]. Therefore, drug combination studies for acarbose and fucoidan are a plausible approach for leveraging each compound's inhibitory potential. This study sought to characterise *Ecklonia radiata* derived fucoidan and demonstrate its potency as an inhibitor of α-amylase and α-glucosidase, significant targets in T2DM therapy. Furthermore, drug combination studies using acarbose and *E. radiata* fucoidan were explored and shown to inhibit α-glucosidase to a significantly higher degree than when the two compounds were used alone.

2. Materials and Methods

The fucoidan was extracted from harvested *E. radiata* seaweed species. Both the fucoidan obtained from *Fucus vesiculosus* (Cat. No. F5631) and acarbose (Cat No A8980) were purchased from Sigma-Aldrich (St. Louis, MO, USA). The two amylolytic enzymes, porcine pancreatic α-amylase (Cat. No. E-PANAA) and *Saccharomyces cerevisiae* α-glucosidase (Cat. No. G5003), were purchased from Megazyme™ (Bray, Ireland) and Sigma-Aldrich, respectively. The rest of the reagents used in the study were of analytical grade and purchased from Sigma-Aldrich, MERCK (Darmstadt, Germany) and Megazyme™.

2.1. Harvesting and Preparation of Seaweeds

The brown seaweed, *E. radiata*, was harvested from the South African Indian Ocean coastline in the Eastern Cape province at the coordinates: 33°36′ 36.8424″ S; 26°53′ 23.4996″ E. The harvested seaweed was stored on ice during transportation to the laboratory. Upon arrival in the laboratory, the seaweed was washed thoroughly with distilled water, cut into small pieces and dried at 40 °C in an oven for 72 h. The dried seaweed samples were subsequently pulverised with a coffee bean grinder and stored at room temperature until required.

2.2. Fucoidan Extraction

The seaweed was defatted by extracting lipids and pigments using a mixture of methanol, dichloromethane and water in a ratio of 4:2:1, as described previously [20,21]. Fucoidan was extracted from *E. radiata* using the hot water extraction method [22] with minor modifications. Briefly, 15 g of dry defatted seaweed powder was suspended in 450 mL of distilled water at a mass loading of 1:30. The mixture was heated to 70 °C with agitation overnight. The extracted fucoidan yield was expressed as a percentage of the dry seaweed weight (% dry wt).

2.3. Structural Analysis of Extracted Fucoidan

2.3.1. Fourier Transform Infrared Spectrometer (FT-IR) Analysis

About 100 mg of ground fucoidan was analysed by Fourier transform infrared spectroscopy (FTIR) using a Model 100 FT-IR spectrometer system (PerkinElmer®, Waltham, MA, USA). The signals were automatically recorded by averaging four scans over the range of 4000–650 cm^{-1}. The Spectrum One software (version 1.2.1) was used to carry out the baseline and ATR corrections for penetration depth and frequency variations.

2.3.2. NMR Spectroscopy Analysis

The *E. radiata* fucoidan extract (10 mg/mL) was dissolved in 1 mL of D_2O, followed by centrifugation at 13,000× g for 2 min. The supernatant was filtered through a 0.45-μm filter to remove any insoluble material. The deuterium-exchanged sample was subjected to ^{1}H-NMR analysis. Spectra were recorded at 23 °C using a 400 MHz spectrometer (Bruker, Fällanden, Switzerland) equipped with Topspin 3.5 software (Bruker, Billerica, MA, USA). The chemical shifts were expressed in ppm.

2.3.3. Thermogravimetric Analysis

Thermogravimetric analysis of *E. radiata* fucoidan was conducted with a Pyris Diamond model thermogravimetric analyser (PerkinElmer®, Shelton, CT, USA). Approximately 4 mg of fucoidan was placed in an alumina crucible. Pure nitrogen (purity of 99.99%), with a 20 mL/min flow rate, was used as a carrier gas to extinguish the mass transfer effect to a minimum level. The fucoidan was heated from 30 °C to 900 °C at a heating rate of 30 °C/min. A separate blank run using an empty pan was conducted for baseline correction. The weight loss relative to the temperature increment was automatically recorded.

2.4. Chemical Composition Analysis of Fucoidan

The total sugar content of fucoidan was analysed using the phenol-sulfuric acid method, with L-fucose as a standard [23]. Partially hydrolysed fucoidan total reducing sugar content was determined according to the previously established dinitrosalicylic acid (DNS) method [24]. Protein content was measured by the Bradford method using bovine serum albumin as the standard [25]. Formic acid (60% *v/v*) was used to desulphurise the fucoidan's sulphate content, which was measured using a modified barium chloride-gelatin method described previously [26]. The total polyphenols were determined using a modified Folin-Ciocalteu method [27]. Also, the ash content in fucoidan was derived from derivative thermogravimetry (DTG) data obtained during TGA analysis. Monosaccharides, including L-fucose, D-glucose, D-galactose, D-mannose, L-arabinose and D-fructose, generated from

2 M trifluoroacetic acid (TFA) hydrolysis of fucoidan were quantified using a Shimadzu (Kyoto, Japan) high-performance liquid chromatography (HPLC) instrument equipped with a refractive index (RID) detector. The neutral sugars were separated using a Fortis Amino analytical column (150 mm × 4.6 mm) with slight modification of the recommended method from the supplier (Fortis Technologies Ltd., Neston, UK). The mobile phase was a mixture of HPLC grade acetonitrile and degassed H_2O (Milli-Q, MERCK) in a ratio of 3:1. The flow rate was 0.8 mL/min, and the column temperature was at 30 °C with a sample injection volume of 20 µL. Finally, uronic acids, D-galacturonic acid and D-glucuronic acid, were quantified enzymatically according to the microtiter plate format described in the Megazyme™ K-URONIC kit.

2.5. Determination of Average Molecular Weight

The molecular weight of the extracted fucoidan was estimated using analytical ultracentrifugation. A volume of 15 mL of *E. radiata* fucoidan stock solution (0.5 mg/mL) prepared in distilled water was transferred into and filtered through 100 K, 50 K, 30 K and 10 K Amicon® ultra-centrifugation filters (MERCK). The retentates were obtained by centrifuging the filters at 4000 g for 20 min. The filtrate and retentate samples from each filtration step were analysed for the presence of fucoidan according to the phenol-sulfuric acid assay.

2.6. Carbohydrate Digesting Enzymes Inhibition Studies

2.6.1. α-Amylase Activity Assay

The inhibition potential of the extracted *E. radiata* fucoidan and the controls (*F. vesiculosus* fucoidan and acarbose) were investigated and compared using the α-amylase activity assay. In brief, 2% (*w/v*) potato starch dissolved in 0.05 M sodium phosphate buffer (pH~7.0) was mixed with variable concentrations of potential inhibitors ranging from 0.01–1 mg/mL. The reaction was started by adding porcine α-amylase (10 units/mL in a total reaction volume of 400 µL). The reaction mixture was incubated at 37 °C for 20 min with gentle agitation at 70 rpm. The amount of reducing sugars produced was measured according to the DNS method [24]. A control reaction was prepared using the same procedure replacing the inhibitor sample with distilled water. Extrapolation of absorbance values was performed against a glucose standard calibration curve. The α-amylase inhibition was expressed as a relative product (reducing sugar (RS)) percentage according to the following formula:

$$Enzyme\ inhibition\ \% = \frac{(RS\ released\ by\ control - RS\ released\ by\ test\ reaction)}{RS\ released\ by\ control} \times 100 \quad (1)$$

The inhibitor concentration resulting in 50% inhibition of α-amylase activity (IC_{50}) was determined graphically using GraphPad Prism Software version 6.0 (GraphPad Inc., San Diego, CA, USA).

2.6.2. α-Glucosidase Activity Assay

The inhibition of α-glucosidase activity was determined by dissolving the substrate, 10 mM *p*-nitrophenyl-α-D-glucopyranoside (*p*NPG), in 0.05 M sodium phosphate buffer (pH~7.0) in the presence of varied concentrations (0.01–1 mg/mL) of potential inhibitors. The reaction was started by adding α-glucosidase (0.1 units/mL) to a total reaction volume of 400 µL. The reaction mixture was incubated at 37 °C for 20 min and terminated by the addition of 400 µL of 2M $NaCO_3$ solution. The reaction mixture with water instead of the inhibitors was used as a negative control. The absorbance of the released *p*-nitrophenol was measured at 405 nm. The amount of *p*-nitrophenol produced was extrapolated from a *p*-nitrophenol standard curve. The α-glucosidase % inhibition was calculated as relative product (*p*-nitrophenol (*p*N)) percentage as follows:

$$Enzyme\ inhibition\ \% = \frac{(pN\ released\ by\ control - pN\ released\ by\ test\ reaction)}{pN\ released\ by\ control} \times 100 \quad (2)$$

The inhibitor concentrations resulting in 50% inhibition of enzyme activity (IC_{50}) was determined graphically using the GraphPad Prism Software version 6.0 (GraphPad Inc., San Diego, CA, USA).

2.7. Interaction between Fucoidan and α-Glucosidase

2.7.1. Tryptophan Fluorescence Analysis of the α-Glucosidase-Fucoidan Interaction

The α-glucosidase-fucoidan interaction was analysed through the intrinsic tryptophan fluorescence of the enzyme [28]. Briefly, 20 µg/mL of α-glucosidase and *E. radiata* fucoidan (0.0625–0.5 mg/mL) in 0.05 M sodium phosphate buffer (pH 7.4) were incubated for 20 min at 37 °C. Thereafter, fluorescence was measured between 300 and 500 nm after initial excitation at 295 nm using a SpectraMax M3 (Separations, Roodeport, South Africa) microplate reader at 25 °C using standard 96-well black microplates with 5 nm increments. The relative fluorescence was calculated as the average value obtained from at least 4 spectrum scans corrected for their baseline (buffer with or without inhibitors in the absence of enzyme) reading.

2.7.2. Determination of Binding Parameters of the α-Glucosidase-Fucoidan Interaction

To further elucidate the quenching mechanism of *E. radiata* fucoidan, the fluorescence data was analysed by constructing a modified Stern-Volmer plot ((F_0/F_0-F) versus 1/[fucoidan]), where F_0 and F are the fluorescence intensities of α-glucosidase in the absence and presence of the quencher (fucoidan), respectively. The values for the association constant (K) and the number of binding sites (n) were obtained from the intercept and slope of the constructed secondary plot, respectively.

2.7.3. Circular Dichroism Analysis of Secondary Structural Changes of α-Glucosidase upon Interaction with *E. radiata* Fucoidan

The secondary structural conformation of the α-glucosidase (upon interacting with fucoidan) was analysed using Far-UV circular dichroism (CD) as previously described [29]. Briefly, 0.2 µM of α-glucosidase was suspended in 0.05 M phosphate buffer, pH 7.0. The analysis was conducted on a Chirascan v.4.4.1 Build spectrometer (Applied Photophysics Ltd., Leatherhead, UK) equipped with a Peltier temperature controller at 19 °C, using a 0.1 cm path-length quartz cuvette (Hellma Analytics, Baden-Württemberg, Germany). The data was analysed and deconvoluted to α-helix, β-sheet, β-turns and unordered regions using the CONTIN program of the Dichroweb online server (accessed 3 October 2019) [30]. The procedure was repeated on α-glucosidase, which was pre-treated with varying concentrations of fucoidan, and a boiled enzyme positive control was included in the assay.

2.7.4. Mode of α-Glucosidase Inhibition

The α-glucosidase activity was assayed in the presence of fucoidan at fixed concentrations between 0 and 1 mg/mL in the presence of varying *p*NPG concentrations (0.25–6.25 mM), according to the protocol previously described in Section 2.6.2. The enzymatic reaction rates (v_0) were calculated using the *p*-nitrophenol released. A Michaelis-Menten curve was constructed, and the V_{max} and K_M values were determined using GraphPad Prism v. 6.0 software (GraphPad Inc., San Diego, CA, USA).

2.8. Investigating the Synergistic Potential of Extracted Fucoidan with Acarbose

The α-glucosidase assay was performed as described in Section 2.6.2 with varying concentrations of inhibitors from 0 to 2 mg/mL in the reaction. The combination (acarbose: fucoidan) was maintained at a constant ratio (1:1) to determine the dose-dependent effect of the inhibitors acting alone or in combination. The % inhibition was represented as normalised enzyme inhibition using GraphPad Prism v.6.0. Also, IC_{25}, IC_{50} and IC_{75} were determined for each inhibitor. The combinations of these inhibitor potencies (non-constant ratios) were investigated as per the α-glucosidase assay protocol. The synergistic effect of inhibitors was calculated by using the combination index (CI). The CI, a quantification of the degree of inhibitor interactions based on the median-effect principle, was calculated as described previously by Chou and colleagues using the CompuSyn software (ComboSyn Inc., Paramus, NJ, USA) [18].

The median-effect equation is stated below:

$$log\left(\frac{fa}{fu}\right) = mlogD - mlogD_m \tag{3}$$

where *fa* is the fraction affected by dose *D*, *fu* is the unaffected fraction *(fu = 1 − fa)*, *m* is the coefficient signifying the shape of the dose-effect curve, *D* is the inhibitor and D_m is the median-effect dose (IC_{50} in this case).

The *CI* was determined through the equation below:

$$CI = \frac{D_1}{(D_x)_1} + \frac{D_2}{(D_x)_2} \tag{4}$$

where D_1 and D_2 are the doses of inhibitors that produce a certain level of inhibition in the combination system and $(D_x)_1$ and $(D_x)_2$ are the doses of inhibitors added alone that lead to the same level of inhibition. CI was calculated from the data as a measure of the interaction among drugs. CI values lower than 1 indicate synergy, CI values equal to 1 indicate an additive effect and CI values higher than 1 indicate antagonism.

2.9. Fucoidan Toxicity Screening

2.9.1. Cell Culture

The HCT116 human colon cancer cell line was purchased from the American Type Culture Collection (ATCC CCL-247). The cell line was cultured in Dulbecco's Modified Eagle's Media (DMEM) with GlutaMAX™-I, 10% *(v/v)* FBS and 1% *(v/v)* sodium pyruvate and was maintained at 37 °C with 9% CO_2 in a humidified atmosphere.

2.9.2. Resazurin Assay

The extracted fucoidan was screened for its potential cytotoxicity on the HCT116 colon cell line using the resazurin assay. Cells were seeded at density of 1×10^5 cells/well in DMEM growth media in a 96 well plate. The cells were allowed to adhere to the plate matrix overnight at 37 °C in 9% CO_2 and then treated with fucoidan extract in the range 0.1 mg/mL to 2.5 mg/mL in the reaction. Also, fluorouracil (5-FU) in the concentration range of 0.0064 μM to 2500 μM was included as a positive control for cytotoxicity. The treated and untreated cultures were incubated for another 72 h. Thereafter, resazurin (0.54 mM) was added to each well and further incubated for 3 h. After the incubation, fluorescence was measured (excitation = 560 nm and emission = 590 nm). The experiment was done in biological triplicates. The half maximum response concentration (IC_{50}) was determined by non-linear regression analysis using GraphPad Prism v. 6.0.

2.10. Statistical Analysis

The experiments were repeated in biological triplicate and the data was expressed as means ± standard deviation (SD) where applicable. One-way analysis of variance (ANOVA) determined significant differences between the enzyme activity in the absence and presence of inhibitors with 95% confidence interval where $p < 0.05$ depicted significant differences. The ANOVA tests were performed using the data analysis features in GraphPad Prism software v. 6.0. (GraphPad Inc., San Diego, CA, USA).

3. Results and Discussion

3.1. Fucoidan Yield

The *E. radiata* fucoidan was extracted using a slightly modified hot water method [31] and a yield of 5.2% *(w/w)* fucoidan was obtained from defatted seaweed. This yield was considered high as most water extraction procedures produce between 1.1 and 4.8% fucoidan [6].

3.2. Structural Validation of E. radiata Fucoidan

3.2.1. FT-IR Spectroscopy Analysis

The FTIR spectra for the extracted *E. radiata* fucoidan (Figure 1) was characteristic of most fucoidan extracts reported in the literature [6,19,32].

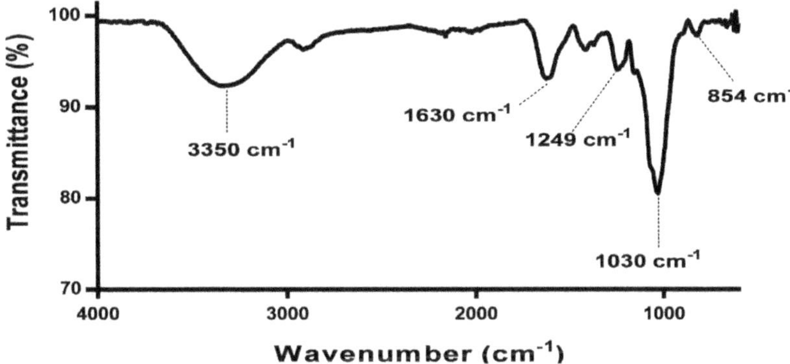

Figure 1. Fourier transform infrared spectra analysis of *E. radiata* fucoidan. The annotated peaks are representative of fucoidan structure reported in the literature.

The broadened peak from wavenumber 3500 to 3000 cm^{-1} signifies the OH group stretching vibrations characteristic of most polysaccharides. The peak at 1650 cm^{-1} represents the carbonyl groups and stretching of O-acetyl groups [32]. The peaks between 1210 cm^{-1} and 1270 cm^{-1} are associated with stretching of S=O bond linked with sulphate groups [6]. Stretching vibrations of the glycosidic C–O bond are represented by the wave number close to 1100 cm^{-1} [33]. The small peak at 854 represents sulphate groups attached to the carbonyl groups of sidechains such as galactose [32]. The absence of peaks in the 1700–1715 cm^{-1} range, 808 cm^{-1} and 822 cm^{-1} (Figure 1) showed the absence or limited amounts of uronic acid contamination in the extracted fucoidan. In addition, IR bands at 940 cm^{-1} were absent, which denote stretching C–O bonds in uronic acids [34,35].

3.2.2. ^{1}H NMR Analysis

The structural backbone of the fucoidan extract was also validated using proton nuclear magnetic resonance. The chemical shifts in the *E. radiata* fucoidan spectra showed visible peaks between 1 ppm and 5.2 ppm (Figure 2), consistent with the NMR spectra of fucoidan reported in the literature. The chemical shifts observed in the range 1.1 to 1.5 ppm (Figure 1) suggest the presence of alternating α-1,3 and α-1,4 linkages of fucose residues (α-L-Fuc, α-L-Fuc (2-SO$_3$$^-$) and α-L-Fuc (2,3-diSO$_3$$^-$)) [36]. Also, vibration bands at 1.45 ppm are assigned to symmetric CH$_3$ deformations emanating from hydrogens on C6 of fucose [37]. The peak at 2.1 ppm assigned to the H-6 methylated protons of L-fucopyranosides [38] was present in the extract. Also, the extracted fucoidan showed characteristic peaks (Figure 2) in the range 3.5–4.5 ppm attributed to the (H2 to H5) ring protons of the L-fucopyranosides. The peaks in the ring proton region also suggest variable fucosal sulphates that are located at variable glycosidic linkages with varying monosaccharide patterns [38]. The prominent peak at 4.7 ppm signals the presence of 3-linked D-galactopyranosyl residues [39].

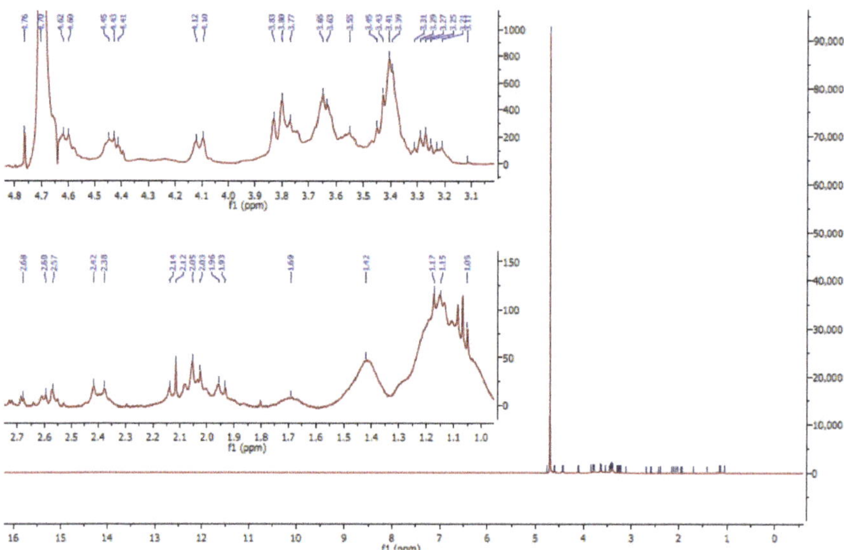

Figure 2. ^{1}H-nuclear magnetic resonance spectrum of *E. radiata* fucoidan. The spectra show peaks from 1 ppm to about 5.1 ppm, with zoom in spectra to show peaks within the regions of interest.

Moreover, definitive peaks were observed close to 3.3 ppm and 3.7 ppm in all extracts (Figure 2), suggesting the presence of hexoses, including, glucose, galactose and mannose [38]. These observations concurred with the chemical characterisation reported in this study, which generally showed variable monosaccharides within the extracted fucoidan (Table 1). Lastly, no chemical shifts in the region around 5.8 ppm were observed (Figure 2). Peaks in this region are representative of uronic acids and the presence of alginate impurities [40].

Table 1. Chemical composition of extracted *E. radiata* fucoidan.

Components	w/w % $\pm$ SD
Total carbohydrate [a]	88.1 $\pm$ 7.4
Total reducing sugars [b]	50.9 $\pm$ 10.3
Total phenolics [c]	1.9 $\pm$ 0.4
Sulphate content [d]	8.8 $\pm$ 1.4
Uronic acids [e]	2.2 $\pm$ 0.7
Total protein [f]	2.3 $\pm$ 0.89
L-fucose [g]	3.7 $\pm$ 0.43
Glucose [g]	7.31 $\pm$ 0.64
Galactose [g]	4.9 $\pm$ 1.2
Mannose [g]	4.23 $\pm$ 0.22
Arabinose [g]	ND
Fructose [g]	ND
Ash content [h]	15.5
Mw (kDa) [i]	>100

Determined by [a] Phenol sulphuric acid method; [b] DNS method; [c] Folin-Ciocalteu method; [d] Barium chloride gelatin method; [e] MegazymeTM uronic acid kit; [f] Bradford's assay; [g] HPLC (RID); [h] TGA; [i] Ultracentrifugation.

3.2.3. Thermogravimetric Analysis

Thermogravimetric analysis of the fucoidan was done primarily to determine the ash content of fucoidan. Furthermore, fucoidan decomposition through heating validated fucoidan as a polysaccharide as its decomposition started just after 200 °C (Figure 3), characteristic of organic polymers [41].

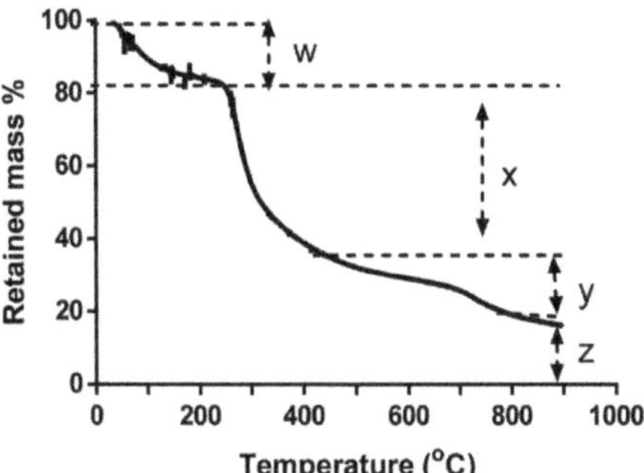

Figure 3. Thermogravimetric analysis (*TGA*) analysis of extracted *E. radiata* fucoidan. W—shows loss of readily volatile material (moisture content), X—represents polymer degradation, Y—combustion of carbon black and Z—ash content.

The TGA plot of *E. radiata* fucoidan shows about 17.4% loss of mass at temperature of 240 °C (Figure 3). This decrease in mass may be due to loss in moisture content through evaporation of water [38] and some volatile matter [39]. The major loss of mass (~45%) occurred between 240 °C and 413 °C, which accounted for the arbitrary depolymerisation/decomposition of organic constituents such as carbohydrates. Above 420 °C, combustion of carbon black occurred. After heating at 900 °C, the residual mass was about 15.5% and this accounted for the ash content, which may contain sulphates, phosphates and carbonates [40].

3.3. Chemical Profiling and Molecular Size Estimation of E. radiata Fucoidan

E. radiata fucoidan was partially chemically characterised by determining total sugar content, monosaccharides and impurities, including protein, phenolics and uronic acids. Moreover, the ash content and size estimation of the fucoidan polymer is also reported in this section (Table 1). *E. radiata* fucoidan had a total carbohydrate content of 88%, of which 51% constituted total reducing sugars (Table 1). These observations showed that the fucoidan had a high carbohydrate content which concurs with findings by Daub et al. [17] who also reported a high carbohydrate content in fucoidan. The *E. radiata* fucoidan contained about 4% L-fucose and 9% sulphate. The amount of sulphate quantified in our experiments was similar to findings by Charoensiddhi and colleagues, who reported about 7% sulphate content in their extracted fucoidan [42]. HPLC was further used to determine the composition of the fucoidan upon TFA hydrolysis; it contained the monosaccharides glucose, fucose, galactose and mannose. The most prominent monosaccharide from the extracted fucoidan was glucose, one of the highest monosaccharides previously detected in fucoidan extracted from the *Ecklonia* species in the literature [19,31]. Also, the extracted fucoidan contained trace amounts of protein and phenolics (Table 1). Moreover, the extracted *E. radiata* fucoidan contained minute amounts of uronic acids (Table 1), which are the most common fucoidan impurities.

This uronic acid determination data also concurred with the NMR spectra (Figure 2), which did not identify any peaks representing uronic acids. Therefore, the extracted *E. radiata* fucoidan was deemed pure. In addition, the molecular weight of the *E. radiata* fucoidan was estimated to be >100 kDa by ultracentrifugation (Table 1). Of note, the molecular weight of *E. radiata* fucoidan extracted by Charoensiddhi et al. [42] was determined by size exclusion chromatography (SEC) to be 339.78 kDa. It is important to note that fucoidan size is an important factor in their observed biological activities, and smaller fractions are deemed better suited for bioaccessibility [43].

3.4. Inhibition of Carbohydrate Digestion Enzymes

3.4.1. Inhibition of α-Amylase by Acarbose

The water extracted *E. radiata* and commercial *F. vesiculosus* fucoidans did not inhibit porcine α-amylase activity (data not shown). However, the acarbose control inhibited α-amylase activity and displayed an IC_{50} of 109.1 µg/mL (Figure 4). These observations were expected for *F. vesiculosus*, as Kim and colleagues demonstrated that seaweed extracted fucoidans did not possess inhibitory activity towards α-amylase [10]. *E. radiata* fucoidan inhibition of porcine α-amylase activity has not been reported in the literature. In addition, most fucoidans reported in the literature do not display any inhibitory potential towards porcine α-amylase [17,42,43].

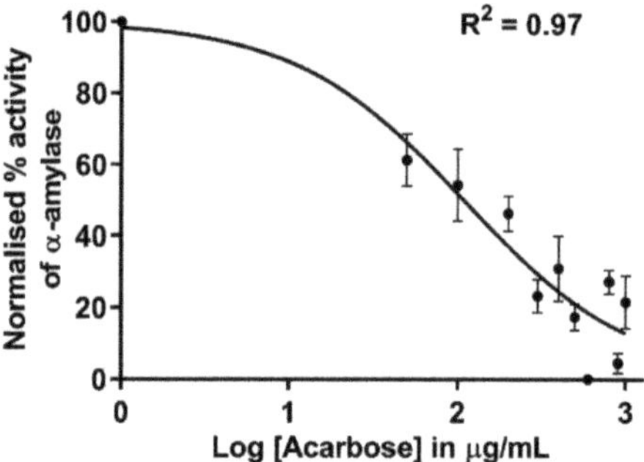

Figure 4. Dose- response curve of inhibition of α-amylase by acarbose. Values are represented as means ± SD ($n = 3$).

3.4.2. Inhibition of α-Glucosidase

The extracted *E. radiata* fucoidan and commercial *F. vesiculosus* fucoidan both exhibited a significant reduction of *Saccharomyces* α-glucosidase activity—in fact, more than acarbose (Table 2). *E. radiata* fucoidan potency (IC_{50}) is also comparable to that of the standard in the field, *F. vesiculosus* fucoidan (Table 2), which suggests that the extracted fucoidan is a powerful inhibitor.

Table 2. Table showing IC_{50} values of *E. radiata* fucoidan, *F. vesiculosus* commercial fucoidan and acarbose obtained using non-linear fit (GraphPad Prism v 6) of normalised experimental α-glucosidase response to the presence of inhibitor. Values are represented as means with a SD < 5%.

Inhibitor	IC_{50} of Fucoidan/Compound (µg/mL)
E. radiata	19
F. vesiculosus control	16
Acarbose control	332

Our data shows that acarbose is a weaker inhibitor of α-glucosidase compared to the fucoidan extract. Moreover, acarbose is currently used as a medication for type 2 diabetes and has well-documented side effects, including flatulence, meteorism, abdominal distension and even diarrhoea [44]. Therefore, the extracted fucoidan may constitute credible alternate sources of potent α-glucosidase inhibitors, presenting fewer side effects as natural bioproducts. Furthermore, literature has shown that most fucoidan extracts are more active in suppressing α-glucosidase activity than α-amylase. Moreover, natural product research contributing to diabetes treatment and prevention has focused more on α-glucosidase inhibition. The enzyme is directly responsible for releasing glucose from maltose and sucrose [10]. Also, the extensive inhibition of α-amylase is not desirable. Undigested carbohydrates may reach the colon, where they ferment due to bacterial degradation, which often results in diarrhoea, abdominal distension and flatulence. Therefore, our extract is a viable candidate as it inhibits α-glucosidase, but not α-amylase.

3.4.3. Fucoidan Directly Interacts with α-Glucosidase

The fucoidan perturbations caused conformational changes in the tertiary structure of α-glucosidase shown by the dose-dependent shifts of relative tryptophan fluorescence of the enzyme with an increase in fucoidan concentration (Figure 5A).

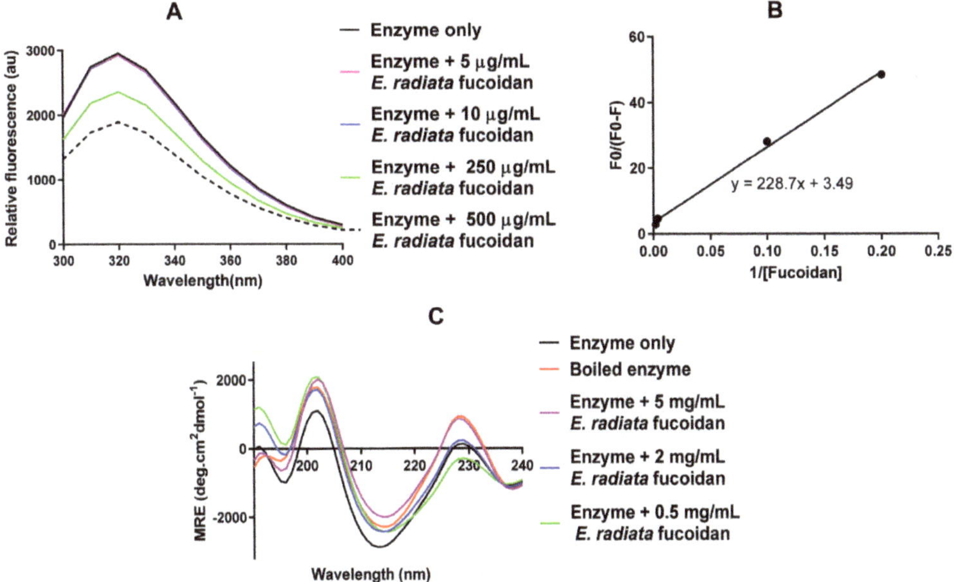

Figure 5. *E. radiata* fucoidan directly interacts with the α-glucosidase enzyme. (**A**) Fluorescence emission spectra of intrinsic fluorescent quenching of α-glucosidase in the presence of *E. radiata* fucoidan. (**B**) F0/(F0−F) plot versus 1/[fucoidan] for the fucoidan-α-glucosidase complex. (**C**) Circular dichroism spectra showing changes in α-glucosidase econdary structure in the presence of *E. radiata* fucoidan. Values are represented as means $\pm$ SD ($n = 3$).

To consolidate the direct interaction of fucoidan and the enzyme shown in Figure 5A, a modified Stren Volmer plot was constructed (Figure 5B). From this plot, fucoidan had an n value of approximately 3.5, indicating that fucoidan has 3 to 4 binding sites to facilitate the interaction with α-glucosidase. Also, the binding constant (K) between fucoidan and α-glucosidase was calculated to be 227.8 µg/mL.

The deconvoluted α glucosidase enzyme showed the β sheet conformation predominantly with 39.6% β strands, 18.9% β turns, 38% unordered and 3.6% α helices. A similar pattern was observed within the secondary structural conformation where the presence of fucoidan shifted the deconvoluted spectrum towards the positive horizontal axis upon the addition of fucoidan (Figure 5C). This observed pattern also consolidated that fucoidan directly interacts with the α glucosidase.

These experiments suggest the direct interaction between the fucoidan and α-glucosidase, concurring with the findings by Daub and colleagues with *E. maxima* fucoidan [17]. Our findings also agree with reports which stated small shifts in the secondary and tertiary structure of α-glucosidase, which were observed in the presence of inhibitors [45,46]. Molecular modelling of *S. cerevisiae* α-glucosidase identified one active site and four allosteric sites [47]. The number of binding sites identified for *E. radiata* fucoidan, suggest that the fucoidan binding possibly occurs at the active site and allosteric sites or exclusively at the allosteric sites of α-glucosidase. The 3.5 identified binding sites do not account for full binding to all allosteric sites, indicating that the large size of fucoidan likely prohibits it from binding to all available sites equally. The mode of enzyme inhibition by fucoidan was then investigated by determining kinetic parameters using Michaelis-Menten modelling.

3.4.4. Mode of Inhibition of α-Glucosidase Activity by *E. radiata* Fucoidan

E. radiata fucoidan was confirmed to inhibit the function of α-glucosidase (Figure 6A) and to be a mixed type inhibitor (Figure 6B).

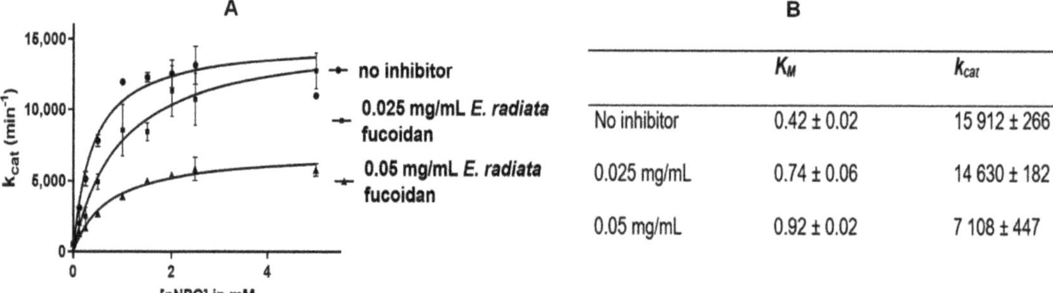

	K_M	k_{cat}
No inhibitor	0.42 ± 0.02	15 912 ± 266
0.025 mg/mL	0.74 ± 0.06	14 630 ± 182
0.05 mg/mL	0.92 ± 0.02	7 108 ± 447

Figure 6. Kinetic parameters for α-glucosidase inhibition. (**A**) Michaelis Menten curve for *E. radiata* fucoidan and (**B**) Kinetic parameters to determine the mode of inhibition of *E. radiata* fucoidan on α-glucosidase. Values are represented as means ± SD (*n* =3).

The kinetic parameters of α-glucosidase determined from Michaelis-Menten modelling (GraphPad Prism v. 6.0) illustrated an increase in the K_M and a decrease in the *kcat* with increasing inhibitor concentration (Figure 6B) which suggested fucoidan is a mixed inhibitor. Mixed inhibitors increase the K_M and decrease the V_{max} of enzymes [48]. Most fucoidans lack efficacy against α-amylase, although they are excellent inhibitors of α glucosidase. However, the vice versa is true for acarbose. The use of both compounds could prove helpful in the management of T2DM.

3.5. A Poly Compound (Acarbose and Fucoidan) Combination Approach in Inhibiting α-Glucosidase

The combination of drugs may have three possible effects: additive or synergistic and/or undesirably antagonistic interaction. The prospect of the polydrug approach was investigated between the extracted fucoidan and acarbose on α-glucosidase. As fucoidan lacked efficacy in inhibiting α-amylase, quantitative synergy by the combination did not suit our intended Compusyn model. Therefore, the drug potencies of the individual compounds and acarbose-fucoidan combinations (at constant ratios) on α-glucosidase were investigated. As shown in Figure 7A, it was evident that fucoidan is a better drug than acarbose. The acarbose-fucoidan combination showed some additive to synergistic effects (line slightly shifted to the left) at some concentration points. The Chou and Talalay method was used to investigate further and quantify the potential synergistic interactions of these compounds [49]. Briefly, the experimental data for the α-glucosidase assay were analysed using Compusyn software. Thereafter, normalised isobolograms were generated using Compusyn software [49]. The isobolograms visually illustrate synergistic combinations (Figure 7B).

All the points below the hypotenuse show synergistic points; those on the line illustrate additivity, and antagonistic combinations are plotted above the line (Figure 8). The further the points are from the line, the stronger the effect, whether synergistic or antagonistic. The best synergistic combinations for the two compounds were at IC_{25}:IC_{50}, IC_{50}:IC_{50} and IC_{25}:IC_{75} [Acarbose]: [*E. radiata*] with total inhibitions at 70.4%, 74.2% and 79.4%, respectively (Table 3).

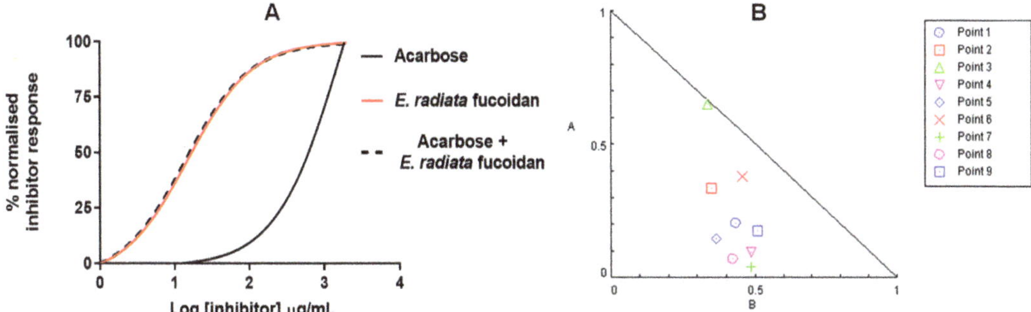

Figure 7. *E. radiata* fucoidan synergistically inhibits the activity of α-glucosidase. (**A**) Dose-response curves of the inhibition potential of compounds on α-glucosidase. (**B**)Normalised isobologram analysis of acarbose and *E. radiata* fucoidan combinations. The axes scales (A-acarbose & B-*E.radiata*) represents the affected fraction (inhibited enzyme). Various combinations of acarbose and *E. radiata* fucoidan based on IC_{75}, IC_{50} and IC_{25} values were tested by the Compusyn software and the combination indexes determined. The points below the hypotenuse line of the triangles indicate synergy at a particular effect.

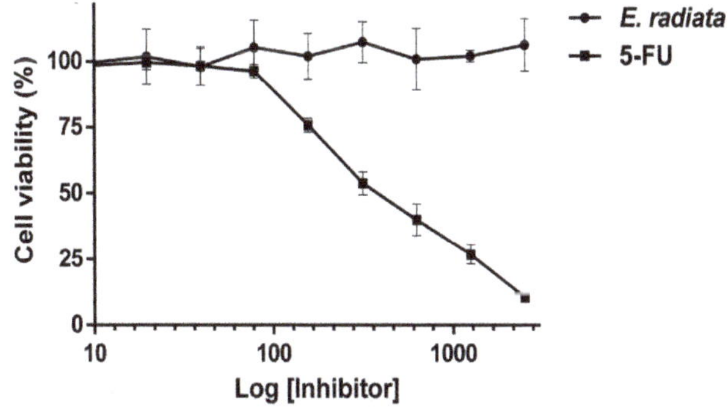

Figure 8. Cytotoxic effects of *E. radiata* fucoidan and 5-fluorouracil (5-FU) on HCT116 cancer cells. Cell viability was assessed by the resazurin assay. The 5-FU concentration is represented as logarithmic concentrations in μM and the *E. radiata* fucoidan as logarithmic concentrations μg/mL. Points represent means of biological replicates (*n* = 3). The error bars shown represent the standard deviations about the mean.

Table 3. Synergistic effects of acarbose and fucoidan extracts on α-glucosidase activity.

Points on Isobologram	Compounds Combinations	Compound Concentration (μg/mL) Acarbose: *E. radiata*		% Residual α-Glucosidase Activity ± SD	CI	Effect
1	IC_{75}-IC_{75}	2223	35.2	18.1 ± 2.35	0.63	Synergistic
2	IC_{75}-IC_{50}	2223	18.6	26.1 ± 3.64	0.68	Synergistic
3	IC_{75}-IC_{25}	2223	9.9	39.8 ± 2.96	0.98	Synergistic
4	IC_{50}-IC_{75}	922	35.2	20.1 ± 2.56	0.58	Synergistic
5	IC_{50}-IC_{50}	922	18.6	26.8 ± 3.4	0.51	Synergistic
6	IC_{50}-IC_{25}	922	9.9	47.7 ± 6.39	0.83	Synergistic
7	IC_{25}-IC_{75}	382	35.2	20.6 ± 3.41	0.52	Synergistic
8	IC_{25}-IC_{50}	382	18.6	29.6 ± 3.72	0.49	Synergistic
9	IC_{25}-IC_{25}	382	9.9	51.4 ± 7.2	0.68	Synergistic

CI = the combination index quantifying the degree of inhibitor interactions. *E. radiata* within the table represents fucoidan extracted from *E. radiata* seaweeds.

These data suggest that combinations of extracted fucoidan with acarbose act synergistically, giving rise to a potential combination therapy for T2DM. The current study has shown that fucoidan lacks efficacy towards α-amylase but is an excellent inhibitor of α-glucosidase. However, acarbose is a good inhibitor of α-amylase (Figure 4) but a less potent inhibitor of α-glucosidase (Table 2). Moreover, their combination illustrated additive or potentially synergistic interactions on α glucosidase (Figure 7 and Table 3). The synergistic potential of fucoidan has been investigated using this approach to suppress the proliferation of the measles virus [15]. More studies may be necessary to investigate fucoidan application within the prospects of polydrug combination approaches for the various bioactivities it possesses. Moreover, the Chou-Talalay method of quantifying synergy is the most used model [17]. Also, considering that fucoidan is a mixed inhibitor, the Chou-Talalay method is best suited to our situation, compared to Bliss models, which quantify synergy among compounds with a different mode of action, and Lewis models quantify synergy of compounds with a similar mode of action [17].

The side effects of acarbose use may be limited at its reduced dose when combined with fucoidan. In addition, a reduction in toxicity and side effects is one of the rationales of using drug combinations [16]. Also, combining compounds with non-overlapping side effects, fucoidan and acarbose, may allow for a more significant total efficacy with possibly fewer side effects. Lastly, considering fucoidan's viscosity, the drug dose may be available for longer before excretion as high viscosity slows gastric emptying.

In summary, the extracted fucoidan showed synergistic interactions in combination with acarbose. This may be important for combination therapy prospects in regulating the activities of carbohydrate digesting enzymes in the quest to manage T2DM.

3.6. Cytotoxicity of Fucoidan

A cytotoxicity assay was performed on HCT116 (a human colon cancer cell line) to evaluate the safety of the extracted fucoidan. The resazurin reduction assay was applied to determine the viability of HCT116 cells treated with *E. radiata* fucoidan in the range of 5 to 2500 μg/mL for 72 h. Fucoidan did not have any cytotoxic effect on the HCT116 colon cells (Figure 8). On the other hand, 5-fluorouracil (5-FU), a known chemotherapeutic drug, was used as a positive control for cytotoxicity and demonstrated this effect on HTC116 cells with an IC_{50} of 9.9 μM ($r^2 = 0.98$). The HCT116 is a relevant model for this study due to the oral administration of these compounds that may remain in the digestive tract for extended periods, especially in the colon, where food takes about 36 h to pass through [50]. No studies have been reported in the cytotoxic screening of fucoidan on "normal" human colorectal cells. Also, most normal cells are derived from cancer cell lines or transformed in some way for them to be able to divide and grow in culture [51]. Based on these facts and availability, the HCT116 colorectal cancer cell line was deemed appropriate for suggesting the lack of fucoidan toxicity to the cells. However, further studies on this aspect using "normal" colon cancer cell lines may be necessary at a later stage.

4. Conclusions

A biologically active non-toxic fucoidan was successfully extracted from *E. radiata* seaweed. The extracted fucoidan inhibited α-glucosidase more strongly than acarbose. Therefore, this fucoidan is an attractive drug candidate for managing post-prandial hyperglycemia, which causes type 2 diabetes mellitus (T2DM). Our data suggested that acarbose can be used in combination with *E. radiata* fucoidan, as they showed synergistic associations at selected concentrations; $IC_{25}:IC_{50}$, $IC_{50}:IC_{50}$ and $IC_{25}:IC_{75}$ [Acarbose]: [*E. radiata*], with more than 70% inhibition. Furthermore, combining acarbose and fucoidan may target both carbohydrates digesting enzymes; α-amylase and α-glucosidase. This combinatorial approach would compensate fucoidan's lack of efficacy against α-amylase when used alone. To conclude, this study highlights the prospect of a polydrug (*E. radiata* fucoidan and acarbose) combination strategy in managing T2DM.

Author Contributions: Conceptualisation, B.M., C.D.D., S.M., A.L.E. and B.I.P.; Investigation, B.M. and C.D.D.; writing—original draft preparation, B.M. and C.D.D.; writing—review and editing, B.M., C.D.D., S.M., A.L.E. and B.I.P.; supervision, S.M., A.L.E. and B.I.P.; project administration, B.I.P.; funding acquisition, B.I.P. and A.L.E. All authors have read and agreed to the published version of the manuscript.

Funding: B.M. was funded by the German Academic Exchange Service (DAAD) In-Region Scholarship (grant no. 57408782). C.D.D. received financial support for this study from the South African National Research Foundation (NRF, grant no. 116951), Henderson Scholarship as well as Pearson-Young Memorial scholarship. A.L.E is supported by the National Research Foundation of South Africa (Grant Numbers 98566, 105829 and 129262), S.M. is supported by the University of Pretoria and both A.L.E. and B.I.P. are supported by Rhodes University. This research was supported in part by KelpX.

Data Availability Statement: Data available upon request.

Acknowledgments: The authors would like to acknowledge Xavier Siwe Noundou for assisting with NMR analysis and some chemistry aspects presented in this paper.

Conflicts of Interest: The corresponding author (Brett I. Pletschke) has received research funding from the kelp industrial agency (KelpX, no grant number assigned) which has partly funded the research conducted in this manuscript. The company had no role in the design of the study; in the collection, analyses, or interpretation of data; in the writing of the manuscript, or in the decision to publish the results. The other authors declare no conflict of interest.

References

1. WHO. World Health Organization. Available online: https://www.who.int/news-room/fact-sheets/detail/diabetes (accessed on 18 March 2021).
2. Moini, J. The Epidemic and Prevalence of Diabetes in the United States. In *Epidemiology of Diabetes*; Elsevier BV: Amsterdam, The Netherlands, 2019; pp. 45–55.
3. Galicia-Garcia, U.; Benito-Vicente, A.; Jebari, S.; Larrea-Sebal, A.; Siddiqi, H.; Uribe, K.B.; Ostolaza, H.; Martín, C. Pathophysiology of Type 2 Diabetes Mellitus. *Int. J. Mol. Sci.* **2020**, *21*, 6275. [CrossRef]
4. Cho, M.; Han, J.H.; You, S. Inhibitory effects of fucan sulfates on enzymatic hydrolysis of starch. *LWT* **2011**, *44*, 1164–1171. [CrossRef]
5. Wilcox, G. Insulin and Insulin Resistance. *Clin. Biochem. Rev.* **2005**, *26*, 19–39.
6. Kumar, T.V.; Lakshmanasenthil, S.; Geetharamani, D.; Marudhupandi, T.; Suja, G.; Suganya, P. Fucoidan. A α-d-glucosidase inhibitor from Sargassum wightii with relevance to type 2 diabetes mellitus therapy. *Int. J. Biol. Macromol.* **2015**, *72*, 1044–1047. [CrossRef]
7. Lopes, G.; Andrade, P.B.; Valentão, P. Phlorotannins: Towards New Pharmacological Interventions for Diabetes Mellitus Type 2. *Molecules* **2016**, *22*, 56. [CrossRef]
8. McIver, L.A.; Tripp, J. StatPearls. Available online: https://www.ncbi.nlm.nih.gov/books/NBK493214/ (accessed on 18 March 2021).
9. Yuan, H.; Ma, Q.; Ye, L.; Piao, G. The Traditional Medicine and Modern Medicine from Natural Products. *Molecules* **2016**, *21*, 559. [CrossRef]
10. Kim, K.-T.; Rioux, L.-E.; Turgeon, S.L. α-amylase and α-glucosidase inhibition is differentially modulated by fucoidan obtained from *Fucus vesiculosus* and *Ascophyllum nodosum*. *Phytochemistry* **2014**, *98*, 27–33. [CrossRef]
11. Mabate, B.; Daub, C.D.; Malgas, S.; Edkins, A.L.; Pletschke, B.I. Fucoidan Structure and Its Impact on Glucose Metabolism: Implications for Diabetes and Cancer Therapy. *Mar. Drugs* **2021**, *19*, 30. [CrossRef]
12. Pozharitskaya, O.N.; Obluchinskaya, E.D.; Shikov, A.N. Mechanisms of Bioactivities of Fucoidan from the Brown Seaweed *Fucus vesiculosus* L. of the Barents Sea. *Mar. Drugs* **2020**, *18*, 275. [CrossRef]
13. Yang, X.-D.; Liu, C.-G.; Tian, Y.-J.; Gao, D.-H.; Li, W.-S.; Ma, H.-L. Inhibitory effect of fucoidan on hypoglycemia in diabetes mellitus anim. *Int J Clin Exp Med.* **2017**, *10*, 8529–8534.
14. Bolton, J.J.; Stegenga, H. Seaweed species diversity in South Africa. *South Afr. J. Mar. Sci.* **2002**, *24*, 9–18. [CrossRef]
15. Morán-Santibañez, K.; Cruz-Suárez, L.E.; Ricque-Marie, D.; Robledo, D.; Freile-Pelegrin, Y.; Peña-Hernández, M.A.; Rodríguez-Padilla, C.; Trejo-Avila, L.M. Synergistic Effects of Sulfated Polysaccharides from Mexican Seaweeds against Measles Virus. *BioMed Res. Int.* **2016**, *2016*, 1–11. [CrossRef]
16. Greco, W.R.; Faessel, H.; Levasseur, L. The Search for Cytotoxic Synergy Between Anticancer Agents: A Case of Dorothy and the Ruby Slippers? *J. Natl. Cancer Inst.* **1996**, *88*, 699–700. [CrossRef]
17. Roell, K.R.; Reif, D.; Motsinger-Reif, A.A. An Introduction to Terminology and Methodology of Chemical Synergy—Perspectives from Across Disciplines. *Front. Pharmacol.* **2017**, *8*, 158. [CrossRef]
18. Chou, T.T.C. Compusyn. Available online: http://www.Combosyn.com (accessed on 10 January 2020).

19. Daub, C.D.; Mabate, B.; Malgas, S.; Pletschke, B.I. Fucoidan from *Ecklonia maxima* is a powerful inhibitor of the diabetes-related enzyme, α-glucosidase. *Int. J. Biol. Macromol.* **2020**, *151*, 412–420. [CrossRef]
20. Suresh, V.; Senthilkumar, N.; Thangam, R.; Rajkumar, M.; Anbazhagan, C.; Rengasamy, R.; Gunasekaran, P.; Kannan, S.; Palani, P. Separation, purification and preliminary characterization of sulfated polysaccharides from *Sargassum plagiophyllum* and its in vitro anticancer and antioxidant activity. *Process. Biochem.* **2013**, *48*, 364–373. [CrossRef]
21. Yuan, Y.; Macquarrie, D. Microwave assisted extraction of sulfated polysaccharides (fucoidan) from *Ascophyllum nodosum* and its antioxidant activity. *Carbohydr. Polym.* **2015**, *129*, 101–107. [CrossRef]
22. Lee, S.-H.; Ko, C.-I.; Ahn, G.; You, S.; Kim, J.-S.; Heu, M.S.; Kim, J.; Jee, Y.; Jeon, Y.-J. Molecular characteristics and anti-inflammatory activity of the fucoidan extracted from *Ecklonia cava. Carbohydr. Polym.* **2012**, *89*, 599–606. [CrossRef]
23. Dubois, M.; Gilles, K.A.; Hamilton, J.K.; Rebers, P.A.; Smith, F. Colorimetric Method for Determination of Sugars and Related Substances. *Anal. Chem.* **1956**, *28*, 350–356. [CrossRef]
24. Miller, G.L. Use of Dinitrosalicylic Acid Reagent for Determination of Reducing Sugars. *Anal. Chem.* **1959**, *31*, 426–428. [CrossRef]
25. Bradford, M.M. A rapid and sensitive method for the quantitation of microgram quantities of protein utilising the principle of protein dye binding. *Anal. Biochem.* **1976**, *72*, 248–254. [CrossRef]
26. Dodgson, K.S.; Price, R. A note on the determination of the ester sulphate content of sulphated polysaccharides. *Biochem. J.* **1962**, *84*, 106–110. [CrossRef]
27. Huang, D.; Ou, B.; Prior, R.L. The Chemistry behind Antioxidant Capacity Assays. *J. Agric. Food Chem.* **2005**, *53*, 1841–1856. [CrossRef] [PubMed]
28. Mabate, B.; Zininga, T.; Ramatsui, L.; Makumire, S.; Achilonu, I.; Dirr, H.; Shonhai, A. Structural and biochemical characterization of *Plasmodium falciparum* Hsp70-x reveals functional versatility of its C-terminal EEVN motif. *Proteins: Struct. Funct. Bioinform.* **2018**, *86*, 1189–1201. [CrossRef] [PubMed]
29. Zininga, T.; Achilonu, I.; Hoppe, H.; Prinsloo, E.; Dirr, H.; Shonhai, A. Overexpression, Purification and Characterisation of the *Plasmodium falciparum* Hsp70-z (PfHsp70-z) Protein. *PLoS ONE* **2015**, *10*, e0129445. [CrossRef]
30. Whitmore, L.; Wallace, B.A. Protein secondary structure analyses from circular dichroism spectroscopy: Methods and reference databases. *Biopolymers* **2007**, *89*, 392–400. [CrossRef]
31. January, G.; Naidoo, R.; Kirby-McCullough, B.; Bauer, R. Assessing methodologies for fucoidan extraction from South African brown algae. *Algal Res.* **2019**, *40*, 101517. [CrossRef]
32. Fernando, S.; Sanjeewa, K.K.A.; Samarakoon, K.W.; Lee, W.W.; Kim, H.-S.; Kim, E.-A.; Gunasekara, U.K.D.S.S.; Abeytunga, D.T.U.; Nanayakkara, C.; De Silva, E.D.; et al. FTIR characterization and antioxidant activity of water soluble crude polysaccharides of Sri Lankan marine algae. *ALGAE* **2017**, *32*, 75–86. [CrossRef]
33. Pereira, L.; Gheda, S.F.; Ribeiro-Claro, P.J.A. Analysis by Vibrational Spectroscopy of Seaweed Polysaccharides with Potential Use in Food, Pharmaceutical, and Cosmetic Industries. *Int. J. Carbohydr. Chem.* **2013**, *2013*, 1–7. [CrossRef]
34. Chandía, N.P.; Matsuhiro, B.; Mejías, E.; Moenne, A. Alginic acids in *Lessonia vadosa*: Partial hydrolysis and elicitor properties of the polymannuronic acid fraction. *Environ. Boil. Fishes* **2004**, *16*, 127–133. [CrossRef]
35. Leal, D.; Matsuhiro, B.; Rossi, M.; Caruso, F. FT-IR spectra of alginic acid block fractions in three species of brown seaweeds. *Carbohydr. Res.* **2008**, *343*, 308–316. [CrossRef]
36. Shan, X.; Liu, X.; Hao, J.; Cai, C.; Fan, F.; Dun, Y.; Zhao, X.; Liu, X.; Li, C.; Yu, G. In vitro and in vivo hypoglycemic effects of brown algal fucoidans. *Int. J. Biol. Macromol.* **2016**, *82*, 249–255. [CrossRef] [PubMed]
37. Kopplin, G.; Rokstad, A.M.; Mélida, H.; Bulone, V.; Skjåk-Bræk, G.; Aachmann, F.L. Structural Characterization of Fucoidan from *Laminaria hyperborea*: Assessment of Coagulation and Inflammatory Properties and Their Structure–Function Relationship. *ACS Appl. Bio Mater.* **2018**, *1*, 1880–1892. [CrossRef]
38. Alwarsamy, M.; Gooneratne, R.; Ravichandran, R. Effect of fucoidan from *Turbinaria conoides* on human lung adenocarcinoma epithelial (A549) cells. *Carbohydr. Polym.* **2016**, *152*, 207–213. [CrossRef]
39. Thangapandi, M.; Kumar, A.T.T. Effect of fucoidan from *Turbinaria ornata* against marine ornamental fish pathogens. *J. Coast. Life Med.* **2013**, *1*, 282–286.
40. Nguyen, T.T.; Mikkelsen, M.D.; Tran, V.H.N.; Trang, V.T.D.; Rhein-Knudsen, N.; Holck, J.; Rasin, A.B.; Cao, H.T.T.; Van, T.T.T.; Meyer, A.S. Enzyme-Assisted Fucoidan Extraction from Brown Macroalgae *Fucus distichus* subsp. evanescens and *Saccharina latissima. Mar. Drugs* **2020**, *18*, 296. [CrossRef]
41. Liu, X.; Yu, W. Evaluating the thermal stability of high performance fibers by TGA. *J. Appl. Polym. Sci.* **2006**, *99*, 937–944. [CrossRef]
42. Charoensiddhi, S.; Conlon, M.A.; Methacanon, P.; Franco, C.; Su, P.; Zhang, W. Gut health benefits of brown seaweed *Ecklonia radiata* and its polysaccharides demonstrated in vivo in a rat model. *J. Funct. Foods* **2017**, *37*, 676–684. [CrossRef]
43. Fitton, J.H.; Stringer, D.N.; Karpiniec, S.S. Therapies from Fucoidan: An Update. *Mar. Drugs* **2015**, *13*, 5920–5946. [CrossRef]
44. Kotowaroo, M.I.; Mahomoodally, M.F.; Gurib-Fakim, A.; Subratty, A.H. Screening of Traditional Anti-diabetic Medicinal Plants of Mauritius for Possible α-Amylase Inhibitory Effects in vitro. *Phytother.Res.* **2006**, *20*, 228–231. [CrossRef]
45. Liu, M.; Zhang, W.; Wei, J.; Lin, X. Synthesis and α-Glucosidase Inhibitory Mechanisms of Bis(2,3-dibromo-4,5-dihydroxybenzyl) Ether, a Potential Marine Bromophenol α-Glucosidase Inhibitor. *Mar. Drugs* **2011**, *9*, 1554–1565. [CrossRef]
46. Ma, H.; Wang, L.; Niesen, D.B.; Cai, A.; Cho, B.P.; Tan, W.; Gu, Q.; Xu, J.; Seeram, N.P. Structure activity related, mechanistic, and modeling studies of gallotannins containing a glucitol-core and α-glucosidase. *RSC Adv.* **2015**, *5*, 107904–107915. [CrossRef]

47. Şöhretoğlu, D.; Sari, S.; Özel, A.; Barut, B. α-Glucosidase inhibitory effect of Potentilla astracanica and some isoflavones: Inhibition kinetics and mechanistic insights through in vitro and in silico studies. *Int. J. Biol. Macromol.* **2017**, *105*, 1062–1070. [CrossRef]
48. Lopina, O.D. Enzyme inhibitors and activators. In *Enzyme Inhibitors and Activators*; Senturk, M., Ed.; IntechOpen: London, UK, 2017; p. 247. [CrossRef]
49. Chou, T.C. Drug Combination Studies and Their Synergy Quantification Using the Chou-Talalay Method. *Cancer Res.* **2010**, *70*, 440–446. [CrossRef]
50. Goldman, L.; Bennett, J.C. Disorders of gastrointestinal motility. In *Cecil Medicine*, 26th ed.; Goldman, L., Schafer, A.I., Eds.; Saunders Co: Philadelphia, PA, USA, 2000.
51. Stauffer, J.S.; Manzano, L.A.; Balch, G.C.; Merriman, R.L.; Tanzer, L.R.; Moyer, M.P. Development and characterization of normal colonic epithelial cell lines derived from normal mucosa of patients with colon cancer. *Am. J. Surg.* **1995**, *169*, 190–196. [CrossRef]

pharmaceutics

Article

Article

The Beneficial Additive Effect of Silymarin in Metformin Therapy of Liver Steatosis in a Pre-Diabetic Model

Martina Hüttl [1,*], Irena Markova [1], Denisa Miklankova [1], Iveta Zapletalova [2], Martin Poruba [2], Zuzana Racova [2], Rostislav Vecera [2] and Hana Malinska [1]

[1] Centre for Experimental Medicine, Institute for Clinical and Experimental Medicine, 14021 Prague, Czech Republic; irena.markova@ikem.cz (I.M.); denisa.miklankova@ikem.cz (D.M.); hana.malinska@ikem.cz (H.M.)

[2] Department of Pharmacology, Faculty of Medicine and Dentistry, Palacky University, 77900 Olomouc, Czech Republic; iveta.zapletalova@upol.cz (I.Z.); martin.poruba@upol.cz (M.P.); zuzu.matuskova@seznam.cz (Z.R.); vecera@seznam.cz (R.V.)

* Correspondence: martina.huttl@ikem.cz; Tel.: +420-261-365-369

Abstract: The combination of plant-derived compounds with anti-diabetic agents to manage hepatic steatosis closely associated with diabetes mellitus may be a new therapeutic approach. Silymarin, a complex of bioactive substances extracted from *Silybum marianum*, evinces an antioxidative, anti-inflammatory, and hepatoprotective activity. In this study, we investigated whether metformin (300 mg/kg/day for four weeks) supplemented with micronized silymarin (600 mg/kg/day) would be effective in mitigating fatty liver disturbances in a pre-diabetic model with dyslipidemia. Compared with metformin monotherapy, the metformin–silymarin combination reduced the content of neutral lipids (TAGs) and lipotoxic intermediates (DAGs). Hepatic gene expression of enzymes and transcription factors involved in lipogenesis (*Scd-1*, *Srebp1*, *Ppar*γ, and *Nr1h*) and fatty acid oxidation (*Ppar*α) were positively affected, with hepatic lipid accumulation reducing as a result. Combination therapy also positively influenced arachidonic acid metabolism, including its metabolites (14,15-EET and 20-HETE), mitigating inflammation and oxidative stress. Changes in the gene expression of cytochrome P450 enzymes, particularly Cyp4A, can improve hepatic lipid metabolism and moderate inflammation. All these effects play a significant role in ameliorating insulin resistance, a principal background of liver steatosis closely linked to T2DM. The additive effect of silymarin in metformin therapy can mitigate fatty liver development in the pre-diabetic state and before the onset of diabetes.

Keywords: metformin; silymarin; combination therapy; liver steatosis; pre-diabetes

Citation: Hüttl, M.; Markova, I.; Miklankova, D.; Zapletalova, I.; Poruba, M.; Racova, Z.; Vecera, R.; Malinska, H. The Beneficial Additive Effect of Silymarin in Metformin Therapy of Liver Steatosis in a Pre-Diabetic Model. *Pharmaceutics* 2022, 14, 45. https://doi.org/10.3390/pharmaceutics14010045

Academic Editors: Diana Marcela Aragon Novoa and Fátima Regina Mena Barreto Silva

Received: 29 November 2021
Accepted: 21 December 2021
Published: 27 December 2021

Publisher's Note: MDPI stays neutral with regard to jurisdictional claims in published maps and institutional affiliations.

1. Introduction

Type 2 diabetes (T2DM) and liver steatosis are closely related diseases. T2DM, a complex disorder primarily affecting glycaemic status, is characterised by insulin resistance and insulin secretory deficiency. The condition develops during pre-diabetes, which is characterised by impaired fasting glucose and glucose tolerance [1]. With prevalence increasing worldwide, pre-diabetes is connected with overweight and impaired insulin resistance and liver lipid storage, and it is also considered a high-risk state for progression to T2DM. There is a strong connection between T2DM and non-alcoholic fatty liver disease (NAFLD). Moreover, liver steatosis may precede the development of T2DM and even hyperglycaemia. Therefore, NAFLD is not simply a consequence but also a causal factor in the pathophysiology of these complications [2] and, as such, a strong independent risk factor for pre-diabetes [3]. The prevalence of NAFLD is reported at 20–30% in the general population of Western countries and 30–50% in patients with T2DM [4]. However, because of difficulties with NAFLD quantification, these figures are most likely underestimated. NAFLD encompasses a range of pathological states and processes, beginning with simple hepatic steatosis (fatty liver), followed by non-alcoholic steatohepatitis (NASH), which is characterised by inflammation and more serious damage to hepatocytes, and then fibrosis and finally cirrhosis, the most severe stage, which can lead to hepatocellular carcinoma. The progression and clinical manifestation of fatty liver to other stages are individual and heterogeneous [5]. The environment, microbiome, metabolism, comorbidities, and genetic risk factors have all been established as causes [6,7]. Fatty liver is usually connected with overweight or obesity [8]. However, this metabolic disturbance is also diagnosed in non-obese and non-overweight individuals. As stated, the prevalence of non-obese NAFLD ranges from 3% to 30% [9], and both obese and non-obese patients have the same risk for adverse metabolic outcomes, including T2DM [10].

Metformin (MET), the first-line medication in the management of T2DM, is a relatively well-tolerated, insulin-sensitising, anti-hyperglycaemic drug with a very low risk of hypoglycaemia. Importantly, it is the only anti-diabetic drug recommended for the prevention of T2DM in people with pre-diabetes [11]. MET reduces hyperglycaemia and alleviates accompanying clinical symptoms by inhibiting hepatic gluconeogenesis, leading to reduced hepatic glucose expenditure, improved insulin signalling, and increased glucose uptake in skeletal muscle [12]. More positively, the glycaemia-lowering properties of MET are connected with a number of pleiotropic effects that are responsible for the glycemia-lowering properties. As it was established in animal models, MET acts as a therapeutic agent by reducing hepatic lipid storage. However, the results of human studies are less conclusive [13].

Phytochemicals isolated from medicinal plants pose an attractive opportunity for the development of new types of therapeutics for T2DM. In the therapy of liver steatosis, for which a license has yet to be granted to a pharmacological agent, the practice of combining commonly prescribed anti-diabetic drugs with phytochemicals has grown in interest. Compared with monotherapy, combination therapy involving drugs and bioactive substances targets key pathways in a characteristically synergistic or additive manner [14], reduces doses of the active substance, retards the elimination rate, can reduce liver load, and provides additional benefits for diabetics compared with standard multidrug therapy [15].

Silymarin (SM) is a complex of biological compounds extracted from the seeds of milk thistle (*Silybum marianum*). Its main bioactive substances include flavonolignans (silybin A and B, isosilybin A and B, and silychristin), flavonoids (taxifolin and quercetin) and polyphenolic compounds [16]. SM's protective effects for different organs have been shown in many human and experimental studies—its cardioprotective, renoprotective, neuroprotective, and strong hepatoprotective activity was confirmed. SM boasts potent antioxidative, radical scavenging, anti-inflammatory, immunomodulatory, antifibrotic, antiviral, and anti-apoptotic properties. Administered in therapeutic doses, SM is also considered a safe herbal product, having no side effects. The efficiency of SM therapy may be limited by its low bioavailability: SM has a short half-life and fast absorption and elimination. Mainly because of its better solubility in water, the micronized form of SM has more pronounced effects than the widely used standardised extract [17]. This was also confirmed in our previous studies [18,19] in which micronized SM more intensively improved lipid and glucose metabolism in the animal model of metabolic syndrome compared with standard forms.

In connection with hepatic lipid metabolism, SM has attracted attention for its ability to ameliorate alcoholic and non-alcoholic liver steatosis as well as liver fibrosis or cirrhosis [20–22] in animal and human studies. Our previous studies revealed its marked hypolipidaemic effect in various experimental models. In rats fed a high-fat diet, SM reduced intestinal cholesterol absorption, decreased serum total-cholesterol levels, and increased HDL-cholesterol concentrations [23]. In the model of metabolic syndrome, we also found a favourable effect of SM on lipid metabolism and oxidative stress parameters; SM administration reduced VLDL cholesterol levels in the bloodstream, elevated concentrations of glutathione (GSH) in the circulation and liver, increased the activity of hepatic superoxide dismutase (SOD), and improved parameters of lipoperoxidation in the liver [24].

The aim of this study was to investigate whether metformin therapy supplemented with silymarin would be effective in mitigating fatty liver disturbances in a pre-diabetes model, the hereditary hypertriglyceridaemic (HHTg) rat strain. Apart from the genetically fixed hypertriglyceridaemia, this non-obese rodent model exhibits insulin resistance in peripheral tissues, liver steatosis, oxidative stress, and low-grade chronic inflammation in the absence of fasting hyperglycaemia [25].

2. Materials and Methods

2.1. Animal Model, Diet and Drugs

Five-month-old male HHTg rats (bred by the Institute for Clinical and Experimental Medicine, Prague, Czech Republic; approved by the research ethics committee—Protocol Number 28/2016) were randomly divided into four experimental groups. The control group (HHTg/C, $n = 8$) was fed a standard diet (Altromin, Maintenance diet for rats and mice, Lage, Germany), the metformin-treated group (HHTg/MET) was fed a standard diet supplemented with metformin (Teva Pharmaceuticals, Brno, Czech Republic) at a dose of 300 mg/kg of body weight per day, and the combination therapy-treated group (HHTg/MET+SM) was fed a standard diet supplemented with a mix of MET and micronized SM with a declared purity of 80% (supplied by Favea, Koprivnice, Czech Republic) at 600 mg/kg body weight per day for four weeks. The quality control analysis of micronized SM is shown in Table 1. Laboratory analyses were performed in a certified and accredited laboratory of the Faculty of Chemical Technology (University of Pardubice). Bacterial, yeast, and fungal strains provided by the Czech Collection of Microorganisms (Brno, Czech Republic) were used for microbiological analysis. A suspension of bacterial and yeast strains was grown into colonies on nutrient agar no. 2 or blood agar and malt agar after 24–48 h incubation at an optimum temperature. Suspensions of fungal spores were prepared from cultures grown on malt agar slants at 24 °C. Upon the completion of the incubations, samples were counted using standard methods. The heavy metal content was measured using atomic absorption spectrophotometry (AAS). For the individual SM component determination, HPLC with UV-VIS and tandem mass spectrometry were used (Agilent LC/MSD Trap SL).

Table 1. Analysis of micronized silymarin.

Microbiological Control		
Yeast and mold (cfu/g)	$<10^2$	
Salmonella spp.	0	
Staphylococcus aureus (cfu/g)	$<10^2$	
Pseudomonas aeruginosa (cfu/g)	$<10^2$	
E. coli (cfu/mL)	$<10^2$	
Heavy metals		
Arsenic (As) (µg/kg)	20–60	
Lead (Pb) (µg/kg)	52 ± 11	
Mercury (Hg) (µg/kg)	0.55 ± 0.11	
Compound	UV/VIS ratio (%)	MS/MS ratio (%)
Silychristin A + B + isomers	31.87	32.02
Silydianin	0.43	0.86
Silybin A + B + isomers	54.39	48.27
Isosilybin A + B + isomers	13.32	18.85

All experiments involving laboratory rats were conducted in compliance with the Animal Protection Law of the Czech Republic (311/1997) and with European Community Council recommendations (86-609/ECC) for the use of laboratory animals and approved by the Ethics Committee of the Institute for Clinical and Experimental Medicine, Prague. Animals were housed in cages in a room with a controlled temperature (22–25 °C), humidity (55–60%), and natural light conditions (12 h light/dark cycle) with free access to chow and drinking water. Daily food consumption and body weight were measured regularly.

Animals were euthanised by anaesthetisation (zoletil 5 mg/kg body weight) in a postprandial state. Tissue samples and aliquots of serum were collected, immediately frozen in liquid nitrogen, and stored at $-80\ ^{\circ}C$ for further analysis.

2.2. Biochemical Analysis of Serum

Plasma levels of glucose, triglycerides (TAG), free fatty acids (FFA), and total cholesterol were measured using commercially available kits (Erba Lachema, Brno, Czech Republic). *Alanine aminotransferase* (ALT) and aspartate aminotransferase (AST) enzyme activity was determined spectrophotometrically using routine clinical biochemistry methods and kits (Roche Diagnostics, Mannheim, Germany). Plasma insulin concentrations were determined using the Rat Insulin ELISA kit (Mercodia AB, Uppsala, Sweden), while monocyte chemoattractant protein-1 (MCP-1) levels were measured using the Rat MCP-1 Instant ELISA kit (eBioscience, Vienna, Austria). The other inflammatory parameters, plasma interleukin 6 (IL-6) and high sensitivity C-reactive protein (hsCRP), were measured using ELISA kits (MyBioSource, San Diego, CA, USA; BioVendor, Brno, Czech Republic).

The homeostasis model assessment of insulin resistance (HOMA-IR) was calculated as follows: HOMA-IR = serum insulin (mmol/L) $\times$ blood glucose (mmol/L)/22.5 [26].

2.3. Biochemical Analysis of Tissues

Adiposity index and relative liver weight were expressed as visceral epididymal adipose tissue and liver weight per 100 g of body weight, respectively.

To determine TAG and DAG in tissues, samples were powdered under liquid N_2 and extracted in a mixture of chloroform/methanol (2:1). A solution of 2% potassium dihydrogenphosphate was then added to the mixture and centrifuged; the organic phase formed from the mixture was evaporated under N_2. The resulting pellet was dissolved in isopropyl alcohol. TAG concentrations were determined by enzymatic assay (TG L 250S, Erba-Lachema, Brno, Czech Republic). The content of 14,15-EET and 20-HETE in the liver was measured using rat ELISA kits (MyBioSource, San Diego, CA, USA).

2.4. Fatty Acid Composition and Fatty Acid Desaturase Activity in the Liver

Total lipids were extracted using dichloromethane/methanol (2:1, v/v) using the Folch method. Individual lipid classes were separated by thin-layer chromatography, converted to fatty acid methyl esters and established by GC using the Hewlett–Packard GC system as previously described [27]. Fatty acid levels in the liver were expressed as a percentage of the total fatty acids. Desaturase activity was estimated based on the product/precursor ratio as follows: D9-desaturase (16:1n7/16:0).

2.5. Tissue Insulin Sensitivity

Tissue insulin sensitivity was measured according to insulin-stimulated incorporation of glucose into skeletal muscle glycogen or visceral adipose tissue lipids. Diaphragm or epididymal adipose tissue was incubated for 2 h in 95% O_2 with 5% CO_2 in Krebs–Ringer bicarbonate buffer (pH 7.4) containing 0.1 µCi/mL of ^{14}C-U glucose, 5 mmol/L of unlabelled glucose, and 2.5 mg/mL of bovine serum albumin (Fraction V, Sigma, Brno, Czech Republic) with or without 250 µU/mL of insulin. Glycogen and lipids were extracted, while insulin-stimulated incorporation of glucose into glycogen or lipids was determined by scintillation counting as previously described [28].

2.6. Oxidative Stress Parameters

The anti-oxidant enzyme activity of superoxide dismutase (SOD), glutathione reductase (GR), glutathione transferase (GT), and glutathione peroxidase (GPx) was measured using commercially available kits (Cayman Chemicals, Ann Arbor, MI, USA). Catalase (CAT) activity was determined on the basis of the ability of H2O2 to form a colour complex with ammonium molybdate and then detected spectrophotometrically. Malondialdehyde (MDA), a parameter of lipid peroxidation, was determined by HPLC with fluorescence detection, with 4-hydroxynonenal (4-HNE), a sensitive product of lipid peroxidation, detected by rat ELISA assay (MyBioSource, San Diego, CA, USA).

2.7. Histological Evaluation

Slices of hepatic tissue were fixed in formaldehyde solution and processed into paraffin blocks using standard techniques. Three-micrometre-thick sections were cut from each sample using a microtome. Samples were stained using routine haematoxylin and eosin (HE) or processed using a Lipid (Oil Red O) Staining Kit (Sigma-Aldrich; St. Luis, MO, USA) to evaluate the neutral hepatic lipid content. The prepared slides were then evaluated by a veterinary histopathologist in a blinded protocol.

2.8. Relative mRNA Expression

Relative gene expression of hepatic enzymes, receptors, and transcriptional factors was determined by quantitative real-time PCR analysis using the TaqMan RNA-to-CT 1-Step Kit and the ViiATM 7 Real Time PCR System (ThermoFisher Scientific, Waltham, MA, USA). TaqMan probes were used to determine the mRNA of *Lpl, Hmgcr, Srebp1, Srebp2, Fas, Nr1h4, Nr1h3, Ppara, Ppary, Abca1, Abcg5, Abcg8, Cyp1a1, Cyp2e1, Cyp2b1, Cyp2c11, Cyp2d1, Cyp3a23, Cyp4a1, Cyp4a2, Cyp4a3, Cyp5a1, Cyp7a1, Scd-1* and *Nrf2* genes (ThermoFisher Scientific, Waltham, MA, USA). Relative expression was determined after normalisation against *Hprt 1* as an internal reference and then calculated using the $2^{-\Delta\Delta Ct}$ method. Results were run in triplicate.

2.9. Statistical Analysis

Data were evaluated on StatSoft® Statistica software (ver. 14, Statsoft CZ, Prague, Czech Republic) using two-way ANOVA for multiple comparisons followed by Fisher's post hoc LSD test. Statistical significance was set at a value of $p < 0.05$. All data were expressed as means ± standard error of the mean (SEM).

3. Results

3.1. Effect of Metformin Monotherapy

As expected, in HHTg rats MET reduced non-fasting glucose levels, favourably influenced serum lipids, decreased circulating TAGs, and elevated HDL cholesterol (Table 2). Although there was no significant reduction in body weight or visceral adiposity, circulating leptin levels decreased. MET monotherapy did not improve insulin sensitivity in skeletal muscle or visceral adipose tissue (Figure 1). Surprisingly, hepatokine fetuin A was adversely elevated in the MET-treated group. MET reduced pro-inflammatory markers (Table 3) as well as resistin content and mitigated hepatic oxidative stress. While the GSH/GSSG ratio and SOD activity increased, the lipid peroxidation-derived aldehydes 4-hydroxynonenal (4-HNE) and malondialdehyde (MDA) both decreased. Compared to untreated HHTg controls, the livers of MET-treated rats contained 9.5% fewer TAGs and 12% less DAG accumulation. MET markedly changed fatty acid composition in the DAG lipid class (Figure 2). Favourable changes in lipid composition were accompanied by decreased activity of D9-desaturase and reduced relative mRNA expression of *Scd-1*. MET reduced hepatic 20-HETE content, an intermediate of arachidonic acid metabolism. MET altered the relative mRNA expression of four genes involved in lipid metabolism, *Srebf-1* and *Fas* were downregulated and *Ldlr* and *Ppara* genes were up-regulated (Figure 2).

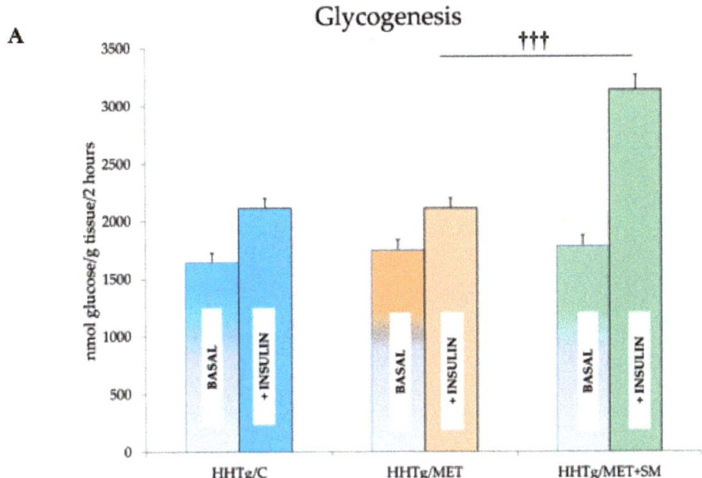

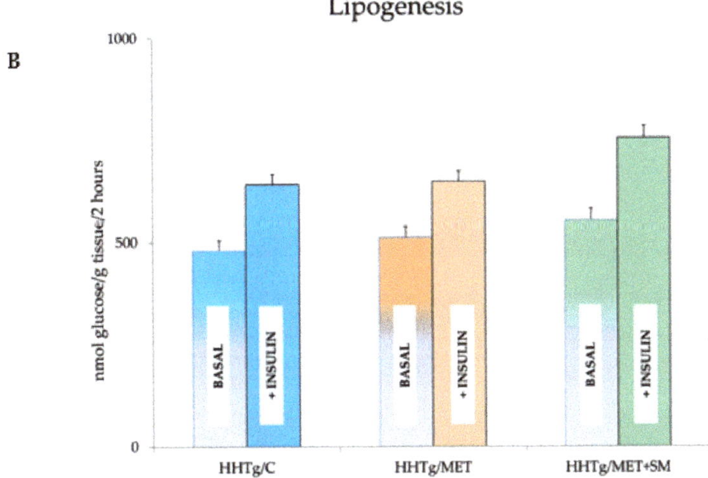

Figure 1. Effect of metformin and silymarin (MET+SM) combination therapy on (**A**) skeletal muscle and (**B**) adipose tissue insulin sensitivity expressed as basal and insulin-stimulated glycogenesis in hereditary hypertriglyceridaemic (HHTg) rats. Values are expressed as mean $\pm$ SEM; $n = 6$ for HHTg/C, $n = 7$ for HHTg/MET, $n = 8$ for HHTg/MET+SM; ††† $p < 0.001$ probability reflecting the effect of MET therapy vs. MET+SM combination therapy.

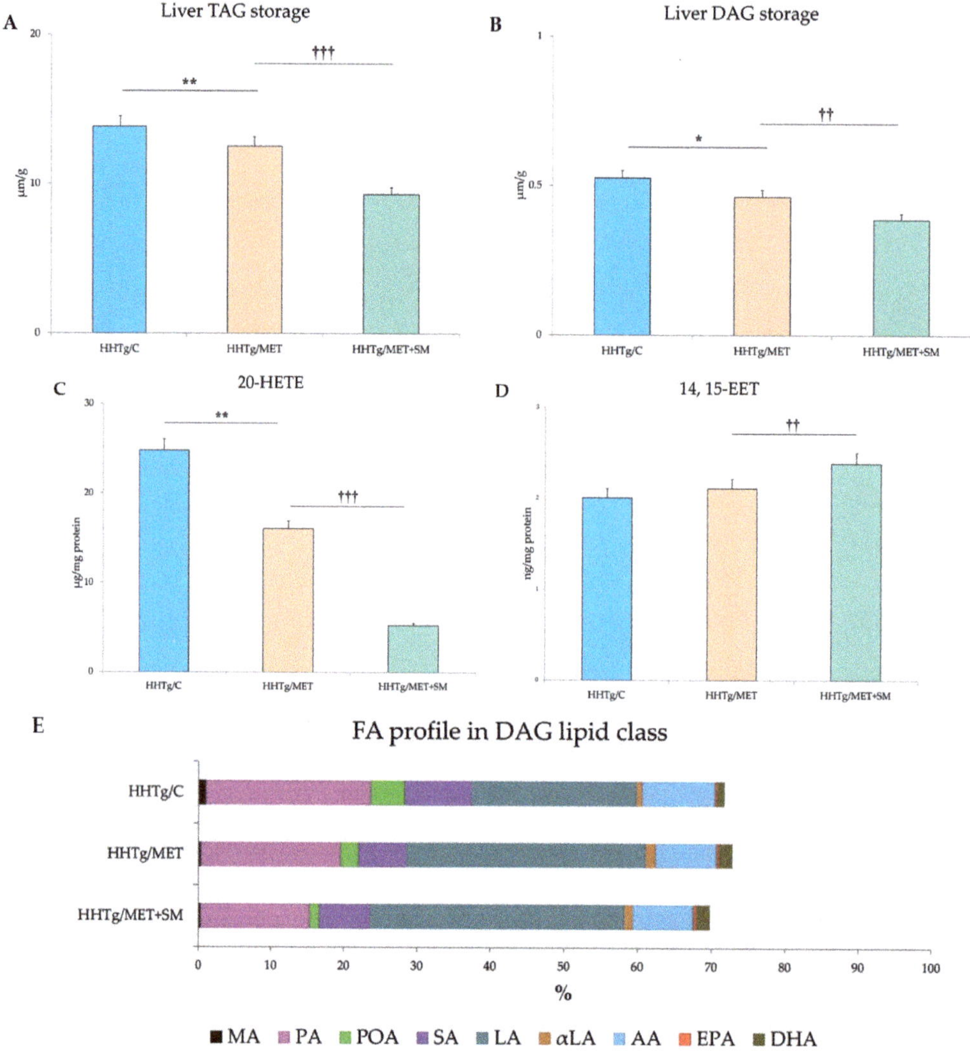

Figure 2. *Cont.*

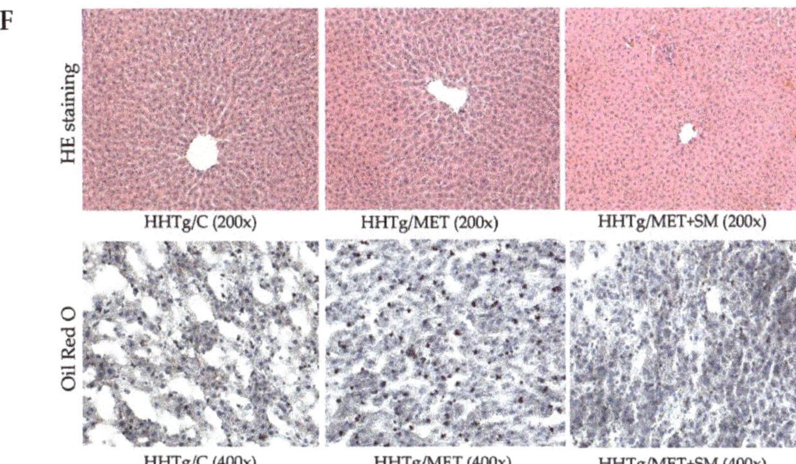

Figure 2. Effect of MET+SM combination therapy on hepatic lipid storage: (**A**) TAG, (**B**) DAG, (**C**) arachidonic acid metabolites 14,15-EET and (**D**) 20-HETE, (**E**) fatty acid composition in liver DAG class (**E,F**) histology of liver tissue: liver sections of all three groups showed minimal alterations of the liver tissue—normally arranged liver with minimal macrophage aggregates and eosinophils in the portal space. Histochemically, in the HHTg/C group, the positivity for cholesterol esters/TAGs was estimated at <5%, and in the HHTg/MET and HHTg/MET+SM group, positivity was <1%; MA—myristic acid, PA—palmitic acid, POA—palmitoleic acid, SA—stearic acid, LA—linoleic acid, αLA—α-linoleic acid, AA—arachidonic acid, EPA—eicosapentaenoic acid, DHA—docosahexaenoic acid. Values are expressed as means ± SEM; $n = 6$ for HHTg/C, $n = 7$ for HHTg/MET, $n = 8$ for HHTg/MET+SM; * $p < 0.05$, ** $p < 0.01$ probability reflecting the effect of MET monotherapy vs. the control group without treatment; †† $p < 0.01$, ††† $p < 0.001$ probability reflecting the effect of MET therapy vs. MET+SM combination therapy.

Table 2. Effect of MET and SM therapy on basal metabolic and morphological parameters in the bloodstream of HHTg rats.

	HHTg/C	HHTg/MET	HHTg/MET+SM	P_{MET}	P_{MET+SM}
Body weight (g)	447.5 ± 3.3	436.7 ± 5.4	414.6 ± 10.1	ns	0.05
Adiposity index (g/100 g b.wt.)	2.094 ± 0.106	1.896 ± 0.044	1.900 ± 0.114	ns	ns
Relative liver weight (g/100 g b.wt.)	3.089 ± 0.033	2.962 ± 0.032	2.687 ± 0.114	ns	0.01
Non-fasting glucose (mmol/L)	9.300 ± 0.379	8.257 ± 0.252 **	7.638 ± 0.173	0.01	ns
Insulin (nmol/L)	0.285 ± 0.047	0.261 ± 0.013	0.237 ± 0.017	ns	ns
HOMA-IR	3.037 ± 0.239	3.369 ± 0.322	2.657 ± 0.119 *	ns	0.01
TAG (mmol/L)	6.667 ± 0.364	5.446 ± 0.352 *	3.350 ± 0.205 *	0.01	0.001
Total cholesterol (mmol/L)	2.040 ± 0.055	2.111 ± 0.057 *	2.001 ± 0.017	ns	ns
HDL cholesterol (mmol/L)	0.813 ± 0.137	1.029 ± 0.118 ***	1.186 ± 0.152	0.01	0.05
FFA (mmol/L)	0.608 ± 0.011	0.621 ± 0.021	0.631 ± 0.064	ns	ns
ALT (μkat/L)	1.242 ± 0.156	1.433 ± 0.249	1.441 ± 0.274	ns	ns
MCP-1 (ng/mL)	4.917 ± 0.348	3.772 ± 0.270	4.818 ± 0.225	ns	ns
TNFα (pg/mL)	11.488 ± 0.917	8.732 ± 0.322 **	10.744 ± 0.523	0.01	0.05
Leptin (pg/mL)	9120 ± 359	6636 ± 278 ***	6023 ± 352 *	0.001	ns
HMW adiponectin (μg/mL)	5.48 ± 0.24	5.79 ± 0.51	5.55 ± 0.15	ns	ns
Fetuin-A (μg/mL)	106.89 ± 15.04	152.69 ± 11.69	104.11 ± 12.88***	0.05	0.01

Values are expressed as means ± SEM; $n = 6$ for HHTg/C, $n = 7$ for HHTg/MET, $n = 8$ for HHTg/MET+SM; P_{MET}—probability reflecting the effect of metformin monotherapy vs. the control group without any treatment, P_{MET+SM}—probability reflecting the effect of metformin therapy vs. metformin + silymarin combination therapy; data analysed by two-way-ANOVA; Fisher's post-hoc LSD test applied for multiple comparisons between groups; * $p < 0.05$, ** $p < 0.01$, *** $p < 0.001$.

Table 3. Effect of MET and SM therapy on hepatic inflammation and oxidative stress parameters in HHTg rats.

	HHTg/C	HHTg/MET	HHTg/MET+SM	P_{MET}	P_{MET+SM}
GSH/GSSG	27.06 ± 2.41	39.44 ± 2.38	44.09 ± 2.64 **	0.01	n.s.
SOD (U/mg protein)	0.127 ± 0.01	0.152 ± 0.01	0.183 ± 0.01	0.05	0.01
CAT (μM H_2O_2 min/mg protein)	1437 ± 80	1311 ± 88	1662 ± 130	n.s.	0.05
GPx (μM NADPH/min/mg protein)	249 ± 14	272 ± 12	349 ± 15	n.s.	0.001
4-HNE (ng/mg protein)	69.1 ± 4.4	47.7 ± 1.6 ***	46.6 ± 1.4	0.001	0.05
MDA (nM/mg protein)	3.43 ± 0.37	2.10 ± 0.29	2.28 ± 0.14 ***	0.01	ns
MCP-1 (pg/mg protein)	27.132 ± 1.494	27.187 ± 2.579 **	13.366 ± 1.074 **	ns	0.001
TNFα (pg/mg protein)	68.615 ± 5.493	54.114 ± 0.994 ***	45.274 ± 0.984	0.01	0.05
hsCRP (ng/mg protein)	93.618 ± 7.824	68.836 ± 2.616 ***	50.240 ± 2.423 *	0.001	0.01
Resistin (pg/mg protein)	4.672 ± 0.315	3.715 ± 0.178 ***	3.767 ± 0.161	0.01	ns

Values are expressed as means $\pm$ SEM; n = 6 for HHTg/C, n = 7 for HHTg/MET, n = 8 for HHTg/MET+SM; P_{MET}—probability reflecting the effect of metformin monotherapy vs. the control group without any treatment, P_{MET+SM}—probability reflecting the effect of metformin therapy vs. metformin + silymarin combination therapy; data analysed by two-way-ANOVA; Fisher's post-hoc LSD test applied for multiple comparisons between groups; * $p < 0.05$, ** $p < 0.01$, *** $p < 0.001$, n.s.—the difference is not significant.

3.2. Effect of Metformin and Silymarin Combination Therapy on Basal Metabolic Parameters

In HHTg rats, MET+SM combination therapy had a greater effect on weight loss than monotherapy but had no influence on visceral adiposity (expressed as the adiposity index) or on the sensitivity of adipose tissue to insulin action (Figure 1). Food intake was 15% less in the MET+SM-treated group than in the MET monotherapy group (27.1 ± 0.09 vs. 23.5 ± 0.07 g/day; $p < 0.01$) and accompanied by a mild decrease in circulating leptin (Table 2). The additive effect of SM manifested in reduced non-fasting glucose levels and a significant improvement in HOMA-IR. In comparison with MET monotherapy, muscle insulin sensitivity (measured based on the incorporation of radioactive-labelled glucose into glycogen in skeletal muscle) significantly improved after MET+SM combination therapy (Figure 1). MET therapy alone ameliorated lipid metabolism disturbances, which manifest as high serum TAG levels in HHTg rats. Compared with monotherapy, however, the addition of SM intensified the effect, increasing TAG, decreasing total cholesterol levels, and improving HDL cholesterol levels. MET+SM therapy had no effect on free fatty acid (FFA) levels or on the concentration of the liver enzyme ALT. In contrast to the observed increase in the circulating hepatokine fetuin-A in MET-treated rats, MET+SM combination therapy ameliorated this undesirable effect, with fetuin-A levels decreasing to control group values.

3.3. Effect of Metformin and Silymarin Combination Therapy on Hepatic Lipid Storage, Lipotoxic Intermediates, and Fatty Acid Profiles

The decreased relative liver weight in the MET+SM-treated group was accompanied by a reduction in neutral lipids (TAG) and in the concentration of the lipotoxic intermediate DAG (Figure 2) in comparison with the MET group. However, none of the follow-up treatments had any effect on cholesterol content (9.61 ± 0.36 vs. 9.62 ± 0.30 vs. 9.76 ± 0.32 vs. 9.33 ± 0.34 mmol/L; n.s.). As shown in Figure 2, neither MET nor MET+SM therapy altered the morphology of the liver tissue, and the histochemical analysis for neutral lipids verification revealed no significant impact of the treatments. Compared with the control group, samples of MET-treated animals showed moderately decreased content of cholesteryl esters and TAGs in hepatocytes, and SM supplementation did not have an additive effect.

These quantitative changes in lipid storage were accompanied by a qualitative improvement in FA composition in the DAG lipid class. As shown in Figure 2, MET reduced the proportional representation of saturated FAs, namely myristic acid (MA), palmitic acid (PA), stearic acid (SA), monounsaturated palmitoleic acid (POA), and the pro-inflammatory omega-6 arachidonic acid (AA). This decreasing tendency was augmented by the addition of SM, especially with regard to PA (P_{MET} < 0.05) and POA (P_{MET+SM} < 0.01). Moreover, compared to monotherapy, MET+SM therapy had a greater effect on lipid metabolites, reducing levels of the pro-inflammatory arachidonic acid metabolite 20-HETE (20-hydroxyeicosatetraenoic acid) and elevating levels of the anti-inflammatory metabolite 14,15-EET (14,15-eicosatetraenoic acid) (Figure 2). Additionally, the MET-treated group had a higher percentage of linoleic acid (LA), α-linoleic acid (αLA), eicosapentaenoic acid (EPA), and docosahexaenoic acid (DHA) compared with only a slight increase in the MET+SM group, indicating an improvement in FA composition in the liver tissue.

3.4. Effect of Metformin and Silymarin Combination Therapy on Hepatic Oxidative Stress

As expected from a potent anti-oxidative compound, SM significantly influenced parameters of oxidative stress in our pre-diabetes HHTg rat model. The addition of SM to MET therapy increased the activity of the superoxide dismutase (SOD) enzyme, which initiates the antioxidant response, and catalase (CAT), and it markedly increased the activity of the glutathione-dependent enzyme GPx (Table 3). SM also elevated relative mRNA expression of *Nrf2*, a transcriptional factor that plays a key role in the response to oxidative stress (Figure 3). Decreased levels of 4-HNE and MDA, reactive markers produced by lipid peroxidation, were evident in the livers of MET-treated models, with the addition of SM leading to a further significant reduction in these products.

A

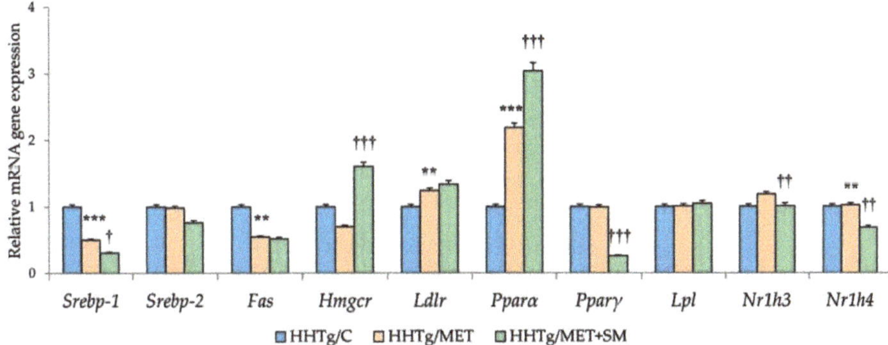

Figure 3. *Cont.*

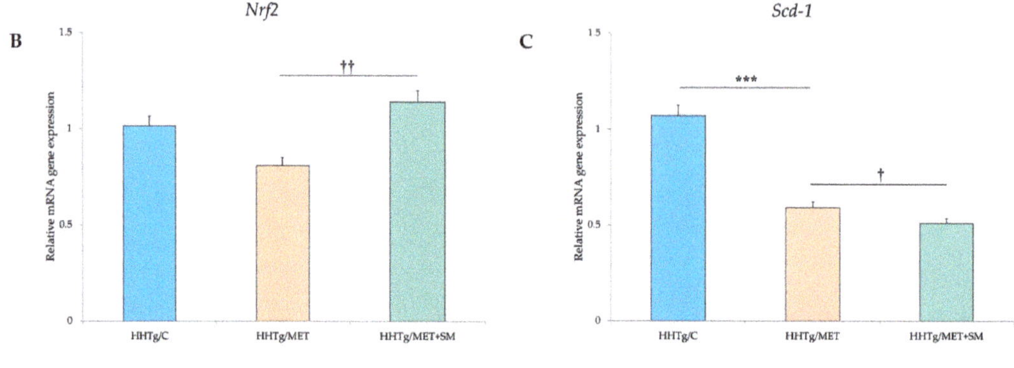

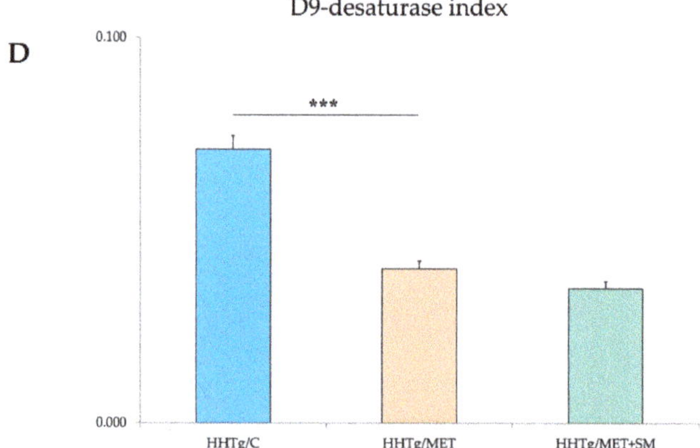

Figure 3. Effect of MET+SM therapy combination on the relative mRNA expression of (**A**) hepatic genes involved in lipid metabolism, (**B**) Nrf2, (**C**) Scd-1, and (**D**) D9-desaturase index genes. Values are expressed as means $\pm$ SEM; $n = 6$ for HHTg/C, $n = 7$ for HHTg/MET, $n = 8$ for HHTg/MET+SM; ** $p < 0.01$, *** $p < 0.001$ probability reflecting the effect of MET monotherapy vs. the control group without treatment; † $p < 0.05$, †† $p < 0.01$, ††† $p < 0.001$ probability reflecting the effect of MET therapy vs. MET+SM combination therapy.

3.5. Effect of Metformin and Silymarin Combination Therapy on Inflammation Parameters

SM supplementation exerted no anti-inflammatory effect on the circulation of HHTg rats (Table 2). However, MET+SM therapy did result in a noticeable improvement in inflammatory markers in liver tissue. Concentrations of MCP-1, TNF-α, and CRP were significantly reduced compared with MET monotherapy (Table 3).

3.6. Effect of Metformin and Silymarin Combination Therapy on Relative mRNA Expression of Genes and Enzymes Involved in Lipid Metabolism

MET+SM combination therapy markedly altered the gene expression of enzymes and transcriptional factors involved in lipid metabolism regulation.

Relative mRNA expression of the *Scd-1* gene, which regulates the expression of other genes involved in the lipogenesis and regulation of mitochondrial FA oxidation, decreased after MET+SM treatment compared with MET monotherapy (Figure 3). The activity index of D9-desaturase in phospholipids reduced significantly after MET, with the addition of SM slightly augmenting the effect. As shown in Figure 3, MET+SM combination therapy decreased mRNA expression of the sterol regulatory element-binding protein (*Srebp1*) gene, a key transcription factor that regulates genes involved in cholesterol biosynthesis and lipid homeostasis. The peroxisome proliferator-activated receptor gamma (*Pparγ*) and nuclear receptor subfamily 1 (*Nr1h3* and *Nr1h4*) genes were significantly downregulated. Compared with monotherapy, MET+SM combination therapy elevated the relative mRNA expression of the rate-limiting enzyme for cholesterol synthesis *Hmgcr* (3-hydroxy-3-methylglutaryl-CoA), *Pparα*, the cholesterol transporters G5 and G8 (*Abcg5* and *Abcg8*), and the membrane transporter *Abca1* (Figure 4). It is likely that MET+SM combination therapy had a greater effect than MET monotherapy on some of the cytochrome P450 family proteins involved in hepatic lipid regulation, with relative mRNA expression of *Cyp4a* (*Cyp4a1*, *Cyp4a2*, and *Cyp4a3*), *Cyp5a1*, and *Cyp2d1* significantly upregulated after SM supplementation.

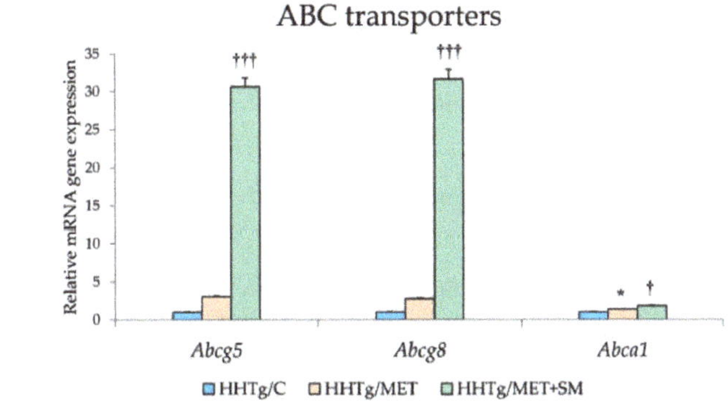

Figure 4. Effect of MET+SM combination therapy on the relative mRNA expression of (**A**) hepatic cytochrome P450 family protein genes and (**B**) ABC transporters. Values are expressed as means ± SEM; $n = 6$ for HHTg/C, $n = 7$ for HHTg/MET, $n = 8$ for HHTg/MET+SM; * $p < 0.05$, ** $p < 0.01$, *** $p < 0.001$ probability reflecting the effect of MET monotherapy vs. the control group without treatment; † $p < 0.05$, †† $p < 0.01$, ††† $p < 0.001$ probability reflecting the effect of MET therapy vs. MET+SM combination therapy.

4. Discussion

Our study examined the potential therapeutic benefits of adding a plant extract SM to traditional MET therapy in a rodent pre-diabetes model with liver steatosis symptoms in the absence of obesity and hyperglycaemia. Liver steatosis is not only a factor in the development of T2DM, but it can be also related to pre-diabetic states regardless of obesity and before hyperglycaemia is increased [3].

In connection with MET and SM in T2DM therapy, current research usually aims to separately evaluate MET compared with various phytochemicals' efficacies in both animal and human studies [29–31]. Only a few studies have investigated the effect of the combination of MET with SM (or another herbal extract) in comparison with monotherapy [32]. Our previous studies with the combined administration of SM with n-3 PUFA [33] or atorvastatin [34] have revealed a significant additive effect in the therapy of hypertriglyceridemia-induced metabolic disorders. SM boosted the hypolipidaemic, antioxidant, and anti-inflammatory effects of n-3 PUFA and statin therapy; moreover, the combination with SM more effectively reduced ectopic lipid storage and improved glucose homeostasis compared with drug monotherapy.

In our non-obese model of prediabetes with genetically fixed hypertriglyceridemia, the four-week combination therapy with SM was more effective in decreasing body weight than MET alone (-5%, $p < 0.05$) (Table 2). In human studies, the enhanced anorectic effect, as well as reduced appetite, are more associated with MET [35]; however, most studies focus on overweight or obese individuals, while our HHTg model involved a non-obese strain of rats. In our study, the mild reduction in food consumption after MET+SM therapy was accompanied by lower circulating levels of leptin (Table 2), a hormone produced by the obese gene that regulates food intake and body mass [36].

Predictably, MET reduced serum non-fasting glucose levels in our pre-diabetes model, with SM exerting a non-significant additive effect. Although the hyperinsulinaemic-euglycaemic glucose clamp is the gold standard for insulin resistance presence, HOMA-IR is an accepted method in both clinical practice and experimental research [37]. In our pre-diabetic HHTg rat model with the HOMA-IR value exceeded (3.037), the SM addition induced a 13% decrease of this marker, highlighting the importance of SM in reducing the risk of diabetes onset. These findings are in accordance with studies with other diabetic models. In high-fat or high-fructose-induced models, the SM/silibinin intervention improved IR, as was shown by the decreased HOMA-IR [38,39] (Table 2).

The pleiotropic effects of MET are widely acknowledged, but its positive impact on lipid metabolism is debatable. While experimental studies have documented an improvement in lipid management across ectopic lipid deposition and circulating lipids [40], clinical studies are less definitive, with some finding no improvement [13]. In our study, SM enhancement reduced serum TAG and elevated HDL cholesterol (Table 2). Given that NAFLD is linked to hypertriglyceridaemia and low levels of HDL cholesterol, these effects seem to highlight an improvement in liver steatosis [41]. ABC transporters, which are proteins responsible for the ATP-driven transfer of substrates across cell membranes, play an important role in cholesterol elimination pathways [42]. In agreement with a study focusing on silibinin [43], the major active constituent of SM, supplementing MET with SM led to upregulation of *Abca1* (Figure 4B), the gene that encodes the membrane transporter ABCA1. Facilitating the transport of cholesterol and phospholipids through the plasma membrane to HDL particles, ABCA1 is a major determinant of HDL levels and also functions as a cholesterol efflux pump [44]. The cholesterol transporters G5 and G8 (*Abcg5* and *Abcg8*) are genes that play an essential role in the selective transport of absorbed cholesterol, returning it to the intestinal lumen [45]. SM supplementation improves cholesterol efflux and transmembrane transport, important processes in multifactorial disorders such as T2DM that require multidrug therapy. Neither MET nor MET+SM therapy was effective in reducing circulating FFAs, which are chronically elevated in HHTg rats because of the higher resistance of this strain to insulin action.

For the management of T2DM as well as pre-diabetes, improving insulin sensitivity is a crucial target. However, the 5% body weight decrease achieved in the MET+SM-treated group did not translate into a reduction in visceral adiposity, and the sensitivity of adipose tissue to insulin action remained unaltered in our insulin-resistant HHTg rat model (Figure 1B). Compared with our previous study with atorvastatin in which the SM potentiated the sensitivity of visceral adipose tissue to insulin action [34], the combination SM with MET mitigated insulin sensitivity. In our model, which exhibits chronically elevated lipid accumulation in skeletal muscles [46], none of the therapies affected lipid storage (data not shown) or serum adiponectin (Table 2), which is positively associated with insulin sensitivity.

The liver, which is understood to be the key driver of insulin resistance, plays a major role in tempering disturbances during the early stages of development. However, the pathogenesis of both NAFLD and simple steatosis is incomplete. According to the 'multiple hit' hypothesis, the most accurate explanation for NAFLD development, a number of insults synergise to induce fatty liver in genetically predisposed individuals. Apart from insulin resistance, these hits are delivered by hormones secreted from adipose tissue, nutritional factors, and the gut microbiota, as well as genetic and epigenetic factors [47].

The non-adipose tissue deposition of lipids and their metabolites—DAGs, ceramides, and fatty acyl-CoA—can be the aggravated background for hepatic steatosis, insulin resistance, and its accompanying metabolic disturbances. Liver steatosis is defined as excessive hepatic lipid content exceeding 5% of total liver weight [48]. In the livers of HHTg rats, where the hepatic TAG content exceeds 12%, adding SM to MET therapy intensified the reduced TAG and DAG storage (Figure 2A,B), providing potential evidence of a mechanism involved in ameliorating hepatic insulin resistance. In the mechanism responsible for MET+SM benefits, the transcription factor Nrf2 can play an important role. Phytochemicals are significant activators of Nrf2, which regulates genes involved in regulating lipid metabolism as well as anti-oxidative and anti-inflammatory responses and enhances insulin signalling [49]. Although biochemically assessed alterations in lipid storage were evident, the histological and histochemical findings were not so convincing in hepatic tissue (Figure 2F). These results correspond with the findings of other animal studies documenting unaltered lipid accumulation in hepatocytes, despite elevated concentrations of neutral lipids and lipotoxic intermediates [13].

Elevated FFAs are closely connected with insulin resistance. It has been reported that FFAs or their metabolites may affect liver damage more than liver TAG accumulation via increased oxidative stress [50]. Deteriorated qualitative changes in serum and tissue FA composition are a typical feature of the HHTg strain [51]. The beneficial alterations in hepatic FA composition observed both in MET and MET+SM treated group (Figure 2E) point to the positive impact on chronic, persistent, low-grade inflammation in HHTg rats.

In the liver after MET and MET+SM administration, the increase in anti-inflammatory n−3 PUFAs (αLA, EPA, DHA) and the reduced saturated FA fraction (PA, SA), deteriorating hepatic insulin sensitivity [52], and decreased monounsaturated POA (reflecting hepatic lipogenesis [53]) highlight the insulin-sensitising effects of both therapies. After four weeks of MET and MET+SM treatment, AA profiles were reduced in the hepatic DAG lipid class. Changes in AA content are understood to be an early indicator of inflammation and NAFLD progression [54], and cytochrome P450 enzymes are supposed to be AA modulators [55]. CYP-AA metabolites HETEs and EETs have different properties and can be stored in tissue lipids. EETs have anti-inflammatory, thrombolytic and angiogenic properties [56], while 20-HETE is a potent vasoconstrictor with pro-inflammatory activity. Alterations in cytochrome P450 enzymes, which are dysregulated in T2DM and insulin resistance, can also contribute to the improvement of lipid metabolism. Compared with monotherapy, MET+SM combination treatment more effectively upregulated Cyp4a isoforms and members of the Cyp2 family (Figure 2A) responsible for catalysing AA metabolism to HETEs and EETs [57]. Altered expressions of other cytochrome P450 enzymes (CYP1A1, CYP2E1, CYP7A1, etc.) connecting with fatty liver and T2DM, were not detected in our study.

Taking a more systemic view, chronic low-grade inflammation of any cause plays a key role in the pathogenesis of insulin resistance and other dyslipidemia-induced disorders. Disturbances connected with both T2DM and NAFLD correlate with increased inflammation states. While most studies of inflammatory markers focus on circulation in the bloodstream or cell cultures, our study examines the concentrations of pro-inflammatory markers in hepatic tissue. In our study, markedly reduced MCP-1, TNF-α, and hsCRP concentrations were observed in the liver tissue of MET+SM-treated HHTg rats (Table 3), while the reduced content of resistin, a novel adipokine and promising marker of inflammation, which is associated with obesity, insulin resistance, NAFLD, and T2DM in animal and human studies [58,59], was not potentiated with SM supplementation. The beneficial effect of SM addition can be mediated through the inhibition of the nuclear transcription factor kappa B (NF-κB) signalling pathway. In vitro and in vivo studies documented that plant-derived polyphenols can suppress NF-κB associated inflammatory pathways and reduce pro-inflammatory markers [60]. As for liver steatosis in human studies, hsCRP is suggested to be a useful marker in differentiating between simple steatosis and NASH [61].

Novel evidence suggests that fatty liver alters the secretion of various factors from hepatocytes such as hepatokines, which affect glucose and lipid metabolism and mediate inter-tissue crosstalk [41]. Epidemiological studies have shown that elevated serum levels of fetuin-A are connected with T2DM, insulin resistance, and NAFLD [62]; however, the relation with liver steatosis is not clear [63]. Fetuin A promotes proinflammatory activation by acting as an endogenous ligand for toll-like receptor 4 (TLR4), which is involved in a lipid-induced pro-inflammatory response [64]. In our study of HHTg rats, MET monotherapy adversely increased circulating fetuin-A levels, a negative effect reversed by the addition of SM (Table 2).

The improved inflammatory state in the livers of MET+SM treated animals was accompanied by ameliorated oxidative stress, which (together with chronic, low-grade inflammation) is a typical feature of the pre-diabetes model with genetically fixed hypertriglyceridemia. There is a strong link between oxidative stress, inflammation, and the development and progression of T2DM [65]. Oxidative stress contributes to insulin resistance and, most importantly, oxidative stress and mitochondrial damage have been reported to be causative in NAFLD initiation and progression [66]. Moreover, in the livers of patients with NAFLD or simple steatosis, increased levels of lipid peroxidation products and reduced levels of SOD, CAT, and GSH-Px have been demonstrated [67]. In accordance with these findings, SM supplementation intensified the activity of liver GSH-Px, CAT, and SOD and increased the concentration of GSH and the GSH/GSSG ratio (Table 3). Furthermore, the MET+SM combination reduced products of lipoperoxidation, namely the MDA and 4-HNE concentrations, more effectively than monotherapy. Activation of the antioxidant system is one of the most hepatoprotective mechanisms of flavonoids such as SM, which could be mediated via free radical scavenging, decreased mitochondrial reactive oxygen species (ROS) production, alterations in cytochrome P450 enzyme activity, and elevated expression of the transcriptional factor *Nrf2*, a key regulator of the cell defence against oxidative damage [68]. In addition to the above positive impacts on liver steatosis mitigation in MET with SM-treated pre-diabetes models, our analysis focused on the hepatic mRNA expressions of genes involved in lipid metabolism (Figure 3).

Compared with monotherapy, MET+SM therapy augmented the upregulation of the *Hmgcr* gene, a rate-limiting enzyme that contributes to cholesterol synthesis; *Srebp-1*, a target gene of lipid synthesis and a key regulator of hepatic lipogenesis; and *Pparα*, a transcription factor responsible for regulating genes involved in FA uptake and metabolism [69]. The upregulation of *Pparα* and *Srebp-1* downregulation can result in reduced FA and TAG synthesis [70]. Supplementation of MET with silybin has been shown to inhibit *Scd-1*, *Fas*, and *Srebp-1* relative mRNA expression in hepatocytes [71], which is in agreement with our results. SCD-1 is an enzyme that modulates lipogenesis and FA oxidation, and decreasing the activity of this mediator can improve insulin action. In an experimental study with high-calorie-diet induced models, SM reduced elevated intraperitoneal fat and mitigated the gene expression of *Pparγ* and the fatty acid synthase (*Fas*) enzyme [72]. SM supplementation enhanced the downregulation of the *Nr1h* family members, key regulators of macrophage function responsible for controlling genes involved in lipid homeostasis and inflammation. Nr1h nuclear receptors affect the expression of *Srebp1* and *Pparγ* and the cholesterol transporters G5 and G8 [73] (Figures 4B and 5).

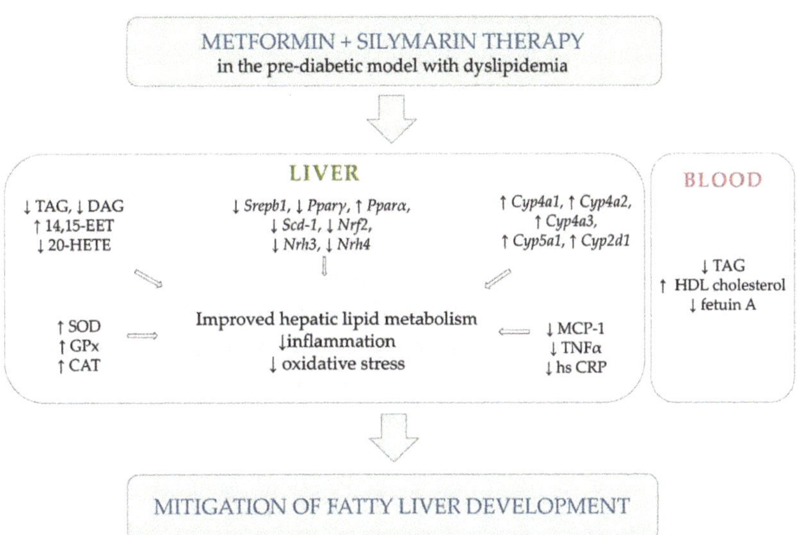

Figure 5. Summary scheme. Diagram of significant results supporting conclusions of the study.

It was reported that plant-derived compounds, such as SM, target various cellular processes at a molecular level [74]. Flavonoids, such as SM, are generally poorly absorbed in the intestine, resulting in relatively low serum and tissue levels. However, even these low concentrations seem to be sufficient for Nrf2 activation and NF-κB-related pathway suppression. Indeed, these mechanisms may be more important in driving the phytochemical-induced health benefits of SM than the direct scavenging of free radicals [75].

A limitation of this study could be the absence of a female rat group. However, the sexual differences are discussed in various hepatic metabolism pathways, and we do not believe that the additive effect of silymarin in metformin therapy is different in male and female animals.

5. Conclusions

Our results provide novel evidence that silymarin supplementation may augment metformin therapy and can modulate the disturbances characteristic of liver steatosis development in a pre-diabetic model. Compared with metformin monotherapy, metformin–silymarin combination therapy improved hepatic lipid metabolism more, reducing the content of neutral lipids and lipotoxic intermediates. In addition, hepatic gene expression of enzymes and transcription factors involved in lipogenesis and fatty acid oxidation can contribute to the hepatic lipid accumulation reducing as a result. In the liver, combination therapy also reduced arachidonic acid metabolism and pro-inflammatory AA metabolites, mitigating inflammation processes and moderating oxidative stress. Changes in the gene expression of the cytochrome P450 family of enzymes, in particular Cyp4A, can improve hepatic lipid metabolism and moderate inflammation. All these metabolic effects significantly participate in insulin resistance, which is the background of liver steatosis and is closely linked to T2DM.

In summary, the additive effect of silymarin in metformin therapy can mitigate and even prevent fatty liver development, particularly in the pre-diabetic state before the onset of diabetes. However, further detailed clinical studies are required.

Author Contributions: M.H. and H.M.—study conception and design, manuscript draft and composition; I.M.—biochemical/molecular analysis and data interpretation; D.M.—statistical and biochemical analysis; I.Z.—molecular analysis; M.P. and Z.R.—cytochrome P450 interpretation; R.V.—study conception. All authors have critically revised the manuscript for intellectual content and have approved the final version for publication. All authors have read and agreed to the published version of the manuscript.

Funding: This work was supported by a grant from the Czech Science Foundation and by the Ministry of Health of the Czech Republic under its programme for the conceptual development of research organisations (Institute for Clinical and Experimental Medicine—IKEM, IN 00023001) and grant no. IGA_LF_2021_013.

Institutional Review Board Statement: Animal Protection Law of the Czech Republic (311/1997) and with European Community Council recommendations (86-609/ECC) for the use of laboratory animals and approved by the Ethics Committee of the Institute for Clinical and Experimental Medicine, Prague.

Informed Consent Statement: Not applicable.

Data Availability Statement: All datasets generated for this study are included in the article.

Acknowledgments: The authors wish to thank Ludmila Kazdova for her expert recommendations, Olena Oliyarnyk for laboratory assistance, Peter Makovicky for histological evaluation, and Michael FitzGerald for English language editing.

Conflicts of Interest: The authors declare that there are no conflicts of interest regarding the publication of this paper.

Abbreviations

AMPK	adenosine monophosphate-activated protein kinase
DAG	diacylglycerol
EET	eicosatetraenoic acid
FAS	fatty acid synthase
GSH	reduced form of glutathione
GSSG	oxidised form of glutathione
HETE	hydroxytetraenoic acid
HHTg	hereditary hypertriglyceridaemia
HMGCR	3-hydroxy-3-methylglutaryl-coenzyme A reductase
HPLC	High Performance Liquid Chromatography
4-HNE	4-hydroxynonenale
hsCRP	high-sensitivity C-reactive protein
IL-6	interleukin 6
MDA	malondialdehyde
MAPK	mitogen-activated protein kinase
MCP-1	monocyte chemoattractant protein-1
NAFLD	non-alcoholic fatty liver disease
FFA	free fatty acid
NF-κB	nuclear factor kappa B
NASH	non-alcoholic steatohepatitis
NRF2	nuclear factor erythroid-2-related factor 2
PPAR	peroxisome proliferator-activated receptor
PUFA	polyunsaturated fatty acid
ROS	reactive oxygen species
SCD	stearoyl-coenzyme A desaturase
SOD	superoxide dismutase
SREBP	sterol regulatory element-binding protein
TAG	triacylglycerol
TNFα	tumour necrosis factor alpha
VLDL	very-low-density lipoprotein

References

1. Buysschaert, M.; Bergman, M. Definition of prediabetes. *Med. Clin. N. Am.* **2011**, *95*, 289–297. [CrossRef] [PubMed]
2. Francque, S.M. The Role of Non-alcoholic Fatty Liver Disease in Cardiovascular Disease. *Eur. Cardiol.* **2014**, *9*, 10–15. [CrossRef] [PubMed]

3. Zelber-Sagi, S.; Lotan, R.; Shibolet, O.; Webb, M.; Buch, A.; Nitzan-Kaluski, D.; Halpern, Z.; Santo, E.; Oren, R. Non-alcoholic fatty liver disease independently predicts prediabetes during a 7-year prospective follow-up. *Liver Int.* **2013**, *33*, 1406–1412. [CrossRef]
4. Bellentani, S.; Scaglioni, F.; Marino, M.; Bedogni, G. Epidemiology of non-alcoholic fatty liver disease. *Dig. Dis.* **2010**, *28*, 155–161. [CrossRef] [PubMed]
5. Friedman, S.L.; Neuschwander-Tetri, B.A.; Rinella, M.; Sanyal, A.J. Mechanisms of NAFLD development and therapeutic strategies. *Nat. Med.* **2018**, *24*, 908–922. [CrossRef] [PubMed]
6. Caussy, C.; Soni, M.; Cui, J.; Bettencourt, R.; Schork, N.; Chen, C.H.; Ikhwan, M.A.; Bassirian, S.; Cepin, S.; Gonzalez, M.P.; et al. Nonalcoholic fatty liver disease with cirrhosis increases familial risk for advanced fibrosis. *J. Clin. Investig.* **2017**, *127*, 2697–2704. [CrossRef]
7. Loomba, R.; Schork, N.; Chen, C.H.; Bettencourt, R.; Bhatt, A.; Ang, B.; Nguyen, P.; Hernandez, C.; Richards, L.; Salotti, J.; et al. Heritability of Hepatic Fibrosis and Steatosis Based on a Prospective Twin Study. *Gastroenterology* **2015**, *149*, 1784–1793. [CrossRef]
8. Albhaisi, S.; Chowdhury, A.; Sanyal, A.J. Non-alcoholic fatty liver disease in lean individuals. *JHEP Rep.* **2019**, *1*, 329–341. [CrossRef]
9. Kim, D.; Kim, W.R. Nonobese Fatty Liver Disease. *Clin. Gastroenterol. Hepatol.* **2017**, *15*, 474–485. [CrossRef]
10. Sookoian, S.; Pirola, C.J. Systematic review with meta-analysis: Risk factors for non-alcoholic fatty liver disease suggest a shared altered metabolic and cardiovascular profile between lean and obese patients. *Aliment. Pharmacol. Ther.* **2017**, *46*, 85–95. [CrossRef]
11. American Diabetes Association. Prevention or Delay of Type 2 Diabetes: Standards of Medical Care in Diabetes-2020. *Diabetes Care* **2020**, *43*, S32–S36. [CrossRef]
12. Nasri, H.; Rafieian-Kopaei, M. Metformin: Current knowledge. *J. Res. Med. Sci.* **2014**, *19*, 658–664. [PubMed]
13. Green, C.J.; Marjot, T.; Tomlinson, J.W.; Hodson, L. Of mice and men: Is there a future for metformin in the treatment of hepatic steatosis? *Diabetes Obes. Metab.* **2019**, *21*, 749–760. [CrossRef] [PubMed]
14. Bayat Mokhtari, R.; Homayouni, T.S.; Baluch, N.; Morgatskaya, E.; Kumar, S.; Das, B.; Yeger, H. Combination therapy in combating cancer. *Oncotarget* **2017**, *8*, 38022–38043. [CrossRef]
15. Zhang, A.; Sun, H.; Yuan, Y.; Sun, W.; Jiao, G.; Wang, X. An in vivo analysis of the therapeutic and synergistic properties of Chinese medicinal formula Yin-Chen-Hao-Tang based on its active constituents. *Fitoterapia* **2011**, *82*, 1160–1168. [CrossRef]
16. Abenavoli, L.; Capasso, R.; Milic, N.; Capasso, F. Milk thistle in liver diseases: Past, present, future. *Phytother. Res.* **2010**, *24*, 1423–1432. [CrossRef] [PubMed]
17. Zhang, Z.B.; Shen, Z.G.; Wang, J.X.; Zhang, H.X.; Zhao, H.; Chen, J.F.; Yun, J. Micronization of silybin by the emulsion solvent diffusion method. *Int. J. Pharm.* **2009**, *376*, 116–122. [CrossRef]
18. Poruba, M.; Kazdova, L.; Oliyarnyk, O.; Malinska, H.; Matuskova, Z.; Tozzi di Angelo, I.; Skop, V.; Vecera, R. Improvement bioavailability of silymarin ameliorates severe dyslipidemia associated with metabolic syndrome. *Xenobiotica* **2015**, *45*, 751–756. [CrossRef]
19. Poruba, M.; Matuskova, Z.; Kazdova, L.; Oliyarnyk, O.; Malinska, H.; Tozzi di Angelo, I.; Vecera, R. Positive effects of different drug forms of silybin in the treatment of metabolic syndrome. *Physiol. Res.* **2015**, *64*, S507–S512. [CrossRef]
20. MacDonald-Ramos, K.; Michan, L.; Martinez-Ibarra, A.; Cerbon, M. Silymarin is an ally against insulin resistance: A review. *Ann. Hepatol.* **2021**, *23*, 100255. [CrossRef]
21. Krecman, V.; Skottova, N.; Walterova, D.; Ulrichova, J.; Simanek, V. Silymarin inhibits the development of diet-induced hypercholesterolemia in rats. *Planta Med.* **1998**, *64*, 138–142. [CrossRef] [PubMed]
22. Wah Kheong, C.; Nik Mustapha, N.R.; Mahadeva, S. A Randomized Trial of Silymarin for the Treatment of Nonalcoholic Steatohepatitis. *Clin. Gastroenterol. Hepatol.* **2017**, *15*, 1940–1949.e8. [CrossRef] [PubMed]
23. Sobolova, L.; Skottova, N.; Vecera, R.; Urbanek, K. Effect of silymarin and its polyphenolic fraction on cholesterol absorption in rats. *Pharmacol. Res.* **2006**, *53*, 104–112. [CrossRef] [PubMed]
24. Skottova, N.; Kazdova, L.; Oliyarnyk, O.; Vecera, R.; Sobolova, L.; Ulrichova, J. Phenolics-rich extracts from Silybum marianum and Prunella vulgaris reduce a high-sucrose diet induced oxidative stress in hereditary hypertriglyceridemic rats. *Pharmacol. Res.* **2004**, *50*, 123–130. [CrossRef]
25. Zicha, J.; Pechanova, O.; Cacanyiova, S.; Cebova, M.; Kristek, F.; Torok, J.; Simko, F.; Dobesova, Z.; Kunes, J. Hereditary hypertriglyceridemic rat: A suitable model of cardiovascular disease and metabolic syndrome? *Physiol. Res.* **2006**, *55*, S49–S63. [PubMed]
26. Hanley, A.J.; Williams, K.; Stern, M.P.; Haffner, S.M. Homeostasis model assessment of insulin resistance in relation to the incidence of cardiovascular disease: The San Antonio Heart Study. *Diabetes Care* **2002**, *25*, 1177–1184. [CrossRef]
27. Miklankova, D.; Markova, I.; Huttl, M.; Zapletalova, I.; Poruba, M.; Malinska, H. Metformin Affects Cardiac Arachidonic Acid Metabolism and Cardiac Lipid Metabolite Storage in a Prediabetic Rat Model. *Int. J. Mol. Sci.* **2021**, *22*, 7680. [CrossRef]
28. Qi, N.; Kazdova, L.; Zidek, V.; Landa, V.; Kren, V.; Pershadsingh, H.A.; Lezin, E.S.; Abumrad, N.A.; Pravenec, M.; Kurtz, T.W. Pharmacogenetic evidence that cd36 is a key determinant of the metabolic effects of pioglitazone. *J. Biol. Chem.* **2002**, *277*, 48501–48507. [CrossRef]
29. Stephen Robert, J.M.; Peddha, M.S.; Srivastava, A.K. Effect of Silymarin and Quercetin in a Miniaturized Scaffold in Wistar Rats against Non-alcoholic Fatty Liver Disease. *ACS Omega* **2021**, *6*, 20735–20745. [CrossRef]

30. Roxo, D.F.; Arcaro, C.A.; Gutierres, V.O.; Costa, M.C.; Oliveira, J.O.; Lima, T.F.O.; Assis, R.P.; Brunetti, I.L.; Baviera, A.M. Curcumin combined with metformin decreases glycemia and dyslipidemia, and increases paraoxonase activity in diabetic rats. *Diabetol. Metab. Syndr.* **2019**, *11*, 33. [CrossRef]

31. Rahimi-Madiseh, M.; Heidarian, E.; Kheiri, S.; Rafieian-Kopaei, M. Effect of hydroalcoholic Allium ampeloprasum extract on oxidative stress, diabetes mellitus and dyslipidemia in alloxan-induced diabetic rats. *Biomed. Pharmacother.* **2017**, *86*, 363–367. [CrossRef]

32. Mohammadi, H.; Manouchehri, H.; Changizi, R.; Bootorabi, F.; Khorramizadeh, M.R. Concurrent metformin and silibinin therapy in diabetes: Assessments in zebrafish (Danio rerio) animal model. *J. Diabetes Metab. Disord.* **2020**, *19*, 1233–1244. [CrossRef] [PubMed]

33. Poruba, M.; Anzenbacher, P.; Racova, Z.; Oliyarnyk, O.; Huttl, M.; Malinska, H.; Markova, I.; Gurska, S.; Kazdova, L.; Vecera, R. The effect of combined diet containing n-3 polyunsaturated fatty acids and silymarin on metabolic syndrome in rats. *Physiol. Res.* **2019**, *68*, S39–S50. [CrossRef]

34. Markova, I.; Malinska, H.; Huttl, M.; Miklankova, D.; Oliyarnyk, O.; Poruba, M.; Racova, Z.; Kazdova, L.; Vecera, R. The combination of atorvastatin with silymarin enhances hypolipidemic, antioxidant and anti-inflammatory effects in a rat model of metabolic syndrome. *Physiol. Res.* **2021**, *70*, 33–43. [CrossRef] [PubMed]

35. Lv, W.S.; Wen, J.P.; Li, L.; Sun, R.X.; Wang, J.; Xian, Y.X.; Cao, C.X.; Wang, Y.L.; Gao, Y.Y. The effect of metformin on food intake and its potential role in hypothalamic regulation in obese diabetic rats. *Brain Res.* **2012**, *1444*, 11–19. [CrossRef]

36. Obradovic, M.; Sudar-Milovanovic, E.; Soskic, S.; Essack, M.; Arya, S.; Stewart, A.J.; Gojobori, T.; Isenovic, E.R. Leptin and Obesity: Role and Clinical Implication. *Front. Endocrinol.* **2021**, *12*, 585887. [CrossRef]

37. Antunes, L.C.; Elkfury, J.L.; Jornada, M.N.; Foletto, K.C.; Bertoluci, M.C. Validation of HOMA-IR in a model of insulin-resistance induced by a high-fat diet in Wistar rats. *Arch. Endocrinol. Metab.* **2016**, *60*, 138–142. [CrossRef] [PubMed]

38. Yao, J.; Zhi, M.; Gao, X.; Hu, P.; Li, C.; Yang, X. Effect and the probable mechanisms of silibinin in regulating insulin resistance in the liver of rats with non-alcoholic fatty liver. *Braz. J. Med. Biol. Res.* **2013**, *46*, 270–277. [CrossRef]

39. Zhang, Y.; Hai, J.; Cao, M.; Zhang, Y.; Pei, S.; Wang, J.; Zhang, Q. Silibinin ameliorates steatosis and insulin resistance during non-alcoholic fatty liver disease development partly through targeting IRS-1/PI3K/Akt pathway. *Int. Immunopharmacol.* **2013**, *17*, 714–720. [CrossRef]

40. Zhang, D.; Ma, Y.; Liu, J.; Deng, Y.; Zhou, B.; Wen, Y.; Li, M.; Wen, D.; Ying, Y.; Luo, S.; et al. Metformin Alleviates Hepatic Steatosis and Insulin Resistance in a Mouse Model of High-Fat Diet-Induced Nonalcoholic Fatty Liver Disease by Promoting Transcription Factor EB-Dependent Autophagy. *Front. Pharmacol.* **2021**, *12*, 689111. [CrossRef]

41. Meex, R.C.R.; Watt, M.J. Hepatokines: Linking nonalcoholic fatty liver disease and insulin resistance. *Nat. Rev. Endocrinol.* **2017**, *13*, 509–520. [CrossRef] [PubMed]

42. Rees, D.C.; Johnson, E.; Lewinson, O. ABC transporters: The power to change. *Nat. Rev. Mol. Cell. Biol.* **2009**, *10*, 218–227. [CrossRef] [PubMed]

43. Hu, W.Y.; Ma, X.H.; Zhou, W.Y.; Li, X.X.; Sun, T.T.; Sun, H. Preventive effect of Silibinin in combination with Pu-erh tea extract on non-alcoholic fatty liver disease in ob/ob mice. *Food Funct.* **2017**, *8*, 1105–1115. [CrossRef] [PubMed]

44. Piehler, A.P.; Haug, K.B.; Wenzel, J.J.; Kierulf, P.B.; Kaminski, W.E. ABCA-transporters: Regulators of cellular lipid transport. *Tidsskr. Nor. Laegeforen.* **2007**, *127*, 2930–2933. [PubMed]

45. Zhao, C.; Dahlman-Wright, K. Liver X receptor in cholesterol metabolism. *J. Endocrinol.* **2010**, *204*, 233–240. [CrossRef]

46. Divisova, J.; Kazdova, L.; Hubova, M.; Meschisvili, E. Relationship between insulin resistance and muscle triglyceride content in nonobese and obese experimental models of insulin resistance syndrome. *Ann. N. Y. Acad. Sci.* **2002**, *967*, 440–445. [CrossRef]

47. Buzzetti, E.; Pinzani, M.; Tsochatzis, E.A. The multiple-hit pathogenesis of non-alcoholic fatty liver disease (NAFLD). *Metabolism* **2016**, *65*, 1038–1048. [CrossRef]

48. Takahashi, Y.; Fukusato, T. Histopathology of nonalcoholic fatty liver disease/nonalcoholic steatohepatitis. *World J. Gastroenterol.* **2014**, *20*, 15539–15548. [CrossRef]

49. Yu, Z.W.; Li, D.; Ling, W.H.; Jin, T.R. Role of nuclear factor (erythroid-derived 2)-like 2 in metabolic homeostasis and insulin action: A novel opportunity for diabetes treatment? *World J. Diabetes* **2012**, *3*, 19–28. [CrossRef]

50. Liu, J.; Han, L.; Zhu, L.; Yu, Y. Free fatty acids, not triglycerides, are associated with non-alcoholic liver injury progression in high fat diet induced obese rats. *Lipids Health Dis.* **2016**, *15*, 27. [CrossRef]

51. Markova, I.; Miklankova, D.; Huttl, M.; Kacer, P.; Skibova, J.; Kucera, J.; Sedlacek, R.; Kacerova, T.; Kazdova, L.; Malinska, H. The Effect of Lipotoxicity on Renal Dysfunction in a Nonobese Rat Model of Metabolic Syndrome: A Urinary Proteomic Approach. *J. Diabetes Res.* **2019**, *2019*, 8712979. [CrossRef] [PubMed]

52. Roumans, K.H.M.; Lindeboom, L.; Veeraiah, P.; Remie, C.M.E.; Phielix, E.; Havekes, B.; Bruls, Y.M.H.; Brouwers, M.; Stahlman, M.; Alssema, M.; et al. Hepatic saturated fatty acid fraction is associated with de novo lipogenesis and hepatic insulin resistance. *Nat. Commun.* **2020**, *11*, 1891. [CrossRef] [PubMed]

53. Lee, J.J.; Lambert, J.E.; Hovhannisyan, Y.; Ramos-Roman, M.A.; Trombold, J.R.; Wagner, D.A.; Parks, E.J. Palmitoleic acid is elevated in fatty liver disease and reflects hepatic lipogenesis. *Am. J. Clin. Nutr.* **2015**, *101*, 34–43. [CrossRef]

54. Sztolsztener, K.; Chabowski, A.; Harasim-Symbor, E.; Bielawiec, P.; Konstantynowicz-Nowicka, K. Arachidonic Acid as an Early Indicator of Inflammation during Non-Alcoholic Fatty Liver Disease Development. *Biomolecules* **2020**, *10*, 1133. [CrossRef] [PubMed]

55. Capdevila, J.H.; Falck, J.R. The CYP P450 arachidonic acid monooxygenases: From cell signaling to blood pressure regulation. *Biochem. Biophys. Res. Commun.* **2001**, *285*, 571–576. [CrossRef] [PubMed]
56. Schuck, R.N.; Zha, W.; Edin, M.L.; Gruzdev, A.; Vendrov, K.C.; Miller, T.M.; Xu, Z.; Lih, F.B.; DeGraff, L.M.; Tomer, K.B.; et al. The cytochrome P450 epoxygenase pathway regulates the hepatic inflammatory response in fatty liver disease. *PLoS ONE* **2014**, *9*, e110162. [CrossRef]
57. Park, S.Y.; Kim, C.H.; Lee, J.Y.; Jeon, J.S.; Kim, M.J.; Chae, S.H.; Kim, H.C.; Oh, S.J.; Kim, S.K. Hepatic expression of cytochrome P450 in Zucker diabetic fatty rats. *Food Chem. Toxicol.* **2016**, *96*, 244–253. [CrossRef] [PubMed]
58. Albracht-Schulte, K.; Rosairo, S.; Ramalingam, L.; Wijetunge, S.; Ratnayake, R.; Kotakadeniya, H.; Dawson, J.A.; Kalupahana, N.S.; Moustaid-Moussa, N. Obesity, adipocyte hypertrophy, fasting glucose, and resistin are potential contributors to nonalcoholic fatty liver disease in South Asian women. *Diabetes Metab. Syndr. Obes.* **2019**, *12*, 863–872. [CrossRef]
59. Park, H.K.; Kwak, M.K.; Kim, H.J.; Ahima, R.S. Linking resistin, inflammation, and cardiometabolic diseases. *Korean J. Intern. Med.* **2017**, *32*, 239–247. [CrossRef]
60. Zhao, S.; Jiang, J.; Jing, Y.; Liu, W.; Yang, X.; Hou, X.; Gao, L.; Wei, L. The concentration of tumor necrosis factor-alpha determines its protective or damaging effect on liver injury by regulating Yap activity. *Cell Death Dis.* **2020**, *11*, 70. [CrossRef]
61. Zimmermann, E.; Anty, R.; Tordjman, J.; Verrijken, A.; Gual, P.; Tran, A.; Iannelli, A.; Gugenheim, J.; Bedossa, P.; Francque, S.; et al. C-reactive protein levels in relation to various features of non-alcoholic fatty liver disease among obese patients. *J. Hepatol.* **2011**, *55*, 660–665. [CrossRef] [PubMed]
62. Haukeland, J.W.; Dahl, T.B.; Yndestad, A.; Gladhaug, I.P.; Loberg, E.M.; Haaland, T.; Konopski, Z.; Wium, C.; Aasheim, E.T.; Johansen, O.E.; et al. Fetuin A in nonalcoholic fatty liver disease: In vivo and in vitro studies. *Eur. J. Endocrinol.* **2012**, *166*, 503–510. [CrossRef] [PubMed]
63. Sato, M.; Kamada, Y.; Takeda, Y.; Kida, S.; Ohara, Y.; Fujii, H.; Akita, M.; Mizutani, K.; Yoshida, Y.; Yamada, M.; et al. Fetuin-A negatively correlates with liver and vascular fibrosis in nonalcoholic fatty liver disease subjects. *Liver Int.* **2015**, *35*, 925–935. [CrossRef] [PubMed]
64. Pal, D.; Dasgupta, S.; Kundu, R.; Maitra, S.; Das, G.; Mukhopadhyay, S.; Ray, S.; Majumdar, S.S.; Bhattacharya, S. Fetuin-A acts as an endogenous ligand of TLR4 to promote lipid-induced insulin resistance. *Nat. Med.* **2012**, *18*, 1279–1285. [CrossRef]
65. Oguntibeju, O.O. Type 2 diabetes mellitus, oxidative stress and inflammation: Examining the links. *Int. J. Physiol. Pathophysiol. Pharmacol.* **2019**, *11*, 45–63.
66. Rives, C.; Fougerat, A.; Ellero-Simatos, S.; Loiseau, N.; Guillou, H.; Gamet-Payrastre, L.; Wahli, W. Oxidative Stress in NAFLD: Role of Nutrients and Food Contaminants. *Biomolecules* **2020**, *10*, 1702. [CrossRef]
67. Swiderska, M.; Maciejczyk, M.; Zalewska, A.; Pogorzelska, J.; Flisiak, R.; Chabowski, A. Oxidative stress biomarkers in the serum and plasma of patients with non-alcoholic fatty liver disease (NAFLD). Can plasma AGE be a marker of NAFLD? Oxidative stress biomarkers in NAFLD patients. *Free Radic Res.* **2019**, *53*, 841–850. [CrossRef]
68. Gillessen, A.; Schmidt, H.H. Silymarin as Supportive Treatment in Liver Diseases: A Narrative Review. *Adv. Ther.* **2020**, *37*, 1279–1301. [CrossRef]
69. Wang, Y.; Nakajima, T.; Gonzalez, F.J.; Tanaka, N. PPARs as Metabolic Regulators in the Liver: Lessons from Liver-Specific PPAR-Null Mice. *Int. J. Mol. Sci.* **2020**, *21*, 2061. [CrossRef]
70. Sampath, S.; Karundevi, B. Effect of troxerutin on insulin signaling molecules in the gastrocnemius muscle of high fat and sucrose-induced type-2 diabetic adult male rat. *Mol. Cell. Biochem.* **2014**, *395*, 11–27. [CrossRef]
71. Zhu, X.; Yan, H.; Xia, M.; Chang, X.; Xu, X.; Wang, L.; Sun, X.; Lu, Y.; Bian, H.; Li, X.; et al. Metformin attenuates triglyceride accumulation in HepG2 cells through decreasing stearyl-coenzyme A desaturase 1 expression. *Lipids Health Dis.* **2018**, *17*, 114. [CrossRef] [PubMed]
72. Xiao, P.; Yang, Z.; Sun, J.; Tian, J.; Chang, Z.; Li, X.; Zhang, B.; Ye, Y.; Ji, H.; Yu, E.; et al. Silymarin inhibits adipogenesis in the adipocytes in grass carp Ctenopharyngodon idellus in vitro and in vivo. *Fish Physiol. Biochem.* **2017**, *43*, 1487–1500. [CrossRef] [PubMed]
73. El Kasmi, K.C.; Anderson, A.L.; Devereaux, M.W.; Balasubramaniyan, N.; Suchy, F.J.; Orlicky, D.J.; Shearn, C.T.; Sokol, R.J. Interrupting tumor necrosis factor-alpha signaling prevents parenteral nutrition-associated cholestasis in mice. *J. Parenter. Enteral Nutr.* **2021**. [CrossRef]
74. Tewari, D.; Nabavi, S.F.; Nabavi, S.M.; Sureda, A.; Farooqi, A.A.; Atanasov, A.G.; Vacca, R.A.; Sethi, G.; Bishayee, A. Targeting activator protein 1 signaling pathway by bioactive natural agents: Possible therapeutic strategy for cancer prevention and intervention. *Pharmacol. Res.* **2018**, *128*, 366–375. [CrossRef] [PubMed]
75. Fallah, M.; Davoodvandi, A.; Nikmanzar, S.; Aghili, S.; Mirazimi, S.M.A.; Aschner, M.; Rashidian, A.; Hamblin, M.R.; Chamanara, M.; Naghsh, N.; et al. Silymarin (milk thistle extract) as a therapeutic agent in gastrointestinal cancer. *Biomed. Pharmacother.* **2021**, *142*, 112024. [CrossRef]

pharmaceutics

Article

Endocrine and Metabolic Impact of Oral Ingestion of a Carob-Pod-Derived Natural-Syrup-Containing D-Pinitol: Potential Use as a Novel Sweetener in Diabetes

Juan A. Navarro [1,2], Juan Decara [1], Dina Medina-Vera [1,2,3,4], Ruben Tovar [1,2], Antonio J. Lopez-Gambero [1,4], Juan Suarez [1,5], Francisco Javier Pavón [1,3], Antonia Serrano [1], Marialuisa de Ceglia [1], Carlos Sanjuan [6], Yolanda Alfonso Baltasar [6], Elena Baixeras [7,*] and Fernando Rodríguez de Fonseca [1,*]

[1] Laboratorio de Medicina Regenerativa, Unidad de Gestión Clínica de Salud Mental, Instituto de Investigación Biomédica de Málaga-IBIMA, Hospital Regional Universitario de Málaga, 29010 Málaga, Spain; juan_naga@hotmail.es (J.A.N.); juandecara@uma.es (J.D.); dina.medina@ibima.eu (D.M.-V.); rubentovar7@hotmail.com (R.T.); antonio.lopez@ibima.eu (A.J.L.-G.); juan.suarez@uma.es (J.S.); javier.pavon@ibima.eu (F.J.P.); antonia.serrano@ibima.eu (A.S.); marialuisa.deceglia@ibima.eu (M.d.C.)
[2] Facultad de Medicina, Campus de Teatinos s/n, Universidad de Málaga, 29010 Málaga, Spain
[3] Unidad de Gestión del Corazón, Hospital Universitario Virgen de la Victoria, 29010 Málaga, Spain
[4] Facultad de Ciencias, Campus de Teatinos s/n, Universidad de Málaga, 29010 Málaga, Spain
[5] Departamento de Anatomía Humana, Medicina Legal e Historia de la Ciencia, Universidad de Málaga, 29010 Málaga, Spain
[6] Euronutra S.L. Calle Johannes Kepler, 3, 29590 Málaga, Spain; euronutra@euronutra.eu (C.S.); lab@euronutra.eu (Y.A.B.)
[7] Departamento de Bioquímica y Biología Molecular, Facultad de Medicina, Universidad de Málaga, 29010 Málaga, Spain
* Correspondence: ebaixeras@uma.es (E.B.); fernando.rodriguez@ibima.eu (F.R.d.F.); Tel.: +34-655373093 (E.B.); +34-669426548 (F.R.d.F.)

Citation: Navarro, J.A.; Decara, J.; Medina-Vera, D.; Tovar, R.; Lopez-Gambero, A.J.; Suarez, J.; Pavón, F.J.; Serrano, A.; de Ceglia, M.; Sanjuan, C.; et al. Endocrine and Metabolic Impact of Oral Ingestion of a Carob-Pod-Derived Natural-Syrup-Containing D-Pinitol: Potential Use as a Novel Sweetener in Diabetes. *Pharmaceutics* 2022, 14, 1594. https://doi.org/10.3390/pharmaceutics14081594

Academic Editors: Diana Marcela Aragon Novoa and Fátima Regina Mena Barreto Silva

Received: 25 June 2022
Accepted: 28 July 2022
Published: 30 July 2022

Publisher's Note: MDPI stays neutral with regard to jurisdictional claims in published maps and institutional affiliations.

Abstract: The widespread use of added sugars or non-nutritive sweeteners in processed foods is a challenge for addressing the therapeutics of obesity and diabetes. Both types of sweeteners generate health problems, and both are being blamed for multiple complications associated with these prevalent diseases. As an example, fructose is proven to contribute to obesity and liver steatosis, while non-nutritive sweeteners generate gut dysbiosis that complicates the metabolic control exerted by the liver. The present work explores an alternative approach for sweetening through the use of a simple carob-pod-derived syrup. This sweetener consists of a balanced mixture of fructose (47%) and glucose (45%), as sweetening sugars, and a functional natural ingredient (D-Pinitol) at a concentration (3%) capable of producing active metabolic effects. The administration of this syrup to healthy volunteers (50 g of total carbohydrates) resulted in less persistent glucose excursions, a lower insulin response to the hyperglycemia produced by its ingestion, and an enhanced glucagon/insulin ratio, compared to that observed after the ingestion of 50 g of glucose. Daily administration of the syrup to Wistar rats for 10 days lowered fat depots in the liver, reduced liver glycogen, promoted fat oxidation, and was devoid of toxic effects. In addition, this repeated administration of the syrup improved glucose handling after a glucose (2 g/kg) load. Overall, this alternative functional sweetener retains the natural palatability of a glucose/fructose syrup while displaying beneficial metabolic effects that might serve to protect against the progression towards complicated obesity, especially the development of liver steatosis.

Keywords: sweeteners; D-Pinitol; carob fruit; insulin; insulin resistance; liver steatosis; pituitary hormones

1. Introduction

One of the major challenges in the current fight against the epidemics of obesity and diabetes type 2 (diabesity) is the correct use of dietary sugars [1]. Additionally, the challenge stands not only in reverting the widespread excessive consumption of added sugars in processed foods, but also in achieving a correct use of sugars or non-caloric compounds as sweeteners [2,3]. As an example concerning added sugars, the use of soft drinks has increased in recent decades, paralleling the rise of metabolic diseases. The consumption of sugary drinks, even in young adults, ranges from 0.24 L/day in Europe to 0.45 L/day in the United States, with glucose or high-fructose corn syrup as the source of carbohydrate added [4,5]. However, beyond the excess of calories provided for these added sugars, the emergence of complicated obesity linked to this dietary pattern has opened the debate of how we can substitute added sugars to improve health [5,6]. Remarkably, the presence of high fructose in processed foods has been identified as an unnoticed source of additional metabolic harm, especially to the liver [7,8]. Intensive research in the last decade has revealed that the metabolic pathway of fructose affects, among others, uric acid production, de novo lipogenesis, gluconeogenesis pathways, intestinal barrier integrity, and the gut–liver axis. All these alterations favor the development of non-alcoholic fatty liver, obesity, dyslipidemia, and insulin resistance, and eventually enhance cardiovascular risk [7–10]. The severity of this situation has prompted the WHO to recommend a reduction in excessive fructose and glucose in the diet and the replacement of fructose with non-caloric substitutes as alternative sweeteners.

However, the use of these non-caloric sweeteners has also become a public health problem [1–4]. Their rapid widespread use has not been followed by a noticeable reduction in the diabesity problem. In fact, multiple studies have demonstrated that non-caloric sweeteners do not contribute either to an improvement in insulin resistance or to a reduction in body weight [11–13]. In addition, important safety issues have been raised, although most of them are surrounded by controversy. Thus, alternative non-caloric sweeteners have been blamed for affecting the intestinal microbiome and disrupting the intestinal barrier [14–16]; they have been found to produce alterations in the gut–brain axis, thus affecting appetite and insulin release modulation [3,17]; their consumption alongside pregnancy has been found to result in obesity and insulin resistance in the adult offspring [18,19]; and finally, they have been questioned because of their potential carcinogenic effects, although systematic reviews are inconclusive [20]. It is true that for each study demonstrating a negative problem derived from alternative sweeteners, there are studies claiming the opposite (for an extensive review see, for instance, [3]). However, after a careful analysis of the multiple preclinical and clinical studies published, we can conclude that these non-caloric sweeteners have a neutral effect on glycemic control and body weight homeostasis, they do produce the dysbiosis of gut microbiota, and they have potential transgenerational effects if taken alongside pregnancy (something proven in preclinical models, raising the need of further human research). From this perspective, they are not the solution for the diabesity epidemic [21].

Thus, the panorama on the use of sweeteners is shadowed by serious concerns on the use of high-fructose syrups as added sugars, and the lack of benefits derived from the use of non-caloric sweeteners. Additionally, this view is the one derived from taking the most conservative option: questioning the toxicity of these sweeteners, but placing a serious concern for their use in pregnancy and gut dysbiosis-related disorders. Thus, the challenge stands at the starting point: how do we control the use of added sugars while providing a safe, palatable, sweet flavor to foods? There are non-explored solutions that come from the observation that maximizing the negative contribution of certain natural nutrients is an incorrect strategy since these nutrients must be incorporated into the diet. Additionally, this strategy affects each macronutrient, from sugars to fats. As an example, the incorporation of saturated fatty acids into diets in obesity clinical trials has improved plasma lipid profiles, resulting in healthy outcomes [22] despite the extensive literature banning saturated fats. Moreover, we cannot forget that these natural nutrients have optimized detection systems in the body that elicit an adequate physiological response. As an example, only sugars, not non-caloric sweeteners, elicit an adequate gut–brain response to adapt behavior and metabolism [23,24]. Thus, we must reconsider the use of physiological mechanisms to achieve the goal of keeping a healthy metabolic balance while preserving palatable properties. Let us translate this idea to the sugar/sweetener dilemma: the use of a monosaccharide-equilibrated food as a sweetener can be healthy if we are able to help the body avoid the transformation of these sugars into fat, leading to bodyweight gain and insulin resistance. This can be achieved by using functional foods that incorporate both an equilibrated composition of glucose and fructose plus a functional natural ingredient capable of improving carbohydrate metabolism.

In the present study, we test this hypothesis by analyzing, in humans and rodents, the physiological impact of the consumption of a natural syrup prepared from the pods of carob trees (*Ceratonia siliqua*). This sweetener contains glucose, fructose and D-Pinitol. This inositol is an insulin sensitizer capable of keeping glycaemia while avoiding both unnecessary insulin secretion and the conversion of carbohydrates into fat depots [25–28].

2. Materials and Methods

A. Human studies

2.1. Carob Syrup

A standardized carob syrup (InnoSweet®) manufactured by Euronutra SL (Málaga, Spain) was used in the experimental approaches designed for the present study. Both human healthy volunteers and laboratory Wistar rats were used. To prepare the syrup, physical process and chromatographic separation technologies without chemical modification were used. Natural raw materials (carob pods from *Ceratonia siliqua*) were milled and mixed with hot water. After filtration and chromatographic separation, a natural syrup containing sugars (including inositols) was obtained and adjusted to 70 Brix degrees (see Table 1 for composition). Considering the sugar fraction, the composition of the lot used for the present experiments was: 45.6% glucose, 47.3% fructose, 0.5% sucrose, and 3.2% D-Pinitol.

Table 1. Composition of the carob syrup (Innosweet) used in the study.

	Value	**Comments**
BRIX at 20 °C	69–71	ISO1743
DENSITY at 20 °C	1.35–1.36	g/cm^3
GLUCOSE	40	% dried matter
FRUCTOSE	45	% dried matter
SUCROSE	<5	% dried matter
D-PINITOL	>3	% dried matter
NON-SUGARS	<2.5	% dried matter

2.2. Study Design

The main aim of the study was first to compare the impact of an oral administration of a 50 g carbohydrate dose (using either glucose, carob syrup or a commercial agave syrup) on plasma glucose concentrations to obtain the glycaemic index of carob syrup with respect to glucose. A secondary aim was to analyse glucose homeostasis by monitoring the plasma concentrations of the endocrine hormones regulating it: insulin, glucagon, ghrelin, circulating free fatty acids, β-hydroxybutirate, and ketone concentrations. To achieve these aims, we designed a study consisting of a randomised trial where subjects were randomly assigned to one of the following treatments: (A) 8 subjects received 100 mL of a water solution containing 50 g of glucose; (B) 9 subjects received 50 g of carbohydrates from a natural carob-pod-derived syrup (InnoSweet®) containing 45.6% glucose, 47.3% fructose, 0.5% sucrose and 3.2% D-Pinitol (50 g of syrup contains a dose of D-pinitol of 1600 mg or 22.8 mg/kg body weight in a person weighing 70 kg), diluted with drinking water to a final volume of 100 ml; and finally (C) 6 subjects received 50 g of carbohydrates from a commercial agave syrup sold as a table sweetener (Mercadona, Spain, containing 77.2 g of carbohydrates/100 g, of which more than 75% was fructose). A third aim was to monitor the impact of carob syrup on metabolically active pituitary hormones (prolactin (PRL), growth hormone (GH) and thyroid-stimulating hormone (TSH)). In addition, taking into consideration that D-Pinitol can be converted to D-Chiroinositol that is capable of modifying the function of the hypothalamus–hypophisis–gonadal axis, we also monitored gonadotrophins (luteinizing hormone (LH) and follicle-stimulating hormone (FSH)). Concerning the sample size, we calculated it using the Gpower program, version 3.1.9.2, Heirich Heine University, Düsseldorf, Germany). The main variable of the study was the area under the curve (AUC) derived of plasma glucose excursions after the oral intake of either glucose solution, carob syrup, or agave syrup. Considering that, compared with the standard glucose solution used (50 g), carob syrup will have around 25 g (50%) and agave syrup around 20 g (1/5) of glucose content, effect sizes of 1.6 and 2 for glucose were considered for carob syrup and agave syrup, respectively. That gave a sample size of 10 subjects per group for comparing glucose versus carob syrup, and a size sample of 7 for comparing glucose versus agave syrup. That means a total of 27 subjects. However, after the recruitment period was closed, we only recruited 23 subjects that were allocated to the 3 groups in a proportional distribution with respect to the calculated sample size.

2.3. Human Volunteers

Twenty-three healthy volunteers were recruited among healthy clinical and laboratory staff of the Laboratory of Neuropsychopharmacology of the Regional University Hospital of Málaga. The inclusion criteria for all subjects were (a) age range of 18–65 years, (b) the presence of a baseline capillary blood glucose < 5.6 nmol/L (100 mg/mL) measured with a glucose oxidase method after overnight fasting, (c) the absence of obesity (BMI > 30) and the absence of a diagnosed/treated metabolic disease. Subjects were interviewed for a present or past diagnosis of diabetes (glucose > 7.8 mmol/L after a standard glucose load), hypertriglyceridemia or hypercholesterolemia under treatment, as well as for a clinical record of past endocrine disorders, including thyroid gland dysfunction or treatment with glucocorticoids. Based on the inclusion criteria, exclusion criteria were pregnancy or lactation, fasting glycaemia > 5.6 mmol/L (100 mg/dL), fasting insulinemia determined by ELISA > 25 mIU/L, diabetes, or using a medication known to interfere with carbohydrate metabolism. All the studies were performed under fasting conditions (overnight fasting for 12 h). Any of the women included used contraceptive medication. The main data regarding the male and female participants can be found in Table 2.

Table 2. Characteristics of human healthy volunteers participating in the present study.

	Males	**Females**
N	13	10
AGE (years)	37.7 + 12.2	34.3 + 11.6
WEIGHT (kg)	84.6 + 9.5	66.6 + 12.4
BODY MASS INDEX	27.2 + 3.7	23.6 + 4.0
BASAL GLUCOSE (mg/dl)	81.8 + 14.8	78.8 + 8.4
BASAL INSULIN (mIU/L)	3.5 + 1.4	2.95 + 1.8
HOMA-IR	0.66 + 0.28	0.58 + 0.36

Data are means + standard deviation.

2.4. Blood Sample Collection, Plasma Preparation and Glucose and Fructose Concentration Monitoring

After 12 h overnight fasting, a catheter was inserted into the antecubital vein of each subject, blood was extracted at baseline (minute 0; while still fasting) and immediately after (within 5 min) the oral intake of the assigned doses of either glucose, the natural carob syrup, or agave syrup. Then, blood samples were collected at 0, 15, 30, 45, 60, 90 and 120 min after intake. Individuals were allowed to eat 6 h after the last blood sample. All-time points were used for glucose and insulin concentrations. Time points in between $0'$ (basal) and $120'$ post intake were used for glucose homeostasis-related measures. Only $0'$, $60'$ and $120'$ time points were used for the analysis of pituitary hormones. All the blood samples were collected in Vacutest tubes (Vacutest Kima S.r.l., Arzergrande, Italy, cat. number: #13560) and centrifuged at $2000 \times g$ for 10 min at 4 °C; the plasma was kept at -80 °C for further biochemical analysis. Right after the blood was extracted from the subjects at each time point, glucose blood concentrations were measured with a commercially available glucometer (AccuCheck, Roche, Germany). For fructose measurements, 0.15 mL of plasma was deproteinized by mixing with an equal volume of ice-cold 20% perchloric acid. Samples were vortexed and centrifuged for 15 min at 2500 rpm in a microfuge. Supernatants were injected with HPLC coupled to a refractive index detector for fructose monitoring using an external standard calibration curve for the calculation of the concentrations. The HPLC system consisted of a µBondapak/carbohydrate column, a solvent system of acetonitrile water (75:25), and a flowrate of 1.8 mL/min.

2.5. Quantitation of Pinitol: Liquid Chromatography–Mass Spectrometry Method

MS method: A quantitation method for D-Pinitol in human plasma using liquid chromatography coupled to mass spectrometry (LC-MS/MS) was validated. Analysis was carried out using an Agilent 1290 (Agilent Technologies, Santa Clara, CA, USA) liquid chromatographer and an Api4000 triple-quadrupole mass spectrometer (SCIEX). Quantitation was obtained using the multiple-reaction monitoring (MRM) mode of the transitions at m/z 195.2→109.0 (quantitative) and 195.2/80.0 (qualitative) for D-Pinitol, both with collision energy to 20 Ev and m/z 261.1→205.1 for the internal standard (IS) (see Figure 1A,B). IS was provided by MEDINA.

Mass spectrometry conditions consisted of: decluster potential (DP): 10 eV; GS1 and GS2: 45 psi; T^a: 600 °C; and ion spray: 5500 Ev. Chromatographic conditions were the following: the mobile phase (MP) consisted of 0.1% formic acid$-$[water:AcN] [90:10] (MP A) and 0.1% formic acid$-$[AcN:water] [90:10] (MP B). The gradient program was applied as follows: t = 0$-$0.5 min 95% MP B; t = 4.50$-$6.70 min 25% MP B; t = 6.80 95% MP B; t = 6.80$-$9:00 min 95% MP B. The flow rate was 0.3 mL/min and the run time was 9 min. The injection volume was 5 µL. The chromatographic column used was × bridge BEH amide (waters) with dimensions of 2.1 × 100 mm and a particle size of 3.5 µm. The oven column was maintained at 30 °C. The method performance was validated according to the FDA recommendations proposed in 2018 for selectivity, sensitivity, matrix effect, linearity, precision, accuracy and the recovery of plasma samples [29].

Sample preparation: An aliquot of 50.0 µL plasma was taken and 2% TFA plus 150 µL of cold methanol containing the internal standard was added. After vortex mixing for one minute, the samples were centrifuged for 15 min at 13,300 rpm. The temperature of the centrifuge was set at 4 °C. An aliquot of 160.0 µL of the supernatant was transferred to a vial for evaporation. The samples were reconstituted in 100 µL water/acetonitrile (80/20) and 0.1% ammonia (20%) for LCMS analysis.

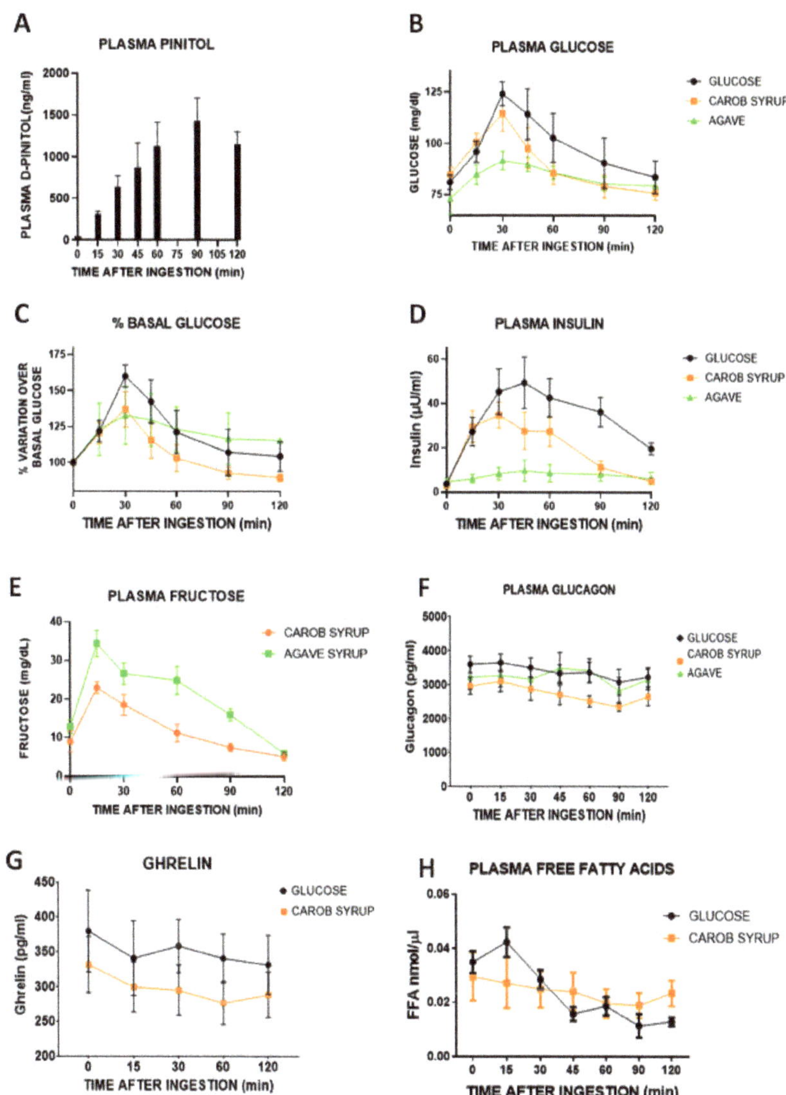

Figure 1. Acute effects of (1) a single oral dose of a glucose solution (50 g in 100 mL of water), (2) a natural carob-pod-derived syrup (Innosweet®, 50 g of carbohydrates in 100 mL water containing equal amounts of glucose and fructose and 1600 mg of D-Pinitol) or 50 g of carbohydrates in 100 mL of water from a commercial agave syrup (containing > 37.5 g of fructose). (**A**) Time-course of D-Pinitol plasma levels after carob syrup ingestion; (**B**) evolution of glucose concentrations after the ingestion; (**C**) percentage of variation of plasma glucose over basal levels previous to ingestion; (**D**) plasma insulin concentrations; (**E**) Insulin Resistance Index (homeostatic model) at different points after the ingestion; (**F**,**G**) evolution of plasma concentration of glucagon (**F**), total ghrelin (**G**) and free fatty acids. (**H**) Data are means ± standard error of the mean of 8 subjects for glucose, 9 subjects for carob syrup, and 6 subjects for agave sweetener.

2.6. Quantitation of Plasma Metabolites and Plasma Hormone Concentrations

Glucose homeostasis-related hormones: plasma levels of hormones regulating glucose homeostasis were determined by the human Enzyme-Linked Immunosorbent Assay (ELISA) method using commercial kits: insulin (EMD Millipore Corporation, Billerica, MA, USA, cat. number: #EZHI-14K), glucagon (Elabscience Biotechnology Inc, Wuhan, Hubei, China, cat. number: #E-EL-H2237), and active ghrelin (Kamiya Biomedical Company, Seattle, WA, USA, cat. number: #KT-364). All samples were assayed in duplicate within one assay, and results were expressed in terms of the particular standard metabolite.

For calculating glycaemic index, area under the curve (AUC) of % of plasma glucose variation over basal levels versus time after ingestion (Graph 1C) was obtained for each experimental subject receiving either glucose, carob syrup or agave syrup. The mean AUC for the glucose group was normalized to 100 and the glycaemic index is expressed of carob and agave syrups obtained as the % over the glucose AUC.

Free fatty acid plasma concentration was measured using a commercial kit (Sigma-Aldrich, Saint Louis, MO, USA, cat. number: MAK044) according to the manufacturer's instructions. All samples were assayed in duplicate within one assay, and results are expressed in terms of the standards provided by the kit.

Insulin resistance was evaluated according to the Homeostatic Model Assessment for Insulin Resistance (HOMA-IR index), calculated using the following formula: HOMA-IR= ((fasting plasma insulin [uIU/mL] $\times$ fasting blood glucose [mmol/L])/22.5) [30].

The plasma levels of beta-hydroxybutyrate were determined using a commercial kit (Sigma Aldrich, Saint Louis, MO, USA, cat. number: MAK041) according to the manufacturer's instructions; results are expressed in terms of the standards provided by the kit.

Plasma levels of β-Ketone were determined using a commercially available meter (GlucoMen® areo 2K, A. Menarini Diagnostics, Badalona, Barcelona, Spain) and the corresponding test strips (GlucoMen® areo β-Ketone Sensor, A. Menarini Diagnostics, Badalona, Barcelona, Spain) according to the manufacturer's instructions. Results are expressed in mmol/L.

Plasma levels of pituitary hormones: follicle-stimulating hormone (FSH, expressed in mU/mL), human growth hormone-1 (GH, expressed in ng/mL), luteinizing hormone (LH, expressed in mU/mL), prolactin (PRL, expressed in ng/mL), and thyroid stimulating hormone (TSH, expressed in µU/mL) were determined by the multiplex immunoassay system using a commercial kit: Bio-Plex Pro™ RBM Human Hormone Panel 1 (Bio-Rad, Hercules, CA, USA, cat. number: #171AHR1CK). The plate was run on a Bio-Plex MAGPIX™ Multiplex Reader with Bio-Plex Manager™ MP Software (Luminex, Austin, TX, USA), as described in [31]. Hormone detection limits were: 0.11 mU/mL (FSH), 0.0076 ng/mL (GH), 0.061 mU/mL (LH), 0.022 ng/mL (PRL) and 0.012 µU/mL (TSH).

B. Preclinical studies in rats

2.7. Statement

Animal experimental procedures were carried out in accordance with the European Communities directive 2010/63/EU and Spanish legislation (Real Decreto 53/2013, BOE 34/11370–11421, 2013) and approved by the Bioethics Committee for Animal Experiments of the University of Malaga, Spain, in accordance with the ARRIVE guidelines [32]. Accordingly, all efforts were made to minimize animal suffering and to reduce the number of animals used.

2.8. Animals

The experiments were performed with 4-to-5-month-old male Wistar rats (Crl:WI(Han)) weighing 400 ± 20 g (Charles River Laboratories, Barcelona, Spain). The animals were kept under a standard condition (light regimen of 12/12 h, day/night) and under temperature and humidity control. The rats were fed on a standard pellet diet (STD) (3.02 Kcal/g with 30 Kcal% protein, 55 Kcal% carbohydrates and 15 Kcal% fat) and were purchased from

Harlam (Tecklad, Madison, WI, USA). Water and food were available ad libitum. Animals were anaesthetized with intraperitoneal (ip) sodium pentobarbital (50 mg/Kg body weight) before being sacrificed by decapitation.

2.9. Drug Preparation and Experimental Design

Carob syrup (Innosweet)® and pure D-Pinitol were generously provided by Euronutra SL. For acute oral administration experiments, they were dissolved in drinking water to an equivalent dose of 2 g glucose/kg body weight/rat for carob syrup, and 100 mg/kg/day of D-Pinitol. For subchronic administration in drinking water, we considered that a rat drinks 12 ml water/100 g body weight/day. Thus, we prepared a water solution containing 10 mg of D-Pinitol/12 ml of drinking water (37.2 g of syrup/L), that would allow us to administer 100 mg/kg of D-Pinitol daily. Acute glucose tolerance experiment: Either glucose (2 g/kg) or carob syrup (equivalent to 2 g/kg glucose) were administered by gavage to 12-hour-fasting rats. Plasma glucose was monitored with a commercially available glucometer (AccuCheck, Roche, Germany) at 0, 5, 10, 15, 30, 45, 60 and 120 min post-administration of the glucose/syrup solutions. Effects of the chronic ingestion of carob syrup: Two groups of male Wistar rats (n = 10) were used. One drank a carob syrup solution (adjusted to a daily dose of 100 mg/kg body weight of D-Pinitol), whereas the second drank tap water. After 8 days, the animals were tested for a glucose tolerance test: Rats were food-deprived for 18 h and given an intraperitoneal injection of 2 g D-glucose/Kg; blood samples were collected from the tail vein at 0 (basal level), 5, 10, 15, 30, 45, 60 and 120 min after injection, and blood glucose concentrations were measured with a commercially available glucometer (AccuCheck, Roche, Germany). Upon completion, the animals resumed drinking and were sacrificed on the morning of the 10th day, not being food-restricted the night before. A third group of rats received only 10 mgkg bw of D-Pinitol for drinking water. This group was designed to be a control for the potential D-Pinitol actions in the liver. They were also sacrificed after 10 days of D-Pinitol drinking.

2.10. Sample Collection

Immediately after animal sacrifice, blood, liver and brain samples were collected. Blood was centrifuged (2100 g for 8 min, 4 °C) and the plasma was kept at −80 °C for biochemical analysis. Liver and brain samples were flash-frozen in liquid nitrogen, then stored at −80 °C until analysis.

2.11. Measurement of Metabolites and Hormones in Plasma and Liver

The following plasma metabolites were measured using commercial kits according to the manufacturer's instructions and a Hitachi 737 Automatic Analyser (Hitachi Ltd., Tokyo, Japan): glucose, uric acid, creatinine, bilirubin, triglycerides, and the hepatic enzymes glutamic oxaloacetic transaminase (GOT) and alanine aminotransferase (ALT). The plasma levels of glucose homeostasis-associated hormones were determined by the Enzyme-Linked Immunosorbent Assay (ELISA) method using commercial kits: leptin, insulin and ghrelin ELISA kits (EMD Millipore Corporation, Billerica, MA, USA, cat. number: #EZRADP-62K, #EZRMI-13K and #EZRGRT-91K, respectively). All plasma samples were assayed in duplicate within one assay, and results were expressed in terms of the particular standard hormone. In order to monitor the presence of oxidative stress, the lipid peroxidation was evaluated measuring thiobarbituric acid reactive substance (TBARS). Using a specific kit (Cell Biolabs Inc., San Diego, CA, USA, cat. number: STA-330) according to the manufacturer's instructions, malondialdehyde (MDA) was measured, which is a split product of an endoperoxide of unsaturated fatty acids resulting from the oxidation of lipid substrates. Plasma levels of beta-hydroxybutyrate were determined using a commercial kit (Sigma Aldrich, Saint Louis, MO, USA, cat. Number: MAK041) and liver glycogen levels were measured by a commercial glycogen assay kit (Sigma-Aldrich, Saint Louis, MO, USA, cat. number: MAK016), both according to the manufacturer's instructions, and results were expressed in terms of the standards provided by the kits.

Fat extraction from liver tissue was performed as previously described [33]. Briefly, according to the (modified) method of Bligh and Dyer [34], total lipids were extracted from frozen liver samples using chloroform–methanol (2:1, *v/v*) and butylated hydroxytoluene (0.025%, *w/v*). Then, two centrifugation steps (2800 g for 10 min, 4 °C) were carried out and the lower phase, containing lipids, was extracted. Each sample was dried using nitrogen and the liver fat content was calculated by subtracting the weight of the empty tube from the weight of the same tube containing the liver lipids. The data are expressed as a percentage of tissue weight.

2.12. Protein Extraction and Western Blot Analysis

Total protein from liver and hypothalamus (17 mg average weight) was extracted using ice-cold cell lysis buffer for 30 min, as previously described [35]. Fifty micrograms of protein were resolved on 4–12% (Bis-Tris) Criterion XT Precast Gels (Bio-Rad, Hercules, CA, USA, cat. number: 3450124) and electroblotted onto nitrocellulose membranes (Bio-Rad, Hercules, CA, USA, cat. number: 1620115). For specific protein detection the membrane was incubated overnight in TBS-T containing 2% BSA and the corresponding primary antibody. Antibodies against GS (#3886), phospho-GS (Ser641) (#3891), GSK-3β (#12456), phospho-GSK-3β (Ser9) (#5558), GSK-3α (#4337), phospho-GSK-3α (Ser21) (#9337), Akt (#9272), phospho-Akt (Ser473) (#9271), mTOR (#2972) and phospho-mTOR (Ser2448) (#2971) were purchased from Cell Signalling Technology Inc. (Danvers, MA, USA); Adaptin (#ab151720) from Abcam (Cambridge, UK); and phospho-GSK-3α/β (Tyr279/Tyr216) (#15648) from Merck Millipore (Burlington, MA, USA). Primary antibodies were detected using anti-rabbit or an anti-mouse HRP-conjugated secondary antibody as appropriate (Promega, Madison, WI, USA, cat. number: W4011 and W4021, respectively). Specific proteins were revealed using the ECL™ Prime Western Blotting System (GE Healthcare, Chicago, IL, USA, cat. number: RPN 2236), in accordance with the manufacturer's instructions. Images were visualized in the ChemiDoc MP Imaging System (Bio-Rad, Hercules, CA, USA). After measuring phosphorylation proteins, the specific antibodies were removed from the membrane by incubation with a stripping buffer (2% SDS, 62.5 mM Tris HCL pH 6.8, 0.8% β-mercaptoethanol) for 30 min at 50 °C. Membranes were extensively washed in ultrapure water and then re-incubated with the corresponding antibody specific for total protein. The quantification of results was performed using ImageJ software (http://imagej.nih.gov/ij (accessed on 4 February 2020), Bethesda, MD, USA). The specific signal level for total proteins was normalized to the signal level of the corresponding Adaptin band of each sample in the same blot. The phosphorylation stage of a protein was expressed as the ratio of the signal obtained with the phospho-specific antibody relative to the appropriate total protein antibody. The amount of the protein of interest in the control samples was arbitrarily set as 1.

2.13. Real-Time qPCR

RNA isolation and cDNA synthesis:

Rat liver sections (50–80 mg) were homogenized on ice and total RNA was extracted from tissue using the Trizol® method according to the manufacturer's instructions (ThermoFisher Scientific, Waltham, MA, USA, EE.UU.). RNA samples were isolated with the RNAeasy MinElute cleanup-kit including digestion with the DNase I column (Qiagen, Hilden, Germany) according to the manufacturers' instructions and quantified using a spectrophotometer Nanodrop TM ND-1000 (Thermo Fisher Scientific, Waltham, MA, USA, EE.UU.) to ensure A260/280 ratios of 1.8 to 2.0. Reverse transcription was carried out from 1 µg of mRNA in a reaction volume of 20 µL total using the qScript XLT cDNA SuperMix (Quantabio, Beverly, MA, USA).

Real-time qPCR and Gene Expression Analysis:

Real-time qPCR reactions were carried out in a CFX96TM Real-Time PCR Detection System (Bio-Rad, Hercules, CA, USA) for each cDNA template, and amplified in 10 µL reaction volume containing 4.5 µL of cDNA (previously diluted 1/100, that is, a total amount of

2.25 ng of cDNA per reaction) and 5.5 µL of PerfeCTa qPCR ToughMix (Quantabio, Beverly, MA, USA, EE.UU.) containing the corresponding primer. The gene-specific primers for the target rat genes: *Fbp1* (fructose 1,6 bisphosphatase 1), *G6pc* (glucose-6-phosphatase catalytic subunit), *Pc* (pyruvate carboxylase), *Pck1* (phosphoenolpyruvate carboxykinase 1), *Pklr* (pyruvate kinase liver/RBC), *Fasn* (fatty acid synthase), *Acox1* (acyl-CoA oxidase 1), *Acaca* (acetyl-CoA carboxylase alpha), *Cox4i1* (cytochrome c oxidase subunit IV isoform 1), *Cox4i2* (cytochrome c oxidase subunit IV isoform 2), *Scd1* (stearoyl-coenzyme A desaturase 1), *Cpt1a* (carnitine palmitoyltransferase 1a), *Actb* (beta actin), are shown in Supplementary Table S1. All primers were obtained based on TaqMan® Gene Expression Assays and the FAM™ dye label format (ThermoFisher Scientific, Waltham, MA, USA, EE.UU.). Each reaction was run in duplicate. Cycling parameters were 50 °C for 2 min to deactivate single- and double-stranded DNA containing dUTPs, 95 °C for 10 min to activate Taq DNA polymerase followed by 44 cycles at 95 °C for 15 s for cDNA melting, and 60 °C for 1 min to allow for annealing and the extension of the primers, during which fluorescence was acquired. We found that a single product was amplified using a melting curve. For the relative quantification the mean of duplicates was used. The expression of *Actb* gene was unaffected during all experimental treatments. The *Actb* gene was chosen as reference gene and all results are normalized with respect to the water group.

2.14. Statistical Analysis

Graph-Pad Prism 8.0 software (GraphPad Software, Inc., San Diego, CA, USA) was used to analyze the data. Values are represented as mean ± standard error of the mean (SEM) for each experimental group, according to the assay. The significance of differences within and between groups was evaluated by a one-way or two-way (depending on the assay) analysis of variance (ANOVA) followed by post hoc test for multiple comparisons. In case data were not distributed normally, the Kruskal–Wallis rank test was performed, as indicated. A p-value ≤ 0.05 was considered to be statistically significant. (* = $p < 0.05$; ** = $p < 0.01$; *** = $p < 0.001$).

3. Results

3.1. Effects of Acute Administration of Carob Syrup on Plasma Levels of D-Pinitol in Humans

The oral administration of 50 gr of carob syrup, containing 1600 mg of D-Pinitol, resulted in a progressive increase in plasma D-Pinitol levels that peaked 60–90 min after the administration (F(5,54) = 3.24, $p < 0.01$) and remained stable for at least one hour. Detection was significant as early as 30 min after the ingestion, suggesting a rapid incorporation of D-Pinitol to the blood stream, and reflecting that the liver passage was not generating a total clearance of the compound once incorporated to the portal circulation (Figure 1A). We performed a correlation analysis of peak concentrations of D-Pinitol versus age and BMI and an analysis of gender differences. D-Pinitol correlates positively with age ($r^2 = 0.45$, p (two-tailed) = 0.047). There were no correlations with D-Pinitol with BMI ($r^2 = 0.44$, p (two-tailed) = 0.05). Mean peak plasma D-Pinitol values were similar in females (1527 ± 589 ng/mL) compared to male subjects (1562 ± 178 ng/mL).

3.2. Comparative Glucose Handling and Insulin and Glucagon Secretion Observed after the Ingestion of Glucose, Carob Syrup or an Agave Syrup Sweetener, and Calculation of the Glycemic Index of Carob Syrup

The oral administration of 50 g glucose generates a rapid excursion of plasma glucose levels that peaked at 30 min after administration, returning to basal levels only at the end of the 120 min test. The administration of 50 g of carbohydrates in the carob syrup also generated glucose excursions that were similar to peak values at 30 min after ingestion, but recovered much more rapidly, returning to normality within the first hour after ingestion. Agave syrup intake produced a lower increase in glycaemia, but this was sustained for the 120 min of the study. A comparative analysis of the area under the curve (AUC) revealed that individuals taking glucose stayed under hyperglycaemia for longer compared to those receiving either carob syrup or agave syrup (F(2,19) = 9.6, $p < 0.002$). Interestingly, despite the peak of glucose observed at 30 min, the AUC of agave and carob syrup was found to be equal, because of the rapid return to basal levels observed in the carob syrup group (Figure 1B). This was confirmed by analyzing the % of change over basal glucose values, a better index for the physiological response to glucose excursions. Again, AUC revealed that subjects receiving glucose had greater long-lasting variations from basal levels of glucose (F(2,19) = 6.2, $p < 0.009$), with carob syrup being the most efficient sweetener to return to basal glucose levels (Figure 1C). We used the AUC of glucose normalized to 100 to calculate the glycemic index of carob and agave syrups. Carob syrup had a glycemic index of 73.4 ± 6.9, whereas agave syrup had a glycemic index of 76.2 ± 3.3. These results indicate that despite having a double amount of glucose compared to agave syrup, the presence of D-pinitol helps to rapidly retire the excess of glucose, resulting in an equal glycemic index. As expected, the glucose excursions caused a rapid insulin response in subjects receiving glucose or carob syrup, but not fructose-based agave sweetener (Figure 1D). AUC analysis again revealed that the insulin response was shorter and less intense in carob syrup-receiving subjects (F(2,19) = 47.4, $p < 0.001$). Concerning plasma fructose, as expected, agave syrup generated a greater and more prolonged rise in plasma fructose levels than carob syrup, which immediately reduced the rise in fructose observed 15 and 10 min after the intake of the syrup (F(1,73) = 29.06, $p < 0.0001$). Area under the curve was clearly different when carob syrup and agave AUC were compared (AUC carob syrup 1471 ± 232; AUC agave 2534 ± 249, t = 8.44, df = 13, $p < 0.001$) (Figure 1E). Considering the glucagon response, the three groups had a similar profile (Figure 1F). However, the glucagon/insulin ratio was higher in the carob syrup group with respect to the glucose group (time $\times$ treatment interaction (F(4,60) = 4.16, $p < 0.005$), revealing a lesser need for insulin and the more active role of glucagon in the carob-syrup-treated individuals (Supplementary Figure S1).

3.3. Effects of the Ingestion of Glucose and Carob Syrup on the Plasma Concentrations of Ghrelin, Free Fatty Acids, β-Hydroxybutirate and Pituitary Hormones

Plasma ghrelin (Figure 1G) concentration was not affected by either glucose or carob syrup. Free fatty acid plasma concentration (Figure 1H) decreased in subjects receiving glucose, an effect that appeared 45 min after the ingestion of this monosaccharide, and lasted up to 120 min (F(6,28) = 8.98 $p < 0.001$)). This effect was not observed in the individuals receiving carob syrup, where the free fatty acid concentrations remained stable (F(6,31) = 0.28 $p = 0.94$, non-significant). Plasma concentrations of β-hydroxybutirate (Supplementary Figure S2) were found to decrease with time equally in both glucose- and carob-syrup-treated individuals (F(2,42) = 10.8 $p < 0.001$). However, there were no differences in between both treatments (F(1,42) = 1.7 $p = 0.2$, non-significant). Subjects did not exhibit elevated levels of plasma ketones despite being fasting, and neither glucose nor carob syrup enhanced the concentration of ketones (Supplementary Figure S2). Finally, none of the pituitary hormones analyzed (prolactin, thyroid-stimulating hormone, and the gonadotropins LH and FSH) were affected by the treatment with carob syrup (Supplementary Figure S3). A sex difference was observed for LH (F(1,21) = 4.62, $p < 0.05$) and FSH (F(1,21) = 4.94, $p < 0.05$).

3.4. Effects of the Acute and Repeated Administration of Carob Syrup on Glucose Handling in Rats

In order to further explore the functional properties of the carob pod syrup, we tested its effects of glucose handling in male Wistar rats. First, we tested if the presence of D-Pinitol affected glucose excursions in a differential way to that observed for glucose alone. Figure 2A shows that the administration of carob syrup at a dose containing 2 g/kg glucose (plus 2.1 g/kg of fructose and 130 mg/kg D-Pinitol) resulted in less intense hyperglycemia (F(1,144) = 69.1 $p < 0.001$). Area under the curve revealed that the time spent under hyperglycemia was less than that spent by the animals receiving the same amount of glucose without D-Pinitol (t = 7.81, df = 18, $p < 0.001$). These results are similar to those observed in humans. Next, we tested if the repeated administration of the carob syrup (at a dose of 100mg/kg/day of D-Pinitol for 10 days) was sufficient to improve glucose handling, after an oral load with 2 g/kg glucose. Figure 2B demonstrates that the intake of this carob syrup reduced the intensity of hyperglycemia when compared with animals fed with the standard chow ($F_{(1,144)}$ = 12.1 $p < 0.001$)). Again, area under the curve analysis showed that the time spent under hyperglycemia in animals fed with the carob syrup was less than the time spent by animals fed with standard chow and water (t = 3.35, df = 18, $p < 0.003$).

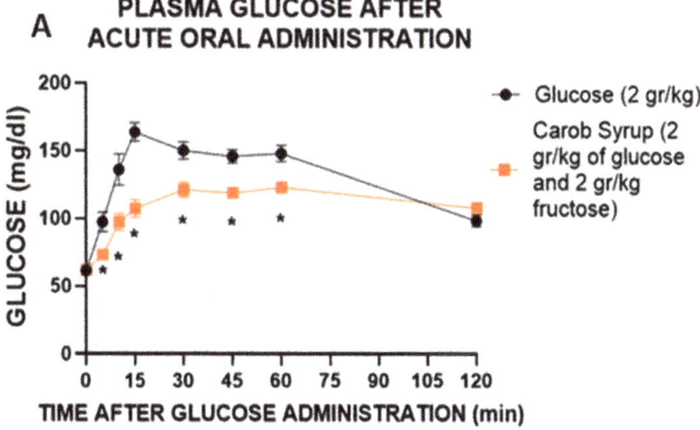

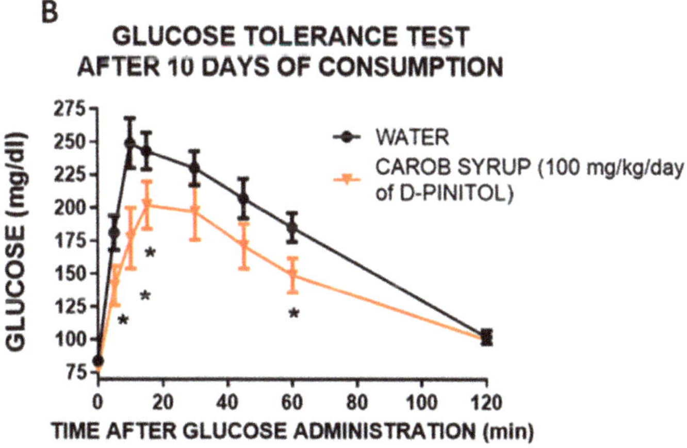

Figure 2. (**A**) Acute effects of (1) a single oral dose of a glucose solution (2 g/kg) or (2) a single oral dose of a carob syrup containing glucose (2 g/kg), fructose (2.1 g/kg) or D-Pinitol (130 mg/kg), on plasma glucose levels in adult male Wistar rats, measured 120 min after administration. (**B**) Time-curse of plasma glucose levels after an oral glucose load (2 g/kg) in both control animals and animals drinking water mixed with carob syrup (up to a 100 mg/kg D-Pinitol per day of study) for 10 days. In both cases, the glycaemic response was better in animals receiving the carob syrup. Data are means ± standard error of the mean for at least 8 subjects per treatment group (* $p < 0.05$ carob syrup vs. control (water) group).

3.5. Metabolic Effects of Repeated Administration of a Carob Syrup in Rats

As described above, excessive glucose consumption has been thought to increase de novo lipogenesis, resulting in liver fat depots while promoting uric acid production and boosting gluconeogenesis, favoring hyperglycemia. In order to control if the consumption of carob syrup results in alterations in both plasma and liver biochemistry, several analytical controls were performed and displayed in Table 3. Carob syrup consumption did not alter the major indexes of liver or kidney function (normal levels of AST, uric acid, bilirubin, creatinine or the oxidative stress indicative compound malonyl dialdehyde). Only plasma urea concentration was increased ($t = 2.98$, df $= 18$, $p < 0.01$), suggesting enhanced urea production by the liver probably derived of the described actions of D-Pinitol as a promoter of glucagon effect, and not derived from kidney functioning, since creatinine clearance was preserved. Neither insulin, glucagon nor leptin hormones involved in glucose and appetite homeostasis were affected by carob pod consumption. However, ghrelin concentrations, a hormone found to be increased after the acute administration of D-Pinitol, were higher in animals receiving carob syrup through drinking water ($t = 3.43$, df $= 18$, $p < 0.01$). The liver content of fat was decreased after carob syrup consumption ($t = 9.6$, df $= 18$, $p < 0.001$). This effect was not linked either to the export of triglycerides or to the production of b-hidroxybutyrate, which were found not to be affected by carob syrup consumption. Finally, liver glycogen contents were found to be reduced in animals consuming carob syrup) ($t = 5.65$, df $= 18$, $p < 0.01$). The effects on urea production and liver fat content were replicated when a dose of 100 mg/kg of D-Pinitol alone was given to the animals through drinking water for 10 days (Supplementary Table S2).

Table 3. Plasma and liver biochemistry parameters after 10 days of drinking **water** or water-diluted carob syrup (equivalent to 100 mg/kg b.w./day of D-Pinitol).

	Water	Carob Syrup
N	10	10
Glucose (mg/dL)	247.0 ± 63.8	269.4 ± 46.3
CCreatinin (mg/dL)	0.57 ± 0.38	0.73 ± 0.15
Urea (mg/dL)	21.6 ± 3.2	**40.7$\pm$ 5.9 (*)**
Bilirubin (mg/dL)	0.10 ± 0.09	0.13 ± 0.08
Uric Acid (mg/dL)	1.67 ± 0.29	1.90 ± 0.30
Triglycerides (mg/dL)	146.9 ± 30.1	156.3 ± 31.24
β-Hydroxy butirate (mg/dL)	1005 ± 86	1000 ± 74
AST (U/L)	152.6 ± 46.5	195.3 ± 97.4
Insulin (ng/mL))	14.9 ± 1.9	15.0 ± 1.6
Glucagon/Insulin ratio	26.6 ± 4.9	35.6 ± 9.4
Leptin (ng/mL)	14.8 ± 4.6	15.3 ± 4.3
Ghrelin (ng/mL)	0.42 ± 0.13	**0.67$\pm$ 0.19 (*)**
TBARS (Malonyl dialdehyde, μM)	11.1 ± 2.7	11.6 ± 3.8
Total Fat in Liver (mg/g)	40.8 ± 1.3	**34.8$\pm$1.5 (*)**
Liver Glycogen (μg/g)	137.7 ± 34.1	**76.1$\pm$ 5.2 (*)**

Data are means $\pm$ standard deviation. (*) indicates $p < 0.05$, ANOVA or Kruskal–Wallis test. AST (aspartate aminotransferase), TBARS (tiobarbituric acid-reactive species).

3.6. Effects of Repeated Administration of a Carob Syrup on the Liver Expression of Key Enzymes for Neoglucogenic and Lipid Metabolism Pathways in Rat Liver

Because of the reduction in liver content for both fat and glycogen, we examined by real-time PCR the expression of the key enzymes of gluconeogenesis, de novo lipid synthesis, and fatty acid oxidation pathways. Figure 3 illustrates how carob syrup consumption increases the expression of pyruvate kinase ($t = 2.2$ df $= 18$, $p < 0.05$), phosphoenolpyruvate carboxykinase ($t = 3.39$, df $= 18$, $p < 0.01$), and the catalytic subunit of the glucose-6 phosphatase ($t = 2.77$, df $= 18$, $p < 0.05$), suggestive of an increased activity of neoglucogenic pathways. Figure 4 depicts that the gene expression of any of the enzymes promoting the de novo synthesis of lipids were affected by the consumption of carob syrup. However, a clear increase in the gene expression of relevant enzymes related to fatty acid oxidation was detected after the administration of this natural sweetener. These enzymes were acyl-coenzymeA oxydase (Mann–Whitney U = 21.5, $p < 0.03$), carnitine palmitoyltransferse 1A (t = 2.2, df = 18, $p < 0.05$) and cytochrome C oxydase isoform 4 (Mann–Whitney U = 20, $p < 0.05$).

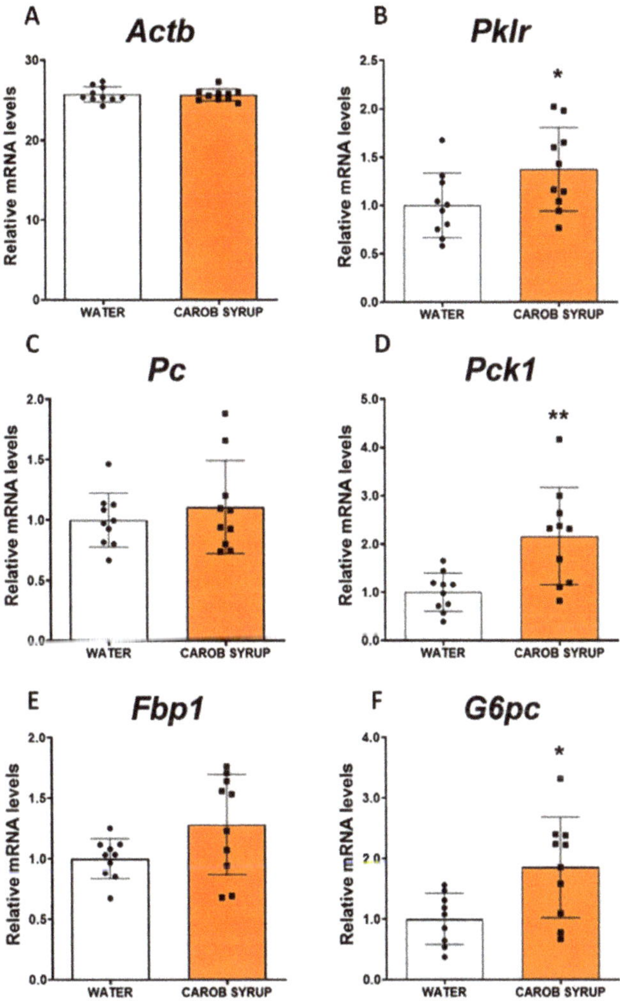

Figure 3. Effects on the liver neoglucogenic pathway of carob syrup mixed in drinking water at a dose of 100 mg/kg D-Pinitol per day of study with repeated administration for 10 days. Panels depict the quantitative expression of mRNA coding for (**A**) actin B (Actb, housekeeping gene), (**B**) pyruvate kinase (Pklr), (**C**) pyruvate carboxylase (PC), (**D**) phosphoenolpyruvate carboxykinase (Pck1) (**E**) fructose bis phosphatase (Fbp1), and (**F**) the catalytic subunit of glucose-6 phosphatase (G6pc). Data are means ± standard errors of the mean of at least 8 determinations per group. (*) $p < 0.05$, (**) $p < 0.01$ carob syrup vs. water drinking control animals.

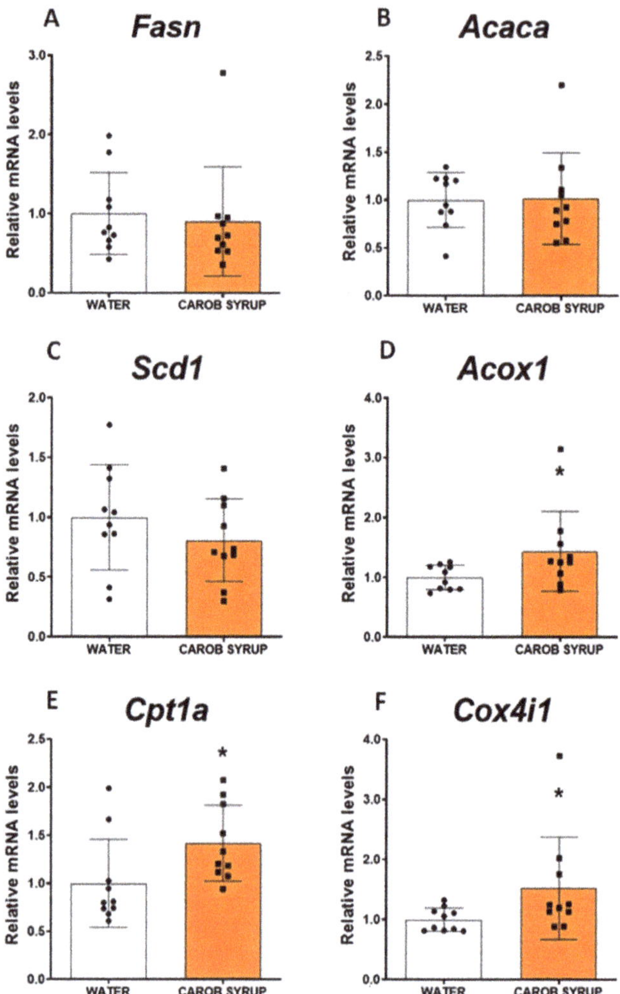

Figure 4. Effects of repeated administration of carob syrup for 10 days on liver lipid synthesis and oxidation pathways. Panels depict the quantitative expression of the mRNA coding for (**A**) fatty acid synthase (Fasn), (**B**) acetyl-coenzyme-A carboxylase (Acaca), (**C**) stearoyl-coenzymeA desaturase-1 (PC), (**D**) acyl-coenzyme-A oxidase-1 (Pck1), (**E**) carnitine palmitoyltransferase 1 (Cpt1), and (**F**) cytochrome C oxydase 4 isoform 1 (Cox4i1). Data are means ± standard errors of the mean of at least 8 determinations per group. (*) $p < 0.05$ carob syrup vs. water drinking control animals.

3.7. Effects of Repeated Administration of Carob Syrup on Insulin Signaling in the Liver, the Hippocampus, and the Hypothalamus of Rats

Since previous works [25] have described that D-pinitol alters the glucagon/insulin ration, promoting an activation of the gluconeogenesis, we explored the status of insulin signaling in the liver and the brain by Western blot analysis. Figure 5 shows that carob pos syrup enhanced the inhibitory phosphorylation of glycogen synthase kinase 3 and isoforms α (t = 2.49, df = 8, $p < 0.05$) and β (t = 2.33, df = 13, $p < 0.05$) without affecting the phosphorylation status of protein kinase B/AKT, glycogen synthase, or the mammalian target of rapamycin (mTOR). Thus, we did not observe a recruitment of the glycogen synthesis pathway at liver tissue. In the hypothalamus (Supplementary Figure S4) we found an activatory phosphorylation of glycogen synthase (t = 3.3, df = 10, $p < 0.01$) and mTOR (t = 3.1, df = 10, $p < 0.01$), and a tendency to display a greater phosphorylation of AKT (t = 2.05, df = 10, $p = 0.06$). In the hippocampus (Figure 6), we found no effects of carob syrup on the insulin signaling chain, with the exception of a decreased activatory phosphorylation of GSK3β (t = 2.6, df = 184, $p < 0.03$) and a clear decrease in the phosphorylation of the microtubule-associated protein tau (t = 3.2, df = 13, $p < 0.01$), a specific pharmacological target of D-Pinitol that we have recently described to be mediated by cyclin-dependent kinase 5 (CDK5).

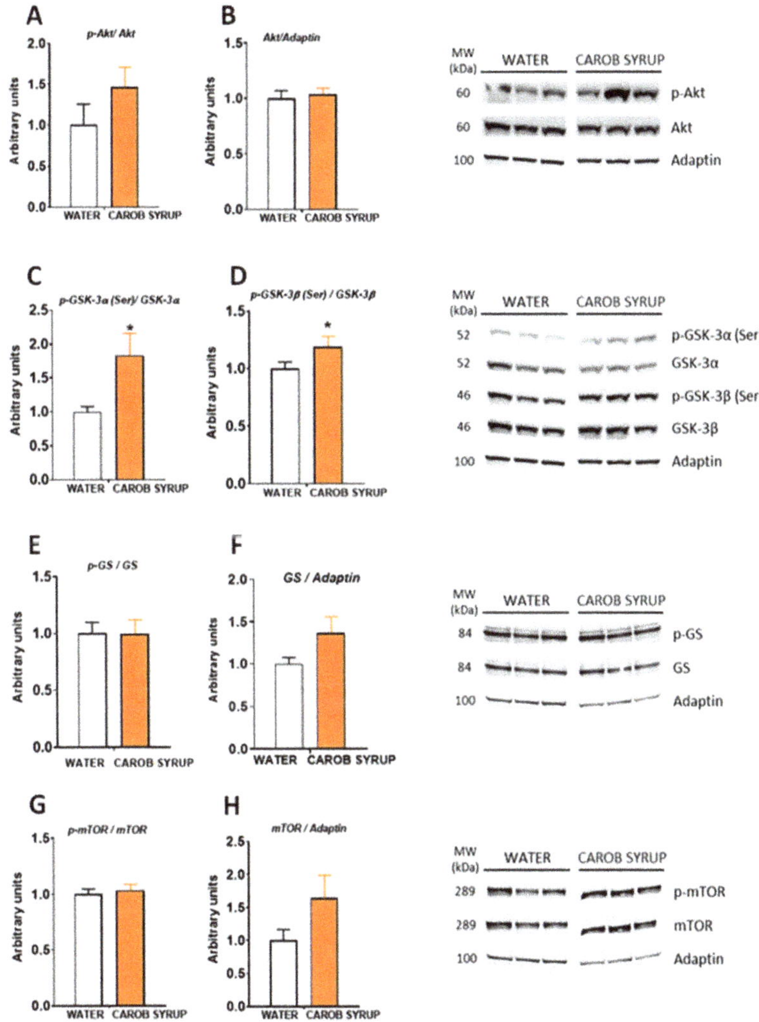

Figure 5. Effects of repeated administration of carob syrup on liver insulin signaling cascade for 10 days measured by Western blot analysis. (**A**) Phospho-protein kinase b/AKT (p-AKT), (**B**) protein kinase b/AKT (AKT), (**C**) phospho-glycogen synthase kinase 3 α (p-GSK3 α), (**D**) phospho-glycogen synthase kinase 3 β (p-GSK3 β), (**E**) phospho-glycogen synthase (pGS), (**F**) glycogen synthase (GS), (**G**) phospho-mammalian target of rapamicin (p-mTOR), and (**H**) mammalian target of rapamicin (mTOR). Data are means or adaptin-normalized band densities ± standard errors of the mean of 5–8 determinations per group. (*) $p < 0.05$ carob syrup vs. water drinking control animals.

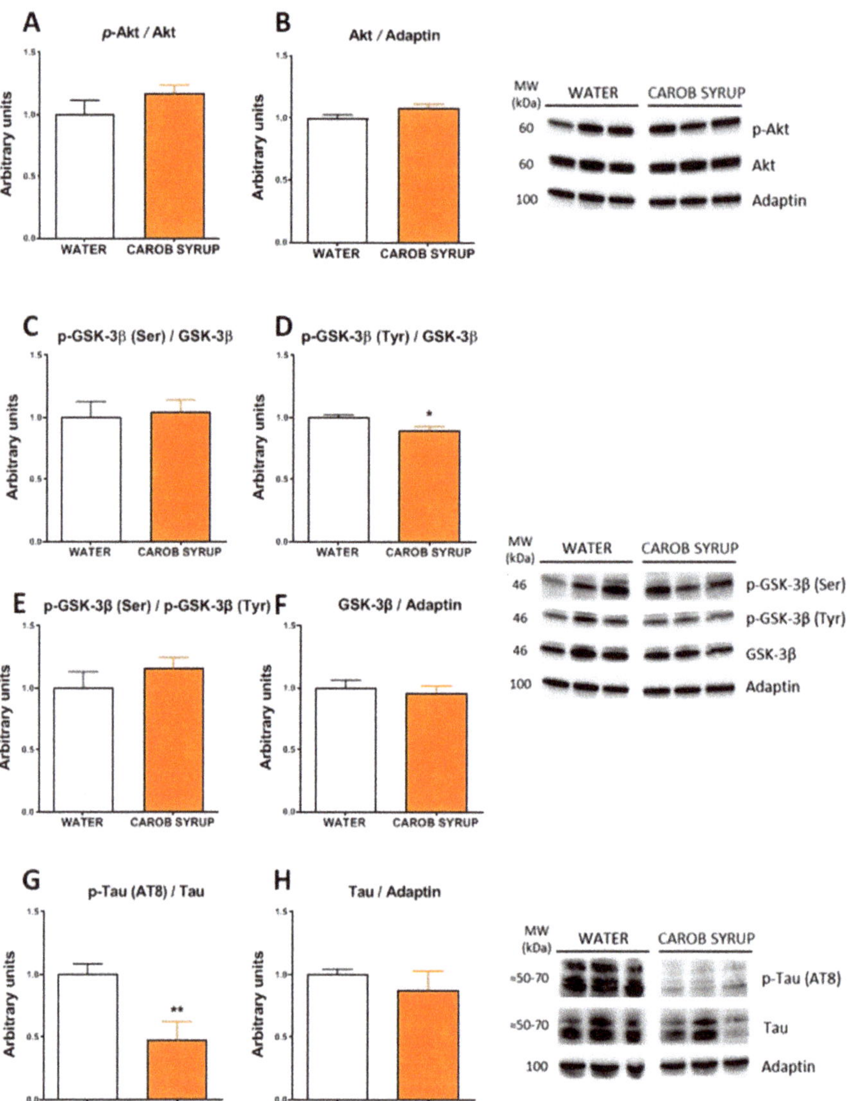

Figure 6. Effects of repeated administration of carob syrup on hippocampus AKT-signaling cascade and tau phosphorylation for 10 days, measured by Western blot analysis. (**A**) Phospho-protein kinase b/AKT (p-AKT), (**B**) protein kinase b/AKT (AKT), (**C**) serine-phosphorylated glycogen synthase kinase 3 β (p-GSK3 β)(Ser), (**D**) tyrosine-phosphorylated glycogen synthase kinase 3 β (p-GSK3 β)(Tyr)), (**E**) ratio p-GSK3 β)(Ser)/p-GSK3 β)(Tyr), (**F**) total GSK3 β, (**G**) phospho-tau (Ser202, Thr205) marked with AT8 antibody (p-TAU(AT8)) and (**H**) total tau. Representative westerns of each target are represented on the right panels. Data are means or adaptin-normalized band densities ± standard errors of the mean of 5–8 determinations per group. (*) $p < 0.05$, (**) $p < 0.01$ carob syrup vs. water drinking control animals.

4. Discussion

The rationale for the present experiment is to provide evidence of the feasibility of improving the normal physiological response to caloric sweet intake by boosting endogenous modulatory mechanisms that help both to keep weight and to sustain energy expenditure. Additionally, we aimed to do so by using a simple design: a balanced composition of fructose and glucose and a natural insulin sensitizer capable of boosting glucagon/ghrelin secretion while lowering insulin requirements: D-Pinitol [25–27,36–39]. A major criticism for the use of complex extracts stands precisely on the unknown composition that makes it impossible to fully understand the pharmacological properties of these extracts. The use of a simple composition such as glucose/fructose/D-pinitol allows us to better understand both pharmacological effects and physiological impact.

Our study in human volunteers reflects clearly that a natural syrup containing only glucose, fructose and D-Pinitol, keeping sweet palatability and the corresponding caloric value, produces glucose excursions that are shorter in time than those observed for glucose intake alone. Moreover, both the percentage of change over basal glucose levels and the insulin release induced by the intake of the carob syrup had less intense and shorter durations than those observed for glucose. It is true that the fructose-based sweetener (agave syrup) had better glucose and insulin-associated responses, but as we have described, fructose has a dark metabolic side that we should avoid [7–10], and thus its reduction is recommended (again, not its complete suppression, unless there is a relevant medical intolerance). The results obtained in humans suggest that by adding D-Pinitol we can modify the glycemic index of glucose/fructose syrup, improving glucose handling and insulin responses. Interestingly, the presence of fructose and glucose in the syrup is not associated with a decrease in circulating free fatty acids, something described for both glucose (see the decrease in circulating FFAs in Figure 1H) and especially fructose [40], suggesting that D-pinitol prevents the antilipolytic-prolipogenic actions of fructose. Nonetheless, plasma fructose concentrations after the intake of carob syrup are time-limited, as is the case with glucose, suggesting a relevant role of D-Pinitol on its regulation. Compared with agave syrup, the duration and intensity of the plasma peak of fructose was much lower. However, the presence of fructose demands more studies to clarify the potential unwanted side effects derived of the repeated consumption of carob syrup. The additional preclinical studies designed have addressed most of these potential problems, as well as added more information regarding the recently described properties of D-Pinitol [25,41,42].

First, we wanted to confirm that the lowering of glucose excursion observed after the administration of the carob syrup in humans was not derived of the different amount of glucose administered. We used Wistar rats to analyze the glucose handling after the administration of glucose (2 g/kg) or an amount of syrup with the same dose of glucose (2 g/kg glucose plus 2.1 g/kg fructose). The results clearly indicate that despite the presence of both glucose and fructose, glycaemia variation was substantially lower in animals receiving the syrup, supporting the insulin sensitizer activity of D-Pinitol. Moreover, the administration of carob syrup at a dose of 100 mg/kg daily (equivalent to supplementation with 1.48 g/kg day of fructose and 1.43 g/kg/day of glucose) clearly improved the glucose handling when an oral glucose load of 2 g/kg was given to the animals. Although, as we will discuss below, experimental data suggest that this effect derives both direct actions of D-pinitol and metabolic adaptions induced by this inositol, and thus we cannot exclude that modifications in metabolically active tissues such as muscle and adipose tissue might contribute to the improvement in glucose handling. We did not measure food intake or energy expenditure in this animal model, although preliminary data (Supplementary Figure S4) suggest that D-Pinitol does not affect feeding behavior.

This modulation of carob syrup could be observed in other metabolic parameters, and was partially replicated in a parallel experiment using only D-Pinitol at the same dose (Supplementary Table S2). Thus, after 10 days of ingestion of the syrup, animals had enhanced ghrelin secretion, and a lowering of liver depots of fat associated with a reduction in glycogen. A potential explanation for the decrease in the liver content of fat is that carob syrup exposure redirects liver metabolism towards glucose export and lipid oxidation. We attribute these effects to D-Pinitol, since previous studies with this inositol [25] have produced the same set of effects. In this study, we observed a decrease in insulin secretion associated with a rise in ghrelin and glucagon. Since D-Pinitol is able to activate both glucose uptake in muscle cells [27] and canonical insulin signaling in the hypothalamus [41], our interpretation is that D-Pinitol favors glucose transfer from liver to other tissues, including muscle. We do not know the impact of D-Pinitol (given alone or under the carob syrup format) on either adipose tissue function or energy expenditure, so future studies must address these important aspects to understand the pharmacological profile of this inositol. In addition, plasma concentrations of urea, but not uric acid, were also increased, suggesting that the glucagon pathway was activated. Glucagon is a hormone secreted by the α-cells of the pancreatic islets that oppose insulin and sustain glycaemia by promoting gluconeogenesis and releasing glucose from liver glycogen deports [43]. In addition, glucagon inhibits the de novo synthesis of lipids, boosting the use of amino acids and enhancing urea production [44,45]. These physiological actions are coincident with the metabolic profile we found in animals receiving the carob syrup, where we found a decrease in both fat and glycogen content in the liver, and an increase in plasma urea. D-Pinitol has been described to enhance the glucagon/insulin ratio [25], a finding we have replicated in humans (see Supplementary Figure S2), although the data obtained in the preclinical study in rats in the present report were not significant (Table 3). Probably, the effects D-pinitol on glucagon secretion are associated with the syrup drinking pattern, so we did not observe it at the moment of sacrifice of the animals. Nonetheless, we did observe a potentiation of the expression of the gluconeogenic pathway (an enhancement of some of the main enzymes for the de novo synthesis of glucose, including pyruvate kinase, phosphoenolpyruvate carboxykinase, and the catalytic subunit of glucose-6 phosphatase), and an enhanced expression of the fatty acid oxidation pathway (acyl-CoenzymeA oxydase, carnitine palmitoyltransferse 1A and cytochrome C oxydase isoform 4). These effects on lipid oxidation were not coupled to the plasma export of b-hydroxybutyrate, nor resulted in ketone production, indicating that fatty acid oxidation was not used to generate ketone bodies as glucose suppliers. In addition, we did not find evidence for oxidative stress, since malonyl dialdehyde in the liver was not enhanced after carob syrup consumption. Thus, the overall profile of carob syrup is that of a safe sweetener in the short term; it is capable of precluding the unwanted effects of fructose, and favors the actions of glucagon, which are seriously compromised in liver steatosis. Further studies using long-term exposure (4 to 8 months) to carob syrup are necessary to fully ensure the lack of fructose-associated toxicity derived of its use.

The present results are in accordance with the described actions of D-Pinitol, that eventually prompted clinical trials for positioning this inositol as a food supplement for the treatment of non-alcoholic steatohepatitis [28]. In addition, the effects observed on the insulin signaling cascade in the liver, after 10 days of consumption of the carob syrup, suggest an inhibitory phosphorylation of GSK3α and GSK3β, thus precluding the activation of glycogen synthase, which was not affected by the treatment. This finding suggests that the syrup does not promote glycogen storage, being in accordance with the reduced liver glycogen found. The present study cannot explain whether the effects observed are solely derived from the presence of D-Pinitol, or if they come from the association of this inositol to glucose and fructose. In a parallel experiment, we found increased urea production and a decrease in the contents of fat in the liver after exposure to D-Pinitol alone for 10 days. In addition, the profile of the effects of the syrup on the insulin signaling cascade found in the hippocampus and the hypothalamus were very similar to those described in experimental conditions where D-Pinitol was administered alone [41,42]. In the hippocampus, the acute and repeated administration of D-Pinitol did not affect insulin signaling, but resulted in the dephosphorylation of tau protein. This is exactly the same profile found in the hippocampus of the animals fed on the carob syrup [42]. The possibility of the neuroprotective actions of this syrup composition in tauopathies will have to be studied to fully confirm this possibility, opening an additional utility for inositols in aging and neurodegeneration [46]. We have observed that the effect on tau phosphorylation could be replicated also in a 3TG transgenic model of Alzheimer's disease. In the case of the hypothalamus, where D-Pinitol administration produces a general activation of AKT signaling cascade, the effects observed in the animals fed with the syrup demonstrated a similar pattern to the phosphorylation of GSK3β, the mammalian target of rapamicine (mTOR), and glycogen synthase [41]. Taken all together, the results suggest that D-Pinitol present in the syrup is bioactive and that the presence of glucose and fructose does not affect its pharmacological properties.

Limitations of the Present Study

The present study explores the acute actions of a carob syrup in a reduced number of human volunteers. Future studies using a within-subject design (the same subject receiving all the treatments) and a higher number of male and female volunteers are needed to better compare the actions of carob syrup with respect to other sweeteners such as agave or high-fructose corn syrup (HFCS). In addition, there is a need for studying the impact of this type of functional sweetener in the diabetic population. The toxicity derived from very long-term exposure to carob syrup in preclinical models must be addressed, especially in comparison with the well-known toxicity derived from chronic exposure to high-fructose-containing sweeteners such as agave or HFCS.

5. Conclusions

The use of natural syrup formulae containing glucose, fructose and D-Pinitol had a safe profile and resulted in a better glycemic index than that observed when glucose was given alone. This was likely caused by the reduction in the insulin response to increased glycaemia and the promotion of glucagon activity. The repeated administration of this syrup to male rats did not produce lipid depots, oxidative stress, or liver/kidney toxicity, as has been reported to occur with fructose when given alone. Overall, the present study is a positive proof of concept of the use of formulae capable of retaining sweetening properties, caloric value, and a safe metabolic profile. Whether the use of this "functional sweetener" is useful in the context of obesity and diabetes will require additional research.

Supplementary Materials: The following supporting information can be downloaded at: https://www.mdpi.com/article/10.3390/pharmaceutics14081594/s1.

Author Contributions: Conceptualization, C.S., E.B. and F.R.d.F.; data curation, J.A.N., J.D., D.M.-V. and J.S.; formal analysis, J.A.N., F.J.P., J.D., D.M.-V., A.J.L.-G., E.B. and F.R.d.F.; funding acquisition, C.S. and F.R.d.F.; investigation, J.A.N., J.D., A.S., F.J.P., A.J.L.-G., R.T., M.d.C., Y.A.B.; methodology, J.A.N., A.S., F.J.P., J.D., Y.A.B.; project administration, F.R.d.F.; writing—original draft, J.A.N. and F.R.d.F.; writing—review and editing, all authors. All authors have read and agreed to the published version of the manuscript.

Funding: The present study has been supported by grants from the following institutions: Ministerio de Economía y Competitividad, Gobierno de España, (Grant RTC-2016-4983-1), EU-ERDF-Instituto de Salud Carlos III (Grant PI19/01577), FIS-Instituto de Salud Carlos III (Grant PI19/00343) and Consejería de Economía, Conocimiento y Universidad, Junta de Andalucía (Grant P18-TP-5194). D.M-V. (FI20/00227) holds a "PFIS" predoctoral contract from the National System of Health, EUERDF-ISCIII. A.J.L.-G. (IFI18/00042) holds an "iPFIS" predoctoral contract from the National System of Health, EU-ERDF-ISCIII. J.D. thanks to Consejería de Salud y Familias (Grant PI-0139-2018) and "Miguel Servet" research contract funded by ISCIII and ERDF-EU (Grant CPII19/00022).

Institutional Review Board Statement: The study was conducted under the guidelines of the Declaration of Helsinki, and the research protocol was approved by the Ethics Committee of the Regional University Hospital of Málaga, under protocol number P18-TP-5194. Informed consent was obtained from all the subjects.

Informed Consent Statement: Informed consent was obtained from all subjects involved in the study. Written informed consent has been obtained from the patient(s) to publish this paper.

Data Availability Statement: All data generated or analysed during this study are available upon request by email to fernando.rodriguez@ibima.eu as a raw data file.

Conflicts of Interest: Carlos San Juan declares he receives salary and has shares from Euronutra company. Yolanda Alfonso Baltasar declares she receives salary from EUronutra company. The remaining authors declare that they have no known competing financial interest or personal relationships that could have appeared to influence the work reported in this paper. The company had no role in the design of the study; in the collection, analyses, or interpretation of data; in the writing of the manuscript, and in the decision to publish the results.

References

1. Schiano, C.; Grimaldi, V.; Scognamiglio, M.; Costa, D.; Soricelli, A.; Nicoletti, G.F.; Napoli, C. Soft drinks and sweeteners intake: Possible contribution to the development of metabolic syndrome and cardiovascular diseases. Beneficial or detrimental action of alternative sweeteners? *Food Res. Int.* **2021**, *142*, 110220. [CrossRef]

2. Czarnecka, K.; Pilarz, A.; Rogut, A.; Maj, P.; Szymańska, J.; Olejnik, Ł.; Szymański, P. Aspartame-True or False? Narrative Review of Safety Analysis of General Use in Products. *Nutrients* **2021**, *13*, 1957. [CrossRef]

3. Pang, M.D.; Goossens, G.H.; Blaak, E.E. The Impact of Artificial Sweeteners on Body Weight Control and Glucose Homeostasis. *Front. Nutr.* **2021**, *7*, 598340. [CrossRef] [PubMed]

4. Greenwood, D.C.; Threapleton, D.E.; Evans, C.E.L.; Cleghorn, C.L.; Nykjaer, C.; Woodhead, C.; Burley, V.J. Association between sugar-sweetened and artificially sweetened soft drinks and type 2 diabetes: Systematic review and dose-response meta-analysis of prospective studies. *Br. J. Nutr.* **2014**, *112*, 725–734. [CrossRef]

5. Azaïs-Braesco, V.; Sluik, D.; Maillot, M.; Kok, F.; Moreno, L.A. A review of total & added sugar intakes and dietary sources in Europe. *Nutr. J.* **2017**, *16*, 6. [CrossRef]

6. Blaak, E.E. Carbohydrate quantity and quality and cardio-metabolic risk. *Curr. Opin. Clin. Nutr. Metab. Care* **2016**, *19*, 289–293. [CrossRef]

7. Febbraio, M.A.; Karin, M. "Sweet death": Fructose as a metabolic toxin that targets the gut-liver axis. *Cell Metab.* **2021**, *33*, 2316–2328. [CrossRef] [PubMed]

8. Ter Horst, K.W.; Serlie, M.J. Fructose Consumption, Lipogenesis, and Non-Alcoholic Fatty Liver Disease. *Nutrients* **2017**, *9*, 981. [CrossRef] [PubMed]

9. Federico, A.; Rosato, V.; Masarone, M.; Torre, P.; Dallio, M.; Romeo, M.; Persico, M. The Role of Fructose in Non-Alcoholic Steatohepatitis: Old Relationship and New Insights. *Nutrients* **2021**, *13*, 1314. [CrossRef]

10. Roeb, E.; Weiskirchen, R. Fructose and Non-Alcoholic Steatohepatitis. *Front. Pharmacol.* **2021**, *12*, 634344. [CrossRef]

11. Azad, M.B.; Abou-Setta, A.M.; Chauhan, B.F.; Rabbani, R.; Lys, J.; Copstein, L.; Mann, A.; Jeyaraman, M.M.; Reid, A.E.; Fiander, M.; et al. Nonnutritive sweeteners and cardiometabolic health: A systematic review and meta-analysis of randomized controlled trials and prospective cohort studies. *CMAJ* **2017**, *189*, E929–E939. [CrossRef] [PubMed]

12. Swithers, S.E. Artificial sweeteners produce the counterintuitive effect of inducing metabolic derangements. *Trends Endocrinol. Metab.* **2013**, *24*, 431–441. [CrossRef] [PubMed]

13. Swithers, S.E.; Shearer, J. Obesity: Sweetener associated with increased adiposity in young adults. *Nat. Rev. Endocrinol.* **2017**, *13*, 443–444. [CrossRef]

14. Farup, P.G.; Lydersen, S.; Valeur, J. Are Nonnutritive Sweeteners Obesogenic? Associations between Diet, Faecal Microbiota, and Short-Chain Fatty Acids in Morbidly Obese Subjects. *J. Obes.* **2019**, *2019*, 4608315. [CrossRef]

15. Bian, X.; Tu, P.; Chi, L.; Gao, B.; Ru, H.; Lu, K. Saccharin induced liver inflammation in mice by altering the gut microbiota and its metabolic functions. *Food Chem. Toxicol.* **2017**, *107*, 530–539. [CrossRef] [PubMed]

16. Bian, X.; Chi, L.; Gao, B.; Tu, P.; Ru, H.; Lu, K. The artificial sweetener acesulfame potassium affects the gut microbiome and body weight gain in CD-1 mice. *PLoS ONE* **2017**, *12*, e0178426. [CrossRef]

17. Lertrit, A.; Srimachai, S.; Saetung, S.; Chanprasertyothin, S.; Chailurkit, L.O.; Areevut, C.; Katekao, P.; Ongphiphadhanakul, B.; Sriphrapradang, C. Effects of sucralose on insulin and glucagon-like peptide-1 secretion in healthy subjects: A randomized, double-blind, placebo-controlled trial. *Nutrition* **2018**, *55–56*, 125–130. [CrossRef] [PubMed]

18. Araújo, J.R.; Martel, F.; Keating, E. Exposure to non-nutritive sweeteners during pregnancy and lactation: Impact in programming of metabolic diseases in the progeny later in life. *Reprod. Toxicol.* **2014**, *49*, 196–201. [CrossRef]

19. Azad, M.B.; Archibald, A.; Tomczyk, M.M.; Head, A.; Cheung, K.G.; de Souza, R.J.; Becker, A.B.; Mandhane, P.J.; Turvey, S.E.; Moraes, T.J.; et al. Nonnutritive sweetener consumption during pregnancy, adiposity, and adipocyte differentiation in offspring: Evidence from humans, mice, and cells. *Int. J. Obes.* **2020**, *44*, 2137–2148. [CrossRef] [PubMed]

20. Mishra, A.; Ahmed, K.; Froghi, S.; Dasgupta, P. Systematic review of the relationship between artificial sweetener consumption and cancer in humans: Analysis of 599,741 participants. *Int. J. Clin. Pract.* **2015**, *69*, 1418–1426. [CrossRef]

21. Toews, I.; Lohner, S.; de Gaudry, D.K.; Sommer, H.; Meerpohl, J.J. Association between intake of non-sugar sweeteners and health outcomes: Systematic review and meta-analyses of randomised and non-randomised controlled trials and observational studies. *BMJ* **2019**, *364*, k4718. [CrossRef] [PubMed]

22. Shih, C.W.; Hauser, M.E.; Aronica, L.; Rigdon, J.; Gardner, C.D. Changes in blood lipid concentrations associated with changes in intake of dietary saturated fat in the context of a healthy low-carbohydrate weight-loss diet: A secondary analysis of the Diet Intervention Examining The Factors Interacting with Treatment Success (DIETFITS) trial. *Am. J. Clin. Nutr.* **2019**, *109*, 433–441. [CrossRef] [PubMed]

23. Smeets, P.A.; de Graaf, C.; Stafleu, A.; van Osch, M.J.; van der Grond, J. Functional magnetic resonance imaging of human hypothalamic responses to sweet taste and calories. *Am. J. Clin. Nutr.* **2005**, *82*, 1011–1016. [CrossRef]

24. Tan, H.E.; Sisti, A.C.; Jin, H.; Vignovich, M.; Villavicencio, M.; Tsang, K.S.; Goffer, Y.; Zuker, C.S. The gut-brain axis mediates sugar preference. *Nature* **2020**, *580*, 511–516. [CrossRef]

25. Navarro, J.A.; Decara, J.; Medina-Vera, D.; Tovar, R.; Suarez, J.; Pavón, J.; Serrano, A.; Vida, M.; Gutierrez-Adan, A.; Sanjuan, C.; et al. D-Pinitol from Ceratonia siliqua Is an Orally Active Natural Inositol That Reduces Pancreas Insulin Secretion and Increases Circulating Ghrelin Levels in Wistar Rats. *Nutrients* **2020**, *12*, 2030. [CrossRef]

26. Kang, M.J.; Kim, J.I.; Yoon, S.Y.; Kim, J.C.; Cha, I.J. Pinitol from soybeans reduces postprandial blood glucose in patients with type 2 diabetes mellitus. *J. Med. Food* **2006**, *9*, 182–186. [CrossRef]

27. Bates, S.H.; Jones, R.B.; Bailey, C.J. Insulin-like effect of pinitol. *Br. J. Pharmacol.* **2000**, *130*, 1944–1948. [CrossRef]

28. Lee, E.; Lim, Y.; Kwon, S.W.; Kwon, O. Pinitol consumption improves liver health status by reducing oxidative stress and fatty acid accumulation in subjects with non-alcoholic fatty liver disease: A randomized, double-blind, placebo-controlled trial. *J. Nutr. Biochem.* **2019**, *68*, 33–41. [CrossRef] [PubMed]

29. U.S. Department of Health and Human Services, Food and Drug Administration, Center for Drug Evaluation and Research (CDER). *Guidance for Industry: Bioanalytical Method Validation*; U.S. Department of Health and Human Services, Food and Drug Administration, Center for Drug Evaluation and Research (CDER), Center for Veterinary Medicine (CV): Laurel, MD, USA, 2018.

30. Matthews, D.R.; Hosker, J.P.; Rudenski, A.S.; Naylor, B.A.; Treacher, D.F.; Turner, R.C. Homeostasis model assessment: Insulin resistance and beta-cell function from fasting plasma glucose and insulin concentrations in man. *Diabetologia* **1985**, *28*, 412–419. [CrossRef]

31. López-Gambero, A.J.; Pacheco-Sánchez, B.; Rosell-Valle, C.; Medina-Vera, D.; Navarro, J.A.; del Mar Fernández-Arjona, M.; de Ceglia, M.; Sanjuan, C.; Simon, V.; Cota, D.; et al. Dietary administration of D-chiro-inositol attenuates sex-specific metabolic imbalances in the 5xFAD mouse model of Alzheimer's disease. *Biomed. Pharmacother.* **2022**, *150*, 112994. [CrossRef]

32. Kilkenny, C.; Browne, W.J.; Cuthill, I.C.; Emerson, M.; Altman, D.G. Improving bioscience research reporting: The ARRIVE guidelines for reporting animal research. *PLoS Biol.* **2010**, *8*, e1000412. [CrossRef] [PubMed]

33. Decara, J.M.; Pavon, F.J.; Suarez, J.; Romero-Cuevas, M.; Baixeras, E.; Vazquez, M.; Rivera, P.; Gavito, A.L.; Almeida, B.; Joglar, J.; et al. Treatment with a novel oleic-acid-dihydroxyamphetamine conjugation ameliorates non-alcoholic fatty liver disease in obese Zucker rats. *Dis. Models Mech.* **2015**, *8*, 1213–1225. [CrossRef] [PubMed]

34. Bligh, E.G.; Dyer, W.J. A rapid method of total lipid extraction and purification. *Can. J. Biochem. Physiol.* **1959**, *37*, 911–917. [CrossRef] [PubMed]

35. Vida, M.; Serrano, A.; Romero-Cuevas, M.; Pavon, F.J.; Gonzalez-Rodriguez, A.; Gavito, A.L.; Cuesta, A.L.; Valverde, A.M.; Rodriguez de Fonseca, F.; Baixeras, E. IL-6 cooperates with peroxisome proliferator-activated receptor-alpha-ligands to induce liver fatty acid binding protein (LFABP) up-regulation. *Liver Int. Off. J. Int. Assoc. Study Liver* **2013**, *33*, 1019–1028. [CrossRef]

36. Gao, Y.; Zhang, M.; Wu, T.; Xu, M.; Cai, H.; Zhang, Z. Effects of D-Pinitol on Insulin Resistance through the PI3K/Akt Signaling Pathway in Type 2 Diabetes Mellitus Rats. *J. Agric. Food Chem.* **2015**, *63*, 6019–6026. [CrossRef]

37. Hernández-Mijares, A.; Bañuls, C.; Peris, J.E.; Monzó, N.; Jover, A.; Bellod, L.; Victor, V.M.; Rocha, M. A single acute dose of pinitol from a naturally-occurring food ingredient decreases hyperglycaemia and circulating insulin levels in healthy subjects. *Food Chem.* **2013**, *141*, 1267–1272. [CrossRef] [PubMed]

38. Narayanan, C.R.; Joshi, D.D.; Mujumdar, A.M.; Dhekne, V.V. Pinitol—A new anti-diabetic compound from the leaves of bougainvillea spectabilis. *Curr. Sci.* **1987**, *56*, 139–141.

39. Carrera-Lanestosa, A.; Moguel-Ordóñez, Y.; Segura-Campos, M. Stevia rebaudiana Bertoni: A Natural Alternative for Treating Diseases Associated with Metabolic Syndrome. *J. Med. Food* **2017**, *20*, 933–943. [CrossRef]

40. Tappy, L.; Randin, J.P.; Felber, J.P.; Chiolero, R.; Simonson, D.C.; Jequier, E.; DeFronzo, R.A. Comparison of thermogenic effect of fructose and glucose in normal humans. *Am. J. Physiol.* **1986**, *250*, E718–E724. [CrossRef]

41. Medina-Vera, D.; Navarro, J.A.; Tovar, R.; Rosell-Valle, C.; Gutiérrez-Adan, A.; Ledesma, J.C.; Sanjuan, C.; Pavón, F.J.; Baixeras, E.; Rodríguez de Fonseca, F.; et al. Activation of PI3K/Akt Signaling Pathway in Rat Hypothalamus Induced by an Acute Oral Administration of D-Pinitol. *Nutrients* **2021**, *13*, 2268. [CrossRef]

42. Medina-Vera, D.; Navarro, J.N.; Rivera, P.; Rosell-Valle, C.; Gutiérrez-Adán, A.; Sanjuan, C.; López-Gambero, A.J.; Tovar, R.; Suarez, J.; Pavón, F.J.; et al. D-Pinitol promotes tau dephosphorylation through a Cyclin-dependent kinase 5 regulation mechanism: A new potential approach for Tauopathies? *Br. J. Pharmacol.* **2022**, *in press*. [CrossRef]

43. Richter, M.M.; Galsgaard, K.D.; Elmelund, E.; Knop, F.K.; Suppli, M.P.; Holst, J.J.; Winther-Sørensen, M.; Kjeldsen, S.A.; Albrechtsen, N.J.W. The Liver-Alpha Cell Axis in Health and in Disease. *Diabetes* **2022**, *3*, dbi220004. [CrossRef]

44. Janah, L.; Kjeldsen, S.; Galsgaard, K.D.; Winther-Sørensen, M.; Stojanovska, E.; Pedersen, J.; Knop, F.K.; Holst, J.J.; Wewer Albrechtsen, N.J. Glucagon Receptor Signaling and Glucagon Resistance. *Int. J. Mol. Sci.* **2019**, *20*, 3314. [CrossRef]

45. Galsgaard, K.D.; Pedersen, J.; Knop, F.K.; Holst, J.J.; Wewer Albrechtsen, N.J. Glucagon Receptor Signaling and Lipid Metabolism. *Front. Physiol.* **2019**, *10*, 413. [CrossRef]
46. López-Gambero, A.J.; Sanjuan, C.; Serrano-Castro, P.J.; Suárez, J.; Rodríguez de Fonseca, F. The Biomedical Uses of Inositols: A Nutraceutical Approach to Metabolic Dysfunction in Aging and Neurodegenerative Diseases. *Biomedicines* **2020**, *8*, 295. [CrossRef]

 *pharmaceutics*

Article

Polydatin Alleviates Diabetes-Induced Hyposalivation through Anti-Glycation Activity in db/db Mouse

Hyung Rae Kim, Woo Kwon Jung, Su-Bin Park, Hwa Young Ryu, Yong Hwan Kim and Junghyun Kim *

Department of Oral Pathology, School of Dentistry, Jeonbuk National University, Jeonju 54896, Korea;
rlagudfo31@gmail.com (H.R.K.); wkjungjbnu@gmail.com (W.K.J.); tnqls309@gmail.com (S.-B.P.);
naive17jbnu@gmail.com (H.Y.R.); kg7229ku@gmail.com (Y.H.K.)
* Correspondence: dvmhyun@jbnu.ac.kr; Tel.: +82-63-270-4032; Fax: +82-63-270-4025

Abstract: Polydatin (resveratrol-3-O-β-mono-D-glucoside) is a polyphenol that can be easily accessed from peanuts, grapes, and red wine, and is known to have antiglycation, antioxidant, and anti-inflammatory effects. Diabetes mellitus is a very common disease, and diabetic complications are very common complications. The dry mouth symptom is one of the most common oral complaints in patients with diabetes mellitus. Diabetes mellitus is thought to promote hyposalivation. In this study, we aimed to investigate the improvement effect of polydatin on diabetes-induced hyposalivation in db/db mouse model of type 2 diabetes. We examined salivary flow rate, TUNEL assay, PAS staining, and immunohistochemical staining for AGEs, RAGE, HMGB1, 8-OHdG, and AQP5 to evaluate the efficacy of polydatin in the submandibular salivary gland. Diabetic db/db mice had a decreased salivary flow rate and salivary gland weight. The salivary gland of the vehicle-treated db/db mice showed an increased apoptotic cell injury. The AGEs were highly accumulated, and its receptor, RAGE expression was also enhanced. Expressions of HMGB1, an oxidative cell damage marker, and 8-OHdG, an oxidative DNA damage marker, increased greatly. However, polydatin ameliorated this hypofunction of the salivary gland and inhibited diabetes-related salivary cell injury. Furthermore, polydatin improved mucin accumulation, which is used as a damage marker for salivary gland acinar cells, and decreased expression of water channel AQP5 was improved by polydatin. In conclusion, polydatin has a potent protective effect on diabetes-related salivary gland hypofunction through its antioxidant and anti-glycation activities, and its AQP5 upregulation. This result suggests the possibility of the use of polydatin as a therapeutic drug to improve hyposalivation caused by diabetes.

Keywords: advanced glycation end products; diabetes; polydatin; salivary gland

Citation: Kim, H.R.; Jung, W.K.; Park, S.-B.; Ryu, H.Y.; Kim, Y.H.; Kim, J. Polydatin Alleviates Diabetes-Induced Hyposalivation through Anti-Glycation Activity in db/db Mouse. *Pharmaceutics* **2022**, *14*, 51. https://doi.org/10.3390/pharmaceutics14010051

Academic Editors: Diana Marcela Aragon Novoa and Fátima Regina Mena Barreto Silva

Received: 18 November 2021
Accepted: 23 December 2021
Published: 27 December 2021

Publisher's Note: MDPI stays neutral with regard to jurisdictional claims in published maps and institutional affiliations.

1. Introduction

Most Saliva is a mixture of fluids secreted by the three major glands (submandibular, sublingual, and parotid) and several minor glands, containing 99% water; electrolytes such as potassium, magnesium, calcium, sodium, and immunoglobulins; enzymes; etc., and protects the oral cavity [1]. It is also necessary for the initiation of digestion and absorption of food and for maintaining the pH, tooth mineralization, oral microbiome, moist oral mucosal surfaces, and infection prevention of teeth and mucosa [2,3]. Therefore, proper salivation is essential for maintaining the oral environment.

Xerostomia is a disease that decreases salivation, can occur even if an objective decrease in salivation is not observed, can be triggered by a change in the ingredients of the saliva, and can be caused by radiotherapy, Sjögren's syndrome, pathologies of the salivary glands, or medications [1,4]. Xerostomia causes diseases such as enamel demineralization, rampant decay, and super-infections by fungal diseases, is associated with oral dryness and decreased or altered taste and lowers the overall quality of life by negatively affecting swallowing and ingestion [5,6].

Diabetes mellitus (DM) is divided into type 1 and prevalent type 2, with its symptoms appearing slowly after the age of 40, which triggers a chain reaction of increased protein glycation due to high blood glucose, causes functional damage generated in many organ systems by promoting the creation of advanced glycation end products (AGEs), and results in ultimate complications such as cardiovascular disease, nephropathy, and retinopathy [7–9]. Oral manifestations due to DM include xerostomia, dental caries, periodontal disease, burning mouth, and poor wound healing, with xerostomia being a very common phenomenon, which 34% to 51% of diabetic patients experience [10–12].

Salivary stimulants such as chewing gum, vitamin C, malic acid, and pilocarpine or salivary substitutes are solutions for xerostomia, which only provide temporary symptom relief [13]. Additionally, there are some medications that cause xerostomia among the medications for the treatment of some existing diseases [14], and there are cases in which xerostomia is caused by xerostomia-containing drugs or the synergistic effect of the drugs if the person takes multiple medications as they get older [15]. Therefore, there is a need for research on therapeutic drugs that can simultaneously improve xerostomia in addition to alleviating existing diseases.

The rhizome and root of *Polygonum cuspidatum* are widely used as traditional herbal medicines for analgesic, diuretic, antipyretic, and expectorant purposes, its main active ingredient being polydatin (resveratrol-3-O-β-mono-D-glucoside) [16]. Polydatin exists in nature in the form of resveratrol and is abundantly present in peanuts, grapes, chocolates, and red wine [17]. Polydatin is known to have anti-diabetic, antioxidant, anti-inflammation, anticancer, and cardiovascular protective effects [18]. In addition, resveratrol is involved in the recovery process in the experimental irradiation-induced xerostomia in vivo model [19,20], and there are studies about resveratrol's improvement on salivary dysfunction in a non-obese diabetic (NOD) mice model [21]. However, there are no studies about polydatin related to salivary secretion.

Therefore, this study aims to investigate the improvement effect of diabetes-induced salivary gland hyposalivation by polydatin in type 2 diabetic db/db mice.

2. Materials and Methods

2.1. Chemicals

Polydatin, with purity greater than 95%, was obtained from Sigma-Aldrich (St. Louis, MA, USA).

2.2. In Vivo Experiment Design in Normoglycemic Mice

To determine the oral dose of polydatin for in vivo experiment, 5-week-old male C57BL/6 mice were purchased from Damul Science (Daejeon, Korea). The mice were adapted to the environment for 1 week and then randomly divided into four groups: normal control ($n = 5$), polydatin 25 mg/kg (PD-25, $n = 5$), polydatin 50 mg/kg (PD-50, $n = 5$), polydatin 100 mg/kg (PD-100, $n = 5$). Polydatin was dissolved in 0.5% (*w/v*) sodium carboxymethylcellulose (Sigma-Aldrich, St. Louis, MA, USA). Polydatin was administered orally to mice on a daily basis for 3 weeks.

2.3. In Vivo Experiment Design in Hyperglycemic Mice

Fourteen male db/db leptin receptor-deficient type 2 diabetic mice aged 5 weeks and seven male wild type C57BL/6 mice aged 5 weeks were supplied by Central Lab Animal Inc. (Seoul, Korea). The mice were adapted to the environment for 1 week and then randomly divided into 3 groups based on the weight and fasting blood glucose levels as follows: normal group (Normal, $n = 7$), diabetic group (DM, $n = 7$), polydatin 100 mg/kg treated group (DM + PD, $n = 7$). Polydatin was dissolved in 0.5% (*w/v*) sodium carboxymethylcellulose and orally administered for 5 weeks. The animals were housed in a barrier system controlled by temperature (20–25 °C) and humidity (40–70%), with a 12:12 h light and dark cycle. Animal experiments were carried out according to procedures approved by the Institutional Animal Care and Use Committee of the Jeonbuk National University Laboratory Animal Center (IACUC, Approved No.: JBNU 2021-087).

2.4. Collection of Saliva

Mice were anesthetized with an intraperitoneal injection of ketamine at a dose of 80 mg/kg. Salivation was induced with an intraperitoneal injection of pilocarpine (Sigma-Aldrich, St. Louis, MA, USA) at a dose of 0.8 mg/kg. The volume of saliva was measured for 10 min after injection of pilocarpine using cotton balls. Cotton balls were kept in the oral cavity for 10 min to absorb saliva and were then immediately weighed on an electronic balance. Total saliva was calculated from the difference of pre and post-collection cotton ball weight in milligrams and converted to microliters. The salivary flow rate was calculated as microliters per minute.

2.5. Hematoxylin and Eosin (H&E) Staining

At necropsy, submandibular gland tissues were isolated. Formalin-fixed tissues were embedded in paraffin and sliced into 4 μm thick sections. The sections were deparaffinized and rehydrated using standard techniques and stained with hematoxylin solution gill no. 3 and eosin. The stained sections were dehydrated and cleared with xylene, before being observed under a light microscope.

2.6. Immunohistochemical Staining

The tissue sections were deparaffinized and rehydrated using standard techniques. The sections were incubated with 3% H_2O_2 for 10 min at room temperature (RT). They were then washed with 1X TBST buffer for 10 min at RT. Primary antibodies were diluted in 2.5% normal horse serum (Vector Laboratory, Bulingame, CA, USA). The primary antibodies were AGE (TransGenic, Kyoto, Japan), the receptor for AGE (RAGE, Santa Cruz, CA, USA), high mobility group box 1 (HMGB1, Abcam, Boston, MA, USA), 8-hydroxydeoxyguanosine (8-OHdG, Abcam, Boston, MA, USA), and aquaporin5 (AQP5, Bioworld Technology, St. Louis Park, MN, USA). Sections were visualized with the VEC-TASTAIN Elite ABC Universal Kit (Vector Laboratory, Bulingame, CA, USA). The staining was observed using a BX51 light microscope (Olympus, Tokyo, Japan). Ten unique fields of view were randomly selected in each slice at $100\times$ magnification. Image-Pro software (Media Cybernetics, Rockville, MD, USA) was used for semi-quantitative analysis of the intensity of the immunohistochemical positive signal. The average optical density per unit area (mm^2) was calculated for the relative amount of protein.

2.7. Oxidative Stress Assay in Salivary Gland

Frozen salivary gland tissues were homogenized in lysis buffer (150 mM NaCl, 1% Triton X-100 and 10 mM Tris, pH 7.4) containing protease inhibitor. The homogenate was centrifuged at $10,000\times g$ for 10 min at 4 °C and the supernatant was collected for measurement of reactive oxygen species (ROS) levels. ROS levels were examined using a Mouse ROS ELISA Kit (MyBioSource, San Diego, CA, USA) according to the manufacturer's instructions.

2.8. Periodic Acid-Schiff (PAS) Staining

The tissue sections were deparaffinized and rehydrated using standard techniques and oxidized in 1% periodic acid solution for 5 min and rinsed in distilled water. The sections were placed in Schiff reagent for 15 min and counterstained with hematoxylin.

2.9. TUNEL Assay

Apoptotic cells in the tissue were detected by an in situ cell death detection kit (Roche, Mannheim, Germany), according to the manufacturer's instructions. The numbers of TUNEL-positive cells were counted under a fluorescence microscope (BX51, Olympus, Tokyo, Japan). Ten fields at $100\times$ magnification were photographed randomly from each salivary gland with a fluorescence microscope. The total number of TUNEL-positive cells per unit area (mm^2) was counted in each field using image analysis software (Image-Pro, Media Cybernetics, Rockville, MD, USA).

2.10. Statistical Analysis

Results are presented as mean ± standard deviation. Significant differences between groups were determined using one-way analysis of variance (ANOVA), followed by Tukey's multiple comparison test. Differences with a *p*-value lower than 0.05 were considered to represent significant differences.

3. Results

3.1. Increased Salivation in Normoglycemic Mice by Polydatin and Selection of the Oral Dose of Polydatin

Changes in salivation by polydatin (Figure 1A) were checked in normoglycemic mice. First, it was confirmed that there was no significant change in body weight (Figure 1B). In addition, it was confirmed that salvation increased depending on the concentration after administration of polydatin for 3 weeks (Figure 1C). H&E staining was performed to confirm histological changes and no compound-related histopathological changes were found (Figure 1D). In addition, the expression level of AQP5, a water channel protein important for salivation, was increased by polydatin treatment (Figure 1E). Therefore, 100 mg/kg of polydatin was selected as the administration concentration for the in vivo test using hyperglycemic db/db mice.

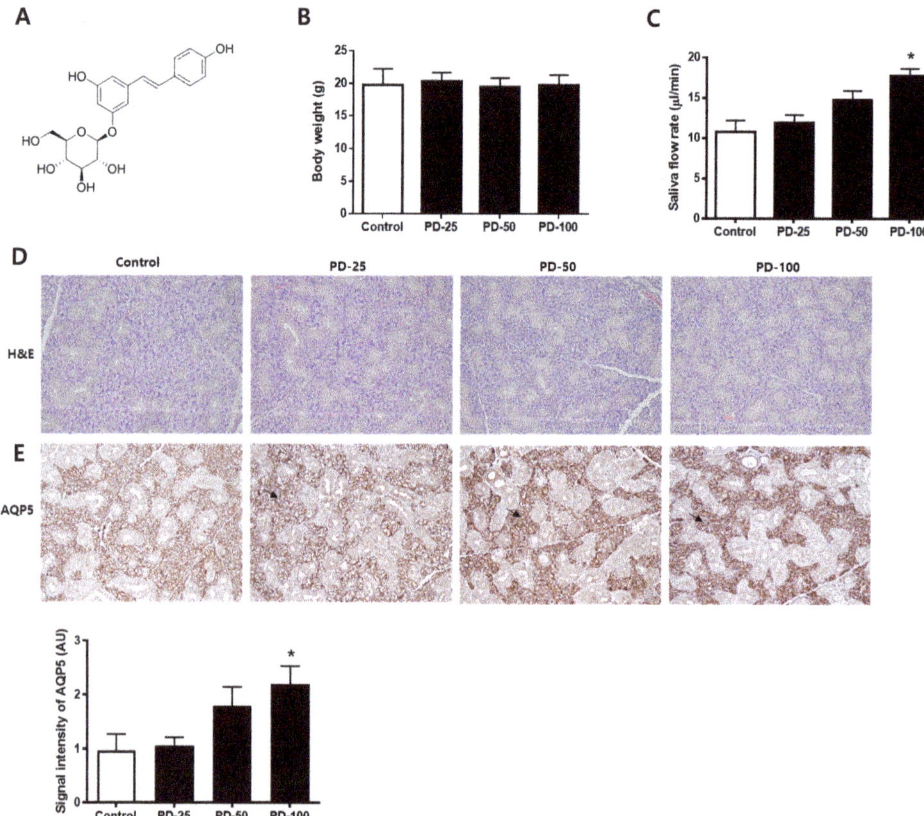

Figure 1. Effect of polydatin in the normoglycemic mice. (**A**) Chemical structure of polydatin. (**B**) Body weight. (**C**) Salivary flow rate. (**D**) Representative submandibular glands were stained with H&E. (**E**) Immunohistochemical staining for AQP5. Quantification of the AQP5 signal intensity. Arrows indicate AQP5 positively stained cells. The values in the bar graphs represent the means ± SD, *n* = 5. * *p* < 0.05 vs. normal control group.

3.2. Polydatin Improves Hyposalivation in Type 2 Diabetic Mice

At the endpoint, the body weight was higher in the DM group with no difference between the DM + PD groups. Additionally, blood glucose levels also increased in the DM group with no difference between the DM + PD groups (Table 1). As a result of the measurement of salivation, it was confirmed that salivation decreased significantly in the DM group and increased in the DM + PD groups (Figure 2A). The weight of the salivary gland decreased significantly in the DM group with no difference between the DM + PD groups (Figure 2B).

Table 1. Body weight and blood glucose.

	Normal	DM	DM + PD
Body weight (g)	26.46 ± 0.37	30.74 ± 1.30 *	30.06 ± 1.86 *
Blood glucose (mg/dL)	179.29 ± 17.61	571.33 ± 70.22 *	562.86 ± 98.27

All data are expressed as the mean ± the SD. * $p < 0.05$ vs. normal control group.

A

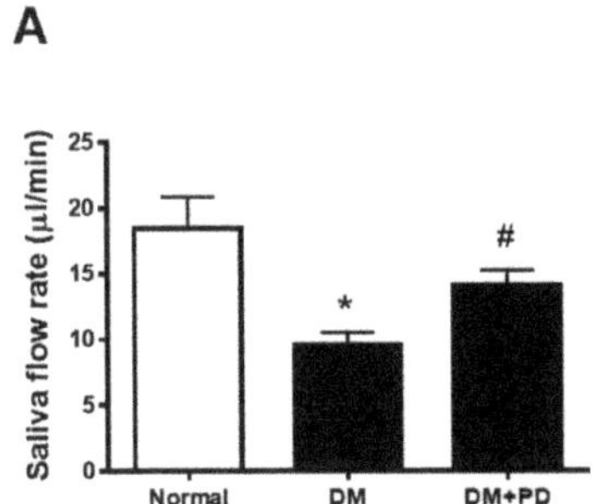

B

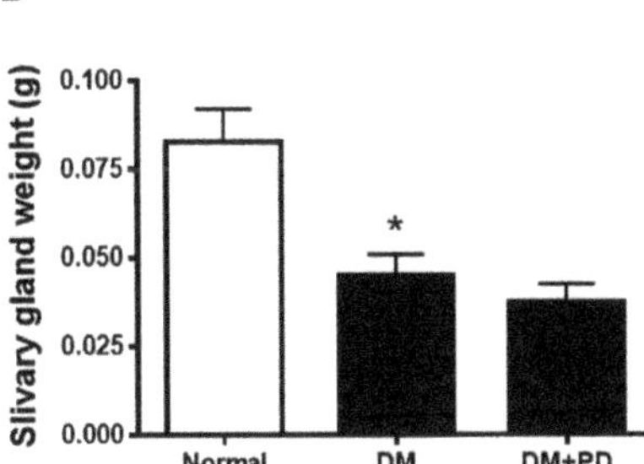

Figure 2. Effect of polydatin on salivation in type 2 diabetic db/db mouse. (**A**) Salivary flow rate. (**B**) Submandibular gland weight. Data are presented as the mean ± standard deviation ($n = 7$). * $p < 0.05$ vs. Normal group; # $p < 0.05$ vs. DM group.

3.3. Polydatin Inhibits Apoptosis in the Submandibular Gland in Type 2 Diabetic Mice

H&E staining was conducted to evaluate the histological change of the submandibular gland. As shown in Figure 3A, the size of duct markedly decreased in the DM group with no difference between the DM + PD groups. To confirm whether this histological change is due to apoptotic injury, a TUNEL assay was conducted in the submandibular gland. The number of TUNEL-positive cells increased significantly in the DM group and decreased significantly in the DM + PD groups (Figure 3B).

3.4. Polydatin Decreases AGEs Accumulations and RAGE Expression in the Submandibular Gland in Type 2 Diabetic Mice

In hyperglycemia, an increase in the glycation process accelerates the accumulation of AGEs [22]. Therefore, the measurement of AGEs in the submandibular gland confirmed the accumulation of AGEs in the DM group and the reduction of AGEs by polydatin treatment (Figure 4A). As for RAGE, the expression level of RAGE increases due to hyperglycemia, resulting in the generation increase in ROS [23]. Therefore, the result of checking the expression level of RAGE confirmed that the expression of RAGE was increased in the DM group and decreased in the DM + PD groups treated with polydatin (Figure 4B).

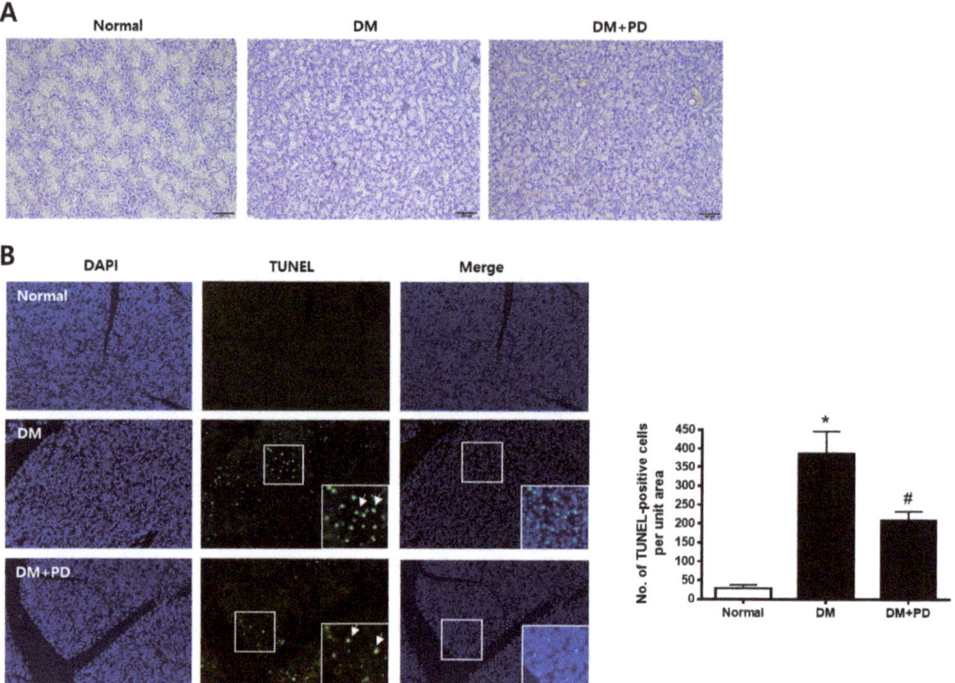

Figure 3. Effect of polydatin on histological changes and apoptosis in submandibular glands. (**A**) Histological analysis of the submandibular gland was performed and images of H&E staining are presented. (**B**) TUNEL staining is presented. Arrows indicate TUNEL-positive cells. Data are presented as the mean ± standard deviation (n = 7). * p < 0.05 vs. Normal group; # p < 0.05 vs. DM group.

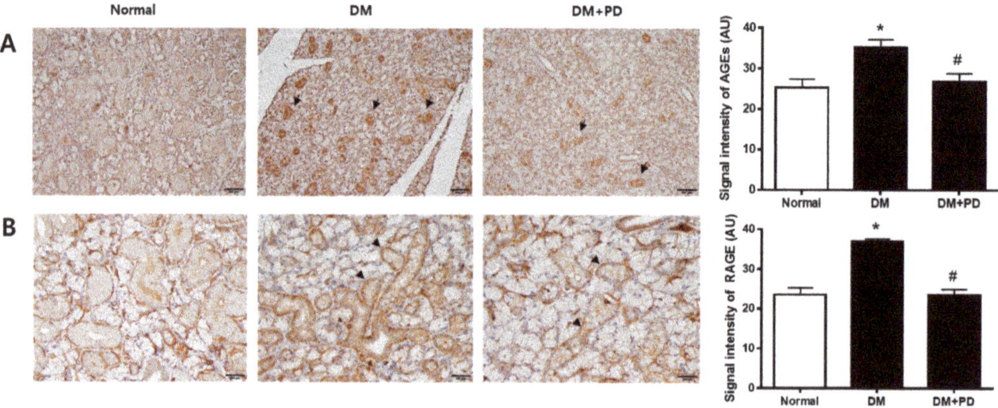

Figure 4. Effect of polydatin on the expression of AGEs and RAGE in submandibular glands. (**A**) Immunohistochemical staining for AGEs. Quantification of the AGEs signal intensity. Arrows indicate AGEs positively stained cells. (**B**) Immunohistochemical staining for RAGE. Quantification of the RAGE signal intensity. Arrowheads indicate RAGE positively stained cells. Data are presented as the mean ± standard deviation (n = 7). * p < 0.05 vs. Normal group; # p < 0.05 vs. DM group.

3.5. Polydatin Decreases the Expressions of HMGB1 and 8-OHdG in the Submandibular Gland in Type 2 Diabetic Mice

To confirm the change caused by oxidative stress-induced cell damage, the accumulation pattern of HMGB1 was checked [24]. In the DM group, the accumulation of HMGB1 and accumulation reduction due to polydatin treatment were confirmed (Figure 5A). To evaluate salivary ROS levels in d-galactose-induced aging rats, we examined ROS ELISA assay and immunohistochemical staining of 8-OHdG, a marker for oxidative DNA damage [25], in salivary gland tissues. As shown in Figure 5B,C, ROS generation and 8-OHdG expression were largely increased in the DM group. Polydatin significantly prevented the generation of ROS and oxidative DNA damage in diabetic mice.

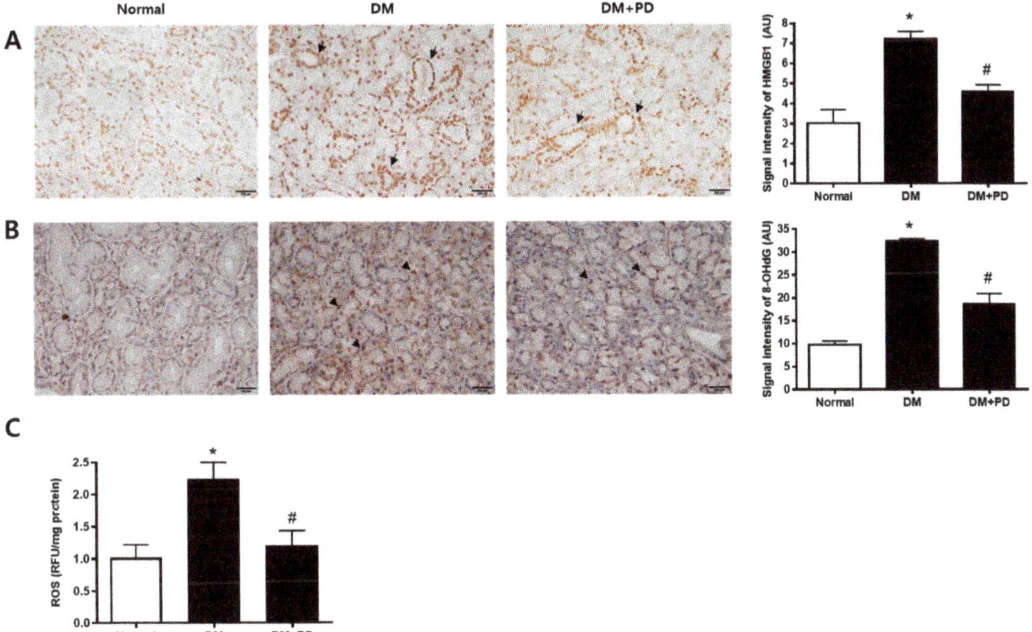

Figure 5. Effect of polydatin on ROS generation in submandibular glands. (**A**) Immunohistochemical staining for HMGB1. Quantification of the HMGB1 signal intensity. Arrows indicate HMGB1 positively stained cells. (**B**) Immunohistochemical staining for 8-OHdG. Quantification of the 8-OHdG signal intensity. Arrowheads indicate 8-OHdG positively stained cells. (**C**) ELISA assay for ROS in the salivary gland. Data are presented as the mean ± standard deviation ($n = 7$). * $p < 0.05$ vs. Normal group; # $p < 0.05$ vs. DM group.

3.6. Polydatin Decreases Mucin Accumulation and the Increased AQP5 Channel in the Submandibular Gland in Type 2 Diabetic Mice

The change aspect of mucin accumulation in salivary gland acinar cells was confirmed through Periodic acid-Schiff (PAS) staining. In the DM group, it was possible to confirm the increase in mucin accumulation and that it decreased due to polydatin treatment (Figure 6A). In addition, the expression level of AQP5 was significantly reduced in the DM group and increased in the DM + PD groups by polydatin treatment (Figure 6B).

3.5. Polydatin Decreases the Expressions of HMGB1 and 8-OHdG in the Submandibular Gland in Type 2 Diabetic Mice

To confirm the change caused by oxidative stress-induced cell damage, the accumulation pattern of HMGB1 was checked [24]. In the DM group, the accumulation of HMGB1 and accumulation reduction due to polydatin treatment were confirmed (Figure 5A). To evaluate salivary ROS levels in d-galactose-induced aging rats, we examined ROS ELISA assay and immunohistochemical staining of 8-OHdG, a marker for oxidative DNA damage [25], in salivary gland tissues. As shown in Figure 5B,C, ROS generation and 8-OHdG expression were largely increased in the DM group. Polydatin significantly prevented the generation of ROS and oxidative DNA damage in diabetic mice.

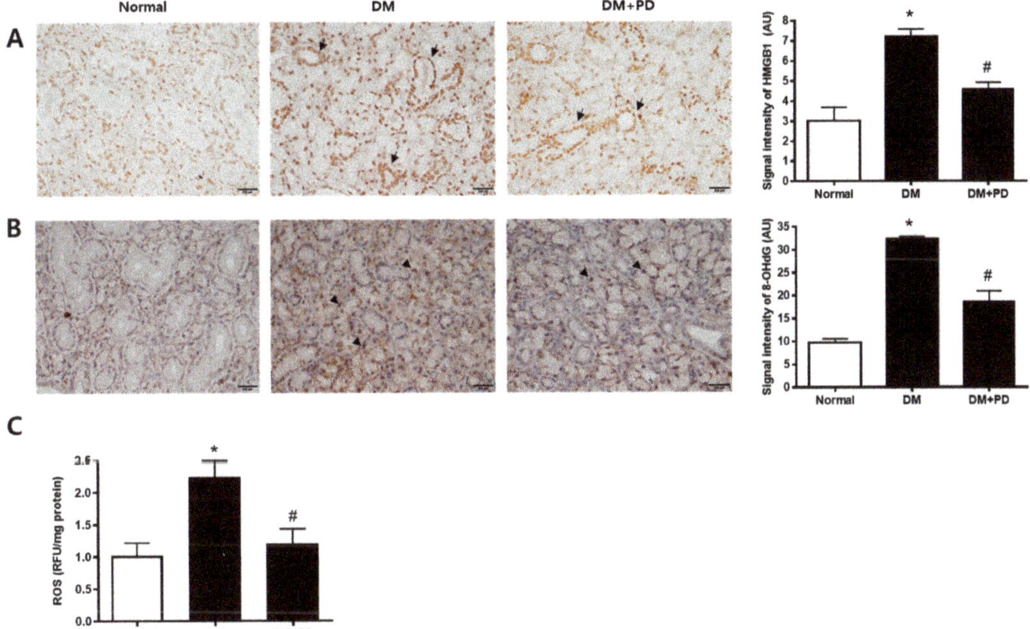

Figure 5. Effect of polydatin on ROS generation in submandibular glands. (**A**) Immunohistochemical staining for HMGB1. Quantification of the HMGB1 signal intensity. Arrows indicate HMGB1 positively stained cells. (**B**) Immunohistochemical staining for 8-OHdG. Quantification of the 8-OHdG signal intensity. Arrowheads indicate 8-OHdG positively stained cells. (**C**) ELISA assay for ROS in the salivary gland. Data are presented as the mean ± standard deviation ($n = 7$). * $p < 0.05$ vs. Normal group; # $p < 0.05$ vs. DM group.

3.6. Polydatin Decreases Mucin Accumulation and the Increased AQP5 Channel in the Submandibular Gland in Type 2 Diabetic Mice

The change aspect of mucin accumulation in salivary gland acinar cells was confirmed through Periodic acid-Schiff (PAS) staining. In the DM group, it was possible to confirm the increase in mucin accumulation and that it decreased due to polydatin treatment (Figure 6A). In addition, the expression level of AQP5 was significantly reduced in the DM group and increased in the DM + PD groups by polydatin treatment (Figure 6B).

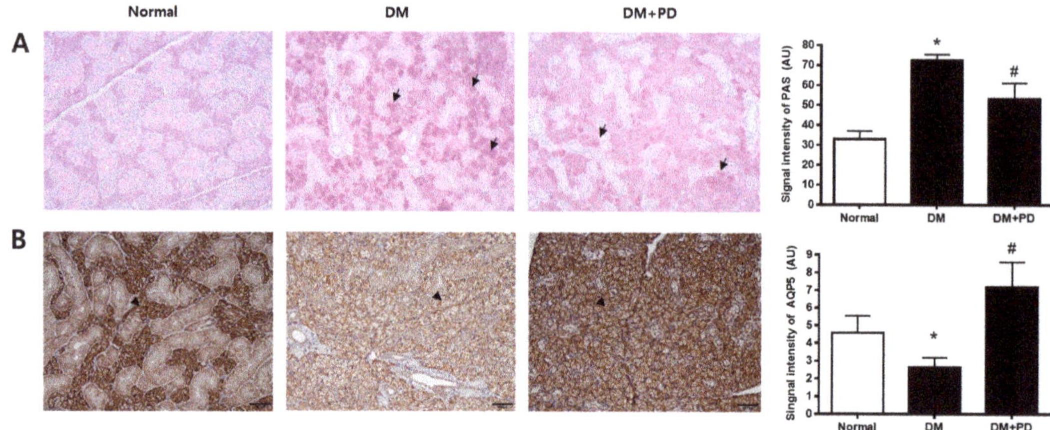

Figure 6. Effect of polydatin on mucin accumulation and AQP5 expression in submandibular glands. (**A**) Periodic acid-Schiff (PAS) staining of submandibular glands. Arrows indicate PAS positively stained cells. (**B**) Immunohistochemical staining for AQP5. Quantification of the AQP5 signal intensity. Arrowheads indicate AQP5 positively stained cells. Data are presented as the mean ± standard deviation ($n = 7$). * $p < 0.05$ vs. Normal group; # $p < 0.05$ vs. DM group.

4. Discussion

Hyperglycemia due to high blood glucose levels is a characteristic of diabetes, which causes many changes in the body through damage to various organs. Hyposalivation and submandibular gland dysfunction have been reported in in vivo models, especially in type 2 diabetic mice [26], non-obese diabetic mice [27], streptozotocin (STZ)-induced type 1 diabetic rats [28], and this study confirmed the improvement effect of polydatin on diabetes-induced hyposalivation in a type 2 diabetic db/db mice model.

This study could not confirm the glycemic control effect of polydatin. Previous studies have confirmed a decrease in glucose in a STZ-induced diabetic mice model [29,30]. A decrease in blood glucose was also confirmed in STZ-induced diabetic male C57BL/6J mice after 8 weeks of polydatin administration [31]. In addition, similar to this study, there is a study showing that blood glucose decreases when 100 mg/kg of polydatin is administered in type 2 diabetic db/db female mice for 4 weeks [18,32]. However, in the oral administration experiment of 100 mg/kg polydatin for 4 weeks in type 2 diabetic male C57BL/6J mice via a high-fat diet, the reduction of blood glucose could not be confirmed [33]. These results are thought to have resulted from the combination of female and male insulin resistance difference [34,35], the difference in pathogenesis for type 1 and 2 diabetes mellitus [36], and the usage, which requires a more detailed study to find the causes. However, in this study, it was possible to confirm the improvement of salivary gland hyposalivation by polydatin without the alleviation of hyperglycemia, which is considered to be a significant finding since there are therapeutic drugs that show a protective effect of diabetic complications without glycemic control [37].

The level of ROS maintained by the antioxidant system of cells in the body increases due to hyperglycemia, disrupting the balance; increased ROS causes apoptosis and inflammation through various signaling paths in cell compositions such as proteins and lipids and organ systems such as the brain, blood vessels, and kidneys, ultimately destroying the systems [38–40].

In addition, the Maillard reaction (non-enzymatic) initiated by the glycation of amino groups and carbonyl groups of proteins causes molecular glycation in circulating proteins, including albumin, lipoprotein, and insulin, and allows for the easy generation of ROS through increased oxidation sensitivity, resulting in the increase in the apoptosis pathway. Hyperglycemia promotes the generation of a substance called AGEs, the final form of products produced by glycation; AGEs produce irreversible cross-links between proteins, transform properties and structure, increase tissue fibrosis and stiffness, cause organic dysfunction, and combine with RAGE to increase oxidative stress and pro-inflammatory signaling pathways. Consequently, it induces vascular complications such as diabetic atherosclerosis, retinopathy, diabetic nephropathy, neuropathy, and impaired wound healing [41–43]. In type 2 DM, hyperglycemia induces salivary gland cell death through the generation of ROS, leading to xerostomia [44], and there is a study in which an increase in RAGE was confirmed in the salivary gland of patients with Sjögren's syndrome [45].

In this study, the accumulation of AGEs and the increased expression of RAGE in the DM group showed that the increase in glycation caused by hyperglycemia was related to salivary gland hyposalivation. There are reports about the accumulation inhibition of AGEs by resveratrol [46], the formation inhibition of polydatin's BSA-MGO reaction induced AGEs [47], and the expression inhibition of RAGE in Schwann cells of diabetic rats [48], and this study confirmed the reduction of AGEs and RAGE by polydatin in the DM + PD group, showing in addition that the improvement of diabetes-induced hyposalivation by polydatin treatment is related to the anti-glycation effect of polydatin.

HMGB1 is a chromatin-binding nuclear protein that acts as an extracellular signal in damaged cells, is related to oxidative stress, and is released in apoptosis [24,49]. In addition, there is a report that the accumulation of HMGB1 is observed in live cells with oxidative-induced damage [50]. There is another report that 8-OHdG is a biomarker of DNA-endogenous oxidative damage, which is mainly used as an indicator of oxidative stress in diabetic nephropathy [25,51]. HMGB1 increased in the submandibular gland of the diabetes-xerostomia rats model [52], and 8-OHdG increased in the saliva of xerostomia patients [53]. In this study, the result of checking DNA damage caused by a hyperglycemia-induced ROS increase in the DM group confirmed the increase in oxidative DNA damage of salivary gland cells by oxidative stress through the increase in HMGB1 and 8-OHdG. This study confirmed that glycation and oxidative stress markers increased by hyperglycemia were decreased by polydatin in the submandibular gland of type 2 diabetic mice, which confirmed that polydatin has antioxidative and anti-glycation effects.

Resveratrol, commonly found in nature in things such as grapes, peanuts, wine, and chocolate products, is a type of polyphenol and is known to have various pharmacological effects, such as anti-inflammation, antioxidation, and anti-tumor effects [54]. There is a report that polydatin, a form in which glucose is combined with resveratrol, is the richest form of resveratrol and metabolizes with resveratrol, which has several pharmacological effects when absorbed in the body, has a higher antioxidation and anti-glycation effect based on better absorption, and has better metabolic stability than resveratrol when administered orally with the same dose as resveratrol [17,47,55,56]. It can be seen that these pharmacological activities of polydatin were involved in the inhibition of hyposalivation and apoptosis in db/db mice.

In the present study, the effective dose of polydatin is 100 mg/kg. Considering an average body weight of an adult of 60 kg, this dose for a 60 kg human is equal to 0.4 g/day. The recommended dose of polydatin as a human food supplement is 160 mg/day [57]. Even if a high dose of polydatin was required to obtain a significant effect in salivary gland dysfunction, polydatin is known to be absent of side effects and toxicity.

Mucin, the main component of saliva, can change in the pathological condition of the salivary gland. In addition, the over-expression of mucin is promoted by pro-inflammatory cytokines, and cytokine production is associated with an increase in ROS [58–60]. Huang et al. reported that db/db mice demonstrated that PAS-positive mucins accumulated in the acini of submandibular glands [58]. Mucin accumulation in salivary glands also suggests that the salivary secretory function was decreased under hyperglycemic conditions. Therefore, as a result of measuring submandibular gland mucin through PAS staining, the damage of salivary gland acinar cells caused by diabetes in the DM group was confirmed through an increase in mucin. The prevention of salivary gland dysfunction by the antioxidant effect of polydatin was confirmed through a decrease in mucin in the DM + PD groups

Saliva is produced and secreted by the acinar cells and duct cells of the salivary gland with its flow initiated by an increase and release of Ca^{2+} in the acinar cells, which in turn creates osmotic pressure, and fluid is secreted by the AQP5 channel in the membrane by the osmotic gradient [61]. AQP is present in most organisms, including mammals, plants, and other microbes. There are 13 AQP gene families in humans, which function as water channels in the eyes, brain, and secretory glands [62]. Among them, AQP5, which is localized to the apical membrane of the salivary gland acinar cells, plays a very important role in the generation of tears, saliva, and exocrine secretions [63], and mice with AQP5 knockout have salivation reduced by more than 60% compared to normal mice [64]. As a result of checking the expression of AQP5, a significant decrease was confirmed, which is presumed to be related to the decrease in AQP5 expression due to the increase in ROS [65], and it is assumed that the reduction of AQP5 expression was alleviated by the antioxidation effect of polydatin.

In the present study, polydatin also promoted saliva secretion in normoglycemic mice. we did not clearly show the mode-of-action of polydatin on salvation in normal mice. However, polydatin is a natural precursor of resveratrol. Polydatin showed better oral absorption and metabolic stability than resveratrol since its serum concentration was 3–4 times higher after oral administration at the same dosage [55]. Polydatin is structurally the same as resveratrol except that it has a glucoside group attached to the C-3 position in place of a hydroxyl group. This substitution makes polydatin more water-soluble and resistant to enzymatic breakdown than resveratrol [16]. Polydatin shares resveratrol's beneficial pharmaceutical properties, with the further advantage of being more abundant than resveratrol [66]. Multiple direct targets of resveratrol, including COX, PPAR, eNOS, and Sirt1, have been identified. Inoue et al. reported that resveratrol increased Sirt1 activity in the salivary gland of NOD mice [21]. This protein regulates a wide range of cellular processes. Forkhead box O (FOXO) transcription factors have been identified as substrates of SIRT1 [67]. SIRT1 regulates the transactivation activity of FOXO by catalyzing its deacetylation [68]. It was recently reported that FOXO1 directly regulates AQP5 expression in salivary gland acinar cells. FOXO1 is a direct regulator of AQP5 expression in salivary gland acinar cells through its interaction with the promoter region of AQP5 [69]. AQP5 is the most important protein for salivation since it directly mediates the transcellular movement of water from the basal to apical or lumen side of acinar cells [70]. Based on the results, we have suggested a possibility that polydatin may promote salivation via the upregulation of AQP5 expression. To identify this additional role of polydatin on AQP5 expression, we performed the immunohistochemical staining for AQP5 in salivary glands. AQP5 was highly expressed in acinar cells of the normoglycemic mice tread with polydatin.

The effects of polydatin on salivation in normoglycemic mice as well as db/db mice were studied in this work, but one of the limitations of this study was that the animal study using the normoglycemic mice was a pilot study to determine the oral dose of polydatin. Therefore, there is not enough analysis data at the molecular level. Another limitation of this study was that multiple oral doses of polydatin were required to determine the effective dose range in the animal study using the db/db mice. The detailed beneficial role of polydatin on salivation under normoglycemic and hyperglycemic conditions needs to be studied further.

In conclusion, an increase in ROS, a glycation product in the submandibular gland; a decrease in AQP5, which plays an important role in salvation caused by hyperglycemia, an apoptotic signal increase; and a salivation decrease in type 2 diabetic mice were confirmed. Polydatin mitigated the generation of excessive ROS due to the reduction of the glycation process initiated by the anti-glycation effect of polydatin and upregulated AQP5 expression, which is presumed to lead to the improvement of diabetic salivary gland dysfunction (Figure 7). Overall, this study shows polydatin's potential for use as a therapeutic drug in diabetic-xerostomia due to its anti-glycation and antioxidant effects.

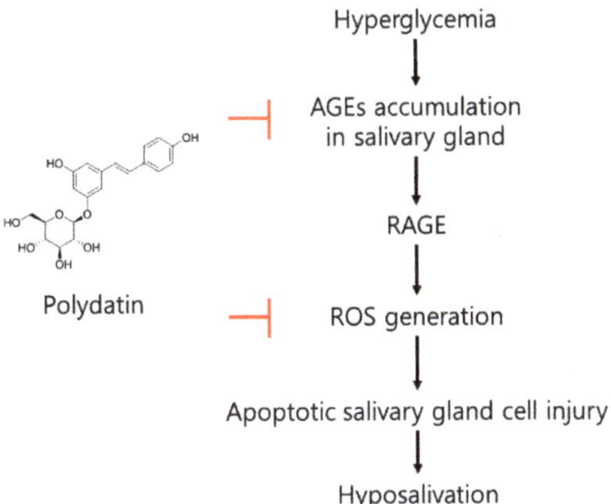

Figure 7. The possible mode of action of polydatin on diabetes-induced hypofunction of salivary glands.

Author Contributions: H.R.K. performed the experiments and wrote the manuscript; W.K.J., S.-B.P., H.Y.R. and Y.H.K. performed the experiments and analyzed the data; J.K. designed and supervised the study. All authors have read and agreed to the published version of the manuscript.

Funding: This study was supported by a National Research Foundation of Korea (NRF) grant funded by the Korea government (MEST) (No. NRF-2019R1A2C1008773).

Institutional Review Board Statement: The animals used in this study were treated according to the protocol approved by Ethics Committee of Animal Experimentation of Jeonbuk National University (IACUC No. 2021-087).

Informed Consent Statement: Not applicable.

Data Availability Statement: The data presented in this study are available on request from the corresponding author.

Conflicts of Interest: The authors declare no conflict of interest.

References

1. Miranda-Rius, J.; Brunet-Llobet, L.; Lahor-Soler, E.; Farré, M. Salivary secretory disorders, inducing drugs, and clinical management. *Int. J. Med. Sci.* **2015**, *12*, 811–824. [CrossRef]
2. Proctor, G.B. The physiology of salivary secretion. *Periodontology 2000* **2016**, *70*, 11–25. [CrossRef]
3. Mese, H.; Matsuo, R. Salivary secretion, taste and hyposalivation. *J. Oral Rehabil.* **2007**, *34*, 711–723. [CrossRef] [PubMed]
4. van der Putten, G.J.; Brand, H.S.; Schols, J.M.; de Baat, C. The diagnostic suitability of a xerostomia questionnaire and the association between xerostomia, hyposalivation and medication use in a group of nursing home residents. *Clin. Oral Investig.* **2011**, *15*, 185–192. [CrossRef] [PubMed]
5. Guggenheimer, J.; Moore, P.A. Xerostomia: Etiology, recognition and treatment. *J. Am. Dent. Assoc.* **2003**, *134*, 61–69. [CrossRef]
6. Atkinson, J.C.; Grisius, M.; Massey, W. Salivary hypofunction and xerostomia: Diagnosis and treatment. *Dent. Clin. N. Am.* **2005**, *49*, 309–326. [CrossRef] [PubMed]
7. Moore, P.A.; Guggenheimer, J.; Etzel, K.R.; Weyant, R.J.; Orchard, T. Type 1 diabetes mellitus, xerostomia, and salivary flow rates. *Oral Surg. Oral Med. Oral Pathol. Oral Radiol. Endod.* **2001**, *92*, 281–291. [CrossRef]
8. Feng, J.K.; Lu, Y.F.; Li, J.; Qi, Y.H.; Yi, M.L.; Ma, D.Y. Upregulation of salivary $\alpha 2$ macroglobulin in patients with type 2 diabetes mellitus. *Genet. Mol. Res.* **2015**, *14*, 2268–2274. [CrossRef]
9. Singh, V.P.; Bali, A.; Singh, N.; Jaggi, A.S. Advanced glycation end products and diabetic complications. *Korean J. Physiol. Pharmacol.* **2014**, *18*, 1–14. [CrossRef]
10. Sreebny, L.M.; Yu, A.; Green, A.; Valdini, A. Xerostomia in diabetes mellitus. *Diabetes Care* **1992**, *15*, 900–904. [CrossRef]
11. Rohani, B. Oral manifestations in patients with diabetes mellitus. *World J. Diabetes* **2019**, *10*, 485–489. [CrossRef] [PubMed]
12. Cicmil, S.; Mladenović, I.; Krunić, J.; Ivanović, D.; Stojanovic, N. Oral alterations in diabetes mdellitus. *Balk. J. Dent. Med.* **2018**, *22*, 7–14. [CrossRef]

13. Visvanathan, V.; Nix, P. Managing the patient presenting with xerostomia: A review. *Int. J. Clin. Pract.* **2010**, *64*, 404–407. [CrossRef]

14. Bascones-Martínez, A.; Muñoz-Corcuera, M.; Bascones-Ilundain, C. Side effects of drugs on the oral cavity. *Med. Clin.* **2015**, *144*, 126–131. [CrossRef]

15. Selvan, S.R.; Venugopalan, S. Effect of oral hypoglycemic drugs on salivary flow—A review. *Int. J. Pharmtech. Res.* **2013**, *5*, 1608–1610.

16. Xie, X.; Peng, J.; Huang, K.; Huang, J.; Shen, X.; Liu, P.; Huang, H. Polydatin ameliorates experimental diabetes-induced fibronectin through inhibiting the activation of NF-κB signaling pathway in rat glomerular mesangial cells. *Mol. Cell Endocrinol.* **2012**, *362*, 183–193. [CrossRef]

17. Du, Q.H.; Peng, C.; Zhang, H. Polydatin: A review of pharmacology and pharmacokinetics. *Pharm. Biol.* **2013**, *51*, 1347–1354. [CrossRef] [PubMed]

18. Wang, Y.; Ye, J.; Li, J.; Chen, C.; Huang, J.; Liu, P.; Huang, H. Polydatin ameliorates lipid and glucose metabolism in type 2 diabetes mellitus by downregulating proprotein convertase subtilisin/kexin type 9 (PCSK9). *Cardiovasc. Diabetol.* **2016**, *15*, 19. [CrossRef] [PubMed]

19. Şimşek, G.; Gürocak, S.; Karadağ, N.; Karabulut, A.B.; Demirtaş, E.; Karataş, E.; Pepele, E. Protective effects of resveratrol on salivary gland damage induced by total body irradiation in rats. *Laryngoscope* **2012**, *122*, 2743–2748. [CrossRef] [PubMed]

20. Xu, L.; Yang, X.; Cai, J.; Ma, J.; Cheng, H.; Zhao, K.; Yang, L.; Cao, Y.; Qin, Q.; Zhang, C.; et al. Resveratrol attenuates radiation-induced salivary gland dysfunction in mice. *Laryngoscope* **2013**, *123*, 23–29. [CrossRef]

21. Inoue, H.; Kishimoto, A.; Ushikoshi-Nakayama, R.; Hasaka, A.; Takahashi, A.; Ryo, K.; Muramatsu, T.; Ide, F.; Mishima, K.; Saito, I. Resveratrol improves salivary dysfunction in a non-obese diabetic (NOD) mouse model of Sjögren's syndrome. *J. Clin. Biochem. Nutr.* **2016**, *59*, 107–112. [CrossRef]

22. Yan, S.F.; Ramasamy, R.; Schmidt, A.M. Mechanisms of disease: Advanced glycation end-products and their receptor in inflammation and diabetes complications. *Nat. Clin. Pract. Endocrinol. Metab.* **2008**, *4*, 285–293. [CrossRef] [PubMed]

23. Yao, D.; Brownlee, M. Hyperglycemia-induced reactive oxygen species increase expression of the receptor for advanced glycation end products (RAGE) and RAGE ligands. *Diabetes* **2010**, *59*, 249–255. [CrossRef] [PubMed]

24. Tang, D.; Kang, R.; Zeh, H.J., 3rd; Lotze, M.T. High-mobility group box 1, oxidative stress, and disease. *Antioxid. Redox. Signal.* **2011**, *14*, 1315–1335. [CrossRef] [PubMed]

25. Valavanidis, A.; Vlachogianni, T.; Fiotakis, C. 8-hydroxy-2′ -deoxyguanosine (8-OHdG): A critical biomarker of oxidative stress and carcinogenesis. *J. Environ. Sci. Health C Environ. Carcinog. Ecotoxicol. Rev.* **2009**, *27*, 120–139. [CrossRef]

26. Xiang, R.-L.; Huang, Y.; Zhang, Y.; Cong, X.; Zhang, Z.-J.; Wu, L.-L.; Yu, G.-Y. Type 2 diabetes-induced hyposalivation of the submandibular gland through PINK1/Parkin-mediated mitophagy. *J. Cell. Physiol.* **2020**, *235*, 232–244. [CrossRef]

27. Allushi, B.; Bagavant, H.; Papinska, J.; Deshmukh, U.S. Hyperglycemia and salivary gland dysfunction in the non-obese diabetic mouse: Caveats for preclinical studies in Sjögren's syndrome. *Sci. Rep.* **2019**, *9*, 17969. [CrossRef]

28. Fedirko, N.V.; Kruglikov, I.A.; Kopach, O.V.; Vats, J.A.; Kostyuk, P.G.; Voitenko, N.V. Changes in functioning of rat submandibular salivary gland under streptozotocin-induced diabetes are associated with alterations of Ca^{2+} signaling and Ca^{2+} transporting pumps. *Biochim. Biophys. Acta* **2006**, *1762*, 294–303. [CrossRef]

29. Wang, L.; Huang, L.; Li, N.; Miao, J.; Liu, W.; Yu, J. Ameliorative effect of polydatin on hyperglycemia and renal injury in streptozotocin-induced diabetic rats. *Cell Mol. Biol.* **2019**, *65*, 55–59. [CrossRef]

30. Yousef, A.; Shawki, H.; El-Shahawy, A.; El-Twab, S.; Abdel Moneim, A.; Oishi, H. Polydatin mitigates pancreatic β-cell damage through its antioxidant activity. *Biomed. Pharmacother.* **2021**, *133*, 111027. [CrossRef]

31. Gong, W.; Li, J.; Chen, Z.; Huang, J.; Chen, Q.; Cai, W.; Liu, P.; Huang, H. Polydatin promotes Nrf2-ARE anti-oxidative pathway through activating CKIP-1 to resist HG-induced up-regulation of FN and ICAM-1 in GMCs and diabetic mice kidneys. *Free Radic. Biol. Med.* **2017**, *106*, 393–405. [CrossRef]

32. Chen, C.; Huang, K.; Hao, J.; Huang, J.; Yang, Z.; Xiong, F.; Liu, P.; Huang, H. Polydatin attenuates AGEs-induced upregulation of fibronectin and ICAM-1 in rat glomerular mesangial cells and db/db diabetic mice kidneys by inhibiting the activation of the SphK1-S1P signaling pathway. *Mol. Cell Endocrinol.* **2016**, *427*, 45–56. [CrossRef]

33. Zheng, L.; Wu, J.; Mo, J.; Guo, L.; Wu, X.; Bao, Y. Polydatin inhibits adipose tissue inflammation and ameliorates lipid metabolism in high-fat-fed mice. *Biomed Res. Int.* **2019**, *2019*, 7196535. [CrossRef]

34. Tramunt, B.; Smati, S.; Grandgeorge, N.; Lenfant, F.; Arnal, J.F.; Montagner, A.; Gourdy, P. Sex differences in metabolic regulation and diabetes susceptibility. *Diabetologia* **2020**, *63*, 453–461. [CrossRef]

35. Elzinga, S.E.; Savelieff, M.G.; O'Brien, P.D.; Mendelson, F.E.; Hayes, J.M.; Feldman, E.L. Sex differences in insulin resistance, but not peripheral neuropathy, in a diet-induced prediabetes mouse model. *Dis. Model Mech.* **2021**, *14*, dmm048909. [CrossRef] [PubMed]

36. Salsali, A.; Nathan, M. A review of types 1 and 2 diabetes mellitus and their treatment with insulin. *Am. J. Ther.* **2006**, *13*, 349–361. [CrossRef] [PubMed]

37. Kang, W.S.; Jung, W.K.; Park, S.-B.; Kim, H.R.; Kim, J. Gemigliptin suppresses salivary dysfunction in streptozotocin-induced diabetic rats. *Biomed. Pharmacother.* **2021**, *137*, 111297. [CrossRef] [PubMed]

38. Du, X.-L.; Edelstein, D.; Rossetti, L.; Fantus, I.G.; Goldberg, H.; Ziyadeh, F.; Wu, J.; Brownlee, M. Hyperglycemia-induced mitochondrial superoxide overproduction activates the hexosamine pathway and induces plasminogen activator inhibitor-1 expression by increasing Sp1 glycosylation. *Proc. Natl. Acad. Sci. USA* **2000**, *97*, 12222–12226. [CrossRef]
39. Yan, L.J. Redox imbalance stress in diabetes mellitus: Role of the polyol pathway. *Anim. Model Exp. Med.* **2018**, *1*, 7–13. [CrossRef] [PubMed]
40. Auten, R.L.; Davis, J.M. Oxygen toxicity and reactive oxygen species: The devil is in the details. *Pediatr. Res.* **2009**, *66*, 121–127. [CrossRef]
41. Kranstuber, A.L.; Del Rio, C.; Biesiadecki, B.J.; Hamlin, R.L.; Ottobre, J.; Gyorke, S.; Lacombe, V.A. Advanced glycation end product cross-link breaker attenuates diabetes-induced cardiac dysfunction by improving sarcoplasmic reticulum calcium handling. *Front. Physiol.* **2012**, *3*, 292. [CrossRef] [PubMed]
42. Sanajou, D.; Ghorbani Haghjo, A.; Argani, H.; Aslani, S. AGE-RAGE axis blockade in diabetic nephropathy: Current status and future directions. *Eur. J. Pharmacol.* **2018**, *833*, 158–164. [CrossRef] [PubMed]
43. Negre-Salvayre, A.; Salvayre, R.; Augé, N.; Pamplona, R.; Portero-Otín, M. Hyperglycemia and glycation in diabetic complications. *Antioxid. Redox. Signal.* **2009**, *11*, 3071–3109. [CrossRef]
44. Matsumoto, N.; Omagari, D.; Ushikoshi-Nakayama, R.; Yamazaki, T.; Inoue, H.; Saito, I. Hyperglycemia induces generation of reactive oxygen species and accelerates apoptotic cell death in salivary gland cells. *Pathobiology* **2021**, *88*, 234–241. [CrossRef]
45. Katz, J.; Stavropoulos, F.; Bhattacharyya, I.; Stewart, C.; Perez, F.M.; Caudle, R.M. Receptor of advanced glycation end product (RAGE) expression in the minor salivary glands of patients with Sjögren's syndrome: A preliminary study. *Scand. J. Rheumatol.* **2004**, *33*, 174–178. [CrossRef] [PubMed]
46. Maleki, V.; Foroumandi, E.; Hajizadeh-Sharafabad, F.; Kheirouri, S.; Alizadeh, M. The effect of resveratrol on advanced glycation end products in diabetes mellitus: A systematic review. *Arch. Physiol. Biochem.* **2020**. [CrossRef]
47. Sheng, Z.; Ai, B.; Zheng, L.; Zheng, X.; Xu, Z.; Shen, Y.; Jin, Z. Inhibitory activities of kaempferol, galangin, carnosic acid and polydatin against glycation and α-amylase and α-glucosidase enzymes. *Int. J. Food Sci. Technol.* **2018**, *53*, 755–766. [CrossRef]
48. Chen, L.; Chen, Z.; Xu, Z.; Feng, W.; Yang, X.; Qi, Z. Polydatin protects Schwann cells from methylglyoxal induced cytotoxicity and promotes crushed sciatic nerves regeneration of diabetic rats. *Phytother. Res.* **2021**, *35*, 4592–4604. [CrossRef]
49. Yu, Y.; Tang, D.; Kang, R. Oxidative stress-mediated HMGB1 biology. *Front. Physiol.* **2015**, *6*, 93. [CrossRef]
50. Prasad, R.; Liu, Y.; Deterding, L.J.; Poltoratsky, V.P.; Kedar, P.S.; Horton, J.K.; Kanno, S.; Asagoshi, K.; Hou, E.W.; Khodyreva, S.N.; et al. HMGB1 is a cofactor in mammalian base excision repair. *Mol. Cell* **2007**, *27*, 829–841. [CrossRef]
51. Kim, C.S.; Jo, K.; Kim, J.S.; Pyo, M.K.; Kim, J. GS-E3D, a new pectin lyase-modified red ginseng extract, inhibited diabetes-related renal dysfunction in streptozotocin-induced diabetic rats. *BMC Complement. Altern. Med.* **2017**, *17*, 430. [CrossRef]
52. Fukuoka, C.Y.; Simões, A.; Uchiyama, T.; Arana-Chavez, V.E.; Abiko, Y.; Kuboyama, N.; Bhawal, U.K. The effects of low-power laser irradiation on inflammation and apoptosis in submandibular glands of diabetes-induced rats. *PLoS ONE* **2017**, *12*, e0169443. [CrossRef]
53. Ryo, K.; Yamada, H.; Nakagawa, Y.; Tai, Y.; Obara, K.; Inoue, H.; Mishima, K.; Saito, I. Possible involvement of oxidative stress in salivary gland of patients with Sjogren's syndrome. *Pathobiology* **2006**, *73*, 252–260. [CrossRef]
54. Martín, A.R.; Villegas, I.; La Casa, C.; de la Lastra, C.A. Resveratrol, a polyphenol found in grapes, suppresses oxidative damage and stimulates apoptosis during early colonic inflammation in rats. *Biochem. Pharmacol.* **2004**, *67*, 1399–1410.
55. Wang, H.L.; Gao, J.P.; Han, Y.L.; Xu, X.; Wu, R.; Gao, Y.; Cui, X.H. Comparative studies of polydatin and resveratrol on mutual transformation and antioxidative effect in vivo. *Phytomedicine* **2015**, *22*, 553–559. [CrossRef] [PubMed]
56. Regev-Shoshani, G.; Shoseyov, O.; Bilkis, I.; Kerem, Z. Glycosylation of resveratrol protects it from enzymic oxidation. *Biochem. J.* **2003**, *374*, 157–163. [CrossRef]
57. Perrella, F.; Coppola, F.; Petrone, A.; Platella, C.; Montesarchio, D.; Stringaro, A.; Ravagnan, G.; Fuggetta, M.P.; Rega, N.; Musumeci, D. Interference of polydatin/resveratrol in the ACE2: Spike recognition during COVID-19 infection. A focus on their potential mechanism of action through computational and biochemical assays. *Biomolecules* **2021**, *11*, 1048. [CrossRef]
58. Huang, Y.; Mao, Q.Y.; Shi, X.J.; Cong, X.; Zhang, Y.; Wu, L.L.; Yu, G.Y.; Xiang, R.L. Disruption of tight junctions contributes to hyposalivation of salivary glands in a mouse model of type 2 diabetes. *J. Anat.* **2020**, *237*, 556–567. [CrossRef] [PubMed]
59. Castro, I.; Barrera, M.-J.; González, S.; Aguilera, S.; Urzúa, U.; Cortés, J.; González, M.-J. Mucins in salivary gland development, regeneration, and disease. In *Salivary Gland Development and Regeneration: Advances in Research and Clinical Approaches to Functional Restoration*; Springer: New York, NY, USA, 2017; pp. 45–71.
60. Maciejczyk, M.; Skutnik-Radziszewska, A.; Zieniewska, I.; Matczuk, J.; Domel, E.; Waszkiel, D.; Żendzian-Piotrowska, M.; Szarmach, I.; Zalewska, A. Antioxidant defense, oxidative modification, and salivary gland function in an early phase of cerulein pancreatitis. *Oxid. Med. Cell. Longev.* **2019**, *2019*, 8403578. [CrossRef]
61. Ambudkar, I.S. Ca^{2+} signaling and regulation of fluid secretion in salivary gland acinar cells. *Cell Calcium* **2014**, *55*, 297–305. [CrossRef] [PubMed]
62. Agre, P.; King, L.S.; Yasui, M.; Guggino, W.B.; Ottersen, O.P.; Fujiyoshi, Y.; Engel, A.; Nielsen, S. Aquaporin water channels–from atomic structure to clinical medicine. *J. Physiol.* **2002**, *542*, 3–16. [CrossRef]
63. Satoh, K.; Narita, T.; Matsuki-Fukushima, M.; Okabayashi, K.; Ito, T.; Senpuku, H.; Sugiya, H. E2f1-deficient NOD/SCID mice have dry mouth due to a change of acinar/duct structure and the down-regulation of AQP5 in the salivary gland. *Pflugers Arch.* **2013**, *465*, 271–281. [CrossRef]

64. Ma, T.; Song, Y.; Gillespie, A.; Carlson, E.J.; Epstein, C.J.; Verkman, A.S. Defective secretion of saliva in transgenic mice lacking aquaporin-5 water channels. *J. Biol. Chem.* **1999**, *274*, 20071–20074. [CrossRef] [PubMed]
65. Saito, K.; Mori, S.; Date, F.; Hong, G. Epigallocatechin gallate stimulates the neuroreactive salivary secretomotor system in autoimmune sialadenitis of MRL-Fas(lpr) mice via activation of cAMP-dependent protein kinase A and inactivation of nuclear factor κB. *Autoimmunity* **2015**, *48*, 379–388. [CrossRef] [PubMed]
66. Di Benedetto, A.; Posa, F.; De Maria, S.; Ravagnan, G.; Ballini, A.; Porro, C.; Trotta, T.; Grano, M.; Muzio, L.L.; Mori, G. Polydatin, natural precursor of resveratrol, promotes osteogenic differentiation of mesenchymal stem cells. *Int. J. Med. Sci.* **2018**, *15*, 944–952. [CrossRef] [PubMed]
67. Huang, H.; Tindall, D.J. Dynamic FoxO transcription factors. *J. Cell Sci.* **2007**, *120*, 2479–2487. [CrossRef]
68. Motta, M.C.; Divecha, N.; Lemieux, M.; Kamel, C.; Chen, D.; Gu, W.; Bultsma, Y.; McBurney, M.; Guarente, L. Mammalian SIRT1 represses forkhead transcription factors. *Cell* **2004**, *116*, 551–563. [CrossRef]
69. Lee, S.M.; Lee, S.W.; Kang, M.; Choi, J.K.; Park, K.; Byun, J.S.; Kim, D.Y. FoxO1 as a Regulator of Aquaporin 5 Expression in the Salivary Gland. *J. Dent. Res.* **2021**, *100*, 1281–1288. [CrossRef]
70. Matsuzaki, T.; Suzuki, T.; Koyama, H.; Tanaka, S.; Takata, K. Aquaporin-5 (AQP5), a water channel protein, in the rat salivary and lacrimal glands: Immunolocalization and effect of secretory stimulation. *Cell Tissue Res.* **1999**, *295*, 513–521. [CrossRef]

pharmaceutics

Article

Cardioprotective Role of BGP-15 in Ageing Zucker Diabetic Fatty Rat (ZDF) Model: Extended Mitochondrial Longevity

Mate Kozma [1], Mariann Bombicz [1], Balazs Varga [1], Daniel Priksz [1], Rudolf Gesztelyi [1], Vera Tarjanyi [1], Rita Kiss [1], Reka Szekeres [1], Barbara Takacs [1], Akos Menes [1], Jozsef Balla [2], Gyorgy Balla [3], Judit Szilvassy [4], Zoltan Szilvassy [1] and Bela Juhasz [1,*]

[1] Department of Pharmacology and Pharmacotherapy, Faculty of Medicine, University of Debrecen, H-4032 Debrecen, Hungary; matekozma@yahoo.com (M.K.); bombicz.mariann@pharm.unideb.hu (M.B.); varga.balazs@pharm.unideb.hu (B.V.); priksz.daniel@pharm.unideb.hu (D.P.); gesztelyi.rudolf@pharm.unideb.hu (R.G.); veratarjanyi@gmail.com (V.T.); kiss.rita@med.unideb.hu (R.K.); szekeres.reka@med.unideb.hu (R.S.); takacs.barbara@pharm.unideb.hu (B.T.); menesaki@hotmail.com (A.M.); szilvassy.zoltan@med.unideb.hu (Z.S.)

[2] Institute of Internal Medicine, Faculty of Medicine, University of Debrecen, H-4032 Debrecen, Hungary; balla.jozsef@med.unideb.hu

[3] Department of Paediatrics, Clinical Centre, University of Debrecen, H-4032 Debrecen, Hungary; balla@med.unideb.hu

[4] Department of Oto-Rhino-Laryngology and Head and Neck Surgery, Faculty of Medicine, University of Debrecen, H-4032 Debrecen, Hungary; szj@med.unideb.hu

* Correspondence: juhasz.bela@med.unideb.hu; Tel.: +36-5242-7899 (ext. 56109)

Citation: Kozma, M.; Bombicz, M.; Varga, B.; Priksz, D.; Gesztelyi, R.; Tarjanyi, V.; Kiss, R.; Szekeres, R.; Takacs, B.; Menes, A.; et al. Cardioprotective Role of BGP-15 in Ageing Zucker Diabetic Fatty Rat (ZDF) Model: Extended Mitochondrial Longevity. *Pharmaceutics* 2022, 14, 226. https://doi.org/10.3390/pharmaceutics14020226

Academic Editors: Fátima Regina Mena Barreto Silva and Diana Marcela Aragon Novoa

Received: 3 December 2021
Accepted: 14 January 2022
Published: 19 January 2022

Publisher's Note: MDPI stays neutral with regard to jurisdictional claims in published maps and institutional affiliations.

Abstract: Impaired mitochondrial function is associated with several metabolic diseases and health conditions, including insulin resistance and type 2 diabetes (T2DM), as well as ageing. The close relationship between the above-mentioned diseases and cardiovascular disease (CVD) (diabetic cardiomyopathy and age-related cardiovascular diseases) has long been known. Mitochondria have a crucial role: they are a primary source of energy produced in the form of ATP via fatty acid oxidation, tricarboxylic acid (TCA) cycle, and electron transport chain (ETC), and ATP synthase acts as a key regulator of cardiomyocyte survival. Mitochondrial medicine has been increasingly discussed as a promising therapeutic approach in the treatment of CVD. It is well known that vitamin B3 as an NAD^+ precursor exists in several forms, e.g., nicotinic acid (niacin) and nicotinamide (NAM). These cofactors are central to cellular homeostasis, mitochondrial respiration, ATP production, and reactive oxygen species generation and inhibition. Increasing evidence suggests that the nicotinic acid derivative BGP-15 ((3-piperidine-2-hydroxy-1-propyl)-nicotinic amidoxime) improves cardiac function by reducing the incidence of arrhythmias and improves diastolic function in different animal models. Our team has valid reasons to assume that these cardioprotective effects of BGP-15 are based on its NAD^+ precursor property. Our hypothesis was supported by an animal experiment where ageing ZDF rats were treated with BGP-15 for one year. Haemodynamic variables were measured with echocardiography to detect diabetic cardiomyopathy (DbCM) and age-related CVD as well. In the ZDF group, advanced HF was diagnosed, whereas the BGP-15-treated ZDF group showed diastolic dysfunction only. The significant difference between the two groups was supported by post-mortem Haematoxylin and eosin (HE) and Masson's trichrome staining of cardiac tissues. Moreover, our hypothesis was further confirmed by the significantly elevated Cytochrome c oxidase (MTCO) and ATP synthase activity and expression detected with ELISA and Western blot analysis. To the best of our knowledge, this is the first study to demonstrate the protective effect of BGP-15 on cardiac mitochondrial respiration in an ageing ZDF model.

Keywords: ageing; diabetic cardiomyopathy; mitochondrial longevity; electron transport chain; BGP-15; nicotinic acid; NAD precursor; bioactive molecule; antioxidants

1. Introduction

The close association between diabetes and cardiovascular disease has long been known. In addition to being a risk factor for coronary atherosclerosis and ischemic heart disease, diabetes also leads directly to myocardial dysfunction (diabetic cardiomyopathy) [1]. Metabolic abnormalities and the production of reactive oxygen species (ROS) can cause pathological activation of several signalling pathways, thus modifying the myocardial expression of various genes [2]. Ageing myocardial oxidative stress, myocardial hypertrophy, and fibrotic remodelling, along with increased apoptosis, all have a critical role in the decreased myocardial ability to contract and relax. Age-related defects in mitochondrial function have been related to normal cardiac ageing. It has been reported that the mitochondrial respiratory chain complexes decline with age in cardiac muscle, particularly in complex I and IV, although complexes II, III, and V are less affected by age in cardiomyocytes. It is widely accepted that nicotinamide adenine dinucleotide (NAD^+) and its reduced and phosphorylated forms such as NADH, NADPH, and $NADP^+$ are crucial regarding mitochondrial. Furthermore, NAD^+ harmonizes the function of several pathways such as glycolysis, TCA cycle, oxidative phosphorylation (OXPHOS), or fatty acid and amino acid metabolism as a cofactor for a wide range of oxidoreductase reactions. NAD^+ is converted to NADH in the mitochondrial compartment, and later it is oxidized by complex I, which is the first part of the mitochondrial electron transport chain [3]. Finally, ATP will be generated from ADP through a series of redox reactions (OXPHOS) [4]. NADP and NADPH have a prominent role in the fight against oxidative damage and the prevention of oxidative stress-induced damage, also under physiological conditions [5]. Vitamin B3 and its derivatives (nicotinic acid, nicotinamide, and nicotinamide riboside) are precursors of nicotinamide adenine dinucleotide (NAD), thus playing a key role in the above-mentioned processes. Several studies confirmed that, in high doses, niacin can be useful in hypercholesterolemia, mitochondrial myopathy or other systemic NAD^+ deficiency [6]. The drug candidate BGP-15, also named N′ -(2-hydroxy-3-(piperidin-1-yl)propoxy)-3-pyridine-carboximidamide, is a nicotinic acid amide derivative, which is a particularly important feature considering the fact that nicotinic acid amide, which can also be called niacin amide or nicotinamide, is the water-soluble active form of vitamin B3 [7,8]. Several studies confirm the beneficial effects of BGP-15, including its impact on mitochondrial and cardiac function especially in animal models [9,10]. There is increasing evidence that BGP-15 improves muscle strength and function and can restore cardiac function in animal models of muscular dystrophy, heart failure, and diabetic cardiomyopathy [11,12]. In a previous study, our research group established that the drug candidate improves diastolic function in Goto–Kakizaki rats suffering from diabetic cardiomyopathy [13]. Additionally, there is further evidence for the beneficial effect of BGP-15 on mitochondrial functions exerted via the reduction of reactive oxygen species (ROS), which indicates the key role of BGP-15 in the prevention of oxidative damage, although these studies are confined to cell cultures [14]. Although there are several animal models that provide a great opportunity for researchers to examine diabetes mellitus or cardiac diseases, we need to emphasize that the Zucker Diabetic Fatty rat is one of the most representative experimental models that can exhibit these two features. The aim of the present study was to confirm the outstanding importance of drug candidate BGP-15 in enhancing mitochondrial function in this ageing ZDF animal model. Last but not least, we also examined the subsequent effects on the heart function in the aforementioned ageing type 2 diabetes mellitus rat model.

2. Materials and Methods

2.1. Animal Model

In the present study, 30 ZDF male rats (8-week-old, 200 g), and 10 LEAN male rats (8-week-old, 200 g) as control group were used, which were purchased from Charles River Laboratories International, Inc. (Wilmington, MA, USA). The ZDF is an obese model, a substrain of the outbreed Zucker rats in the laboratory of Dr. Walter Shaw at Eli Lilly Research Laboratories in Indianapolis, which enables us to conduct research with them in the field of diabetology and obesity. All of the methods used in this present study were approved by the local Ethics Committee of University of Debrecen (25/2013DEMÁB; 29.04.2014). The animals received proper care and were caged on the basis of the "Principles of Laboratory Animal Care" by EU Directive 2010/63/EU. They were fed with Purina 5008 diet and had free access to water.

The present study began with two weeks of acclimatization. After this period of time, the animals were randomly divided into three groups, a healthy control group (LEAN, $n = 10$), a diseased group modelled by ZDF rats ($n = 15$), and a group of ZDF rats that were administered BGP-15 ($n = 15$). Each animal received the same dose (10 mg/kg/daily) of either vehicle (Hydroxyethyl cellulose: distilled water; 1:5 mixture) or drug candidate BGP-15 for 52 weeks by oral gavage. BGP-15 was purchased from Sigma-Aldrich-Merck KGaA (Darmstadt, Germany). The dosage was defined based on previous studies [13].

2.2. Chemicals

All chemicals, reagents, and buffer solutions used for isolation, Western blot, Cytochrome c oxidase and ATP synthase assay were obtained from Sigma-Aldrich–Merck KGaA (Darmstadt, Germany), and Abcam Plc. (Cambridge, UK). Ca^{2+}-Containing Modified Krebs Solution (in mmol/L): NaCl: 118, KCl: 4.7, $CaCl_2$: 2.5, NaH_2PO_4: 1, $MgCl_2$: 1.2, $NaHCO_3$: 24.9, glucose: 11.5, and ascorbic acid: 0.1, dissolved in redistilled water. Ca^{2+}-Containing Modified Krebs buffer was used to wash isolated organs for further studies.

2.3. Echocardiographic Studies

Transthoracic echocardiography of rats was carried out by a Vivid E9 sonographic equipped with an i13L linear-array probe (GE Healthcare, New York, NY, USA). All of the animals were anaesthetized intramuscularly with ketamine/xylazine combination (75/5 mg/kg, respectively), then the chest hair was removed and animals were placed on a heated table. Data acquisition was performed in 2D-, M-, and Doppler modes, from parasternal short (SAX)- and long axis (PLAX, respectively), as well as apical axis (APLAX). Cardiac function was evaluated in accordance with the guideline of American Society of Echocardiography [15,16]. Wall thickness and chamber diameters were measured in M-mode, at the mid-papillary level. Diameter of the left atrium (LA) was normalized to aortic root (Ao) diameter (LA/Ao). Left ventricle internal diameter in diastole (LVIDd) and in systole (LVIDs), wall thickness of the septal, and lateral walls were measured by standard methods. Fractional shortening (FS), Ejection fraction (EF) mitral, and tricuspid plane systolic excursion (MAPSE and TAPSE, respectively) were calculated from M-mode traces. Diastolic function was assessed by Doppler (pulsed wave, PW) and Tissue Doppler imaging (TDI) from apical 4-chamber views. The ratio of peak early (E) and atrial (A) transmitral flow velocities were assessed (E/A ratio), and deceleration time of the E wave (DecT) was measured. Left ventricle outflow tract (LVOT) parameters (velocity: V; and pressure gradient: PG) were evaluated using PW mode from the apical 5-chamber view. Early (e′) and atrial (a′) diastolic myocardial relaxation velocity waves, systolic myocardial velocity (s′), mitral valve closure to opening time (MCOT), isovolumic contraction time (IVCT), ejection time (ET), and isovolumic relaxation time (IVRT) were defined in tissue Doppler (TDI) imaging at the septal and mitral annulus. The ratio of E/e′ was calculated offline. Myocardial performance index (Tei-index) was determined as a sum of the isovolumic contraction and relaxation times (ICT and IRT) divided by the ejection time (ET). Images were analysed by a single reader in a blinded fashion using the EchoPAC PC software ver. 112, (GE Healthcare, New York, NY, USA). To assess the quality of measurements, three cardiac cycles were averaged for each parameter at each animal, and data are presented as mean $\pm$ SD.

2.4. Analysis of Serum Parameters

Blood samples were collected in Vacutainer Plast SSTII tubes (BD Vacutainer, Bergen County, NJ, USA), using a 23-gauge needle from the lateral saphenous veins of each animal after 12 h fasting. The samples were handled aseptically to minimize haemolytic activity. Insulin was measured on the Liaison XL DiaSorin platform (DiaSorin Inc., Stillwater, MN, USA). All other serum parameters were measured on the Roche Cobas Integrated platform (Roche Diagnostics GmbH, Mannheim, Germany). Lipid parameters included total cholesterol, low-density lipoprotein (LDLc), and high-density lipoprotein (HDLc). Specific cardiac biomarkers were troponin T, creatine kinase MB isoform (CK-MB), and lactate dehydrogenase (LDH). For blood sugar levels, related measurements blood was collected from the tail vein. Fasting blood glucose concentration was calculated by an Accu-Check glucose meter (Roche Diagnostics, Mannheim, Germany). For quantifying insulin resistance and ß-cell function, the homeostasis model assessment (HOMA) was used for calculation with the following formulas, where HOMA-IR corresponds to insulin resistance and HOMA-B to ß-cell function: HOMA-IR: fasting glucose $\times$ insulin/22,5; HOMA-B: 20 $\times$ insulin/fasting glucose–3,5 [13].

2.5. Morphometry

At the endpoint of the study, all animals were weighed and anaesthetized deeply with ketamine-xylazine (75/5 mg/kg) intramuscular injection, and thoracotomy was performed. Organ samples (heart, lung, kidney, liver, and right tibia) were excised, washed in modified Krebs buffer, and weighed using a milligram scale. Left ventricle and septum of the hearts were isolated, and weights were measured and normalized to tibia length. After weighing, kidney and lung samples were kept overnight at 60 °C and wet-to-dry tissue ratios were determined. Cardiac samples that remained were stored in 4% formalin solution or were rapidly deep-frozen in N2 and stored at −80 °C for further analyses.

2.6. Histology

Hearts were dissected transversely at mid-LV level, and samples were fixed for 24 h in 4% neutral buffered formalin (pH = 7.4). From the formalin-fixed, paraffin-embedded (FFPE) blocks, 4 µm thick sections were created and stained with haematoxylin-eosin (HE). To estimate the degree of interstitial fibrosis Masson's trichrome staining was performed too on the sections, based on the protocol provided by the manufacturer (Sigma-Aldrich Co., St. Louis, MO, USA). Image acquisition was performed using Nikon Eclipse 80i microscope. On the other hand, in order to determine the extent of each cardiomyocyte, transnuclear transversal area was measured. Seven animals from the control group, six from the ZDF group, and seven from the ZDF+BGP-15 group were selected, and 11 sections from the control and ZDF+BGP-15 groups and 12 sections from the ZDF group were created from their dissected heart. One hundred longitudinally oriented cardiomyocytes from LV were examined on each section (magnification 100$\times$), and the diameters at the transnuclear position were defined with Image J Software. According to statistical analysis the mean values of 100 measurements represent one section, and the mean values of sections originating from the same animal were adjusted (mean of the mean).

2.7. Western Blot

To identify proteins from myocardial tissue of the left ventricle, Western blot technique was used. At the beginning of the procedure, deep frozen samples (300 mg at –80 °C) were treated with liquid nitrogen and they were milled for protein extraction. The powder arising from the tissue of the heart was homogenized by a Poltroon-homogenizer (IKA-WERKE, Staufen, Germany) in 800 µL Buffer (25 mM Tris-HCl, pH = 8, 25 mM-NaCl, 4 mMNa-orthovanadate, 10 mMNaF, 10 mMNa-pyrophosphate, 10 nM okadaic acid, 0.5 mM EDTA, 1 mM PMSF and protease inhibitor cocktail (Sigma-Aldrich, St. Louis, MO, USA)). Total protein concentration was determined from the supernatant after centrifugation (at 10,000$\times$ *g* for 20 min). For this part of the analysis, Bicinchoninic Acid (BCA) Protein Assay Kit (QuantiPro™ BCA Assay Kit, Sigma-Aldrich-Merck KGaA, Darmstadt, Germany) was used. Samples containing the proper amount of total protein (20 µg) were separated by Gel electrophoresis (using 10, 12 or 18% gel), since this method enables us to divide proteins examined according to their molecular weight. The procedure was conducted at 40 mA for 100–120 min. After SDS-Polyacrylamide gel electrophoresis, electro-blotting (at 25 V for 90 min) made it possible to transfer proteins onto a nitrocellulose membrane (GE Healthcare, New York, NY, USA). After a period of blocking (1 h at room temperature) in TBS-T, which contained 3% BSA the membranes were incubated overnight with the following antibodies: anti-Beta actin (Beta actin), anti-superoxide-dismutase 1 (SOD1), anti-superoxide-dismutase 2 (SOD2), anti-heat-shock-protein 32 (HSP32), anti-Ubiquinol-Cytochrome-C-Reductase-Binding-Protein (UQCRB), anti-cyclooxygenase 2(MTCO), and anti-ATP-synthase (ATPS). All of the primary and secondary antibodies (anti-Rabbit and anti-Mouse) were purchased from Sigma-Aldrich (Sigma-Aldrich-Merck KGaA) and Abcam (Abcam Plc., Cambridge, UK) and were applied after the recommendation of the manufacturer. The second incubation was conducted with secondary antibodies followed by labelling with horseradish peroxidase. For detection of the proteins, enhanced chemiluminescence reagent was used. Then, the membranes were scanned with a C-Digit© blot scanner with Image Studio Digits ver. 5.2. software (LI-COR Inc., Lincoln, NE, USA). In all cases, the background was normalized, and a standardization was conducted to a housekeeping protein (Beta-actin); then, the statistical analysis was carried out from the average of three independent experiment.

2.8. Cytochrome C Oxidase Enzyme Activity Microplate Assay

To measure the activity of Cytochrome c oxidase enzyme (CYTOCOX), an optimized colorimetric assay was used. The method was carried out according to the instructions of manufacturer. Briefly, the Kit is a colorimetric assay based on the observation of the decrease in absorbance of Ferro cytochrome c measured at 550 nm, which is caused by its oxidation to Ferri cytochrome c by cytochrome c oxidase (Sigma-Aldrich-Merck KGaA, Darmstadt, Germany).

2.9. ATP Synthase Enzyme Activity Microplate Assay

For examination of the activity of complex V, ATP synthase Enzyme Activity Microplate Assay kit was used, which was obtained from Abcam (Abcam Plc., Cambridge, UK). After preparing samples from left ventricle myocardial tissue following the instruction of the manufacturer, the plates included in the kit were loaded and incubated at room temperature for 3 h. The previously prepared samples were rinsed with solution 1 used during the preparation, lipid mix, and reagent mix were added as recommended. The activity of the enzyme was measured on OD340 in 1 min intervals for 1–2 h.

2.10. Statistical Analyses

All data are presented as mean $\pm$ standard deviation (SD). D'Agostino and Pearson normality test was used to estimate normal distribution. Then, statistical analysis was performed using one-way analysis of variance (ANOVA) followed by Tukey post hoc analysis (only if F achieved $p < 0.05$, normality test was passed, and there was no significant variance inhomogeneity), or Kruskal–Wallis test followed by Dunn's post-test (when normality test was not passed). Analyses were carried out using GraphPad Prism software for Windows, version 7.00 (GraphPad Software Inc., La Jolla, CA, USA). $p < 0.05$ was considered as statistically significant.

3. Results

3.1. BGP-15 Prevented Ageing and Type 2 Diabetes Mellitus-Associated Cardiac Dysfunction In Vivo

ZDF rats showed signs of both systolic and diastolic dysfunction. Ejection fraction of ZDF group decreased in comparison to LEAN (66.22 ± 1.949 vs. $80.63 \pm 1.711\%$, $p < 0.0001$), similarly to fractional shortening (32.67 ± 1.374 vs. $44.75 \pm 1.78\%$, $p < 0.0001$). MAPSE was diminished in ZDF group compared to LEAN (1.777 ± 0.0577 mm vs. 2.335 ± 0.098, $p = 0.0022$) (Figure 1d). Systolic mitral annular velocity (s') decreased in the ZDF rats vs.

LEAN (30.5 ± 2.885 vs. 41.88 ± 2.248 mm/s, $p = 0.0105$). Right ventricle systolic function, described by TAPSE, was deteriorated in the ZDF group vs. LEAN (2.281 ± 0.0167 vs. 3.596 ± 0.0242 mm, $p = 0.0001$). ZDF rats showed impaired diastolic function compared to controls. LA/Ao increased in ZDF (1.283 ± 0.052 vs. 0.9697 ± 0.0351, $p < 0.0001$), e′/a′ ratio decreased (0.776 ± 0.054 vs. 1.298 ± 0.65, $p = 0.0495$), and E/e′ ratio increased (26.03 ± 2.525 vs. 18.20 ± 1.246, $p = 0.0088$) (Figure 1h). IVRT lengthened in the ZDF group compared to LEAN (55.0 ± 3.674 vs. 33.13 ± 1.093, $p < 0.0001$) and Tei-index worsened (0.643 ± 0.0297 vs. 0.5091 ± 0.0164) (Figure 1e). E/A ratio and DecT did not show significant changes; however, we note that the animals in the ZDF group showed either extremely high or extremely low E/A values, and together with the above parameters (E/e′, IVRT), we showed that all animals in the ZDF group were in different stages of DD at the age of 52 weeks. Echocardiographic parameters were significantly improved in the BGP-15 treated group in comparison to ZDF. Systolic function of ZDF+BGP-15 rats were preserved in comparison to ZDF, as treated rats showed improved EF: (82.20 ± 0.509, $p < 0.0001$), FS (45.93 ± 0.502, $p < 0.0001$), MAPSE (2.354 ± 0.1089, $p = 0.0007$), and TAPSE (3.658 ± 0.081, $p < 0.0001$) values (Figure 1c,d). Left atrial enlargement was attenuated in the BGP-15 treated group vs. ZDF (LA/Ao: 0.951 ± 0.021, $p < 0.0001$), E/A ratio was normal (1.598 ± 0.073), E/e′ was improved (18.01 ± 1.018, $p = 0.0022$), IVRT was normalized (44.6 ± 1.704, $p = 0.0068$), and Tei-index improved (0.516 ± 0.016, $p = 0.0003$). No significant differences were found in LVOT parameters (outflow velocity and pressure gradients).

3.2. Severely Diabetic Rats Demonstrated Decreased Body Weight

Throughout the time of the study, the bodyweight of the animals was measured monthly after the sixth month. At the endpoint, the weight of the animals in the control group was almost twice as much as in the other groups, and the difference was significant (Control vs. ZDF and Control vs. ZDF+BGP-15 $p < 0.0001$). The heart weight was higher in the control group, but the difference was only significant between the control and the treated group (Control vs. ZDF+BGP $p = 0.031$). There was no significant difference in the heart weight–tibial length quotient, but for both the bodyweight–tibial length and heart weight–bodyweight quotient, there was a recognizable difference between the control, the diseased, and the treated group (BW/TL: Control vs. ZDF and Control vs. ZDF+BGP $p < 0.0001$; HW/BW) (Table 1).

Table 1. Bodyweight, heart weight (BW, HW) to tibia length (TL) ratios, lung and kidney wet (w) to dry (d) tissue ratios at the endpoint of the study.

		LEAN	**ZDF**	**ZDF+BGP-15**
N		8	7	9
Parameter	Unit	Mean ± SD	Mean ± SD	Mean ± SD
BW	g	453.6 ± 14.76	# 279.3 ± 53.76	# 279.9 ± 40.84
HW	g	1.221 ± 0.15	1.09 ± 0.12	# 1.02 ± 0.18
HW/TL	g/cm	0.25 ± 0.05	0.24 ± 0.03	0.23 ± 0.04
BW/TL	g/cm	91.14 ± 6.79	# 60.13 ± 11.2	# 58.86 ± 9.59
HW/BW	g/kg	2.69 ± 0.36	# 4 ± 0.7	# 3.76 ± 0.95
lung w/d		0.19 ± 0.02	0.19 ± 0.01	0.18 ± 0.01
kidney w/d		0.26 ± 0.03	0.23 ± 0.07	0.25 ± 0.03

Weight of the animals and organs in the endpoint of the study. All data are presented as mean ± standard deviation (SD). $p < 0.05$. #: significant difference compared to control group. Statistical analysis was carried out by GraphPad Prism 7.00: D'Agostino and Pearson normality test was used to estimate Gaussian distribution than data were analysed with ordinary one-way ANOVA or Kruskal–Wallis test. The groups were matched by Tukey's multiple comparisons test.

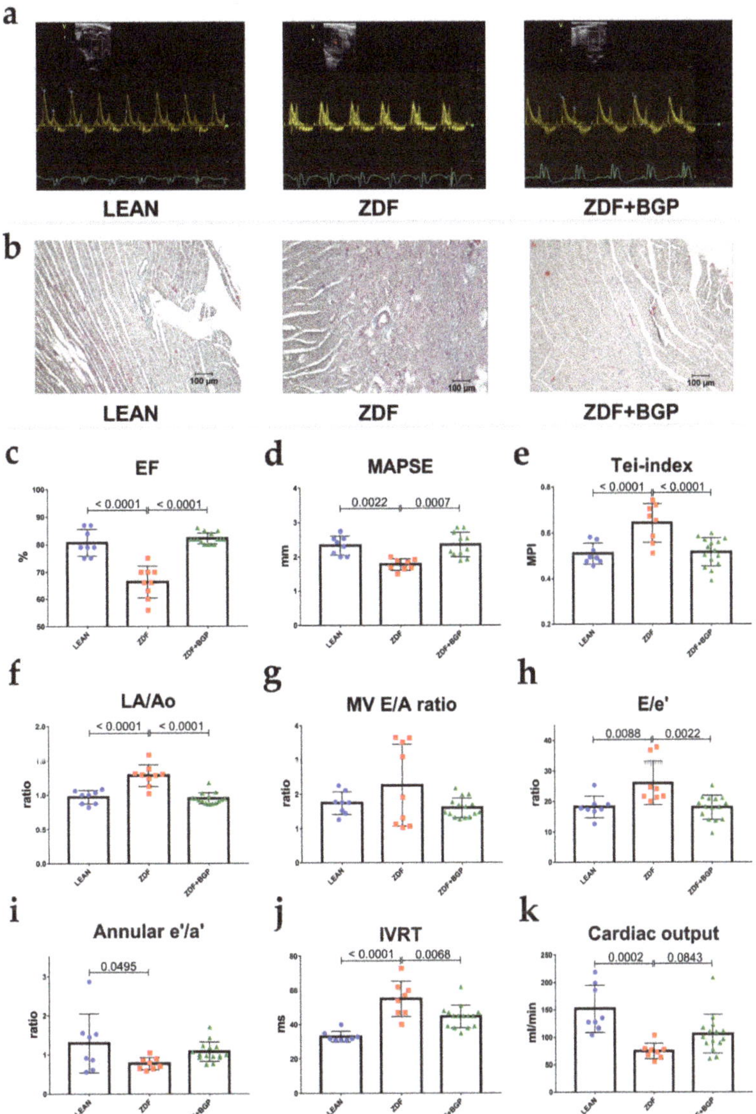

Figure 1. Echocardiographic parameters of rats at the endpoint of the study. BGP-15 treatment improved the echocardiographic parameters of aged ZDF rats. (**a**) Representative Doppler traces of the mitral inflow (E and A waves) in LEAN, ZDF, and ZDF+BGP groups, respectively. (**b**) Histological samples of the dissected heart tissues stained with Masson's trichrome in LEAN, ZDF, and ZDF+BGP groups, respectively. Fibrotic tissue dyes with blue colour. (**c**) Ejection fraction (EF) in the three endpoints groups. EF was deteriorated in the ZDF, but was normal in the ZDF+BGP group. (**d**) Mitral annular plain systolic excursion (MAPSE) improved in the treated animals. (**e**) Tei-index (Myocardial Performance Index, MPI) worsened in the ZDF but improved in the ZDF+BGP group. (**f**) Left atrial enlargement (LA/Ao) improved in the treated animals. (**g**) Mitral valve inflow E/A ratio (MV E/A) in the endpoint groups. (**h**) E/e' ratio, indicating left ventricle filling pressure was elevated in the ZDF but was restored in the ZDF+BGP group. (**i**) Septal annular wall motion e'/a' ratio worsened in

the ZDF animals. (**j**) Isovolumic relaxation time (IVRT) lengthened in the ZDF, but was restored in the ZDF+BGP group. (**k**) Cardiac output (calculated by echocardiography) reduced in the ZDF but increased in the treated group. Normal data distribution, data presented as mean $\pm$ SD, one-way ANOVA, and Tukey post-test.

3.3. Ageing and T2DM Elevates Serum Parameters

Values of serum lipid parameters and cardiac biomarkers are shown in Table 2. The level of total cholesterol was the highest in the ZDF group, but there was a significant difference observable between the treated and control groups as well (Control vs. ZDF $p = 0.0018$; Control vs. ZDF+BGP-15 $p = 0.0484$). The results for the LDL levels are similar; the highest levels were measured in the ZDF group (Control vs. ZDF $p = 0.0028$; ZDF vs. ZDF+BGP-15 $p = 0.043$). There was no significant difference between the HDL levels. Creatinine levels ought to be similar to the total cholesterol and LDL levels (Control vs. ZDF $p = 0.0249$; Control vs. ZDF+BGP-15 $p = 0.0348$). The creatine-kinase level of ZDF rats was significantly higher than measured in the treated group (ZDF vs. ZDF+BGP-15 $p = 0.04$). Although the levels of LDH and Troponin T were higher in the diseased group than either in the control or the treated group, but the difference were not showing any significance.

Table 2. Values of cardiac- and lipid homeostasis biomarkers.

		LEAN	**ZDF**	**ZDF+BGP-15**
N		8	6	8
Parameter	Unit	Mean $\pm$ SD	Mean $\pm$ SD	Mean $\pm$ SD
Total Cholesterol	mmol/L	2.19 $\pm$ 0.36	# 8.09 $\pm$ 4.21	# 5.62 $\pm$ 2.59
LDLc	mmol/L	0.23 $\pm$ 0.08	# 1.08 $\pm$ 0.68	* 0.51 $\pm$ 0.33
HDLc	mmol/L	0.49 $\pm$ 0.15	1.16 $\pm$ 0.68	0.69 $\pm$ 0.58
Creatinine	µmol/L	26.7 $\pm$ 12.85	# 141 $\pm$ 221.1	# 66.88 $\pm$ 53.83
CK	U/L	380.8 $\pm$ 304.6	548.8 $\pm$ 312.4	* 182.1 $\pm$ 124.9
LDH	U/L	1143 $\pm$ 1559	1871 $\pm$ 1256	1114 $\pm$ 856.8
Troponin T	ng/L	3263 $\pm$ 2356	4162 $\pm$ 3293	3349 $\pm$ 2276

All data are presented as mean $\pm$ standard deviation (SD). $p < 0.05$. #: significant difference compared to control group; *: significant difference compared to ZDF group. Statistical analysis was carried out by GraphPad Prism 7.0s0: D'Agostino and Pearson normality test was used to estimate Gaussian distribution than data were analysed with ordinary one-way ANOVA or Kruskal–Wallis test. The groups were matched by Tukey's multiple comparisons test. Units of data are represented according to the recommendation of local Department of Laboratory Medicine.

3.4. BGP-15 Has Mild Influence on Glucose Homeostasis

Fasting glucose levels were higher in both the ZDF and the treated groups, but the difference was not significant between the control and ZDF+BGP 15 groups (Control vs. ZDF $p = 0.0005$; ZDF vs. ZDF+BGP-15 $p = 0.0148$). The insulin levels resulted in a significant difference in the same groups (Control vs. ZDF $p = 0.0076$; ZDF vs. ZDF+BGP-15 $p = 0.0076$). In the point of HOMA-IR a significant difference was observable between the control and treated groups (Control vs. ZDF+ZDFBGP-15 $p = 0.0259$). According to HOMA-B index the ß-cell function was in the ZDF group the worst, the difference was significant in relation to the other two groups (Control vs. ZDF $p = 0.0016$; ZDF vs. ZDF+BGP-15 $p = 0.0146$) (Table 3).

Table 3. Glucose metabolism related values.

		Control	**ZDF**	**ZDF+BGP-15**
N		8	6	8
Parameter	Unit	Mean $\pm$ SD	Mean $\pm$ SD	Mean $\pm$ SD
Glucose	mmol/litre	7.05 $\pm$ 0.59	# 18.42 $\pm$ 7.93	* 10.74 $\pm$ 3.34
Insulin	mU/litre	5.81 $\pm$ 1.48	# 2.05 $\pm$ 1.02	* 5.81 $\pm$ 1.48

Table 3. *Cont.*

	Control	ZDF	ZDF+BGP-15
HOMA-IR	1.79 ± 0.40	1.87 ± 1.50	# 3.19 ± 0.88
HOMA-B	13.3 ± 5.15	# −0.89 ± 1.54	* 11.92 ± 10.57

All data are presented as mean ± standard deviation (SD). $p < 0.05$. #: significant difference compared to control group; *: significant difference compared to ZDF group. Statistical analysis was carried out by GraphPad Prism 7.00. D'Agostino and Pearson normality test was used to estimate Gaussian distribution than data were analysed with ordinary one-way ANOVA or Kruskal–Wallis test. The groups were matched by Tukey's multiple comparisons test. Unit of data were represented according to the recommendation of local Department of Laboratory Medicine. HOMA-IR corresponds to homeostasis model assessment of insulin resistance and HOMA-B to homeostasis model assessment of B-cell function.

3.5. BPG-15 Decrease Myocardial Hypertrophy in Ageing Zucker Diabetic Fatty Rats

A significant difference was observed in point of the cardiomyocyte diameters between the control and diseased group ($p = 0.041$). Although the extent of cardiomyocytes ought to be higher in the diseased group (ZDF) then in than in the treated group (ZDF+BGP-15), the difference was not significant ($p = 0.213$) (Figure 2).

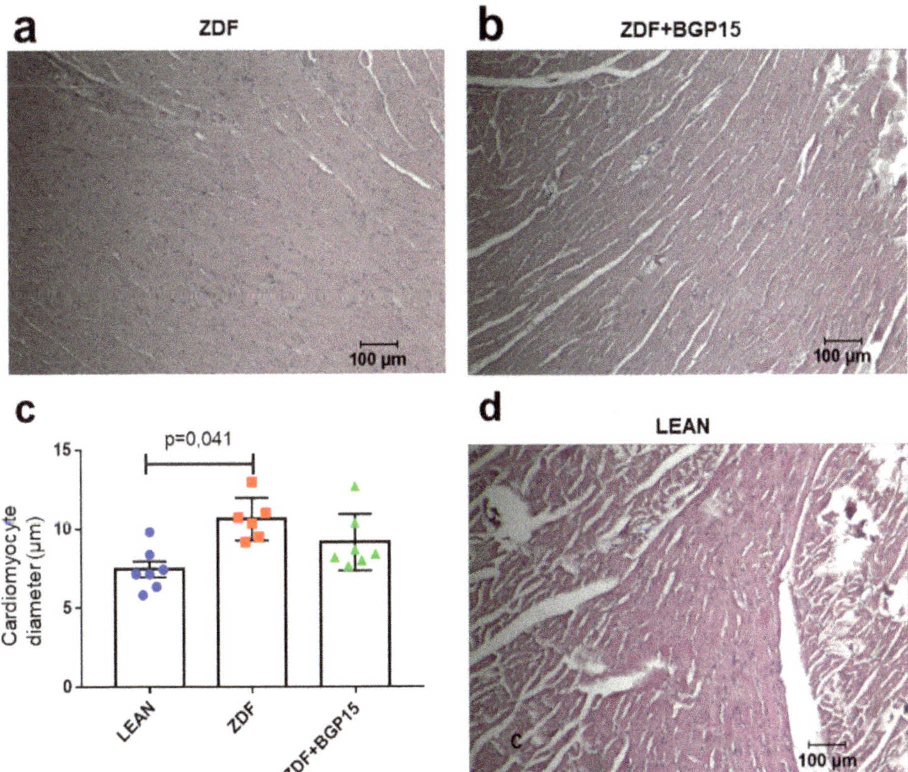

Figure 2. Histological samples stained with haematoxylin and eosin and their quantification. (**a**) Histological sample from the diseased group (ZDF), (**b**) histological sample from the BGP-15-treated ZDF group, (**c**) cardiomyocyte diameters among groups. All data are presented as mean ± standard deviation (SD, (**d**) histological sample from the control (LEAN) group, and). $p < 0.05$. Statistical analysis was carried out by GraphPad Prism 7.00: D'Agostino and Pearson normality test was used to estimate Gaussian distribution, and then data were analysed with ordinary one-way ANOVA or Kruskal–Wallis test. The groups were matched by Tukey's multiple comparisons test.

3.6. BGP-15 Improves Mitochondrial Function in Ageing Zucker Diabetic Fatty Rat Hearts

In an attempt to further confirm the increase in mitochondrial oxidative capacity, we analysed the subunits of the electron transport chain by Western blotting. Most of the subunits of oxidative phosphorylation were increased in mitochondria from ZDF BGP-15-treated rats compared to the ZDF non-treated group. The relative expression of NADH dehydrogenase (NDFUS4) was the highest in the control group, and there was a significant difference observable between the diseased and the healthy group (Control vs. ZDF $p = 0.0111$). A high relative expression was recognized in Cytochrome c oxidase (MTCO) in BGP-15 treated groups compared to ZDF and LEAN groups (Control vs. ZDF+BGP-15 $p = 0.0007$; ZDF vs. ZDF+BGP-15 $p = 0.0007$). Concerning the relative expression of ATPS, similar results were inspected (Control vs. ZDF+BGP-15 $p = 0.0031$; ZDF vs. ZDF+BGP-15 $p = 0.002$). Further analysis of the enzymatic activity of the mitochondrial ETC complexes (Complex IV-V) exhibited the same tendencies (Figure 3).

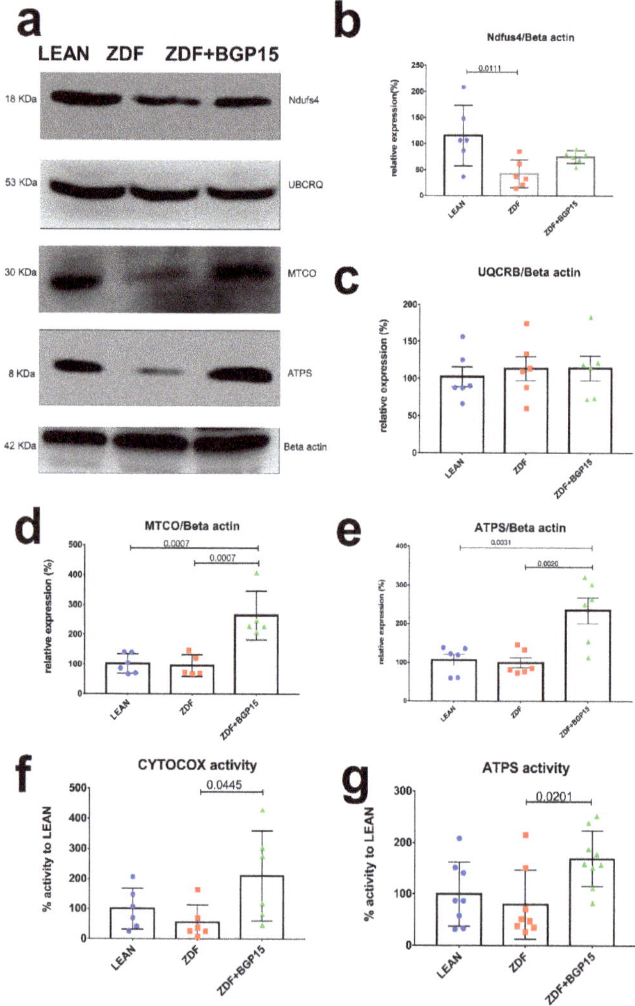

Figure 3. Protein expression and activity of electron transport chain complexes. BGP-15 treatment protects mitochondrial function against diabetes-induced oxidative stress in ageing ZDF rat myocardium.

(**a**) A representative Western blot for electric transport chain complex I, III, IV, and V in ZDF rat myocardium treated with BGP-15. (**b–e**) Quantitative band density analyses of the ETC proteins normalized to Beta actin and expressed in the percentage of LEAN (that was considered 100%). (**f,g**) Graphs show MTCO (aka CYTOCOX) and ATP synthase enzymes activity analysed by microplate assays. Results are expressed as percentage of LEAN (which was considered 100%). All data are presented as the average outcome in a group (mean) ± standard deviation (SD). Statistical analysis was carried out by GraphPad Prism 7.00: D'Agostino and Pearson normality test was used to estimate Gaussian distribution, and then data were analysed with ordinary one-way ANOVA or Kruskal–Wallis test. The groups were matched by Tukey's multiple comparisons test.

3.7. BGP-15 Treatment Induces MTCO and ATP Synthase Activity in Ageing Zucker Diabetic Fatty Rats

Cardiac muscle mitochondrial content was determined via MTCO (aka CYTOCOX) enzyme activity analysis, an established marker of mitochondrial biogenesis. A remarkable increase was observed in BGP-15 treated group in MTCO activity compared with ZDF group (209.8 ± 150.3 vs. 54.9 ± 57.11; $p = 0.0445$). In addition, BGP-15 group showed no significant differences compared to LEAN group (209.8 ± 150.3 vs. 100 ± 68.53, $p = 0.1773$). Moreover, ATP synthase activity was significantly increased in BGP-15 group compared to ZDF as well (169.1 ± 54.23 vs. 79.84 ± 67.42, $p = 0.0201$) (Figure 3).

3.8. Ageing and T2DM Exerts an Antioxidant Response on ZDF Rat Hearts by HO-1 Dependent Activation

To assess the antioxidant response in cardiac cytoplasm and mitochondria, we examined HO-1 (aka Hsp32), SOD1 (aka Cu/ZnSOD) and SOD2 (aka MnSOD) expression by Western blot technique as well. SOD1 and SOD2 proteins expression did not showed any differences among groups, in contrast HO-1 level was significantly elevated in ZDF group, compared with LEAN and ZDF BGP-15 groups (Control vs. ZDF $p = 0.0001$; ZDF vs. ZDF+BGP-15 $p = 0.0011$) (Figure 4).

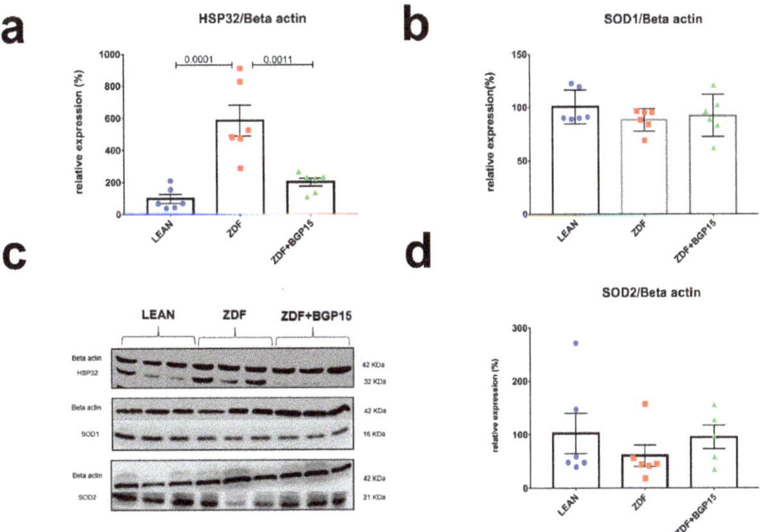

Figure 4. Expressions of antioxidant proteins from rat myocardium. HO-1 protein expression elevated significantly to diabetes induced oxidative stress in aged rats. (**a,b,d**) quantitative band density analyses of HO-1, SOD1, and SOD2 proteins normalized to Beta actin and expressed in the percentage of LEAN (that was considered 100%). (**c**) A representative Western blot for Beta actin,

Hsp32, SOD1, and SOD2 in ZDF rat myocardium treated with BGP-15. All data are presented as the average outcome in a group (mean) ± standard deviation (SD). Statistical analysis was carried out by GraphPad Prism 7.00: D'Agostino and Pearson normality test was used to estimate Gaussian distribution than data were analysed with ordinary one-way ANOVA or Kruskal–Wallis test. The groups were matched by Tukey's multiple comparisons test.

4. Discussion

4.1. General Characteristics of the ZDF Rats: Glucose and Lipid Homeostasis

It is a well-known phenomenon that diabetes mellitus (DM) increases the risk of cardiac dysfunction independently of coronary artery disease or even blood pressure. The main problem is that uncontrolled diabetes acts in the affected non-healthy patient as a sneaking silent killer; therefore, it is crucial to reveal all signal transduction pathways of its pathomechanism. In addition, having a mutation in leptin receptor, like in Zucker diabetic fatty (ZDF) rat is a well-established animal model of type 2 DM. Regarding cardiac dysfunction, earlier studies reported highly divergent results in relation to depressed diastolic function alone or in combination with systolic LV function [17,18]. Since the association between increasing age and T2DM is strong, studying aged rats is more relevant to the clinical situation. To dissolve this discrepancy, in our present study, we conducted multiple investigations and only terminated them after 52 weeks. Taking into account that ZDF rats have limited lifespan, the animals involved in this study should be considered as aged animals. Additionally, to the best of our knowledge, this long-term follow-up and treatment is unique so far [19]. Based on our blood parameters of the serum, morphometric examination, aged ZDF rats showed representative features of severe and prolonged T2DM. T2DM was exemplified by strongly elevated fasting glucose, low insulin level, and extremely low body weight compared to the healthy LEAN group [20]. Considering the glucose-metabolism-related values, in BGP-15-treated ZDF animals, only insulin resistance was observed due to the higher insulin level compared to the ZDF group. These findings suggest that even though BGP-15 therapy fails to reduce blood sugar level effectively, it clearly slows down the progression of T2DM. The ZDF rat fed with lipid-rich diet closely mimics human adult onset of metabolic syndrome with age-dependent evolution from insulin resistance to type-II diabetes mellitus, hypertriglyceridemia, and hypercholesterolemia. In our model, all of the lipid homeostasis markers were elevated in both ZDF groups, although LDLc was significantly lower in BGP-15 treated group compared to ZDF rats. To our knowledge, this is the first study describing BGP-15 LDL cholesterol lowering effect. Peterson et al. define, that serum lipids rises after 22–42 weeks of age, when insulin resistance develops into severe T2DM [21]. These data support the hypothesis that BGP-15 delays the onset of metabolic syndrome in ZDF rats.

4.2. Assessment of Cardiac Dysfunction and Remodelling

During haemodynamic measurements, signs of diabetes-induced myocardial structural and metabolic remodelling were clearly presented. In this current study, we observed that severe T2DM in ZDF rats leads to significantly depressed cardiac haemodynamic performance with depressed LV diastolic and systolic function as well. A study on 45-week-old ZDF rats even reported similar results. Daniels et al. observed that long-term severe diabetes and ageing leads to mild impairment of diastolic LV function, whereas systolic function was well preserved [19]. Interestingly, in our study, animals in the untreated ZDF group were in different stages of diastolic dysfunction with extremely low and extremely high E/A values, based on the diastolic dysfunction stages/grades. The cardioprotective effects of BGP-15 were seen even in diastolic as well as in systolic parameters. This finding is in line with investigations of Spontaneously Hypertensive Rat (SHR) rats. BGP-15 treatment results considerably decrease the worsening in systolic (EF) and in diastolic (E/E′) LV function. Moreover, only mild LV hypertrophy was seen in the BGP-15-treated group [22]. In contrast, our findings showed preserved cardiomyocyte diameters in BGP-15 treated group compared to LEAN healthy control group, and it was only the ZDF group that showed significant difference in terms of cardiac hypertrophy. Recent findings suggest a so-called reverse remodelling phenomenon induced by BGP-15 long-term treatment. Cardiac biomarkers (e.g., CK including CK-MB, LDH, Troponin T) which leak out from damaged tissues into the blood stream serve as diagnostic markers of myocardial injury. Some of these, e.g., CK, LDH, and Troponin T, showed slight elevation in the ZDF group compared to the control group; nevertheless, they did not reach the level of significance. In addition, creatinine levels were elevated significantly in both ZDF group, but only CK level was significantly decreased in BPG15 compared to ZDF rats. These results indicate the obvious detrimental effects of diabetes on cardiac tissues [23]. Additionally, the elevated level of these markers correlate with the disease severity and with the cardiac dysfunction detected with the help of echocardiographic measurements.

4.3. ROS Capacity

To our knowledge, increased intramitochondrial haem and subsequent ROS generation may be the driving force for mobilizing HO-1 in mitochondria. Under oxidative injury provoked by diabetes or ageing in some tissues, haem-derived Fe and CO may exacerbate intracellular oxidative stress and cellular injury by promoting ROS generation in mitochondria. HO-1 enzyme is a well-known cytoprotective stress protein (Hsp32); it provides defence against oxidative stress by catalysing the degradation of pro-oxidant haem to biliverdin and bilirubin. At the same time, acting as a double-edged sword, the overexpression of HO-1 promotes mitochondrial sequestration and macroautophagy [24–27]. In turn, the overexpression of HO-1 drives the overproduction of CO, which leads to tissue hypoxia and direct CO-mediated cell damage. Due to the short half-life, mitochondria and especially complex IV (also known as cytochrome c oxidase) are crucial members of the electron transport chain and seem to be the directly inhibited target by CO [28,29]. In our ageing ZDF model, a noticeable HO-1 overexpression was observed, supporting the abovementioned correlation with oxidative stress and diabetes. However, mitochondrial and cytoplasmic superoxide dismutases (SOD1 and SOD2) were also detected by Western blot technique, but significant elevation in protein expression in our hyperglycaemic tissue was not seen. The limitation of this study is that SOD activity was not measured, although based on the literature, a decreased activity is expressed in patients with DM. The inactivity of SOD might be due to the hyperglycaemia, which leads to glucose auto oxidation and non-enzymatic glycosylation of numerous proteins, such as SOD [30].

4.4. Upregulation of Cardiac ATP Synthase and Cytochrome C Oxidase

As was noted above, the heart has a high and continuous demand for oxidative metabolism to maintain ATP production. Accordingly, being the centre of fatty acid and glucose metabolism, a cardiomyocyte has a relatively high number of mitochondria compared to other tissues. In physiologic demand, the heart preferentially uses fatty acids, but in pathologic conditions, such as hypertrophy, heart failure, and myocardial infarction, the fuel preference is switched toward glucose oxidation with less oxygen consumption. This means that a diabetic heart has an increased rate of fatty acid oxidation and decreased rate of glucose oxidation. In addition, development of advanced glycation end (AGE) products and the modulation of NADPH oxidase and Superoxide contribute an increased ROS level as well. Due to their short half-life, mitochondrial ROS production may play important role regarding the mitochondrial damage during diabetes [31,32]. Considering the above, metabolic remodelling and a defect in mitochondrial bioenergetics lead to cardiac dysfunction in diabetes. In our study, the decrease in mitochondrial function was revealed by a decline in oxidative phosphorylation at the level of Complex I and II. Under normal conditions, both complexes generate NADH and Dihydroflavine-adenine dinucleotide (FADH$_2$) during the oxidization of substrates, pass electrons to coenzyme Q, and further reduce Complex III, cytochrome c, and Complex IV [32]. Our study identifies the decreased ETC function on the level of protein expression of Complex I, IV, and V in diabetes. Defects of Complex IV and V were proved by decreased activity of enzymes as well. We showed that LV dysfunction and remodelling are associated with Complex I, IV, and V defects. At the same time, we found a dramatic increase in the activity of Cytochrome c oxidase and ATP synthase with the treatment of aged rats that suffered from T2DM with the pharmacological agent BGP-15.

4.5. Translational Aspects and Future Directions

According to these findings, we strongly hypothesize that BGP-15 protects mitochondrial function against diabetes- and ageing-derived oxidative stress. Long-term treatment with this nicotinic acid (Vitamin B3) derivative delays the onset of severe T2DM and metabolic changes. Based on our and previous results, the cardioprotective effect of BGP-15 is indisputable. In the future, it would be worth considering the BGP-15 drug candidate as a NAD precursor since it may provide sound basis for a promising strategy in the treatment of multiple pathologic conditions related to NAD deficiency.

5. Conclusions

To the best of our knowledge, this is the first study that demonstrates BGP-15's protective effect on cardiac mitochondrial respiration in Zucker Diabetic Fatty Rat model at the age of 52 weeks. Due to a strong association between increasing age and T2DM, examination of aged rats is more relevant to the clinical situation. At the end of our study, age-dependent aggravation of diabetes, cardiac dysfunction, and cardiac remodelling were all clearly seen in diabetic rats, although in BPG15-treated animals, this metabolic features did not reach this life-threatening level. We assume that as a nicotinic acid derivative, BGP-15 may act as a NAD precursor and have all the beneficial effects on mitochondrial energy metabolism on subcellular level. In spite of our statement, further investigations are needed, and these observations undoubtedly place BPG-15 into a new focus of therapeutic approach.

Author Contributions: Conceptualization, M.K., M.B., D.P., J.B., J.S., G.B., Z.S. and B.J.; Data curation, A.M. and Z.S.; Formal analysis, M.K., D.P., M.B., A.M., R.S., B.T. and R.G.; Funding acquisition, Z.S. and B.J.; Investigation, M.K., D.P., M.B., V.T., R.K., R.S., B.T., A.M., B.V., R.G. and B.J.; Methodology, M.K., M.B., D.P., V.T., B.V., R.S., R.K. and A.M.; Project administration, B.J.; Supervision, G.B., J.S., J.B., Z.S. and B.J.; Validation, J.B.; Visualization, M.K., M.B. and R.K.; Writing—Original draft, M.K., M.B., V.T., D.P., A.M. and B.J. All authors have read and agreed to the published version of the manuscript.

Funding: The present research was supported by the European Union, the State of Hungary, and the Uni-versity of Debrecen under grant number GINOP-2.3.4-15-2020-00008 (Mate Kozma, Daniel Priksz, Mariann Bombicz, Reka Szekeres, Rudolf Gesztelyi, Balazs Varga, Bela Juhasz, Zoltan Szilvassy). The project is co-financed by the European Union and the European Regional Devel-opment Fund. Project no. TKP2020-IKA-04 has been implemented with the support provided from the National Research, Development and Innovation Fund of Hungary, financed under the 2020-4.1.1-TKP2020 funding scheme, and Ministry of Human Capacities, Hungary (grant 20391-3/2018/FEKUSTRAT (A.P.)). Project no. TKP2021-EGA-18 has been implemented with the support provided from the National Research, Development and Innovation Fund of Hungary, financed under the TKP2021-EGA funding scheme. The research was financed by the Thematic Excellence Programme of the Ministry for Innovation and Technology in Hungary (TKP2020-NKA-04_), within the framework of the Space Sciences thematic program of the University of Debrecen.

Institutional Review Board Statement: The animal study protocol was approved by the Institutional Ethics Committee of University of Debrecen (protocol code 25/2013DEMÁB; date of approval: 29 April 2014).

Informed Consent Statement: Not applicable.

Data Availability Statement: The data that support the findings of this study are available from the corresponding author upon reasonable request. Some data may not be made available because of privacy or ethical restrictions.

Conflicts of Interest: The authors declare no conflict of interest.

References

1. Maayah, Z.H.; McGinn, E.; Al Batran, R.; Gopal, K.; Ussher, J.R.; El-Kadi, A.O.S. Role of Cytochrome p450 and Soluble Epoxide Hydrolase Enzymes and Their Associated Metabolites in the Pathogenesis of Diabetic Cardiomyopathy. *J. Cardiovasc. Pharmacol.* **2019**, *74*, 235–245. [CrossRef] [PubMed]
2. Abushouk, A.I.; El-Husseny, M.W.A.; Bahbah, E.; Elmaraezy, A.; Ali, A.A.; Ashraf, A.; Abdel-Daim, M.M. Peroxisome proliferator-activated receptors as therapeutic targets for heart failure. *Biomed. Pharmacother.* **2017**, *95*, 692–700. [CrossRef] [PubMed]
3. Duncan, J.G. Mitochondrial dysfunction in diabetic cardiomyopathy. *Biochim. Biophys. Acta* **2011**, *1813*, 1351–1359. [CrossRef]
4. Srivastava, S. Emerging therapeutic roles for NAD + metabolism in mitochondrial and age-related disorders. *Clin. Transl. Med.* **2016**, *5*, 25. [CrossRef] [PubMed]
5. Pollak, N.; Dölle, C.; Ziegler, M. The power to reduce: Pyridine nucleotides-small molecules with a multitude of functions. *Biochem. J.* **2007**, *402*, 205–218. [CrossRef]
6. Pirinen, E.; Auranen, M.; Khan, N.A.; Brilhante, V.; Urho, N.; Pessia, A.; Hakkarainen, A.; Kuula, J.; Heinonen, U.; Schmidt, M.S.; et al. Niacin Cures Systemic NAD+ Deficiency and Improves Muscle Performance in Adult-Onset Mitochondrial Myopathy. *Cell Metab.* **2020**, *31*, 1078–1090.e5. [CrossRef]
7. Pető, Á.; Kósa, D.; Fehér, P.; Ujhelyi, Z.; Sinka, D.; Vecsernyés, M.; Szilvássy, Z.; Juhász, B.; Csanádi, Z.; Vígh, L.; et al. Pharmacological Overview of the BGP-15 Chemical Agent as a New Drug Candidate for the Treatment of Symptoms of Metabolic Syndrome. *Molecules* **2020**, *25*, 429. [CrossRef] [PubMed]
8. Literati-Nagy, B.; Kulcsar, E.; Literati-Nagy, Z.; Buday, B.; Peterfai, E.; Horvath, T.; Tory, K.; Kolonics, A.; Fleming, A.; Mandl, J.; et al. Improvement of Insulin Sensitivity by a Novel Drug, BGP-15, in Insulin-resistant Patients: A Proof of Concept Randomized Double-blind Clinical Trial. *Horm. Metab. Res.* **2009**, *41*, 374–380. [CrossRef]
9. Szabados, E.; Literati-Nagy, P.; Farkas, B.; Sumegi, B. BGP-15, a nicotinic amidoxime derivate protecting heart from ischemia reperfusion injury through modulation of poly(ADP-ribose) polymerase. *Biochem. Pharmacol.* **2000**, *59*, 937–945. [CrossRef]
10. Halmosi, R.; Berente, Z.; Osz, E.; Toth, K.; Literati-Nagy, P.; Sümegi, B. Effect of Poly(ADP-Ribose) Polymerase Inhibitors on the Ischemia-Reperfusion-Induced Oxidative Cell Damage and Mitochondrial Metabolism in Langendorff Heart Perfusion System. *Mol. Pharmacol.* **2001**, *59*, 1497–1505. [CrossRef]
11. Kennedy, T.L.; Swiderski, K.; Murphy, K.T.; Gehrig, S.M.; Curl, C.L.; Chandramouli, C.; Febbraio, M.A.; Delbridge, L.M.; Koopman, R.; Lynch, G.S. BGP-15 Improves Aspects of the Dystrophic Pathology in mdx and dko Mice with Differing Efficacies in Heart and Skeletal Muscle. *Am. J. Pathol.* **2016**, *186*, 3246–3260. [CrossRef]
12. Sapra, G.; Tham, Y.K.; Cemerlang, N.; Matsumoto, A.; Kiriazis, H.; Bernardo, B.C.; Henstridge, D.C.; Ooi, J.Y.Y.; Pretorius, L.; Boey, E.J.H.; et al. The small-molecule BGP-15 protects against heart failure and atrial fibrillation in mice. *Nat. Commun.* **2014**, *5*, 5705. [CrossRef] [PubMed]
13. Bombicz, M.; Priksz, D.; Gesztelyi, R.; Kiss, R.; Hollos, N.; Varga, B.; Nemeth, J.; Toth, A.; Papp, Z.; Szilvassy, Z.; et al. The Drug Candidate BGP-15 Delays the Onset of Diastolic Dysfunction in the Goto-Kakizaki Rat Model of Diabetic Cardiomyopathy. *Molecules* **2019**, *24*, 586. [CrossRef]
14. Sumegi, K.; Fekete, K.; Antus, C.; Debreceni, B.; Gallyas, F., Jr.; Sumegi, B.; Szabo, A. BGP-15 Protects against Oxidative Stress- or Lipopolysaccharide-Induced Mitochondrial Destabilization and Reduces Mitochondrial Production of Reactive Oxygen Species. *PLoS ONE* **2017**, *12*, e0169372. [CrossRef] [PubMed]
15. Lang, R.M.; Badano, L.P.; Mor-Avi, V.; Afilalo, J.; Armstrong, A.; Ernande, L.; Flachskampf, F.A.; Foster, E.; Goldstein, S.A.; Kuznetsova, T.; et al. Recommendations for Cardiac Chamber Quantification by Echocardiography in Adults: An Update from the American Society of Echocardiography and the European Association of Cardiovascular Imaging. *J. Am. Soc. Echocardiogr.* **2015**, *28*, 1–39.e14. [CrossRef]
16. Nagueh, S.F.; Smiseth, O.A.; Appleton, C.P.; Byrd, B.F.; Dokainish, H.; Edvardsen, T.; Flachskampf, F.A.; Gillebert, T.C.; Klein, A.L.; Lancellotti, P.; et al. Recommendations for the Evaluation of Left Ventricular Diastolic Function by Echocardiography: An Update from the American Society of Echocardiography and the European Association of Cardiovascular Imaging. *Eur. Hear. J. Cardiovasc. Imaging* **2016**, *17*, 1321–1360. [CrossRef]
17. Marsh, S.A.; Powell, P.C.; Agarwal, A.; Dell'Italia, L.J.; Chatham, J.C. Cardiovascular dysfunction in Zucker obese and Zucker diabetic fatty rats: Role of hydronephrosis. *Am. J. Physiol. Circ. Physiol.* **2007**, *293*, H292–H298. [CrossRef]
18. Brom, C.E.V.D.; Huisman, M.C.; Vlasblom, R.; Boontje, N.M.; Duijst, S.; Lubberink, M.; Molthoff, C.F.; Lammertsma, A.A.; van der Velden, J.; Boer, C.; et al. Altered myocardial substrate metabolism is associated with myocardial dysfunction in early diabetic cardiomyopathy in rats: Studies using positron emission tomography. *Cardiovasc. Diabetol.* **2009**, *8*, 39. [CrossRef] [PubMed]
19. Daniels, A.; Linz, D.; van Bilsen, M.; Rütten, H.; Sadowski, T.; Ruf, S.; Juretschke, H.-P.; Neumann-Haefelin, C.; Munts, C.; van der Vusse, G.J.; et al. Long-term severe diabetes only leads to mild cardiac diastolic dysfunction in Zucker diabetic fatty rats. *Eur. J. Heart Fail.* **2012**, *14*, 193–201. [CrossRef] [PubMed]
20. Siwy, J.; Zoja, C.; Klein, J.; Benigni, A.; Mullen, W.; Mayer, B.; Mischak, H.; Jankowski, J.; Stevens, R.; Vlahou, A.; et al. Evaluation of the Zucker Diabetic Fatty (ZDF) Rat as a Model for Human Disease Based on Urinary Peptidomic Profiles. *PLoS ONE* **2012**, *7*, e51334. [CrossRef]
21. Peterson, R.G.; Shaw, W.N.; Neel, M.-A.; Little, L.A.; Eichberg, J. Zucker Diabetic Fatty Rat as a Model for Non-insulin-dependent Diabetes Mellitus. *ILAR J.* **1990**, *32*, 16–19. [CrossRef]

22. Horvath, O.; Ordog, K.; Bruszt, K.; Deres, L.; Gallyas, F.; Sumegi, B.; Toth, K.; Halmosi, R. BGP-15 Protects against Heart Failure by Enhanced Mitochondrial Biogenesis and Decreased Fibrotic Remodelling in Spontaneously Hypertensive Rats. *Oxidative Med. Cell. Longev.* **2021**, *2021*, 1250858. [CrossRef]

23. Huang, E.-J.; Kuo, W.-W.; Chen, Y.-J.; Chen, T.-H.; Chang, M.-H.; Lu, M.-C.; Tzang, B.-S.; Hsu, H.-H.; Huang, C.-Y.; Lee, S.-D. Homocysteine and other biochemical parameters in Type 2 diabetes mellitus with different diabetic duration or diabetic retinopathy. *Clin. Chim. Acta* **2006**, *366*, 293–298. [CrossRef]

24. Bansal, S.; Biswas, G.; Avadhani, N.G. Mitochondria-targeted heme oxygenase-1 induces oxidative stress and mitochondrial dysfunction in macrophages, kidney fibroblasts and in chronic alcohol hepatotoxicity. *Redox Biol.* **2013**, *2*, 273–283. [CrossRef]

25. Abraham, N.; Lin, J.-C.; Schwartzman, M.; Levere, R.; Shibahara, S. The physiological significance of heme oxygenase. *Int. J. Biochem.* **1988**, *20*, 543–558. [CrossRef]

26. Llesuy, S.F.; Tomaro, M.L. Heme oxygenase and oxidative stress. Evidence of involvement of bilirubin as physiological protector against oxidative damage. *Biochim. Biophys. Acta (BBA) Bioenerg.* **1994**, *1223*, 9–14. [CrossRef]

27. Di Noia, M.A.; Van Driesche, S.; Palmieri, F.; Yang, L.-M.; Quan, S.; Goodman, A.I.; Abraham, N.G. Heme Oxygenase-1 Enhances Renal Mitochondrial Transport Carriers and Cytochrome c Oxidase Activity in Experimental Diabetes. *J. Biol. Chem.* **2006**, *281*, 15687–15693. [CrossRef]

28. Alonso, J.-R.; Cardellach, F.; López, S.; Casademont, J.; Miró, Ò. Carbon Monoxide Specifically Inhibits Cytochrome C Oxidase of Human Mitochondrial Respiratory Chain. *Pharmacol. Toxicol.* **2003**, *93*, 142–146. [CrossRef] [PubMed]

29. Srinivasan, S.; Avadhani, N.G. Cytochrome c oxidase dysfunction in oxidative stress. *Free Radic. Biol. Med.* **2012**, *53*, 1252–1263. [CrossRef]

30. Goodarzi, M.T.; Varmaziar, L.; Navidi, A.A.; Parivar, K. Study of oxidative stress in type 2 diabetic patients and its relationship with glycated hemoglobin. *Saudi Med. J.* **2008**, *29*, 503–506. [PubMed]

31. Sivitz, W.I.; Yorek, M.A. Mitochondrial Dysfunction in Diabetes: From Molecular Mechanisms to Functional Significance and Therapeutic Opportunities. *Antioxid. Redox Signal.* **2010**, *12*, 537–577. [CrossRef] [PubMed]

32. Vazquez, E.J.; Berthiaume, J.M.; Kamath, V.; Achike, O.; Buchanan, E.; Montano, M.M.; Chandler, M.P.; Miyagi, M.; Rosca, M.G. Mitochondrial complex I defect and increased fatty acid oxidation enhance protein lysine acetylation in the diabetic heart. *Cardiovasc. Res.* **2015**, *107*, 453–465. [CrossRef] [PubMed]

pharmaceutics

Article

The Antidiabetic Effects and Modes of Action of the *Balanites aegyptiaca* Fruit and Seed Aqueous Extracts in NA/STZ-Induced Diabetic Rats

Asmaa S. Zaky [1], Mohamed Kandeil [2], Mohamed Abdel-Gabbar [1], Eman M. Fahmy [3], Mazen M. Almehmadi [4], Tarek M. Ali [5] and Osama M. Ahmed [6,*]

[1] Biochemistry Department, Faculty of Science, Beni-Suef University, Beni-Suef P.O. Box 62521, Egypt; asmaa_zaky221@yahoo.com (A.S.Z.); hhmgabar@yahoo.com (M.A.-G.)
[2] Biochemistry Department, Faculty of Veterinary Medicine, Beni-Suef University, Beni-Suef P.O. Box 62521, Egypt; mohamed.kandeal@vet.bsu.edu.eg
[3] Department of Internal Medicine, Faculty of Medicine, Helwan University, Helwan 11795, Egypt; dremf@hotmail.com
[4] Department of Clinical Laboratory Sciences, College of Applied Medical Sciences, Taif University, P.O. Box 11099, Taif 21944, Saudi Arabia; dr.mazen.ma@gmail.com
[5] Department of Physiology, College of Medicine, Taif University, P.O. Box 11099, Taif 21944, Saudi Arabia; tarek70ali@gmail.com
[6] Physiology Division, Department of Zoology, Faculty of Science, Beni-Suef University, Beni-Suef P.O. Box 62521, Egypt
* Correspondence: osamamoha@yahoo.com or osama.ahmed@science.bsu.edu.eg

Citation: Zaky, A.S.; Kandeil, M.; Abdel-Gabbar, M.; Fahmy, E.M.; Almehmadi, M.M.; Ali, T.M.; Ahmed, O.M. The Antidiabetic Effects and Modes of Action of the *Balanites aegyptiaca* Fruit and Seed Aqueous Extracts in NA/STZ-Induced Diabetic Rats. *Pharmaceutics* **2022**, 14, 263. https://doi.org/10.3390/pharmaceutics14020263

Academic Editors: Diana Marcela Aragon Novoa and Fátima Regina Mena Barreto Silva

Received: 11 December 2021
Accepted: 19 January 2022
Published: 22 January 2022

Publisher's Note: MDPI stays neutral with regard to jurisdictional claims in published maps and institutional affiliations.

Abstract: Diabetes mellitus (DM) is a chronic metabolic disorder that threatens human health. Medicinal plants have been a source of wide varieties of pharmacologically active constituents and used extensively as crude extracts or as pure compounds for treating various disease conditions. Thus, the aim of this study is to assess the anti-hyperglycemic and anti-hyperlipidemic effects and the modes of action of the aqueous extracts of the fruits and seeds of *Balanites aegyptiaca* (*B. aegyptiaca*) in nicotinamide (NA)/streptozotocin (STZ)-induced diabetic rats. Gas chromatography–mass spectrometry analysis indicated that 3,4,6-tri-O-methyl-d-glucose and 9,12-octadecadienoic acid (Z,Z)- were the major components of the *B. aegyptiaca* fruit and seed extracts, respectively. A single intraperitoneal injection of STZ (60 mg/kg body weight (b.w.)) 15 min after intraperitoneal NA injection (60 mg/kg b.w.) was administered to induce type 2 DM. After induction was established, the diabetic rats were treated with the *B. aegyptiaca* fruit and seed aqueous extracts (200 mg/kg b.w./day) via oral gavage for 4 weeks. As a result of the treatments with the *B. aegyptiaca* fruit and seed extracts, the treated diabetic-treated rats exhibited a significant improvement in the deleterious effects on oral glucose tolerance; serum insulin, and C-peptide levels; liver glycogen content; liver glucose-6-phosphatase and glycogen phosphorylase activities; serum lipid profile; serum free fatty acid level; liver lipid peroxidation; glutathione content and anti-oxidant enzyme (glutathione peroxidase, glutathione-S-transferase, and superoxide dismutase) activities; and the mRNA expression of the adipose tissue expression of the insulin receptor β-subunit. Moreover, the treatment with fruit and seed extracts also produced a remarkable improvement of the pancreatic islet architecture and integrity and increased the islet size and islet cell number. In conclusion, the *B. aegyptiaca* fruit and seed aqueous extracts exhibit potential anti-hyperglycemic and anti-hyperlipidemic effects, which may be mediated by increasing the serum insulin levels, decreasing insulin resistance, and enhancing the anti-oxidant defense system in diabetic rats.

Keywords: NA/STZ-induced diabetes mellitus; *Balanitis aegyptiaca*; fruit; seed; aqueous extracts

1. Introduction

Diabetes mellitus (DM) is a chronic metabolic syndrome with a number of different etiologies. It severely affects the life of patients and heightens the risk of developing other diseases [1]. It is characterized by abnormal carbohydrate, lipid, and protein catabolism and anabolism due to insulin resistance or hypoinsulinism [2]. The recent statistics from the International Diabetes Federation (IDF) indicated that approximately 463 million adults between the ages of 20 and 79 years have diabetes, most of whom live in poor and developing countries, and this is expected to increase to 700 million by 2045 [3]. Many factors contribute to this increasing prevalence of DM, including population growth, urbanization, nutritional transition, physical inactivity, and dietary change [4,5].

Although the existing antidiabetic synthetic drugs have several benefits, they are accompanied by many adverse side effects [6]. Thus, alternative antidiabetic agents with less or no hazardous side effects are needed [6–8]. Recently, new active medicines have been extracted from plants and have antidiabetic activity with more effectiveness than oral chemical hypoglycemic drugs used in proven therapy [9]. Medicinal plants contain various bioactive compounds that have multiple activities in insulin production, insulin action, or both [10].

Eskander and WonJun described several types of Egyptian plant and herb prescriptions for the treatment of DM, and these belong to various families [11]. *Balanites aegyptiaca* (L.) Delile, which belongs to the Zygophyllaceae family, is used traditionally in African countries as an antihelmintic and in the treatment of jaundice [12,13]. In Egyptian folkloric medicine, the fruit is used as an oral anti-hyperglycemic agent [14], and herbalists in the Egyptian market sell the fruits as an antidiabetic agent. Nevertheless, the quality control of such herbal products remains a great challenge. The aqueous extract of the mesocarp of the *B. aegyptiaca* fruit exhibited antidiabetic activities in streptozotocin (STZ)-induced diabetic mice and rats [14,15], and several saponins were isolated from the mesocarp [14,16,17]. Moreover, the *B. aegyptiaca* seed kernel contains a high amount of oil and protein, which differs from one source to another [18].

Therefore, the aims of this study are to evaluate the effects of aqueous extracts of the *B. aegyptiaca* fruit and seed on the glycemic state and lipid profile and to indicate their probable modes of action in nicotinamide (NA)/STZ-induced diabetic rats.

2. Materials and Methods

2.1. Chemicals

NA, STZ [2-deoxy-2-(3-methyl-3nitrosoureido)-D-glycopyranoside], glucose-6-phosphate, glucose-1-phosphate, anthrone, reduced glutathione (GSH), malondialdehyde (MDA), and 1-Chloro-2,4-dinitrobenzene were purchased from Sigma-Aldrich Chemical Co., St Louis, MO, USA. All other chemicals were of analytical grade and were obtained from standard commercial supplies.

2.2. Experimental Animals

Male Wistar rats weighing approximately 110–140 g were used as experimental animals in the present study. The rats were housed in standard polypropylene cages and placed under a regulated room temperature of $22 \pm 2\,°C$ and humidity of $55 \pm 5\%$ with a 12:12 light–dark cycle. They were fed with a standard diet of known composition and water *ad libitum*. All animal procedures were in accordance with the ethical guidelines for the use and care of animals of the Experimental Animal Ethics Committee, Faculty of Science, Beni-Suef University, Egypt (Ethical Approval Number: BSU/FS/2015/17). All attempts were made to minimize the number and pain of used animals.

2.3. Induction of DM

After fasting for 16 h, DM was experimentally induced in male Wistar rats via an intraperitoneal (IP) injection of 60 mg NA/kg body weight (b.w.) to 16-h fasted rats before the IP injection of 60 STZ mg/kg b.w. [19]. The rats were tested for serum glucose levels 10 days after STZ was injected. The overnight-fasted (10–12 h) animals were given glucose (3 g/kg b.w) via an intragastric tube. The blood samples were taken from the lateral tail vein after 2 h of oral administration, left to coagulate, and centrifuged. The serum glucose level was then measured. The experiment included rats with a serum glucose level between 180 and 300 mg/dL, after 2 h of glucose intake, whereas the others were excluded.

2.4. Preparation of the B. aegyptiaca Fruit and Seed Aqueous Extracts

The *B. aegyptiaca* fruits and seeds were powdered using an electrical grinder. The fruit or seed powders were infused in boiled distilled water (200 mg/10 mL) for 15 min. The obtained extracts were filtered pending their use via oral gavage.

2.5. Gas Chromatography–Mass Spectrometry (GC-MS) Analysis

Both the *B. aegyptiaca* fruit and seed aqueous extracts were phytochemically analyzed via GC-MS (Producer, City, Country) according to the method described in our previous publication [20].

2.6. Experimental Design and Blood and Tissue Sampling

The rats were allocated into four groups of six rats (Scheme 1):

Scheme 1. Schematic figure showing the experimental design and animal grouping. NA: nicotinamide; STZ: streptozotocin; *B. aegptiaca*: *Balanites aegptiaca*; and ip: intraperitoneal.

Group I (Normal group): This group was assigned as the normal control group, and rats included in this group was given distilled water daily (5 mL/kg b.w./day) via oral gavage for 4 weeks.

Group II (Diabetic control): This group was assigned as the diabetic control group, and the diabetic rats within this group were given distilled water daily (5 mL/kg b.w./day) via oral gavage for 4 weeks.

Group III (Diabetic rats treated with the *B. aegyptiaca* fruit extract): This group consisted of diabetic rats that were treated daily with *B. aegyptiaca* fruit extract at a dose level of 200 mg/kg b.w./day via oral gavage for 4 weeks.

Group IV (diabetic rats treated with the *B. aegyptiaca* seed extract): This group consisted of diabetic rats that were treated daily with *B. aegyptiaca* seed extract at a dose level of 200 mg/kg b.w./day via oral gavage for 4 weeks.

At the day before sacrifice, oral glucose tolerance (OGT) test (OGTT) was performed by administering glucose solution (3 g/kg b.w.) to overnight-fasted rats via oral gavage. Successive blood samples were then obtained at 0, 30, 60, 90, and 120 min. Blood samples were left to coagulate and centrifuged. Sera were separated via centrifugation at 3000 rpm for 15 min, and the serum glucose levels were determined. One day after the end of the experiment, blood samples were collected from the jugular vein under diethyl ether inhalation. Moreover, the rats were euthanized and dissected for the excision of various organs.

2.7. Blood Sampling and Tissue Sampling

The blood obtained from the jugular vein of each rat was left to coagulate at room temperature. Serum was separated for 15 min via centrifugation at 3000 rpm and stored at $-20\,^{\circ}C$ pending its use for the determination of insulin and C-peptide levels as well as other biochemical parameters. The rats were rapidly dissected. Visceral adipose tissues were excised and kept at $-70\,^{\circ}C$ until use for ribonucleic acid (RNA) extraction and the detection of the messenger RNA (mRNA) of adiponectin and resistin via reverse transcription-polymerase chain reaction (RT-PCR). The liver was excised for the determination of oxidative stress parameters and glycogen content and glycogen metabolizing enzymes. The pancreas was also excised for histological investigation.

2.8. Biochemical Analysis

The serum glucose levels were measured using the reagent kits purchased from Spinreact Company (Spain) using the method of Trinder et al. [21]. The serum insulin level was measured using sandwich enzyme-linked immunosorbent assay (ELISA) using kits purchased from Linco Research, USA, in accordance with the manufacturer's instructions. Similarly, the serum C-peptide level was measured using the ELISA kits purchased from Linco Research, USA, in accordance with the manufacturer's instructions. Homeostatic model assessment (HOMA)-insulin resistance (IR), HOMA-insulin sensitivity (IS) [22], and HOMA-β cell function [23] were calculated using the following formulas, respectively:

$$\text{HOMA-IR} = (\text{fasting insulin } [\mu IU/mL] \times \text{fasting glucose } [mg/dL])/405$$

$$\text{HOMA-IS} = 10{,}000/(\text{fasting insulin } [\mu IU/mL] \times \text{fasting glucose } [mg/dL])$$

$$\text{HOMA-}\beta \text{ cell function} = (20 \times \text{fasting insulin } [\mu IU/mL])/(\text{fasting glucose } [mg/dL/18] - 3.5).$$

The liver glycogen content was measured using the method of Seifter et al. [24]. The liver glucose-6-phosphatase and glycogen phosphorylase activities were measured using laboratory-prepared chemicals and the methods of Begum et al. [25] and Stallman and Hers, respectively [26]. The serum cholesterol level was assayed using the method of Allain et al. and the reagent kits purchased from Spinreact Company (Spain) [27].

The serum triglyceride level was determined using the reagent kit purchased from Reactivos Spinreact Company (Girona, Spain) and Fossati and Prencipe's method [28]. The serum high density lipoprotein (HDL)-cholesterol level was measured using the method of Allain et al. (1974) and the reagent kit obtained from Spinreact Company, Spain [27]. The serum low density lipoprotein (LDL) cholesterol level was calculated using the formula of Friendewald et al. [29]:

$$\text{LDL cholesterol} = \text{total cholesterol} - \text{triglycerides}/5 - \text{HDL cholesterol}$$

Serum very low density lipoprotein (vLDL)-cholesterol was calculated using Norbert's formula [30]:

$$\text{vLDL cholesterol conc.} = \text{triglycerides}/5$$

The serum free fatty acid (FFA) level was determined using Duncombe's method [31].

2.9. RNA Isolation and RT-PCR

RNA was isolated from visceral adipose tissue using the GeneJet RNA purification kit produced by Thermo Fisher Scientific Inc., Branchburg, NJ 08876, USA, according to the procedures of Chomzynski and Sacchi [32] and Boom et al. [33]. The isolated RNA was quantified and qualified [34,35]. Thermo Scientific Verso 1-Step RT-PCR ReddyMix was used to produce cloned DNA that was amplified in the presence of specific forward and reverse primers using a Techne thermal cycler, Cole-Parmer, IL 60061, USA [35]. The primer pair sequences for the insulin receptor β-subunit were F: 5′CTGGAGAACTGCTCGGTCATT3′ and R: 5′GGCCA-TAGACACGGAAAAGAAG3′ [36], and those for β-actin re F: 5′TCACCCTGAAGTACCCCATGGAG3′ and R: 5′TTGGCCTTGG GGTTCAGGGGG3′ [37,38].

2.10. Determination of Oxidative Stress and Anti-Oxidant Defense Parameters

The glutathione content (GSH) in the liver was determined [39]. Moreover, glutathione-S-transferase (GST) activity in the liver was measured using Mannervik and Gutenberg's method [40]. The liver glutathione peroxidase (GPx) activity was determined using the method of Matkovics et al. [41]. The superoxide dismutase (SOD) and lipid peroxidation (LPO) were determined using the methods of Marklund and Marklund [42] and Preuss et al. (1998), respectively [43].

2.11. Histological Investigation

The pancreas from each rat was rapidly excised after dissection and then fixed in 10% neutral buffered formalin for 24 h. The organs were routinely processed and sectioned at a thickness of 4 to 5 μm. The sections of the pancreas were stained with hematoxylin and eosin [44,45].

2.12. Statistical Analysis

The results were analyzed using the PC-STAT Program [46]. One-way analysis of variance (ANOVA) was followed by the least significant difference (LSD) test to compare various groups. Data were described as the mean $\pm$ SE. A p value of >0.05 was considered nonsignificantly different, whereas p values of <0.05 and <0.01 were considered significant and highly significant, respectively.

3. Results

3.1. GC-MS Analysis

The GC-MS analysis of the *B. aegyptiaca* fruit and seed extracts showed the presence of several phytocomponents. Tables 1 and 2 and Figures 1 and 2 show the identified phyto-components with their retention time, which was expressed as the peak area %. In the fruit extract, compounds 3,4,6-tri-O-methyl-d-glucose (52.55%) and triethylphosphine (9.31%) were the most abundant. Conversely, in the seed extract, compounds 9,12-octadecadienoic acid (Z,Z)- (38.27%), 8-dodecen-1-ol, (Z)- (15.09%), 2,3-dihydroxypropyl ester (11.47%), and H-cyclopenta [b]quinoxaline-1,2,3trione (11.39%) were the most abundant.

Table 1. Chemical groups and compounds present in the *B. aegyptiaca* fruit aqueous extract.

Number	(RT) Retention Time	Compound Name	Area% (Higher than 1%)
1	2.123	Hydrazine, 1,1-dimethyl-	4.31
2	3.008	Butanal, 2-methyl-	3.04
3	3.043	Butanal, 2-methyl-	1.89
4	3.363	Propanoic acid, propyl ester	3.41
5	3.907	Glyceraldehyde	5.39
6	4.250	2-Furanmethanol	2.55
7	4.645	Isopropyl isothiocyanate	2.64
8	4.889	Triethylphosphine	9.31
9	9.454	2,4(3H,5H)-Furandione	1.46
10	9.744	Pyridine, 4-chloro-2,6-dimethyl-	0.96
11	18.561	3,4,6-Tri-O-methyl-d-glucose	52.55
12	21.667	Perhydrohistrionicotoxin-2-thione, 2-depentyl-	2.07
13	23.758	Propyl 11,12-methylene-octadecanoate	3.06

Table 2. Chemical groups and compounds present in the *B. aegyptiaca* seed aqueous extract.

Number	Retention Time	Compound Name	Area% (Higher than 1%)
1	4.263	Butane, 2-methyl-	1.67
2	20.538	-3,4,6Tri-O-methyl-d-glucose	2.59
3	24.018	H-Cyclopenta [b]quinoxaline-1,2,3trione	11.39
4	25.505	-9,12Octadecadienoic acid (Z,Z)-	2.47
5	25.568	13-Octadecenoic acid, methyl ester	1.62
6	26.152	9,12-Octadecadienoic acid (Z,Z)-	38.27
7	26.192	8-Dodecen-1-ol, (Z)-	15.09
8	26.281	9,12-Octadecadienoic acid (Z,Z)-, 2,3-dihydroxypropyl ester	11.47
9	26.339	Octadec-9-enoic acid	6.06
10	26.371	Benzoic acid, 4-(4-hydroxybenzylidenamino)-, propyl ester	6.79

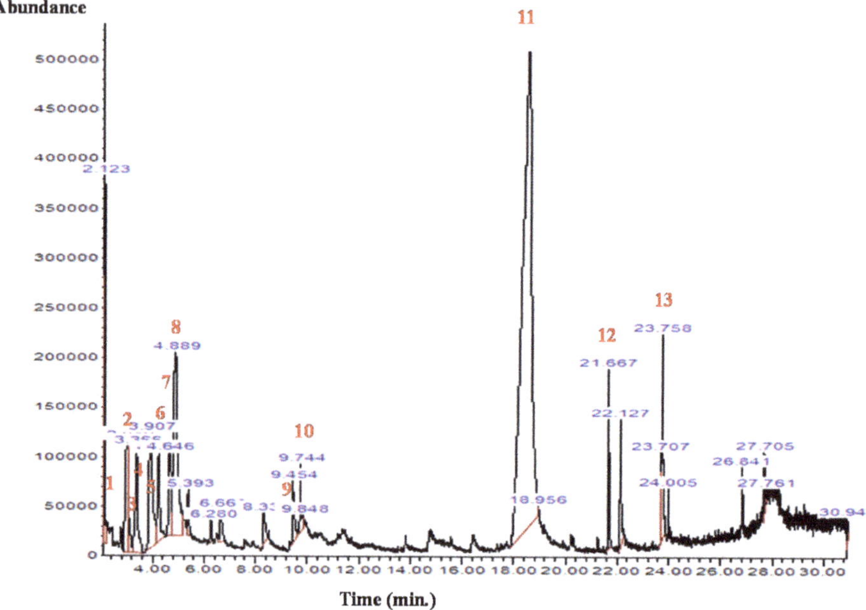

Figure 1. GC-MS chromatogram of the *B. aegyptiaca* fruit aqueous extract. 1: Hydrazine, 1,1-dimethyl-; 2 and 3: Butanal, 2-methyl-; 4: Propanoic acid, propyl ester; 5: Glyceraldehyde; 6: 2-Furanmethanol; 7: Isopropyl isothiocyanate; 8: Triethylphosphine; 9: 2,4(3H,5H)-Furandione; 10: Pyridine, 4-chloro-2,6-dimethyl-; 11: 3,4,6-Tri-O-methyl-d-glucose; 12: Perhydrohistrionicotoxin-2-thione, 2-depentyl-; and 13: Propyl 11,12-methylene-octadecanoate.

3.2. Effects on OGT

The OGT curve of the diabetic rats exhibited a significant ($p < 0.01$; LSD) elevation at all tested periods (0, 30, 60, 90, and 120 min) after oral glucose intake compared with that of normal animals. The oral administration of *B. aegyptiaca* fruit and seed extracts to diabetic rats induced a potential amelioration of elevated values at all tested points of the OGT curve. However, the seed extract was more potent at 30 and 60 min after oral glucose intake (Figure 3). The F-probability of OGTT data indicated that the effect between groups was very highly significant ($p < 0.01$).

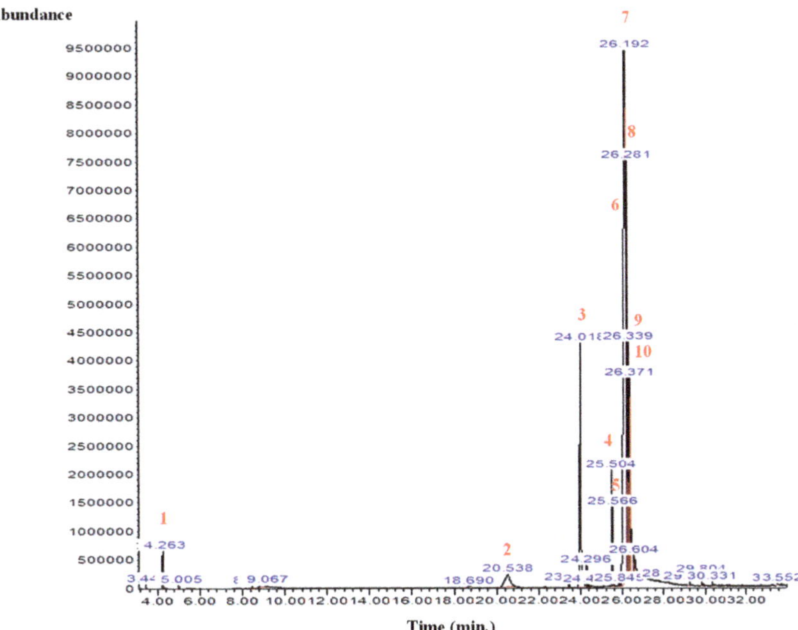

Figure 2. GC-MS chromatogram of the *B. aegyptiaca* seed aqueous extract. 1: Butane, 2-methyl-; 2: 3,4,6-Tri-O-methyl-d-glucose; 3: 3,4,6-Tri-O-methyl-d-glucose; 4: 9,12-Octadecadienoic acid (Z,Z)-; 5: 13-Octadecenoic acid, methyl ester; 6: 9,12-Octadecadienoic acid (Z,Z)-; 7: 8-Dodecen-1-ol, (Z)-; 8: 9,12-Octadecadienoic acid (Z,Z)-, 2,3-dihydroxypropyl ester; 9: Octadec-9-enoic acid; and 10: Benzoic acid, 4-(4-hydroxybenzylidenamino)-, propyl ester.

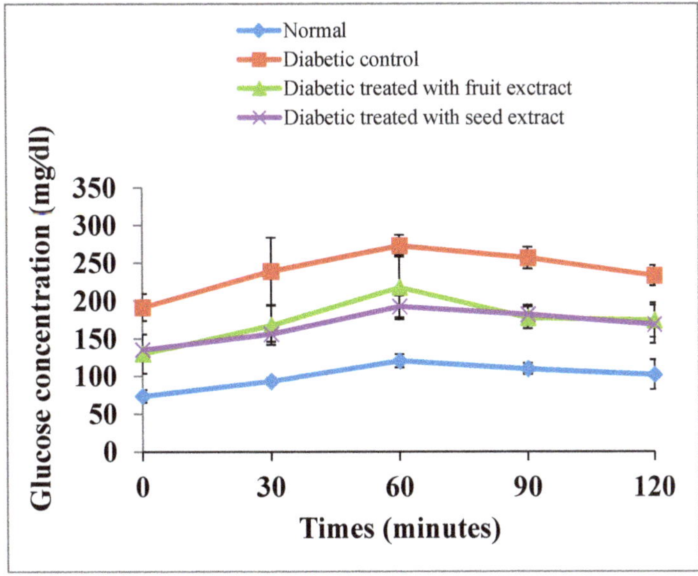

Figure 3. Effects of the *B. aegyptiaca* fruit and seed extracts on OGTT in NA/STZ-induced diabetic rats.

3.3. Effects on Serum Insulin and C-Peptide Levels

As indicated in Table 3, the diabetic rats showed a highly significant decrease ($p < 0.01$; LSD) in the insulin and C-peptide levels in serum. The treatments of the diabetic rats with the fruit and seed extracts caused a highly significant increase ($p < 0.01$; LSD) in these lowered levels. The diabetic rats treated with the *B. aegyptiaca* fruit extract exhibited no significant effects on the insulin and C-peptide levels in comparison with the diabetic rats treated with the *B. aegyptiaca* seed extract. However, the effects of the fruit extract was more potent in increasing the serum insulin levels. The F-probability indicated that the general effect between groups was very highly significant ($p < 0.01$).

Table 3. Effects of the *B. aegyptiaca* fruit and seed extracts on the serum insulin and C-peptide levels in NA/STZ-induced diabetic rats.

Parameter Group	Insulin (µIU/mL)	C-Peptide (pg/mL)
Normal	2.56 ± 0.51 [a]	4.7 ± 0.13 [a]
Diabetic control	1.00 ± 0.78 [c]	1.19 ± 0.13 [c]
Diabetic treated with fruit extract	1.71 ± 0.11 [b]	3.47 ± 0.44 [b]
Diabetic treated with seed extract	1.56 ± 0.11 [b]	3.50 ± 0.22 [b]

- Data were expressed as the mean $\pm$ SE. The number of animals in each group was six. - Means, which have different superscript symbols (a, b, and c), were significantly different at $p < 0.05$.

3.4. Effect on HOMA-IR Cell Function, HOMA-IS, and HOMA-β Cell Function

In diabetic rats, the HOMA-IS and HOMA-β cell functions were highly significantly ($p < 0.01$; LSD) decreased, whereas HOMA-IR was highly significantly ($p < 0.01$; LSD) increased. The treatment of diabetic rats with *B. aegyptiaca* fruit and seed extracts induced a highly significant increase in HOMA-β cell function and HOMA-IS. In contrast, HOMA-IR was highly significantly decreased after the treatments with the *B. aegyptiaca* fruit and seed extracts.

Although the effects of the *B. aegyptiaca* fruit and seed extracts on HOMA-IR and HOMA-IS were more or less similar, the effects of the fruit extract on HOMA-β cell function was more potent than that of the seed extract (Table 4). One-way ANOVA revealed that the effect between groups on the HOMA-IR, HOMA-IS, and HOMA-β cell function was very highly significant ($p < 0.01$; F-probability).

Table 4. Effects of the *B. aegyptiaca* fruit and seed extracts on HOMA-IR cell function, HOMA-IS, and HOMA-β cell function in NA/STZ-induced diabetic rats.

Parameter Group	HOMA-IR	HOMA-IS	HOMA-β Cell Function
Normal	0.37 ± 0.03 [c]	55.48 ± 2.23 [a]	55.50 ± 1.42 [a]
Diabetic control	0.51 ± 0.04 [a]	45.50 ± 0.66 [b]	2.71 ± 0.17 [d]
Diabetic treated with fruit extract	0.40 ± 0.03 [b]	55.00 ± 1.23 [a]	11.20 ± 0.90 [b]
Diabetic treated with seed extract	0.45 ± 0.03 [b]	53.33 ± 1.03 [a]	7.80 ± 0.39 [c]

- Data were expressed as the mean $\pm$ SE. The number of animals in each group was six. - Means, which have different superscript symbols (a, b, c, and d c), were significantly different at $p < 0.05$.

3.5. Effects on Liver Glycogen Content and Glucose-6-phospatase and Glycogen Phosphorylase Activities

The diabetic rats exhibited a highly significant ($p < 0.01$; LSD) depletion in liver glycogen content and a highly significant ($p < 0.01$; LSD) elevation in liver glucose-6-phosphatase and glycogen phosphorylase activities. The treatments with the *B. aegyptiaca* fruit and seed extracts highly significantly improved ($p < 0.01$; LSD) the reduced liver glycogen content of the diabetic rats and the elevated glucose-6-phosphatase and glycogen phosphorylase activities (Table 5). One-way ANOVA revealed that the effect between groups on liver glycogen content and glucose-6-phosphatase and glycogen phosphorylase activities was very highly significant ($p < 0.01$; F-probability).

Table 5. Effects of the *B. aegyptiaca* fruit and seed extracts on liver glycogen content and glucose-6-phosphatase and glycogen phosphorylase activities in NA/STZ-induced diabetic rats.

Group \ Parameter	Liver Glycogen (mg P liberated/g Tissue/h)	Glucose-6-phosphatase (mg Pi liberated/g Tissue/h)	Glycogen Phosphorylase (mg Pi liberated/g Tissue/h)
Normal	54.55 ± 3.60 [a]	25.79 ± 1.53 [b]	22.78 ± 3.09 [b]
Diabetic control	8.10 ± 1.66 [b]	67.42 ± 3.38 [a]	41.85 ± 6.08 [a]
Diabetic treated with fruit extract	42.75 ± 2.91 [a]	28.02 ± 0.41 [b]	25.45 ± 4.95 [b]
Diabetic treated with seed extract	50.51 ± 3.60 [a]	31.45 ± 2.28 [b]	15.22 ± 2.20 [b]

- Data were expressed as the mean ± SE. The number of animals in each group was six. - Means, which have different superscript symbols (a, b, and c), were significantly different at $p < 0.05$.

3.6. Effects on Serum Lipid Profile

The total cholesterol, triglycerides, LDL-cholesterol, vLDL- cholesterol, and FFA levels in serum exhibited a highly significant elevation ($p < 0.01$; LSD) in diabetic rats compared with those in the normal group. The treatment of diabetic rats with the *B. aegyptiaca* fruit and seed extracts produced a highly significant improvement in the altered lipid profile in the serum. Moreover, the HDL- cholesterol level was affected in an inverse pattern, as it was highly significantly decreased ($p < 0.01$; LSD) in diabetic rats.

Conversely, the treatment with the fruit extract induced a significant increase ($p < 0.01$; LSD) compared with the diabetic control, whereas the treatment with the seed extract had no significant effect ($p > 0.05$; LSD) (Table 6). The seed extract was more effective in decreasing the elevated total cholesterol and triglyceride levels in diabetic rats than the fruit extract, whereas the fruit extract was more potent in decreasing the elevated LDL-cholesterol, vLDL-cholesterol, and FFA levels and increasing the lowered HDL-cholesterol level (Table 6). The F-probability revealed that the effect on serum lipid profile between groups was very highly significant ($p < 0.01$).

Table 6. Effects of the *B. aegyptiaca* fruit and seed extracts on serum lipid profile and FFA level in NA/STZ-induced diabetic rats.

Group \ Parameter	Total Cholesterol (mg/dL)	Triglycerides (mg/dL)	LDL-Cholesterol (mg/dL)	HDL-Cholesterol (mg/dL)	vLDL-Cholesterol (mg/dL)	FFAs (mg/dL)
Normal	42.70 ± 1.34 [c]	35.66 ± 2.25 [c]	5.10 ± 2.10 [c]	31.30 ± 2.30 [a]	7.11 ± 0.33 [c]	5.10 ± 0.86 [b]
Diabetic control	79.40 ± 9.48 [a]	80.25 ± 15.30 [a]	37.48 ± 6.03 [a]	24.80 ± 2.80 [b]	16.08 ± 3.00 [a]	18.70 ± 2.7 [a]
Diabetic treated with fruit extract	59.70 ± 9.30 [b]	53.70 ± 2.83 [b]	13.18 ± 4.76 [b]	31.00 ± 2.10 [a]	8.11 ± 1.05 [c]	5.10 ± 0.70 [b]
Diabetic treated with seed extract	51.80 ± 9.70 [bc]	41.25 ± 5.34 [c]	15.94 ± 7.90 [b]	27.10 ± 0.80 [b]	10.71 ± 0.056 [b]	6.75 ± 1.80 [b]

- Data were expressed as the mean ± SE. The number of animals in each group was six. - Means, which have different superscript symbols (a, b, and c), were significantly different at $p < 0.05$.

3.7. Effect on Insulin Receptor β-Subunit

The densitometric analysis of the electrophoretogram showed a highly significant ($p < 0.01$; LSD) decrease in the mRNA expression of the adipose tissue insulin receptor β-subunit in diabetic rats compared with that in the normal group. The treatment of diabetic rats with the *B. aegyptiaca* fruit extract produced a highly significant ($p < 0.01$; LSD) amelioration of the insulin receptor β-subunit mRNA expression (Figure 4), whereas the treatment with the *B. aegyptiaca* seed extract did not show a significant effect ($p > 0.05$; LSD). One-ANOVA indicated that the effect between groups on the mRNA expression of insulin receptor β-subunit was highly significant ($p < 0.01$; F-probability).

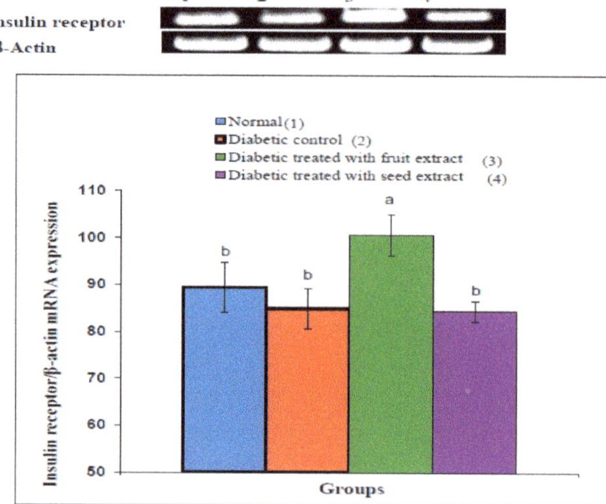

Figure 4. Effects of the *B. aegyptiaca* fruit and seed extracts on the adipose mRNA expression of insulin receptor β-subunit relative to β-actin in NA/STZ-administered rats. The means, which have different superscript symbols (a and b), were significantly different at $p < 0.05$. The number of detected samples in each group was three.

3.8. Effect on Oxidative Stress and Anti-Oxidant Defense Parameters

The liver LPO exhibited a highly significant ($p < 0.01$; LSD) increase in diabetic rats compared with that in normal rats. The treatment with the *B. aegyptiaca* fruit and seed extracts in diabetic rats resulted in a highly significant ($p < 0.01$; LSD) amelioration in LPO; the effects of the fruit extract were the most potent.

The GSH content as well as the GPx, GST, and SOD activities showed a highly significant ($p < 0.01$; LSD) decline in diabetic control rats compared with those in the normal rats. The treatments with the *B. aegyptiaca* fruit and seed extracts successfully improved the GSH content and GPx, GST ($p < 0.01$; LSD), and SOD activities ($p < 0.05$; LSD) (Table 7). While the effects of the fruit and seed extracts on GSH content and anti-oxidant enzyme activities were more or less similar, the seed extract was more potent in decreasing the elevated LPO.

Table 7. Effects of the *B. aegyptiaca* fruit and seed extracts on liver LPO, GSH content, and GPx, GST, and SOD activities in NASTZ-induced diabetic rats.

Group \ Parameter	LPO (nmole MDA/100 mg Tissue/h)	GSH (nmole/100 mg Tissue)	GPx (mU/100 mg Tissue)	GST (U/100mg Tissue)	SOD (U/g Tissue)
Normal	67.17 ± 1.25 [bc]	13.30 ± 1.63 [a]	42.60 ± 5.1 [a]	12.65 ± 3.6 [b]	16.78 ± 1.59 [a]
Diabetic control	111.39 ± 7.57 [a]	6.10 ± 1.60 [c]	17.20 ± 5.1 [c]	6.57 ± 1.65 [c]	10.31 ± 1.58 [c]
Diabetic treated with fruit extract	68.75 ± 3.80 [b]	12.10 ± 1.54 [ab]	30.80 ± 4.6 [b]	16.54 ± 0.86 [a]	13.46 ± 2.55 [b]
Diabetic treated with seed extract	52.95 ± 5.821 [c]	11.20 ± 1.5 [b]	31.00 ± 2.5 [b]	16.43 ± 0.85 [a]	13.26 ± 2.08 [b]

- Data were expressed as the mean ± SE. The number of animals in each group was six. - Means, which have different superscript symbols (a, b, and c), were significantly different at $p < 0.05$.

The F-probability indicated that the effect on the liver LPO, GSH content, and GPx, GST, and SOD activities between groups was very highly significant ($p < 0.01$; F-probability)

3.9. Histological Changes in the Pancreas

The histopathological examination of the control pancreas sections showed closely packed lobules of the pancreatic acini. The islet of Langerhans is composed of numerous compactly arranged cells (alpha, beta, and delta cells) occurring as dense cords (Figure 5; Photomicrographs A and B). The diabetic control revealed the histopathological changes of endocrine portions represented by a marked decrease in the size of the islets of pancreas (pancreatic shrinkage) and decreased number of the cells.

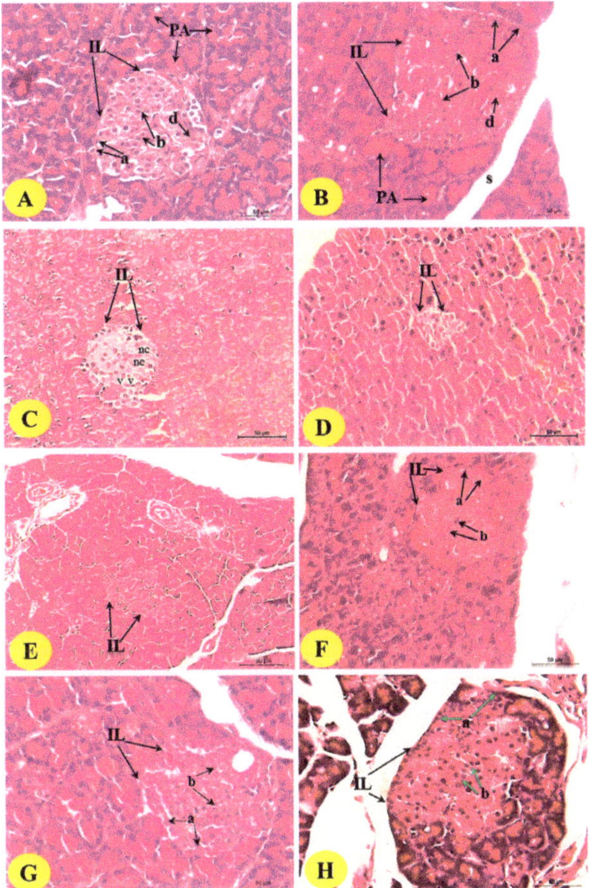

Figure 5. Photomicrographs of the H&E-stained pancreas of the normal (**A,B**), diabetic control (**C,D**), and diabetic rats treated with the *B. aegyptiaca* fruit (**E,F**) and seed extracts (**G,H**). Photomicrographs A and B show normal pancreatic lobules consisting of pancreatic acini and intact islets of Langerhans (IL) with alpha (a), beta (b), and delta cells. Photomicrographs C and D show the islets of Langerhans with the reduced size and smaller number of the cells; the islets exhibited necrosis (nc) and vacuolations (v) (H&E stain: Scale bar = 50 μm). Photomicrographs E–H show considerable improvements in the islets of Langerhans with a greater increase in the islet size and the number of islet cells.

The islets also exhibited cytoplasmic vacuolations (v) and necrosis (nc) (Figure 5; Photomicrographs C and D). The islets of Langerhans exhibited nearly normal and organized architecture, and the necrotic and degenerative changes were markedly improved in rats treated with the *B. aegyptiaca* fruit (Figure 5; Photomicrographs E and F) and seed extracts (Figure 5; Photomicrographs G and H) compared with those in the diabetic control rats; the seed extract was found to be the most potent.

4. Discussion

DM is a group of metabolic diseases characterized by chronic hyperglycemia due to defects in insulin secretion, insulin action, or both [47]. STZ is known for its selective pancreatic islet β-cell cytotoxicity and has been frequently used to induce DM in animals. NA/STZ-induced DM is selected to be a model of type 2 DM (T2DM) to assess the effects of the *B. aegyptiaca* fruit and seed aqueous extracts. In this model, when NA is injected prior to the administration of STZ, the severity of DM will be reduced to a certain extent, leading to a T2DM-like condition with deteriorated IS [20,35,48–50].

Several traditional medicines have been discovered for DM. Isolated substances and extracts isolated from various natural resources particularly plants have always been a rich arsenal for the control and treatment of DM problems and complications [9]. Plants have rich sources of antidiabetic as well as anti-hyperlipidemic and anti-oxidant substances, such as flavonoids, amino acids, gallotannins, and many other related polyphenols [51]. *B. aegyptiaca* is a plant widely used as a hypoglycemic agent in Egyptian folkloric medicine [15]. However, only few studies have investigated the antidiabetic effects and the mechanisms of action of *B. aegyptiaca*, especially its seeds.

In the present study, GC-MS data presented in Tables 1 and 2 revealed the presence of several active ingredients. In the *B. aegyptiaca* fruit aqueous extract, 3,4,6-tri-O-methyl-d-glucose (52.55%) and triethylphosphine (9.31%) were the most abundant, whereas in the seed aqueous extract, 9,12-octadecadienoic acid (Z,Z)- (38.27%), 8-dodecen-1-ol, (Z)-(15.09%), 2,3-dihydroxypropyl ester (11.47%), and H-cyclopenta [b]quinoxaline-1,2,3trione (11.39%) were the major components.

Many of the constituting ingredients of the *B. aegyptiaca* fruit and seed aqueous extracts exhibit many biological activities. Of these, 9,12-octadecadienoic acid (Z,Z)-have anti-inflammatory, hepatoprotective, cancer preventive, and hypocholesterolemic effects [20,52,53]. The derivative 9-octadecenoic acid was reported to have antitumor and anti-inflammatory activities [53]. Isothiocyanates have been demonstrated to have both anti-inflammatory and antioxidant activities [54].

OGTT is a well-known and common assay used to determine the anti-hyperglycemic activity of antidiabetic agents [55]. OGTT is considered the gold standard test for the diagnosis of DM by the World Health Organization [56]. The data in the present study revealed a marked increase in the serum glucose levels of the diabetic groups compared with those of the normal rats. These findings are consistent with those of Akhani et al. [57], Ahmed [58], Schaalan et al. [59], Ahmed et al. [60], Ahmed et al. [61], and Ali et al. [62].

The increase in glucose level may be due to the decreased glucose consumption in the peripheral, muscle, and adipose tissues [63] and increased glycogen breakdown [64], gluconeogenesis, and production of hepatic glucose [65]. Furthermore, Powers confirmed that IR in T2DM induces an increase in blood glucose because of the same causes [66]. In the present study, the treatment of diabetic rats with the *B. aegyptiaca* fruit or seed extracts caused a potential improvement in OGT. The decrease in the elevated serum glucose levels is in accordance with the results of Zaahkouk et al. [67], Helal et al. [68], and Al-Malki et al. [69] who verified the anti-hyperglycemic effects of the *B. aegyptiaca* fruit.

The ameliorative effects of the *B. aegyptiaca* extracts may be associated with insulinomimetic activities [70], stimulation and potentiation of insulin secretion, increased affinity of insulin receptors [71], improved concentration of hepatic glycogen, accelerated glucose metabolism, reduced production of intestinal glucosidase, and decreased gluconeogenesis of the liver [15]. These actions may be attributed to the active constituting ingredients found in the *B. aegyptiaca* fruit and seed extracts. In this regard, a previous publication found that *B. aegyptiaca* may contain interketones, organic constituents, rutin, and oils (fatty acids and volatile oils) present in the internal kernel according to the phytochemical investigation [72].

Furthermore, Baragob et al. attributed the hypoglycemic effects of the aqueous extract to its constituents, such as saponins, rutin, and organic constituents [73]. In the present study, the GC-MS analysis indicated the presence of many organic ingredients, which have several biological activities, including antidiabetic potencies.

In the present study, the NA/STZ-induced diabetic rats exhibited a significant decrease in serum insulin and C-peptide levels. It is important to note the relationship between decreased insulin and C-peptide levels in diabetic rats and decreased size and number of the islets of Langerhans that have a decreased number of β-cells, necrosis, and vacuolations. These decreases are also correlated with the calculated HOMA-β cell function.

The administration of the *B. aegyptiaca* fruit and seed aqueous extracts produced a significant increase in serum insulin and C-peptide levels of diabetic rats, and this finding is consistent with that of Abou Khalil et al. [74], El-Bayomy et al. [75], and Hassan [76]. In this regard, Abdel-Moneim [70] hypothesized that the hypoglycemic action of the *B. aegyptiaca* aqueous extract stimulated the β-cells of the pancreatic islets to secrete insulin, potentiate glucose-stimulated insulin secretion, and increase the number and sensitivity of insulin receptors and post-receptor effects in peripheral tissues.

Furthermore, the *B. aegyptiaca* seeds contain diosgenin [77], which may be useful in ameliorating the glucose metabolic disorder related with DM and obesity as reported by Ulbricht et al. [78]. C-peptide is formed during insulin biosynthesis, and the two peptides, insulin and C-peptide, are then released to the circulation in equal amounts [79]. An increase in C-peptide levels in diabetic rats treated with *B. aegyptiaca* corresponds well with the increase in insulin secretion (endogenous secretion), which is possibly due to the regeneration of the β-cells of the islets of Langerhans.

This association is demonstrated in the present study by the significant increase in HOMA-β cell function and marked improvement in the histological architecture and number of β cells of the pancreatic islets as a result of the treatment with the *B. aegyptiaca* fruit and seed extracts.

The level of liver glycogen may be regarded as the best marker for evaluating the anti-hyperglycemic activity of any drug [80]. The present study showed that the diabetic rats exhibited a significant depletion of liver glycogen content correlated with a marked increase of glucose 6 phosphatase and glycogen phosphorylase activities. These results are consistent with those of Sundaram et al. [81] and Mahmoud et al. [82]. In the present study, the treatment with the *B. aegyptiaca* fruit and seed extracts significantly improved the lowered liver glycogen content and the elevated hepatic glucose-6-phosphatase and glycogen phosphorylase activities.

These ameliorations may be secondary to the increase in the insulin levels in the blood and the enhanced IS. These results are consistent with those of Helal et al. [68] who found that the regeneration of β-cells led to an increase in the insulin level and improvement in IS, both of which can lower the glucose levels in the blood. The present results support this finding as the treatment of diabetic rats with the *B. aegyptiaca* fruit and seed extracts significantly increased the HOMA-β cell function and HOMA-IS but significantly decreased HOMA-IR (Figure 6).

Hypoinsulinemia and IR clearly shown in untreated diabetic rats are considered the main cause of the recorded dyslipidemia represented by hypertriglyceridemia and hypercholesterolemia associated with the increased production of vLDL cholesterol and LDL cholesterol and decreased HDL cholesterol level. These findings are consistent with those of Abdel-Moneim et al. [83] who reported a marked increase in the levels of serum triglycerides, cholesterol, and LDL cholesterol of diabetic rats. This increase may be due to a decrease in lipoprotein lipase function caused by insulin deficiency [84].

In addition, Goodman and Gilman [85] reached the same results, which were explained by inhibition of lipoprotein lipase transcription inside the capillary endothelium as a result of insulin deficiency. The data of the present study are consistent with those of Harvey and Ferrier [86] who reported that the metabolic abnormalities of T2DM as a result of IR lead to dyslipidemia in the liver where fatty acids are converted into triacylglycerol, which, in turn, are packaged and secreted in vLDL. Both accumulations of lipids, especially triglycerides, and reduced anti-oxidant activity contribute to the development of oxidative stress in diabetic rats [87].

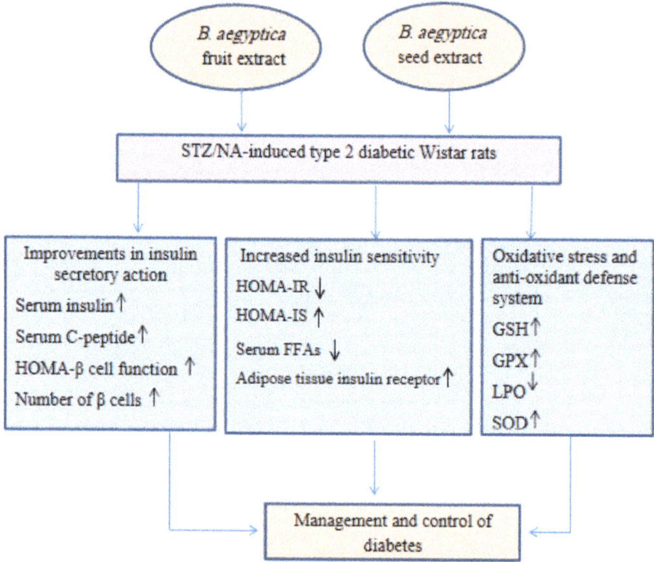

Figure 6. Schematic illustrative figure depicting the suggested hypothesis of the modes of action of the *B. aegyptiaca* fruit and seed aqueous extracts.

The results of the present study are consistent with those of previous studies [88–90], which revealed a reduction in the HDL cholesterol level in the diabetic control group. The present results of the serum lipid profile are consistent with the findings of Stanfield [91] who stated that DM increases the number of LDL particles that transport lipids, including cholesterol to peripheral tissues, and decreases the number of HDL particles, which transport lipids and cholesterol to the liver.

Simultaneously, the elevated serum triglyceride level in the diabetic group of the present study may be related to the decreased clearance and increased production of endogenously synthesized main triglyceride transporters [92]. Moreover, the expansion of the cholesterol pool in DM could be explained by the increased input into the system by accelerating the synthesis of intestinal cholesterol or by increasing the rate of absorption of intestinal cholesterol [93].

Indeed, the improvement in dyslipidemia through *B. aegyptiaca* treatment may be related to the increased level and sensitivity of insulin. The present results are not in concordance with those of Matter and Helal [94] who reported that the level of serum triglycerides and cholesterol was not significantly different when compared with the control group after treatment with the *B. aegyptiaca* seed extract.

In accordance with the findings of Abd El-Rahman and Al-ahmari [95], the improvement of lipid profile may be due to the presence of saponins in its extract, indicating antihypercholesterolemic and hypoglycemic activities. Moreover, diosgenin in the *B. aegyptiaca* seed kernels plays an important role in the regulation of cholesterol metabolism [96]. In the present study, the GC-MS analysis indicated the presence of many organic ingredients, which have several biological activities, including antihypercholesterolemic properties.

The elevated serum FFA level recorded in diabetic rats in the present study is consistent with that estimated in many preceding studies [35,62,97,98]. The elevated release of FFAs from the adipose tissue can be attributed to the lipolysis of visceral adipose depots; this effect can result in IR, excessive endogenous glucose formation, and progression to

T2DM [99]. Thus, decreasing the plasma FFA level is recommended as a method for IR prevention and treatment [100].

During the treatment of diabetic animals with *B. aegyptiaca* aqueous extracts, a reduction in the amount of serum FFA levels that could be associated with the insulin-sensitizing activity of the extract was observed [101]. Furthermore, several studies have found significantly low levels of resistin mRNA in the adipose tissue in various obese mouse models, such as db/db or high-fat diet–induced obesity, as well as in rat models with IR [102].

The present study showed a significant reduction in the mRNA expression of insulin receptor β-subunit in the adipose tissues of the diabetic group in comparison with that of the normal group. This result is consistent with that of Ali et al. [62] and Abdel Aziz et al. [51] who demonstrated that the mRNA expression of insulin receptor β-subunit was significantly decreased in NA/STZ-induced diabetic rats. This effect provides evidence of the presence of IR and impaired IS in such animal models, which in turn is a suitable model of T2DM. In the present study, the treatment with the *B. aegyptiaca* fruit aqueous extract produced a significant increase in the mRNA expression of insulin receptor β-subunit reflecting the ability of this extract to reduce IR and enhance IS in the adipose tissues (Figure 6).

Oxidative stress is an important factor in DM etiology and pathogenesis, causing interactions with polyunsaturated fatty acids that contribute to LPO [103]. According to Randle's theory on glucose-fatty acids [104], the excessive release of free fatty acids from the adipose tissue for oxidation induces the production of metabolites that prevent tissue use of glucose. Such metabolites are reactive oxygen species and hydrogen peroxide, which are involved with the glucose-fatty acid process [105].

Belfort et al. [106] showed that the increase in plasma FFA caused a dose-dependent inhibition of insulin-stimulated glucose disposal and insulin signaling in the skeletal muscle of lean healthy individuals. The present findings showed a significant increase in LPO in the liver. Additionally, the GSH level and anti-oxidant enzyme defenses decreased simultaneously in the liver of diabetic rats. These findings are consistent with those of several studies [35,89,89,107,108]. GSH plays a multifaceted role in the defense against anti-oxidants. It actively scavenges free radicals or indirectly detoxifies reactive species via GST and GPx [109].

The treatment of diabetic rats with *B. aegyptiaca* extracts significantly decreased MDA, which is attributed to the increased levels of anti-oxidants that fight free radicals [110] and markedly increased GSH level and SOD and GPx activities. Thus, it is worth noting that the improvement in the glycemic state, lipid profile, and insulinotropic and insulin-sensitizing effects is associated with the suppression of oxidative stress and enhancement of the anti-oxidant defense system.

This indicated that the decrease in oxidative stress and enhancement of the anti-oxidant defense system may have an important role in the improvement of the architecture and tissue IS of the pancreatic islets, which in turn result in the effective management of diabetes. These findings are consistent with those of Hassanin et al. [111] who indicated that *B. aegyptiaca* exerted hypoglycemic, hypolipidemic, and insulinotropic actions associated with the reduction in oxidative stress, enhancement in the anti-oxidant defense system, and reduced apoptosis in pancreatic β-cells.

In conclusion, the *B. aegyptiaca* fruit and seed aqueous extracts have potent antidiabetic potencies, which may be mediated via improvements in the insulin secretory response, β-cell function, tissue IS, and anti-oxidant defense system (Figure 6).

Author Contributions: Conceptualization, O.M.A. M.K. and M.A.-G.; project administration, M.A.-G., M.K., M.A.-G., E.M.F., M.M.A., T.M.A. and O.M.A.; supervision, M.A.-G., M.K., M.A.-G. and O.M.A.; funding acquisition, A.S.Z., M.M.A. and T.M.A.; methodology, A.S.Z., M.K. and O.M.A.; data curation, A.S.Z., M.A.-G., E.M.F. and O.M.A.; statistical analysis, A.S.Z. and O.M.A.; software, A.S.Z.; validation and visualization, M.K., M.A.-G., E.M.F., M.M.A., T.M.A. and O.M.A.; formal analysis, A.S.Z. and O.M.A.; writing the original draft, A.S.Z.; revising and editing, M.K., M.A.-G.,

E.M.F., M.M.A., T.M.A. and O.M.A. All authors have read and agreed to the published version of the manuscript.

Funding: This work was supported by Taif University, Taif, Saudi Arabia (Taif University Researchers supporting project number: TURSP-2020/80).

Institutional Review Board Statement: The animal experiments were approved by Experimental Animal Ethics Committee, Faculty of Science, Beni-Suef University, Egypt (Ethical Approval Number: BSU/FS/2015/17).

Informed Consent Statement: Not applicable.

Data Availability Statement: The data presented in this study are available in this manuscript.

Acknowledgments: The authors acknowledged Taif University, Taif, Saudi Arabia (Taif University Researchers supporting project number: TURSP-2020/80) for supporting and funding.

Conflicts of Interest: The authors declare no conflict of interest.

Abbreviations

B. aegyptiaca: *Balanites aegyptiaca*; b.w.: body weight; DM: Diabetes mellitus; ELISA: sandwich enzyme-linked immunosorbent assay; FFA: free fatty acid; GC-MS: Gas chromatography–mass spectrometry; GPx: glutathione peroxidase; GST: glutathione-S-transferase; H&E: haematoxylin and eosin; HDL: high density lipoprotein; HOMA-IR: homeostatic model assessment-insulin resistance; HOMA-IS: homeostatic model assessment- insulin sensitivity; HOMA-β cell function: homeostatic model assessment-β cell function; IDF: International Diabetes Federation; GSH: reduced glutathione; IP: intraperitoneal; IR: insulin resistance; IS: insulin sensitivity; LDL: low density lipoprotein; LPO: lipid peroxidation; LSD: least significant difference; MDA: malondialdehyde; mRNA: messenger RNA; OGT: oral glucose tolerance; OGT: oral glucose tolerance; OGTT: oral glucose tolerance test; RNA: Ribonucleic acid; RT-PCR: reverse transcription-polymerase chain reaction; SOD: superoxide dismutase; STZ: streptozotocin; T2DM: Type 2 DM; and vLDL: very low density lipoprotein.

References

1. He, J.H.; Chen, L.X.; Li, H. Progress in the discovery of naturally occurring anti-diabetic drugs and in the identification of their molecular targets. *Fitoterapia* **2019**, *134*, 270–289. [CrossRef] [PubMed]
2. Reach, G.; Pechtner, V.; Gentilella, R.; Corcos, A.; Ceriello, A. Clinical inertia and its impact on treatment intensification in people with type 2 diabetes mellitus. *Diabetes Metab.* **2017**, *43*, 501–511. [CrossRef]
3. *Diabetes Atlas, International Diabetes Federation (IDF)*, 9th ed.; International Diabetes Federation: Brussels, Belgium, 2019. Available online: https://www.diabetesatlas.org/en/ (accessed on 16 January 2019).
4. Hirst, M. Diabetes in 2013, The new figures. *Diabetes Res. Clin. Pract.* **2013**, *102*, 265. [CrossRef]
5. Guariguata, L.; Whiting, D.R.; Hambleton, I.; Beagley, J.; Linnenkamp, U.; Shaw, J.E. Global estimates of diabetes prevalence for 2013 and projections for 2035. *Diabetes Res. Clin. Pract.* **2014**, *103*, 137–149. [CrossRef]
6. Osadebe, P.O.; Odoh, E.U.; Uzor, P.F. Natural Products as Potential Sources of Antidiabetic Drugs. *Br. J. Pharmaceut. Res.* **2014**, *4*, 2075–2095. [CrossRef]
7. Abd El Mgeed, A.; Bstawi, M.; Mohamed, U.; Gabbar, M.A. Histopathological and biochemical effects of green tea and/or licorice aqueous extracts on thyroid functions in male albino rats intoxicated with dimethylnitrosamine. *Nutr. Metab.* **2009**, *6*, 2. [CrossRef] [PubMed]
8. Ahmed, O.M.; Hassan, M.A.; Saleh, A.S. Combinatory effect of hesperetin and mesenchymal stem cells on the deteriorated lipid profile, heart and kidney functions band antioxidant activity in STZ-induced diabetic rats. *Biocell* **2020**, *44*, 27–29. [CrossRef]
9. Verma, S.; Gupta, M.; Popli, H.; Aggarwal, G. Diabetes Mellitus Treatment Using Herbal Drugs. *Int. J. Phytomed.* **2018**, *10*, 1–10. [CrossRef]
10. Chang, C.L.; Lin, Y.; Bartolome, A.P.; Chen, Y.C.; Chiu, S.C.; Yang, W.C. Herbal therapies for type 2 diabetes mellitus: Chemistry, biology, and potential application of selected plants and compounds. *Evid. Based Complement. Altern. Med.* **2013**, *2013*, 378657. [CrossRef] [PubMed]
11. Eskander, E.F.; Jun, H.W. Hypoglycaemic and hyperinsulinemic effects of some egyptian herbs used forthe treatment of diabetes mellitus (type II) in rats. *Egypt. J. Pharm. Sci.* **1995**, *36*, 331–342.
12. Koko, W.S.; Galal, M.; Khalid, H.S. Fasciolicidal efficacy of Albizia anthelmintica and Balanites aegyptiaca compared with albendazole. *J. Ethnopharmacol.* **2000**, *71*, 247–252. [CrossRef]
13. Sarker, S.D.; Bartholomew, B.; Nash, R.J. Alkaloids from Balanites aegyptiaca. *Fitoterapia* **2000**, *71*, 328–330. [CrossRef]

14. Kamel, M.S.; Ohtani, K.; Kurokawa, T.; Assaf, M.H.; El-Shanawany, M.A.; Ali, A.A.; Tanaka, O. Studies on *Balanites Aegyptiaca* fruits, an antidiabetic Egyptian folk medicine. *Chem. Pharm. Bull.* **1991**, *39*, 1229–1233. [CrossRef] [PubMed]
15. Gad, M.Z.; El-Sawalhi, M.M.; Ismail, M.F.; El-Tanbouly, N.D. Biochemical study of the anti-diabetic action of the Egyptian plants Fenugreek and Balanites. *Mol. Cell. Biochem.* **2006**, *281*, 173–183. [CrossRef] [PubMed]
16. Hosny, M.; Khalifa, T.; Caliġ, I.; Wright, A.D.; Sticher, O. Balanitoside, a furostanol glycoside, and 6-methyldiosgenin from *Balanites aegyptiaca*. *Phytochemistry* **1992**, *31*, 3565–3569. [CrossRef]
17. Staerk, D.; Chapagain, B.P.; Lindin, T.; Wiesman, Z.; Jaroszewski, J.W. Structural analysis of complex saponins of *Balanites aegyptiaca* by 800 MHz 1H NMR spectroscopy. *Magn. Reson. Chem.* **2006**, *44*, 923–928. [CrossRef] [PubMed]
18. Elfeel, A.A.; Warrag, E.I. Uses and Conservation Status of *Balanites aegyptiaca* (L.) Del. (Hegleig Tree) in Sudan: Local People Perspective. *Asian J. Agric. Sci.* **2011**, *3*, 286–290.
19. Aboonabi, A.; Rahmat, A.; Othman, F. Antioxidant effect of pomegranate against streptozotocin-nicotinamide generated oxidative stress induced diabetic rats. *Toxicol. Rep.* **2014**, *1*, 915–922. [CrossRef]
20. Aziz, S.M.A.; Ahmed, O.M.; EL-Twab, S.M.A.; Al-Muzafar, H.M.; Amin, K.A.; Abdel-Gabbar, M. Antihyperglycemic effects and mode of actions of Musa paradisiaca leaf and fruit peel hydroethanolic extracts in nicotinamide/streptozotocin-induced diabetic rats. *Evid. Based Complement. Altern. Med.* **2020**, *2020*, 9276343.
21. Trinder, P. Determination of glucose in blood using glucose oxidase with an alternative oxygen acceptor. *Ann. Clin. Biochem.* **1969**, *6*, 24–27. [CrossRef]
22. Mishra, J.S.; More, A.S.; Kumar, S. Elevated androgen levels induce hyperinsulinemia through increase in Ins1 transcription in pancreatic beta cells in female rats. *Biol. Reprod.* **2018**, *98*, 520–531. [CrossRef]
23. Kuang, L.; Huang, Z.; Hong, Z.; Chen, A.; Li, Y. Predictability of 1-h postload plasma glucose concentration: A 10-year retrospective cohort study. *J. Diabetes Investig.* **2015**, *6*, 648–654. [CrossRef]
24. Seifer, S.; Dayton, S.; Novic, B.; Muntwyler, E. The estimation of glycogen with anthrone reagent. *Arch. Biochem.* **1950**, *25*, 191–200.
25. Begum, N.; Moses, S.G.; Shanmugasundaram, K.R. Serum enzymes in human and experimental diabetes mellitus. *Indian J. Med. Res.* **1978**, *68*, 774–784.
26. Stallman, W.; Hers, H.G. The stimulation of liver phosphorylase b by AMP, fluoride and sulfate. A technical note of the specific determination of the a and b forms of liver glycogen phosphorylase. *Eur. J. Biochem.* **1975**, *54*, 341–350. [CrossRef] [PubMed]
27. Allain, C.C.; Poon, L.S.; Chan, C.S.; Richmond, W.; Fu, P.C. Enzymatic determination of total serum cholesterol. *Clin. Chem.* **1974**, *20*, 470–475. [CrossRef]
28. Fossati, P.; Prenicpe, L. Serum triglycerides determined colorimetrically with an enzyme that produces hydrogen peroxide. *Clin. Chem.* **1982**, *28*, 2077–2080. [CrossRef] [PubMed]
29. Friedewald, W.T.; Levy, R.I.; Fredrickson, D.S. Estimation of the concentration of low-density lipoprotein cholesterol in plasma, without use of the preparative ultracentrifuge. *Clin. Chem.* **1972**, *18*, 499–502. [CrossRef]
30. Norbert, W.T. *Clinical Guide to Laboratory Tests*, 3rd ed.; W.B. Saunders Company: Philadilphia, PA, USA, 1995.
31. Duncombe, W.G. The coloremitric Micro-determination of long chain fatty acids. *Biochem. J.* **1964**, *88*, 7–10. [CrossRef] [PubMed]
32. Chomzynski, P.; Sacchi, N. Single-step method of RNA isolation by acid guanidinium thiocyanate-phenol-chloroform extraction. *Ann. Rev. Biochem.* **1987**, *162*, 156–159. [CrossRef]
33. Boom, R.C.J.A.; Sol, C.J.; Salimans, M.M.; Jansen, C.L.; Dillen, P.M.W.-V.; Van der Noordaa, J.P.M.E. Rapid and simple method for purification of nucleic acids. *J. Clin. Microbiol.* **1990**, *28*, 495–503. [CrossRef] [PubMed]
34. Sambrook, J.; Russell, D.W. *Molecular Cloning Laboratory Manual*, 5th ed.; Chemical Industry Press: Beijing, China, 2008.
35. Ahmed, O.M.; Hassan, M.A.; Abdel-Twab, S.M.; Azeem, M.N.A. Navel orange peel hydroethanolic extract, naringin and naringenin have anti-diabetic potentials in type 2 diabetic rats. *Biomed. Pharmacother.* **2017**, *94*, 197–205. [CrossRef] [PubMed]
36. Limin, T.; Hou, X.; Liu, J.; Zhang, X.; Sun, N.; Gao, L.; Zhao, J. Chronic ethanol consumption resulting in the downregulation of insulin receptor-β subunit, insulin receptor substrate-1, and glucose transporter 4 expression in rat cardiac muscles. *Alcohol* **2009**, *43*, 51–58. [CrossRef]
37. Shaker, O.G.; Sadik, N.A.H. Vaspin gene in rat adipose tissue: Relation to obesity-induced insulin resistance. *Mol. Cell. Biochem.* **2013**, *373*, 229–239. [CrossRef]
38. Ahmed, O.M.; Ahmed, A.A.; Fahim, H.I.; Zaky, M.Y. Quercetin and naringenin abate diethylnitrosamine/acetylaminofluorene-induced hepatocarcinogenesis in Wistar rats: The roles of oxidative stress, inflammation and cell apoptosis. *Drug Chem. Toxicol.* **2019**, *42*, 1–12, Erratum in *Drug Chem. Toxicol.* **2020**, *43*, 112. [CrossRef]
39. Beutler, E.; Duron, O.; Kelly, B.M. Improved method for the determination of blood glutathione. *J. Lab. Clin. Med.* **1963**, *61*, 882–888.
40. Mannervik, B.; Guthenberg, C. Glutathione transferase (human placenta). *Methods Enzymol.* **1981**, *77*, 231–235.
41. MacDougald, O.A.; Burant, C.F. The rapidly expanding family of adipokines. *Cell Metab.* **2007**, *6*, 159–161. [CrossRef]
42. Marklund, S.; Marklund, G. Involvement of the superoxide anion radical in the autoxidation of pyrogallol and a convenient assay for superoxide dismutase. *Eur. J. Biochem.* **1974**, *47*, 469–474. [CrossRef] [PubMed]
43. Preuss, H.G.; Jarrell, S.T.; Scheckenbach, R.; Lieberman, S.; Anderson, R.A. Comparative effect of chromium vanadium and Gymnema sylvestre on sugar-induced blood pressure elevation in SHR. *J. Am. Coll. Nutr.* **1998**, *17*, 116–123. [CrossRef]
44. Bancroft, J.; Gamble, M. *Throry and Practice of Histological Techniques*; Churchill Livingstone: London, UK, 2002; pp. 172–175.

45. Suvarna, S.K.; Layton, C.; Bancroft, J.D. *Bancroft Theory and Practice of Histological Techniques*, 7th ed.; Elsevier, Churchilli Livingstone: London, UK, 2013.
46. Roa, M.; Blane, K.; Zonneberg, M. *One Way Analysis of Variance, Version IA (C), PC-STAT, Program Coded by University of Georgia*; University of Georgia: Athens, GA, USA, 1985.
47. Trikkalinou, A.; Papazafropoulou, A.K.; Melidonis, A. Type 2 diabetes and quality of life. *World J. Diabetes* **2017**, *8*, 120–129. [CrossRef]
48. Masiello, P. Animal models of type 2 diabetes with reduced pancreatic β-cell mass. *Int. J. Biochem. Cell Biol.* **2006**, *38*, 873–893. [CrossRef] [PubMed]
49. Punitha, I.R.; Rajendran, K.; Shirwaikar, A.; Shirwaikar, A. Alcoholic stem extract of Coscinium fenestratum regulates carbohydrate metabolism and improves antioxidant status in streptozotocin–nicotinamide induced diabetic rats. *Evid. Based Complement. Altern. Med.* **2005**, *2*, 375–381. [CrossRef] [PubMed]
50. Veerapur, V.P.; Prabhakar, K.R.; Kandadi, M.R.; Srinivasan, K.K.; Unnikrishnan, M.K. Antidiabetic effect of *Dodonaea viscosa* aerial parts in high fat diet and low dose streptozotocin-induced type 2 diabetic rats: A mechanistic approach. *Pharm. Biol.* **2010**, *48*, 1137–1148. [CrossRef] [PubMed]
51. Ashok-Kumar, B.S.; Lakshman, K.; Jayaveea, K.N.; SheshadriShekar, D.; Saleemulla Khan, B.S.; Veeresh, T.; Veerapur, P. Antidiabetic, antihyperlipidemic and antioxidant activities of methanolic extract of *Amaran* thus *viridis* Linn in alloxan induced diabetic rats. *J. Exp. Toxicol. Pathol.* **2012**, *64*, 75–79. [CrossRef]
52. Salem, M.Z.; Zayed, M.Z.; Ali, H.M.; Abd El-Kareem, M.S. Chemical composition, antioxidant and antibacterial activities of extracts from *Schinus molle* wood branch growing in Egypt. *J. Wood Sci.* **2016**, *62*, 548–561. [CrossRef]
53. Ahmed, O.M.; Fahim, H.I.; Ahmed, H.Y.; Al-Muzafar, H.M.; Ahmed, R.R.; Amin, K.A.; Abdelazeem, W.H. The preventive effects and the mechanisms of action of navel orange peel hydroethanolic extract, naringin, and naringenin in N-acetyl-p-aminophenol-induced liver injury in Wistar rats. *Oxid. Med. Cell. Longev.* **2019**, *2019*, 2745352. [CrossRef] [PubMed]
54. Burčul, F.; Mekinić, I.G.; Radan, M.; Rollin, P.; Blažević, I. Isothiocyanates: Cholinesterase inhibiting, antioxidant, and anti-inflammatory activity. *J. Enzym. Inhib. Med. Chem.* **2018**, *33*, 577–582. [CrossRef]
55. Alberti, K.G.; Zimmet, P.Z. Definition, diagnosis and classification of diabetes mellitus and its complications. Part 1: Diagnosis and classification of diabetes mellitus. Provisional report of a WHO Consultation. *Diabetes Med.* **1998**, *15*, 539–553. [CrossRef]
56. Katulanda, G.W.; Katulanda, P.; Dematapitiya, C.; Dissanayake, H.A.; Wijeratne, S.; Sheriff, M.H.R.; Matthews, D.R. Plasma glucose in screening for diabetes and pre-diabetes: How much is too much? Analysis of fasting plasma glucose and oral glucose tolerance test in Sri Lankans. *BMC Endocr. Disord.* **2019**, *19*, 11. [CrossRef]
57. Akhani, S.P.; Vishwakarma, S.L.; Goyal, R.K. Anti-diabetic activity of Zingiber officinale in streptozotocin-induced type I diabetic rats. *J. Pharm. Pharmacol.* **2004**, *56*, 101–105. [CrossRef] [PubMed]
58. Ahmed, O.M. Evaluation of the antihyperglycemic, antihyperlipidemic and myocardial enhancing properties of pioglitazone in diabetic and hyperthyroid rats. *J. Egypt. Ger. Soc. Zool.* **2006**, *51*, 253–278.
59. Schaalan, M.; El-Abhar, H.S.; Barakat, M.; El-Denshary, E.S. Westernized-like-diet-fed rats: Effect on glucose homeostasis, lipid profile, and adipocyte hormones and their modulation by rosiglitazone and glimepiride. *J. Diabetes Complicat.* **2009**, *23*, 199–208. [CrossRef] [PubMed]
60. Ahmed, O.M.; Moneim, A.A.; Yazid, I.A.; Mahmoud, A.M. Antihyperglycemic, antihyperlipidemic and antioxidant effects and the probable mechanisms of action of Ruta graveolens infusion and rutin in nicotinamide-streptozotocin-induced diabetic rats. *Diabetol. Croat.* **2010**, *39*, 15–32.
61. Ahmed, O.M.; Hozayen, W.G.M.; Bastawy, M.; Hamed, M.Z. Biochemical effects of Cichorium intybus and Sonchus oleraceus infusions and esculetin on streptozotocin-induced diabetic albino rats. *J. Am. Sci.* **2011**, *7*, 1124–1138.
62. Ali, A.M.; Gabbar, M.A.; Abdel-Twab, S.M.; Fahmy, E.M.; Ebaid, H.; Alhazza, I.M.; Ahmed, O.M. Antidiabetic Potency, Antioxidant Effects, and Mode of Actions of Citrus reticulata Fruit Peel Hydroethanolic Extract, Hesperidin, and Quercetin in Nicotinamide/Streptozotocin-Induced Wistar Diabetic Rats. *Oxid. Med. Cell. Longev.* **2020**, *2020*, 1730492. [CrossRef]
63. Beck-Nielsen, H. Insulin resistance: Organ manifestations and cellular mechanisms. *Ugeskr. Laeg.* **2002**, *164*, 2130–2135.
64. Gold, A.H. The effect of diabetes and insulin on liver glycogen synthetase activation. *J. Biol. Chem.* **1970**, *245*, 903–905. [CrossRef]
65. Raju, J.; Gupta, D.; Rao, A.R.; Yadava, P.K.; Baquer, N.Z. Trigonella foenum graecum (fenugreek) seed powder improves glucose homeostasis in alloxan diabetic rat tissues by reversing the altered glycolytic, gluconeogenic and lipogenic enzymes. *Mol. Cell. Biochem.* **2001**, *224*, 45–51. [CrossRef]
66. Powers, A. Diabetes Mellitus. In *Harrison's Principles of Internal Medicine*; Kasper, D.L., Fauci, A.S., Longo, D., Braunwald, E., Hauser, S., Jameson, J.L., Eds.; Mcgraw-Hill: New York, NY, USA, 2005.
67. Zaahkouk, S.A.; Rashid, S.Z.; Mattar, A.F. Anti-diabetic properties of water and ethanolic extracts of *Balanites aegyptiaca* fruits flesh in senile diabetic rats. *Egypt. J. Hosp. Med.* **2003**, *10*, 90–108. [CrossRef]
68. Helal, E.G.; El-Wahab, A.; Samia, M.; El Refaey, H.; Mohammad, A.A. Antidiabetic and antihyperlipidemic effect of Balanites aegyptiaca seeds (aqueous extract) on diabetic rats. *Egypt. J. Hosp. Med.* **2013**, *52*, 725–739. [CrossRef]
69. Al-Malki, A.L.; Barbour, E.K.; Abulnaja, K.O.; Moselhy, S.S. Management of hyperglycaemia by ethyl acetate extract of Balanites aegyptiaca (desert date). *Molecules* **2015**, *20*, 14425–14434. [CrossRef]
70. Motaal, A.A.; Shaker, S.; Haddad, P.S. Antidiabetic activity of standardized extracts of Balanites aegyptiaca fruits using cell-based bioassays. *Pharmacol. J.* **2012**, *4*, 20–24. [CrossRef]

71. Abdel-Moneim, A. Effect of some medicinal plants and gliciazide on insulin release in vitro. *J. Egypt. Ger. Soc. Zool.* **1998**, *25*, 423–445.

72. Nour, M.E.; Khalifa, E.L.; Massimo, K.; Hassen, B. Preliminary study on seed pregermination treatments and vegetative propagation of *Balanites aegyptiaca* (L). *Physiology* **1991**, *4*, 413–415.

73. Baragob, A.E.A.; AlMalki, W.H.; Shahid, I.; Bakhdhar, F.A.; Bafhaid, H.S.; Eldeen, O.M.I. The hypoglycemic effects of the aqueous extract of the fruits of *Balanites aegypticea* in Alloxan-induced diabetic rats. *Pharmacogn. Res.* **2014**, *6*, 1–5.

74. Khalil, N.S.A.; Abou-Elhamd, A.S.; Wasfy, S.I.; El Mileegy, I.M.; Hamed, M.Y.; Ageely, H.M. Antidiabetic and antioxidant impacts of desert date (*Balanites aegyptiaca*) and parsley (*Petroselinum sativum*) aqueous extracts: Lessons from experimental rats. *J. Diabetes Res.* **2016**, *2016*, 8408326.

75. El-Bayomy, M.M.; El-Mously, M.; El-Stoohy, F.; Mehanna, S.S. Effect of oral administration of the aqueous extract of *Balanites aegyptiaca* dates on blood glucose, serum insulin and lipids in streptozotocin-induced diabetes in rats. *J. Biomed. Ther. Sci.* **1992**, *8*, 82–89.

76. Hassan, M. *Physiological and Herpbiochemical Studies on the Effects of Balanites Aegyptiaca in Albino Ratápdez*; Department of Botany, Faculty of Science, Assiut University: Asyut, Egypt, 2000.

77. Chapagain, B.; Wiesman, Z. Variation in diosgenin level in seed kernels among different provenances of *Balanites aegyptiaca* Del (Zygophyllaceae) and its correlation with oil content. *Afr. J. Biotechnol.* **2005**, *4*, 1209–1213.

78. Ulbricht, C.; Basch, E.; Burke, D.; Cheung, L.; Ernst, E.; Giese, N.; Weissner, W. Fenugreek (*Trigonella foenum-graecum* L. Leguminosae):anevidence-based systematic review by the natural standard research collaboration. *J. Herb. Pharmacother.* **2008**, *7*, 143–177. [CrossRef]

79. Wahren, J.; Ekberg, K.; Samnegard, B.; Johansson, B.L. C peptide: A new potential in the treatment of diabetic nephro pathy. *Curr. Diabetese Rep.* **2001**, *1*, 261–266. [CrossRef]

80. Grover, J.K.; Vats, V.; Rathi, S.S. Antihyperglycemic effect of Eugenia jambolana and Tinospora cordifolia in experimental diabetes and their effects on key metabolic enzymes involved in carbohydrate metabolism. *J. Ethnopharmacol.* **2000**, *73*, 461–470. [CrossRef]

81. Sundaram, R.; Shanthi, P.; Sachdanandam, P. Effect of tangeretin, a polymethoxylated flavone on glucose metabolism in streptozotocin-induced diabetic rats. *Phytomedicine* **2014**, *21*, 793–799. [CrossRef] [PubMed]

82. Mahmoud, A.M.; Ahmed, O.M.; Ashour, M.B.; Abdel-Moneim, A. In vivo and in vitro antidiabetic effects of citrus flavonoids; a study on the mechanism of action. *Int. J. Diabetes Dev. Ctries.* **2015**, *35*, 250–263. [CrossRef]

83. Abdel-Moneim, A.; Al-Zayat, E.; Mahmoud, S. Effect of some antioxidants on streptozotocin diabetic rat's comparative physiology. *J. Egypt. Ger. Soc. Zool.* **2002**, *38*, 213–245.

84. Minnich, A.; Zilversmit, D.B. Impaired triacylglycerol catabolism in hypertriglyceridemia of the diabetic, cholesterol fed rabbit: A possible mechanism for protection from atherosclerosis. *Biochim. Biophys. Acta (BBA) Lipids Lipid Metab.* **1989**, *1002*, 324–332. [CrossRef]

85. Goodman, L.S. *Goodman Gilman's the Pharmacological Basis of Therapeutics*, 11th ed.; Brunton, L.L., Lazo, J.S., Parker, K.L., Eds.; Mcgraw-Hill: New York, NY, USA, 2006; pp. 1613–1644.

86. Harvey, R.A.; Ferrier, D.R. Lippincott's Illustrated Reviews. In *Biochemistry*, 5th ed.; Lippincott Williams & Wilkins: Philadelphia, PA, USA, 307, 2011.

87. Budin, S.B.; Othman, F.; Louis, S.R.; Bakar, M.A.; Radzi, M.; Osman, K.; Das, S.; Mohamed, J. Effect of alpha lipoic acid on oxidative stress and vascular wall of diabetic rats. *Rom. J. Morphol. Embryol.* **2009**, *50*, 23–30.

88. Pottathil, S.; Nain, P.; Morsy, M.A.; Kaur, J.; Al-Dhubiab, B.E.; Jaiswal, S.; Nair, A.B. Mechanisms of Antidiabetic Activity of Methanolic Extract of Punica granatum Leaves in Nicotinamide/Streptozotocin-Induced Type 2 Diabetes in Rats. *Plants* **2020**, *9*, 1609. [CrossRef] [PubMed]

89. Morakinyo, A.O.; Samuel, T.A.; Adekunbi, D.A. Magnesium upregulates insulin receptor and glucose transporter-4 in streptozotocin-nicotinamide-induced type-2 diabetic rats. *Endocr. Regul.* **2018**, *52*, 6–16. [CrossRef] [PubMed]

90. Omolola, S.O.; Bukoye, O.H. Improvement of Oral Glucose Tolerance and Total Lipid Profile of Diabetic Rats Treated with Ficus exasperata Leaf-Based Diet. *Covenant J. Phys. Life Sci.* **2019**, *7*, 982–993.

91. Stanfield, C.L. *Principles of Human Physiology*, 4th ed.; Publishing as Benjamin Cummings; Manufactured in the United States of America; Pearson Education, Inc.: London, UK, 2011.

92. Rawi, S.M.; Abdel-Moneim, A.; Ahmed, O.M. Studies on the effect of garlic oil and glibenclamide on alloxan diabetic rats. *Egypt J. Zool.* **1998**, *30*, 211–228.

93. Mathe, D. Dyslipidemia and diabetes: Animal models. *Diabetes Metab.* **1995**, *21*, 106–111.

94. Matter, F.E.; Helal, E.G.E. Effect of *Blanitis aegyptiaca* seed extract on factors affecting serum glucose level in hyperglycemic senile albino rats. *Al-Azhar Bull. Sci.* **2001**, *12*, 231–245.

95. El-Rahman, S.N.A.; Al-ahmari, H. Evaluation of fertility potential of *Balanites aegyptiaca* sapogenin extract in male rats. *Int. J. Sudan Res.* **2013**, *3*, 15–33. [CrossRef]

96. Roman, I.D.; Thewles, A.; Coleman, R. Fractionation of livers following diosgenin treatment to elevate biliary cholesterol. *Biochim. Biophys. Acta* **1995**, *1255*, 77–81. [CrossRef]

97. Abdin, A.A.; Baalash, A.A.; Hamooda, H.E. Effects of rosiglitazone and aspirin on experimental model of induced type 2 diabetes in rats: Focus on insulin resistance and inflammatory markers. *J. Diab. Complicat.* **2010**, *24*, 168–178. [CrossRef]

98. Ahmed, O.M.; Mahmoud, A.M.; Abdel-Moneim, A.; Ashour, M.B. Antidiabetic effects of hesperidin and naringin in type 2 diabetic rats. *Diabetol. Croat.* **2012**, *41*, 53–67.

99. Lewis, G.F.; Carpentier, A.; Adeli, K.; Giacca, A. Disordered fat storage and mobilization in the pathogenesis of insulin resistance and type 2 diabetes. *Endocr. Rev.* **2002**, *23*, 201–229. [CrossRef]

100. Na, L.X.; Zhang, Y.L.; Li, Y.; Liu, L.Y.; Li, R.; Kong, T.; Sun, C.H. Curcumin improves insulin resistance in skeletal muscle of rats. *Nutr. Metab. Cardiovasc. Diseases.* **2011**, *21*, 526–533. [CrossRef]

101. Qi, Y.; Nie, Z.; Lee, Y.S.; Singhal, N.S.; Scherer, P.E.; Lazar, M.A.; Ahima, R.S. Loss of resistin improves glucose homeostasis in leptin deficiency. *Diabetes* **2006**, *55*, 3083–3090. [CrossRef]

102. Way., J.M.; Görgün, C.Z.; Tong, Q.; Uysal, K.T.; Brown, K.K.; Harrington, W.W.; Oliver, W.R.; Willson, T.M.; Kliewer, S.A.; Hotamisligil, G.S. Adipose tissue resistin expression is severely suppressed in obesity and stimulated by peroxisome proliferator-activated receptor gamma agonists. *J. Biol. Chem.* **2001**, *276*, 25651–25653. [CrossRef] [PubMed]

103. Karthikesan, K.; Pari, L.; Menon, V.P. Combined treatment of tetra hyd rocu cumin and chlorogenic acid exerts potential antihyperglycemic effect on streptozotocin -nicotinamide-induced diabetic rats. *Gen. Physiol. Biophys.* **2010**, *29*, 23–30. [CrossRef]

104. Randle, P.J.; Garland, C.N.; Hales, E.A. Newsholme.The glucose fatty-acid cycle. Its role in insulin sensitivity and the metabolic disturbances of diabetes mellitus. *Lancet* **1963**, *281*, 785–789. [CrossRef]

105. Brøns, C.; Grunnet, L.G. Mechanisms in endocrinology: Skeletal muscle lipotoxicity in insulin resistance and type 2 diabetes: A causal mechanism or an innocent bystander. *Eur. J. Endocr.* **2017**, *176*, 67–78. [CrossRef]

106. Belfort, R.; Mandarino, L.; Kashyap, S.; Wirfel, K.; Pratipanawatr, T.; Berria, R.; Defronzo, R.A.; Cusi, K. Dose-response effect of elevated plasma free fatty acid on insulin signaling. *Diabetes* **2005**, *54*, 1640–1648. [CrossRef]

107. Iranloye, B. Anti-diabetic and antioxidant effects of virgin coconut oil in alloxan induced diabetic male Sprague Dawley rats. *J. Diabetes Mellit.* **2013**, *3*, 221. [CrossRef]

108. Ahmed, O.M.; El-Twab, S.M.A.; Al-Muzafar, H.M.; Amin, K.A.; Aziz, S.M.A.; Abdel-Gabbar, M. Musa paradisiaca L. leaf and fruit peel hydroethanolic extracts improved the lipid profile, glycemic index and oxidative stress in nicotinamide/streptozotocin-induced diabetic rats. *Vet. Med. Sci.* **2021**, *7*, 500–511. [CrossRef]

109. Valko, M.; Leibfritz, D.; Moncol, J.; Cronin, M.T.D.; Mazur, M.; Telser, J. Free radicals and antioxidants in normal physiological functions and human disease. *Int. J. Biochem. Cell Biol.* **2007**, *9*, 44–84. [CrossRef] [PubMed]

110. Çelik, N.; Vurmaz, A.; Kahraman, A. Protective effect of quercetin on homocysteine-induced oxidative stress. *Nutrition* **2017**, *33*, 291–296. [CrossRef]

111. Hassanin, K.M.; Mahmoud, M.O.; Hassan, H.M.; Abdel-Razik, A.R.H.; Aziz, L.N.; Rateb, M.E. *Balanites aegyptiaca* ameliorates insulin secretion and decreases pancreatic apoptosis in diabetic rats: Role of SAPK/JNK pathway. *Biomed. Pharmacother.* **2018**, *102*, 1084–1091. [CrossRef] [PubMed]

 *pharmaceutics*

Article

Artemisia absinthium L. Aqueous and Ethyl Acetate Extracts: Antioxidant Effect and Potential Activity In Vitro and In Vivo against Pancreatic α-Amylase and Intestinal α-Glucosidase

Asmae Hbika [1], Nour Elhouda Daoudi [2], Abdelhamid Bouyanzer [1], Mohamed Bouhrim [2], Hicham Mohti [3], El Hassania Loukili [1], Hamza Mechchate [4,*], Rashad Al-Salahi [5], Fahd A. Nasr [6], Mohamed Bnouham [2] and Abdelhamid Zaid [3]

[1] Laboratory of Applied Chemistry and Environment, Team Applied Analytical Chemistry of Materials and Environment Faculty of Sciences, Mohammed First University, Oujda 60000, Morocco; asmaae.hbika@gmail.com (A.H.); bouyanzer@yahoo.fr (A.B.); e.loukili@ump.ac.ma (E.H.L.)

[2] Laboratory of Bioresources, Biotechnology, Ethnopharmacology and Health, Department of Biology, Faculty of Sciences, University Mohamed First, Boulevard Mohamed VI, Oujda 60000, Morocco; nourelhoudada95@gmail.com (N.E.D.); mohamed.bouhrim@gmail.com (M.B.); m.bnouham@yahoo.fr (M.B.)

[3] Laboratory of Management and Valorization of Natural Resources, Department of Biology, Faculty of Sciences, Moulay Ismail University, BP 11201 Zitoune, Meknes 50070, Morocco; hicham.mohti@gmail.com (H.M.); a.zaid@umi.ac.ma (A.Z.)

[4] Laboratory of Inorganic Chemistry, Department of Chemistry, University of Helsinki, FI-00014 Helsinki, Finland

[5] Department of Pharmaceutical Chemistry, College of Pharmacy, King Saud University, Riyadh 11451, Saudi Arabia; ralsalahi@ksu.edu.sa

[6] Department of Pharmacognosy, College of Pharmacy, King Saud University, Riyadh 11451, Saudi Arabia; fnasr@ksu.edu.sa

* Correspondence: hamza.mechchate@helsinki.fi

Citation: Hbika, A.; Daoudi, N.E.; Bouyanzer, A.; Bouhrim, M.; Mohti, H.; Loukili, E.H.; Mechchate, H.; Al-Salahi, R.; Nasr, F.A.; Bnouham, M.; et al. *Artemisia absinthium* L. Aqueous and Ethyl Acetate Extracts: Antioxidant Effect and Potential Activity In Vitro and In Vivo against Pancreatic α-Amylase and Intestinal α-Glucosidase. *Pharmaceutics* **2022**, *14*, 481. https://doi.org/10.3390/pharmaceutics14030481

Academic Editors: Fátima Regina Mena Barreto Silva and Diana Marcela Aragon Novoa

Received: 15 January 2022
Accepted: 18 February 2022
Published: 22 February 2022

Publisher's Note: MDPI stays neutral with regard to jurisdictional claims in published maps and institutional affiliations.

Abstract: *Artemisia absinthium* L. is one of the plants which has been used in folk medicine for many diseases over many centuries. This study aims to analyze the chemical composition of the *Artemisia absinthium* ethyl acetate and its aqueous extracts and to evaluate their effect on the pancreatic α-amylase enzyme and the intestinal α-glucosidase enzyme. In this study, the total contents of phenolic compounds, flavonoids, and condensed tannins in ethyl acetate and the aqueous extracts of *Artemisia absinthium* leaves were determined by using spectrophotometric techniques, then the antioxidant capacity of these extracts was examined using three methods, namely, the DPPH (2, 2-diphenyl-1picrylhydrazyl) free radical scavenging method, the iron reduction method FRAP, and the β-carotene bleaching method. The determination of the chemical composition of the extracts was carried out using high-performance liquid chromatography—the photodiode array detector (HPLC-DAD). These extracts were also evaluated for their ability to inhibit the activity of the pancreatic α-amylase enzyme, as well as the intestinal α-glucosidase enzyme, in vitro and in vivo, thus causing the reduction of blood glucose. The results of this study showed that high polyphenol and flavonoid contents were obtained in ethyl acetate extract with values of 60.34 ± 0.43 mg GAE/g and 25.842 ± 0.241 mg QE/g, respectively, compared to the aqueous extract. The results indicated that the aqueous extract had a higher condensed tannin content (3.070 ± 0.022 mg EC/g) than the ethyl acetate extract (0.987 ± 0.078 mg EC/g). Ethyl acetate extract showed good DPPH radical scavenging and iron reduction FRAP activity, with an IC_{50} of 0.167 ± 0.004 mg/mL and 0.923 ± 0.0283 mg/mL, respectively. The β-carotene test indicated that the aqueous and ethyl acetate extracts were able to delay the decoloration of β-carotene with an inhibition of 48.7% and 48.3%, respectively, which may mean that the extracts have antioxidant activity. HPLC analysis revealed the presence of naringenin and caffeic acid as major products in AQE and EAE, respectively. Indeed, this study showed that the aqueous and ethyl acetate extracts significantly inhibited the pancreatic α-amylase and intestinal α-glucosidase, in vitro. To confirm this result, the inhibitory effect of these plant extracts on the enzymes has been evaluated in vivo. Oral intake of the aqueous extract significantly attenuated starch- and sucrose-induced hyperglycemia in normal rats, and evidently, in STZ-diabetic rats as well.

The ethyl acetate extract had no inhibitory activity against the intestinal α-glucosidase enzyme in vivo. The antioxidant and the enzyme inhibitory effects may be related to the presence of naringenin and caffeic acid or their synergistic effect with the other compounds in the extracts.

Keywords: *Artemisia absinthium*; antioxidant activity; phenolic compounds; hyperglycemia; pancreatic α-amylase; intestinal α-glucosidase

1. Introduction

The variety of molecules constituting aromatic and medicinal plants raises interest in several different fields, especially in the pharmaceutical industry; in fact, numerous food supplements and drugs are manufactured from plants. For this reason, various studies have been oriented towards the characterization and identification of new bioactive substances that can improve pharmaceutical production. *Artemisia absinthium* L. (*A. absinthium*) is among the plants with several pharmaceutical properties. This plant belongs to the *Asteraceae* family, and it is popularly known as wormwood in the United Kingdom and absinthe in France; it has been utilized as a medicinal plant in Europe, Asia, the Middle East, and North of Africa [1]. Well known for its many biological properties, including as an antispasmodic [2], antimalarial [3] antipyretic [4], antitrypanosomal [5], acaricidal [6] antibacterial [7], anti-inflammatory [8], antioxidant [9], antidepressant [10], and also as a treatment for chronic fever [11]. *A. absinthium* is being used in face serums, essences, masks, shampoos, and other cosmetology products [12]. It is used in the food industry, and it is also widely used as a key aromatic ingredient [13] in the manufacture of some alcoholic drinks and also as an ingredient in absinthe [14].

The Egyptian Ebers Papyrus from about 1552 BC and the Old Testament of the Bible both contain references to *A. absinthium* [15]. Archaeologists have found tablets with cuneiform characters referring to *A. absinthium* that probably originated in the Babylonian civilizations [16]. The Greek mathematician and philosopher Pythagoras of Samos recommended the use of absinthe as a treatment for the pains of childbirth (569–475 BC). Hippocrates (460–377 BC) used *A. absinthium* extracts for the treatment of rheumatism and menstrual pain, and he also recommended it for jaundice [16]. In Ethiopia, *A. absinthium* is used in rituals called Atete; it was known as ariti, and was also used for the treatment of non-infectious and infectious diseases such as malaria, helminths, and animal injuries [10,17]. *A. absinthium* has been used as a treatment for gastric irritation since the Middle Ages, since it acts as an aromatic bitter and is believed to stimulate the acid secretion and bile production in small doses. In the Middle Ages, *A. absinthium* was employed as both a vermifuge and a purge, and it became known as a "general remedy for all diseases". Pure essential oil is very toxic and can lead to the death of those who consume it. *A. absinthium* has been used in the treatment of swelling, chronic fever, inflammation of the liver, and also as a stimulant and tonic [18]. For centuries, it has been used in the treatment of intermittent and chronic fevers in indigenous medicine [19]. For centuries, this plant has been used infused with tea by the Moroccan people in winter.

Diabetes mellitus is among the most common diseases in the world. Indeed, in 2017, 325 million people suffered from type 2 diabetes, and the number of cases is progressively increasing [20]. In 2006, type 2 diabetes was responsible for about 5% of global deaths, according to World Health Organization. The incidence of diabetes is increasing daily, and it is proposed that the number of cases will around 552 million by the year 2030 [21]. Diabetes mellitus is known as a chronic metabolic disorder defined by hyperglycemia, with disturbances in the metabolism of carbohydrates, proteins, and lipids. This disturbance is either due to an extreme deficiency in insulin synthesis, which is called type 1 diabetes, or primarily by insulin resistance, which is called type 2 diabetes. [22]. To achieve normal blood glucose levels in people with type 2 diabetes mellitus, oral hypoglycemic agents or insulin are required. However, the use of these drugs has been shown to have limited efficacy and is combined with undesirable side effects, which has led to an increasing focus on the use of herbs to limit these effects [23,24]. Pancreatic α-amylase and intestinal α-glucosidase are enzymes that decompose the long-chain carbohydrates that catalyze the cleavage of the starch to the disaccharide and the disaccharide to the glucose, respectively, and are therefore effective in retarding the absorption of glucose into the bloodstream, resulting in a hypoglycemic effect [25]. Pancreatic α-amylase is an enzyme that comes from the salivary glands and the pancreas. It is a species that is responsible for the digestion of carbohydrates, thanks to its ability to catalyze the initial hydrolysis of starch by acting on the internal bonds of carbohydrates α-D1,4. Pancreatic α-amylase is an enzyme that can break down starch into dextrins, maltotriose, and maltose [26]. Intestinal α-glucosidase is an enzyme of the intestinal brush border. It catalyzes the release of absorbable monosaccharides, such as glucose, from the substrate, ultimately facilitating absorption by the small intestine [27]. Natural phenolics have been shown to reduce the activity of enzymes such as pancreatic α-amylase and intestinal α-glucosidase [28]. It is estimated that about 800 plants may have anti-diabetic properties [29]. In the traditional treatment of diabetes, the plants belonging to the *Asteraceae* family are the most widely documented [22]. Li et al. [30] showed that *A. absinthium* exhibits antidiabetic activity in diabetic humans, with no significant effect on lipid profiles. Additionally, Daradka et al. [31] showed that the ethanolic extract of *A. absinthium* has a hypoglycemic activity in rats with alloxan-induced diabetes, exhibiting biosafety, with the improvement of associated biochemical parameters and the prevention of serious reduction in body weight. These studies led us to the study of the antidiabetic effect of this plant, using other methods and other solvents of extraction, as well as other techniques, and also to elucidate some mechanisms of action that explain the antidiabetic effect of this plant, targeting the digestive enzymes related to the digestion of carbohydrates (α-amylase and α-glucosidase). The analysis of *A. Absinthium* showed the presence of chemical compounds such as polyphenolic compounds [32] and flavonoids [33]. In studies on the bioactive compounds of *A. Absinthium*, flavonoids, terpenoids, coumarins, sterols, tannins, carotenoid glycosides, and bitter principles have been cited. [10,34]. In this work, we determined the phenolic compounds, condensed tannins, and flavonoids, as well as the chemical compounds in the extracts by HPLC-DAD; we also evaluated the extract antioxidant activities using DPPH, FRAP, and β-carotene bleaching methods. In addition, we studied the inhibitory activity of the extract against pancreatic α-amylase and intestinal α-glucosidase by using various approaches in vivo and in vitro.

2. Materials and Methods

2.1. Chemicals and Reagents

Phlorizin dehydrates and the pancreatic α-amylase enzyme were obtained from Sigma-Aldrich, USA; intestinal α-glucosidase, acarbose, streptozotocin, and dinitrosalicylic acid were obtained from Sigma-Aldrich in China. Anhydrous D (+) glucose was sourced from Riedel-de Haen, Germany. The solvents used in this work are of analytical grade and have been provided from Honeywell Riedel-de Haen. All chemicals were high-quality analytical chemicals and were used exactly as they were obtained from Merck.

2.2. Plant Material and Extraction

Artemisia absinthium L. (Figure 1) was collected in December 2020 in Zkara, a rural Moroccan commune of Mestferki, in the prefecture of Oujda Angad, in the eastern region of Morocco. It is located approximately 25 km from the city of Oujda. The leaves of *A. absinthium* were collected fresh and dried in a dry, airy place, away from light and sunlight. To obtain the ethyl acetate extract, the leaves (100 g) were extracted using a mixture of acetone/water (70/30) under reflux for 2 h, and the resulting solution was then filtered and concentrated by evaporation to 1/4 of the initial volume by using a rotary evaporator under reduced pressure at 40 °C. The extract obtained was fractionated with solvents of increasing polarity: hexane, diethyl ether, and finally, with an extraction of ethyl acetate to obtain the ethyl acetate extract (EAE). The aqueous extract was obtained by using five successive extractions of *A. absinthium* leaves with a Soxhlet extractor using five solvents (hexane, dichloromethane, ethyl acetate, acetone, ethanol); the sixth extraction is removed by distilled water to obtain the aqueous fraction (AQE); this technique was used to allow a minimum of molecules in this extract.

Figure 1. *A. absinthium* herb.

Extract yields (Y) were determined as the ratio between the mass of the extract obtained and the initial mass of the dried plant.

$$Y(\%) = \frac{Mext}{Mdp} \times 100 \tag{1}$$

where Mext and Mdp are extract mass and dry plant mass, respectively.

2.3. Content of Phenolic Compounds, Flavonoids, and Tannins

2.3.1. The Total Polyphenol Quantification

The total polyphenol content in the *A. absinthium* extracts was determined according to the Folin–Ciocalteu method [35]. Indeed, 100 µL of extract solution with a concentration of 2 mg/mL was then mixed with 200 µL of Folin–Ciocalteu reagent in 2 mL of distilled water, and finally, 1 mL of sodium carbonate was added. (15%); the resulting mixture was incubated in the dark for 2 h at room temperature. Then the absorbance was measured at 765 nm using a spectrophotometer. The calibration curve is generated by using gallic acid according to a concentration range of (0–0.1 mg/mL). All experiments were performed in triplicate to take the mean of the experiments with the standard deviation. The amount of total phenolic compounds was expressed in mg gallic acid equivalents per gram of dry extract (mg GAE/g DE).

2.3.2. The Flavonoids Quantification

The total flavonoid content of the *A. absinthium* extracts was obtained by using a colorimetric test using aluminum chloride (AlCl$_3$), according to the method of Natalizia Miceli et al. [35]. Next, 500 µL of each extract (2 mg/mL) was blended with 1.5 mL of MeOH, and then 100 µL of AlCl$_3$ (10%) was added, as well as 100 µL of potassium acetate (1 M) with 2.8 mL distilled water. The absorbance measurement was performed after 30 min of incubation at room temperature in the dark against a white solution at 415 nm. To determine the calibration curve, quercetin was used; a concentration range was established separately with quercetin (0–0.1 mg/mL). The results were expressed in mg quercetin equivalent per gram of dry extract (mg QE/g DE). All measurements were determined in three independent experiments to ensure the reproducibility of the results.

2.3.3. The Condensed Tannin Quantification

The content of condensed tannin in the *A. absinthium* extracts was determined by the vanillin method according to the instructions of Mohti et al. [36]. First, 50 μL of the extract solution was mixed with 1.5 mL vanillin (MeOH, 4%) and 750 μL of concentrated acid (HCl) was added. The absorbance of the resulting mixture was measured at 500 nm after 20 min of incubation in the dark and at room temperature. Catechin was used to establish the calibration curve. The content of the condensed tannins was expressed in mg catechin equivalents per gram of dry extract (mg CE/g DE) and each extract was analyzed in triplicate.

2.4. High-Performance Liquid Chromatography Analysis

Our HPLC/DAD equipment comes from the Waters Corporation in the United States, the analysis of the extracts was performed using a liquid chromatography separation module (Waters e2695) coupled with a diode array detector (Waters 2998 PDA) and data were processed using Empower data processing software. The chromatogram was recorded at a wavelength of 254 nm to 300 nm. AC18 column (4.6 mm × 250 mm, 5 μm), a gradient mode of the mobile phase, was used. Solvent A is a mixture of ultrapure water/acetic acid (2% *v/v*) and solvent B is acetonitrile: 0 to 5 min: 95% A and 5% B, 25 to 30 min; 65% A and 35% B, 35 to 40 min; 30% A and 70% B, 40 to 45 min; 95% A and 5% B. The flow rate is 0.9 mL/min, and the injection volume is 20 μL. Peaks were identified by comparing their retention times and UV spectra with the standards used. The standard polyphenolic compounds used were caffeic acid, gallic acid, catechin, *p*-hydroxy benzoic acid, vanillin, naringenin, *p*-coumaric acid, ferulic acid, rosmarinic acid, rutin, vanillic acid, ascorbic acid, quercetin, trans-chalcone, malic acid, syringic acid, and kaempferol, which have all been previously found in *A. absinthium*.

2.5. Antioxidant Activity

2.5.1. Scavenging 2, 2-Diphenyl-1-picrylhydrazyl Radical Test

The antioxidant activity of *A. absinthium* leaf extracts was determined by the method based on the decoloration of the DPPH (1,1-diphenyl-2-picrylhydrazyl) radical according to the protocol described by Miceli et al. [37]. A volume of 500 μL of each extract solution at different concentrations (from 0.0625 to 2 mg/mL) was added to 3 mL of the freshly prepared DPPH methanolic solution (0.1 mM). The negative control was made by mixing 500 μL of the solvent used for the solubilization of the extracts with a 3 mL solution of DPPH prepared in methanol. The absorbance value of the decolorization of the DPPH solution was read against a blank solution at a wavelength of 517 nm after an incubation time of 20 min at room temperature and in the dark using a UV/visible spectrophotometer. The positive control was represented by ascorbic acid, whose absorbance was determined under the same circumstances and with the same procedure as for the samples and for each concentration, and the assay was repeated three times. The scavenging activity was determined by the formula below:

$$\text{Radical scavenging activity (\%)} = \frac{\text{Adpph} - \text{As}}{\text{As}} \times 100 \qquad (2)$$

where Adpph is the absorbance of the negative control, and As is the absorbance of the solution which contains the extract.

2.5.2. The Ferric Reducing Power Assay (FRAP)

The reducing power of *A. absinthium* extracts was determined by the spectrophotometric method described by Miceli et al. [37], with modifications. Next, 0.5 mL of the different concentrations of the extracts solubilized in the suitable solvent (0.0625–2 mg/mL) were added to 1.25 mL of the phosphate buffer solution (0.2 M, pH 6.6) and to 1.25 mL of solution of potassium ferricyanide [$K_3Fe(CN)_6$] (1%). The mixtures were incubated at a temperature of 50 °C for 20 min. Afterward, 1.25 mL trichloroacetic acid solution (10%)

was then added. The resulting mixture was then centrifuged at 3000 rpm for 10 min. In the end, 1.25 mL of the supernatant of each of the resulting solutions was added to the solution of 1.25 mL of distilled water and 0.25 mL of $FeCl_3$ (0.1%). The absorbance was determined at a wavelength of 700 nm at room temperature after 10 min of incubation. The increase in the reaction medium absorbance indicated an increase in iron reduction. The positive control used was ascorbic acid. To ensure reproducibility, the results are the mean of three separate experiments.

2.5.3. β-Carotene Bleaching Test

The antioxidant capacity of the plant extracts using the carotene bleaching test was measured using the following procedure from Kartal et al. [38]. The emulsion of β-carotene/linoleic acid was prepared by solubilizing 6 mg of β carotene in 1.5 mL of chloroform, 30 μL of linoleic acid, and 250 mg of Tween 80. The chloroform was evaporated using a rotary evaporator, and then 100 mL of distilled water saturated with oxygen H_2O_2 (30%) was added. The emulsion obtained was stirred to homogenize it. Next, 175 μL of extract solution or a reference antioxidant solution (BHA) at a concentration of 2mg/mL were added to 1.25 mL of the previous emulsion. The decoloration kinetics of the emulsion of the negative control and the extracts, or the BHA was followed at 490 nm for 120 min at regular time intervals. The relative antioxidant activity of the extracts (RAA) was determined according to the formula below:

$$RAA(\%) = \frac{A120 - C120}{C0 - C120} \times 100 \tag{3}$$

where A120 is the absorbance of the solution which contains the extract after 120 min of incubation; C120 and C0 are the absorbances of the control after and before 120 min of incubation, respectively.

2.6. Inhibition of Carbohydrates Hydrolase Enzymes, In Vitro
2.6.1. Pancreatic α-Amylase

The inhibition of the activity of the pancreatic α-amylase by the *A. absinthium* extracts was studied following the experimental protocol reported by Nour Elhouda Daoudi et al. [39]. The mixtures tested contained 200 mL of pancreatic α-amylase enzyme solution (13 IU), with 200 mL of the phosphate buffer (0.02 M; pH = 6.9), and an additional 200 mL of extracts of *A. absinthium* or acarbose for concentrations of 0.45 mg/mL and 0.9 mg/mL. The resulting mixtures were pre-incubated for 10 min at 37 °C. Then 200 mL of starch (1%) dissolved in the phosphate buffer was added and incubated at 37 °C for 20 min. To stop the enzymatic reaction, 600 mL of a colored DNSA reagent was added. Immediately after the addition of this reagent, the tubes were incubated a third time at 100 °C for 8 min and then placed in an ice-water bath for a few minutes. Finally, 1 mL of distilled water was added to dilute the resulting solution and the absorbance was measured at a wavelength of 540 nm. The percentage of pancreatic α-amylase enzyme inhibition was determined using the formula below:

$$\text{Inhibitory activity percentage} = \frac{\text{Atest 540 nm} - \text{Acontrol 540}}{\text{Atest 540nm}} \times 100 \tag{4}$$

where (Atest 540 nm) and (Acontrol 540 nm) were the absorbances of the solutions containing the extracts and without the extracts, respectively, at a wavelength of 540 nm.

2.6.2. Intestinal α-Glucosidase

A. absinthium extracts have been tested for their ability to inhibit intestinal α-glucosidase activity according to the protocol described by Nour Elhouda Daoudi et al. [39], which consists of controlling the release of glucose obtained from sucrose degradation. The test solutions contained 100 mL of sucrose (50 mM) in addition to 1000 mL of the phosphate buffer (50 mM; pH = 7.5), and 100 mL of the intestinal α-glucosidase enzyme solution

(10 IU). Next, 10 mL of the control (distilled water), positive control (acarbose), or solutions containing the *A. absinthium* extract at the concentrations of 165 and 328 mg/mL, respectively, were added to the previous mixture. Then the whole was incubated in a water bath at 37 °C for 25 min. The resulting solution was then heated for 5 min at 100 °C to stop the enzyme reaction. Finally, the glucose liberation was estimated using the glucose oxidase method with an automatically accessible kit (Glucose oxidase peroxidase). The absorbance of the final resulting solution was measured at 500 nm. The percentage of inhibition was determined by the formula given below:

$$\text{Inhibitory activity percentage} = \frac{\text{Acontrol 500 nm} - \text{Atest 500}}{\text{Acontrol 540 nm}} \times 100 \tag{5}$$

where (Atest 500) and (Acontrol 500) were the absorbances of the solutions containing the extracts and without the extracts, respectively, at a wavelength of 500 nm.

2.7. Inhibition of Carbohydrates Hydrolase Enzymes, In Vivo

2.7.1. Animals

The Wistar rats, weighing between 150 g and 250 g, used in this work were taken from the animal house of the Department of Biology, Faculty of Science, Mohammed First University, Oujda, Morocco. The rats were housed in animal cages, with access to water and food. They were kept in a well-ventilated room at a temperature of 24 ± 2 °C with a 12 h light/12 h dark cycle. The rats were treated and cared for according to the guide published by the United States National Institutes of Health (1985), which is internationally recognized for the proper treatment and care of laboratory animals.

2.7.2. Induction of Diabetes

Diabetes was provoked in the rats by following the procedure described by [39]. The animals fasted for 16 h with ad libitum accessibility to water. They were then given an intraperitoneal injection of a unique dose of alloxan (140 mg/Kg) dissolved in a sodium-citrate buffer (pH = 3), cold and freshly prepared. One week later, rats with blood glucose levels above 1.5 g/L were chosen for use in this study.

2.7.3. Pancreatic α-Amylase

Healthy and diabetic Wistar rats that had fasted for 16 h were used in this study. The animals tested were separated into four groups (n = 6; male/female = 1). The rats in the control group were treated with distilled water only (10 mL/Kg; po). The positive group or the rats were received the Acarbose (10 mg/Kg; po). Two treated groups received EAE or AQE for a dose of (200 mg/K; po). Then, 30 min of the administration of the solution, the rats of the different groups were orally loaded with starch (2 g/Kg). Afterward, by employing the glucose-peroxidase method, the blood glucose levels of the rats were estimated at 0, 30, 60 and 120 min.

2.7.4. Intestinal α-Glucosidase

Healthy and diabetic Wistar rats used in this study were treated in the same way [40]. After fasting for 16 h, they were divided into four groups (n = 6; male/female = 1). The rats in the control group were treated with distilled water only (10 mL/Kg; po). The positive group received acarbose (10 mg/Kg; po). Group 1 received EAE (250 mg/Kg; po) and group 2 received AQE (250 mg/Kg; po). Then, 30 min of oral administration of the test substances, the rats of the different groups were loaded orally with sucrose (2 g/Kg). Blood samples were taken from the rats' tails at different times under light anesthesia. Finally, the blood glucose levels of the rats were estimated by employing the glucose-peroxidase method for t = 0, 30, 60, and 120 min.

2.8. Statistical Analysis of Results

The results carried out are presented in the form of means ± standard errors, then they were statistically analyzed using computer software (Graph Pad Prism 5.04-San Diego, CA, USA). Comparative analysis between multiple groups was done by one-way variance analysis (ANOVA) and the statistical significance was accepted as $p \leq 0.05$.

3. Results and Discussion

3.1. Yields, Phenols, Flavonoids, and Tannins Contents

Polyphenols are widely present in almost all medicinal and aromatic plant species; they are therefore an indispensable part of the human diet because of their health-promoting and antioxidant properties [41]. The total phenolic, flavonoid, and condensed tannin content, along with the extraction yields of EAE and AQE of *A. absinthium*, were examined and the results are presented in Table 1. The polyphenol content was calculated from the calibration curve of gallic acid (R = 0.998). EAE had the highest concentration of phenolic compounds (69 mg AGE/g DE), compared to AQE (31.534 ± 0.408 mg AGE/g DE). We obtained extraction yields of 15.95% for EAE and 0.672% for AQE. The flavonoid content of the *A. absinthium* extracts calculated from the quercetin calibration curve (R = 0.998), gives the highest content observed for EAE, with a value of 69.02 ± 0.33 mg QE/g DE vs. 31.534 ± 0.408 mg QE/g DE for AQE. For the condensed tannin (another class of bioactive compounds) [42], the results were calculated from the catechin calibration curve (R = 0.995) and contrarily to the previous results, AQE recorded the highest tannin content (3.070 ± 0.022) compared to EAE (0.987 ± 0.078). EAE showed a high content of polyphenols and flavonoids, while the AQE had a higher content of condensed tannin.

Table 1. The yield of extraction and the contents of the total phenolic compounds, flavonoids, and condensed tannins in *A. absinthium* extracts.

	Yield (%)	Total Phenolic (mg GAE/g DE)	Flavonoid (mg QE/g DE)	Condensed Tannins (mg CE/g DE)
AQE	0.672 ± 0.088	31.534 ± 0.408	11.246 ± 0.184	3.070 ± 0.022
EAE	15.95 ± 0.252	69.013 ± 0.249	25.842 ± 0.241	0.987 ± 0.078

EAE—ethyl acetate extract; AQE—aqueous extract.

The values of the phenolic compound contents found in our extracts are lower than those obtained by Boudjelal et al. [43] (180.33 ± 16.25 mg GAE/g DE). On the other hand, the values obtained from the present study are higher than those found by Kružinauskaitė et al. [44], whose phenolic compound content of *A. absinthium* extracts varies from 21.19–24.74 mg GAE/g ED.

Polyphenols are one of the most common classes of secondary metabolites in flowering plants, as well as in fruits and flowers. These compounds are considered essential in the defense of plants against interferences and predators [45]. They are composed of one or more aromatic rings with one or more hydroxyl groups, along with other substituents. The presence of hydroxyl groups makes polyphenols very reactive in the neutralization of free radicals by giving a hydrogen atom or an electron, as well as chelate metal ions, thus reducing their pro-oxidant activity [46,47]. They have received the greatest attention for their effects in the treatment and prevention of several diseases. It has been proved that this type of molecule has numerous cardio-protective functions [48]; they can also prevent or delay depression, anxiety, and other diseases related to oxidative stress [49]. Several studies have indicated that polyphenols and their derivatives have shown anticancer capacities and an antioxidant potential for animal and human cervical cancer cells [50]. These compounds also can react with free radicals by stopping their activity, modulating the expression of genes involved in metabolism, thus repairing and protecting DNA damage, in addition to their ability to act as signaling molecules enhancing antioxidant defense [51].

Flavonoids are a class of phenolic compounds that are widespread in the plant kingdom [45]; many flavonoids are responsible for the attractive colors of leaves, fruits, and flowers [52]. Flavonoids are more appropriately called "nutraceuticals" because of the diversity of their pharmacological activities in the body. These elements have a powerful free radical scavenging capacity, as well as the advantage of being easily absorbed in the intestine after ingestion with minimal side effects and low toxicity in animals [53]. Flavonoids have many health benefits, thanks to the strength of their in vivo and in vitro antioxidant capacity [54,55]. Tannins are high molecular weight polyphenols that react with proteins when not oxidized [56]. These compounds are, in many cases, bioactive in plants [57]. They exist mainly in two forms: hydrolyzable or condensed [56]. Tannins, in turn, are endowed with antioxidant power. Thus, they inhibit superoxide formation, and hydrolyzable tannins inhibit lipid peroxidation [58]. Proanthocyanidins, or condensed tannins, are polymeric and oligomeric derivatives of the flavonoid biosynthetic process; they are increasingly recognized for their beneficial effects on health [59]. The quantity of tannin contained in the extracts is lower than that of the other bioactive components; however, their presence in the extracts cannot be ignored [60]. The degradation products of condensed tannins are potentially toxic to ruminants [61] and are absorbed by the small intestine of animals [62].

Based on our study outcome, it has been shown that polyphenols and flavonoids are most likely to be extracted using a solvent of average polarities, such as ethyl acetate, and that tannins can be obtained using a solvent with high polarity, such as water.

3.2. High-Performance Liquid Chromatography HPLC

The chemical components contained in the *A. absinthium* extracts were analyzed using high-performance liquid chromatography coupled with a diode array detector (HPLC-DAD) by comparing their retention times and UV spectra to those of the standards. The HPLC chromatograms for the identified polyphenolics are presented in Figure 2. Figure 3 shows the chemical structure of the majority compounds found in AQE and EAE. The most abundant phenolic compounds were caffeic acid (21.49%) in EAE and naringenin (58.76%) in AQE. Other compounds were identified in these two extracts of *A. absinthium*, but at low percentages, namely *p*-hydroxybenzoic acid (5.44%) in AQE and quercetin (1.29%), *p*-hydroxybenzoic acid (0.21%), *p*-coumaric acid (0.77%) and naringenin (5.85%) in EAE. Similar findings were reported in the study by Lee et al. [63], with a total of 20 polyphenolic compounds in *A. absinthium* leaves, including hydroxybenzoic acids, hydroxycinnamic acids, flavanols, and caffeic acid.

The use of multiple successive extractions causes the AQE to be of low chemical composition, which is justified by a limited number of peaks; the solvents used successively in the extractions are of different polarity, and each solvent can selectively extract certain compounds depending on specific parameters, which therefore leads to the weakening of the ability of water, which is the sixth solvent used, to extract the previously extracted compounds; for EQA, the use of three liquid-liquid extractions decreases the extraction selectivity of the solvent ethyl acetate. These methods, therefore, consist of using extracts or fractions with a limited number of chemical compounds, making it possible to identify the element that is likely responsible for the observed activity.

Moacă et al. [64] indicated the presence of chlorogenic acid as the major product; quercitrin, rutin and isoquercitrin were also detected in lower concentrations, while *p*-coumaric acid, luteolin, gentisic acid, caffeic acid, and apigenin were also detected in trace amounts. Ivanescu et al. [65] detected the presence of ferulic acid and kaempferol in the unhydrolyzed extract, while the presence of *p*-coumaric acid, ferulic acid, fisetin, and patuletin was found in the extract of *A. Absinthium* after acid hydrolysis. Lee et al. [63] revealed the predominance of salicylic acid in the leaf extract, along with myricetin, caffeic acid, gallic acid and ferulic acid. The methanolic extract revealed the presence of chlorogenic acid as the main product [66]. Some works have reported the presence of gallic acid, caffeic acid [67], vanillic acid [68], vanillin, and naringenin [63] in *A. absinthium*. The studies of the chemical composition of *A. absinthium* yield different results, as the composition varies both qualitatively and quantitatively according to intrinsic factors related to the plant, the type of soil and climate, the periods of harvest, and the maturity of the plant, along with extrinsic factors related to the technique, duration, and temperature of the extraction, as well as the environment.

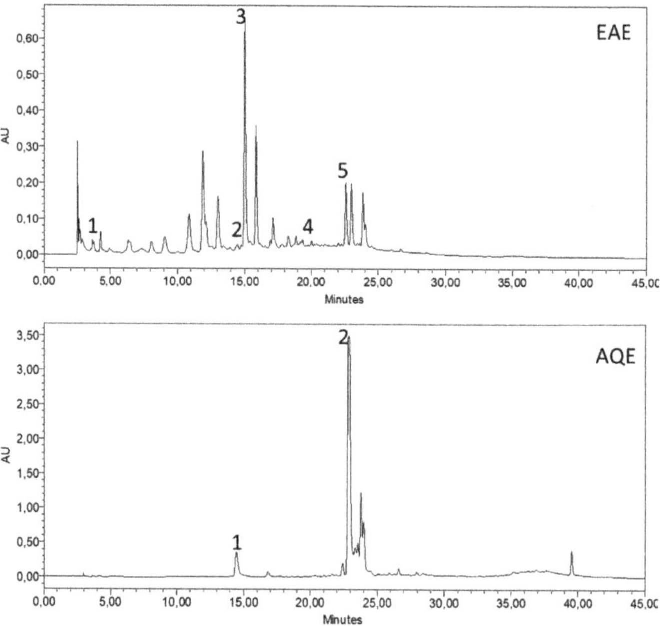

Figure 2. The HPLC chromatogram patterns of extracts from of *A. absinthium*: EAE (1) quercetin (1.29%), (2) *p*-hydroxybenzoic acid (0.21%), (3) caffeic acid (21.49%), (4) *p*-coumaric acid (0.77%), and (5) naringenin (5.85%); AQE (1) *p*-hydroxybenzoic acid (5.44%) and (2) naringenin (58.76%). EAE— ethyl acetate extract; AQE—aqueous extract.

Figure 3. The structures of caffeic acid (**A**) and naringenin (**B**).

Caffeic acid is a hydroxycinnamic acid derived from plants, and it is produced by various plant species as a secondary metabolite [69]. It is well known for its antioxidant activity; it has a direct primary function of scavenging harmful reactive species by electron transfer or hydrogen donation [70,71], a second function of chelating transition metals, especially iron, and copper [72] and an additional action of activating the redox-sensitive enzymes, improving the cytoprotective response [73]. This molecule has anti-diabetic properties [74].

Naringenin (flavonoid), a molecule that is considered biologically active, has positive bioactivities in diabetes, including hypoglycemia and antioxidant properties [75,76]. Indeed, naringenin may reduce renal glucose reabsorption and glucose adsorption by the brush border of the intestine, as well as increase glucose absorption and utilization by muscle and fat tissues [77], making it an excellent candidate for utilization in medicine as a treatment for type 2 diabetes in the fight against of insulin resistance. This molecule also has the advantage of low toxicity [77]. In addition, it is known for its antioxidant activity [78]. Naringenin reduces blood glucose and restores body weight, while normalizing serum lipid concentrations and the biomarkers of oxidative stress in the pancreas and liver, thus granting it potential as an anti-diabetic compound for future drug production [79].

3.3. Antioxidant Activity of A. absinthium Extracts

Phenolic compounds work as antioxidants; they have a strong ability to trap free radicals through their hydroxyl groups, which could serve as a basis for screening for antioxidant activity [80]. Flavonoids and condensed tannins also act as antioxidants as a result of their free hydroxyl groups.

In this study, we tested the DPPH free radical scavenging activities of EAE and AQE of *A. absinthium*. The free radical scavenging activity of the tested extracts is expressed as IC_{50} value (mg/mL) and the results are given in Table 2. The results indicate that EAE illustrates higher antioxidant activity (IC_{50} = 0.167 ± 0.004 mg/mL) than AQE (IC_{50} = 0.352 ± 0.019 mg/mL). The IC_{50} value of EAE is very close to that of ascorbic acid (IC_{50} = 0.158 ± 0.003 mg/mL).

Table 2. The IC_{50} values were obtained for the DPPH and FRAP tests, and the RAA% for the β-carotene bleaching test.

	IC_{50} (mg/mL)		RAA %
	DPPH	**FRAP**	**β-Carotene**
BHA	-	-	73.4
Ascorbic acid	0.158 ± 0.003	0.137 ± 0.077	-
AQE	0.352 ± 0.019	3.361 ± 0.043	48.7
EAE	0.167 ± 0.004	0.923 ± 0.028	48.3

RAA—relative antioxidant activity; BHA—butylated hydroxyanisole; EAE—ethyl acetate extract; AQE—aqueous extract.

In the iron-reducing power test, the plant extracts induced a reduction of Fe^{3+} to Fe^{2+}, which can be observed by appearance, as well as by measuring the absorbance of the blue coloration formed at 700 nm. Figure 4A shows the curves presenting the reducing powers of the *A. absinthium* extracts, and IC_{50} values are presented in Table 3. The reducing power sequence occurs in the following order: ascorbic acid > EAE > AQE. The reducing power of EAE and AQE increased from 0.249 ± 0.005 and 0.137 ± 0.007, respectively, at 0.062 mg/mL to 3.461 ± 0.041 and 1.107 ± 0.05 at 2 mg/mL, respectively.

Table 3. The IC_{50} values of *A. absinthium* extracts and acarbose in pancreatic α-amylase and intestinal α-glucosidase inhibition.

	IC_{50} (mg/mL)	
	Pancreatic α-Amylase	**Intestinal α-Glucosidase**
Acarbose	0.58 ± 0.003	0.148 ± 0.002
EAE	0.68 ± 0.010	0.155 ± 0.0009
AQE	0.76 ± 0.064	0.170 ± 0.002

EAE—ethyl acetate extract; AQE—aqueous extract.

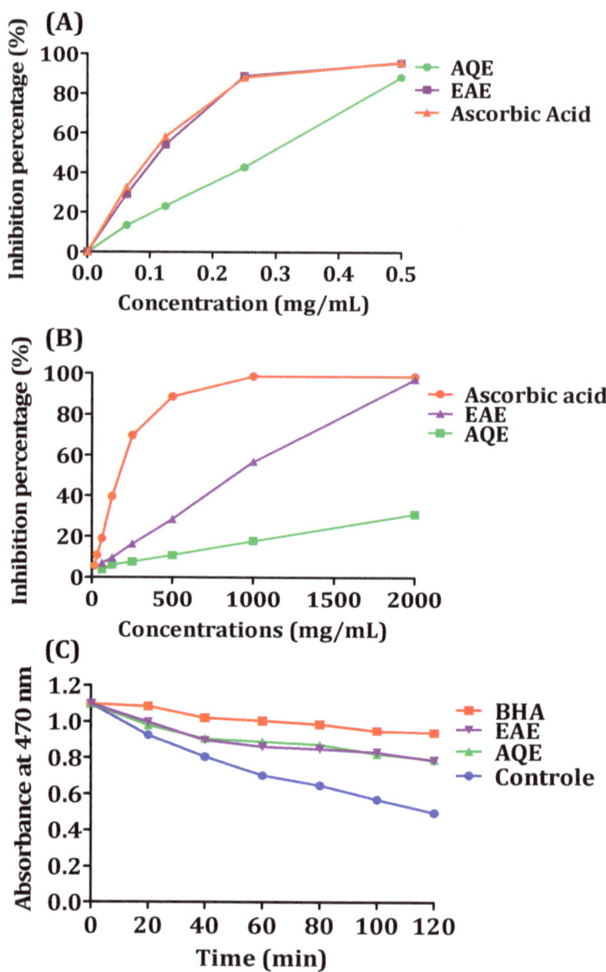

Figure 4. The activity of the scavenging DPPH radical, (**A**) iron reduction, (**B**) and bleaching kinetics of β-carotene, (**C**) in the presence and absence of EAE, AQE, and reference. EAQ—ethyl acetate extract; AQE—aqueous extract.

The capacity of *A. absinthium* extracts to stop or retard lipid peroxidation was evaluated using the bleaching method of the β-carotene molecule, and the results are shown in Figure 4B. The results of relative antioxidant activity are shown in Table 3, and these results indicate that AQE and EAE were able to delay the decoloration of beta carotene by a similar inhibition of 48.7% and 48.3%, respectively.

The antioxidant capacity of *A. absinthium* extracts using the DPPH test is higher than that obtained by Moacă et al. [64] ($IC_{50} = 0.4993 \pm 0.0201$ mg/mL), and by Craciunescu et al. [67] ($IC_{50} = 0.57 \pm 0.05$ mg/mL), and lower than the results given by Msaada et al. [81], with IC_{50} values ranging from 9.38 ± 0.82 to 0.044 ± 1.92 mg/mL, contrastingly. The results for EAE are in accordance with those of Sidaoui et al. [82], who demonstrated that methanol–water extracts have powerful antioxidant capacities ($IC_{50} = 0.118$ mg/mL). It should be noted that the chemical characteristics of the solvents used during extraction have a remarkable effect on the organic type of the agent's antioxidants, resulting from the extraction process [83].

This difference in antioxidant activity may also be related to the differences in the content of the phenolic compounds, which differ according to several parameters.

From our results, we found a considerable correlation between phenolic compounds in both the DPPH and FRAP assays; these two techniques allow us to conclude that EAE presents a higher antioxidant activity than AQE. The antioxidant activity of EAE can be justified by the existence of polyphenolic compounds, as reflected by the high phenolic content. It is known that polyphenols exhibit considerable antioxidant activity. Nabavi et al. have found that increasing flavonoid content in the diet may reduce some human diseases [84].

The EAE and AQE of *A. absinthium* leaves showed similar and promising results in the β-carotene test, which was related to the polyphenols and tannins content in the two extracts. The reducing capacity of the plant's extracts may indicate their potentially significant antioxidant power, which is usually related to the presence of reductones [85]. The antioxidant function prevents the formation of peroxide due to the presence of reductones that have the power to break the chain of free radicals by donating a hydrogen atom, or by reacting with certain peroxide precursors [86,87]. The free radical of linoleic acid is attacked by β-carotene so that it undergoes rapid bleaching as it loses the double bonds, and therefore, its orange color [88]; the existence of a molecule that plays the role of an antioxidant can inhibit or delay the destruction of the β-carotene molecule by neutralizing the radicals of the linoleic acid formed.

The presence of caffeic acid and naringenin as the major products in EAE and AQE, respectively, may be responsible for the observed antioxidant activities. Caffeic acid was found to have excellent antioxidant activity [89]; this activity was also shown to be stronger than that of Trolox and ascorbic acid [90], in a dose-dependent manner with an IC_{50} of 0.038 mg/mL [91]. Naringenin is a flavanone of natural origin, known for its effect on health, thanks to its antioxidant activity and its free radical scavenging capacity [92]. These two molecules may be the source of the antioxidant activity observed in EAE and AQE.

An excess of oxidants in the body can induce a phenomenon called oxidative stress, which is linked to the pathogenesis of several chronic diseases that can cause oxidative harm to proteins, DNA, and lipids. This process plays a key role in the development of several diseases that are the cause of most deaths today, including diabetes, cancer, atherosclerosis, eye disease, Alzheimer's and Parkinson's diseases [93], as well as other chronic diseases [93,94]. The intervention of antioxidants consists of the neutralization of the excess free radicals found in the organism, and therefore, an absence of oxidative stress responsible for the diseases. In fact, antioxidant substances have a strong capacity to trap free radicals. Antioxidants can be used in the cosmetic industry due to their ability to reduce oxidative damage, making them a good therapeutic alternative to prevent the premature onset of diseases [95]. The protective role of phytochemicals may be related to their antioxidant activity. The use of plant extracts rich in phenolic compounds appears to be a sustainable choice for cosmetic applications, ensuring a commitment to durability. Antioxidant compounds are used to inhibit the oxidation of the oil portion of cosmetic preparations and are also employed to reduce or prevent the oxidative deterioration of the active components of the product [96]. These compounds play a photoprotective role that helps in the treatment of sun-stressed or sensitive skin [97]. *A. Absinthium* has exhibited cytotoxic action on human colon and endometrial cancer cells, which may be due to its antioxidant power [66]. The methanolic extract of *A. absinthium* leaves was able to actively inhibit the proliferation of breast cancer cells with an IC_{50} value of 80.96 ± 3.94 µg/mL [98]. These studies suggest the use of this plant to develop new agents that can be used industrially or pharmaceutically, especially as anti-cancer agents.

3.4. Inhibitory Effect of Pancreatic α-Amylase Enzyme and Intestinal α-Glucosidase Enzyme, In Vitro

Due to the side effects and toxicity of the drugs currently used to control hyperglycemia, research has been directed at discovering new pancreatic α-amylase and intestinal α-glucosidase inhibitors from natural sources, especially plants that show a hypoglycemic effect with no or fewer side effects. Various plants that act as enzyme inhibitors have been tested for the management of diabetes. We evaluated the pancreatic α-amylase and intestinal α-glucosidase inhibitory activities of *A. absinthium* extracts. Acarbose, an oral hypoglycemic drug employed for treating diabetes mellitus, was used as a positive control [99].

Figure 5A showed that both extracts of *A. absinthium* significantly inhibited the activity of pancreatic α-amylase enzyme, in vitro, in a nearly similar manner to that of the control. The concentration of 0.9 mg/mL showed a more potent inhibitory effect than 0.45 mg/mL concentration, with an inhibitory percentage of $58.14 \pm 4.57\%$ for EAE and $72.06 \pm 1.17\%$ for AQE. AQE and EAE inhibit the enzyme, as shown by the IC_{50} values of 0.68 ± 0.01 mg/mL and 0.76 ± 0.064 mg/mL, respectively (Table 3). Moreover, the statistical analyses showed that AQE possesses the same effect as EAE ($p > 0.05$).

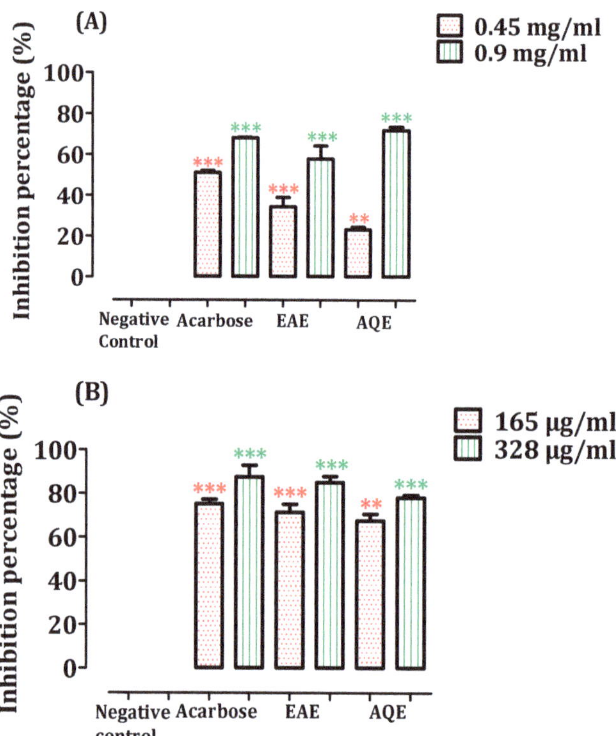

Figure 5. The effect of AQE and EAE of *A. absinthium* on pancreatic α-amylase enzyme (**A**), and intestinal α-glucosidase enzyme (**B**), inhibition in vitro ($n = 3$). ** $p < 0.01$; *** $p < 0.001$ as compared to the control. EAE—ethyl acetate extract; AQE—aqueous extract.

The results of the inhibitory activity of the intestinal α-glucosidase enzyme in the extracts of *A. absinthium* are shown in Figure 5, revealing a considerable inhibition against this enzyme compared to the positive control. The results indicated that both EAE and AQE significantly inhibited the intestinal α-glucosidase enzyme activity ($p < 0.001$), in vitro, in comparison with the control, which shows a value of 0.155 ± 0.0009 mg/mL for EAE and 0.170 ± 0.002 mg/mL for AQE (Table 3). As compared with the positive control (acarbose), *A. absinthium* revealed statistically the same activity as the positive control at the concentrations of 0.165 mg/mL and 0.328 mg/mL. The ethyl acetate extract EAE appeared to be the most active extract, with an inhibition percentage of 85.1%, close to that of the positive control (acarbose), which has an inhibitory activity of 87.7 for a concentration of 0.328 mg/mL.

Oligosaccharides and disaccharides resulting after the hydrolysis of polysaccharides by pancreatic α-amylase are transformed into monosaccharides in the presence of the enzyme α-glycosidase, which are absorbed in the hepatic portal vein by the small intestine and subsequently, increase postprandial glycemia [22]. Therefore, the use of these extracts inhibits the formation of monosaccharides; consequently, there will not be an increase in blood glucose levels, indicating a hypoglycemic effect.

The activity of AQE shown for the inhibition of these two enzymes may be related to the presence of caffeic acid, which presents as the majority product after extraction. Indeed, Ganiyu Oboh et al. showed that caffeic acid displays an inhibitory effect on pancreatic α-amylase (IC_{50} = 0.003 mg/mL) and intestinal α-glucosidase (IC_{50} = 0.004 mg/mL) [91]. These IC_{50} values remain smaller than those of our extract, and we can attribute this result to the antagonism effect of caffeic acid in the presence of other molecules.

For many years, plants were considered as the main source of drugs and oral hypo-glycemic agents, so research has been directed towards the discovery of those plants possessing pancreatic α-amylase and intestinal α-glucosidase inhibitory power, and thus with potential for use as hypoglycemic agents. Among the plants used as inhibitors of pancreatic α-amylase and intestinal α-glucosidase are *Andrographis paniculata* [100], *Hibiscus sabdariffa* Linn. [101], *Morinda lucida* Benth [102], *Telfairia occidentalis* [103], *Phaseolus vulgaris* L. [104] and *Andromachia igniaria* [105].

The *Asteraceae* family is considered one of the most common anti-diabetic plant sources in traditional Moroccan medicine [106], and artemisia is a large and diverse genus of plants in the *Asteraceae* family. In the last few years, there has been a growing focus on the research of bioactive components of artemisia. Some plants belonging to this genus have shown a capacity for the inhibition of the pancreatic α-amylase enzyme and the intestinal α-glucosidase enzyme, among these plants we find Artemisia anethifolia, Artemisia desertorum, Artemisia latifolia, Artemisia umbrosa, Artemisia tanacetifolia, Artemisia palustris, Artemisia leucophylla and Artemisia commutata [107].

3.5. Inhibitory Effect of Pancreatic α-Amylase Enzyme and Intestinal α-Glucosidase Enzyme, In Vivo

To confirm the inhibitory activity of AQE and EAE extracts of *A. absinthium* on pancreatic the α-amylase enzyme and intestinal α-glucosidase enzyme, in vivo, the test was performed and the results are presented in Figure 6. Figure 6A represents the effect of *A. absinthium* extracts on blood. After oral starch overload in normal rats, the results indicated that the blood glucose level in the control group increased from 0.81 g/L to 1.32 g/L after 30 and 120 min of starch administration, respectively. While in the presence of AQE and EAE, the postprandial blood glucose decreased compared to the control from 1.84 g/L to 0.60 g/L and from 0.82 g/L to 0.62 g/L, respectively, after 120 min. However, EAE and AQE showed significantly higher activity than acarbose at a dose of 10 mg/Kg.

Figure 6B demonstrates the results of glucose level monitoring in normal rats. The blood glucose level increased to 1.65 g/L and 1.22 g/L during 120 min in the control group, whereas in the presence of the aqueous extract, blood glucose levels decreased from 1.26 g/L to 0.99 g/L 120 min after sucrose administration, respectively. EAE showed no significant anti-hyperglycemic effect of this enzyme in the normal Wistar rats; in fact, it shows a similar effect to that of the control.

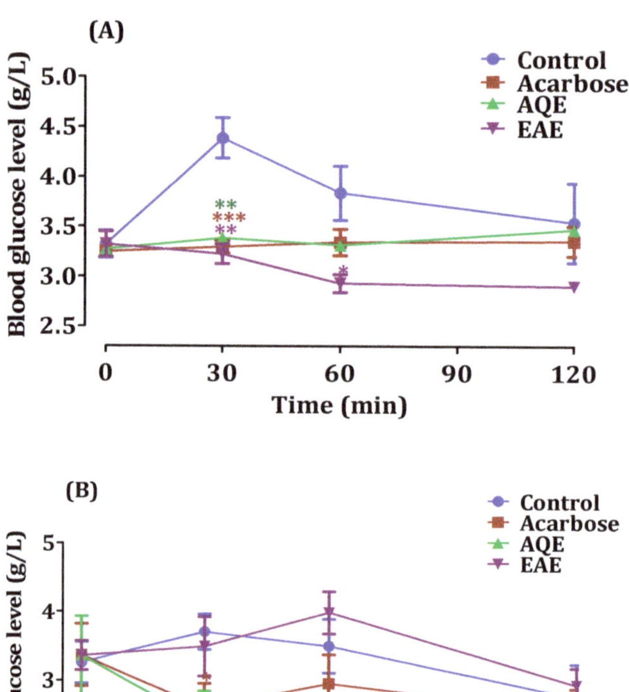

Figure 6. The effects of *A. absinthium* extracts and acarbose on glycemia in normal rats after starch overload (**A**), and after sucrose overload (**B**) ($n = 5$) ((* $p < 0.05$), (** $p < 0.01$) and (*** $p < 0.001$)).

The results in Figure 7 correspond to the similar oral starch and sucrose tolerance test conducted previously, but this time on diabetic rats. The effect of *A. absinthium* extracts on blood glucose after oral starch overload is shown in Figure 7A. The results show that in the presence of EAE, postprandial blood glucose decreased remarkably 30 min after oral starch overload, decreasing to 3.22 and 2.9 g/L at 60 and 120 min, respectively. On the other hand, blood glucose levels increased by 3.67 g/L and 4.52 g/L in the control non-treated group. The group treated with AQE showed a decrease in blood glucose at both 30 and 120 min, with levels of 4.38 and 3.47 g/L, respectively. It can be seen that the blood glucose value in diabetic rats treated with AQE showed similar levels to those treated with acarbose after the oral starch overload; however, EAE showed a remarkably higher hypoglycemic effect than acarbose.

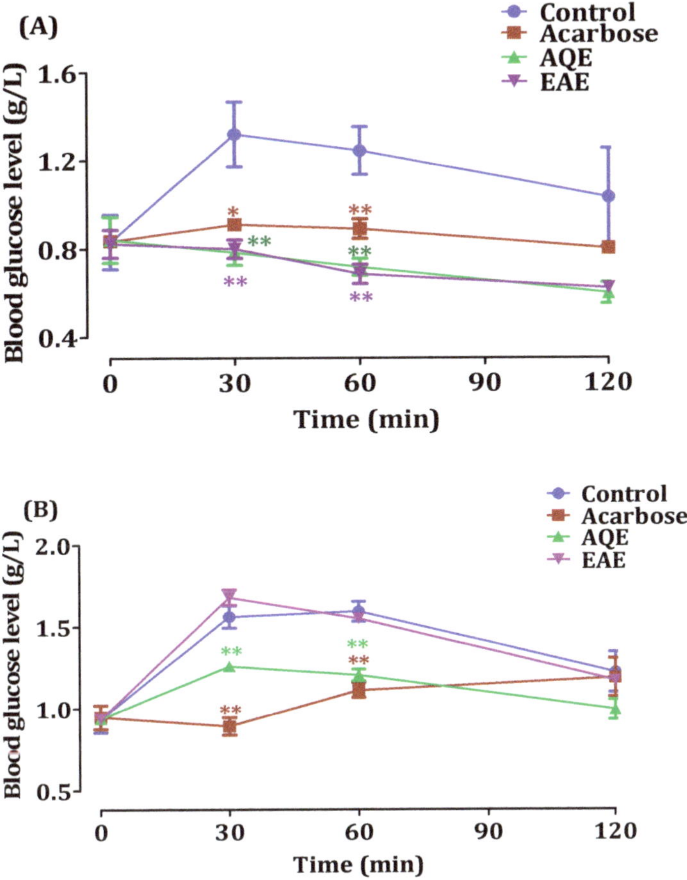

Figure 7. The effect *A. absinthium* extracts and acarbose on glycemia in diabetic rats after starch overload (**A**), and after sucrose overload (**B**) ($n = 5$) ((* $p < 0.05$) and (** $p < 0.01$)).

Figure 7B demonstrated the results of the oral sucrose tolerance test of *A. absinthium* extracts on streptozotocin-induced diabetic rats. In the control group, the level of glycemia increased from 3.26 to 3.71 g/l 30 min after sucrose overload, while, in presence of 250 mg/Kg of EAE, the glycemia level decreased ($p < 0.01$) from 3.36 to 2.39 g/L and from 3.36 to 2.024 g/L 30 and 60 min after sucrose overload, respectively. After 120 min, we found that the blood glucose level changed and became equal to the control values for all the tested groups. In the presence of EAE, the blood glucose levels revealed no significant variation from the control group ($p > 0.05$). Indeed, the blood glucose level increased from 3.36 g/L to 3.5 g/L 30 min after sucrose overload and to 3.99 g/L 60 min after; the blood glucose level decreased to 2.94 after 120 min. Li et al. [30] showed that *A. absinthium* exhibits antidiabetic activity in diabetic humans, with no significant effect on lipid profiles, while, Daradka et al. [31] showed that the ethanolic extract of *A. absinthium* has a hypoglycemic activity in rats with alloxan-induced diabetes. These results are in agreement with our work, which has confirmed the antidiabetic power of this plant.

α-amylase is one of the main enzymes in the human body, identified as either salivary amylase or pancreatic amylase, responsible for the degradation of starch into simple sugars. In general, this enzyme hydrolyzes polysaccharides to produce disaccharides and oligosaccharides. The inhibition of this enzyme by our extracts resulted in a decrease in the formation of oligosaccharides and disaccharides, causing a decrease in the absorption of glucose into the blood. The inhibition of pancreatic α-amylase and intestinal α-glucosidase enzymes has the potential to significantly decrease the postprandial rise in blood glucose; therefore, it may be an important management strategy for type 2 diabetic and borderline diabetic patients [108,109].

The in vivo antidiabetic activity found for AQE may be due to the main by-product after extraction, which is caffeic acid. Feng-Lin Hsu et al. demonstrated that caffeic acid reduced the elevation of plasma glucose levels in insulin-resistant rats subjected to a glucose challenge test. In addition, this molecule increased the glucose uptake in isolated adipocytes [74]. The antidiabetic activity found in AQE may be due to the presence of a high tannin content; indeed, tannins have shown positive effects in the treatment of diabetes mellitus type 2, and this effect is probably due to the inhibition of alpha-amylase thanks to its ability to bind carbohydrates and proteins and possibly also due to the inhibition of alpha-glucosidase [110,111].

The activity found in EAE of pancreatic α-amylase can be linked to the presence of naringenin, with a percentage of 58% in this extract. In fact, Osama M. Ahmed et al. have shown that naringenin has a potent antidiabetic effect in NA/STZ-induced type 2 diabetic rats, thanks to its insulin-enhancing property and its insulinotropic effect, which in turn may be facilitated by adiponectin expression in adipose tissue and the enhancement of the insulin receptor GLUT4 [112]. The activity of this extract in vivo on intestinal α-glucosidase did not show any hypoglycemic effect on rats; this result confirms that the in vitro inhibitory effect is not necessarily the same as the in vivo effect.

4. Conclusions

The abundance of *A. absinthium* in Morocco has led us to search for ways to utilize it. Our work has shown that the ethyl acetate extract and the aqueous form of *A. absinthium* leaves have antioxidant properties, in addition to being able to inhibit the pancreatic α-amylase enzyme and the intestinal α-glucosidase enzyme involved in the degradation of sugars, thus causing a hypoglycemic effect. This result is justified by the high content of polyphenols and by the high antioxidant activity found in the extracts. Ethyl acetate extract does not show any inhibitory activity on intestinal α-glucosidase in vivo, but it exhibits remarkable activity in vitro; hence, the importance of in vivo studies. This result is not justified, but it may be linked to the presence of one or more molecules which do not react with the animal's enzymes. The presence of naringenin in the acetate extract and of caffeic acid in the aqueous fraction of *A. absinthium* could explain the inhibition activity of α-amylase pancreatic and the antioxidant properties found.

Author Contributions: Conceptualization, A.H. and A.Z.; methodology, N.E.D., E.H.L. and A.B.; validation, H.M. (Hicham Mohti); formal analysis, H.M. (Hicham Mohti); investigation, E.H.L.; data curation, H.M. (Hamza Mechchate); writing—original draft preparation, A.H. and M.B. (Mohamed Bouhrim); writing—review and editing, F.A.N. and R.A.-S.; supervision, M.B. (Mohamed Bnouham) and A.Z. All authors have read and agreed to the published version of the manuscript.

Funding: This study was funded by Researchers Supporting Project, King Saud University, Riyadh, Saudi Arabia, through grant no. RSP-2021/353.

Institutional Review Board Statement: The study was conducted according to the guidelines of the Declaration of Helsinki and approved by the Institutional Review Board at the Faculty of Sciences, Mohammed First University, Oujda, Morocco (09/2021/LBBEH-12 and 17/01/2021).

Informed Consent Statement: Not applicable.

Data Availability Statement: Data are available upon request.

Acknowledgments: The authors extend their appreciation to the Researchers Supporting Project, King Saud University, Riyadh, Saudi Arabia, for funding this work through grant no. RSP-2021/353. Open access funding provided by University of Helsinki.

Conflicts of Interest: The authors declare no conflict of interest.

References

1. Sharopov, F.S.; Sulaimonova, V.A.; Setzer, W.N. Composition of the Essential oil of *Artemisia absinthium* from Tajikistan. *Rec. Nat. Prod.* **2012**, *6*, 127–134.
2. Zafar, M.; Hamdard, M.; Hameed, A. Screening of *Artemisia absinthium* for antimalarial effects on *Plasmodium berghei* in mice: A preliminary report. *J. Ethnopharmacol.* **1990**, *30*, 223–226. [PubMed]
3. Irshad, S.; Butt, M.; Younus, H. In-vitro antibacterial activity of two medicinal plants neem (*Azadirachta indica*) and peppermint. *Int Res. J. Pharm.* **2011**, *1*, 9–14.
4. Khattak, S.G.; Gilani, S.N.; Ikram, M. Antipyretic studies on some indigenous Pakistani medicinal plants. *J. Ethnopharmacol.* **1985**, *14*, 45–51. [CrossRef]
5. Nibret, E.; Wink, M. Volatile components of four Ethiopian Artemisia species extracts and their in vitro antitrypanosomal and cytotoxic activities. *Phytomedicine* **2010**, *17*, 369–374. [CrossRef]
6. Chiasson, H.; Bélanger, A.; Bostanian, N.; Vincent, C.; Poliquin, A. Acaricidal properties of *Artemisia absinthium* and *Tanacetum vulgare* (Asteraceae) essential oils obtained by three methods of extraction. *J. Econ. Entomol.* **2001**, *94*, 167–171. [CrossRef]
7. Chopra, C.; Bhatia, M.; Chopra, I. In vitro antibacterial activity of oils from Indian Medicinal Plants I. *J. Am. Pharm. Assoc.* **1960**, *49*, 780–781. [CrossRef]
8. Ahmad, F.; Khan, R.A.; Rasheed, S. Study of analgesic and anti-inflammatory activity from plant extracts of *Lactuca scariola* and Artemisia absinthium. *J. Islamic Acad. Sci.* **1992**, *5*, 111–114.
9. Mahmoudi, M.; Ebrahimzadeh, M.; Ansaroudi, F.; Nabavi, S.; Nabavi, S. Antidepressant and antioxidant activities of *Artemisia absinthium* L. at flowering stage. *Afr. J. Biotechnol.* **2009**, *8*, 7170–7175.
10. Bora, K.S.; Sharma, A. Phytochemical and pharmacological potential of *Artemisia absinthium* Linn. and *Artemisia asiatica* Nakai: A review. *J. Pharm. Res.* **2010**, *3*, 325–328.
11. Gilani, A.-U.H.; Janbaz, K.H. Preventive and curative effects of *Artemisia absinthium* on acetaminophen and CCl4-induced hepatotoxicity. *Gen. Pharmacol. Vasc. Syst.* **1995**, *26*, 309–315. [CrossRef]
12. Ivanov, M.; Gašić, U.; Stojković, D.; Kostić, M.; Mišić, D.; Soković, M. New Evidence for *Artemisia absinthium* L. Application in Gastrointestinal Ailments: Ethnopharmacology, Antimicrobial Capacity, Cytotoxicity, and Phenolic Profile. *Evid.-Based Complement. Altern. Med.* **2021**, *2021*. [CrossRef] [PubMed]
13. Szopa, A.; Pajor, J.; Klin, P.; Rzepiela, A.; Elansary, H.O.; Al-Mana, F.A.; Mattar, M.A.; Ekiert, H. *Artemisia absinthium* L.—Importance in the history of medicine, the latest advances in phytochemistry and therapeutical, cosmetological and culinary uses. *Plants* **2020**, *9*, 1063. [CrossRef]
14. Bhat, R.R.; Rehman, M.U.; Shabir, A.; Mir, M.U.R.; Ahmad, A.; Khan, R.; Masoodi, M.H.; Madkhali, H.; Ganaie, M.A. Chemical composition and biological uses of *Artemisia absinthium* (wormwood). In *Plant and Human Health*; Springer: Cham, Switzerland, 2019; Volume 3, pp. 37–63.
15. Haines, J. Absinthe-return of the Green Fairy. *J. Okla. State Med. Assoc.* **1998**, *91*, 406–407. [PubMed]
16. Padosch, S.A.; Lachenmeier, D.W.; Kröner, L.U. Absinthism: A fictitious 19th century syndrome with present impact. *Subst. Abus. Treat. Prev. Policy* **2006**, *1*, 14. [CrossRef] [PubMed]
17. Yineger, H.; Kelbessa, E.; Bekele, T.; Lulekal, E. Ethnoveterinary medicinal plants at bale mountains national park, Ethiopia. *J. Ethnopharmacol.* **2007**, *112*, 55–70. [CrossRef]
18. Ambasta, S.; Ramchandran, K. The useful plants of India, publication and information directorate. *CSIR New Delhi* **1986**, *109*.
19. Nadkarni, K.; Nadkarni, A. Indian Materia Medica, Popular Prakashan Pvt. Ltd. *Bombay* **1976**, *1*, 799.
20. Solomon, S.D.; Chew, E.; Duh, E.J.; Sobrin, L.; Sun, J.K.; VanderBeek, B.L.; Wykoff, C.C.; Gardner, T.W. Diabetic retinopathy: A position statement by the American Diabetes Association. *Diabetes Care* **2017**, *40*, 412–418. [CrossRef]
21. Gowd, V.; Bao, T.; Wang, L.; Huang, Y.; Chen, S.; Zheng, X.; Cui, S.; Chen, W. Antioxidant and antidiabetic activity of blackberry after gastrointestinal digestion and human gut microbiota fermentation. *Food Chem.* **2018**, *269*, 618–627. [CrossRef]
22. Uddin, N.; Hasan, M.R.; Hossain, M.M.; Sarker, A.; Hasan, A.N.; Islam, A.M.; Chowdhury, M.M.H.; Rana, M.S. In vitro α–amylase inhibitory activity and in vivo hypoglycemic effect of methanol extract of *Citrus macroptera* Montr. fruit. *Asian Pac. J. Trop. Biomed.* **2014**, *4*, 473–479. [CrossRef]
23. Harrower, A. Comparison of efficacy, secondary failure rate, and complications of sulfonylureas. *J. Diabetes Complicat.* **1994**, *8*, 201–203. [CrossRef]
24. Campbell, R.K.; White, J.R., Jr.; Saulie, B.A. Metformin: A new oral biguanide. *Clin. Ther.* **1996**, *18*, 360–371. [CrossRef]
25. Yu, Z.; Yin, Y.; Zhao, W.; Liu, J.; Chen, F. Anti-diabetic activity peptides from albumin against α-glucosidase and α-amylase. *Food Chem.* **2012**, *135*, 2078–2085. [CrossRef] [PubMed]
26. Sales, P.M.; Souza, P.M.; Simeoni, L.A.; Magalhães, P.O.; Silveira, D. α-Amylase inhibitors: A review of raw material and isolated compounds from plant source. *J. Pharm. Pharm. Sci.* **2012**, *15*, 141–183. [CrossRef] [PubMed]
27. Kumar, S.; Narwal, S.; Kumar, V.; Prakash, O. α-glucosidase inhibitors from plants: A natural approach to treat diabetes. *Pharmacogn. Rev.* **2011**, *5*, 19. [CrossRef] [PubMed]
28. Tundis, R.; Loizzo, M.; Menichini, F. Natural products as α-amylase and α-glucosidase inhibitors and their hypoglycaemic potential in the treatment of diabetes: An update. *Mini Rev. Med. Chem.* **2010**, *10*, 315–331. [CrossRef]
29. Grover, J.; Yadav, S.; Vats, V. Medicinal plants of India with anti-diabetic potential. *J. Ethnopharmacol.* **2002**, *81*, 81–100. [CrossRef]

30. Li, Y.; Zheng, M.; Zhai, X.; Huang, Y.; Khalid, A.; Malik, A.; Shah, P.; Karim, S.; Azhar, S.; Hou, X. Effect of *Gymnema sylvestre*, Citrullus colocynthis and *Artemisia absinthium* on blood glucose and lipid profile in diabetic human. *Acta Pol. Pharm.* **2015**, *72*, 981–985.

31. Daradka, H.M.; Abas, M.M.; Mohammad, M.A.; Jaffar, M.M. Antidiabetic effect of *Artemisia absinthium* extracts on alloxan-induced diabetic rats. *Comp. Clin. Pathol.* **2014**, *23*, 1733–1742. [CrossRef]

32. Slepetys, J. Biology and biochemistry of wormwood. VIII. Accumulation dynamics of tannins, ascorbic acid and carotene (Russian). Труды Академии наук Литовской ССР Серия С Биологические науки **1975**, *1*, 43–48.

33. Canadanovic-Brunet, J.M.; Djilas, S.M.; Cetkovic, G.S.; Tumbas, V.T. Free-radical scavenging activity of wormwood (*Artemisia absinthium* L) extracts. *J. Sci. Food Agric.* **2005**, *85*, 265–272. [CrossRef]

34. Da Silva, J.A.T. Mining the essential oils of the Anthemideae. *Afr. J. Biotechnol.* **2004**, *3*, 706–720.

35. Miceli, N.; Buongiorno, L.P.; Celi, M.G.; Cacciola, F.; Dugo, P.; Donato, P.; Mondello, L.; Bonaccorsi, I.; Taviano, M.F. Role of the flavonoid-rich fraction in the antioxidant and cytotoxic activities of *Bauhinia forficata* Link.(Fabaceae) leaves extract. *Nat. Prod. Res.* **2016**, *30*, 1229–1239. [CrossRef] [PubMed]

36. Mohti, H.; Taviano, M.F.; Cacciola, F.; Dugo, P.; Mondello, L.; Zaid, A.; Cavò, E.; Miceli, N. *Silene vulgaris* subsp. macrocarpa leaves and roots from morocco: Assessment of the efficiency of different extraction techniques and solvents on their antioxidant capacity, brine shrimp toxicity and phenolic characterization. *Plant. Biosyst.-Int. J. Deal. All Asp. Plant. Biol.* **2020**, *154*, 692–699. [CrossRef]

37. Miceli, N.; Filocamo, A.; Ragusa, S.; Cacciola, F.; Dugo, P.; Mondello, L.; Celano, M.; Maggisano, V.; Taviano, M.F. Chemical characterization and biological activities of phenolic-rich fraction from cauline leaves of *Isatis tinctoria* L.(Brassicaceae) growing in Sicily, Italy. *Chem. Biodivers.* **2017**, *14*, e1700073. [CrossRef]

38. Kartal, N.; Sokmen, M.; Tepe, B.; Daferera, D.; Polissiou, M.; Sokmen, A. Investigation of the antioxidant properties of *Ferula orientalis* L. using a suitable extraction procedure. *Food Chem.* **2007**, *100*, 584–589. [CrossRef]

39. Daoudi, N.E.; Bouhrim, M.; Ouassou, H.; Legssyer, A.; Mekhfi, H.; Ziyyat, A.; Aziz, M.; Bnouham, M. Inhibitory effect of roasted/unroasted *Argania spinosa* seeds oil on α-glucosidase, α-amylase and intestinal glucose absorption activities. *South. Afr. J. Bot.* **2020**, *135*, 413–420. [CrossRef]

40. Ortiz-Andrade, R.; Rodríguez-López, V.; Garduño-Ramírez, M.; Castillo-España, P.; Estrada-Soto, S. Anti-diabetic effect on alloxanized and normoglycemic rats and some pharmacological evaluations of *Tournefortia hartwegiana*. *J. Ethnopharmacol.* **2005**, *101*, 37–42. [CrossRef]

41. Balasundram, N.; Sundram, K.; Samman, S. Phenolic compounds in plants and agri-industrial by-products: Antioxidant activity, occurrence, and potential uses. *Food Chem.* **2006**, *99*, 191–203. [CrossRef]

42. Salar, R.K.; Purewal, S.S. Phenolic content, antioxidant potential and DNA damage protection of pearl millet (*Pennisetum glaucum*) cultivars of North Indian region. *J. Food Meas. Charact.* **2017**, *11*, 126–133. [CrossRef]

43. Boudjelal, A.; Smeriglio, A.; Ginestra, G.; Denaro, M.; Trombetta, D. Phytochemical Profile, Safety Assessment and Wound Healing Activity of *Artemisia absinthium* L. *Plants* **2020**, *9*, 1744. [CrossRef]

44. Kružinauskaitė, J.; Raudoné, L. Determination of phenolic compounds content and antiradical activity in Artemisia absinthium l. During different vegetation periods. *PLANTA Sci. Pract. Educ.* **2021**, *2021*, 23–24.

45. Ferrazzano, G.F.; Amato, I.; Ingenito, A.; Zarrelli, A.; Pinto, G.; Pollio, A. Plant polyphenols and their anti-cariogenic properties: A review. *Molecules* **2011**, *16*, 1486–1507. [CrossRef]

46. Zhu, M.; Phillipson, J.D.; Greengrass, P.M.; Bowery, N.E.; Cai, Y. Plant polyphenols: Biologically active compounds or non-selective binders to protein? *Phytochemistry* **1997**, *44*, 441–447. [CrossRef]

47. Charlton, A.J.; Baxter, N.J.; Khan, M.L.; Moir, A.J.; Haslam, E.; Davies, A.P.; Williamson, M.P. Polyphenol/peptide binding and precipitation. *J. Agric. Food Chem.* **2002**, *50*, 1593–1601. [CrossRef]

48. Giglio, R.V.; Patti, A.M.; Cicero, A.F.; Lippi, G.; Rizzo, M.; Toth, P.P.; Banach, M. Polyphenols: Potential use in the prevention and treatment of cardiovascular diseases. *Curr. Pharm. Des.* **2018**, *24*, 239–258. [CrossRef]

49. Bouayed, J. Polyphenols: A potential new strategy for the prevention and treatment of anxiety and depression. *Curr. Nutr. Food Sci.* **2010**, *6*, 13–18. [CrossRef]

50. Moga, M.A.; Dimienescu, O.G.; Arvatescu, C.A.; Mironescu, A.; Dracea, L.; Ples, L. The role of natural polyphenols in the prevention and treatment of cervical cancer—An overview. *Molecules* **2016**, *21*, 1055. [CrossRef]

51. Hügel, H.M.; Jackson, N. Polyphenols for the prevention and treatment of dementia diseases. *Neural Regen. Res.* **2015**, *10*, 1756. [CrossRef]

52. De Groot, H.d.; Rauen, U. Tissue injury by reactive oxygen species and the protective effects of flavonoids. *Fundam. Clin. Pharmacol.* **1998**, *12*, 249–255. [CrossRef]

53. Agrawal, A. Pharmacological activities of flavonoids: A review. *Int. J. Pharm. Sci. Nanotechnol.* **2011**, *4*, 1394–1398. [CrossRef]

54. Cook, N.C.; Samman, S. Flavonoids—Chemistry, metabolism, cardioprotective effects, and dietary sources. *J. Nutr. Biochem.* **1996**, *7*, 66–76. [CrossRef]

55. Rice-evans, C.A.; Miller, N.J.; Bolwell, P.G.; Bramley, P.M.; Pridham, J.B. The relative antioxidant activities of plant-derived polyphenolic flavonoids. *Free Radic. Res.* **1995**, *22*, 375–383. [CrossRef]

56. Haslam, E. *Plant. Polyphenols: Vegetable Tannins Revisited*; Cambridge University Press: Cambridge, UK, 1989.

57. Battestin, V.; Matsuda, L.K.; Macedo, G.A. Fontes e aplicações de taninos e tanases em alimentos. *Aliment. E Nutr. Araraquara* **2008**, *15*, 63–72.
58. Lavid, N.; Schwartz, A.; Yarden, O.; Tel-Or, E. The involvement of polyphenols and peroxidase activities in heavy-metal accumulation by epidermal glands of the waterlily (Nymphaeaceae). *Planta* **2001**, *212*, 323–331. [CrossRef]
59. Lutgen, P. Tannins in Artemisia: The hidden treasure of prophylaxis. *Pharm. Pharmacol. Int. J.* **2018**, *6*, 176–181. [CrossRef]
60. Dhull, S.B.; Kaur, P.; Purewal, S.S. Phytochemical analysis, phenolic compounds, condensed tannin content and antioxidant potential in Marwa (*Origanum majorana*) seed extracts. *Resour. Effic. Technol.* **2016**, *2*, 168–174. [CrossRef]
61. Dollahite, J.; Pigeon, R.; Camp, B. The toxicity of gallic acid, pyrogallol, tannic acid, and *Quercus havardi* in the rabbit. *Am. J. Vet. Res.* **1962**, *23*, 1264–1267.
62. McLeod, M.N. Plant tannins-their role in forage quality. *Nutr. Abstr. Rev.* **1974**, *44*, 803–815.
63. Lee, Y.-J.; Thiruvengadam, M.; Chung, I.-M.; Nagella, P. Polyphenol composition and antioxidant activity from the vegetable plant'Artemisia absinthium'L. *Aust. J. Crop. Sci.* **2013**, *7*, 1921–1926.
64. Moacă, E.-A.; Pavel, I.Z.; Danciu, C.; Crăiniceanu, Z.; Minda, D.; Ardelean, F.; Antal, D.S.; Ghiulai, R.; Cioca, A.; Derban, M. Romanian wormwood (*Artemisia absinthium* L.): Physicochemical and nutraceutical screening. *Molecules* **2019**, *24*, 3087. [CrossRef]
65. Ivanescu, B.; Vlase, L.; Corciova, A.; Lazar, M. HPLC-DAD-MS study of polyphenols from *Artemisia absinthium, A. annua*, and *A. vulgaris*. *Chem. Nat. Compd.* **2010**, *46*, 468–470. [CrossRef]
66. Koyuncu, I. Evaluation of anticancer, antioxidant activity and phenolic compounds of *Artemisia absinthium* L. extract. *Cell. Mol. Biol.* **2018**, *64*, 25–34. [CrossRef]
67. Craciunescu, O.; Constantin, D.; Gaspar, A.; Toma, L.; Utoiu, E.; Moldovan, L. Evaluation of antioxidant and cytoprotective activities of *Arnica montana* L. and *Artemisia absinthium* L. ethanolic extracts. *Chem. Cent. J.* **2012**, *6*, 97. [CrossRef]
68. Kordali, S.; Kotan, R.; Mavi, A.; Cakir, A.; Ala, A.; Yildirim, A. Determination of the chemical composition and antioxidant activity of the essential oil of Artemisia dracunculus and of the antifungal and antibacterial activities of Turkish *Artemisia absinthium, A. dracunculus, Artemisia santonicum*, and *Artemisia spicigera* essential oils. *J. Agric. Food Chem.* **2005**, *53*, 9452–9458.
69. Razzaghi-Asl, N.; Garrido, J.; Khazraei, H.; Borges, F.; Firuzi, O. Antioxidant properties of hydroxycinnamic acids: A review of structure-activity relationships. *Curr. Med. Chem.* **2013**, *20*, 4436–4450. [CrossRef]
70. Silva, F.A.; Borges, F.; Guimarães, C.; Lima, J.L.; Matos, C.; Reis, S. Phenolic acids and derivatives: Studies on the relationship among structure, radical scavenging activity, and physicochemical parameters. *J. Agric. Food Chem.* **2000**, *48*, 2122–2126. [CrossRef]
71. Sroka, Z.; Cisowski, W. Hydrogen peroxide scavenging, antioxidant and anti-radical activity of some phenolic acids. *Food Chem. Toxicol.* **2003**, *41*, 753–758. [CrossRef]
72. Borges, F.; Guimaraes, C.; Lima, J.L.; Pinto, I.; Reis, S. Potentiometric studies on the complexation of copper (II) by phenolic acids as discrete ligand models of humic substances. *Talanta* **2005**, *66*, 670–673. [CrossRef]
73. Yeh, C.-T.; Ching, L.-C.; Yen, G.-C. Inducing gene expression of cardiac antioxidant enzymes by dietary phenolic acids in rats. *J. Nutr. Biochem.* **2009**, *20*, 163–171. [CrossRef] [PubMed]
74. Hsu, F.-L.; Chen, Y.-C.; Cheng, J.-T. Caffeic acid as active principle from the fruit of xanthiumstrumarium to lower plasma glucose in diabetic rats. *Planta Med.* **2000**, *66*, 228–230. [CrossRef] [PubMed]
75. O'Connor, A.B.; Dworkin, R.H. Treatment of neuropathic pain: An overview of recent guidelines. *Am. J. Med.* **2009**, *122*, S22–S32. [CrossRef] [PubMed]
76. Park, H.Y.; Kim, G.-Y.; Choi, Y.H. Naringenin attenuates the release of pro-inflammatory mediators from lipopolysaccharide-stimulated BV2 microglia by inactivating nuclear factor-κB and inhibiting mitogen-activated protein kinases. *Int. J. Mol. Med.* **2012**, *30*, 204–210.
77. Den Hartogh, D.J.; Tsiani, E. Antidiabetic properties of naringenin: A citrus fruit polyphenol. *Biomolecules* **2019**, *9*, 99. [CrossRef]
78. Kumar, S.; Pandey, A.K. Chemistry and biological activities of flavonoids: An overview. *Sci. World J.* **2013**, *2013*, 162750. [CrossRef]
79. Singh, A.K.; Raj, V.; Keshari, A.K.; Rai, A.; Kumar, P.; Rawat, A.; Maity, B.; Kumar, D.; Prakash, A.; De, A. Isolated mangiferin and naringenin exert antidiabetic effect via PPARγ/GLUT4 dual agonistic action with strong metabolic regulation. *Chem. Biol. Interact.* **2018**, *280*, 33–44. [CrossRef]
80. Ali, A.M.A.; El-Nour, M.E.M.; Yagi, S.M. Total phenolic and flavonoid contents and antioxidant activity of ginger (*Zingiber officinale* Rosc.) rhizome, callus and callus treated with some elicitors. *J. Genet. Eng. Biotechnol.* **2018**, *16*, 677–682. [CrossRef]
81. Msaada, K.; Salem, N.; Bachrouch, O.; Bousselmi, S.; Tammar, S.; Alfaify, A.; Al Sane, K.; Ben Ammar, W.; Azeiz, S.; Haj Brahim, A. Chemical composition and antioxidant and antimicrobial activities of wormwood (*Artemisia absinthium* L.) essential oils and phenolics. *J. Chem.* **2015**, *2015*. [CrossRef]
82. Sidaoui, F.; Igueld, S.B.; Yemmen, M.; Mraihi, F.; Barth, D.; Trabelsi-Ayadi, M.; Cherif, J.K. Chemical and functional characterization of Tunisian *Artemisia absinthium* volatiles and non-volatile extracts obtained by supercritical fluid procedure. *Int. J. Pharm. Clin. Res.* **2016**, *8*, 1178–1185.
83. Kocak, M.; Uren, M.; Calapoglu, M.; Tepe, A.S.; Mocan, A.; Rengasamy, K.; Sarikurkcu, C. Phenolic profile, antioxidant and enzyme inhibitory activities of *Stachys annua* subsp. annua var. annua. *South. Afr. J. Bot.* **2017**, *113*, 128–132. [CrossRef]
84. Nabavi, S.; Ebrahimzadeh, M.; Nabavi, S.; Hamidinia, A.; Bekhradnia, A. Determination of antioxidant activity, phenol and flavonoids content of *Parrotia persica* Mey. *Pharmacologyonline* **2008**, *2*, 560–567.
85. Sen, S.; De, B.; Devanna, N.; Chakraborty, R. Total phenolic, total flavonoid content, and antioxidant capacity of the leaves of *Meyna spinosa* Roxb., an Indian medicinal plant. *Chin. J. Nat. Med.* **2013**, *11*, 149–157. [CrossRef]

86. Gülçin, İ.; Huyut, Z.; Elmastaş, M.; Aboul-Enein, H.Y. Radical scavenging and antioxidant activity of tannic acid. *Arab. J. Chem.* **2010**, *3*, 43–53. [CrossRef]
87. Liu, Q.; Yao, H. Antioxidant activities of barley seeds extracts. *Food Chem.* **2007**, *102*, 732–737. [CrossRef]
88. Benchikh, F.; Amira, S.; Benabdallah, H. The evaluation of antioxidant capacity of different fractions of *Myrtus communis* L. leaves. *Annu. Res. Rev. Biol.* **2018**, *22*, 1–14. [CrossRef]
89. Mohammed, F.Z.; El-Shehabi, M. Antidiabetic activity of caffeic acid and 18β-glycyrrhetinic acid and its relationship with the antioxidant property. *Asian J. Pharm. Clin. Res.* **2015**, *8*, 229–235.
90. Spagnol, C.M.; Assis, R.P.; Brunetti, I.L.; Isaac, V.L.B.; Salgado, H.R.N.; Corrêa, M.A. In vitro methods to determine the antioxidant activity of caffeic acid. *Spectrochim. Acta Part A Mol. Biomol. Spectrosc.* **2019**, *219*, 358–366. [CrossRef]
91. Oboh, G.; Agunloye, O.M.; Adefegha, S.A.; Akinyemi, A.J.; Ademiluyi, A.O. Caffeic and chlorogenic acids inhibit key enzymes linked to type 2 diabetes (in vitro): A comparative study. *J. Basic Clin. Physiol. Pharmacol.* **2015**, *26*, 165–170. [CrossRef]
92. Venkateswara Rao, P.; Kiran, S.; Rohini, P.; Bhagyasree, P. Flavonoid: A review on Naringenin. *J. Pharmacogn. Phytochem.* **2017**, *6*, 2778–2783.
93. Hoffman, F. Antioxidant vitamins newsletter. *Nutr Rev.* **1997**, *14*, 234–236.
94. Cookson, M.R.; Shaw, P.J. Oxidative stress and motor neurone disease. *Brain Pathol.* **1999**, *9*, 165–186. [CrossRef] [PubMed]
95. De Lima Cherubim, D.J.; Buzanello Martins, C.V.; Oliveira Fariña, L.; da Silva de Lucca, R.A. Polyphenols as natural antioxidants in cosmetics applications. *J. Cosmet. Dermatol.* **2020**, *19*, 33–37. [CrossRef] [PubMed]
96. Arct, J.; Pytkowska, K. Flavonoids as components of biologically active cosmeceuticals. *Clin. Dermatol.* **2008**, *26*, 347–357. [CrossRef]
97. Monteiro, S.A.; Valarini, M.F.C.; Chorilli, M.; Venturini, A.; Leonardi, G.R. Atividade antioxidante do extrato seco de cacau orgânico (*Theobroma cacao*)-estudo de estabilidade e teste de aceitação de cremes acrescidos deste extrato. *Rev. De Ciências Farm. Básica E Apl.* **2013**, *34*, 493–501.
98. Sultan, M.H.; Zuwaiel, A.A.; Moni, S.S.; Alshahrani, S.; Alqahtani, S.S.; Madkhali, O.; Elmobark, M.E. Bioactive principles and potentiality of hot methanolic extract of the leaves from *Artemisia absinthium* L. "in vitro cytotoxicity against human MCF-7 breast cancer cells, antibacterial study and wound healing activity". *Curr. Pharm. Biotechnol.* **2020**, *21*, 1711–1721. [CrossRef]
99. Yin, Z.; Zhang, W.; Feng, F.; Zhang, Y.; Kang, W. α-Glucosidase inhibitors isolated from medicinal plants. *Food Sci. Hum. Wellness* **2014**, *3*, 136–174. [CrossRef]
100. Subramanian, R.; Asmawi, M.Z.; Sadikun, A. In vitro alpha-glucosidase and alpha-amylase enzyme inhibitory effects of Andrographis paniculata extract and andrographolide. *Acta Biochim. Pol.* **2008**, *55*, 391–398. [CrossRef]
101. Ademiluyi, A.O.; Oboh, G. Aqueous extracts of Roselle (*Hibiscus sabdariffa* Linn.) varieties inhibit α-amylase and α-glucosidase activities in vitro. *J. Med. Food* **2013**, *16*, 88–93. [CrossRef]
102. Kazeem, M.; Adamson, J.; Ogunwande, I. Modes of inhibition of α-amylase and α-glucosidase by aqueous extract of *Morinda lucida* Benth leaf. *BioMed Res. Int.* **2013**, *2013*. [CrossRef]
103. Oboh, G.; Akinyemi, A.; Ademiluyi, A. Inhibition of α-amylase and α-glucosidase activities by ethanolic extract of *Telfairia occidentalis* (fluted pumpkin) leaf. *Asian Pac. J. Trop. Biomed.* **2012**, *2*, 733–738. [CrossRef]
104. Mojica, L.; Meyer, A.; Berhow, M.A.; de Mejía, E.G. Bean cultivars (*Phaseolus vulgaris* L.) have similar high antioxidant capacity, in vitro inhibition of α-amylase and α-glucosidase while diverse phenolic composition and concentration. *Food Res. Int.* **2015**, *69*, 38–48. [CrossRef]
105. Saltos, M.B.V.; Puente, B.F.N.; Faraone, I.; Milella, L.; De Tommasi, N.; Braca, A. Inhibitors of α-amylase and α-glucosidase from Andromachia igniaria Humb. & Bonpl. *Phytochem. Lett.* **2015**, *14*, 45–50.
106. Bouyahya, A.; El Omari, N.; Elmenyiy, N.; Guaouguaou, F.-E.; Balahbib, A.; Belmehdi, O.; Salhi, N.; Imtara, H.; Mrabti, H.N.; El-Shazly, M. Moroccan antidiabetic medicinal plants: Ethnobotanical studies, phytochemical bioactive compounds, preclinical investigations, toxicological validations and clinical evidences; challenges, guidance and perspectives for future management of diabetes worldwide. *Trends Food Sci. Technol.* **2021**, *115*, 147–254.
107. Olennikov, D.N.; Chirikova, N.K.; Kashchenko, N.I.; Nikolaev, V.M.; Kim, S.-W.; Vennos, C. Bioactive phenolics of the genus Artemisia (Asteraceae): HPLC-DAD-ESI-TQ-MS/MS profile of the Siberian species and their inhibitory potential against α-amylase and α-glucosidase. *Front. Pharmacol.* **2018**, *9*, 756. [CrossRef]
108. Dwek, R.A.; Butters, T.D.; Platt, F.M.; Zitzmann, N. Targeting glycosylation as a therapeutic approach. *Nat. Rev. Drug Discov.* **2002**, *1*, 65–75. [CrossRef]
109. Baron, A.D. Postprandial hyperglycaemia and α-glucosidase inhibitors. *Diabetes Res. Clin. Pract.* **1998**, *40*, S51–S55. [CrossRef]
110. Thinkratok, A.; Supkamonseni, N.; Srisawat, R. Inhibitory potential of the rambutan rind extract and tannin against alpha-amylase and alpha-glucosidase activities in vitro. *In Proceedings of the International Conference on Food, Biological and Medical Sciences, 28-29th January 2014, Bangkok, Thailand.* **2014**, *2014*, 44–48.
111. Chelladurai, G.R.M.; Chinnachamy, C. Alpha amylase and Alpha glucosidase inhibitory effects of aqueous stem extract of Salacia oblonga and its GC-MS analysis. *Braz. J. Pharm. Sci.* **2018**, *54*. [CrossRef]
112. Ahmed, O.M.; Hassan, M.A.; Abdel-Twab, S.M.; Azeem, M.N.A. Navel orange peel hydroethanolic extract, naringin and naringenin have anti-diabetic potentials in type 2 diabetic rats. *Biomed. Pharmacother.* **2017**, *94*, 197–205. [CrossRef]

 *pharmaceutics*

Article

Rosinidin Flavonoid Ameliorates Hyperglycemia, Lipid Pathways and Proinflammatory Cytokines in Streptozotocin-Induced Diabetic Rats

Sadaf Jamal Gilani [1], May Nasser Bin-Jumah [2,3], Fahad A. Al-Abbasi [4], Muhammad Shahid Nadeem [4], Syed Sarim Imam [5], Sultan Alshehri [5], Mohammed M. Ghoneim [6], Muhammad Afzal [7], Sami I. Alzarea [7], Nadeem Sayyed [8] and Imran Kazmi [4,*]

[1] Department of Basic Health Sciences, Preparatory Year, Princess Nourah Bint Abdulrahman University, Riyadh 11671, Saudi Arabia; sjglani@pnu.edu.sa
[2] Biology Department, College of Science, Princess Nourah Bint Abdulrahman University, Riyadh 11671, Saudi Arabia; mnbinjumah@pnu.edu.sa
[3] Environment and Biomaterial Unit, Health Sciences Research Center, Princess Nourah Bint Abdulrahman University, Riyadh 11671, Saudi Arabia
[4] Department of Biochemistry, Faculty of Science, King Abdulaziz University Jeddah, Jeddah 21589, Saudi Arabia; fabbasi@kau.edu.sa (F.A.A.-A.); mhalim@kau.edu.sa (M.S.N.)
[5] Department of Pharmaceutics, College of Pharmacy, King Saud University, Riyadh 11451, Saudi Arabia; simam@ksu.edu.sa (S.S.I.); salshehri1@ksu.edu.sa (S.A.)
[6] Department of Pharmacy Practice, College of Pharmacy, AlMaarefa University, Ad Diriyah 13713, Saudi Arabia; mghoneim@mcst.edu.sa
[7] Department of Pharmacology, College of Pharmacy, Jouf University, Aljouf, Sakaka 72341, Saudi Arabia; afzalgufran@ju.edu.sa (M.A.); samisz@ju.edu.sa (S.I.A.)
[8] School of Pharmacy, Glocal University, Saharanpur 247121, India; snadeem.pharma@gmail.com
* Correspondence: ikazmi@kau.edu.sa

Citation: Gilani, S.J.; Bin-Jumah, M.N.; Al-Abbasi, F.A.; Nadeem, M.S.; Imam, S.S.; Alshehri, S.; Ghoneim, M.M.; Afzal, M.; Alzarea, S.I.; Sayyed, N.; et al. Rosinidin Flavonoid Ameliorates Hyperglycemia, Lipid Pathways and Proinflammatory Cytokines in Streptozotocin-Induced Diabetic Rats. *Pharmaceutics* **2022**, *14*, 547. https://doi.org/10.3390/pharmaceutics14030547

Academic Editors: Diana Marcela Aragon Novoa, Fátima Regina Mena Barreto Silva and Montse Mitjans Arnal

Received: 20 December 2021
Accepted: 22 February 2022
Published: 28 February 2022

Publisher's Note: MDPI stays neutral with regard to jurisdictional claims in published maps and institutional affiliations.

Abstract: Diabetes is one of the world's most important public health issues, impacting both public health and socioeconomic advancement; moreover, current pharmacotherapy is still insufficient. The natural flavonoid rosinidin has a long history of use in pharmaceuticals and nutritional supplements, but its role in diabetes has been unknown. The current study was intended to confirm the anti-diabetic activity of rosinidin in our laboratory setting, along with its mechanism. Streptozotocin (60 mg/kg, ip) treatment used to induce type II diabetes in rats and the test medication rosinidin was then administered orally (at doses of 10 mg/kg and 20 mg/kg) for biochemical and histopathological analysis. Treatment with rosinidin reduced negative consequences of diabetes. Rosinidin exerted a protective effect on a number of characteristics, including anti-diabetic responses (lower blood glucose, higher serum insulin and improved pancreatic function) and molecular mechanisms (favorable effects on lipid profiles, total protein, albumin, liver glycogen, proinflammatory cytokine, antioxidant and oxidative stress markers, AST, ALT and urea). Furthermore, the improved pancreatic architecture observed in tissues substantiated the favourable actions of rosinidin in STZ-induced diabetic rats.

Keywords: rosinidin; streptozotocin; diabetes; lipid profile; proinflammatory cytokine

1. Introduction

Diabetes mellitus is indeed one of the world's biggest public health problems, impacting both public health and socioeconomic advancement. It has a devastating economic impact, accounting for approximately 10% of global health expenditures [1]. Diabetes mellitus is a long-lasting endocrine ailment triggered by abnormalities in the production of insulin by pancreatic cells and its actions in outlying tissues, resulting in metabolic abnormalities [2]. Despite substantial advancements in diabetes management over the last decades, patient outcomes from diabetes medications are still far from ideal.

Natural herbal plants have been a rich source of polyphenolic compounds for numerous centuries, some of which have been extensively applied to avert diabetes as a result of their medicinal qualities, multi-targeting abilities and low toxicities. Furthermore, recent studies reveal that their potential as a revolutionary medicine is promising. Several studies have suggested that some dietary bioactive substances, such as flavonoids, may influence cellular pathways [3]. Flavonoids are polyphenolic compounds having a benzo-pyrone structure. Flavonoids are plant secondary metabolites, hence they are also ingested through food [4]. Over 5000 distinct natural flavonoid compounds have been found, making flavonoids widely available. These natural chemicals have a lot of pharmacological power. This subclass was chosen since numerous studies have revealed that such compounds possess a variety of beneficial properties, such as anti-diabetic, anti-oxidant, antitumor, cardioprotective and anti-inflammatory activities, in addition to anti-cancer, anti-allergy, anti-mutagenesis, antimicrobial and antiviral activities [5]. Flavonoids' potential to alter the insulin signaling pathway in typical target tissues such as muscle, liver and adipose tissue has been shown in abundant scientific studies to help prevent or alleviate insulin resistance in diabetics [6].

Rosinidin is a anthocyanidin flavonoid pigment found in flowers such as *Catharanthus roseus* and *Primula rosea* [7]. Rosinidin is a benzopyrylium compound with hydroxy substituents at positions 3 and 5, a methoxy substituent at position 7 and a 4-hydroxy-3-methoxyphenyl substitution at position 2. Rosinidin has a long history of use in pharmaceuticals and nutritional additions for its anti-oxidant and anti-inflammatory properties [8], all of which play prominent roles in the evolution of diabetes mellitus, yet its involvement in type II diabetes mellitus remains unknown. Molecular docking and silico target investigations demonstrated that rosinidin has the requisite structural features and has the potential to be a therapeutic molecule for neurodegenerative treatment and Parkinson's disease neuroprotection [9].

Several new glucose-lowering synthetic medications have been approved for clinical use, each with a unique set of rewards. In this context, novel flavonoid-based compounds that favorably modify a variety of microvascular and macrovascular hazards, that function in tandem with the body's natural defenses, and that present pharmacoeconomically favorable yet minimal side effects, are desirable. In light of this, the goal of this current study was to confirm the anti-diabetic efficacy of rosinidin in a laboratory setting. Furthermore, the impacts of these drugs on a range of targets (lipid profiles, inflammation, antioxidative-oxidant balance, as well as liver, renal and pancreatic function) were investigated in order to decipher their mechanisms. The outcomes of this study could be crucial in providing a lead for the creation of a unique flavonoid-based anti-diabetic medicine for drug companies.

2. Methods

2.1. Chemicals and Drugs for Testing

STZ (Streptozotocin) was procured from Sigma-Aldrich, St. Louis, MO, USA, analytical quality. Rosinidin (isolated compound from *Catharanthus roseus*) was obtained as a gift sample from SRL, India. Other reagents, chemicals and kits used were available locally in India.

2.2. Animals and STZ Paradigm

STZ's diabetogenic action was discovered by Rakieten et al. (1963) [10]. It is especially cytotoxic to pancreatic beta-cells. Wistar rats (weighing 150–200 gm) were acclimatised for 7 days to a regular laboratory setting which included a schedule of 12-hour light and dark cycles, and standard temperature and humidity. The institute's animal ethics committee authorized the study protocol (TRS/PT/021/006).

In rats subjected to overnight fasting, type II diabetes was produced using a 60 mg/kg STZ intraperitoneal injection prepared in a 0.1 M cold citrate buffer (pH 4.5). In order to avoid hypoglycemia-related deaths, the animals were fed a 5% glucose solution instead of water for the next 24 h. Blood samples were obtained from tail veins 48 h after STZ or vehicle injection in order to assess blood glucose levels. Diabetic rats were recognized as those with glucose levels $\geq$250 mg/dL, and were chosen for further pharmacological research. Rats were sacrificed at the termination of the experiment. Following termination, blood samples were obtained for various biochemical investigations and pancreases were removed and fixed in 10% buffered formalin for histopathological exploration.

2.3. Research Protocol Demonstration

To compare the anti-diabetic effects of the test drug rosinidin, rats were arbitrarily assigned into four experimental groups of six animals each (normal control, diabetic control, diabetic + rosinidin-10 and diabetic + rosinidin-20). The test treatment rosinidin doses (10 mg/kg and 20 mg/kg) were administered orally via a serving cannula for thirty days. While the normal control group received saline, the other treatment groups received test dosing from the first to the thirtieth day. At the beginning and completion of the study, blood samples were collected in order to determine glucose levels, and rat weights were measured.

2.4. Parameters
Changes in Body Weight

Every week, the body weights of all group animals were recorded and changes in body weight were estimated after 30 days of nourishment. At the same time, any deaths that occurred throughout the experimental period with the respective drugs were tracked.

2.5. Fasting Blood Glucose

To avoid the impact of homeostatic changes in blood glucose metabolism, the animals were fasted overnight and their blood glucose levels were measured at morning time. Blood samples were obtained by pricking the tips of their tails. On days 1, 4, 7 and 37, glucose was measured using a glucometer [11,12].

2.6. Estimation of Biochemical Parameters

The biochemical indicator evaluation was based on previously published research with slight medication. Under light anaesthesia, blood samples from all experimental rats were taken from the retro-orbital plexus at the end of study period and centrifuged in order to obtain the sera. Blood samples and homogenates were investigated in order to determine the following anti-diabetic parameters: insulin, amylase and liver glycogen. The following lipid parameters were also investigated: total cholesterol (TC), triglyceride (TG), high-density lipoprotein (HDL-C), low density lipoprotein (LDL-C), total protein and albumin. The following proinflammatory cytokines were also analysed: interleukin IL-1β and IL-6, and tumor necrosis factor (TNF)-α. The following antioxidant and oxidative stress markers were also analysed: MDA, superoxide dismutase (SOD) and catalase (CAT). Finally, the following liver and renal function markers were investigated: aspartate aminotransferase (AST), alanine amino transferase (ALT) and urea.

2.7. Determination of Serum Lipid and Total Protein and Albumin

Estimations of serum TC, TG, HDL-C, total protein (TP) and albumin were carried out using biochemical kits (ACCUREX, Biomedical Pvt. Ltd., Maharashtra, India). For the determination of fluorescent AGE, a 96-plate spectrophotofluorimeter was employed. The fluorescence intensity was measured at 440 nm. The Friedewald formula was used in order to compute serum LDL-C.

2.8. Assay of Serum Insulin

Assays were based on a sandwich ELISA kit which contained a target-specific antibody that is pre-coated in the walls of the microplate. Serum insulin was estimated using a commercial rat ELISA insulin kit (Sigma-Aldrich, St. Louis, MO, USA) as per standard provided procedure.

2.9. Determination of Liver Glycogen

Liver glycogen was assayed in the liver homogenate as per the anthrone reagent method [13]. In brief, 500 mg of liver tissue was homogenized in 1.5 mL of 5% potassium hydroxide and boiled for 30 min. Later, 5 mL of 95% ethanol were added to precipitate glycogen. After precipitation, test tubes were centrifuged and the supernatant decanted. Finally, the pellet was dissolved in anthrone reagent and the change in green color was measured at 620 nm.

2.10. Determination of α-Amylase

Amylase is an enzyme that catalyzes the hydrolysis of starch into simple sugar. Amylase concentration was estimated using a commercially available rat amylase ELISA kit using an autoanalyzer device (Meril Life Sciences Pvt. Ltd., Gujarat, India).

2.11. Determination of Proinflammatory Cytokines

The ELISA kit uses the sandwich-ELISA principle which includes a pre-coated antibody and microwell plate. The optical density was measured spectrophotometrically. Proinflammatory cytokines such as TNF-α, IL-1β and IL-6 were assayed using ELISA kits (Merck Millipore; Billerica, MA, USA) by following the assay protocol.

2.12. Determination of Lipid Peroxidation, Superoxide and Catalase

To measure antioxidant enzymes, the liver was dissected and washed in ice-cold saline, then homogenized in a 0.1 M Tris–HCl buffer (pH 7.4). The homogenate was centrifuged and supernatant was used for the assay of antioxidant enzymes, namely superoxide dismutase (SOD) and catalase (CAT). Lipid peroxidation was estimated using OxiSelect kits.

2.13. Determination of Aspartate Aminotransferase (AST), Alanine Aminotransferase (ALT), Urea

Colorimetric assays were used in order to assess serum concentrations of liver enzymes (AST, ALT) in addition to the renal parameter (urea) using commercially available kits according to the manufacturer's instructions (Modern Lab., Maharashtra, India). Colorimetric analysis is based on the interaction of an enzyme with a substrate, and then to estimate colorimetrically a coloured, light-absorbing complex formed by adding another reagent after the enzyme reaction has stopped.

2.14. Histopathological Studies of the Pancreas

The rats were sacrificed at the termination of the experiment (37 days). Pancreatic tissues were immediately fixed in a 10% buffered neutral formalin solution. The tissues were meticulously embedded in molten paraffin using metallic blocks, then covered with flexible plastic castings and arranged under freezing plates in order to solidify the paraffin. The fixed tissues were sliced into 3–5 µm thick cross sections. In order to analyse the microscopic architecture of the pancreas, these tissue sections were stained with haematoxylin and eosin before examination under a light microscope. The sections were then observed independently by two blinded investigators.

2.15. Statistical Analysis

A one-way ANOVA statistical test was used to measure the differences among groups using GraphPad Prism®version 5 statistical software, San Diego, CA, USA, and $p < 0.05$ defined the statistical significance.

3. Results

3.1. Assessment of Body Weight Changes in Diabetic Rats

In general, diabetic rats displayed overt diabetes symptoms such as polyphagia, polydipsia and polyurea. The diabetes was proven by reduced body weight in conjunction with biochemical indicator changes (e.g., increasing blood glucose). Rats were weighed twice, at the outset of the experiment (initial weight) and again at 24 h once the last dosage of either medication was administered (final weight). When compared to normal control rats, diabetic rats exhibited a lower body weight after 30 days. The effect of rosinidin was to sustain body weight in STZ-induced diabetic rats. Figure 1 depicts no significant difference observed between treatment groups (rosinidin at different doses) as compared to diabetic control group rats.

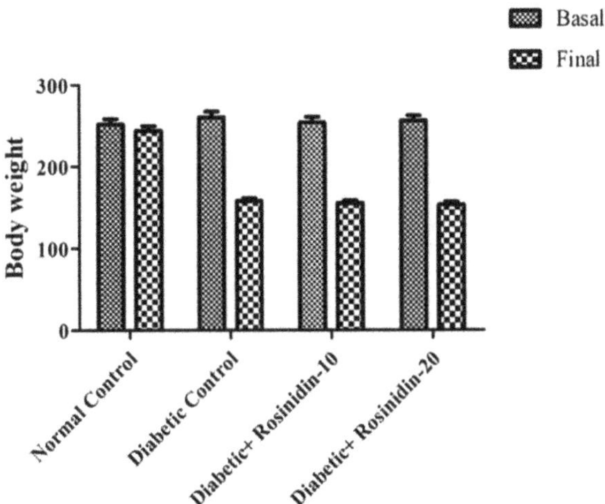

Figure 1. Effect of rosinidin treatments on body weights against those of STZ-induced diabetic rats ($n = 6$).

3.2. Experimental Evidence for Proposed Place of Rosinidin as an Anti-diabetic Therapy

3.2.1. Rosinidin Effect on Glucose in Diabetic Rats

Figure 2 depicts fasting blood glucose levels in rats from various experimental groups. Blood glucose levels in the diabetic control group were higher ($p < 0.05$) after STZ 60 mg/kg intraperitoneal injection than in normal animals. However, in comparison to the diabetic control group, oral administration of rosinidin 10 mg/kg and 20 mg/kg restored the hyperglycemia ($p < 0.001$).

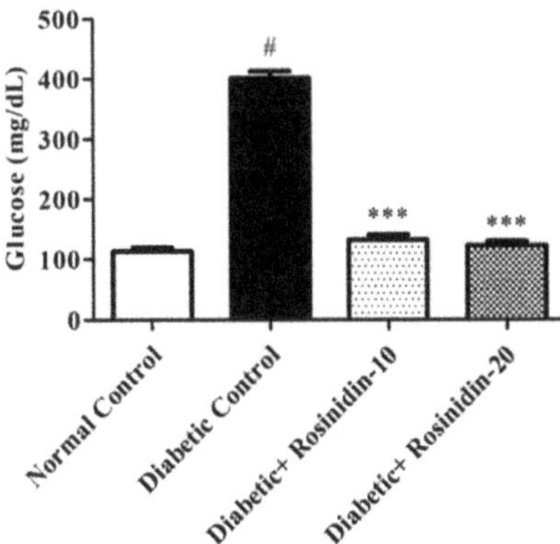

Figure 2. Effect of rosinidin on blood glucose against that of STZ-induced diabetic rats ($n = 6$). Values are expressed as mean $\pm$ SEM. # $p < 0.05$, normal control vs. diabetic control; *** $p < 0.001$, diabetic control vs. rosinidin-10, rosinidin-20.

3.2.2. Rosinidin Effect on Insulin in Diabetic Rats

The serum insulin levels in preventive and corrective strategies are depicted in Figure 3. In comparison to the normal group, the diabetic control group has a substantial drop ($p < 0.05$) in insulin serum levels. Both treatments with rosinidin 10 mg/kg and 20 mg/kg boosted serum insulin levels as compared to the diabetic control group ($p < 0.001$).

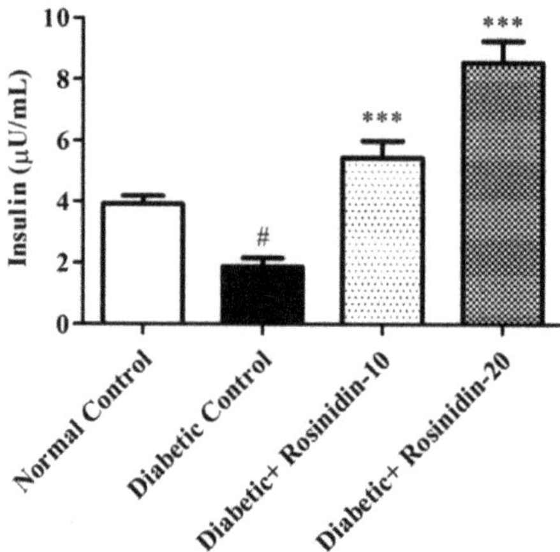

Figure 3. Effect of rosinidin on insulin against that of STZ-induced diabetic rats ($n = 6$). Values are expressed as mean ± SEM. # $p < 0.05$, normal control vs. diabetic control; *** $p < 0.001$, diabetic control vs. rosinidin-10, rosinidin-20.

3.2.3. Rosinidin Effect on Liver Glycogen in Diabetic Rats

Figure 4 reveals that diabetic animals had significantly ($p < 0.05$) reduced levels of liver glycogen when compared with the normal group. The administration of rosinidin for 30 days enhanced the levels of liver glycogen in diabetic rats expressively ($p < 0.001$).

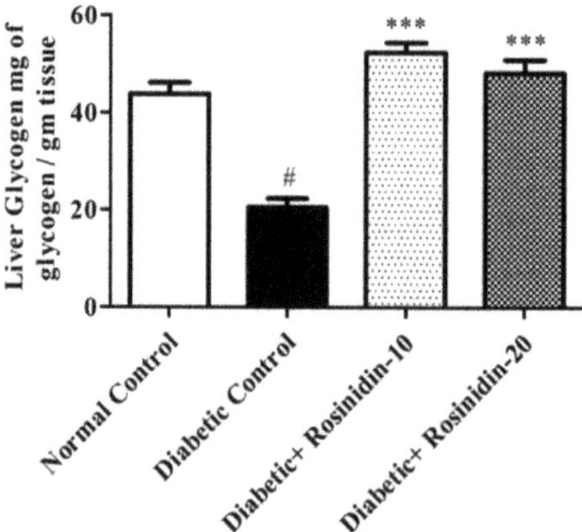

Figure 4. Effect of rosinidin on liver glycogen against that of STZ-induced diabetic rats ($n = 6$). Values are expressed as mean ± SEM. # $p < 0.05$, normal control vs. diabetic control; *** $p < 0.001$, diabetic control vs. rosinidin-10, rosinidin-20.

3.2.4. Rosinidin Effect on Lipid Profiles in Diabetic Rats

Figure 5A–D depict serum levels of lipids TC, TG, LDL-C and HDL-C in various groups. When compared to normal animals, diabetic animals had a substantial ($p < 0.05$)

increase in serum TG, TC and LDL-C. The Friedewald formula was used in order to compute the serum LDL-C level. The administration of rosinidin 10 mg/kg and 20 mg/kg was improved significantly ($p < 0.001$) in TC, TG and LDL-C levels of diabetic animals. In addition, diabetic rats revealed a significant ($p < 0.05$) drop in serum HDL-C levels which was prominently increased after receiving rosinidin at both dosages when compared to diabetic control rats. The administration of rosinidin at doses of 10 mg/kg and 20 mg/kg resulted in very marked improvements in all indices of altered lipid contours.

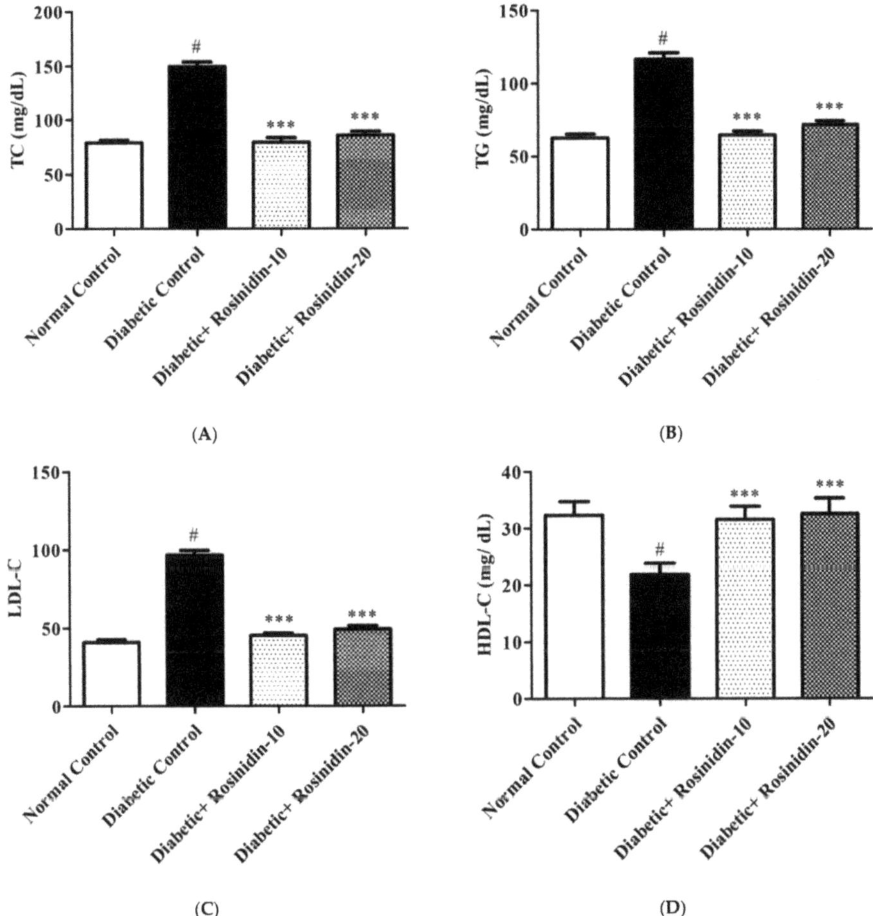

Figure 5. (**A**) Effect of rosinidin on TC against that of STZ-induced diabetic rats ($n = 6$). Values are expressed as mean ± SEM. # $p < 0.05$, normal control vs. diabetic control; *** $p < 0.001$, diabetic control vs. rosinidin-10, rosinidin-20. (**B**) Effect of rosinidin on TG against that of STZ-induced diabetic rats ($n = 6$). Values are expressed as mean ± SEM. # $p < 0.05$, normal control vs. diabetic control; *** $p < 0.001$, diabetic control vs. rosinidin-10, rosinidin-20. (**C**) Effect of rosinidin on LDL-C against that of STZ-induced diabetic rats ($n = 6$). Values are expressed as mean ± SEM. # $p < 0.05$, normal control vs. diabetic control; *** $p < 0.001$, diabetic control vs. rosinidin-10, rosinidin-20. (**D**) Effect of rosinidin on HDL-C against that of STZ-induced diabetic rats ($n = 6$). Values are expressed as mean ± SEM. # $p < 0.05$, normal control vs. diabetic control; *** $p < 0.001$, diabetic control vs. rosinidin-10, rosinidin-20.

3.2.5. Rosinidin Effect on Total Protein and Albumin in Diabetic Rats

In serum levels of total protein, albumin represents the condition of protein absorption and metabolism. The total protein and albumin levels of STZ-induced diabetic rats were considerably lower than those of the normal group rats. In the treatment group rats, rosinidin administration dramatically improved total protein and albumin levels as shown in Figures 6 and 7.

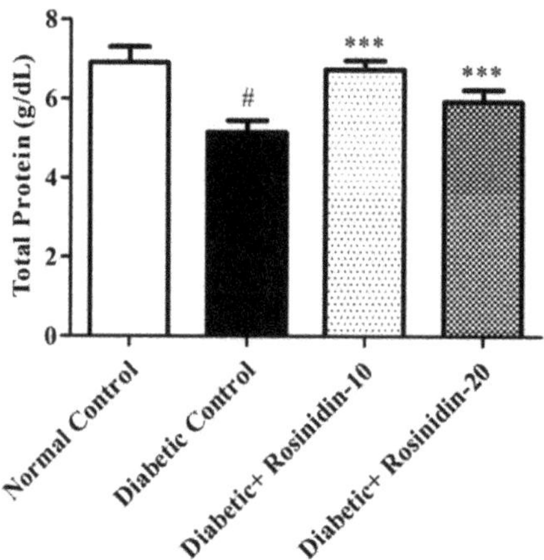

Figure 6. Effect of rosinidin on total protein against that of STZ-induced diabetic rats ($n = 6$). Values are expressed as mean $\pm$ SEM. # $p < 0.05$, normal control vs. diabetic control; *** $p < 0.001$, diabetic control vs. rosinidin-10, rosinidin-20.

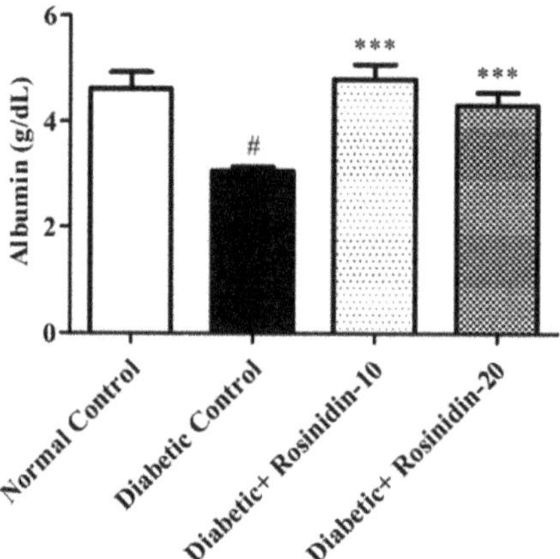

Figure 7. Effect of rosinidin on albumin against that of STZ-induced diabetic rats ($n = 6$). Values are expressed as mean $\pm$ SEM. # $p < 0.05$, normal control vs. diabetic control; *** $p < 0.001$, diabetic control vs. rosinidin-10, rosinidin-20.

3.2.6. Rosinidin Effect on Proinflammatory Cytokines in Diabetic Rats

Diabetes induction resulted in an increase in proinflammatory parameters TNF-α, IL-1β and IL-6, which were estimated using an ELISA technique in this study. Figure 8A–C showed that STZ administration drastically increased the production of proinflammatory markers when compared to normal control rats ($p < 0.05$). Meanwhile, rosinidin therapy significantly brought down the levels of inflammatory mediators when compared to the diabetic control group ($p < 0.001$). Rosinidin implantation exerts protective effects via the inhibition of proinflammatory cytokines (Figure 8A–C).

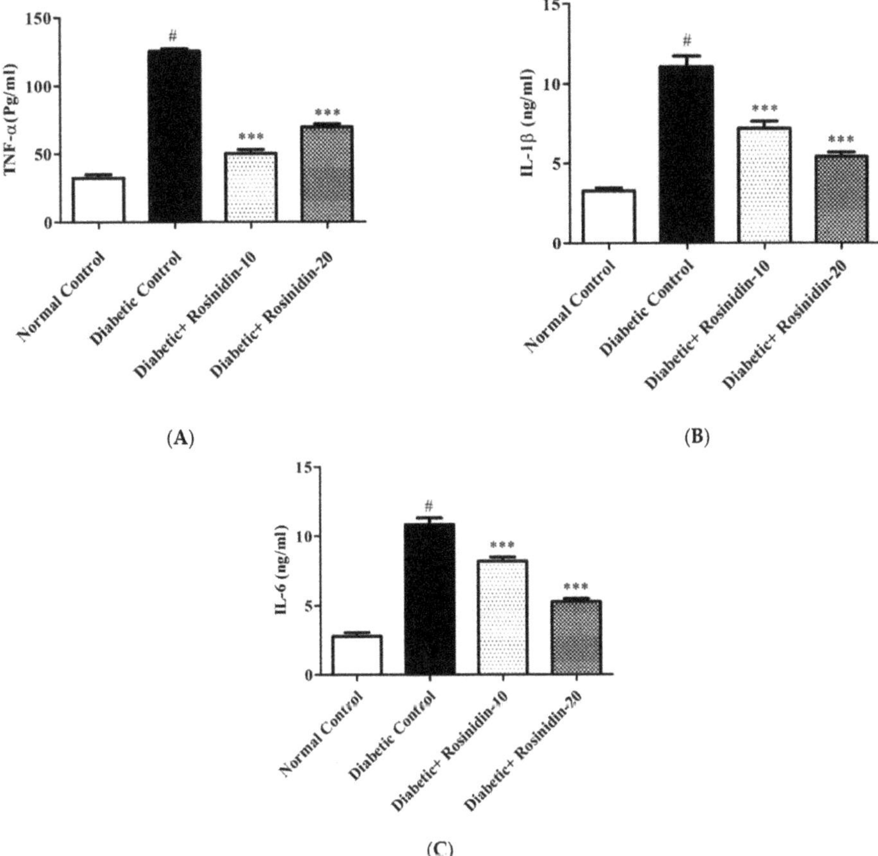

(A)

(B)

(C)

Figure 8. (**A**) Effect of rosinidin on TNF-α against that of STZ-induced diabetic rats ($n = 6$). Values are expressed as mean ± SEM. # $p < 0.05$, normal control vs. diabetic control; *** $p < 0.001$, diabetic control vs. rosinidin-10, rosinidin-20. (**B**) Effect of rosinidin on IL-1β against that of STZ-induced diabetic rats ($n = 6$). Values are expressed as mean ± SEM. # $p < 0.05$, normal control vs. diabetic control; *** $p < 0.001$, diabetic control vs. rosinidin-10, rosinidin-20. (**C**) Effect of rosinidin on IL-6 against that of STZ-induced diabetic rats ($n = 6$). Values are expressed as mean ± SEM. # $p < 0.05$, normal control vs. diabetic control; *** $p < 0.001$, diabetic control vs. rosinidin-10, rosinidin-20.

3.2.7. Rosinidin Effects on Antioxidant and Oxidative Stress Markers in Diabetic Rats

Figure 9A–C illustrate changes in lipid peroxidation and antioxidant defense system indicators such as MDA, SOD and CAT activity in experimental rats. In diabetic control rats, SOD and CAT activities were severely decreased, yet there was a significant increase in MDA levels when compared to normal animals ($p < 0.05$). Increased levels of SOD and CAT in addition to lower levels of MDA in the treatment group were shown to have a substantial upgrading in the activities of these enzymes following rosinidin 10 mg/kg and 20 mg/kg treatments as compared to diabetic control animals ($p < 0.001$). Treatment with rosinidin improved exacerbated alterations in oxidative stress markers and antioxidant enzyme activities in diabetic rats.

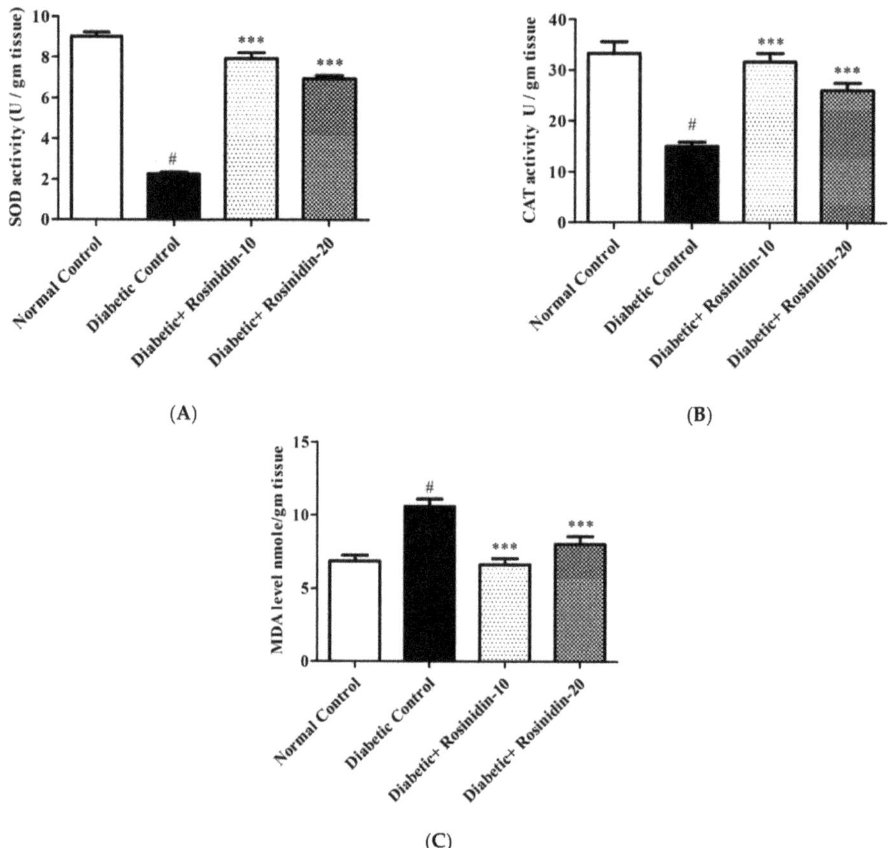

Figure 9. (A) Effect of rosinidin on SOD against that of STZ-induced diabetic rats ($n = 6$). Values are expressed as mean $\pm$ SEM. # $p < 0.05$, normal control vs. diabetic control; *** $p < 0.001$, diabetic control vs. rosinidin-10, rosinidin-20. **(B)** Effect of rosinidin on CAT against that of STZ-induced diabetic rats ($n = 6$). Values are expressed as mean $\pm$ SEM. # $p < 0.05$, normal control vs. diabetic control; *** $p < 0.001$, diabetic control vs. rosinidin-10, rosinidin-20. **(C)** Effect of rosinidin on MDA against that of STZ-induced diabetic rats ($n = 6$). Values are expressed as mean $\pm$ SEM. # $p < 0.05$, normal control vs. diabetic control; *** $p < 0.001$, diabetic control vs. rosinidin-10, rosinidin-20.

3.2.8. Rosinidin Effect on Hepatic and Renal Function in Diabetic Rats

When diabetic control rats were compared to the normal group, serum biomarkers ALT, AST and urea levels increased significantly ($p < 0.05$). Administering rosinidin doses at 10 mg/kg and 20 mg/kg, on the other hand, resulted in noteworthy declines in the elevated levels ($p < 0.001$) of these hepatic and renal functions, as shown in Figure 10A–C.

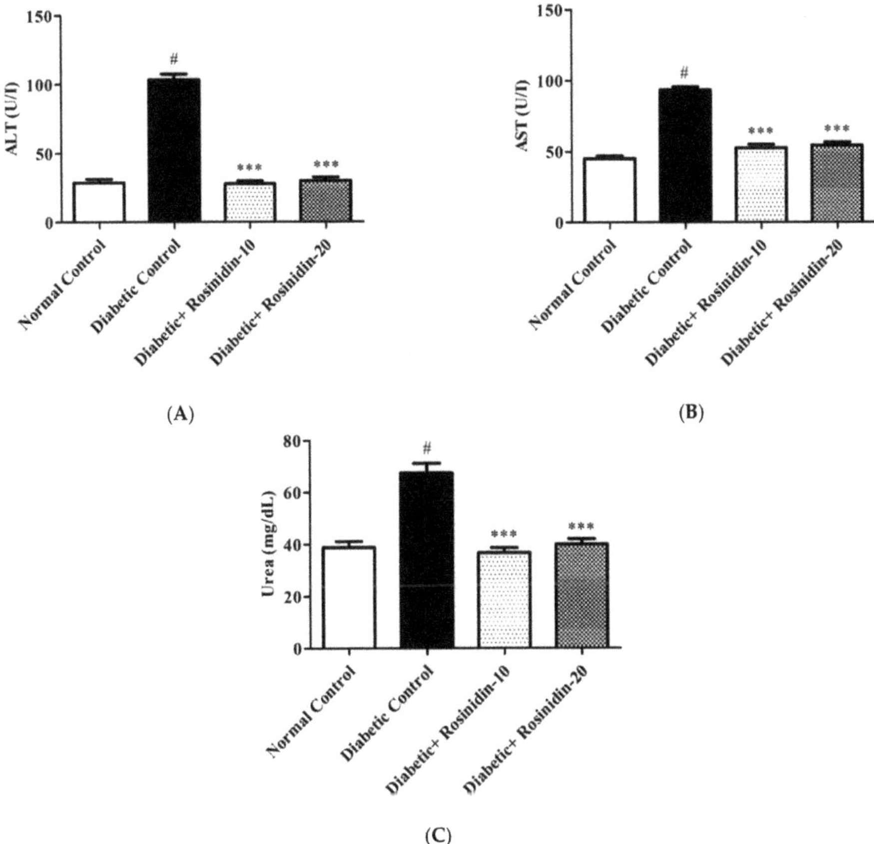

(A)

(B)

(C)

Figure 10. (**A**) Effect of rosinidin on ALT against that of STZ-induced diabetic rats ($n = 6$). Values are expressed as mean ± SEM. # $p < 0.05$, normal control vs. diabetic control; *** $p < 0.001$, diabetic control vs. rosinidin-10, rosinidin-20. (**B**) Effect of rosinidin on AST against that of STZ-induced diabetic rats ($n = 6$). Values are expressed as mean ± SEM. # $p < 0.05$, normal control vs. diabetic control; *** $p < 0.001$, diabetic control vs. rosinidin-10, rosinidin-20. (**C**) Effect of rosinidin on urea against that of STZ-induced diabetic rats ($n = 6$). Values are expressed as mean ± SEM. # $p < 0.05$, normal control vs. diabetic control; *** $p < 0.001$, diabetic control vs. rosinidin-10, rosinidin-20.

3.2.9. Rosinidin Effect on α-Amylase Level in Diabetic Rats

Regarding α-amylase levels, a significant elevation ($p < 0.05$) in its levels was noted in diabetic rats compared to the normal rats; meanwhile, diabetic rats treated with rosinidin doses at 10 mg/kg and 20 mg/kg exhibited a substantial decrease ($p < 0.001$) in serum amylase levels compared to the diabetic rats (Figure 11).

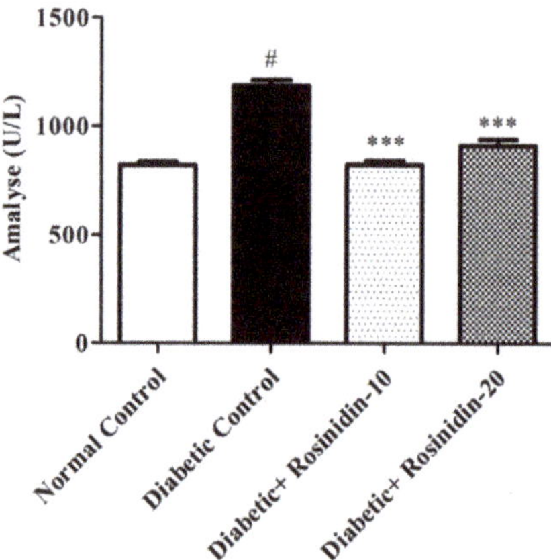

Figure 11. Effect of rosinidin on α-amylase against that of STZ-induced diabetic rats ($n = 6$). Values are expressed as mean $\pm$ SEM. # $p < 0.05$, normal control vs. diabetic control; *** $p < 0.001$, diabetic control vs. rosinidin-10, rosinidin-20.

3.2.10. Rosinidin Effect on Pancreatic Function and Photomicrographs of Pancreatic Tissue

In order to investigate therapeutic effects of rosinidin on the improvement of pancreatic performance in the STZ-induced diabetic model, histopathological modifications of the pancreas were measured. In control rats the pancreatic architecture was observed to be normal, with intact exocrine tissue and islet cells enclosed by a delicate capsule. Diabetic animals showed considerable degeneration in the pancreas portion, according to scientific evaluations. Rosinidin at the 10 mg/kg dose resulted in pancreatic histology that was analogous to that of the control rats, but with slight pancreatic cell crowding or occlusion. The rosinidin 20 mg/kg treatment group, on the other hand, showed pancreatic cell regeneration to a level similar to that of a normal rat pancreas (Figure 12).

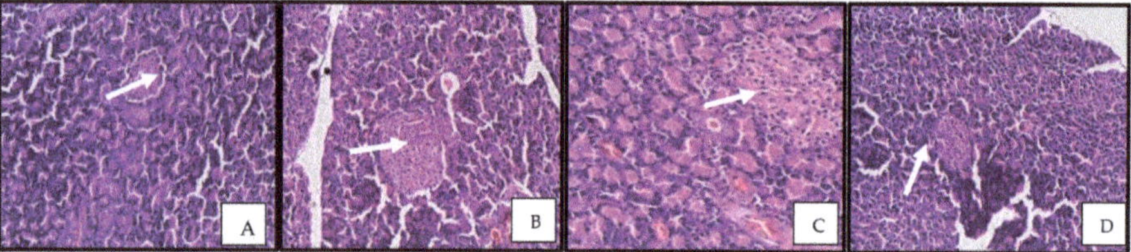

Figure 12. Representative photomicrographs illlustrating pancreatic tissue sections stained with H and E among various experimental groups. (**A**) Normal control: organized pattern and normal architecture of islet cell. (**B**) Diabetic control: damaged islets of Langerhans and degeneration in the pancreas portion. (**C**) Rosinidin-10: improved pancreatic histology, but minor pancreatic cell congestion. (**D**) Rosinidin-20: marked improvement in pancreatic histology. Arrows indicate islets of Langerhans.

4. Discussion

The problems associated with insulin and oral hypoglycemic medications have sparked interest in finding natural substances with anti-diabetic properties. As a result, safer and more effective anti-diabetic medicines in the form of flavonoid-based regimens are required. Moreover, numerous studies demonstrating flavonoids' salutary effects on metabolic abnormalities such as diabetes mellitus and obesity have piqued scientists' interest in flavonoids in recent decades [14]. However, the anti-diabetic action of rosinidin in STZ-induced diabetic rats has yet to be investigated. The goal of this research study was to learn more about the processes by which rosinidin protects against type II diabetes mellitus, in order to offer fresh light on type II diabetes mellitus prevention using flavonoid-based alternatives. Rosinidin at 10 mg/kg and 20 mg/kg doses ameliorated hyperglycemia, lipid pathways and proinflammatory cytokines, in addition to antioxidant and oxidative stress markers in STZ-induced diabetic rats. In addition, improvements to liver, renal and pancreatic function were also reported.

STZ is transported into pancreatic cells by the glucose transporter GLUT2. The nitrosourea moiety of this chemical alkylates DNA, which is the primary basis of STZ-induced cell loss. ROS generation, on the other hand, may cause DNA damage and other STZ-related adverse effects [15]. In the current study, STZ resulted in significant hyperglycemia, hypoinsulinemia and significant proinflammatory cytokinin, in addition to antioxidant and oxidative stress with a noticeable upsurge in levels of TC, TG, LDL-C, amylase, ALT, AST and urea; meanwhile, HDL-C, total protein, albumin and liver glycogen markedly decreased. A variety of studies have provided equivalent results [16].

The biological actions of these naturally occurring flavonoids and their metabolites provide cells with a variety of health-promoting benefits. Flavonoids can also interact directly with proteins such as strategic cellular receptors or signaling pathways, affecting a wide range of functions in many cells and tissues [17]. As a result, flavonoids can diminish insulin resistance in insulin-sensitive tissues in a variety of ways, including via modulating the insulin signaling pathway. An outcome in treating diabetic rats with rosinidin at 10 mg/kg and 20 mg/kg doses was a significant decline in blood glucose. This finding was in line with prior research conducted on flavonoids. Anthocyanins have been presented to have anti-obesity and anti-diabetic properties in a variety of animal models [18]. From in vitro and in vivo investigations, many flavonoid components, particularly auxiliaries, have demonstrated hypoglycemic effects. Apigenin, baicalein and catechin flavonoids reduce blood glucose. Hesperidin is beneficial for diabetic neuropathy, while glycyrrhiza flavonoid has a substantial effect on gestational diabetes mellitus. Some compounds, such as kaempferol and puerarin, are also favorable to cardiomyopathy [19].

Flavonoids help people with diabetes mellitus in a variety of ways. The most prevalent symptom is a decrease in glucose absorption. Flavonoids also function by reducing the action of the enzyme glucosidase in the small intestine. When utilized to explore their functions in glycoside absorption and metabolism, galangin, kaempferol and luteolin demonstrated α-glucosidase inhibitory action for both in vivo and in vitro circumstances [20]. As a result, the flavonoid rosinidin's potential to enhance plasma insulin concentration, as established in this work, could offer a foundation for its hypoglycemic activity in diabetic rats. This outcome is in line with expectations [21]. The capacity of rosinidin to preserve body weight loss is a result of an escalation in insulin secretion and subsequent hypoglycemia reaction. Diabetes impairs glucose production in the liver and skeletal muscles of rats. The diabetic animals in this study had significantly lower hepatic glycogen levels, which were proportional to insulin insufficiency. Liver glycogen content of diabetic rats was considerably raised by rosinidin, possibly resulting from enhanced insulin output. These findings are consistent with earlier research. Li C. et al. demonstrated the instability of the diabetic hepatic glycogen structure [22].

Dyslipidemia is well-thought-out as the fundamental module of metabolic disorder that is characterized by increased levels of TC, TG and LDL-C, in addition to reduced levels of HDL [23]. The analysis of lipid profile levels in this study specified a promising effect of rosinidin for both dosage regimens. These results are in partial agreement with previous results [24]. Flavonoids are also regarded as safe for use in the treatment of diabetes-related lipid abnormalities as well as other metabolic illnesses, according to clinical evidence [25]. Furthermore, rosinidin administration resulted in an increase in total protein and albumin levels, which contributed to the test drug's sound hypoglycemic effects.

Another important point to consider is the substantial rise in proinflammatory cytokines TNF-α, IL-1β and IL-6 in diabetic animals. Excess free radical production promotes the triggering of nuclear factor-kappa B, which in turn promotes the transcription of genes encrypting inflammatory proteins. Previously, it was reported that STZ-induced diabetes in rats promoted NF-B-mediated inflammation [26]. Rosinidin, on the other hand, was effective in lowering TNF-α, IL-1β and IL-6 in diabetic rats, implying that rosinidin may possibly play a defensive starring role against inflammatory responses in diabetic rats. Previous reports revealed similar findings [27].

The progression of diabetes, as well as diabetes-related problems, can be attributed to oxidative stress. Our study disclosed striking elevations in antioxidant enzymes (SOD, CAT) and lipid peroxidation by reduced levels of MDA, which specify the antioxidant strength of the flavonoid found in rosinidin. Rutin, luteolin and quercetin flavonoids employed patent antioxidant events that protected liver and kidney tissues against various damages, consistent with other investigations [28].

Furthermore, the diabetic group in this study had significant increases in ALT, AST and urea levels in their blood. Impairment of hepatic cells is indicated by the leaking of hepatic enzymes from the hepatic cytosol into circulation. The elevated levels of serum urea in diabetic rats suggested reduced renal function, since urea is a sensitive marker for kidney injury. Treatment of diabetic rats with rosinidin, on the other hand, resulted in substantial declines in the actions of these enzymes in addition to decreases in urea, indicating that rosinidin has a hepatorenal protective effect. These findings are consistent with prior research that found flavonoids to have hepatorenal protective properties [29]. Rosinidin molecules significantly suppress amylase enzymes, signifying anti-diabetic effects; this is consistent with prior studies on other flavonoids [30–32]. Furthermore, rosinidin protects vital tissues including the pancreas, thereby lowering risks of diabetes-related complications. Histological considerations of the pancreas revealed substantial abnormalities in untreated diabetic pancreatic tissue and demonstrated degeneration at the level of the islets of Langerhans when compared to healthy rats. These conclusions imply that rosinidin's anti-diabetic properties are linked to its ability to improve pancreatic histological properties.

5. Conclusions

Flavonoid-based therapies could be a new target for researchers. The flavonoid rosinidin was found to control hyperglycemia and lipid metabolism, and moreover restored proinflammatory cytokine levels in STZ-induced diabetic animals. As a consequence, rosinidin's anti-diabetic properties appear to be linked to its ability to achieve improvements in pancreatic histology. More research is desired to confirm these findings and to determine other salutary effects of rosinidin, a flavonoid-based medication, in standard anti-diabetic therapy.

Author Contributions: I.K.: Conceptualization; S.J.G. and M.N.B.-J.: Critical revision of manuscript; F.A.A.-A.: Critical revision of manuscript; S.A.: Data interpretation; M.M.G.: Data interpretation; S.S.I.: Project Administration; M.A. and S.I.A.: Data analysis; M.S.N.: Supervision; N.S.: Methodology and Writing original draft. All authors have read and agreed to the published version of the manuscript.

Funding: This research project is supported by Princess Nourah bint Abdulrahman University Researchers Supporting Project number (PNURSP2022R108), Princess Nourah bint Abdulrahman University, Riyadh, Saudi Arabia.

Institutional Review Board Statement: The institute's animal ethics committee authorized the study protocol (TRS/PT/021/006), Trans-Genica Services, approval date: 19 September 2021.

Informed Consent Statement: Not applicable.

Acknowledgments: This research project is supported by Princess Nourah bint Abdulrahman University Researchers Supporting Project number (PNURSP2022R108), Princess Nourah bint Abdulrahman University, Riyadh, Saudi Arabia.

Conflicts of Interest: The authors declare no conflict of interest.

References

1. Kovács, N.; Nagy, A.; Dombrádi, V.; Bíró, K. Inequalities in the Global Burden of Chronic Kidney Disease Due to Type 2 Diabetes Mellitus: An Analysis of Trends from 1990 to 2019. *Int. J. Environ. Res. Public Health* **2021**, *18*, 4723. [CrossRef]
2. Daryabor, G.; Atashzar, M.R.; Kabelitz, D.; Meri, S.; Kalantar, K. The Effects of Type 2Diabetes Mellitus on Organ Metabolism and the ImmuneSystem. *Front. Immunol.* **2020**, *11*, 1582. [CrossRef] [PubMed]
3. Choudhury, H.; Pandey, M.; Hua, C.K.; Mun, C.S.; Jing, J.K.; Kong, L.; Ern, L.Y.; Ashraf, N.A.; Kit, S.W.; Yee, T.S.; et al. An update on natural compounds in the remedy of diabetes mellitus: A systematic review. *J. Tradit. Complement. Med.* **2017**, *8*, 361–376. [CrossRef]
4. Mutha, R.E.; Tatiya, A.U.; Surana, S.J. Flavonoids as natural phenolic compounds and their role in therapeutics: An overview. *Futur. J. Pharm. Sci.* **2021**, *791*, 25. [CrossRef]
5. Lim, S.H.; Yu, J.S.; Lee, H.S.; Choi, C.-I.; Kim, K.H. Antidiabetic Flavonoids from Fruits of Morus alba Promoting Insulin-Stimulated Glucose Uptake via Akt and AMP-Activated Protein Kinase Activation in 3T3-L1 Adipocytes. *Pharmaceutics* **2021**, *13*, 526. [CrossRef]
6. Martín, M.Á.; Ramos, S. Dietary Flavonoids and Insulin Signaling in Diabetes and Obesity. *Cells* **2021**, *10*, 1474. [CrossRef]
7. Lang, X.; Li, N.; Li, L.; Zhang, S. Integrated Metabolome and Transcriptome Analysis Uncovers the Role of Anthocyanin Metabolism in Micheliamaudiae. *Int. J. Genom.* **2019**, *3*, 4393905.
8. Khoo, H.E.; Azlan, A.; Tang, S.T.; Lim, S.M. Anthocyanidins and anthocyanins: Colored pigments as food, pharmaceutical ingredients, and the potential health benefits. *Food Nutr. Res.* **2017**, *61*, 1361779. [CrossRef]
9. Monteiro, A.F.; Viana, J.D.; Nayarisseri, A.; Zondegoumba, E.N.; Mendonça Junior, F.J.; Scotti, M.T.; Scotti, L. Computation-alStudies Applied to Flavonoids against Alzheimer's and Parkinson's Diseases. *Oxid. Med. Cell. Longev.* **2018**, *2018*, 7912765. [CrossRef]
10. Arokoyo, D.S.; Oyeyipo, I.P.; Du Plessis, S.S.; Chegouand, N.N.; Aboua, Y.G. Modulation of Inflammatory Cytokines and Islet Morphology as Therapeutic Mechanisms of Basella alba in Streptozotocin-Induced Diabetic Rats. *Toxicol. Res.* **2018**, *34*, 325–332. [CrossRef]
11. Fathiazad, F.; Hamedeyazdan, S.; Khosropanah, M.K.; Khaki, A. Hypoglycemic activity of Fumaria parviflora in streptozotocin-induced diabetic rats. *Adv. Pharm. Bull.* **2013**, *3*, 207.
12. Ali, A.A.; Essawy, E.A.; Hamed, H.S.; Abdel Moneim, A.E.; Attaby, F.A. The ameliorative role of Physalis pubescens L. against neurological impairment associated with streptozotocin induced diabetes in rats. *Metab. Brain Dis.* **2021**, *36*, 1191–1200. [CrossRef]
13. Soni, L.K.; Dobhal, M.P.; Arya, D.; Bhagour, K.; Parasher, P.; Gupta, R.S. In vitro and in vivo antidiabetic activity of isolated fraction of Prosopis cineraria against streptozotocin-induced experimental diabetes: A mechanistic study. *Biomed. Pharmacother.* **2018**, *108*, 1015–1021. [CrossRef]
14. Al-Ishaq, R.K.; Abotaleb, M.; Kubatka, P.; Kajo, K.; Büsselberg, D. Flavonoids and Their Anti Diabetic Effects: Cellular Mechanisms and Effects to Improve Blood Sugar Levels. *Biomolecules* **2019**, *9*, 430. [CrossRef]
15. Elshemy, M.M.; Asem, M.; Allemailem, K.S.; Uto, K.; Ebara, M.; Nabil, A. Antioxidative Capacity of Liver- and Adipose-Derived Mesenchymal Stem Cell-Conditioned Media and Their Applicability in Treatment of Type 2 Diabetic Rats. *Oxid. Med. Cell. Longev.* **2021**, *2021*, 8833467. [CrossRef]
16. Rezaei, S.; Ashkar, F.; Koohpeyma, F.; Mahmoodi, M.; Gholamalizadeh, M.; Mazloom, Z.; Doaei, S. Hydroalcoholic extract of Achilleamillefolium improved blood glucose, liver enzymes and lipid profile compared to metformin in streptozotocin-induced diabetic rats. *Lipids Health Dis.* **2020**, *19*, 81. [CrossRef]
17. Hussain, T.; Tan, B.; Murtaza, G.; Liu, G.; Rahu, N.; Saleem-Kalhoro, M.; Hussain-Kalhoro, D.; Adebowale, T.O.; Usman-Mazhar, M.; Rehman, Z.U.; et al. Flavonoids and type 2 diabetes: Evidence of efficacy in clinical and animal studies and delivery strategies to enhance their therapeutic efficacy. *Pharmacol. Res.* **2020**, *152*, 104629. [CrossRef]
18. Bai, L.; Li, X.; He, L.; Zheng, Y.; Lu, H.; Li, J.; Zhong, L.; Tong, R.; Jiang, Z.; Shi, J.; et al. Antidiabetic Potential of Flavonoids from Traditional Chinese Medicine: A Review. *Am. J. Chin. Med.* **2019**, *47*, 933–957. [CrossRef]
19. Othman, M.S.; Khaled, A.M.; Al-Bagawi, A.H.; Fareid, M.A.; Ghany, R.A.; Habotta, O.A.; Moneim AE, A. Hepatorenal protective efficacy of flavonoids from Ocimumbasilicum extract in diabetic albino rats: A focus on hypoglycemic, antioxidant, anti-inflammatory and anti-apoptotic activities. *Biomed. Pharmacother.* **2021**, *144*, 112287. [CrossRef]
20. Li, C.; Gan, H.; Tan, X.; Hu, Z.; Deng, B.; Sullivan, M.A.; Gilbert, R.G. Effects of active ingredients from traditional Chinese medicines on glycogen molecular structure in diabetic mice. *Eur. Polym. J.* **2019**, *112*, 67–72. [CrossRef]
21. Copeland, L.A.; Swendsen, C.S.; Sears, D.M.; MacCarthy, A.A.; McNeal, C.J. Association between triglyceride levels and cardiovascular disease in patients with acute pancreatitis. *PLoS ONE* **2018**, *13*, e0179998. [CrossRef] [PubMed]
22. El-Ouady, F.; Bachir, F.; Eddouks, M. Flavonoids Extracted from Asteris cusgraveolens improve glucose metabolism and lipid profile in diabetic Rats. *Endocr. Metab. Immune Disord. Drug Targets* **2021**, *21*, 895–904. [CrossRef]
23. Zhang, H.; Xu, Z.; Zhao, H.; Wang, X.; Pang, J.; Li, Q.; Yang, Y.; Ling, W. Anthocyanin supplementation improves anti-oxidative and anti-inflammatory capacity in a dose-response manner in subjects with dyslipidemia. *Redox Biol.* **2020**, *32*, 101474. [CrossRef] [PubMed]
24. Kassab, R.B.; Lokman, M.S.; Daabo, H.M.; Gaber, D.A.; Habotta, O.A.; Hafez, M.M.; Zhery, A.S.; Moneim, A.E.A.; Fouda, M.S. Ferulic acid influences Nrf2 activation to restore testicular tissue from cadmium-induced oxidative challenge, inflammation, and apoptosis in rats. *J. Food Biochem.* **2020**, *44*, e13505. [CrossRef]

25. Zou, J.; Sui, D.; Fu, W.; Li, Y.; Yu, P.; Yu, X.; Xu, H. Total flavonoids extracted from the leaves of *Murraya paniculata* (L.) Jack alleviate oxidative stress, inflammation and apoptosis in a rat model of diabetic cardiomyopathy. *J. Funct. Foods* **2021**, *76*, 104319. [CrossRef]

26. Ajayi, A.M.; Umukoro, S.; Ben-Azu, B.; Adzu, B.; Ademowo, O.G. Toxicity and protective effect of phenolic-enriched ethylacetate fraction of *Ocimum gratissimum* (Linn.) leaf against acute inflammation and oxidative stress in rats. *Drug Dev. Res.* **2017**, *78*, 135–145. [CrossRef] [PubMed]

27. Liu, X.L.; Liu, W.P.; Wang, L.L.; Feng, L. Effects of flavonoids from Pyrrosiae folium on pathological changesand inflammatory response of diabetic nephropathy. *China J. Chin. Mater. Med.* **2018**, *43*, 2352–2357.

28. Nazir, N.; Zahoor, M.; Ullah, R.; Ezzeldin, E.; Mostafa, G.A.E. Curative Effect of Catechin Isolated from Elaeagnus Umbellata Thunb. Berries for Diabetes and Related complications in Streptozotocin-Induced Diabetic Rats Model. *Molecules* **2020**, *26*, 137. [CrossRef]

29. Zaky, A.S.; Kandeil, M.; Abdel-Gabbar, M.; Fahmy, E.M.; Almehmadi, M.M.; Ali, T.M.; Ahmed, O.M. The Antidiabetic Effects and Modes of Action of the Balanites aegyptiaca Fruit and Seed Aqueous Extracts in NA/STZ-Induced Diabetic Rats. *Pharmaceutics* **2022**, *14*, 263. [CrossRef]

30. Arigela, C.S.; Nelli, G.; Gan, S.H.; Sirajudeen, K.N.S.; Krishnan, K.; Abdul Rahman, N.; Pasupuleti, V.R. Bitter Gourd Honey Ameliorates Hepatic and Renal Diabetic Complications on Type 2 Diabetes Rat Models by Antioxidant, Anti-Inflammatory, and Anti-Apoptotic Mechanisms. *Foods* **2021**, *10*, 2872. [CrossRef]

31. Ogunlana, O.O.; Adetuyi, B.O.; Esalomi, E.F.; Rotimi, M.I.; Popoola, J.O.; Ogunlana, O.E.; Adetuyi, O.A. Antidiabetic and Antioxidant Activities of the Twigs of Andrograhis paniculata on Streptozotocin-Induced Diabetic Male Rats. *BioChem* **2021**, *1*, 238–249. [CrossRef]

32. Esquivel-Gutiérrez, E.R.; Manzo-Avalos, S.; Peña-Montes, D.J.; Saavedra-Molina, A.; Morreeuw, Z.P.; Reyes, A.G. Hypolipidemic and Antioxidant Effects of Guishe Extract from *Agave lechuguilla*, a Mexican Plant with Biotechnological Potential, on Streptozotocin-Induced Diabetic Male Rats. *Plants* **2021**, *10*, 2492. [CrossRef]

pharmaceutics

Article

Resveratrol Inhibited ADAM10 Mediated CXCL16-Cleavage and T-Cells Recruitment to Pancreatic β-Cells in Type 1 Diabetes Mellitus in Mice

Mohamed S. Abdel-Bakky [1,2], Abdulmajeed Alqasoumi [3], Waleed M. Altowayan [3], Elham Amin [4,5] and Mostafa A. Darwish [6,*]

[1] Department of Pharmacology and Toxicology, College of Pharmacy, Qassim University, Buraydah 52471, Saudi Arabia; m.abdelbakky@qu.edu.sa
[2] Department of Pharmacology and Toxicology, Faculty of Pharmacy, Al-Azhar University, Cairo 11884, Egypt
[3] Department of Pharmacy Practice, College of Pharmacy, Qassim University, Buraydah 52471, Saudi Arabia; a.alqasomi@qu.edu.sa (A.A.); w.altowayan@qu.edu.sa (W.M.A.)
[4] Department of Pharmacognosy, Faculty of Pharmacy, Beni-Suef University, Beni Suef 62514, Egypt; el.saleh@qu.edu.sa
[5] Department of Medicinal Chemistry and Pharmacognosy, College of Pharmacy, Qassim University, Buraydah 52471, Saudi Arabia
[6] Department of Pharmacology and Toxicology, Faculty of Pharmacy, Nahda University, Beni Suef 11787, Egypt
* Correspondence: mostafa.darwish@nub.edu.eg or mostafaassemmostafa@yahoo.com; Tel.: +20-1011672367

Citation: Abdel-Bakky, M.S.; Alqasoumi, A.; Altowayan, W.M.; Amin, E.; Darwish, M.A. Resveratrol Inhibited ADAM10 Mediated CXCL16-Cleavage and T-Cells Recruitment to Pancreatic β-Cells in Type 1 Diabetes Mellitus in Mice. *Pharmaceutics* 2022, 14, 594. https://doi.org/10.3390/pharmaceutics14030594

Academic Editors: Diana Marcela Aragon Novoa and Fátima Regina Mena Barreto Silva

Received: 17 January 2022
Accepted: 23 February 2022
Published: 9 March 2022

Publisher's Note: MDPI stays neutral with regard to jurisdictional claims in published maps and institutional affiliations.

Abstract: Background: CXCL16 attracts T-cells to the site of inflammation after cleaving by A Disintegrin and Metalloproteinase (ADAM10). **Aim:** The current study explored the role of ADAM10/CXCL16/T-cell/NF-κB in the initiation of type 1 diabetes (T1D) with special reference to the potential protecting role of resveratrol (RES). **Methods:** Four sets of Balb/c mice were created: a diabetes mellitus (DM) group (streptozotocin (STZ) 55 mg/kg, i.p.], a control group administered buffer, a RES group [RES, 50 mg/kg, i.p.), and a DM + RES group (RES (50 mg/kg, i.p.) and STZ (55 mg/kg, i.p.) administered daily for 12 days commencing from the fourth day of STZ injection). Histopathological changes, fasting blood insulin (FBI), glucose (FBG), serum and pancreatic ADAM10, CXCL16, NF-κB, T-cells pancreatic expression, inflammatory, and apoptotic markers were analyzed. **Results:** FBG, inflammatory and apoptotic markers, serum TNF-α, cellular CXCL16 and ADAM10 protein expression, pancreatic T-cell migration and NF-κB were significantly increased in diabetic mice compared to normal mice. RES significantly improved the biochemical and inflammatory parameters distorted in STZ-treated mice. **Conclusions:** ADAM10 promotes the cleaved form of CXCL16 driving T-cells into the islets of the pancreatic in T1D. RES successfully prevented the deleterious effect caused by STZ. ADAM10 and CXCL16 may serve as novel therapeutic targets for T1D.

Keywords: CXCL16; ADAM10; NF-κβ; apoptosis; pancreatic islets; resveratrol; T1D

1. Introduction

There are over 37 members of the group of A Disintegrin and Metallopeptidases (ADAMs); among them, ADAM10 and ADAM17 are functionally and structurally similar [1,2]. ADAM10 has an important role in the signaling of Notch and shedding amyloid precursor protein (APP) accompanying Alzheimer's disease pathophysiology. ADAM10 has a central role in cell migration, cell adhesion, and the regulation of immunity responses, in addition to controlling apoptotic process [3,4]. C-X-C motif chemokine ligand 16 (CXCL16) is an exceptional chemokine existing in two different forms: the soluble CXCL16 binds to immune cells that express its receptor, CXCR6, to drive them into the site of inflammation [5], and the cell membrane form that acts as a receptor then internalizes oxidized low-density lipoprotein (ox-LDL) [6]. ADAM10 and/or ADAM17 can cleave the membranous CXCL16 to its soluble form [7].

Our earlier work has revealed that both proteins (CXCL16 and ADAM10) are expressed in the renal glomerular podocytes, principle cells of the distal tubules, collecting duct, and early part of Henle's loop [8]. In addition, recent work has reported that CXCL16 may have an essential role in the promotion of ovarian cancer [9]. Also, serum levels of CXCL16 were increased in the case of gout, in addition to chronic kidney diseases (CKD), and is accompanied by deterioration of renal function [10,11]. Notably, CXCL16 has an important role in NK and T-cell recruitment [12]. Moreover, it is enhanced by TNF-alpha and IFN-gamma inflammatory cytokines [13].

Tawfik et al., (2021) has shown in a recent study that the CXCL16 serum level was elevated significantly in patients with Type 2 DM, relative to that of normal individuals [14]. Similarly, CXCL16 serum levels in T2DM patients, with or without coronary artery disease, were increased if compared with healthy individuals [15]. Likewise, serum CXCL16 level was found to increase in gestational diabetes mellitus [16]. Interestingly, TNF-α induced apoptosis is enhanced by serum CXCL16 in the DLBCL cell-type; this may comprise serum CXCL16, ADAM10, and the TNF-α mediated NF-κB pathway. However, few records concerning the role of ADAM10 and CXCL16 in the progression of T1D were found. Indeed, the role of ADAM10 in diabetes is still questionable. Abdel-Bakky et al., 2022 found that simvastatin treatment reduced CXCL16 expression in the islets of the pancreas in STZ-treated mice [17]. An earlier report demonstrated that treating 5XFAD mice with STZ did not change ADAM10 levels [18]. Furthermore, an additional report demonstrated significant elevation in ADAM10 protein expression in diabetic minipigs after vascular injury [19]. This work aimed to investigate the role of ADAM10 and CXCL16 in STZ-induced T1D in mice.

Resveratrol (RES) is a phytoalexin polyphenolic compound which exists in berries, grapes, and peanuts. It has anti-inflammatory, anti-oxidant, cardioprotective, antidiabetic, anticancer, and neuroprotective activities [20–22]. It may protect against the consequences of diabetes, including nephropathy or cataracts [23,24]. In addition, RES reduces pancreatic inflammatory factors and glucose levels in the serum of cardiovascular-complicated diabetic models of rats [25]. Also, RES showed antidiabetic effects experimentally and clinically through several mechanistic pathways and targets [26]. The aim of the current study is to explore the possible antidiabetic mechanism influencing RES against T1D induced by STZ. Moreover, the interdependence between ADAM10 and CXCL16 in STZ-induced T1D will also be explored.

2. Materials and Methods

2.1. Work Design

The experimental design and work involved were in agreement with the Institutional Animal Care and Use Committee of the University of Nahda in Egypt (NUB-002-021). BALB/c male albino mice (weighing 23–29 g; supplied by the Nahda animal facility) were given standard lab chow (El-Nasr Company, Cairo, Egypt) and water ad libitum. Mice were accommodated at 22 °C $\pm$ 2 °C with 50% $\pm$ 10 humidification percentage with a 12 h day-night cycle. Animals were randomly classified into 4 groups (n = 10). Normal mice received i.p. sodium citrate buffer. RES animals were administered RES alone (50 mg/kg, i.p.) for 12 days commencing from day 4 until day 15. Diabetic mice (DM) received STZ (55 mg/kg, i.p.), dissolved in a sodium citrate buffer (pH 4.5, 50 mmol/L) once daily for 5 consecutive days [27]. DM + RES group animals received i.p., 55 mg/kg STZ, once daily for 5 consecutive days and RES (50 mg/kg, i.p. dissolved in 25% dimethylsulfoxide, DMSO) for 12 days beginning on day 4 after injecting STZ until the day 15 of the experiment. In the DM + RES group, RES was administered after STZ treatment by 30 min on days 4 and 5 of the experiment. Doses used of RES and STZ injection were guided by our previous work [28,29].

2.2. Tissue Samples and Serum Collection

Diethyl ether was used as anesthesia, and mice were euthanized by cervical dislocation on day 15 of the experiment. Collected pancreatic tissues were fixed in Davidson buffer to perform immunofluorescence analysis. Blood samples were collected by small capillary tubes from the retro-orbital plexus before sacrifice. The preparation of the serum was performed by centrifugation of the collected blood at 4000 rpm for 15 min, and the parameters were analyzed using the ELISA technique and blood analyzer.

2.3. Pancreatic Islet Isolation

The pancreatic islets were isolated in a separate experiment as previously described [30]. In brief, isolated pancreas was cut into tiny slices and collected in cold collagenase (2 mL, 1.0 mg/mL) in Hank's balanced salt solution (HBSS; Invitrogen, Waltham, CA, USA) for digestion. Samples were digested for 15 min at 37 °C in an incubator. Then, ice-cold HBSS was mixed into the digested samples, and the islets were separated by Ficoll gradient (Ficoll–Paque Plus; GE Healthcare, Chicago, IL, USA). After a gentle vortex of the suspended mixture using high speed for 10 s, the islets were separated and picked up, aided by an inverted Leica DMIL LED microscope. The separated pancreatic islets were then subjected directly to Western blot analysis.

2.4. Estimation of Blood Glucose, Insulin, WBC Count, and Body Weight Change

Blood glucose, white blood cell (WBC) count, and serum insulin were measured in blood samples collected from all groups at the end of the experiment. Direct measurement of blood glucose levels was determined in all groups using a glucometer (Uright, New Taipei City, Taiwan). An ELISA kit (Biovision Inc., Milpitas, CA, USA) was used to determine serum insulin. Furthermore, WBCs were counted using an ABX Micros 60 Analyzer auto blood analyzer (ABX Micros 60, Montpellier, France). The following equation was used to determine changes in body weight: Δ body weight = (weight of animals on day 15 − weight of animals on day 0).

2.5. Serum NO Estimation

Estimation of serum NO was accomplished according to the previous method [31]. In brief, 0.5 mL of cold absolute ethanol was mixed with 250 µL serum samples, and the mixture was lifted for 2 days at +4 °C in order to achieve full precipitation of the protein. Samples were then centrifuged at 15,000 rpm for 60 min, 250 µL of the obtained supernatant was mixed with 250 µL of VCl3 and 125 µL of sulfanilamide, and 125 µL of *N*-(1-Naphtyl) ethylenediamine dihydrochloride (NEDD) was rapidly added. Samples were incubated at 37 °C for 30 min and the resulting pink chromophore was then analyzed using a double-beam spectrophotometer (UV-150-02, Shimadzu, Toyko, Japan) at 540 nm.

2.6. Immunofluorescence Analysis

Paraffin-fixed tissue samples were deparaffinized by dipping the slides in xylene and then in series of ethyl alcohol concentrations for rehydration. After washing the sections in 0.05% Tween 20 in 10 mM phosphate-buffered saline (pH 7.4), antigen retrieval was performed by dipping the slides in sodium citrate buffer (0.01 M, pH 6.0) and boiling them for 15 min in a microwave oven (500 W). To fix or permeabilize tissues, sections were covered for 20 min by drops of 100% methyl alcohol. The blocking step was performed by incubation of the sections with a blocking buffer (BPS containing 1% bovine serum albumin in 10% horse serum) for 60 min. Slides were then washed and incubated with the primary Abs (polyclonal goat anti-mouse CXCL16 or mouse monoclonal ADAM10) overnight in a refrigerator. After washing, the sections were incubated with secondary Abs (rabbit anti-goat Alexa 488 or goat anti-mouse Cy3) for 30 min. Nuclei were counterstained by the incubation of the slides in 4′,6′-diamidino-2-phenylindole (DAPI) for 1 min. Tissue sections were then washed for 30 min and mounted using fluoromount G and analyzed using fluorescence microscopy (Leica DM5000 B, Wetzlar, Germany) [32]. Quantification of the florescence intensities was analyzed by measuring 4 to 6 fields to obtain the mean intensity for each tissue section by ImageJ/NIH software 1.51 (MD, USA).

2.7. Immunohistochemistry of Pancreatic Tissue Sections

Tissue samples fixed in paraffin were deparaffinized by dipping the slides in xylene and in a series of ethyl alcohol concentrations for rehydration. After washing sections in buffer (10 mM phosphate-buffered saline containing 0.05% tween 20, pH 7.4), slides were boiled for 20 min in an oven (500 W) in 0.01 M sodium citrate buffer (pH 6.0) for the antigen retrieval step. Blocking endogenous peroxidase was performed by incubating the sections with 3% H_2O_2/methanol. The sections were blocked by BPS (containing 1% bovine serum albumin in 10% horse serum) for 60 min and kept with avidin/biotin-blocking kits for 15 min. After incubating the slides with the primary antibody (mouse monoclonal CD3, 1:200 diluted in blocking buffer overnight at 4 °C, Universal Quick kits were used, followed by an AEC Substrate kit, to develop the red color. Nuclei were counterstained with hematoxylin and finally mounted with mounting solution. The pictures were captured by light microscope (Leica DM5000 B, Wetzlar, Germany)

2.8. Western Blot Analyses

Western blot analysis was performed in accordance to the method previously described [33]. Ready Prep™ buffer for the extraction of the protein (Bio-Rad Inc., catalog #163-2086, Hercules, CA, USA) was used to determine protein content by Bradford protein assay kit in the pancreatic isolated islet tissue lysates of all the groups. Subsequently, equal concentrations of protein from all samples were mixed with loading buffer (Laemmli) and separated by 10% sodium dodecyl sulfate (SDS)-polyacrylamide gels. Separated protein contents were blotted to nitrocellulose membrane (Millipore, Burlington, MA, USA). The membrane was then blocked using 5% skim milk for 1 hr and incubated with the primary Abs against β-actin, NF-κB, and ADAM10, and then incubated with the secondary Ab (HRP-conjugated goat IgG). Visualization of the proteins was performed using the enhanced chemiluminescent kit. For equal loading, the membrane was incubated with β-actin Ab. Densitometry was performed to semi-quantify the protein bands of β-actin and presented as a bar graph.

2.9. Kits, Chemicals, and Antibodies

A mouse insulin ELISA kit was procured from Biovision (Milpitas, CA, USA). Mouse ADAM10, CD3, insulin, cleaved caspases-3, and NF-κB were purchased from Santa Cruz Biotechnology (Dallas, TX, USA). Goat anti-mouse Cy3 and rabbit anti-goat Alexafluor 488 secondary Abs were obtained from Invitrogen (Waltham, CA, USA). The mouse CXCL16 ELISA kit and TUNEL assay kit—HRP-DAB were obtained from Abcam (Cambridge, UK). Goat anti-rabbit polyclonal CXCL16 Ab was obtained from Peprotech (London, UK). STZ, monoclonal mouse against β-actin Ab, RES, and collagenase were purchased from Sigma-Aldrich (St. Louis, MO, USA). RES was purchased from Colchester (Colchester, UK). The Bradford kit for protein assay was purchased from Bio Basic Inc. (Burlington, ON, Canada). Goat IgG secondary antibody-conjugated horseradish peroxidase (HRP) was obtained from Novus Biologicals (Abingdon, UK). ECL substrate (Clarity™) for Western was purchased from Bio-Rad (Hercules, CA, USA).

2.10. Statistical Analysis

Data were presented as mean ± SEM and statistically evaluated using one-way ANOVA for multiple group comparisons and Tukey–Kramer as a post-ANOVA test using GraphPad Prism, version 6 (San Diego, CA, USA). The difference between groups was deemed statistically significant at $p < 0.05$.

3. Results

3.1. Effect of RES on Fasting Blood Glucose, Serum Insulin and Body Weight

STZ-injected mice had significantly lower body weights and serum insulin levels (Figure 1A,B, respectively) compared to the control group. However, STZ-injected mice demonstrated significant elevation in the level of fasting blood glucose (Figure 1C) compared to normal mice. On the other hand, RES-injected animals exhibited non-significant differences in serum insulin, body weight, and fasting blood glucose when compared to the control group. Remarkably, significant improvement in the above-mentioned parameters was found in treatment mice with RES in STZ-treated mice (Figure 1A–C).

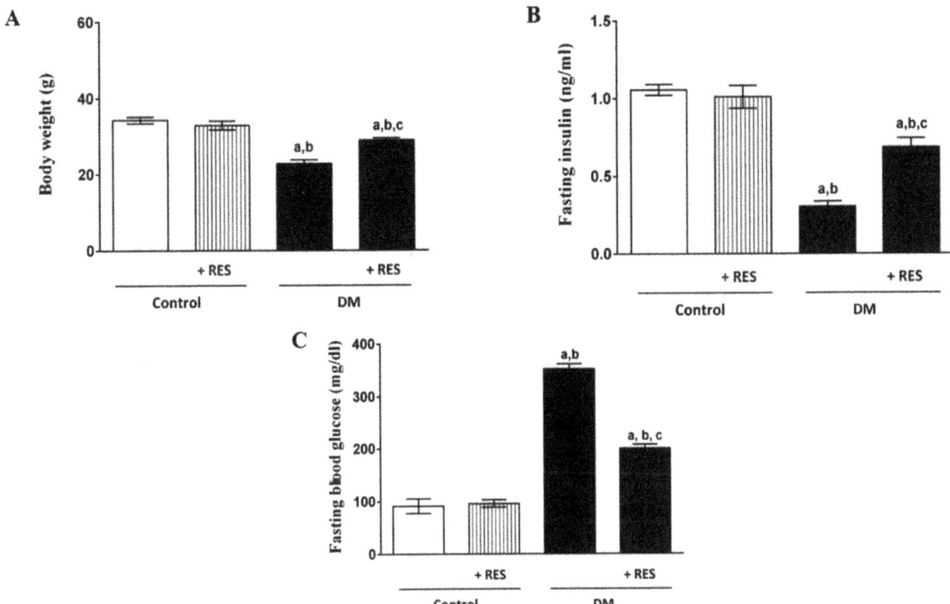

Figure 1. STZ-injected mice showed significant decreases in body weight (**A**) and fasting serum insulin (**B**) compared to control. STZ-injected mice revealed significant elevation in fasting blood glucose level (**C**) when compared to control mice. RES-injected animals showed no difference in fasting blood glucose, serum insulin, or body weight compared to control group. Remarkably, significant improvement was seen in levels of fasting blood glucose, body weight, and fasting serum insulin in RES-treated animals in the presence of STZ. Data represent mean ± SEM: [a] difference is significant compared to normal group; [b] difference is significant compared to RES group; [c] difference is significant compared to diabetic group at $p < 0.05$.

3.2. Effect of RES with or without STZ on WBCs, NO or TNF-α

Increased (by approximately 3-fold) levels of WBCs, NO, and TNF-α were observed in mice injected with STZ compared with normal controls. Mice injected with RES displayed no significant changes in the level of WBCs, TNF-α, or NO when compared to the control group. RES treatment significantly reduced WBC count, serum levels of NO, and TNF-α when compared to diabetic mice (Figure 2A–C).

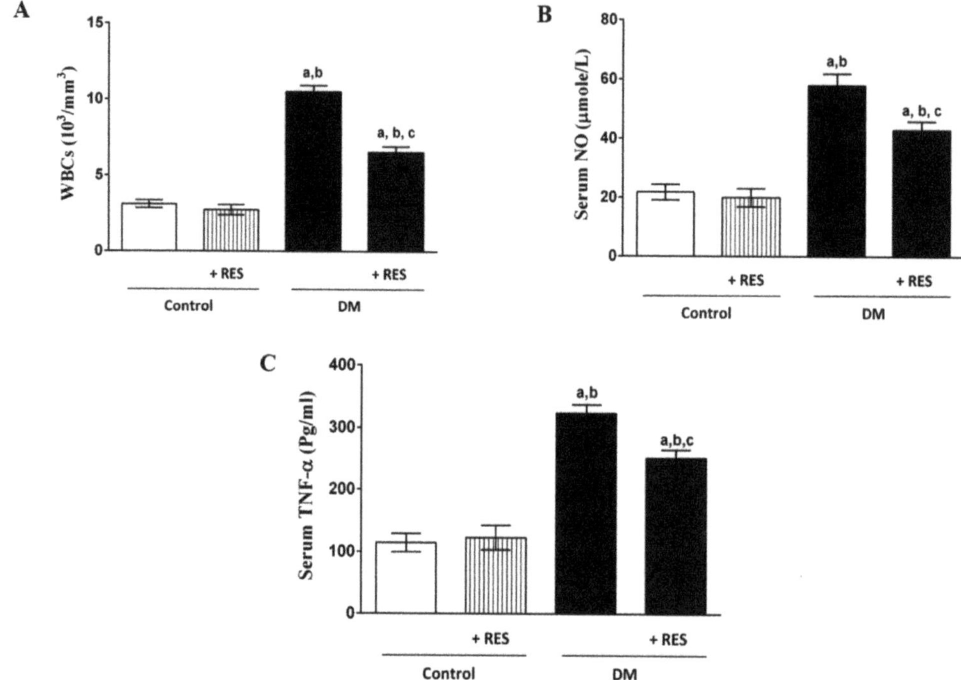

Figure 2. Levels of WBCs (**A**), NO (**B**), and TNF-α (**C**) are significantly increased (approximately 3-fold) in mice injected with STZ compared with normal or RES treatment mice. STZ-treated group injected with RES display significant decreases in WBC count, serum level of NO, and serum level of TNF-α when compared to STZ-treated mice. No significant change in WBC count, serum level of NO, and serum level of TNF-α was seen in RES group compared to control group. Data represent mean ± SEM: [a] difference is significant compared to control group; [b] difference is significant compared to RES group; [c] difference is significant compared to diabetic group at $p < 0.05$.

3.3. Effect of RES on STZ-Induced CXCL16 Pancreatic Expression

Using double immunofluorescence staining for insulin, pancreatic β cell marker, and CXCL16, we observed basal expression of CXCL16 in normal control mouse tissues. Treating the mice with RES showed a non-significant difference in the expression of CXCL16 protein in comparison with control mice. Increased pancreatic β cell expression of CXCL16 was seen confirmed by CXCL16 co-localization with insulin (β cell marker). Significant reduction in CXCL16 expression was found in mice pancreatic β cells of diabetic mice treated with RES (Figure 3A). Fluorescence intensity of CXCL16 protein expression for all groups was quantified and blotted, as shown in Figure 3B. Similar findings were found with serum CXCL16 as measured by ELISA (Figure 3C).

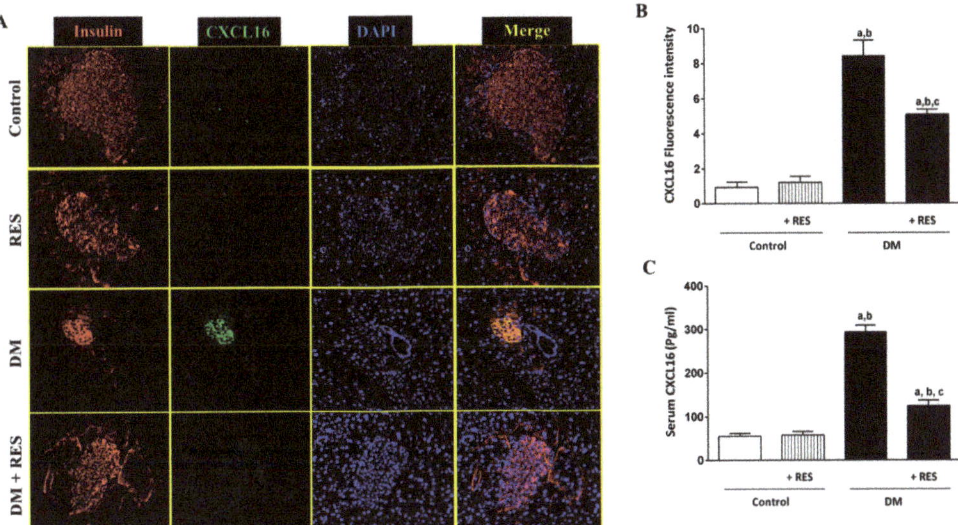

Figure 3. (**A**) Double immunofluorescence staining for insulin (red), a pancreatic β cell marker, and CXCL16 (green) was performed; basal expression of pancreatic β cell CXCL16 in normal control and RES-treated mouse tissues. Increased CXCL16 expression in pancreatic β cells was confirmed by the co-localization of CXCL16 with insulin. Marked reduction of CXCL16 expression in mouse pancreatic β cells of RES-treated mice with STZ was seen; (**B**) Fluorescence intensity of CXCL16 protein expression for all groups was quantified and blotted; (**C**) Serum levels of CXCL16 using ELISA in all treated groups show similar findings of immunofluorescence. Data represent mean ± SEM: [a] difference is significant compared to control group; [b] difference is significant compared to RES group; [c] difference is significant compared to diabetic group at $p < 0.05$.

3.4. Effect of RES on STZ-Induced Pancreatic ADAM10 Expression

Consistent with CXCL16 expression in β cell islets, ADAM10 upregulation in the pancreatic islets was found in animals treated with STZ in comparison with control animals. However, RES decreased pancreatic β cell expression of ADAM10 increased by STZ-injection (Figure 4A). Fluorescence intensity analysis of ADAM10 protein expression from immunofluorescence was quantified and blotted, as shown in Figure 4B. For more confirmation, expression of ADAM10 protein was analyzed by Western blot in all treated animals. As shown in Figure 4C, no difference was found between control and RES-treated mouse tissues. On the other hand, STZ-treated mice demonstrated a marked elevation in ADAM10 protein expression as compared to control mice. Reduced ADAM10 protein expression was detected in the [RES + STZ] group in comparison with diabetic mice. Densiometry for Western blotting bands for ADAM10 was performed using NIH ImageJ software (Figure 4D)

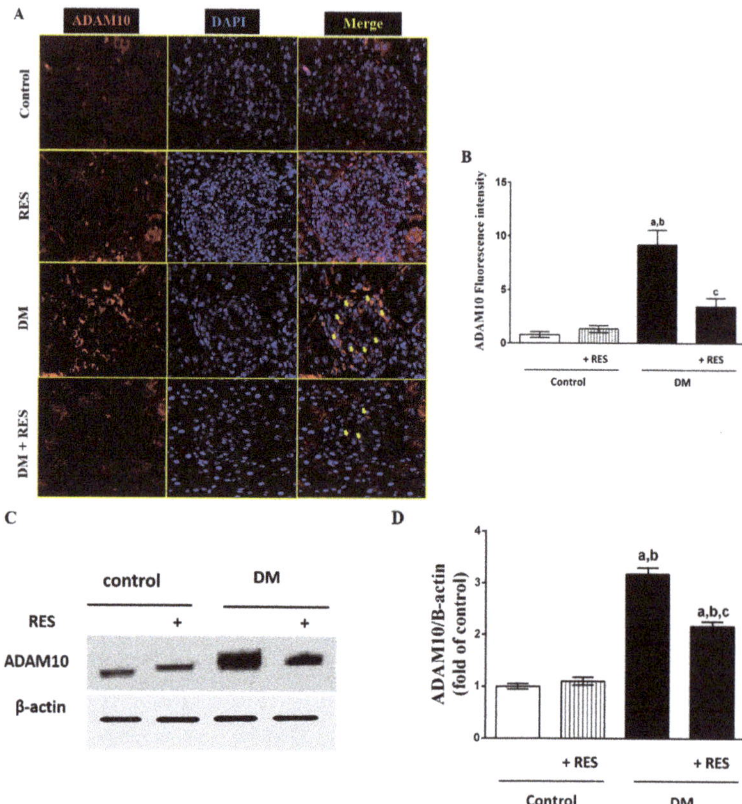

Figure 4. (**A**) Increased expression of ADAM10 protein in the islets of pancreases of STZ-injected mice compared to control group. Decreased expression of ADAM10 in pancreatic β cells after treatment with RES increased by STZ-injection; (**B**) Fluorescence intensity analysis of ADAM10 protein expression from immunofluorescence was quantified and blotted in all experimental groups; (**C**) Expression of ADAM10 protein was analyzed using Western blot in all treated animals. No difference was seen between control and RES-treated groups. STZ-injected animals showed marked increases in ADAM10 protein expression in comparison with the normal group. Animals injected with RES and STZ showed marked reduction in the expression of ADAM10 in islets of pancreas compared to STZ-injected group; (**D**) Densiometry for Western blot bands for ADAM10 in all groups was performed using NIH ImageJ software (MD, USA). Data represent mean ± SEM: [a] difference is significant compared to control group; [b] difference is significant compared to RES group; [c] difference is significant compared to diabetic group at $p < 0.05$.

3.5. Expression of NF-κB Protein in the Islets of Pancreas in RES-Injected Mice with or without STZ

Increased NF-κB expression in the islets of the pancreas was found in STZ-injected animals in comparison with normal mice, as shown by Western blot analyses. Treatment with RES in the STZ group revealed a marked reduction in the expression of NF-κB protein in comparison with the STZ-injected mice. No marked changes in NF-κB protein expression were found in the RES-injected group in comparison with normal animals (Figure 5A,B).

3.6. Effect of RES on STZ-Induced Apoptosis

Pancreatic sections of the control group showed negative reactivity in the nucleus for TUNEL in the cells of the pancreatic islets while in the RES group; strong reactivity in the nucleus in a few cells/ islets for TUNEL was seen. On the other hand, the islets of the pancreas of STZ-injected mice displayed a strong reactivity in the nucleus for TUNEL that was higher in number of cells in comparison with the normal islets. Animals injected with RES with STZ displayed mild reactivity in the nucleus for TUNEL in few cells in comparison with the STZ-treated mice (Figure 6A,B).

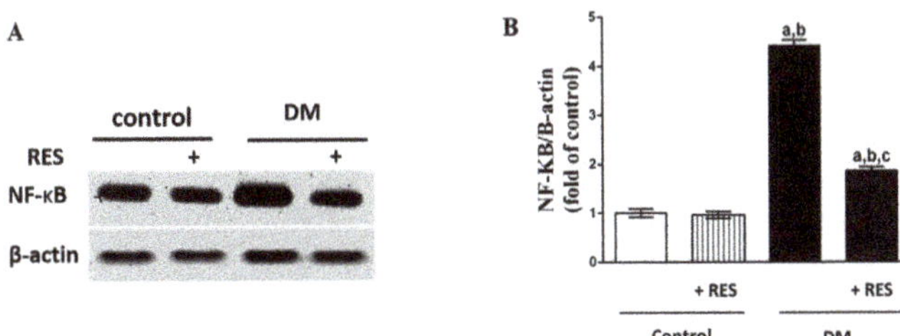

Figure 5. (**A,B**) Expression of NF-κB in β cells was significantly increased in STZ-injected animals in comparison with normal mice. RES treatment significantly reduced the expression level of NF-κB of STZ-treated group in comparison with the STZ-injected group. No change was found in NF-κB expression in RES-injected mice in comparison with the normal group. Data represent mean ± SEM: [a] difference is significant compared to control group; [b] difference is significant compared to RES group; [c] difference is significant compared to diabetic group at $p < 0.05$.

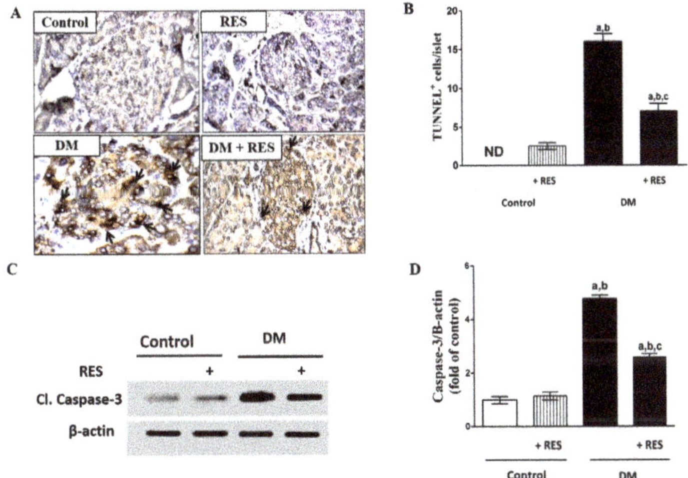

Figure 6. (**A,B**) Pancreas of normal mice demonstrates negative TUNEL reactivity in the nucleus in islet cells. In RES group, strong reactivity in the nucleus in few cells/ islets for TUNEL was seen. On the other hand, islets of pancreas of STZ-injected mice revealed strong reactivity in the nucleus for TUNEL that was higher in number of cells in comparison with the normal islets. RES and STZ-injected group displayed mild reactivity in the nucleus for TUNEL in limited cell numbers when compared with mice treated with STZ; (**C,D**) Expression of pro-apoptotic protein, cleaved caspase-3, increased significantly in STZ treatment in pancreatic islets of mice using Western blot analysis compared to normal mice. Animals injected with RES with STZ displayed mild reactivity in the nucleus for TUNEL in few cells in comparison with the STZ-treated mice. No marked change in the expression of caspase-3 protein in the RES treatment group was noted compared to the normal group. Data represent mean ± SEM: [a] difference is significant compared to control group; [b] difference is significant compared to RES group; [c] difference is significant compared to diabetic group at $p < 0.05$.

For more confirmation of the apoptotic pathway, expression of caspase-3 was also tested in isolated islets. The expression of pro-apoptotic protein, cleaved caspase-3, increased significantly in STZ treatment in the pancreatic islets of mice using Western blot analysis compared with normal mice, while mice injected with RES together with STZ displayed a reduction in the level of expression of cleaved caspase-3 in the pancreatic islets as compared to diabetic mice. RES-treated mice showed a similar expression of caspase-3 to that of the control group (Figure 6C,D).

3.7. Effect of RES with or without STZ on Splenic and Pancreatic T-Cell Protein Expression

The spleens of diabetic mice showed marked increases in CD3 positive cells in the peri-arteriolar area in some follicles and high reactivity in the red bulb area in comparison with the average positive cells for CD3 in the peri-arteriolar area (T-cell area) with mild reactivity in the red bulb in the normal spleen of the control or RES-treated groups. However, the STZ-injected group in the presence of RES displayed spleens with average CD3 positive cells in the peri-arteriolar area and mild reactivity in the red bulb area (Figure 7A).

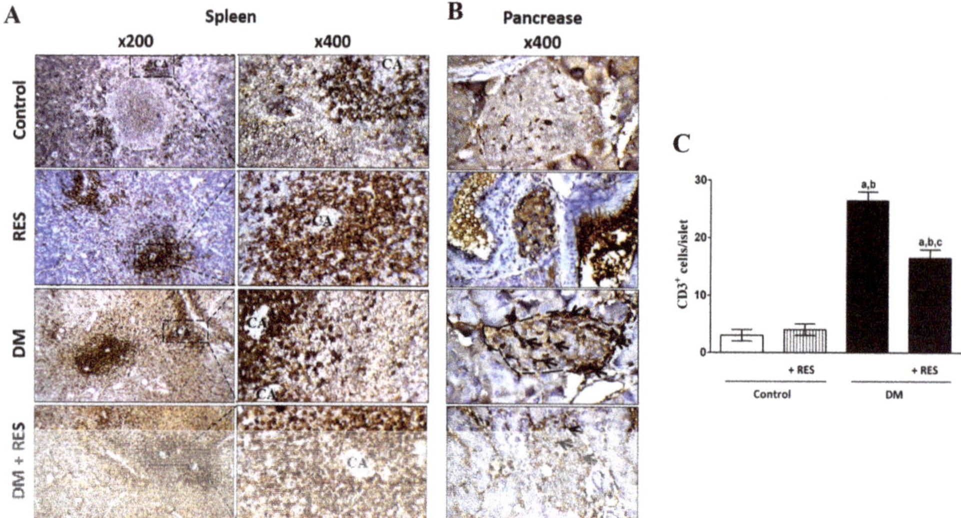

Figure 7. (**A**) Spleen of STZ-treated mice shows marked increase in CD3 positive cells in peri-arteriolar area in certain follicles and high reactivity in red bulb area in comparison with the average positive cells for CD3 in peri-arteriolar area (T-cell area) with mild reactivity in red bulb in normal spleen of the control or RES groups. However, STZ-injected group in the presence of RES displayed spleen with average positive cells for CD3 in peri-arteriolar area and mild reactivity in red bulb area (400× and 200×) compared to STZ-treated group; (**B,C**) Normal and RES-injected animals show normal number of positive cells for CD3/ islet. Pancreatic islets from STZ-injected mice display marked increase in CD3 positive cells/islets in comparison with normal animals. Diabetic group injected with RES reveals marked decrease in positive cells for CD3/islets in comparison with STZ-injected animals, as shown by immunohistochemistry and the average count of T-cells/islets for all treated groups (400×). Data represent mean ± SEM; [a] difference is significant compared to control group; [b] difference is significant compared to RES group; [c] difference is significant compared to diabetic group at $p < 0.05$.

In addition, the pancreases of the control animals revealed normal positive numbers of CD3 cells/islet. Moreover, no significant difference in CD3 positive cells numbers in the islets of pancreas of RES treated animals in comparison with an increase in normal islets. In contrast, islets of pancreas of STZ-injected mice displayed marked in CD3 positive cells/islets in comparison with normal animals. Diabetic mice injected with RES showed marked decrease in the positive CD3 cells/islets in comparison with STZ-treated mice as shown by CD3 positive T-cell expression using immunohistochemistry (Figure 7A) and the average number of T-cells/islets for all treated groups (Figure 7B).

3.8. RES and/or STZ Effect on Histopathological Features

The islets of the pancreas from the normal group showed normal islets with normal β cells, a low number of α cells with pink cytoplasm in the border (blue arrows), pale blue cytoplasm in the center (black arrows), normal dominant thin-wall capillaries (green arrow), and normal exocrine regions (yellow arrows). RES-injected mice exhibited relatively small-size islets with low distributed apoptotic β cells (black arrows) with mildly dilated intervening capillaries (blue arrows), and normal exocrine zones (yellow arrows). Animals injected with STZ demonstrated noticeable small-size, hypo-cellular islets and clear apoptotic β cells (black arrows), apparent dilated dominant blood capillaries (blue arrows), inflammatory infiltrate (green arrows), and interstitial congested blood vessels (red arrows) with normal exocrine spaces (yellow arrows). However, the RES and STZ-injected group displayed islets with a normal size with a low number of dispersed apoptotic β cells (black arrows), normal interstitial blood vessels (red arrows) and average exocrine parts (yellow arrows) (H&E × 400) (Figure 8).

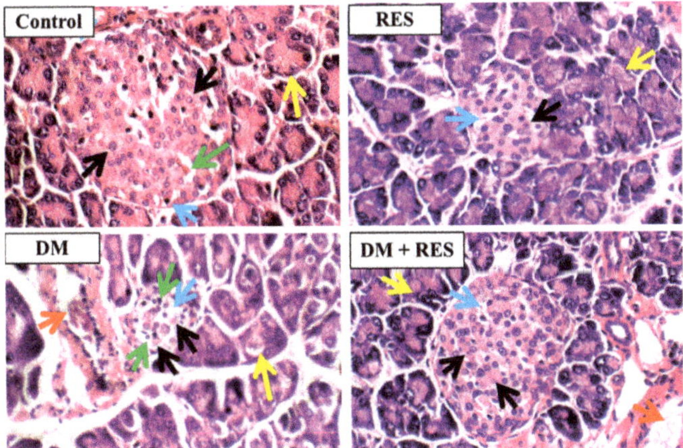

Figure 8. Islets of pancreas from control and RES-injected mice showed average islets, fewer α cells, marginal pink cytoplasm (blue arrows), average dominant thin-wall blood capillaries (green arrows), normal exocrine spaces (yellow arrows), and predominating β-cells with middle-pale-blue cytoplasm (black arrows). Pancreatic tissue of mice treated with STZ shows obvious apoptotic β-cells (black arrows); noticeable dilated principal blood capillaries (blue arrows); inflammatory infiltrate (green arrows); noticeable small-sized hypo-cellular islets, average exocrine zones (yellow arrows); and congested interstitial blood vessels (red arrows). Section from pancreas of mice injected with RES and STZ displays normal-sized islets and a low number of dispersed apoptotic β-cells (black arrows) and normal exocrine spaces (yellow arrows), normal prevailing blood capillaries (blue arrows), and normal blood vessels (red arrows) (H&E × 400).

4. Discussion

Diabetes mellitus is a common metabolic disease characterized by an imbalance between the resource and consumption of glucose at the cellular level [34]. The exact mechanism underlying the development of this disease is still unclear. Increased oxidative stress, inflammation, and apoptosis during T1D may lead to pancreatic β-cell loss [35,36]. In addition, T-cell infiltration into pancreatic β-cells mediates inflammatory cascade and consequently β-cell loss [37]. A previous study reported that weight loss is a common finding in diabetes induced by STZ injection that may be due to the degeneration of adipocyte and muscle tissues in order to recompense energy loss [38]. Furthermore, STZ in a high, single dose causes β-cell death due to necrosis and consequently T1D [39]. However, STZ with low multiple doses causes partial β-cell destruction and generates an inflammatory process [40]. The current work was designed to investigate the possible protecting effect of RES in T1D and the possible mechanism(s) underlying this protection, in addition to determining the role of CXCL16 and ADAM10 in the initiation of T1D. Recorded results demonstrate that RES prevents T1D in mice through the attenuation of CXCL16/ADAM10/TNF-α/NF-κB pathway, T-cell infiltration, and caspase-3 mediated apoptosis in the pancreatic islets. To our knowledge, this is the first work exploring the protective effect of RES in term of its effects on CXCL16/ADAM10-mediated pancreatic islets apoptosis, inflammation, and T-cell infiltration in T1D pathogenesis.

In T1D, the increased fasting blood glucose level, reduced serum insulin level, and reduction of animal weight are essential markers of diabetes [41]. Accordingly, the recorded significant weight loss, increased fasting blood glucose level, decreased fasting blood insulin level, obvious small-size hypo-cellular islets, and clear apoptotic β-cell destruction confirmed the successful induction of diabetes with low multiple doses of STZ (55 mg/kg i.p. for 5 consecutive days) in mice. The current findings reveal that RES prevented weight loss, which could be due to the glucose-lowering effect of RES. In addition, RES ameliorated the reduced serum level of insulin caused by STZ injection and prevented β-cell damage, as shown by histopathological results with subsequently increased insulin secretion and glucose utilization.

TNF-α is one of the important serum pro-inflammatory cytokines [42]. Increased serum levels of TNF-α have been accompanied by diabetes development, as it is known to interfere with signal transduction mediated by insulin receptor that leads to the inhibition of the insulin effect [43,44]. Oxidative stress and metabolic disturbance have been reported in T1D [45,46]. In addition, increased oxidative stress in the pancreatic islets is known to be an important pathological mechanism of β-cell damage in an STZ animal model [47,48]. The results of the current study reveal that RES inhibits TNF-α, WBC count, and serum NO-induced by STZ injection. Lin et al., (2019) reported that RES downregulated the expression of monocyte chemoattractant protein-1 (MCP-1) via TNF-α- inhibition in rats with acute pulmonary thromboembolism [49]. In addition, RES-attenuated renal injury resulted from ischemia/reperfusion (I/R) in diabetic rats through TNF-α-stimulated inflammation and oxidative stress [50]. Additionally, RES decreased hepatic I/R injuries associated with increased liver tissue levels of TNF-α and oxidative stress in diabetic rats [51].

Being an exceptional chemokine, CXCL16 exists in two forms: the transmembrane that acts as a receptor for ox-LDL [52] and the soluble form resulting from cleavage of its transmembrane form by ADAM (ADAM10 or ADAM17), and consequently, the released part can attract CXCR6-expressing T-cells to the site of injury [53]. On this basis, we reported the involvement and the role of the chemokine CXCL16 with its processing enzyme, ADAM10, for the first time, as well as its associated signaling mechanism in the development of T1D. For the pancreatic localization of CXCL16, double immunofluorescence of CXCL16 and insulin (β-cell marker) was performed. In addition, ADAM10 protein expression using immunofluorescence and Western blot analysis of isolated pancreatic islets were also accomplished. Our findings display a marked CXCL16 protein upregulation in β-cells of the pancreas and increase of its serum level in T1D induced by STZ. Similarly, increased expression of ADAM10 protein in β-cells of the pancreas was confirmed, adopting immunofluorescence and Western blot analysis. Our previous work confirmed that the activation of CXCL16/ox-LDL pathway in beta cells is a possible cause of TF activation and autophagy of the pancreatic beta cell in T1D [54]. Similarly, CXCL16 promotes inflammatory markers and infiltrated cell factors, and reduces antioxidants, subsequently accelerating diabetic nephropathy development [55]. Also, simvastatin alleviated STZ-induced ADAM10 and CXCL16 in diabetic mice [17].

NF-κB mediates β-cell inflammation resulting in the recruitment of immune cells and insulitis development that mediates T1D development [56]. Abu El-Asrar et al. (2021) suggested that CXCL16/CXCR6 and the processing enzyme ADAM10 play a role in the initiation and progression of proliferative diabetic retinopathy [57]. The author mentioned that the proliferative diabetic retinopathy mediated through increased NF-κB, VEGF and ICAM-1 in the retina. Similarly, regulation of CXCL16 and ADAM10 expression is considered to be an early response in diabetic nephropathy development [52]. Interestingly, a recent report showed that the CXCL16-regulated feedback mechanism is critical for the activation of NF-κB in prostate cancer cells [58]. The role of CXCL16 in the recruitment of T-cells has been described in inflammatory valvular heart disease [59], psoriasis pathogenesis [60], rheumatoid joints [61], colorectal cancer [62], hepatic ischemia-reperfusion injury [63], and interstitial rejection of renal allograft [8]. In the current study, we observed a significant increase in T-cell recruitment to the pancreatic islets as compared with control mice. The recruitment of T-cells in our study is accompanied by ADAM10 upregulation in the pancreatic islets, indicating that ADAM10 may cleave CXCL16 into its soluble form, resulting in an accumulation of T-cells. Notably, the current results show that RES treatment significantly reduces ADAM10 and CXCL16 expressions in the pancreatic β-cells. This inhibition may be explained by the fact that RES may inhibit the membranous level and consequently the ability of ADAM10 to cleave the cellular CXCL16, therefore reducing T-cell recruitment from the blood vessels into the pancreatic β-cells.

Remarkably, an earlier report suggested that apoptosis is induced by CXCL16 in colorectal cancer liver metastasis [64]. Furthermore, ROCK1 activated by CXCL16 causes upregulation of caspase-3 in injured HK-2 cells [6]. Similarly, CXCL16 attracts circulating inflammatory cells to the injured kidney, leading to tubular epithelial-cell apoptosis after cisplatin injection [65]. In support of these results, our findings demonstrate that STZ-induced T1D increased the cellular and soluble form of CXCL16, which mediates apoptosis, as confirmed by cleaved caspases-3 induction, and further confirmed by increased DNA fragmentation shown by TUNNEL assay. Treatment of mice with RES ameliorated the effect of STZ on apoptosis consequences.

5. Conclusions

The current study demonstrated for the first time that ADAM10/CXCL16/NF-κB mediated STZ-induced diabetes. Additionally, cleaved CXCL16 mediated T-cell recruitment into β-cells, and therefore enhanced oxidative stress, inflammatory response, and apoptosis. RES successfully mitigated CXCL16 cleavage by ADAM10 and markedly improved the deleterious effects induced by STZ treatment.

Author Contributions: M.A.D. collected data; carried out the mouse model, in vivo experiments, and biochemical assessments; analysed the results; and drafted the manuscript. M.S.A.-B. participated in the study's design, supervision of the practical study, overall article review, and immunofluorescence imaging and analysis. E.A. contributed to the manuscript editing and thorough revision. A.A. and W.M.A. conceptualized the study, participated in its design, and revised it overall. All authors have read and agreed to the published version of the manuscript.

Funding: The authors gratefully acknowledge Qassim University, represented by the Deanship of Scientific Research, on the financial support for this research under the number (10103-pharmacy-2020-1-3-1) during the academic year 1441AH/2020AD.

Institutional Review Board Statement: The experimental design and work involved were in agreement with the Institutional Animal Care and by NADA University Scientific Research Ethical Committee under the number (NUB-002-021).

Informed Consent Statement: Not applicable.

Data Availability Statement: Not applicable.

Conflicts of Interest: The authors declare no conflict of interest.

Abbreviations

ADAM10	A Disintegrin and Metallopeptidase 10
ANOVA	Analysis of variance
CKD	chronic kidney diseases
CXCL16	CXC chemokine ligand 16
DAPI	4′,6′-diamidino-2-phenylindole
DMSO	Dimethylsulfoxide
ELISA	Enzyme-linked immunosorbent assay
HBSS	Hank's balanced salt solution
NF-KB	Nuclear factor-kappa B
PBS	Phosphate-buffered saline
RES	Resveratrol
STZ	Streptozotocin
TNF-alpha	Tumor necrosis factor-alpha
T1D	Type 1 diabetes
WBC	White blood cell

References

1. Fagerberg, L.; Hallström, B.M.; Oksvold, P.; Kampf, C.; Djureinovic, D.; Odeberg, J.; Habuka, M.; Tahmasebpoor, S.; Danielsson, A.; Edlund, K. Analysis of the human tissue-specific expression by genome-wide integration of transcriptomics and antibody-based proteomics. *Mol. Cell. Proteom.* **2014**, *13*, 397–406. [CrossRef]
2. Weber, S.; Saftig, P. Ectodomain shedding and ADAMs in development. *Development* **2012**, *139*, 3693–3709. [CrossRef]
3. Pruessmeyer, J.; Ludwig, A. The good, the bad and the ugly substrates for ADAM10 and ADAM17 in brain pathology, inflammation and cancer. *Semin. Cell Dev. Biol.* **2009**, *20*, 164–174. [CrossRef]
4. Dreymueller, D.; Pruessmeyer, J.; Groth, E.; Ludwig, A. The role of ADAM-mediated shedding in vascular biology. *Eur. J. Cell Biol.* **2012**, *91*, 472–485. [CrossRef]
5. Fang, Y.; Henderson, F.C.; Yi, Q.; Lei, Q.; Li, Y.; Chen, N. Chemokine CXCL16 expression suppresses migration and invasiveness and induces apoptosis in breast cancer cells. *Mediat. Inflamm.* **2014**, *2014*. [CrossRef]
6. Liang, H.; Liao, M.; Zhao, W.; Zheng, X.; Xu, F.; Wang, H.; Huang, J. CXCL16/ROCK1 signaling pathway exacerbates acute kidney injury induced by ischemia-reperfusion. *Biomed. Pharmacother.* **2018**, *98*, 347–356. [CrossRef]
7. Kato, T.; Hagiyama, M.; Ito, A. Renal ADAM10 and 17: Their physiological and medical meanings. *Front. Cell Dev. Biol.* **2018**, *6*, 153. [CrossRef]
8. Schramme, A.; Abdel-Bakky, M.S.; Gutwein, P.; Obermüller, N.; Baer, P.C.; Hauser, I.A.; Ludwig, A.; Gauer, S.; Schäfer, L.; Sobkowiak, E. Characterization of CXCL16 and ADAM10 in the normal and transplanted kidney. *Kidney Int.* **2008**, *74*, 328–338. [CrossRef]
9. Mir, H.; Kaur, G.; Kapur, N.; Bae, S.; Lillard, J.W.; Singh, S. Higher CXCL16 exodomain is associated with aggressive ovarian cancer and promotes the disease by CXCR6 activation and MMP modulation. *Sci. Rep.* **2019**, *9*, 2527. [CrossRef]
10. Gong, Q.; Wu, F.; Pan, X.; Yu, J.; Li, Y.; Lu, T.; Li, X.; Lin, Z. Soluble CXC chemokine ligand 16 levels are increased in gout patients. *Clin. Biochem.* **2012**, *45*, 1368–1373. [CrossRef]
11. Lin, Z.; Gong, Q.; Zhou, Z.; Zhang, W.; Liao, S.; Liu, Y.; Yan, X.; Pan, X.; Lin, S.; Li, X. Increased plasma CXCL16 levels in patients with chronic kidney diseases. *Eur. J. Clin. Investig.* **2011**, *41*, 836–845. [CrossRef] [PubMed]
12. Chandrasekar, B.; Bysani, S.; Mummidi, S. CXCL16 signals via Gi, phosphatidylinositol 3-kinase, Akt, IκB kinase, and nuclear factor-κB and induces cell-cell adhesion and aortic smooth muscle cell proliferation. *J. Biol. Chem.* **2004**, *279*, 3188–3196. [CrossRef]
13. Abel, S.; Hundhausen, C.; Mentlein, R.; Schulte, A.; Berkhout, T.A.; Broadway, N.; Hartmann, D.; Sedlacek, R.; Dietrich, S.; Muetze, B. The transmembrane CXC-chemokine ligand 16 is induced by IFN-γ and TNF-α and shed by the activity of the disintegrin-like metalloproteinase ADAM10. *J. Immunol.* **2004**, *172*, 6362–6372. [CrossRef] [PubMed]
14. Tawfik, M.S.; Abdel-Messeih, P.L.; Nosseir, N.M.; Mansour, H.H. Circulating CXCL16 in type 2 diabetes mellitus Egyptian patients. *J. Radiat. Res. Appl. Sci.* **2021**, *14*, 9–15. [CrossRef]
15. Zhou, F.; Wang, J.; Wang, K.; Zhu, X.; Pang, R.; Li, X.; Zhu, G.; Pan, X. Serum CXCL16 as a novel biomarker of coronary artery disease in type 2 diabetes mellitus: A pilot study. *Ann. Clin. Lab. Sci.* **2016**, *46*, 184–189.
16. Lekva, T.; Michelsen, A.E.; Aukrust, P.; Roland, M.C.P.; Henriksen, T.; Bollerslev, J.; Ueland, T. CXC chemokine ligand 16 is increased in gestational diabetes mellitus and preeclampsia and associated with lipoproteins in gestational diabetes mellitus at 5 years follow-up. *Diabetes Vasc. Dis. Res.* **2017**, *14*, 525–533. [CrossRef]
17. Devi, L.; Alldred, M.J.; Ginsberg, S.D.; Ohno, M. Mechanisms underlying insulin deficiency-induced acceleration of β-amyloidosis in a mouse model of Alzheimer's disease. *PLoS ONE* **2012**, *7*, e32792. [CrossRef]
18. Yang, K.; Lu, L.; Liu, Y.; Zhang, Q.; Pu, L.J.; Wang, L.J.; Zhu, Z.B.; Wang, Y.N.; Meng, H.; Zhang, X.J. Increase of ADAM10 level in coronary artery in-stent restenosis segments in diabetic minipigs: High ADAM10 expression promoting growth and migration in human vascular smooth muscle cells via Notch 1 and 3. *PLoS ONE* **2013**, *8*, e83853. [CrossRef]

19. Allagnat, F.; Fukaya, M.; Nogueira, T.C.; Delaroche, D.; Welsh, N.; Marselli, L.; Marchetti, P.; Haefliger, J.A.; Eizirik, D.L.; Cardozo, A.K. C/EBP homologous protein contributes to cytokine-induced pro-inflammatory responses and apoptosis in β-cells. *Cell Death Differ.* **2012**, *19*, 1836–1846. [CrossRef]

20. Garcia, G.E.; Truong, L.D.; Li, P.; Zhang, P.; Johnson, R.J.; Wilson, C.B.; Feng, L. Inhibition of CXCL16 attenuates inflammatory and progressive phases of anti-glomerular basement membrane antibody-associated glomerulonephritis. *Am. J. Pathol.* **2007**, *170*, 1485–1496. [CrossRef]

21. Darwish, M.A.; Abo-Youssef, A.M.; Messiha, B.A.S.; Abo-Saif, A.A.; Abdel-Bakky, M.S. Resveratrol inhibits macrophage infiltration of pancreatic islets in streptozotocin-induced type 1 diabetic mice via attenuation of the CXCL16/NF-κB p65 signaling pathway. *Life Sci.* **2021**, *272*, 119250. [CrossRef] [PubMed]

22. Ciddi, V.; Dodda, D. Therapeutic potential of resveratrol in diabetic complications: In vitro and in vivo studies. *Pharmacol. Rep.* **2014**, *66*, 799–803. [CrossRef] [PubMed]

23. Hussein, M.M.A.; Mahfouz, M.K. Effect of resveratrol and rosuvastatin on experimental diabetic nephropathy in rats. *Biomed. Pharmacother.* **2016**, *82*, 685–692. [CrossRef] [PubMed]

24. Huo, X.; Zhang, T.; Meng, Q.; Li, C.; You, B. Resveratrol effects on a diabetic rat model with coronary heart disease. *Med. Sci. Monit. Int. Med. J. Exp. Clin. Res.* **2019**, *25*, 540. [CrossRef] [PubMed]

25. Oyenihi, O.R.; Oyenihi, A.B.; Adeyanju, A.A.; Oguntibeju, O.O. Antidiabetic effects of resveratrol: The way forward in its clinical utility. *J. Diabetes Res.* **2016**, *2016*, 9737483. [CrossRef]

26. Kim, Y.H.; Kim, Y.S.; Kang, S.S.; Cho, G.J.; Choi, W.S. Resveratrol inhibits neuronal apoptosis and elevated Ca^{2+}/ calmodulin-dependent protein kinase II activity in diabetic mouse retina. *Diabetes* **2010**, *59*, 1825–1835. [CrossRef]

27. Blanchet, J.; Longpré, F.; Bureau, G.; Morissette, M.; DiPaolo, T.; Bronchti, G.; Martinoli, M.-G. Resveratrol, a red wine polyphenol, protects dopaminergic neurons in MPTP-treated mice. *Prog. Neuro-Psychopharmacol. Biol. Psychiatry* **2008**, *32*, 1243–1250. [CrossRef]

28. Rydgren, T.; Vaarala, O.; Sandler, S. Simvastatin protects against multiple low-dose streptozotocin-induced type 1 diabetes in CD-1 mice and recurrence of disease in nonobese diabetic mice. *J. Pharmacol. Exp. Ther.* **2007**, *323*, 180–185. [CrossRef]

29. Bender, C.; Christen, S.; Scholich, K.; Bayer, M.; Pfeilschifter, J.M.; Hintermann, E.; Christen, U. Islet-expressed CXCL10 promotes autoimmune destruction of islet isografts in mice with type 1 diabetes. *Diabetes* **2017**, *66*, 113–126. [CrossRef]

30. Miranda, K.M.; Espey, M.G.; Wink, D.A. A rapid, simple spectrophotometric method for simultaneous detection of nitrate and nitrite. *Nitric Oxide* **2001**, *5*, 62–71. [CrossRef]

31. Abdel-bakky, M.S.; Hammad, M.A.; Walker, L.A.; Ashfaq, M.K. Silencing of tissue factor by antisense deoxyoligonucleotide prevents monocrotaline / LPS renal injury in mice. *Arch. Toxicol.* **2011**, 1245–1256. [CrossRef] [PubMed]

32. Martin, A.P.; Alexander-Brett, J.M.; Canasto-Chibuque, C.; Garin, A.; Bromberg, J.S.; Fremont, D.H.; Lira, S.A. The Chemokine Binding Protein M3 Prevents Diabetes Induced by Multiple Low Doses of Streptozotocin. *J. Immunol.* **2007**, *178*, 4623–4631. [CrossRef] [PubMed]

33. Wild, S.; Roglic, G.; Green, A.; Sicree, R.; King, H. Global prevalence of diabetes: Estimates for the year 2000 and projections for 2030. *Diabetes Care* **2004**, *27*, 1047–1053. [CrossRef] [PubMed]

34. Gerber, P.A.; Rutter, G.A. The role of oxidative stress and hypoxia in pancreatic beta-cell dysfunction in diabetes mellitus. *Antioxid. Redox Signal.* **2017**, *26*, 501–518. [CrossRef] [PubMed]

35. Baynes, J.W. Role of oxidative stress in development of complications in diabetes. *Diabetes* **1991**, *40*, 405–412. [CrossRef] [PubMed]

36. Eizirik, D.L.; Colli, M.L.; Ortis, F. The role of inflammation in insulitis and β-cell loss in type 1 diabetes. *Nat. Rev. Endocrinol.* **2009**, *5*, 219. [CrossRef]

37. Ye, M.; Qiu, T.; Peng, W.; Chen, W.; Ye, Y.; Lin, Y. Purification, characterization and hypoglycemic activity of extracellular polysaccharides from Lachnum calyculiforme. *Carbohydr. Polym.* **2011**, *86*, 285–290. [CrossRef]

38. Furman, B.L. Streptozotocin-induced diabetic models in mice and rats. *Curr. Protoc. Pharmacol.* **2015**, *70*, 5–47. [CrossRef]

39. Kolb, H. Mouse models of insulin dependent diabetes: Low-dose streptozocin-induced diabetes and nonobese diabetic (NOD) mice. *Diabetes. Metab. Rev.* **1987**, *3*, 751–778. [CrossRef]

40. Tang, T.; Duan, X.; Ke, Y.; Zhang, L.; Shen, Y.; Hu, B.; Liu, A.; Chen, H.; Li, C.; Wu, W. Antidiabetic activities of polysaccharides from Anoectochilus roxburghii and Anoectochilus formosanus in STZ-induced diabetic mice. *Int. J. Biol. Macromol.* **2018**, *112*, 882–888. [CrossRef]

41. Suzuki, M.; Mihara, M. Adiponectin induces CCL20 expression synergistically with IL-6 and TNF-α in THP-1 macrophages. *Cytokine* **2012**, *58*, 344–350. [CrossRef] [PubMed]

42. De Luca, C.; Olefsky, J.M. Inflammation and insulin resistance. *FEBS Lett.* **2008**, *582*, 97–105. [CrossRef] [PubMed]

43. Ruge, T.; Lockton, J.A.; Renstrom, F.; Lystig, T.; Sukonina, V.; Svensson, M.K.; Eriksson, J.W. Acute hyperinsulinemia raises plasma interleukin-6 in both nondiabetic and type 2 diabetes mellitus subjects, and this effect is inversely associated with body mass index. *Metabolism* **2009**, *58*, 860–866. [CrossRef] [PubMed]

44. Gulcubuk, A.; Haktanir, D.; Cakiris, A.; Ustek, D.; Guzel, O.; Erturk, M.; Yildirim, F.; Akyazi, I.; Cicekci, H.; Durak, M.H. The effects of resveratrol on tissue injury, oxidative damage, and pro-inflammatory cytokines in an experimental model of acute pancreatitis. *J. Physiol. Biochem.* **2014**, *70*, 397–406. [CrossRef] [PubMed]

45. Yao, L.; Wan, J.; Li, H.; Ding, J.; Wang, Y.; Wang, X.; Li, M. Resveratrol relieves gestational diabetes mellitus in mice through activating AMPK. *Reprod. Biol. Endocrinol.* **2015**, *13*, 118. [CrossRef] [PubMed]

46. Kwon, N.S.; Lee, S.H.; Choi, C.S.; Kho, T.; Lee, H.S. Nitric oxide generation from streptozotocin 1. *FASEB J.* **1994**, *8*, 529–533. [CrossRef]

47. Haluzik, M.; Nedvidkova, J. The role of nitric oxide in the development of streptozotocin-induced diabetes mellitus: Experimental and clinical implications. *Physiol. Res.* **2000**, *49*, S37–S42.

48. Lin, J.-W.; Yang, L.-H.; Ren, Z.-C.; Mu, D.-G.; Li, Y.-Q.; Yan, J.-P.; Wang, L.-X.; Chen, C. Resveratrol downregulates TNF-α-induced monocyte chemoattractant protein-1 in primary rat pulmonary artery endothelial cells by P38 mitogen-activated protein kinase signaling. *Drug Des. Devel. Ther.* **2019**, *13*, 1843. [CrossRef]

49. Wang, M.; Weng, X.; Chen, H.; Chen, Z.; Liu, X. Resveratrol inhibits TNF-α-induced inflammation to protect against renal ischemia/reperfusion injury in diabetic rats. *Acta Cir. Bras.* **2020**, *35*, e202000506. [CrossRef]

50. Aktaş, H.S.; Ozel, Y.; Ahmad, S.; Pençe, H.H.; Ayaz-Adakul, B.; Kudas, I.; Tetik, S.; Şekerler, T.; Canbey-Göret, C.; Kabasakal, L. Protective effects of resveratrol on hepatic ischemia reperfusion injury in streptozotocin-induced diabetic rats. *Mol. Cell. Biochem.* **2019**, *460*, 217–224. [CrossRef]

51. Gutwein, P.; Abdel-bakky, M.S.; Doberstein, K.; Schramme, A.; Beckmann, J.; Schaefer, L.; Amann, K.; Doller, A.; Kämpfer-kolb, N.; Sayed, E.; et al. CXCL16 and oxLDL are induced in the onset of diabetic nephropathy. *Mol. Med.* **2009**, *13*, 3809–3825. [CrossRef]

52. Schramme, A.; Abdel-bakky, M.S.; Kämpfer-kolb, N.; Pfeilschifter, J.; Gutwein, P. The role of CXCL16 and its processing metalloproteinases ADAM10 and ADAM17 in the proliferation and migration of human mesangial cells. *Biochem. Biophys. Res. Commun.* **2008**, *370*, 311–316. [CrossRef] [PubMed]

53. Darwish, M.A.; Abdel-Bakky, M.S.; Messiha, B.A.S.; Abo-Saif, A.A.; Abo-Youssef, A.M. Resveratrol mitigates pancreatic TF activation and autophagy-mediated beta cell death via inhibition of CXCL16/ox-LDL pathway: A novel protective mechanism against type 1 diabetes mellitus in mice. *Eur. J. Pharmacol.* **2021**, *901*, 174059. [CrossRef]

54. Zhao, L.; Wu, F.; Jin, L.; Lu, T.; Yang, L.; Pan, X.; Shao, C.; Li, X.; Lin, Z. Serum CXCL16 as a novel marker of renal injury in type 2 diabetes mellitus. *PLoS ONE* **2014**, *9*, e87786. [CrossRef]

55. Abdel-Bakky, M.S.; Alqasoumi, A.; Altowayan, W.M.; Amin, E.; Darwish, M.A. Simvastatin mitigates streptozotocin-induced type 1 diabetes in mice through downregulation of ADAM10 and ADAM17. *Life Sci.* **2022**, *289*, 120224. [CrossRef] [PubMed]

56. Salem, H.H.; Trojanowski, B.; Fiedler, K.; Maier, H.J.; Schirmbeck, R.; Wagner, M.; Boehm, B.O.; Wirth, T.; Baumann, B. Long-term IKK2/NF-κB signaling in pancreatic β-cells induces immune-mediated diabetes. *Diabetes* **2014**, *63*, 960–975. [CrossRef] [PubMed]

57. El-Asrar, A.M.A.; Nawaz, M.I.; Ahmad, A.; De Zutter, A.; Siddiquei, M.M.; Blanter, M.; Allegaert, E.; Gikandi, P.W.; De Hertogh, G.; Van Damme, J. Evaluation of Proteoforms of the Transmembrane Chemokines CXCL16 and CX3CL1, Their Receptors, and Their Processing Metalloproteinases ADAM10 and ADAM17 in Proliferative Diabetic Retinopathy. *Front. Immunol.* **2020**, *11*. [CrossRef]

58. Kapur, N.; Mir, H.; Sonpavde, G.P.; Jain, S.; Bae, S.; Lillard, J.W., Jr.; Singh, S. Prostate cancer cells hyper-activate CXCR6 signaling by cleaving CXCL16 to overcome effect of docetaxel. *Cancer Lett.* **2019**, *454*, 1–13. [CrossRef]

59. Yamauchi, R.; Tanaka, M.; Kume, N.; Minami, M.; Kawamoto, T.; Togi, K.; Shimaoka, T.; Takahashi, S.; Yamaguchi, J.; Nishina, T. Upregulation of SR-PSOX/CXCL16 and recruitment of CD8+ T cells in cardiac valves during inflammatory valvular heart disease. *Arterioscler. Thromb. Vasc. Biol.* **2004**, *24*, 282–287. [CrossRef]

60. Günther, C.; Carballido-Perrig, N.; Kaesler, S.; Carballido, J.M.; Biedermann, T. CXCL16 and CXCR6 are upregulated in psoriasis and mediate cutaneous recruitment of human CD8+ T cells. *J. Investig. Dermatol.* **2012**, *132*, 626–634. [CrossRef]

61. Van Der Voort, R.; Van Lieshout, A.W.T.; Toonen, L.W.J.; Slöetjes, A.W.; Van Den Berg, W.B.; Figdor, C.G.; Radstake, T.R.D.J.; Adema, G.J. Elevated CXCL16 expression by synovial macrophages recruits memory T cells into rheumatoid joints. *Arthritis Rheum.* **2005**, *52*, 1381–1391. [CrossRef] [PubMed]

62. Kee, J.-Y.; Ito, A.; Hojo, S.; Hashimoto, I.; Igarashi, Y.; Tsukada, K.; Irimura, T.; Shibahara, N.; Nakayama, T.; Yoshie, O. Chemokine CXCL16 suppresses liver metastasis of colorectal cancer via augmentation of tumor-infiltrating natural killer T cells in a murine model. *Oncol. Rep.* **2013**, *29*, 975–982. [CrossRef] [PubMed]

63. Zhu, H.; Zhang, Q.; Chen, G. CXCR6 deficiency ameliorates ischemia-reperfusion injury by reducing the recruitment and cytokine production of hepatic NKT cells in a mouse model of non-alcoholic fatty liver disease. *Int. Immunopharmacol.* **2019**, *72*, 224–234. [CrossRef] [PubMed]

64. Kee, J.-Y.; Ito, A.; Hojo, S.; Hashimoto, I.; Igarashi, Y.; Tsuneyama, K.; Tsukada, K.; Irimura, T.; Shibahara, N.; Takasaki, I. CXCL16 suppresses liver metastasis of colorectal cancer by promoting TNF-α-induced apoptosis by tumor-associated macrophages. *BMC Cancer* **2014**, *14*, 949. [CrossRef] [PubMed]

65. Liang, H.; Zhang, Z.; He, L.; Wang, Y. CXCL16 regulates cisplatin-induced acute kidney injury. *Oncotarget* **2016**, *7*, 31652. [CrossRef]

 *pharmaceutics*

Article

Citrus Flavanone Narirutin, In Vitro and In Silico Mechanistic Antidiabetic Potential

Ashraf Ahmed Qurtam [1,†], Hamza Mechchate [2,†], Imane Es-safi [2,*], Mohammed Al-zharani [1], Fahd A. Nasr [3], Omar M. Noman [3], Mohammed Aleissa [1], Hamada Imtara [4], Abdulmalik M. Aleissa [5], Mohamed Bouhrim [6] and Ali S. Alqahtani [3]

[1] Biology Department, College of Science, Imam Mohammad Ibn Saud Islamic University (IMSIU), Riyadh 11623, Saudi Arabia; AAQURTAM@imamu.edu.sa (A.A.Q.); MMyAlzahrani@imamu.edu.sa (M.A.-z.); msaleissa@imamu.edu.sa (M.A.)
[2] Laboratory of Biotechnology, Environment, Agrifood and Health, Faculty of Sciences Dhar El Mahraz, University of Sidi Mohamed Ben Abdellah, Fez 30000, Morocco; Hamza.mechchate@usmba.ac.ma
[3] Department of Pharmacognosy, College of Pharmacy, King Saud University, Riyadh 11451, Saudi Arabia; fnasr@ksu.edu.sa (F.A.N.); onoman@ksu.edu.sa (O.M.N.); alalqahtani@ksu.edu.sa (A.S.A.)
[4] Faculty of Arts and Sciences, Arab American University Palestine, Jenin 240, Palestine; Hamada.tarayrah@gmail.com
[5] King Khaled Eye Specialist Hospital (KKESH), Riyadh 11462, Saudi Arabia; aaleissa@kkesh.med.sa
[6] Laboratory of Bioresources, Biotechnology, Ethnopharmacology and Health, Faculty of Sciences, Mohammed First University, Oujda B.P. 717, Morocco; mohamed.bouhrim@gmail.com
* Correspondence: Imane.essafi1@usmba.ac.ma
† These authors contributed equally to this work.

Citation: Qurtam, A.A.; Mechchate, H.; Es-safi, I.; Al-zharani, M.; Nasr, F.A.; Noman, O.M.; Aleissa, M.; Imtara, H.; Aleissa, A.M.; Bouhrim, M.; et al. Citrus Flavanone Narirutin, In Vitro and In Silico Mechanistic Antidiabetic Potential. *Pharmaceutics* 2021, 13, 1818. https://doi.org/10.3390/pharmaceutics13111818

Academic Editors: Diana Marcela Aragon Novoa and Fátima Regina Mena Barreto Silva

Received: 9 September 2021
Accepted: 21 October 2021
Published: 31 October 2021

Publisher's Note: MDPI stays neutral with regard to jurisdictional claims in published maps and institutional affiliations.

Abstract: Citrus fruits and juices have been studied extensively for their potential involvement in the prevention of various diseases. Flavanones, the characteristic polyphenols of citrus species, are the primarily compounds responsible for these studied health benefits. Using in silico and in vitro methods, we are exploring the possible antidiabetic action of narirutin, a flavanone family member. The goal of the in silico research was to anticipate how narirutin would interact with eight distinct receptors implicated in diabetes control and complications, namely, dipeptidyl-peptidase 4 (DPP4), protein tyrosine phosphatase 1B (PTP1B), free fatty acid receptor 1 (FFAR1), aldose reductase (AldR), glycogen phosphorylase (GP), alpha-amylase (AAM), peroxisome proliferator-activated receptor gamma (PPAR-γ), alpha-glucosidase (AGL), while the in vitro study looked into narirutin's possible inhibitory impact on alpha-amylase and alpha-glucosidase. The results indicate that the studied citrus flavanone interacted remarkably with most of the receptors and had an excellent inhibitory activity during the in vitro tests suggesting its potent role among the different constituent of the citrus compounds in the management of diabetes and also its complications.

Keywords: narirutin; naringenin rutinoside; isonaringin; molecular docking; mechanism of action; enzyme; receptors

1. Introduction

Diabetes is a severe, long-term disease that has a significant effect on the lives and well-being of people, families, and communities all over the globe. It is among the top ten causes of mortality in adults, with an estimated four million fatalities worldwide in 2017 [1]. The International Diabetes Federation (IDF) has been tracking diabetes on national, regional, and worldwide scales since 2000. In 2009, 285 million individuals were projected to have diabetes (including T1D and T2D), 366 million in 2011, 382 million in 2013, 415 million in 2015, 425 million in 2017, 463 million in 2019 and the global projection is 578 million by 2030, and 700 million by 2045 [1–5]. Diabetes has a tremendous social cost in terms of increased medical expenses, lost productivity, early death, and intangible costs such as decreased quality of life. Diabetes cost the global health system 727 billion USD in 2017 [1]. Furthermore, changes in the demography of the diabetic community, healthcare use and delivery patterns, technology, medical expenses, insurance coverage, and economic circumstances continue to have an impact on the economic burden of diabetes [6]. While several synthetic oral medicines have been established to treat diabetes, it is still a struggle to control diabetes without any side effects [7]. A new surge of scientific interest in conventional practices has been stimulated by increased research about complementary drugs and natural therapies, and the desire to look for more powerful agents with less side effects [8–11]. It is commonly known that low intake of unhealthy foods, daily physical exercise and strong consumption of plant-based products help to maintain a stable state of health [12]. In fact, an elevated dietary amount of fruit and vegetables is correlated with a decreased risk of some life-threatening diseases, such as cardiovascular disease and cancer [13]. Many phenolic secondary metabolites (flavonoids) found in plant-derived foods are now increasingly recognized to have beneficial health effects on the prevention and treatment of diseases [14,15]. Flavonoids are plant secondary metabolites belonging to polyphenolic family, and constitute a very interesting and important class of natural products commonly present in fruits, vegetables and some beverages [16]. They are an essential component of a variety of nutraceuticals, pharmaceutical, medicinal and cosmetic uses due to their broad spectrum of health-promoting properties [17,18]. This is due to their ability to regulate critical cellular enzyme activities, as well as their antioxidative, anti-inflammatory, antimutagenic, and anticarcinogenic characteristics [19,20]. They are phenolic chemicals of low molecular weight and one of the most distinctive groups of chemicals found in higher plants.

Flavanones, which are found in all citrus fruits such as oranges, lemons and grapes, are one of the most significant subgroups of flavonoids [21,22]. Because of their free radical-scavenging characteristics, they are linked to a variety of health advantages. Citrus fruit juice and peel contain these chemicals, which give them a bitter flavor. Citrus flavonoids have pharmacological actions against reactive oxygen species, inflammation, and are involved in lowering blood lipids and cholesterol [23]. Different studies have investigated their role (i.e., luteolin, apigenin) in epigenetic therapy and cell gene expression by regulating HDAC1, HDAC2 [24] and MMP-2 [25].

In this study, we looked into the antidiabetic potential of narirutin (flavanone-7-O-glycoside), a flavanone family representative, consisting of the flavanone naringenin bonded with the disaccharide rutinose (Figure 1), through different approaches to assess its mechanistic proprieties as a preventive and curative natural alternative. Molecular docking was used to evaluate molecular interaction with the following receptors: dipeptidyl-peptidase 4 (DPP4), protein tyrosine phosphatase 1B (PTP1B), free fatty acid receptor 1 (FFAR1), aldose reductase (AldR), glycogen phosphorylase (GP), alpha-amylase (AAM), peroxisome proliferator-activated receptor gamma (PPAR-γ), and alpha-glucosidase (AGL). The alpha-amylase and alpha-glucosidase enzymes were tested in vitro to see whether the compound has any inhibitory action.

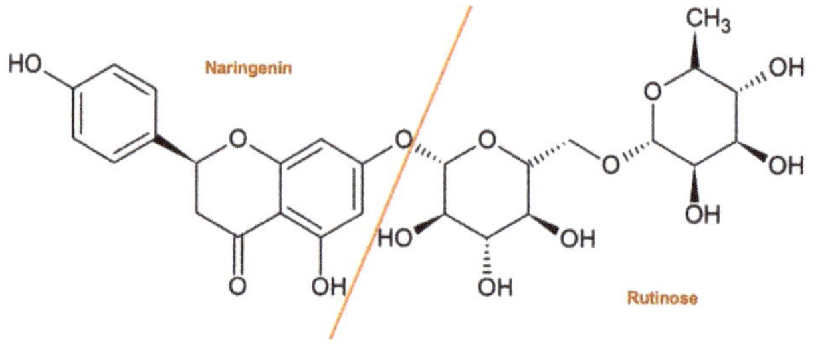

Figure 1. Narirutin chemical structure, formula and molar mass.

2. Results and Discussion

2.1. Molecular Docking

To confirm the interaction of the receptor with the ligand, literature screening was performed to identify the active site of the receptor and the amino acids composing it. Table 1 shows the result of the in silico simulation of the studied molecule and the selected receptors (affinity), the active site description and the amino acids that interact with a hydrogen bond in the site.

Table 1. Summary of narirutin/receptors affinities.

Receptor	Affinity (kcal/mol)	Active Site Described in Literature	Interaction Confirmed with the Active Site	H-Bonds
PTP1B	−8.5	Trp179, Pro180, Asp181 [26], His214, Ser216, Ala217, Gly218, Ile219, Gly220, and Arg221 [27,28] and the active site on Cys 215 (catalytic loop)	Yes	Arg221, Arg24, Ser216
DPP4	−10.4	DPP4 active site (α/β-hydrolase domain) is identified by the residues from 39 to 51 and from 501 to 706 [29,30].	Yes	Tyr662, Ser209, Ser630, Arg125, His740, Trp629,
FFAR1	−8.3	Active site includes Arg183, Arg258 and Tyr2240 [31]. Binding pocket is on Glu172, Arg183, Ser187, Tyr240, Asn241, Asn244, Arg258 and Tyr91 [32,33].	Yes	Tyr44
Alpha amylase	−9.9	Active site:Asp197, Glu233 and Asp300 and other important AA: Arg337, Arg195, Asn298, Phe265, Phe295, His201, Ala307, Gly306, Trp203, Trp284, Trp59, Tyr62, Trp58, His299 and His101 [34–37].	Yes	Gly306, Asp197, Gln63, Trp69, Lys352, Asp352
PPAR gamma		PPARγ ligand-binding domain: Ser289, His323, Tyr473, and His449 [38].	No	
Alpha glucosidase	−8.7	The amino acids involved in the α-Glucosidase activity are Asp404, Asp518, Arg600, Asp616, and His674 Trp376, Ile441, Trp516, Met519, Trp613, and Phe649 Leu405, Trp481, Asp645, and Arg672 [39,40].	Yes	Asp616, Ala284, Arg281, Asp282, Ser523
Aldose reductase	−9.3	The active site is located in the barrel core clearly seen in the 3D structure [41]	Yes	Val47, Gln49, Lys21, Ser32
Glycogen phosphorylase	−8.3	Active site on amino acids 280–288 (The 280's loop) [42,43].	Yes	Asp283, Glu382, leu384

2.1.1. PTP1B

Resistance to insulin's cellular action, a basic pathophysiological flaw associated with the global obesity pandemic, is linked to the development of type 2 diabetes and metabolic syndrome, a collection of cardiovascular risk factors. The discovery of new pharmacological medicines that help alleviate insulin resistance may be crucial not only for the prevention and treatment of diabetes but also for lowering the cardiovascular risk profile associated with it [44,45].

Although that PTP1B direct action to control the cardiovascular function is still unclear, there is more evidence linking this receptor with the inhibition of the regulation of metabolic function and cardiovascular function because of its direct interaction with various receptor tyrosine kinase signaling pathways [46]. Overall, these investigations have opened the way for the commercialization of PTP1B inhibitors, which may be used as a new kind of "insulin sensitizer" in the treatment of type 2 diabetes and cardiovascular/metabolic syndromes [47].

Figure 2 shows that narirutin had a strong contact with the catalytic site and the active site, with a docking affinity of 8.5 kcal/mol and three hydrogen bonds formed.

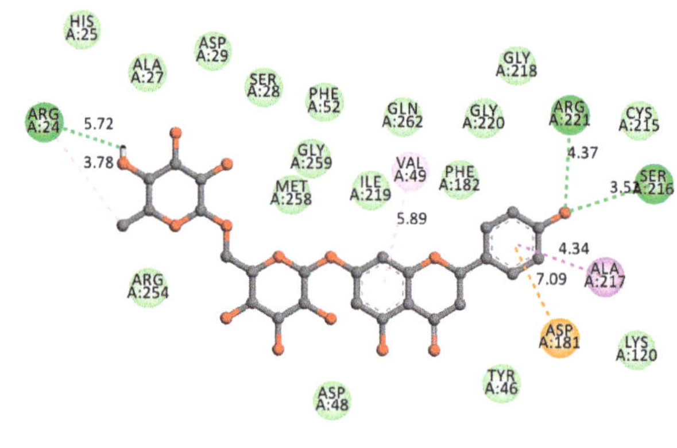

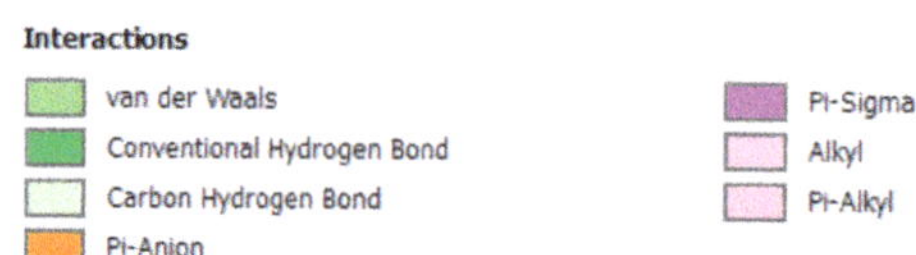

Figure 2. Two-dimensional scheme of the narirutin interactions with PTP1B.

2.1.2. DPP4

DPP4 inhibitors have been linked to better blood glucose management and lower fasting and postprandial blood glucose levels while avoiding weight gain [48]. DPP4 inhibition by a chemotherapeutic agent may raise the levels of circulating endogenous GLP-1 by prolonging its half-life, thus increasing GLP-1's beneficial effects in glucose-dependent insulin production and cell restoration. DPP-4 belongs to the serine protease family, which also includes the fibroblast activation protein (FAP), DPP-8, and DPP-9 [49]. DPP-4 inhibitors provide many benefits, including a reduced risk of hypoglycemia, minimal weight gain, and the possibility for pancreatic -cell repair and separation [50].

In the α/β-hydrolase domain, excellent interactions were established between narirutin and DPP4 receptor, with a docking affinity of 10.4 kcal/mol and 6 hydrogen bonds formed, as shown in Figure 3.

2.1.3. FFAR1

Free fatty acid receptor 1 (FFAR1) agonists have recently been discovered as a promising anti-diabetic target, since they regulate secretion stimulated by glucose in the pancreatic cells without causing hypoglycemia [51]. The FFAR1 mode of action is via interaction with a G protein. The coupling mechanism activates phospholipase-C, which regulates inositol triphosphate and the endoplasmic reticulum's release of intracellular Ca^{2+}. As a result, insulin secretion improves in a glucose concentration-dependent manner [52].

The Docking results demonstrated good interactions between narirutin and the FFAR1 receptor (Affinity of -8.3 kcal/mol) with one single hydrogen bond formed (Figure 4).

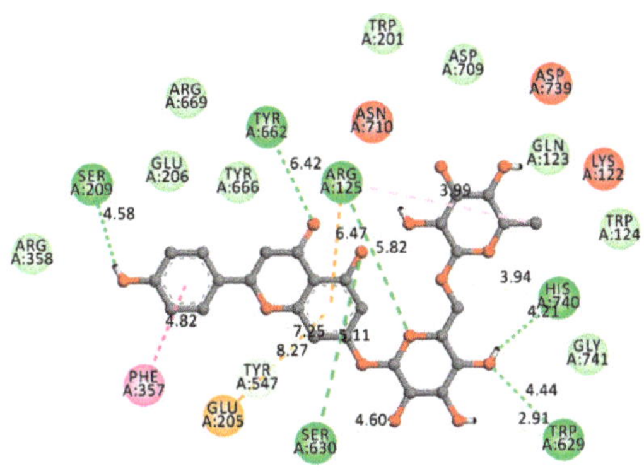

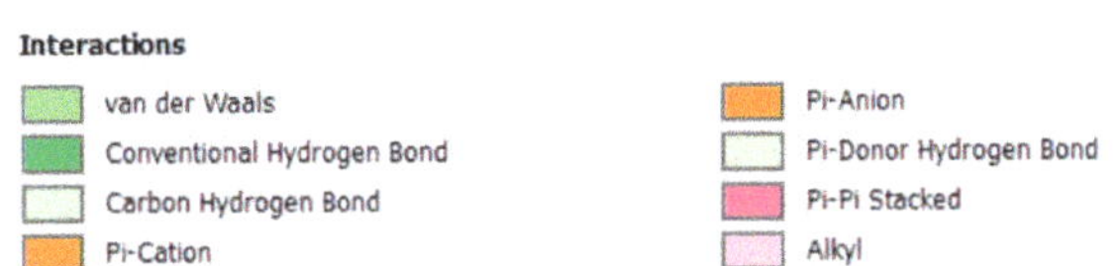

Interactions

van der Waals		Pi-Anion	
Conventional Hydrogen Bond		Pi-Donor Hydrogen Bond	
Carbon Hydrogen Bond		Pi-Pi Stacked	
Pi-Cation		Alkyl	

Figure 3. Two-dimensional scheme of the narirutin interactions with DPP4.

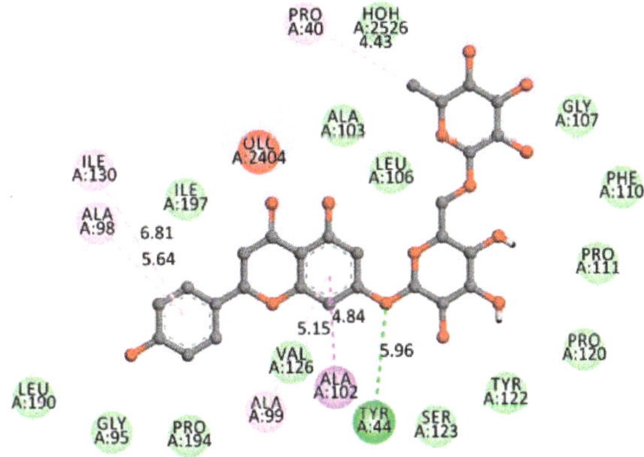

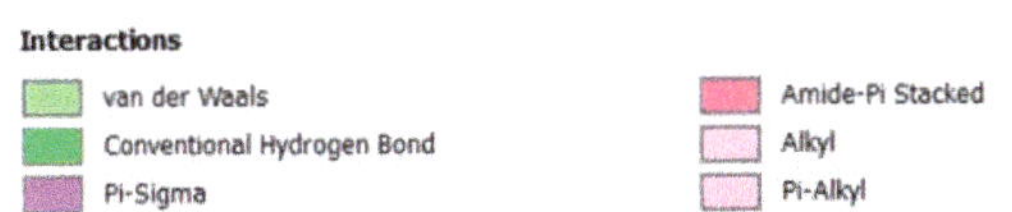

Interactions

van der Waals		Amide-Pi Stacked	
Conventional Hydrogen Bond		Alkyl	
Pi-Sigma		Pi-Alkyl	

Figure 4. Two-dimensional scheme of the narirutin interactions with FFAR1.

2.1.4. Alpha-Amylase

Alpha-amylases are present in a broad range of living creatures (plants, animals, bacteria, and fungi). They are also present in human salivary glands, and the pancreas secretes them into the small intestine on a regular basis during digestion. They belong to the amylase family of enzymes whose main role in the body is to catalyze the hydrolysis of complex polysaccharides into small digestible mono, di and trisaccharide, converted afterwards to glucose as the fuel for energy production [53,54].

Figure 5 shows the interactions between narirutin with the alpha-amylase receptor on the catalytic residues (affinity of −9.9 kcal/mol) with the formation of six hydrogen bonds.

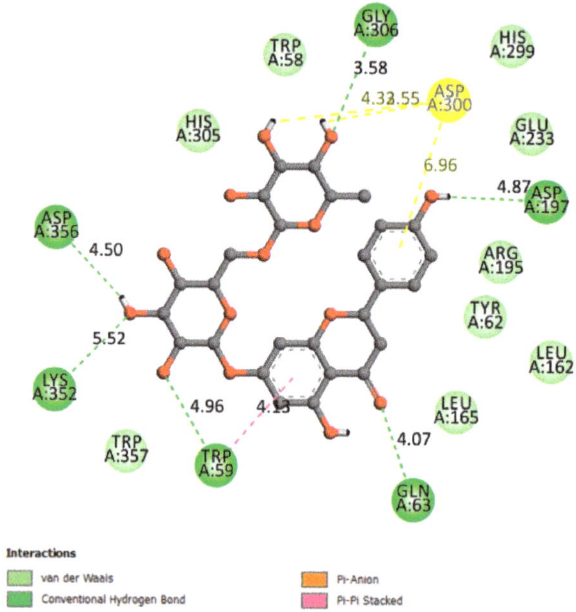

Figure 5. Two-dimensional scheme of the narirutin interactions with AAM.

2.1.5. PPARγ

Peroxisome proliferator-activated receptor gamma (PPARγ) is a protein that controls gene expression and is found in key reproductive components [55]. PPARs are important regulators of cellular differentiation, cellular growth, and carbohydrate, lipid, and protein metabolism in humans. PPARγ has been linked to a key role in glucose homeostasis [56], making it a promising target for T2DM treatment. It is highly expressed in adipose tissue and, for optimum DNA binding and transcriptional activity, needs heterodimerization with the retinoid X-receptor (RXR) [56]. PPARγ stimulates the expression of glucose transporters (GLUT4) and subsequent translocation, leading to decreased levels of blood glucose, blood gluconeogenesis repression and enhanced lipide storage and glucose intake into muscles, and performs a number of psychological activities [57]. Increased insulin sensitizing effect occurs when adipose genes are activated.

Docking findings and various poses analysis revealed no established interaction between narirutin and the PPARγ receptor. The results were confirmed when compared with the default ligand cocrystalized with the receptor PDB format (Rosiglitazone), which binds perfectly into the receptor active site.

2.1.6. Alpha-Glucosidase

Alpha-glucosidase is one of the digestive enzymes, its main function is to accelerate the hydrolysis of polysaccharides (starch) to glucose (acts on (14) bonds) in order to promote glucose absorption and, therefore, raise blood glucose levels [58]. This helps to prevent hyperglycemia and maintain normal blood sugar levels by slowing down the digestion of starch and extra dietary carbohydrates [58].

Figure 6 indicates that narirutin had a good interaction with the receptor active site with 4 hydrogen bonds formed and an affinity of −8.7 kcal/mol.

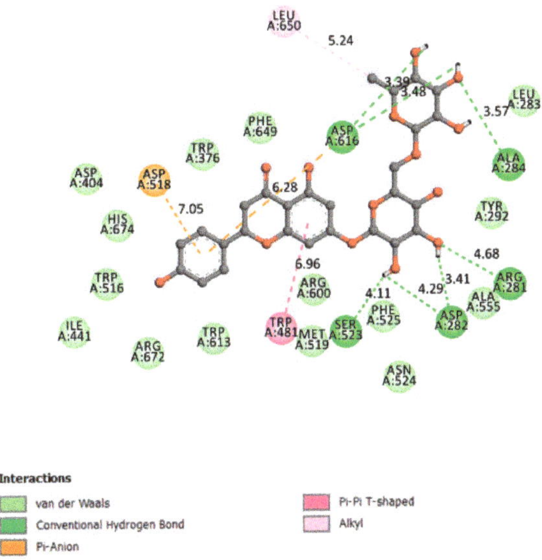

Figure 6. Two-dimensional scheme of the narirutin interactions with AGL.

2.1.7. Aldose Reductase

Aldose reductase is an important enzyme that catalyzes the NAD (P) H-dependent reduction of glucose to sorbitol through the sorbitol-aldose reductase pathway. It is found in numerous organs throughout the body, including red blood cells, retina, Schwann cells, and others. As a consequence of it action, intracellular reactive oxygen species (ROS) accumulate excessively in different organs [59], leading to various health complications [60]. Excessive sorbitol buildup in tissues occurs in T2DM patients with poor sorbitol penetration and metabolism, leading to DM-related problems such as cataracts and glaucoma.

As a result, aldose reductase inhibition has been proposed as a possible treatment for T2DM problems.

The interaction between narirutin and AldR demonstrated one of the best affinities (−9.3 kcal/mol) with the ligand fitting right into the receptor barrel core (active site [41]) with five hydrogen bonds formed (Figure 7).

2.1.8. Glycogen Phosphorylase

Glycogen phosphorylase catalyzes glycogen hydrolysis to produce glucose-1-phosphate. Allosteric effectors and reversible phosphorylation regulate the activity of GP [61]. Phosphorylation control is a widespread intracellular mechanism that regulates, controls, and participates in signal transmission [61]. The liver may store glucose as glycogen and then generate and release glucose into the bloodstream through a reverse mechanism. As a result, the liver is closely connected to glycemic regulation, and hepatic metabolism provides numerous therapeutic options in the setting of T2DM [62].

Figure 8 displays the interaction of narirutin with GP's 280's loop (active site), with an affinity of −8.3 kcal/mol and three hydrogen bonds formed.

The excellent affinities of the receptors/ligand interactions ranged between −8.3 and −10.4 kcal/mol, and narirutin interaction with all receptors demonstrate that this molecule can act as a multitarget medicine to treat diabetes, targeting multiple receptors involved directly and indirectly in overall diabetic status. The in silico study concluded that the predicted mechanism of action exhibited by Narirutin is the inhibition of PTP1B, DPP4, AAM, AGL, AldR, GP, and the activation of FFAR1.

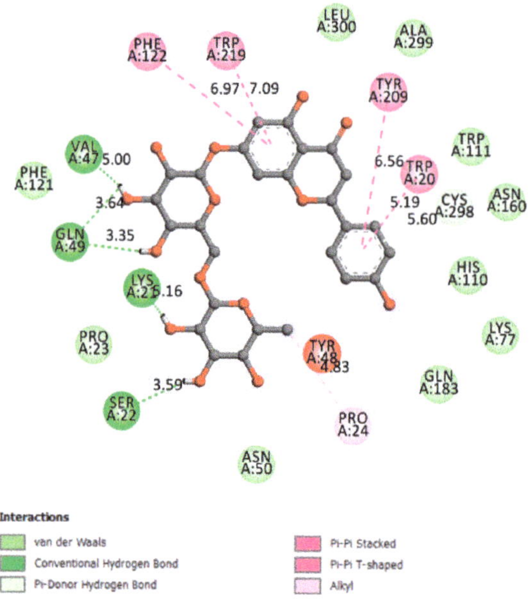

Figure 7. Two-dimensional scheme of the narirutin interactions with AldR.

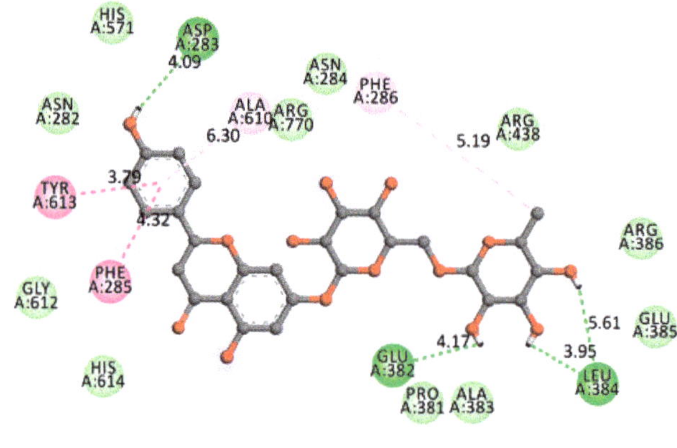

Figure 8. Two-dimensional scheme of the narirutin interactions with Glycogen Phosphorylase receptor.

To have a better insight concerning the obtained results, we compared the obtained results with those obtained with other molecules such as gentisic acid [8] and amentoflavone [63] using the same methodology and against similar receptors. Gentisic acid demonstrated poor affinities ranging between −5.6 and −6.9 kcal/mol, which were partially confirmed with an in vitro investigation in which the molecule also demonstrated low activity. On the other hand, amentoflavone showed excellent affinities ranging between −8.8 and

−11.3 kcal/mol, which was in total accordance with the molecule being well known with antidiabetic potential and tested in vivo [64].

Narirutin in vitro testing was done to partially confirm the results obtained above.

2.2. In Vitro Assays

One of the most effective methods for treating diabetes is to prevent glucose absorption. By blocking the digestive enzymes that hydrolyze polysaccharides into small absorbable pieces, postprandial high blood glucose is avoided. Alpha-glucosidase and alpha-amylase are two of these enzymes. The inhibitory impact of narirutin on these two enzymes were investigated to identify one or more of this plant's mechanisms of action.

2.2.1. Alpha-Amylase Inhibitory Effect

Alpha-amylase is considered as one of the most essential enzymes in the digestive process [34]. because of its critical involvement in the breakdown of polysaccharides. Saliva and pancreatic juice are the two most common places to find it. One approach for avoiding increasing postprandial blood glucose is to target and inhibit this enzyme [39].

Figure 9 shows the ability of narirutin to inhibit alpha-amylase. Since the concentration of narirutin obviously affects the quantity of enzyme inhibited, the inhibition of the enzyme seems to be linked to the dose. The estimated IC50 revealed that acarbose (positive control) had a lower inhibitory potential than narirutin, with an IC50 of 1.012 mg/mL for acarbose compared to 0.0066 mg/mL for narirutin. In comparison to acarbose, narirutin showed outstanding activity, which was consistent with the in silico findings.

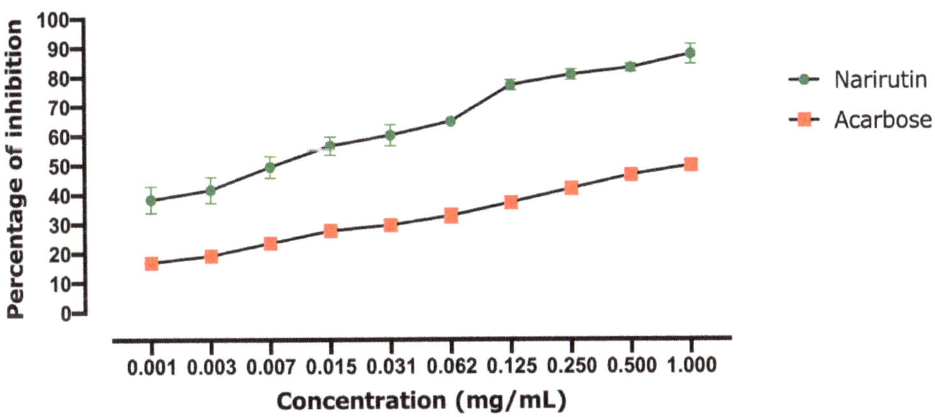

Figure 9. Alpha-amylase inhibitory effect results. Values are expressed as mean ± SD (*n* = 3).

2.2.2. Alpha-Glucosidase Inhibitory Effect

One of the major digesting enzymes is alpha-glucosidase, which is located in the mucosal brush border of the small intestine. Its job is to break down and convert complicated carbohydrates into short, fast, and absorbable sugars. Inhibition is a key strategy for slowing glucose absorption and avoiding high postprandial blood glucose levels, both of which may delay the onset of diabetes.

Figure 10 shows narirutin's ability to block alpha-glucosidase. The inhibitory impact is linked to the narirutin concentration since the highest concentrations showed the greatest inhibition activity. The estimated IC50 revealed that acarbose has a more powerful inhibitory action (IC50 = 0.00035 mg/mL) than that observed with narirutin (IC50 = 0.00091 mg/mL). Nevertheless, the obtained results for narirutin are still considered powerful, and consistent with the docking findings.

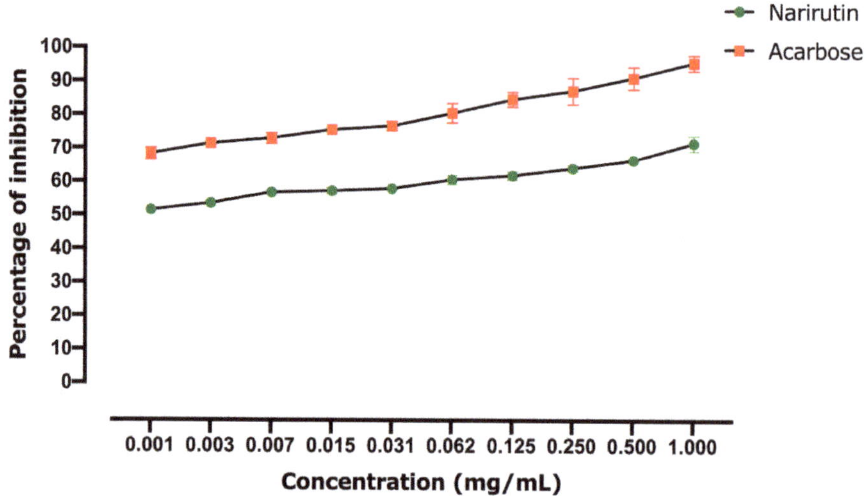

Figure 10. Alpha-glucosidase inhibitory effect results. Values are expressed as mean $\pm$ SD ($n = 3$).

As can be seen from Table 2, the results of the in vitro testing correlate perfectly with those obtained in silico. The difference in the affinity score of narirutin and acarbose against alpha-amylase in silico (-9.9 and -8.1 respectively) was also seen in the in vitro test (0.0066 and 1.012, respectively). Against alpha-glucosidase, almost similar results were obtained in both tests, confirming the accuracy of the performed analysis.

Table 2. Summary of narirutin and acarbose in vitro and in silico activities against alpha-amylase and alpha-glucosidase.

	Alpha-Amylase In Vitro IC50 (mg/mL)	Alpha-Glucosidase In Vitro IC50 (mg/mL)	Alpha-Amylase In Silico Affinity (kcal/mol)	Alpha-Glucosidase In Silico Affinity (kcal/mol)
Narirutin	0.0066	0.00091	-9.9	-8.7
Acarbose	1.012	0.00035	-8.1	-8.4

3. Materials and Methods

3.1. Chemicals and Reagents

Chemicals and reagents used in this study were analytical grade (obtained from Sigma Aldrich, St. Louis, MO, USA). (Narirutin CAS: 14259-46-2; MW: 580.53, alpha-amylase CAS: 9000-85-5, alpha-glucosidase CAS: 9001-42-7).

3.2. Molecular Docking

3.2.1. Preparation of the Ligand

The SDF format of Narirutin (Figure 1) was obtained from PubChem (CID: 442431). The SDF file was converted to the PDBQT format using AutoDock Tools. For the final ligand preparation Gasteiger partial charges were added, rotatable bonds were defined and the nonpolar hydrogen atoms were merged.

3.2.2. Preparation of the Receptors

Each receptor's PDB file was obtained from the protein data bank website [65]. The receptors' X-ray crystal structures were selected for their completeness, resolution, and compatibility with our study goal. Details of the selected receptors are described in Table 3.

Table 3. Selected receptors parameters.

Receptor	PID	Resolution (Å)	Classification
Protein tyrosine phosphatase 1B (PTP1B)	1c83	1.80	Hydrolase
Glycogen phosphorylase (GP)	1l5q	2.25	Transferase
Free fatty acid receptor 1 (FFAR1)	4phu	2.33	Hydrolase
Peroxisome proliferator-activated receptor gamma (PPAR gamma)	5ycp	2.00	Transcription
Alpha-amylase (AAM)	1smd	1.60	Hydrolase
Alpha-glucosidase (AGL)	5nn5	2.00	Hydrolase
Aldose reductase (AR)	2hv5	1.59	Oxidoreductase
dipeptidyl peptidase IV (DPP4)	2p8s	2.20	Hydrolase

For the receptors, Discovery Studio Visualizer v 19.1.0 (BIOVIA, San Diego, CA, USA (windows software)) was used first to begin the preparation of the receptors before running the analysis by deleting water molecules and heteroatoms. AutoDock Tools v1.5.6 (Scripps Research, San Diego, CA, USA) were used later on to add Gasteiger charges and polar hydrogen atoms. The final file was converted to the PDBQT format to prepare for molecular docking.

3.2.3. Simulation

AutoDock Tools were used to establish the grid box size for each receptor, and AutoDock Vina was used to run docking simulations for the narirutin (Ligand) and the eight receptors [66]. The exhaustiveness of the simulation was set to 24. Discovery Studio Visualizer was used to visualize the protein-ligand complexes.

3.3. Narirutin In-Vitro Inhibition Potential on Digestive Enzymes

3.3.1. Alpha-Amylase Inhibition Assay

The test was carried out according to Mechchate procedure [67]. A solution of 0.2 mL of 0.5 M Tris–HCl buffer (pH 6.9) containing 0.01 M $CaCl_2$ was combined with 2 mg of starch to make the substrate solution. The substrate solution was divided into test tubes, boiled for 5 min, and then preincubated for 5 min at 37 °C. Narirutin was dissolved in DMSO and produced at doses ranging from 1 to 1000 g/mL. Porcine pancreatic amylase (0.1 mL in Tris-HCl buffer (2 units/mL) was added to the test tube holding the substrate solution after the narirutin solution (0.2 mL) was added at various concentrations. The reaction was carried out for 10 min at 37 °C before being stopped by adding 0.5 mL of 50% acetic acid to each test tube. The supernatant optical density was measured using a spectrophotometer at 595 nm after centrifugation (3000 rpm for 5 min at 4 °C). Acarbose was utilized as a positive control (alpha-amylase inhibitor) in this test. For each concentration, the tests were performed three times.

The inhibitory activity of alpha-amylase was determined using the formula:

$$\text{alpha-amylase inhibitory activity} = [(A - B)/A] \times 100 \tag{1}$$

where the absorbance of the negative control (solution with only DMSO) is A, whereas that of the sample (Narirutin and acarbose) is B.

After evaluating the inhibitory activity of various concentrations against alpha-amylase, the IC50 values for acarbose and narirutin (concentration needed to inhibit 50% of the enzyme) were calculated.

3.3.2. Alpha-Glucosidase Inhibitory Assay

The test was carried out according to the procedure of Pistia Brueggeman and Hollingsworth [68]. A quantity of 50 microliters of narirutin (concentration ranging between 1–1000 g/mL) was made and incubated for 20 min at 37 °C with a solution comprising 10 microliters of alpha-glucosidase 1 U/mL and 125 µL of 0.1 M phosphate buffer (pH 6.8) A solution of 20 μL of 1 M pNPG (substrate) was added to initiate the reaction, which was then incubated for half an hour. To stop the reaction, 50 microliters of 0.1 N Na_2CO_3 were added. A spectrophotometer was used to measure the optical density at 405 nm. Acarbose was utilized as a positive control in this test. For each concentration, the tests were performed three times.

The inhibitory activity of alpha-glucosidase was determined using the formula:

$$\text{Alpha-glucosidase inhibitory activity} = [(A - B)/A] \times 100 \tag{2}$$

where the absorbance of the negative control (solution with only DMSO) is A, and that of the sample (Narirutin and acarbose) is B After evaluating the inhibitory activity of alpha-glucosidase at various concentrations, the IC50 values for acarbose and narirutin (concentration needed to inhibit 50% of alpha-glucosidase) were calculated.

3.4. Statistical Analysis

Statistical analysis was performed using GraphPad prism version 7 for Windows (GraphPad Software, CA, USA). In vitro analysis was done in three replicates and presented as mean ± SD.

4. Conclusions

The antidiabetic activity of narirutin was remarkable and indicated strong and real potential by inhibition of PTP1B, DPP4, AAM, AGL, AldR, GP, and the activation of FFAR1 with excellent affinities. This conclusion was partially confirmed by in vitro tests, which confirmed the in silico results of the alpha-amylase and alpha-glucosidase experiments. The results were compared to those obtained with other antidiabetic molecules with strong and well-studied mechanism of action. This citrus flavanone deserves further standardization and complete pharmacological study to confirm its overall impact on diabetes prevention, management and also to prevent the complications.

Author Contributions: Conceptualization, A.A.Q., H.M. and I.E.-s.; methodology, M.A.-z., O.M.N. and F.A.N.; software, H.M. and I.E.-s.; validation, M.A., H.I. and M.B. data curation, A.A.Q.; writing—original draft preparation, H.M.; writing—review and editing, A.S.A. and A.M.A.; supervision, A.S.A. All authors have read and agreed to the published version of the manuscript.

Funding: This study was funded by the Deanship of Scientific Research at Imam Mohammad Ibn Saud Islamic University for funding this work through Research Group no. RG-21-09-86.

Institutional Review Board Statement: Not applicable.

Informed Consent Statement: Not applicable.

Data Availability Statement: Data are available upon request.

Acknowledgments: The authors extend their appreciation to the Deanship of Scientific Research at Imam Mohammad Ibn Saud Islamic University for funding this work through Research Group no. RG-21-09-86.

Conflicts of Interest: The authors declare no conflict of interest.

References

1. IDF International Diabetes Federation. *IDF Diabetes Atlas*, 8th ed.; IDF: Brussels, Belgium, 2017.
2. International Diabetes Federation. *IDF Diabetes Atlas*; International Diabetes Federation: Brussels, Belgium, 2015; ISBN 978-2-930229-81-2.
3. Whiting, D.R.; Guariguata, L.; Weil, C.; Shaw, J. IDF Diabetes Atlas: Global Estimates of the Prevalence of Diabetes for 2011 and 2030. *Diabetes Res. Clin. Pract.* **2011**, *94*, 311–321. [CrossRef]
4. IDF. *IDF Diabetes Atlas*, 6th ed.; International Diabetes Federation: Brussels, Belgium, 2013.
5. Saeedi, P.; Petersohn, I.; Salpea, P.; Malanda, B.; Karuranga, S.; Unwin, N.; Colagiuri, S.; Guariguata, L.; Motala, A.A.; Ogurtsova, K.; et al. Global and Regional Diabetes Prevalence Estimates for 2019 and Projections for 2030 and 2045: Results from the International Diabetes Federation Diabetes Atlas, 9th Edition. *Diabetes Res. Clin. Pract.* **2019**, *157*, 107843. [CrossRef] [PubMed]
6. American Diabetes Association. Economic Costs of Diabetes in the US in 2017. *Diabetes Care* **2018**, *41*, 917–928. [CrossRef]

7. Zhou, X.; Ding, L.; Liu, J.; Su, L.; Dong, J.; Liao, L. Efficacy and Short-Term Side Effects of Sitagliptin, Vildagliptin and Saxagliptin in Chinese Diabetes: A Randomized Clinical Trial. *Endocr. Connect.* **2019**, *8*, 318–325. [CrossRef]

8. Mechchate, H.; Es-safi, I.; Mohamed Al kamaly, O.; Bousta, D. Insight into Gentisic Acid Antidiabetic Potential Using In Vitro and In Silico Approaches. *Molecules* **2021**, *26*, 1932. [CrossRef] [PubMed]

9. Khan, M.F.; Kader, F.B.; Arman, M.; Ahmed, S.; Lyzu, C.; Sakib, S.A.; Tanzil, S.M.; Zim, A.I.U.; Imran, M.A.S.; Venneri, T. Pharmacological Insights and Prediction of Lead Bioactive Isolates of Dita Bark through Experimental and Computer-Aided Mechanism. *Biomed. Pharmacother.* **2020**, *131*, 110774. [CrossRef] [PubMed]

10. Es-safi, I.; Mechchate, H.; Amaghnouje, A.; Elbouzidi, A.; Bouhrim, M.; Bencheikh, N.; Hano, C.; Bousta, D. Assessment of Antidepressant-Like, Anxiolytic Effects and Impact on Memory of Pimpinella Anisum L. Total Extract on Swiss Albino Mice. *Plants* **2021**, *10*, 1573. [CrossRef] [PubMed]

11. Iqbal, J.; Abbasi, B.A.; Ahmad, R.; Mahmoodi, M.; Munir, A.; Zahra, S.A.; Shahbaz, A.; Shaukat, M.; Kanwal, S.; Uddin, S. Phytogenic Synthesis of Nickel Oxide Nanoparticles (NiO) Using Fresh Leaves Extract of Rhamnus Triquetra (Wall.) and Investigation of Its Multiple in Vitro Biological Potentials. *Biomedicines* **2020**, *8*, 117. [CrossRef]

12. Wirnitzer, K.C. Vegan nutrition: Latest boom in health and exercise. In *Therapeutic, Probiotic, and Unconventional Foods*; Elsevier: Amsterdam, The Netherlands, 2018; pp. 387–453.

13. Alissa, E.M.; Ferns, G.A. Dietary Fruits and Vegetables and Cardiovascular Diseases Risk. *Crit. Rev. Food Sci. Nutr.* **2017**, *57*, 1950–1962. [CrossRef]

14. Mechchate, H.; Costa de Oliveira, R.; Es-safi, I.; Vasconcelos Mourão, E.M.; Bouhrim, M.; Kyrylchuk, A.; Soares Pontes, G.; Bousta, D.; Grafov, A. Antileukemic Activity and Molecular Docking Study of a Polyphenolic Extract from Coriander Seeds. *Pharmaceuticals* **2021**, *14*, 770. [CrossRef]

15. Al-Ishaq, R.K.; Abotaleb, M.; Kubatka, P.; Kajo, K.; Büsselberg, D. Büsselberg Flavonoids and Their Anti-Diabetic Effects: Cellular Mechanisms and Effects to Improve Blood Sugar Levels. *Biomolecules* **2019**, *9*, 430. [CrossRef] [PubMed]

16. Brodowska, K.M. Natural Flavonoids: Classification, Potential Role, and Application of Flavonoid Analogues. *Eur. J. Biol. Res.* **2017**, *7*, 108–123.

17. Ghasemzadeh, A. Flavonoids and Phenolic Acids: Role and Biochemical Activity in Plants and Human. *J. Med. Plants Res.* **2011**, *5*. [CrossRef]

18. Li, Q.; Wang, Y.; Mai, Y.; Li, H.; Wang, Z.; Xu, J.; He, X. Health Benefits of the Flavonoids from Onion: Constituents and Their Pronounced Antioxidant and Anti-Neuroinflammatory Capacities. *J. Agric. Food Chem.* **2020**, *68*, 799–807. [CrossRef]

19. Abotaleb, M.; Samuel, S.; Varghese, E.; Varghese, S.; Kubatka, P.; Liskova, A.; Büsselberg, D. Flavonoids in Cancer and Apoptosis. *Cancers* **2018**, *11*, 28. [CrossRef] [PubMed]

20. Cushnie, T.P.T.; Lamb, A.J. Antimicrobial Activity of Flavonoids. *Int. J. Antimicrob. Agents* **2005**, *26*, 343–356. [CrossRef]

21. Koolaji, N.; Shanmugasamy, B.; Schindeler, A.; Dong, Q.; Dehghani, F.; Valtchev, P. Citrus Peel Flavonoids as Potential Cancer Prevention Agents. *Curr. Dev. Nutr.* **2020**, *4*, nzaa025. [CrossRef]

22. Graf, B.A.; Milbury, P.E.; Blumberg, J.B. Flavonols, Flavones, Flavanones, and Human Health: Epidemiological Evidence. *J. Med. Food* **2005**, *8*, 281–290. [CrossRef]

23. Barreca, D.; Gattuso, G.; Bellocco, E.; Calderaro, A.; Trombetta, D.; Smeriglio, A.; Laganà, G.; Daglia, M.; Meneghini, S.; Nabavi, S.M. Flavanones: Citrus Phytochemical with Health-promoting Properties. *BioFactors* **2017**, *43*, 495–506. [CrossRef] [PubMed]

24. Scafuri, B.; Bontempo, P.; Altucci, L.; De Masi, L.; Facchiano, A. Molecular Docking Simulations on Histone Deacetylases (HDAC)-1 and -2 to Investigate the Flavone Binding. *Biomedicines* **2020**, *8*, 568. [CrossRef]

25. Velmurugan, B.K.; Lin, J.-T.; Mahalakshmi, B.; Chuang, Y.-C.; Lin, C.-C.; Lo, Y.-S.; Hsieh, M.-J.; Chen, M.-K. Luteolin-7-O-Glucoside Inhibits Oral Cancer Cell Migration and Invasion by Regulating Matrix Metalloproteinase-2 Expression and Extracellular Signal-Regulated Kinase Pathway. *Biomolecules* **2020**, *10*, 502. [CrossRef]

26. Jia, Z.; Barford, D.; Flint, A.; Tonks, N. Structural Basis for Phosphotyrosine Peptide Recognition by Protein Tyrosine Phosphatase 1B. *Science* **1995**, *268*, 1754–1758. [CrossRef] [PubMed]

27. Johnson, T.O.; Ermolieff, J.; Jirousek, M.R. Protein Tyrosine Phosphatase 1B Inhibitors for Diabetes. *Nat. Rev. Drug Discov.* **2002**, *1*, 696–709. [CrossRef]

28. Wiesmann, C.; Barr, K.J.; Kung, J.; Zhu, J.; Erlanson, D.A.; Shen, W.; Fahr, B.J.; Zhong, M.; Taylor, L.; Randal, M.; et al. Allosteric Inhibition of Protein Tyrosine Phosphatase 1B. *Nat. Struct. Mol. Biol.* **2004**, *11*, 730–737. [CrossRef] [PubMed]

29. Bjelke, J.R.; Christensen, J.; Branner, S.; Wagtmann, N.; Olsen, C.; Kanstrup, A.B.; Rasmussen, H.B. Tyrosine 547 Constitutes an Essential Part of the Catalytic Mechanism of Dipeptidyl Peptidase IV. *J. Biol. Chem.* **2004**, *279*, 34691–34697. [CrossRef] [PubMed]

30. Chien, C.-H.; Tsai, C.-H.; Lin, C.-H.; Chou, C.-Y.; Chen, X. Identification of Hydrophobic Residues Critical for DPP-IV Dimerization [†]. *Biochemistry* **2006**, *45*, 7006–7012. [CrossRef]

31. Morgan, N.G.; Dhayal, S. G-Protein Coupled Receptors Mediating Long Chain Fatty Acid Signalling in the Pancreatic Beta-Cell. *Biochem. Pharmacol.* **2009**, *78*, 1419–1427. [CrossRef]

32. Sum, C.S.; Tikhonova, I.G.; Costanzi, S.; Gershengorn, M.C. Two Arginine-Glutamate Ionic Locks Near the Extracellular Surface of FFAR1 Gate Receptor Activation. *J. Biol. Chem.* **2009**, *284*, 3529–3536. [CrossRef] [PubMed]

33. Sum, C.S.; Tikhonova, I.G.; Neumann, S.; Engel, S.; Raaka, B.M.; Costanzi, S.; Gershengorn, M.C. Identification of Residues Important for Agonist Recognition and Activation in GPR40. *J. Biol. Chem.* **2007**, *282*, 29248–29255. [CrossRef]

34. Hsiu, J.; Fischer, E.H.; Stein, E.A. Alpha-amylases as calcium-metalloenzymes. II. Calcium and the catalytic activity. *Biochemistry* **1964**, *3*, 61–66. [CrossRef] [PubMed]

35. Ragunath, C.; Manuel, S.G.A.; Venkataraman, V.; Sait, H.B.R.; Kasinathan, C.; Ramasubbu, N. Probing the Role of Aromatic Residues at the Secondary Saccharide-Binding Sites of Human Salivary α-Amylase in Substrate Hydrolysis and Bacterial Binding. *J. Mol. Biol.* **2008**, *384*, 1232–1248. [CrossRef] [PubMed]

36. Ramasubbu, N.; Ragunath, C.; Sundar, K.; Mishra, P.J.; Gyémánt, G.; Kandra, L. Structure-Function Relationships in Human Salivary α-Amylase: Role of Aromatic Residues. *Biol.-Sect. Cell. Mol. Biol.* **2005**, *60*, 47–56.

37. Ramasubbu, N.; Ragunath, C.; Mishra, P.J.; Thomas, L.M.; Gyemant, G.; Kandra, L. Human Salivary Alpha-Amylase Trp58 Situated at Subsite -2 Is Critical for Enzyme Activity. *Eur. J. Biochem.* **2004**, *271*, 2517–2529. [CrossRef]

38. Pochetti, G.; Godio, C.; Mitro, N.; Caruso, D.; Galmozzi, A.; Scurati, S.; Loiodice, F.; Fracchiolla, G.; Tortorella, P.; Laghezza, A.; et al. Insights into the Mechanism of Partial Agonism: Crystal Structures Of The Peroxisome Proliferator-Activated Receptor Γ Ligand-Binding Domain In The Complex With Two Enantiomeric Ligands. *J. Biol. Chem.* **2007**, *282*, 17314–17324. [CrossRef] [PubMed]

39. Alqahtani, A.S.; Hidayathulla, S.; Rehman, M.T.; ElGamal, A.A.; Al-Massarani, S.; Razmovski-Naumovski, V.; Alqahtani, M.S.; El Dib, R.A.; AlAjmi, M.F. Alpha-Amylase and Alpha-Glucosidase Enzyme Inhibition and Antioxidant Potential of 3-Oxolupenal and Katononic Acid Isolated from Nuxia Oppositifolia. *Biomolecules* **2019**, *10*, 61. [CrossRef] [PubMed]

40. Hermans, M.M.; Kroos, M.A.; van Beeumen, J.; Oostra, B.A.; Reuser, A.J. Human Lysosomal Alpha-Glucosidase. Characterization of the Catalytic Site. *J. Biol. Chem.* **1991**, *266*, 13507–13512. [CrossRef]

41. Wilson, D.K.; Bohren, K.M.; Gabbay, K.H.; Quiocho, F.A. An Unlikely Sugar Substrate Site in the 1.65 A Structure of the Human Aldose Reductase Holoenzyme Implicated in Diabetic Complications. *Science* **1992**, *257*, 81–84. [CrossRef]

42. Barford, D.; Johnson, L.N. The Allosteric Transition of Glycogen Phosphorylase. *Nature* **1989**, *340*, 609–616. [CrossRef] [PubMed]

43. Goldsmith, E.; Sprang, S.R.; Hamlin, R.; Xuong, N.; Fletterick, R. Domain Separation in the Activation of Glycogen Phosphorylase a. *Science* **1989**, *245*, 528–532. [CrossRef]

44. Elchebly, M. Increased Insulin Sensitivity and Obesity Resistance in Mice Lacking the Protein Tyrosine Phosphatase-1B Gene. *Science* **1999**, *283*, 1544–1548. [CrossRef]

45. Goldstein, B.J. Protein-Tyrosine Phosphatase 1B (PTP1B): A Novel Therapeutic Target for Type 2 Diabetes Mellitus, Obesity and Related States of Insulin Resistance. *Curr. Drug Targets Immune Endocr. Metab. Disord.* **2001**, *1*, 265–275. [CrossRef] [PubMed]

46. Ketsawatsomkron, P.; Stepp, D.W.; de Chantemèle, E.J.B. PTP1B in Obesity-Related Cardiovascular Function. In *Protein Tyrosine Phosphatase Control of Metabolism*; Bence, K.K., Ed.; Springer: New York, NY, USA, 2013; pp. 129–145. ISBN 978-1-4614-7855-3.

47. Zhang, Z.-Y.; Lee, S.-Y. PTP1B Inhibitors as Potential Therapeutics in the Treatment of Type 2 Diabetes and Obesity. *Expert Opin. Investig. Drugs* **2003**, *12*, 223–233. [CrossRef]

48. Matteucci, E.; Giampietro, O. Dipeptidyl Peptidase-4 (CD26): Knowing the Function before Inhibiting the Enzyme. *Curr. Med. Chem.* **2009**, *16*, 2943–2951. [CrossRef] [PubMed]

49. Kim, S.-H.; Jung, E.; Yoon, M.K.; Kwon, O.H.; Hwang, D.-M.; Kim, D.-W.; Kim, J.; Lee, S.-M.; Yim, H.J. Pharmacological Profiles of Gemigliptin (LC15-0444), a Novel Dipeptidyl Peptidase-4 Inhibitor, in Vitro and in Vivo. *Eur. J. Pharmacol.* **2016**, *788*, 54–64. [CrossRef]

50. Abd El-Karim, S.S.; Anwar, M.M.; Syam, Y.M.; Nael, M.A.; Ali, H.F.; Motaleb, M.A. Rational Design and Synthesis of New Tetralin-Sulfonamide Derivatives as Potent Anti-Diabetics and DPP-4 Inhibitors: 2D & 3D QSAR, in Vivo Radiolabeling and Bio Distribution Studies. *Bioorg. Chem.* **2018**, *81*, 481–493. [CrossRef]

51. Ren, X.-M.; Cao, L.-Y.; Zhang, J.; Qin, W.-P.; Yang, Y.; Wan, B.; Guo, L.-H. Investigation of the Binding Interaction of Fatty Acids with Human G Protein-Coupled Receptor 40 Using a Site-Specific Fluorescence Probe by Flow Cytometry. *Biochemistry* **2016**, *55*, 1989–1996. [CrossRef]

52. Srivastava, A.; Yano, J.; Hirozane, Y.; Kefala, G.; Gruswitz, F.; Snell, G.; Lane, W.; Ivetac, A.; Aertgeerts, K.; Nguyen, J.; et al. High-Resolution Structure of the Human GPR40 Receptor Bound to Allosteric Agonist TAK-875. *Nature* **2014**, *513*, 124–127. [CrossRef] [PubMed]

53. Suvd, D.; Fujimoto, Z.; Takase, K.; Matsumura, M.; Mizuno, H. Crystal Structure of Bacillus Stearothermophilus Alpha-Amylase: Possible Factors Determining the Thermostability. *J. Biochem. (Tokyo)* **2001**, *129*, 461–468. [CrossRef]

54. Aghajari, N.; Feller, G.; Gerday, C.; Haser, R. Crystal Structures of the Psychrophilic Alpha-Amylase from Alteromonas Haloplanctis in Its Native Form and Complexed with an Inhibitor. *Protein Sci. Publ. Protein Soc.* **1998**, *7*, 564–572. [CrossRef]

55. Ahmadian, M.; Suh, J.M.; Hah, N.; Liddle, C.; Atkins, A.R.; Downes, M.; Evans, R.M. PPARγ Signaling and Metabolism: The Good, the Bad and the Future. *Nat. Med.* **2013**, *19*, 557–566. [CrossRef]

56. Rees, W.D.; McNeil, C.J.; Maloney, C.A. The Roles of PPARs in the Fetal Origins of Metabolic Health and Disease. *PPAR Res.* **2008**, *2008*, 1–8. [CrossRef]

57. Variya, B.C.; Bakrania, A.K.; Patel, S.S. Antidiabetic Potential of Gallic Acid from Emblica Officinalis: Improved Glucose Transporters and Insulin Sensitivity through PPAR-γ and Akt Signaling. *Phytomedicine* **2020**, *73*, 152906. [CrossRef]

58. Attjioui, M.; Ryan, S.; Ristic, A.K.; Higgins, T.; Goñi, O.; Gibney, E.R.; Tierney, J.; O'Connell, S. Comparison of Edible Brown Algae Extracts for the Inhibition of Intestinal Carbohydrate Digestive Enzymes Involved in Glucose Release from the Diet. *J. Nutr. Sci.* **2021**, *10*, e5. [CrossRef]

59. Tang, W.H.; Martin, K.A.; Hwa, J. Aldose Reductase, Oxidative Stress, and Diabetic Mellitus. *Front. Pharmacol.* **2012**, *3*. [CrossRef]

60. Heather, L.C.; Clarke, K. Metabolism, Hypoxia and the Diabetic Heart. *J. Mol. Cell. Cardiol.* **2011**, *50*, 598–605. [CrossRef]
61. Livanova, N.B.; Chebotareva, N.A.; Eronina, T.B.; Kurganov, B.I. Pyridoxal 5'-Phosphate as a Catalytic and Conformational Cofactor of Muscle Glycogen Phosphorylase B. *Biochem. Biokhimiia* **2002**, *67*, 1089–1098. [CrossRef]
62. Ashworth, W.B.; Davies, N.A.; Bogle, I.D.L. A Computational Model of Hepatic Energy Metabolism: Understanding Zonated Damage and Steatosis in NAFLD. *PLoS Comput. Biol.* **2016**, *12*, e1005105. [CrossRef]
63. Mechchate, H.; Es-Safi, I.; Bourhia, M.; Kyrylchuk, A.; El Moussaoui, A.; Conte, R.; Ullah, R.; Ezzeldin, E.; Mostafa, G.A.; Grafov, A.; et al. In-Vivo Antidiabetic Activity and In-Silico Mode of Action of LC/MS-MS Identified Flavonoids in Oleaster Leaves. *Molecules* **2020**, *25*, 5073. [CrossRef]
64. Su, C.; Yang, C.; Gong, M.; Ke, Y.; Yuan, P.; Wang, X.; Li, M.; Zheng, X.; Feng, W. Antidiabetic Activity and Potential Mechanism of Amentoflavone in Diabetic Mice. *Molecules* **2019**, *24*, 2184. [CrossRef]
65. Berman, H.M.; Battistuz, T.; Bhat, T.N.; Bluhm, W.F.; Bourne, P.E.; Burkhardt, K.; Feng, Z.; Gilliland, G.L.; Iype, L.; Jain, S.; et al. The Protein Data Bank. *Acta Crystallogr. D Biol. Crystallogr.* **2002**, *58*, 899–907. [CrossRef]
66. Trott, O.; Olson, A.J. AutoDock Vina: Improving the Speed and Accuracy of Docking with a New Scoring Function, Efficient Optimization, and Multithreading. *J. Comput. Chem.* **2010**, *31*, 455–461. [CrossRef]
67. Mechchate, H.; Es-Safi, I.; Louba, A.; Alqahtani, A.S.; Nasr, F.A.; Noman, O.M.; Farooq, M.; Alharbi, M.S.; Alqahtani, A.; Bari, A.; et al. In Vitro Alpha-Amylase and Alpha-Glucosidase Inhibitory Activity and In Vivo Antidiabetic Activity of Withania Frutescens L. Foliar Extract. *Molecules* **2021**, *26*, 293. [CrossRef]
68. Pistia-Brueggeman, G.; Hollingsworth, R.I. A Preparation and Screening Strategy for Glycosidase Inhibitors. *Tetrahedron* **2001**, *57*, 8773–8778. [CrossRef]

 *pharmaceutics*

Article

Evidence of Insulin-Sensitizing and Mimetic Activity of the Sesquiterpene Quinone Avarone, a Protein Tyrosine Phosphatase 1B and Aldose Reductase Dual Targeting Agent from the Marine Sponge *Dysidea avara*

Marcello Casertano [1,†], Massimo Genovese [2,†], Alice Santi [2], Erica Pranzini [2], Francesco Balestri [3,4], Lucia Piazza [3], Antonella Del Corso [3,4], Sibel Avunduk [5], Concetta Imperatore [1], Marialuisa Menna [1,*] and Paolo Paoli [2,*]

[1] Department of Pharmacy, University of Naples "Federico II", Via D. Montesano 49, 80131 Naples, Italy
[2] Department of Experimental and Clinical Biomedical Sciences, University of Florence, Viale Morgagni 50, 50134 Florence, Italy
[3] Biochemistry Unit, Department of Biology, University of Pisa, Via S. Zeno 51, 56123 Pisa, Italy
[4] Interdepartmental Research Center for Marine Pharmacology, Via Bonanno 6, 56126 Pisa, Italy
[5] Medical Laboratory Programme, Vocational School of Health Care, Mugla University, Marmaris 48187, Turkey
[*] Correspondence: mlmenna@unina.it (M.M.); paolo.paoli@unifi.it (P.P.); Tel.: +39-081678518 (M.M.); +39-0552751248 (P.P.)
[†] These authors contributed equally to this work.

Citation: Casertano, M.; Genovese, M.; Santi, A.; Pranzini, E.; Balestri, F.; Piazza, L.; Del Corso, A.; Avunduk, S.; Imperatore, C.; Menna, M.; et al. Evidence of Insulin-Sensitizing and Mimetic Activity of the Sesquiterpene Quinone Avarone, a Protein Tyrosine Phosphatase 1B and Aldose Reductase Dual Targeting Agent from the Marine Sponge *Dysidea avara*. Pharmaceutics **2023**, 15, 528. https://doi.org/10.3390/pharmaceutics15020528

Academic Editors: Diana Marcela Aragon Novoa and Fátima Regina Mena Barreto Silva

Received: 15 December 2022
Revised: 18 January 2023
Accepted: 31 January 2023
Published: 4 February 2023

Abstract: Type 2 diabetes mellitus (T2DM) is a complex disease characterized by impaired glucose homeostasis and serious long-term complications. First-line therapeutic options for T2DM treatment are monodrug therapies, often replaced by multidrug therapies to ensure that non-responding patients maintain target glycemia levels. The use of multitarget drugs instead of mono- or multidrug therapies has been emerging as a main strategy to treat multifactorial diseases, including T2DM. Therefore, modern drug discovery in its early stages aims to identify potential modulators for multiple targets; for this purpose, exploration of the chemical space of natural products represents a powerful tool. Our study demonstrates that avarone, a sesquiterpene quinone obtained from the sponge *Dysidea avara*, is capable of inhibiting in vitro PTP1B, the main negative regulator of the insulin receptor, while it improves insulin sensitivity, and mitochondria activity in C2C12 cells. We observe that when avarone is administered alone, it acts as an insulin-mimetic agent. In addition, we show that avarone acts as a tight binding inhibitor of aldose reductase (AKR1B1), the enzyme involved in the development of diabetic complications. Overall, avarone could be proposed as a novel natural hit to be developed as a multitarget drug for diabetes and its pathological complications.

Keywords: marine natural products; *Dysidea avara*; sesquiterpene-type quinones; avarone; aldose reductase; protein tyrosine phosphatase 1B; dual target inhibitors; diabetes mellitus

1. Introduction

Insulin, a peptide hormone secreted by β-cells of the pancreas, represents the principal regulator of glycemic homeostasis. Diabetes mellitus (DM) is characterized by hyperglycemia that results from decreased tissue response to insulin and/or inadequate insulin secretion in the complex pathways of hormone action. The traditional classification of diabetes comprises four categories: gestational diabetes, Type 1, Type 2, and specific types of diabetes caused by other factors. Long-term diabetic complications are related to chronic hyperglycemia and consist of cataracts, cardiovascular complications, nephropathy, retinopathy, and neuropathy [1]. According to the World Health Organization (WHO) data, today, more than 420 million people are living with diabetes worldwide. This number is estimated to rise to 570 million by 2030 and 700 million by 2045 [2] and, contrary to the other main non-communicable diseases, the premature mortality for diabetes has increased by 5% from 2000 to 2016 [3]. Furthermore, it is thought that all these predictions could be negatively altered by COVID-19 [4]. The WHO and the International Diabetes Federation are working together to warrant the best quality of life possible for people with diabetes worldwide. In this context, there is a steadily developing need to increase and improve DM treatment approaches by discovering new, more effective and safer therapeutic options. This is relevant today more than ever, as unfortunately, the poor outcome from COVID-19 of people affected by DM has been shown dramatically.

Developing new therapeutics for DM treatment/prevention strongly relies on a detailed knowledge of the molecular mechanisms and cellular pathways underlying the onset of insulin resistance. The insulin signal pathway determines the activation of the insulin receptor (IR)/insulin receptor substrate (IRS)/phosphatidylinositol 3-kinase (PI3K)/protein kinase B (Akt) promoting the uptake of peripheral glucose through GLUT transporters, glycogen synthesis, and other metabolic effects [1]. Among organs that are targets of insulin, adipose tissue responds to endocrine and metabolic signals by modulating a variety of adipokines and cytokines that can affect locally and systemically [5]. Dysfunction in this crosstalk may become deleterious, causing low-grade inflammation and insulin resistance [5]. Aldose Reductase (AKR1B1) and Protein Tyrosine Phosphatase 1B (PTP1B) represent two insulin-resistance-related targets involved in the development of type 2 diabetes mellitus (T2DM) and its chronic complications. AKR1B1 is a cytosolic enzyme that catalyzes the reduction of aldoses to the corresponding alcohols [6], thus being the rate-limiting step of the polyol pathway. This leads to several metabolic alterations related to an increase in NF-kB expression and ROS generation, resulting in oxidative stress, induction of an inflammatory response, and the impairment of the glutathione reductase-dependent recovery of reduced glutathione. Under these conditions, a ROS increase induces lipid peroxidation with the consequent formation of cytotoxic aldehydes, such as 4-hydroxy-trans-2-nonenal (HNE) and its glutathione conjugate. AKR1B1 catalyzes reduction of this last metabolite to the corresponding alcohol (GS-DHN) [7], which also promotes the expression of NF-κB. Thus, the enhanced activity of AKR1B1 promotes not only the onset of oxidative stress linked to insulin resistance but also the long-term complications linked to T2DM [6,8,9]. PTP1B is one of the most important enzymes involved in the regulation of the insulin signaling pathway as it catalyzes the dephosphorylation of phosphotyrosine residues of IR and IRS, thus damping the receptor response following insulin binding [10]. PTP1B overexpression has been related to the onset of insulin resistance, and knockout mice for PTP1B do not develop insulin resistance even if they are fed a high-fat diet [11]. Moreover, inside neurons of the arcuate nucleus of the hypothalamic region, PTP1B catalyzes dephosphorylation of the JAK2 tyrosine kinase associated with the leptin (an adipokine that promotes anorexigenic effects) receptor. Increased PTP1B activity in the hypothalamic region is responsible for the upregulation of orexigenic signals, thereby promoting hyperphagia that, in turn, causes obesity. Both PTP1B and AKR1B1 are, thus, currently emerging as suitable targets to be modulated with drug-like small molecules to discover new therapeutics for insulin resistance and obesity [12,13]. Moreover, in a polypharmacology strategy, dual inhibitors of PTP1B and AKR1B1 could be used as multitarget drugs to treat both conditions of insulin resistance and chronic complications associated with T2DM [14].

Natural Products (NPs) are an excellent resource of chemical diversity for drug discovery [15,16]. In addition to providing bioactive chemical scaffolds for medicinal chemistry optimization, NPs are ideal chemical probes to determine the biochemical pathways underlying a given disease and to discover novel drug targets. Natural quinone/hydroquinone (Q/HQ) compounds are privileged scaffolds, due to their capability to participate in important biological redox processes and to affect radical production, resulting in a wide variety of pharmacological effects [17–24], as well as due to their high prevalence in the environment [17,18]. Recently, we undertook research to identify new chemotypes active as dual inhibitors of PTP1B and AKR1B1 using NPs of marine origin [25,26]. A significant inhibitory activity against several phosphatases, including PTP1B, by sesquiterpenes quinones of marine origin was reported [24,27]. On the other hand, antidiabetic properties due to aldose reductase inhibition were evidenced for benzoquinones, anthraquinones, and naphthoquinones, which were shown to all be noncompetitive inhibitors of the enzyme [17]. We had the opportunity to investigate a sample of the marine sponge *Dysidea avara* (Schmidt, 1862) collected along the coast of Turkey (İzmir Bay, Aegean Sea) and we isolated the sesquiterpene Qs/HQs avarol (**1**), its oxidized form avarone (**2**), and methylamine derivatives of avarone (**3** and **4**, Figure 1) [19]. The great potential for new drug discovery of these compounds as well as other structurally related secondary metabolites (e.g., ilimaquinone [28,29]) is evidenced by the many biological effects reported, including anti-tumor, anti-inflammatory, anti-mutagenic, anti-bacterial, anti-viral, anti-psoriatic, anti-parasitic, radical scavenger, inhibition of the lipid peroxidation, and anti-biofouling activities [19–22,30–37]. In the present study, we evaluated the inhibition effects of the sesquiterpene quinones **1–4** against PTP1B and AKR1B1 enzymes. The obtained results are encouraging for the possibility of optimizing and developing the sesquiterpene quinone scaffold of avarone (**2**) for the discovery of newly designed multiple ligands (DMLs) for the treatment of T2DM.

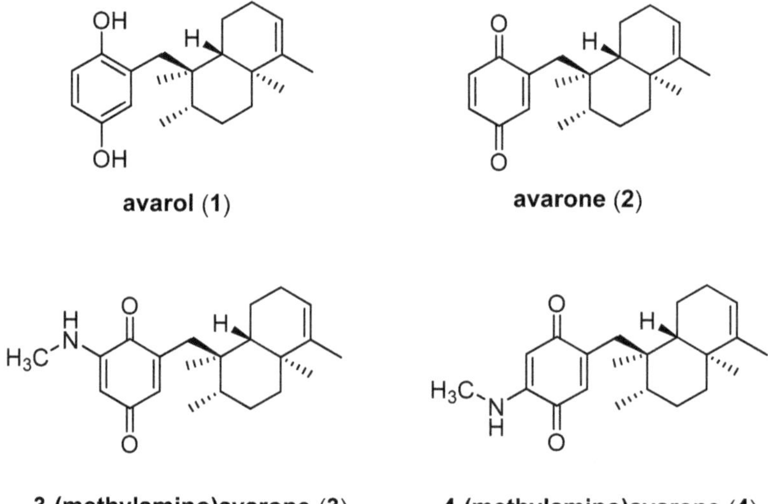

Figure 1. Structures of the marine-derived hydroquinone avarol (**1**), its oxidized form avarone (**2**), and the methylamine derivatives of avarone (**3** and **4**) isolated from the sponge *D. avara*.

2. Materials and Methods

2.1. Materials

The commercial solvents and deuterated solvents used for extraction and HPLC purification were purchased from Sigma-Aldrich (Saint Louis, MO, USA) and were used without further purification. TLC (Silica Gel 60 F254, plates 5 × 20, 0.25 mm) were purchased from Merck (Kenilworth, NJ, USA). NADPH and L-idose were from Carbosynth (Compton, UK). P-nitrophenylphosphate, pNPP and p-IR β subunit Y1162/1163 (sc-25103-R), β-actin clone C-4 (sc-47778), and anti-rabbit and anti-mouse secondary antibodies were from Santa Cruz Biotechnology (Dallas, TX, USA); IRβ subunit antibody clone CT-3 (MABS65) was from Merck-Millipore (Burlington, MA, USA); p-protein kinase B (pAkt) (9271S) and Akt (9272S) antibodies were from Cell Signaling Technology (Danvers, Massachusetts, USA). All other reagents, unless otherwise specified, were obtained from Merck Life Science S.r.l. (St. Louis, MO, USA). High-resolution electrospray ionization−mass spectrometry (ESI-MS) was performed on a Thermo LTQ Orbitrap XL spectrometer (Thermo-Fisher, San Josè,

CA, USA). The MS spectra of compounds **1–4** were recorded by infusion into the ESI source using methanol (MeOH) as a solvent. The ^{1}H (700 MHz) and ^{13}C (175 MHz) NMR experiments were carried out on a Bruker Avance Neo spectrometer (Bruker BioSpin Corporation, Billerica, MA, USA); chemical shifts were reported in parts per million (ppm) and referenced to the residual solvent signal (CDCl$_3$: $\delta_H = 7.26$; $\delta_C = 77.0$) [38,39]. High-performance liquid chromatography (HPLC) separation was achieved on Knauer K-501 and Shimadzu LC-10AT (Shimadzu, Milan, Italy) apparatuses both equipped with a Knauer K-2301 RI detector (LabService Analytica s.r.l., Anzola dell'Emilia, Italy).

2.2. Isolation of Sesquiterpenes **1–4** from the Sponge D. avara

The hydroquinone avarol (**1**) and its oxidized form, the quinone avarone (**2**), were obtained in pure form by extraction of a sample of the Aegean sponge *D. avara* available at the Department of Pharmacy of the University of Naples Federico II. Indeed, a sample of the sponge (collected in Narlidere, Bay of Izmir, Turkey, 38°24′45 N 27°8′18 E) was taken from the existing collection of natural sources, and it was extracted according to our previously reported procedure [19,25,40,41]. Indeed, specimens of *D. avara* were thawed, homogenized, and first extracted with methanol (3 × 500 mL) at room temperature and, subsequently, with chloroform (3 × 500 mL), providing different extracts that were combined and concentrated in vacuo. The obtained suspension was dissolved in water, partitioned first with ethyl acetate and then with n-BuOH. The latter extract (800 mg after evaporation of the solvent as dark brown oil) was chromatographed on a RP-18 silica gel flash column, as reported in [19]. The most interesting fraction, labeled as A (23.2 mg), was eluted with MeOH/H$_2$O 8:2 (v/v), showing the presence of diagnostic NMR signals related to hydroquinone/quinone derivatives. Accordingly, fraction A was subjected to HPLC purification on an RP-18 column (Luna, 5 μm C-18, 250 × 4.6 mm, flow rate = 1.00 mL/min), using MeOH/H$_2$O 95:5 (*v/v*) as the mobile phase. This analysis afforded avarol (**1**, 6.2 mg, $t_R = 5.1$ min), and avarone (**2**, 7.7 mg, $t_R = 9.0$ min) in pure form. As for the ethyl acetate soluble material (550 mg), it was chromatographed by the open column on silica gel with the elution starting from 100% n-hexane and gradually increasing to 100% EtOAc. Then, it was followed by a step gradient of CH$_2$Cl$_2$:MeOH in different concentrations, and increased up to 100% MeOH. The fraction eluted with CH$_2$Cl$_2$:MeOH 8:2 (*v/v*), labeled as fraction B (13.6 mg), was chromatographed on an RP-18 column (Luna, 5 μm C-18, 250 × 4.6 mm, flow rate = 1.00 mL/min), with a mixture MeOH/H$_2$O 9:1 (*v/v*), affording compounds **3** (2.2 mg, $t_R = 14.4$ min) and **4** (2.8 mg, $t_R = 16.6$ min) in the pure state. The identity of the isolated compounds **1–4** was confirmed by comparison of their spectroscopic properties (^{1}H and HRESI-MS) with those reported in the literature [19,36,37,42]. Moreover, their purity (≥99.7%) was confirmed by HPLC analyses.

Avarol (**1**): yellow powder; ^{1}H NMR (CDCl$_3$) and HRESI-MS spectra are reported in Supplementary Materials (Figures S1 and S2); HRESI-MS *m/z* 315.2320 [M+H]$^+$ (calcd. for C$_{21}$H$_{31}$O$_2$: 315.2319).

Avarone (**2**): yellow powder; ^{1}H NMR (CDCl$_3$) and HRESI-MS spectra are reported in Supplementary Materials (Figures S3 and S4); HRESI-MS *m/z* 313.2166 [M+H]$^+$ (calcd. for C$_{21}$H$_{29}$O$_2$: 313.2162).

3-(methylamino)avarone (**3**): red powder; ^{1}H NMR (CDCl$_3$) and HRESI-MS spectra are reported in Supplementary Materials (Figures S5 and S6); HRESI-MS *m/z* 342.2428 [M+H]$^+$, and *m/z* 364.2247 [M+Na]$^+$ (calcd. for C$_{22}$H$_{32}$NO$_2$: 342.2428).

4-(methylamino)avarone (**4**): red powder; ^{1}H NMR (CDCl$_3$) and HRESI-MS spectra are reported in Supplementary Materials (Figures S7 and S8); HRESI-MS *m/z* 342.2426 [M+H]$^+$, and *m/z* 364.2245 [M+Na]$^+$ (calcd. for C$_{22}$H$_{32}$NO$_2$: 342.2428).

2.3. Enzymes Expression and Purification

The purification of human recombinant AKR1B1 was performed as previously described [43]. The specific activity of the purified enzyme was 5.3 U/mg. The purified enzyme, stored at −80 °C in 10 mM sodium phosphate buffer pH 7.0 containing 2 mM DTT and 30% (*w/v*) glycerol, was extensively dialyzed against 10 mM sodium phosphate buffer pH 7.0 before use. Human PTP1B coding sequence (1-302 aa) was cloned into the bacterial expression vector downstream and in frame with the polyHis tag. After expression, recombinant protein was purified using a chromatographic column loaded with Ni-NTA Resin (Thermo Fischer Thermo Fisher Scientific 168 Third Avenue Waltham, MA, USA). The fractions containing PTP1B were collected, concentrated, and injected into a Superdex G75 column equilibrated with Tris-HCl buffer pH 8.0 containing 150 mM NaCl and 1 mM mercaptoethanol. The fractions containing PTP1B were collected and concentrated; the purity of preparation was evaluated using SDS-PAGE. The PTP1B preparation was split into fractions of 0.5 mL volume; each fraction was stored at −80 °C.

2.4. Determination of Enzymatic Activity

The AKR1B1 activity was determined at 37 °C, as previously described [44], by following the decrease in absorbance at 340 nm linked to NADPH oxidation ($\varepsilon 340 = 6.22$ mM^{-1}.cm^{-1}) through a Biochrom Libra S60 spectrophotometer. The standard assay mixture (final volume 0.7 mL) was contained in 0.25 M sodium phosphate buffer pH 6.8, 0.18 mM NADPH, 0.4 M ammonium sulfate, 0.5 mM EDTA, and 4.7 mM GAL. One unit of enzyme activity is the amount of enzyme that catalyzes the conversion of 1 µmol of substrate/min in the above assay conditions. The same assay conditions were adopted in the inhibition studies using L-idose as the substrate at the indicated concentrations. PTP1B activity was determined using p-nitrophenylphosphate (pNPP) as the synthetic substrate. The pNPP was dissolved in 0.075 M β β-dimethylglutarate buffer, pH 7.0, containing 1 mM dithiothreitol. Each assay was carried out at 37 °C in 1 mL of assay buffer. The reactions were started by diluting an aliquot of PTP1B preparation in the assay buffer; after 30 min, the reactions were stopped by adding 2 mL of 0.2 M KOH solution. The amount of p-nitrophenol released was determined spectrophotometrically, reading the absorbance of solutions at 400 nm ($\varepsilon 400$nm $= 18$ mM^{-1}cm^{-1}).

2.5. Determination of IC$_{50}$ Values

Inhibitors were dissolved in DMSO; the solvent concentration in both the assay solution for AKR1B1 and PTP1B was kept constant at 0.5% (*v/v*) to avoid negative effects on enzymatic activity [45]. For IC$_{50}$ evaluation, 10 mU of AKR1B1 in the presence of 1.6 mM L-idose was used. For each compound, at least five different concentrations (each at least in triplicate) were used. IC$_{50}$ values were determined by non-linear regression analysis of experimental data using GraphPad Prism version 7.04 (GraphPad Software, San Diego, CA, USA). The IC$_{50}$ values with respect to PTP1B were determined by measuring the enzymatic activity in the presence of an increasing concentration of compounds and a fixed substrate concentration (2.5 mM pNPP) corresponding to the K$_M$ of PTP1B. For each compound, 12–15 different concentrations of each inhibitor were used. Each test was carried out in triplicate. Data obtained were normalized respect to the control sample. Experimental points obtained were fitted using the following equation:

$$\frac{V_i}{V_0} = \frac{Max - Min}{1 + \left(\frac{x}{IC_{50}}\right)^{slope}} + Min$$

where V_i/V_0 represents the ratio between the enzymatic activity calculated in the presence of the inhibitor (Vi) and the activity of the control sample (V$_0$); "Max" and "Min" indicated the maximum and minimum of activity, respectively; IC$_{50}$ represents the concentration of the inhibitor able to reduce the enzyme activity up to 50% with respect to the control, while the term "slope" represents the slope of the curve in the transition zone. The IC$_{50}$ value was calculated by fitting experimental data with OriginPro 2021 software (OriginLab Corporation, One Roundhouse Plaza, Suite 303 Northampton, MA 01060, USA).

2.6. Kinetic Analyses

The kinetic characterization of avarone was performed considering the molecule as a "tight binding inhibitor". AKR1B1 activity measurements obtained at each substrate concentration were fitted by nonlinear regression analysis to the Morrison equation reported below (Equation (1)) [46].

$$v_i = 1 - \frac{\left([E_T] + [I] + K_i^{app}\right) - \sqrt{\left([E_T] + [I] + K_i^{app}\right)^2 - 4[E_T][I]}}{2[E_T]} \tag{1}$$

This allowed the evaluation of K_i^{app}. In order to obtain the inhibition constants K_i (dissociation constant of the EI complex) and K_i', (dissociation constant of the EIS complex), the K_i^{app} values were plotted against substrate concentration and fitted by nonlinear regression analysis to the following equation:

$$K_i^{app} = ([S] + K_M)/(K_M/K_i + [S]/K_i') \tag{2}$$

Equation (2) refers to a general case of a tight binding non-competitive inhibition model. The K_M value for L-idose was 3.9 mM [43]. The ability of different compounds to act as reversible AKR1B1 inhibitors was evaluated as described in [47]. The ability of avarone to act as a reversible or as a non-reversible inhibitor was evaluated by dilution assay [47]. The mechanism of action of avarone was evaluated by analyzing the effect of increasing avarone concentrations on the main kinetic parameters (K_M and V_{max}) of PTP1B. Data obtained were analyzed using the Lineweaver–Burk plot. The value of K_i was determined using the appropriate equations.

2.7. Molecular Docking

Docking simulations were carried out with AutoDock Tools and AutoDock Vina (version 1.5.6rc3) [48] using the 2HNP structure of PTP1B and 7f5n of TC-PTP obtained from Protein Data Bank. Missing residues were identified and restored using MODELLER 10.2 [49]. Ions, co-crystallized ligan, water molecules, and co-factors were removed from the structure. Polar hydrogen and charges were added to the protein and missing atoms were restored; then, it was saved in PDBQT format. AutoGrid software was used for calculating the grid map of the interaction energies for various atoms of the ligand with the macromolecule. The grid box center and dimension were set at center_x: 46.969, center_y: 15.762, center_z: 1.839 and size_x: 64, size_y: 60, size_z: 56 for PTP1B. For TC-PTP, such data were set at center_x = −51.862, center_y = −51.862, center_z = 31.56 and size_x = 56, size_y = 60, size_z = 58. The dockings were performed using AutoDock Vina with an exhaustiveness of 30. The ligand structure was prepared and energy minimization was performed using DS viewer pro 6.0 (Accelrys, San Diego, CA, USA); the structure was saved in PDB format. Graphical representations of docked structures were performed using UCSF Chimera [50]. Two-dimensional representations of interactions were performed using Ligplot plus (version 2.2.5, Velizy-Villacoublay, Cambridge, United Kingdom) [51] and using Biovia Discovery Studio Visualizer [52].

2.8. Cell Cultures

The C2C12 cell line was grown in DMEM medium (high glucose, 4500 mg/L) supplemented with 10% FBS (Fetal Bovine Serum (FBS, Euroclone, Milan, Italy)), 2 mM glutamine, 100 U/mL of penicillin, and 100 μg/mL of streptomycin, under controlled temperature (37 °C), humidity, and CO_2 concentration (5% CO_2) conditions. When cells reached 70–80% confluence, cells were detached and part of these transferred to a new plate to favor their propagation.

2.9. Ex Vivo Assays

To evaluate the impact of avarone (2) on the insulin signaling pathway, C2C12 cells were plated on P35 dishes and grown until 70% confluence. Then, cells were washed with phosphate-buffered saline (PBS) and incubated for 24 h in the presence of a starvation medium (DMEM high glucose containing 2 mM glutamine, 100 U/mL of penicillin, and 100 µg/mL of streptomycin). After this time, cells were stimulated for 30 min with 10 nM insulin (Humulin R, Ely Lilly, Italy) with avarone (2) or the insulin-avarone combination. After this time, cells were washed with cold PBS and lysed using 100 µL of 1X Laemmli sample buffer solution (50 mM Tris-HCl pH 6.8, 2% SDS, 10% glycerol, 1% β-mercaptoethanol, 12.5 mM EDTA, 0.02% bromophenol blue). Cell proteins were withdrawn, transferred in an Eppendorf tube and boiled for 5 min. All samples were stored at $-20\,^\circ$C before the analysis. Analysis of samples was carried out using SDS-PAGE; after electrophoresis, proteins were transferred on a PVDF membrane by Western blot. Phosphorylation levels of the insulin receptor β chain and of the kinase Akt were evaluated using specific antibodies: IR β subunit, clone CT-3 (MABS65) was from Santa Cruz Biotechnology (Santa Cruz, CA, USA); pIR β subunit, Y1162/1163 (sc-25103-R) was from Merck-Millipore (Burlington, MA, USA). Akt (9272S) and p-Akt (9271S) antibodies were from Cell Signaling Technology (Danvers, MA, USA). Finally, β-actin and clone C-4 (sc-47778) were from Merck-Millipore (Burlington, MA, USA). Anti-rabbit and anti-mouse secondary antibodies were from Santa Cruz Biotechnology (Santa Cruz, CA, USA). Detection was performed using Clarity western ECL substrates (BioRad Laboratories, Inc. 1000 Alfred Nobel DriveHercules, CA, USA). Quantitation of the images obtained was carried out using iBright Analysis Software (Thermo Fisher Scientific 168 Third Avenue Waltham, MA, USA, 02451). Glucose uptake was evaluated as previously described [26]. Briefly, 70% confluent C2C12 cells were starved and then stimulated for 30 min with 10 nM insulin, avarone, or the insulin-avarone combination. After this time, cells were washed with PBS and incubated for 3 h with starvation medium containing 40 µM 2-NBDG (Invitrogen). Then, cells were washed with PBS, trypsinized, pelleted by centrifugation ($1000\times g$ for 5 min), and resuspended in 500 µL of PBS before the analysis. Intrinsic and 2-NBDG-derived cell fluorescences were determined using a flow cytometer (FACSCanto II, BD Biosciences, San Jose, CA, USA). For each sample, 1×104 events were acquired. Quantitative analysis was carried out using FlowJo software (FlowJo™ v10. BD FlowJo™ Software for Windows Version 10. Ashland, OR: Becton, Dickinson and Company; 2021).

2.10. Seahorse XFe96 Metabolic Assays

C2C12 cells (3×104 cells/well) were placed in XFe96 cell culture plates. After 24 h of incubation at 37 °C, cell growth medium was substituted with an XF base medium supplemented with 2 mM glutamine, 1 mM sodium pyruvate, and 10 mM glucose, pH 7.4. The plate was then transferred in a non-CO_2 incubator for 1 h at 37 °C to pre-equilibrate the cells before analysis. An XF Mito Stress Test was performed to assay the cells' ability to exploit mitochondrial oxidative metabolism, according to the manufacturer's instructions [53]. The oxygen consumption rate (OCR) was evaluated after the injection of a sequence of compounds that interfere with the electron transport chain: oligomycin (1 µM), carbonyl cyanide-4 (trifluoromethoxy) phenylhydrazone (FCCP) (1 µM), and Rotenone/Antimycin A (0.5 µM). All data were normalized on the protein content of each well.

3. Results

3.1. Bioprospecting of the Sponge D. avara for the Isolation of Compounds 1–4

The redox couple avarol (**1**) and avarone (**2**) as well as its methylamino derivatives, compounds **3** and **4**, were obtained in pure form by extraction of a sample of the Aegean sponge *D. avara* from the existing biological samples collection available at the Department of Pharmacy of University of Naples Federico II. Compounds **1–4** were isolated and purified according to our previously described procedure [25,40], and unequivocally identified by comparison of their HRMS and NMR data with those reported in the literature

(Figures S1–S8). Solvents (ethyl acetate, n-butanol, and water) with different polarity were used for the extraction of the available sample of the fresh thawed sponge, providing three different dry organic extracts. The n-butanol and ethyl acetate soluble material were separately subjected to chromatographic separation on an RP-18 and SiO_2 open column, respectively, providing two different interesting fractions (A and B) as confirmed by a preliminary ^{1}H NMR analysis. Indeed, the diagnostic protons' signals related to hydroquinone and/or quinone-related compounds were recorded and, accordingly, HPLC purification followed by the comparison of our experimental spectroscopic data with those reported in literature were performed, affording compounds 1–4 (Figure 1) in pure form (see f Materials and Methods) [19,36,37,42].

3.2. Evaluation of Inhibitory Effects against PTP1B and AKR1B1 of Compounds 1–4, and Kinetic Analyses of Avarone (2)

The four sesquiterpene-quinones (1–4, Figure 1) were analyzed to evaluate their ability to inhibit both PTP1B and AKR1B1. For each compound, the IC_{50} values were determined on both PTP1B and AKR1B1. The calculated values are reported in Table 1, together with the IC_{50} values of *p*-benzoquinone as the control [17,24], and dysidine (5, Figure 2), a natural sesquiterpene quinone PTP1B inhibitor [27]. The relative inhibition curves are reported in Figure S9. Vanadate and epalrestat were used as the reference inhibitor on PTP1B and AKR1B1, respectively.

Table 1. Calculated IC_{50} values for PTP1B and AKR1B1.

Compound	IC_{50} (µM) [a]	
	PTP1B	**AKR1B1**
avarol (1)	42.2 ± 18	0.52 ± 0.19
avarone (2)	6.7 ± 0.6	0.078 ± 0.017
3-methylaminoavarone (3)	15.2 ± 2.1	73 ± 15
4-methylaminoavarone (4)	21.6 ± 1.0	62 ± 8
p-benzoquinone	38.9 ± 5.2 [¥]	1.93 [‡]
dysidine (5)	6.7 ± 0.1 [§]	— —-
vanadate	0.4 ± 0.01 [#]	— —-
epalrestat	— —-	0.13 ± 0.03

[a] Data were expressed as $IC_{50} \pm$ SD. [¥] data from Zhang et al. [24]; [‡] data from Demir et al. [17]; [§] data from Li et al. [27]; [#] data from Maccari et al. [54].

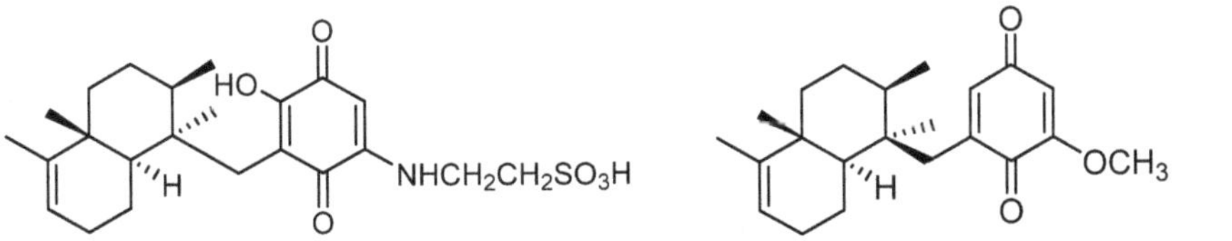

Figure 2. Structure of the marine sesquiterpene quinones dysidine (5) and 21-dehydroxybolinaquinone (6).

As far AKR1B1 is concerned, compounds 1–4 result as reversible inhibitors. Avarone (2) is the most potent in the series and even more active than *p*-benzoquinone, the simplest natural quinone [17] (Table 1). As the IC_{50} value measured for avarone is of the same order of magnitude as the AKR1B1 concentration in the assay (78 nM, calculated on the basis of a molecular weight of 34 KDa), this compound is considered a "tight binding" inhibitor. Among the compounds analyzed, avarone (2) is also confirmed as the most powerful inhibitor of PTP1B, showing an IC_{50} value of 6.7 µM. Considering the high molar ratio (IC_{50}/PTP1B concentration), avarone cannot be considered a tight binding inhibitor for PTP1B.

The kinetic characterization carried out using recombinant AKR1B1 enzyme (see Methods for details) allowed the evaluation of K_i and K_i' values of 410 ± 94 and 55 ± 8 nM, respectively (Figure 3). Thus, avarone presents as a non-competitive mixed-type inhibitor of AKR1B1, with a preferential binding to the ES complex rather than the free enzyme.

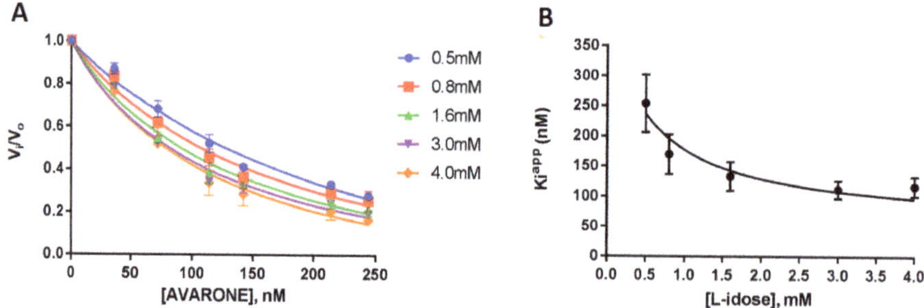

Figure 3. Kinetic characterization of avarone as AKR1B1 inhibitor. Panel A: the activity of the purified enzyme (10 mU) was measured at the indicated concentrations of avarone in the presence of the indicated L-idose concentrations. (**B**) Experimental data reported in (**A**) were fitted to the Morrison equation (see Materials and methods, Equation (1)). The obtained apparent inhibition constants, Ki^{app}, were plotted against substrate concentration and fitted by nonlinear regression analysis to equation 2 (see Materials and methods) relative to a general case of tight binding non-competitive inhibition. Data were plotted using GraphPad 7.04 version.

Kinetic analyses were performed to define the mechanism of PTP1B inhibition by avarone, too. By dilution assay, we confirmed that avarone behaves as a reversible inhibitor of PTP1B (Figure S10). Moreover, to dissect the mechanism of inhibition of avarone, we analyzed the dependence of $^{app}K_M$ and $^{app}V_{max}$ from the concentration of the inhibitor. We observed that as the avarone concentration increases, the value of K_M increases, but the V_{max} substantially does not change (Figure S11); furthermore, the double reciprocal plot showed that the experimental points describe right lines intersecting one another in a point on the abscissa axis, suggesting that avarone behaves as a competitive inhibitor of PTP1B (Figure 4).

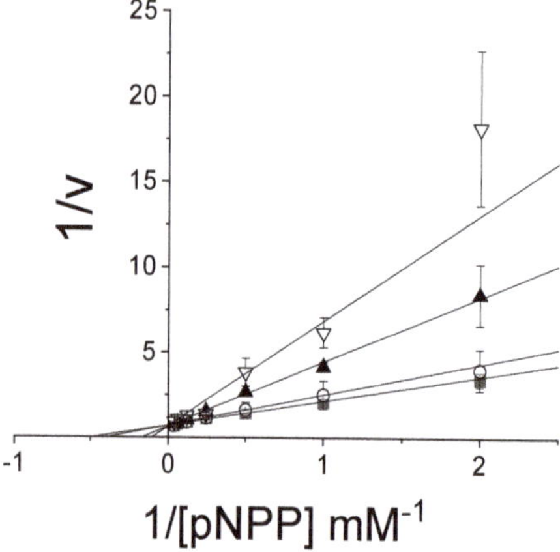

Figure 4. Impact of avarone on PTP1B-catalyzed hydrolysis of pNPP: the Lineweaver–Burk plot. Data reported in the figure were obtained by measuring the initial hydrolysis rate of pNPP at seven different concentrations, in the presence of increasing concentration of avarone. The concentrations of avarone used were: 0 μM, (●); 4 μM, (□); 5.5 μM, (▲); 7 μM (▽). Each test was carried out in triplicate. Data reported in the figure represent the mean value ± SD.

However, it is evident that K_M values increase non-linearly with increasing avarone concentration (Figure S11), raising the suspicion that the inhibition is determined by the binding of more than one avarone molecule within the enzyme site. Based on this evidence, we describe the mechanism of inhibition of avarone for PTP1B as a "cooperative pure competitive inhibition by two non-exclusive inhibitor molecules" (see Scheme 1).

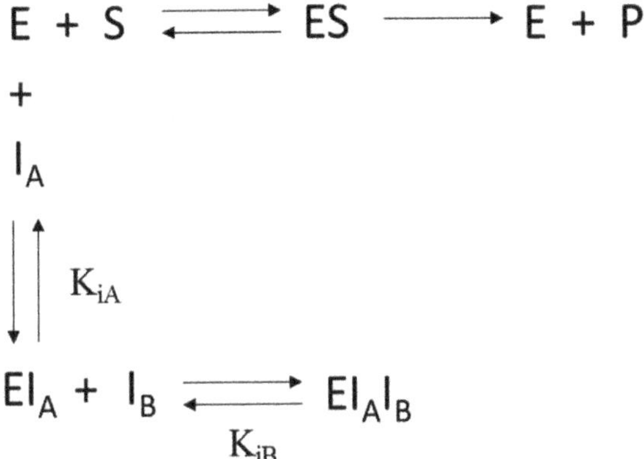

Scheme 1. Mechanism of PTP1B inhibition by avarone. E: free enzyme; ES: Michaelis–Menten complex; P: product of reaction; I_A and I_B: the two molecules of the inhibitor; EI_A and EI_AI_B: enzyme-inhibitor complexes; K_{iA} and K_{iB}: enzyme inhibition constants for I_A and I_B, respectively, i.e., the dissociation constants of the two corresponding enzyme-inhibitor/s complexes.

Such an inhibition model foresees that the first inhibitor molecule (I_A) binds to the active site of the free enzyme and not to the ES complex. The second inhibitor molecule (I_B) can only bind to EI_A, generating the ternary complex (EI_AI_B). Considering the prevailing rapid equilibrium conditions, it was possible to derive Equation (3), which describes the dependence of K_M on the avarone concentration. Therefore, using $^{app}K_M$ values determined at increasing avarone concentrations and Equation (1), we calculated both Ki_A (15.1 ± 1.4 µM) and Ki_B (1.0 ± 0.1 µM) values (Figure 5; details of the kinetic model are reported in Appendix A).

$$K_{M,app} = K_M \times \left\{ 1 + \frac{[I]}{Ki_A} + \frac{[I]^2}{Ki_A \times Ki_B} \right\} \tag{3}$$

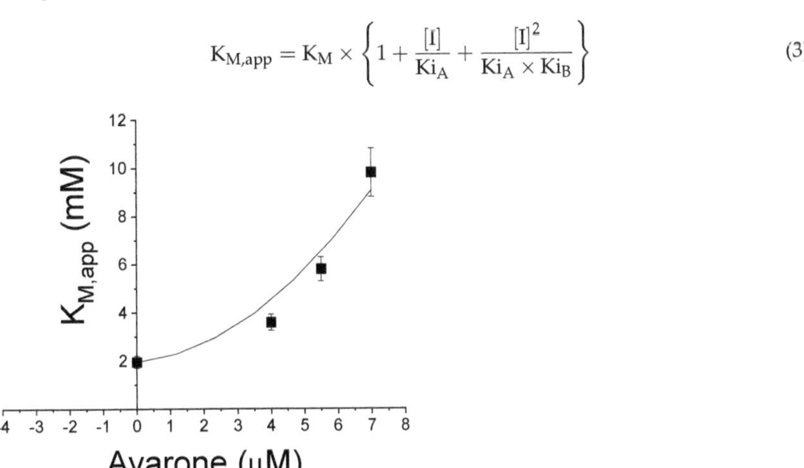

Figure 5. Determination of Ki_A and Ki_B for avarone. Dependence of $K_{M,app}$ from the avarone concentration. Data obtained were fitted using Equation (3). Data fitting was obtained using OriginPlus 2021 software.

3.3. PTP1B Docking Experiments

To better evaluate the interaction modality of avarone with PTP1B, in silico docking analyses were carried out with both compounds **1–5** and *p*-benzoquinone (Figure S12). In accordance with the kinetic analyses, we found that among all the compounds, avarone is the molecule that binds more tightly to the active site of the enzyme, showing a binding free energy of −7.8 kcal/mol (Table S1). The strength of the binding can be explained by the fact that the benzoquinone group of the avarone penetrates deeply into the active site of PTP1B and forms a hydrogen bond with one of the nitrogen atoms of the guanidine group of Arg 221. At the same time, the hexahydronaphthalene group is projected toward the outermost portion of the active site where it forms bonds with Tyr46, Val49, Glu115, Lys116, Lys120, Gly 220, and Gln262 (first pose), or with Trp179, Gly183, Ala217, Gly220, Gln262, and Gln266 (third pose) (Figure 6). None of other compounds can penetrate as deeply into the active site as avarone (Figure S12). Moreover, results of docking show that avarone can interact with the active site of the enzyme in multiple ways, also interacting with residues placed on the mouth of the active site (Figure S13). These results suggest that the active site of the enzyme can accommodate more than one molecule of avarone simultaneously, thus confirming the results obtained from kinetic tests, which show that the inhibitory activity of avarone is the result of a cooperative mechanism.

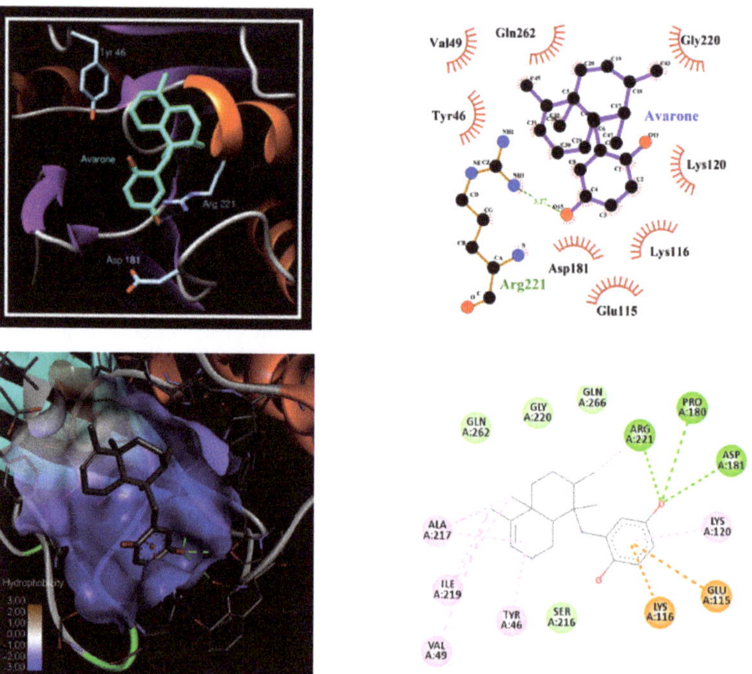

Figure 6. Docking of the avarone into the active site of PTP1B. The calculated free energy of binding was −7.8 kcal/mol. Images of avarone docked into active site of PTP1B were obtained using UCSF Chimera. Moreover, 2D representations of interactions were performed using Ligplot plus.

3.4. Specificity of PTP1B Inhibitory Activity of Avarone

To evaluate the capacity of avarone to specifically target the PTP1B enzyme, we determined the IC$_{50}$ value for TC-PTP, another member of the tyrosine phosphatase family (Figure S14). The results of the analysis demonstrated that the affinity of avarone for TC-PTP is lower (5.8-fold) than that for PTP1B. This result is quite surprising if we consider the high similarity in the primary and tertiary structure of the two enzymes, and suggests that avarone interacts differently with the active sites of the two enzymes [55]. To confirm this hypothesis, the interaction mode of avarone with TC-PTP was analyzed by in silico docking analysis carried out using the crystallographic structure (7F5N) deposited in the PDB database [56]. The result of docking showed that avarone binds into the active site of TC-PTP assuming a different position with respect to that observed in the simulation carried out using the PTP1B enzyme (Figure S15).

3.5. Ex Vivo Assay

To evaluate the impact of avarone on the insulin signaling pathway, further tests were carried out using the myoblast cell line C2C12. Muscle cells were treated with insulin, avarone, or a combination of both. After 30 min, cells were lysed, and cells extracts were analyzed by Western blot to evaluate the phosphorylation levels of Akt (Figures 7 and S16).

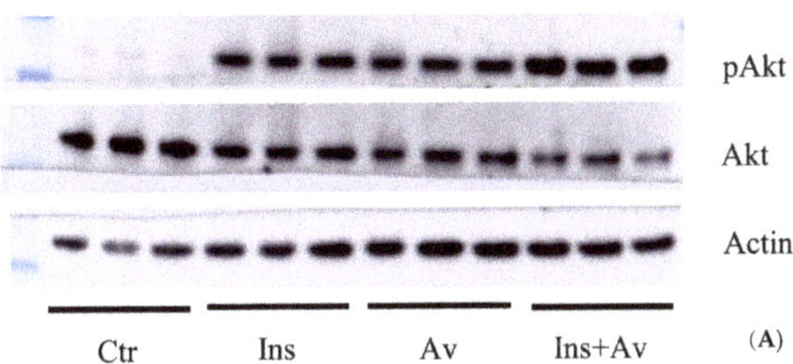

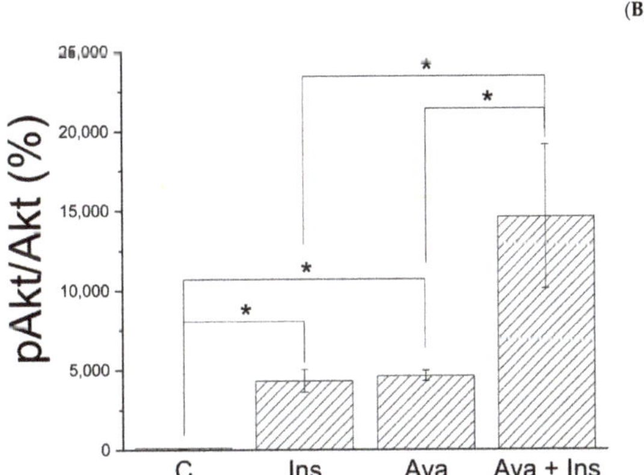

Figure 7. Effects of avarone on insulin signaling pathway. C2C12 cells were serum-starved and stimulated with 10 nM insulin (Ins), 25 μM of avarone (Ava), or with a combination of both (Ins + Ava) for 30 min at 37 °C. After, cells were washed, lysed, and analyzed to evaluate phosphorylation levels of Akt. Each test was carried out in triplicate. (**A**) Representative results of Western blot analysis; (**B**) quantitation of Western blot. Data were normalized with respect to control samples. Data reported in the figure represent the mean value ± SD. Student's *t*-test was used to assess differences in experimental features. Statistical significance is annotated as follows: * $p \leq 0.05$.

We observed that avarone causes an increase in Akt phosphorylation levels similar to that induced by insulin, while treatment with the insulin-avarone combination determines an enhanced effect, raising the levels of Akt phosphorylation to values higher than those observed after treatment with insulin alone. Such data suggest that avarone possesses both insulin-mimetic and insulin-sensitizing activity.

To confirm these results, we evaluated whether avarone can stimulate extracellular glucose uptake in C2C12 cells by incubating them with the fluorescent D-glucose analog 2-NBDG for 3 h and subsequently quantifying levels of intracellular fluorescence by flow cytometry analysis (Figure 8).

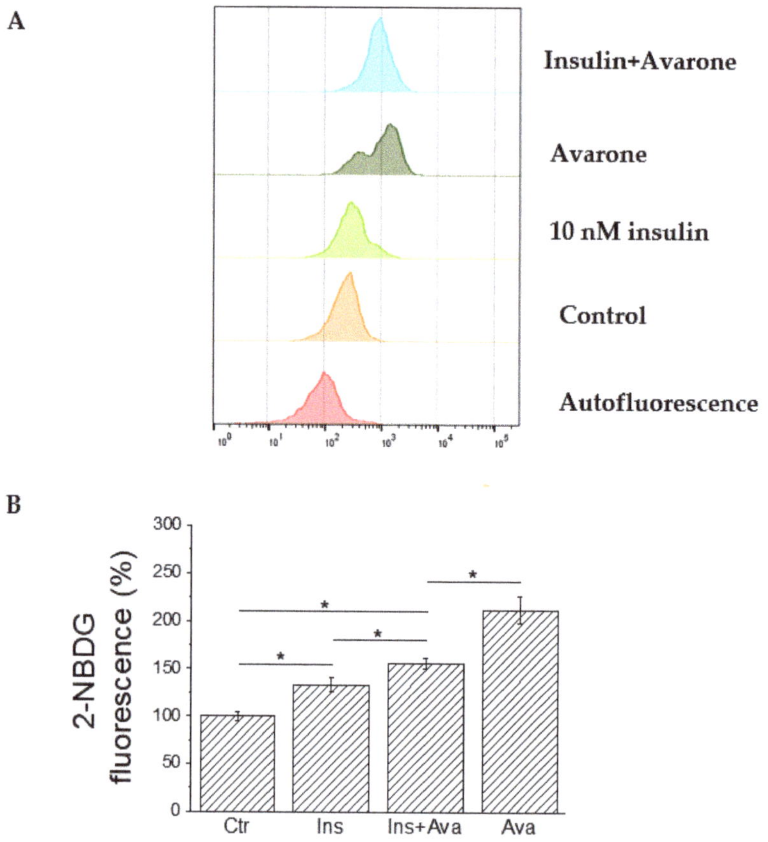

Figure 8. Glucose uptake assay. C2C12 cells were starved for 24 h and then treated with 10 nM insulin, 25 µM avarone, or a combination of both for 30 min at 37 °C. After, cells were washed and incubated for 3 h with 40 µM 2-NBDG. After, cells were washed with PBS, detached using trypsin, collected by centrifugation, and analyzed using a flow cytometer (FACSCanto II, BD Biosciences). (**A**) Flow cytometer analysis. For each experiment, 10,000 events were analyzed. Each test was carried out in quadruplicate. (**B**) Quantitation of fluorescence levels. The data shown in the figure represent the mean value ± SD. Student's *t*-test was used to assess differences in experimental features. Statistical significance is annotated as follows: * $p \leq 0.05$.

We observed that the levels of incorporated glucose significantly increase in cells treated with the insulin-avarone combination with respect to cells treated with insulin alone, confirming that avarone acts as an insulin-sensitizing agent. However, higher amounts of fluorescent glucose were also detected in cells treated with avarone alone (Figure 8), indicating that it is able to act also as an insulin-mimetic compound when administered alone. It is well known that insulin promotes glucose uptake, stimulating GLUT4 translocation to the plasma membrane of C2C12 cells [57]. Therefore, we speculated that avarone could favor the translocation/activation of GLUT4. To confirm this hypothesis, we evaluated the levels of fluorescent glucose incorporated in C2C12 cells pretreated or not with cytochalasin B, a potent inhibitor of GLUT4 activity (Figure 9) [58]. We observed that preincubation with Cytochalasin B strongly affects insulin-stimulated glucose uptake and abolishes the avarone-mediated glucose uptake. Taken together, these data confirm that avarone promotes GLUT4 translocation/activation.

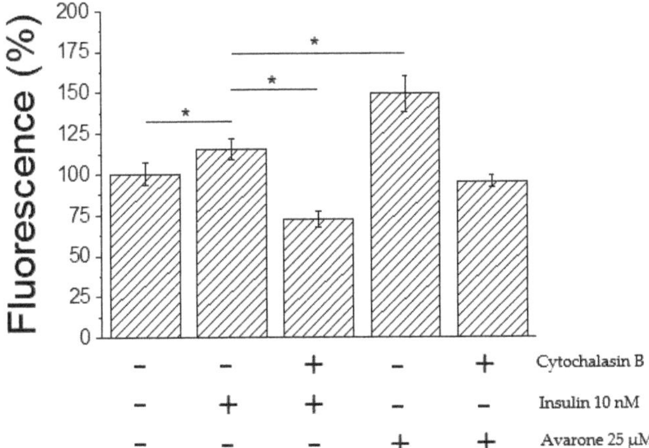

Figure 9. Impact of cytochalasin B on avarone-induced glucose uptake. C2C12 cells were starved for 24 h and then incubated for 2 h in the presence of 10 mM cytochalasin B. Then, cells were treated with 25 µM avarone or 10 nM insulin for 30 min at 37 °C. After, cells were washed and incubated for 3 h with 40 µM 2-NBDG and then analyzed using a flow cytometer (FACSCanto II, BD Biosciences). For each experiment, 10,000 events were analyzed. Each test was carried out in quadruplicate. The data shown in the figure represent the mean value + SD. Student's *t*-test was used to assess differences in experimental features. Statistical significance is annotated as follows: * $p \leq 0.05$.

3.6. Bioenergetics Analyses of C2C12 Cells Treated with Avarone

To evaluate whether avarone may affect mitochondrial metabolism, C2C12 cells were treated for 30 min with avarone and then analyzed for mitochondrial functionality by using the Seahorse XFe96 Analyzer (Figure 10). This approach allowed us to investigate the impact of this molecule on basal respiration, ATP production-related oxygen consumption, maximal respiration, spare respiratory capacity, non-mitochondrial respiration, and mitochondrial proton leak. Meanwhile, a comparative analysis was carried out treating C2C12 cells with UK5099, an inhibitor of the mitochondrial pyruvate carrier [59]. Interestingly, we found that treatment with avarone does not alter basal respiration, while UK5099 leads to a significant reduction in mitochondrial respiration, specifically affecting maximal respiration and ATP production-linked oxygen consumption (Figure 10A). This result is expected as UK5099 inhibits the mitochondrial pyruvate carrier, reducing mitochondrial pyruvate levels, NADH and $FADH_2$ production, the flux of electrons through the mitochondrial electron transport chain, and consequently also the oxygen consumption. This finding also indicates that, unlike UK5099, the treatment with avarone does not impair the oxygen-dependent ATP production. Surprisingly, we found that treatment with avarone slightly increases both spare capacity and maximal respiration, without enhancing the proton leak rate (Figure 10B).

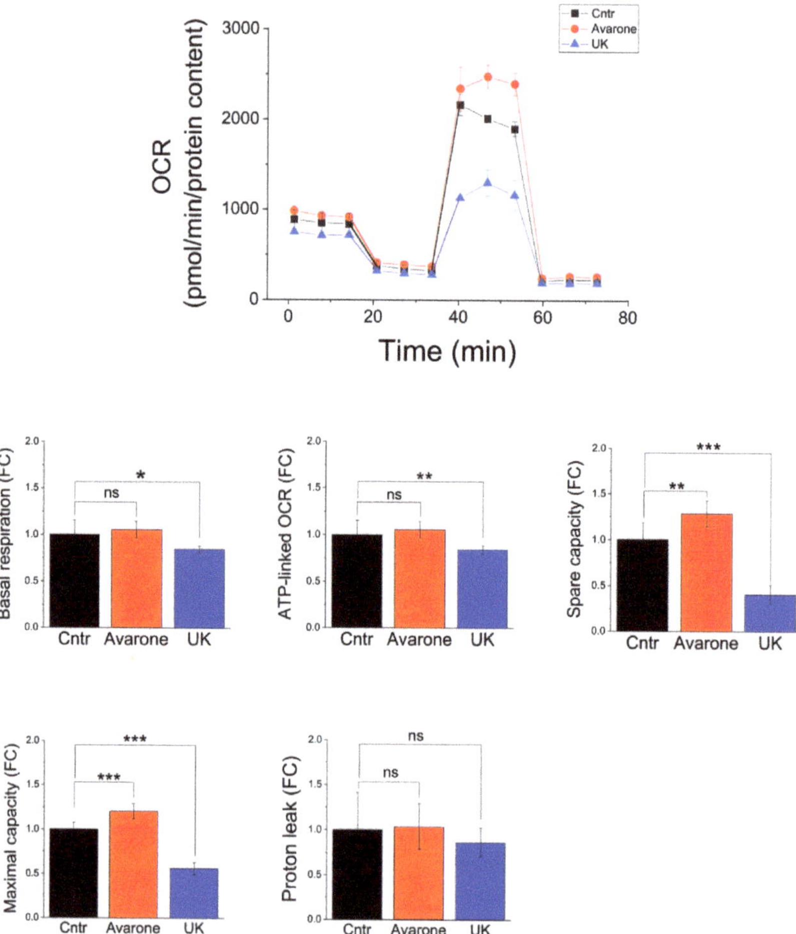

Figure 10. Results of Mito Stress Test assay. (**A**) C2C12 cells were subjected to Seahorse XFe96-Mito Stress Test analysis to evaluate oxygen consumption rate (OCR) in real time. OCR values were normalized on protein content. (**B**) Basal respiration, ATP-linked OCR, spare capacity, maximal respiration, and proton leak were calculated using data obtained from OCR plot after the administration of the adenosine triphosphate ATP synthase inhibitor oligomycin, the proton uncoupler carbonylcyanide *p*-trifluoromethoxyphenylhydrazone (FCCP), and the combination of the respiratory complex I inhibitor rotenone and the respiratory complex III inhibitor antimycin A. Each point reported in the OCR plot represents the medium value ± SEM obtained from independent replicates (n = 3 or 4). One-way ANOVA with Dunnett's correction was used to assess differences in experimental features by using Prism 8 software (GraphPad). Statistical significance is annotated as follows: * $p \leq 0.01$, ** $p \leq 0.001$, *** $p \leq 0.0001$, (*t*-test); ns, not significant; FC, fold change.

4. Discussion

A multitude of risk factors strongly influenced and determined T2DM as a complex disease that has reached alarming rates across the globe; some of these factors are modifiable by lifestyle changes (such as diet, physical activity levels, tobacco use), whereas others, such as age, ethnicity, and genetics, are nonmodifiable and adequate drug treatments are a key issue [60]. The failure of single-drug therapy against T2DM has encouraged the adoption of a multitarget treatment, mainly known as polypharmacology, switching from a one-target–one-drug model to a multiple-target approach. In this regard, searching for chemical scaffolds with a balanced activity toward different and selected targets is a great challenge, as identifying appropriate multifunctional molecules is quite difficult. Therefore, many researchers are concretely exploring the possibility of exploiting natural molecules, including those of marine origin, to discover novel DMLs for the treatment of T2DM. In the present study, we investigated the inhibitory effects against both PTP1B and AKR1B1 of four natural sesquiterpene Q/HQs (compounds **1–4**) isolated from the marine sponge *D. avara*.

Previously, the PTP1B inhibiting activity of related NPs has been reported; the two sesquiterpene quinones dysidine (**5**) and 21-dehydroxybolinaquinone (**6**) were shown to be PTP1B inhibitors with IC_{50} values of 6.7 μM and 39.5 μM, respectively (Figure 2) [27]. More in-depth studies revealed that the most active dysidine (**5**) was a reversible slow-binding PTP1B inhibitor, with moderate selectivity over other PTPs (TC-PTP and CD45). In addition, it has been shown that dysidine could activate the insulin signaling pathway, promote membrane translocation of the glucose transporter 4 (GLUT4) in CHO-K1 and 3T3-L1 cells, and increase glucose uptake in 3T3-L1 cells [24,27].

Thus, we tested the marine-derived compounds **1–4** against PTP1B; all compounds were found to be significantly active, although with different potency. The less active compound was the reduced form avarol (**1**), whereas avarone (**2**) resulted as the most active in the series, with an IC_{50} value of 6.7 μM. The methylamino group, either in 3 or 4 positions, present in compounds **3** and **4**, decreases its potency (Table 1). Moreover, avarone resulted as more active than both the simplest natural quinone, *p*-benzoquinone [24], and dysidine (**5**), a sesquiterpene hydroquinone acting as a competitive slow binding inhibitor of PTP1B [27]. The reason for the elevated inhibitory activity of avarone probably arises from its peculiar chemical structure, which allows it to assume several poses when it binds to the active site of PTP1B. It can penetrate deeply or interact with amino acids that form the active site crevice, suggesting that more than one molecule can bind the active site of PTP1B in the same moment. This hypothesis was also confirmed by results of kinetic analyses showing that two molecules bind to the active site of the enzyme with a different affinity. This led to the hypothesis that the first molecule, upon binding, causes a slight change in the structure of the enzyme's active site that allows the second molecule to interact and bind more effectively to the enzyme's active site. We can hypothesize that avarone is the only molecule capable of doing this among those analyzed.

Moreover, we observed that avarone shows greater specificity for PTP1B than TC-PTP. This is quite surprising considering that the two phosphatases share a high degree of similarity of the primary and tertiary structure of the active site [55]. The observed significant differences in the IC_{50} values of avarone against PTP1B (6.7 μM) over TC-PTP (39.2 μM), indicating a higher affinity for the former, could be attributed to slight differences in the conformation of the active site of the two enzymes that can influence the positioning of the inhibitor. Docking analyses carried out with both enzymes supported this hypothesis, indicating that the position of avarone can acquire different positions in the active site; we cannot exclude the possibility that the active site of PTP1B can host two molecules of avarone simultaneously. Furthermore, the free energy values calculated for the PTP1B-inhibitor complex are lower on average than those calculated for the TC-PTP inhibitor complex.

It is worth noting that avarone's structure, as opposed to many competitive inhibitors identified and/or synthesized in recent years, lacks negative charged groups, such as phosphonates, sulphonates, or carboxylates. Such groups generally favor the interaction of inhibitors with the positive charged arginine, present at the bottom of the active site pocket of the PTPs. This may be considered a favorable feature of the structure of the new inhibitor; the presence of such negatively charged groups could indeed both act as an obstacle to the penetration of inhibitors into target cells and increase the probability that the inhibitors inhibit other members of the PTP family, generating severe side-effects. In fact, dysidine (**6**), featuring the ionizable sulfonic group in its structure, was shown to be a potent PTP1B inhibitor as avarone but with lower selectivity against PTP1B over TC-PTP [24]. Therefore, avarone represents an interesting prototype of a non-charged molecule that could be used as a model to design novel potent and bioavailable competitive inhibitors of PTP1B, thus helping to eliminate the concept that PTPs are undruggable enzymes [61].

Cellular assays revealed that avarone can activate the kinase Akt and stimulate glucose uptake in C2C12 cells either when administered alone or in combination with insulin. The kinase Akt acts downstream in the insulin receptor and is activated following insulin binding to its cognate receptor [62]. However, the degree of activation of the insulin receptor also depends on PTP1B activity, which dephosphorylates the insulin receptor, contributing to switch off the signal activated by insulin. Therefore, PTP1B is a key regulator of the insulin signaling pathway, and its overexpression, or unusual activation, contributes to the onset of insulin resistance [63]. Thus, by treating diabetic mice with PTP1B inhibitors, it is possible to boost insulin receptor activation, increase insulin sensitivity, and reduce blood glucose levels both in fasting conditions and after a meal [64]. The results of the tests on C2C12 cells (Figure 7) demonstrated that avarone, besides acting as an insulin sensitizer, also possesses insulin-mimetic activity, as suggested by its ability to induce the phosphorylation of Akt with the same intensity of insulin when used alone. Thus, we speculated that avarone could determine a strong activation of the Akt/PI3K pathway, triggering GLUT4 translocation to the plasma membrane also in the absence of insulin. Pre-treatment of C2C12 cells with cytochalasin B, a potent inhibitor of GLUT4 translocation, abolishes glucose uptake (Figure 9), confirming that avarone alone can boost migration of GLUT4 to the plasma membrane. However, it is interesting to note that avarone alone stimulates glucose uptake more efficiently than insulin itself; this evidence suggests that avarone acts through a complex mechanism that could also include the direct activation of GLUT4. In accordance with this hypothesis, previous studies reported that GLUT4 undergoes activation after its translocation to the plasma membrane of cells, and that both mechanisms contribute to the full stimulation of glucose uptake by insulin [65]. Therefore, to date, we cannot exclude that avarone may also act as a GLUT4 activator, enhancing, after exposure to the plasma membrane, the transporter's ability to import glucose into muscle cells.

Bioenergetics analyses of C2C12 cells treated with avarone evidenced that the treatment increases the mitochondrial spare respiratory capacity (SRC) and the maximum respiration (MR). The increase in these parameters represents positive events because they witness the increased ability of cells to respond to stress stimuli, increasing the ATP production [66]. Based on literature data, we suppose that this effect can be another consequence of Akt activation as both SRC and MR are modulated by the PI3K/AKT/mTOR signaling pathway. According to this evidence, it has been reported that pharmacological inhibition of this pathway leads to a significant reduction in SRC; vice versa, the activation of the PI3K/AKT/mTOR signaling pathway, triggered by stimulation with different growth factors or through PTEN knockdown, results in an enhancement of glycolytic flux and of the SRC [67]. Therefore, we hypothesize that treatment of C2C12 cells with avarone could not only improve insulin sensitivity, and decrease blood glucose levels, but also positively impact mitochondrial activity. It is well known that the mitochondrial activity of muscle cells resulted as impaired in diabetic patients, leading to a reduction in fatty acid oxidation, promoting lipid accumulation in muscle cells, and favoring the development of the insulin resistance [68]. According to this finding, we hypothesize that avarone could contribute to revert the insulin resistance condition, improving mitochondrial activity in muscle cells of diabetic patients.

Based on the previously reported effects on aldose reductase of several quinone compounds [17], we tested compounds **1–4** against AKR1B1, the enzyme that is responsible for the conversion of glucose into sorbitol through the polyol pathway. Considering this action, AKR1B1 is considered a critical target to be inhibited to prevent the complications linked to an abnormal increase in the flux through the polyol pathway, as it occurs in hyperglycemic conditions [9]. Indeed, the overexpression of AKR1B1 in transgenic mice has been shown to accelerate the onset of diabetic cataracts [69] and exacerbate diabetic cardiomyopathy [70]. Blood glucose fluctuations and aldose reductase expression favor GSH depletion and apoptosis, thereby promoting the onset of retinopathy, and other diabetes-related complications [71]. All compounds **1–4** result in reversible inhibition of the AKR1B1 enzyme, and avarone (**2**) is the most potent in the series, again, with an IC_{50} value in the nanomolar range (0.078 μM). It shows a high affinity for AKR1B1 and can be considered a tight binding inhibitor of this enzyme. Interestingly, 3- or 4-substitution with the methylamino group in compounds **3** and **4** dramatically decreases the inhibition potency (73 and 62 μM). The kinetic characterization of **2** (Figure 3) allowed evaluation of Ki and Ki' values (410 ± 94 and 55 ± 8 nM, respectively), highlighting avarone as a non-competitive mixed-type inhibitor of AKR1B1, with a preferential binding to the ES complex with respect to the free enzyme.

Definitively, we demonstrated that avarone is a potent inhibitor of both PTP1B and AKR1B1 enzymes, and, therefore, it can be considered as a multitarget agent.

A schematic graphic representation of the putative molecular mechanism of avarone, which highlights its ability to interfere with the cellular events related to the activation of the insulin signaling pathway, is reported in Figure 11.

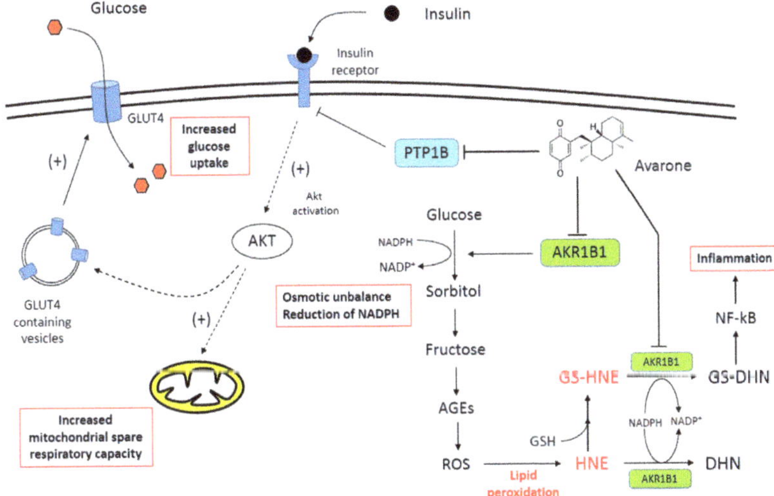

Figure 11. Graphical representation of avarone's mechanism of action.

Furthermore, the potential of avarone as a lead compound is strongly enhanced by its very low cytotoxic effects; in fact, avarone exhibits an IC_{50} value of 62.19 μM when tested on human microvascular endothelial cells (HMEC-1) and results as not cytotoxic against human leukemia monocytic cells (THP-1) [19].

5. Conclusions

The thorough pharmacological characterization we conducted on avarone revealed that it is a potent and selective PTP1B inhibitor with both insulin-sensitizing and mimetic activity and that it displays positive effects on muscle cell mitochondria. These results reveal this natural hit as a promising candidate to be developed as an antidiabetic molecule. On the other hand, the intrinsic ability of avarone to act as a potent AKR1B1 inhibitor makes it an interesting scaffold molecule for developing a new series of potent and safe aldose reductase inhibitors. For these reasons, avarone could be proposed as a novel nontoxic natural hit to be developed for designing a multitarget drug for diabetes and its pathological complications. Our data confirm and reinforce the idea that the use of natural molecules, with their unique chemical structures and the inherent ability to act against multiple targets, play a crucial role in accelerating the discovery of new multitarget drugs.

Supplementary Materials: The following supporting information can be downloaded at: https://www.mdpi.com/article/10.3390/pharmaceutics15020528/s1, Figure S1: HRESI-MS spectrum of avarol (**1**); Figure S2: ^{1}H NMR spectrum of avarol (**1**) in CDCl$_3$ (600 MHz); Figure S3: HRESI-MS spectrum of avarone (**2**); Figure S4: ^{1}H NMR spectrum of avarone (**2**) in CDCl$_3$ (600 MHz); Figure S5: HRESI-MS spectrum of 3-methylaminoavarone (**3**); Figure S6: ^{1}H NMR spectrum of 3-methylaminoavarone (**3**) in CDCl$_3$ (600 MHz); Figure S7: HRESI-MS spectrum of 4-methylaminoavarone (**4**); Figure S8: ^{1}H NMR spectrum of 4-methylaminoavarone (**4**) in CDCl$_3$ (600 MHz); Figure S9: Determination of the IC$_{50}$ values for PTP1B and for AKR1B1; Figure S10: Dilution assay; Figure S11: Dependence of K$_M$ and V$_{max}$ from avarone (**2**) concentration; Figure S12: Docking of compounds **1**–**5** and of *p*-benzoquinone; Figure S13: Binding sites of avarone (**2**) identified by docking analyses; Figure S14. Determination of the IC$_{50}$ values of avarone for PTP1B (A) and TC-PTP (B); Figure S15. Binding sites of avarone (**2**) identified by docking analyses with TC-PTP; Figure S16: Full Western blot of avarone on insulin signaling pathway; Table S1: Binding free energies of compounds **1**–**5** and of *p*-benzoquinone to PTP1B and TC-PTP.

Author Contributions: Conceptualization, P.P. and M.M.; methodology, M.C., A.S. and C.I.; software, M.G.; validation, A.D.C., M.M. and P.P; formal analysis, M.C., M.G., A.S., E.P., F.B., L.P. and C.I.; investigation, M.C., M.G., E.P., F.B. and L.P.; resources, C.I. and S.A.; data curation, M.C., M.G., A.D.C., C.I., M.M. and P.P.; writing—original draft preparation, M.C., A.D.C., C.I., M.M. and P.P.; writing—review and editing, C.I., M.M. and P.P.; visualization, M.C. and M.G.; supervision, M.M. and P.P.; funding acquisition, A.D.C., M.M. and P.P. All authors have read and agreed to the published version of the manuscript.

Funding: This research was funded by a grant from the Department of Pharmacy, University of Naples Federico II, Bando Contributo alla Ricerca Anno 2021 "Targeting PTP1B and/or AR enzymes with marine-derived small molecules and foodstuffs constituents as natural inhibitor in search for novel therapeutics against type 2 diabetes mellitus (T2DM)" (project acronym PTP1B-AR EnNatIn). This research was also funded in part by the University of Florence ("Fondi di Ateneo 2021") and in part by the University of Pisa ("Fondi di Ateneo 2021").

Institutional Review Board Statement: Not applicable.

Informed Consent Statement: Not applicable.

Data Availability Statement: The tested compounds are available at the laboratory of the corresponding author.

Acknowledgments: We thank Asli Kacar and Burcu Omuzbuken for sample collection, and Arturo Facente for identifying the organism.

Conflicts of Interest: The authors declare no conflict of interest.

Appendix A

Steady-state kinetic analysis of the inhibition mechanism of avarone.

Based on our experimental data reported in the Section 3, we can hypothesize that PTP1B is able to bind two avarone molecules in the active site simultaneously. Furthermore, the hyperbolic dependence of KM on the avarone concentration suggests that the binding of the first molecule causes a distortion of the active site of the enzyme, thus facilitating the binding of the second avarone molecule. This model predicts that the two molecules bind with different affinity to the active site of the enzyme, with KiA > KiB, indicating that avarone shows a higher affinity for the EIA complex than the free enzyme (E).

The catalysed p-nitrophenylphosphate (pNPP) hydrolysis can be described as follow:

$$
\begin{array}{c}
\text{pNP} \qquad \text{H}_2\text{O} \\
\uparrow k_2 \qquad \downarrow \\
E + \text{pNPP} \ \rightleftarrows \ E \bullet \text{pNPP} \ \underset{k_{-2}}{\overset{}{\rightleftarrows}} \ E\text{-}P \ \xrightarrow{k_3} \ E + \text{Pi} \\
Ks \\
+ \\
I_A \\
\updownarrow \ Ki_A \\
EI_A + I_B \ \underset{Ki_B}{\rightleftarrows} \ EI_AI_B
\end{array}
$$

where E•pNPP is the Michaelis-Menten complex, and E-P is the covalent phosphoenzyme intermediate. By applying the steady state assumption to [E-P], we can assume:

$$
k_2 \cdot [E \cdot pNPP] = k_3 \cdot [E - P][E \cdot pNPP] = \frac{k_3}{k_2}\,[E - P] \tag{A1}
$$

By the max conservation lay

$$
[E]_T = [E] + [E \cdot pNPP] + [E - P] + [EI_A] + [EI_AI_B] \tag{A2}
$$

Taking into account that:

$$
Ks = \frac{[E] \cdot [pNPP]}{[E \bullet pNPP]} \tag{A3}
$$

$$
K_{iA} = \frac{[E] \cdot [I_A]}{[EI_A]} \tag{A4}
$$

$$
K_{iB} = \frac{[EI_A][I_B]}{[EI_AI_B]} \tag{A5}
$$

Substituting the equations (4) and (5) into (2), and assuming initial velocity condition ([pNPP]$_0$>>[E]$_0$, so [pNPP]$_0$ = [pNPP], we obtained:

$$
[E] = \frac{[E]_T - [E \cdot pNPP] \cdot \left\{ 1 + \frac{k_2}{k_3} \right\}}{\left\{ 1 + \frac{[I_A]}{K_{iA}} + \frac{[I_A] \cdot [I_B]}{K_{iA} \cdot K_{iB}} \right\}} \tag{A6}
$$

Then, substituting (5) into (3), we obtained:

$$
[E \cdot pNPP] = \frac{\dfrac{k_3 \cdot [E]_T \cdot [pNPP]}{k_2 + k_3}}{\dfrac{Ks \cdot k_3}{k_2 + k_3} \cdot \left\{ 1 + \frac{[I_A]}{K_{IA}} + \frac{[I_A][I_B]}{[K_{IA} \cdot K_{IB}]} \right\} + [pNPP]} \tag{A7}
$$

Rate measured as production of [E-P]

$$
v = \frac{\dfrac{k_2 \cdot k_3}{k_2 + k_3} \cdot \left\{ [E]_T \cdot [pNPP] \right\}}{\dfrac{Ks \cdot k_3}{k_2 + k_3} \cdot \left\{ 1 + \frac{[I_A]}{K_{IA}} + \frac{[I_A][I_B]}{[K_{IA} \cdot K_{IB}]} \right\} + [pNPP]}
$$

where

$$
K_M = \left(\frac{Ks \cdot k_3}{k_2 + k_3} \right)
$$

So, we obtained:

$$
K_{M,app} = K_M \cdot \left(1 + \frac{[I_A]}{K_{IA}} + \frac{[I_A][I_B]}{[K_{IA} \cdot K_{IB}]} \right)
$$

Taking into account that

$$[I_A] = [I_B] \mathrm{K_{M,app}} = K_M \cdot \left(1 + \frac{[I]}{K_{IA}} + \frac{[I]^2}{[K_{IA} \cdot K_{IB}]}\right)$$

References

1. Alim, Z.; Kilinc, N.; Sengul, B.; Beydemir, S. Mechanism of Capsaicin Inhibition of Aldose Reductase Activity. *J. Biochem. Mol. Toxicol.* **2017**, *31*, e21898. [CrossRef]
2. Saeedi, P.; Petersohn, I.; Salpea, P.; Malanda, B.; Karuranga, S.; Unwin, N.; Colagiuri, S.; Guariguata, L.; Motala, A.A.; Ogurtsova, K.; et al. Global and Regional Diabetes Prevalence Estimates for 2019 and Projections for 2030 and 2045: Results from the International Diabetes Federation Diabetes Atlas, 9th Edition. *Diabetes Res. Clin. Pract.* **2019**, *157*, 107843. [CrossRef]
3. *World Health Statistics 2020: Monitoring Health for the SDGs, Sustainable Development Goals*; World Health Or-Ganization: Geneva, Switzerland, 2020.
4. Hasanzad, M.; Larijani, B.; Aghaei Meybodi, H.R. Diabetes and COVID-19: A Bitter Nightmare. *J. Diabetes Metab. Disord.* **2022**, *21*, 1191–1193. [CrossRef] [PubMed]
5. Vázquez-Vela, M.E.F.; Torres, N.; Tovar, A.R. White Adipose Tissue as Endocrine Organ and Its Role in Obesity. *Arch. Med. Res.* **2008**, *39*, 715–728. [CrossRef] [PubMed]
6. Thakur, S.; Gupta, S.K.; Ali, V.; Singh, P.; Verma, M. Aldose Reductase: A Cause and a Potential Target for the Treatment of Diabetic Complications. *Arch. Pharm. Res.* **2021**, *44*, 655–667. [CrossRef] [PubMed]
7. Balestri, F.; Barracco, V.; Renzone, G.; Tuccinardi, T.; Pomelli, C.S.; Cappiello, M.; Lessi, M.; Rotondo, R.; Bellina, F.; Scaloni, A.; et al. Stereoselectivity of Aldose Reductase in the Reduction of Glutathionyl-Hydroxynonanal Adduct. *Antioxidants* **2019**, *8*, 502. [CrossRef]
8. del Corso, A.; Cappiello, M.; Mura, U. From a Dull Enzyme to Something Else: Facts and Perspectives Regarding Aldose Reductase. *Curr. Med. Chem.* **2008**, *15*, 1452–1461. [CrossRef]
9. Maccari, R.; Ottanà, R. Targeting Aldose Reductase for the Treatment of Diabetes Complications and Inflammatory Diseases: New Insights and Future Directions. *J. Med. Chem.* **2015**, *58*, 2047–2067. [CrossRef]
10. Goldstein, B.J. Protein-Tyrosine Phosphatases: Emerging Targets for Therapeutic Intervention in Type 2 Diabetes and Related States of Insulin Resistance. *J. Clin. Endocrinol. Metab.* **2002**, *87*, 2474–2480. [CrossRef]
11. Elchebly, M.; Payette, P.; Michaliszyn, E.; Cromlish, W.; Collins, S.; Loy, A.L.; Normandin, D.; Cheng, A.; Himms-Hagen, J.; Chan, C.-C.; et al. Increased Insulin Sensitivity and Obesity Resistance in Mice Lacking the Protein Tyrosine Phosphatase-1B Gene. *Science* **1999**, *283*, 1544–1548. [CrossRef]
12. Pardella, E.; Pranzini, E.; Leo, A.; Taddei, M.L.; Paoli, P.; Raugei, G. Oncogenic Tyrosine Phosphatases: Novel Therapeutic Targets for Melanoma Treatment. *Cancers* **2020**, *12*, 2799. [CrossRef]
13. Lori, G.; Paoli, P.; Femia, A.P.; Pranzini, E.; Caselli, A.; Tortora, K.; Romagnoli, A.; Raugei, G.; Caderni, G. Morin-dependent Inhibition of Low Molecular Weight Protein Tyrosine Phosphatase (LMW-PTP) Restores Sensitivity to Apoptosis during Colon Carcinogenesis: Studies in Vitro and in Vivo, in an *Apc* -driven Model of Colon Cancer. *Mol. Carcinog.* **2019**, *58*, 686–698. [CrossRef] [PubMed]
14. Artasensi, A.; Pedretti, A.; Vistoli, G.; Fumagalli, L. Type 2 Diabetes Mellitus: A Review of Multi-Target Drugs. *Molecules* **2020**, *25*, 1987. [CrossRef] [PubMed]
15. Saldívar-González, F.I.; Aldas-Bulos, V.D.; Medina-Franco, J.L.; Plisson, F. Natural Product Drug Discovery in the Artificial Intelligence Era. *Chem. Sci.* **2022**, *13*, 1526–1546. [CrossRef]
16. Atanasov, A.G.; Zotchev, S.B.; Dirsch, V.M.; Supuran, C.T. Natural Products in Drug Discovery: Advances and Opportunities. *Nat. Rev. Drug Discov.* **2021**, *20*, 200–216. [CrossRef]
17. Demir, Y.; Özaslan, M.S.; Duran, H.E.; Küfrevioğlu, Ö.İ.; Beydemir, Ş. Inhibition Effects of Quinones on Aldose Reductase: Antidiabetic Properties. *Environ. Toxicol. Pharmacol.* **2019**, *70*, 103195. [CrossRef] [PubMed]
18. Menna, M.; Imperatore, C.; D'Aniello, F.; Aiello, A. Meroterpenes from Marine Invertebrates: Structures, Occurrence, and Ecological Implications. *Mar. Drugs* **2013**, *11*, 1602–1643. [CrossRef] [PubMed]
19. Imperatore, C.; Gimmelli, R.; Persico, M.; Casertano, M.; Guidi, A.; Saccoccia, F.; Ruberti, G.; Luciano, P.; Aiello, A.; Parapini, S.; et al. Investigating the Antiparasitic Potential of the Marine Sesquiterpene Avarone, Its Reduced Form Avarol, and the Novel Semisynthetic Thiazinoquinone Analogue Thiazoavarone. *Mar. Drugs* **2020**, *18*, 112. [CrossRef]
20. *Natural Products in the New Millennium: Prospects and Industrial Application*; Rauter, A.P.; Palma, F.B.; Justino, J.; Araújo, M.E.; dos Santos, S.P. (Eds.) Springer Netherlands: Dordrecht, The Netherlands, 2002; ISBN 978-90-481-6186-7.
21. Shen, Y.-C.; Lu, C.-H.; Chakraborty, R.; Kuo, Y.-H. Isolation of Sesquiterpenoids from Sponge Dysidea Avara and Chemical Modification of Avarol as Potential Antitumor Agents. *Nat. Prod. Res.* **2003**, *17*, 83–89. [CrossRef]
22. Sarin, P.S.; Sun, D.; Thornton, A.; Muller, W.E. Inhibition of Replication of the Etiologic Agent of Acquired Immune Deficiency Syndrome (Human T-Lymphotropic Retrovirus/Lymphadenopathy-Associated Virus) by Avarol and Avarone. *J. Natl. Cancer Inst.* **1987**, *78*, 663–666.

23. Gimmelli, R.; Persico, M.; Imperatore, C.; Saccoccia, F.; Guidi, A.; Casertano, M.; Luciano, P.; Pietrantoni, A.; Bertuccini, L.; Paladino, A.; et al. Thiazinoquinones as New Promising Multistage Schistosomicidal Compounds Impacting *Schistosoma Mansoni* and Egg Viability. *ACS Infect. Dis.* **2020**, *6*, 124–137. [CrossRef]

24. Zhang, Y.; Li, Y.; Guo, Y.-W.; Jiang, H.-L.; Shen, X. A Sesquiterpene Quinone, Dysidine, from the Sponge Dysidea Villosa, Activates the Insulin Pathway through Inhibition of PTPases. *Acta Pharmacol. Sin.* **2009**, *30*, 333–345. [CrossRef] [PubMed]

25. Genovese, M.; Imperatore, C.; Casertano, M.; Aiello, A.; Balestri, F.; Piazza, L.; Menna, M.; del Corso, A.; Paoli, P. Dual Targeting of PTP1B and Aldose Reductase with Marine Drug Phosphoeleganin: A Promising Strategy for Treatment of Type 2 Diabetes. *Mar. Drugs* **2021**, *19*, 535. [CrossRef] [PubMed]

26. Casertano, M.; Genovese, M.; Piazza, L.; Balestri, F.; del Corso, A.; Vito, A.; Paoli, P.; Santi, A.; Imperatore, C.; Menna, M. Identifying Human PTP1B Enzyme Inhibitors from Marine Natural Products: Perspectives for Developing of Novel Insulin-Mimetic Drugs. *Pharmaceuticals* **2022**, *15*, 325. [CrossRef] [PubMed]

27. Li, Y.; Zhang, Y.; Shen, X.; Guo, Y.-W. A Novel Sesquiterpene Quinone from Hainan Sponge Dysidea Villosa. *Bioorganic Med. Chem. Lett.* **2009**, *19*, 390–392. [CrossRef]

28. Surti, M.; Patel, M.; Adnan, M.; Moin, A.; Ashraf, S.A.; Siddiqui, A.J.; Snoussi, M.; Deshpande, S.; Reddy, M.N. Ilimaquinone (Marine Sponge Metabolite) as a Novel Inhibitor of SARS-CoV-2 Key Target Proteins in Comparison with Suggested COVID-19 Drugs: Designing, Docking and Molecular Dynamics Simulation Study. *RSC Adv.* **2020**, *10*, 37707–37720. [CrossRef]

29. Casaubon, R.L.; Snapper, M.L. S-Adenosylmethionine Reverses Ilimaquinone's Vesiculation of the Golgi Apparatus. *Bioorganic Med. Chem. Lett.* **2001**, *11*, 133–136. [CrossRef] [PubMed]

30. Kurelec, B.; Zhan, R.K.; Gasić, M.J.; Britvić, S.; Lucić, D.; Müller, W.E.G. Antimutagenic Activity of the Novel Antileukemic Agents, Avarone and Avarol. *Mutat. Res. Lett.* **1985**, *144*, 63–66. [CrossRef]

31. Loya, S.; Hizi, A. The Inhibition of Human Immunodeficiency Virus Type 1 Reverse Transcriptase by Avarol and Avarone Derivatives. *FEBS Lett.* **1990**, *269*, 131–134. [CrossRef]

32. Amigó, M.; Payá, M.; Braza-Boïls, A.; de Rosa, S.; Terencio, M.C. Avarol Inhibits TNF-α Generation and NF-KB Activation in Human Cells and in Animal Models. *Life Sci.* **2008**, *82*, 256–264. [CrossRef]

33. Tsoukatou, M.; Maréchal, J.; Hellio, C.; Novaković, I.; Tufegdzic, S.; Sladić, D.; Gašić, M.; Clare, A.; Vagias, C.; Roussis, V. Evaluation of the Activity of the Sponge Metabolites Avarol and Avarone and Their Synthetic Derivatives Against Fouling Micro- and Macroorganisms. *Molecules* **2007**, *12*, 1022–1034. [CrossRef]

34. Vilipić, J.; Novaković, I.; Stanojković, T.; Matić, I.; Šegan, D.; Kljajić, Z.; Sladić, D. Synthesis and Biological Activity of Amino Acid Derivatives of Avarone and Its Model Compound. *Bioorganic Med. Chem.* **2015**, *23*, 6930–6942. [CrossRef]

35. Belisario, M.A.; Maturo, M.; Avagnale, G.; de Rosa, S.; Scopacasa, F.; Caterina, M. de In Vitro Effect of Avarone and Avarol, a Quinone/Hydroquinone Couple of Marine Origin, on Platelet Aggregation. *Pharmacol. Toxicol.* **1996**, *79*, 300–304. [CrossRef] [PubMed]

36. Minale, L.; Riccio, R.; Sodano, G. Avarol a Novel Sesquiterpenoid Hydroquinone with a Rearranged Drimane Skeleton from the Sponge. *Tetrahedron Lett.* **1974**, *15*, 3401–3404. [CrossRef]

37. Cozzolino, R.; de Giulio, A.; de Rosa, S.; Strazzullo, G.; Gašič, M.J.; Sladić, D.; Zlatović, M. Biological Activities of Avarol Derivatives, 1. Amino Derivatives. *J. Nat. Prod.* **1990**, *53*, 699–702. [CrossRef]

38. Imperatore, C.; Valadan, M.; Tartaglione, L.; Persico, M.; Ramunno, A.; Menna, M.; Casertano, M.; Dell'Aversano, C.; Singh, M.; d'Aulisio Garigliota, M.L.; et al. Exploring the Photodynamic Properties of Two Antiproliferative Benzodiazopyrrole Derivatives. *Int. J. Mol. Sci.* **2020**, *21*, 1246. [CrossRef] [PubMed]

39. Imperatore, C.; Varriale, A.; Rivieccio, E.; Pennacchio, A.; Staiano, M.; D'Auria, S.; Casertano, M.; Altucci, C.; Valadan, M.; Singh, M.; et al. Spectroscopic Properties of Two 5'-(4-Dimethylamino)Azobenzene Conjugated G-Quadruplex Forming Oligonucleotides. *Int. J. Mol. Sci.* **2020**, *21*, 7103. [CrossRef]

40. Casertano, M.; Genovese, M.; Paoli, P.; Santi, A.; Aiello, A.; Menna, M.; Imperatore, C. Insights into Cytotoxic Behavior of Lepadins and Structure Elucidation of the New Alkaloid Lepadin L from the Mediterranean Ascidian Clavelina Lepadiformis. *Mar. Drugs* **2022**, *20*, 65. [CrossRef]

41. Aiello, A.; Fattorusso, E.; Imperatore, C.; Luciano, P.; Menna, M.; Vitalone, R. Aplisulfamines, New Sulfoxide-Containing Metabolites from an Aplidium Tunicate: Absolute Stereochemistry at Chiral Sulfur and Carbon Atoms Assigned Through an Original Combination of Spectroscopic and Computational Methods. *Mar. Drugs* **2012**, *10*, 51–63. [CrossRef]

42. Hamed, A.N.E.-S.; Wätjen, W.; Schmitz, R.; Chovolou, Y.; Edrada-Ebel, R.; Youssef, D.T.A.; Kamel, M.S.; Proksch, P. A New Bioactive Sesquiterpenoid Quinone from the Mediterranean Sea Marine Sponge Dysidea Avara. *Nat. Prod. Commun.* **2013**, *8*, 289–292.

43. Balestri, F.; Cappiello, M.; Moschini, R.; Rotondo, R.; Buggiani, I.; Pelosi, P.; Mura, U.; Del-Corso, A. L-Idose: An Attractive Substrate Alternative to d-Glucose for Measuring Aldose Reductase Activity. *Biochem. Biophys. Res. Commun.* **2015**, *456*, 891–895. [CrossRef]

44. Balestri, F.; Poli, G.; Pineschi, C.; Moschini, R.; Cappiello, M.; Mura, U.; Tuccinardi, T.; del Corso, A. Aldose Reductase Differential Inhibitors in Green Tea. *Biomolecules* **2020**, *10*, 1003. [CrossRef] [PubMed]

45. Misuri, L.; Cappiello, M.; Balestri, F.; Moschini, R.; Barracco, V.; Mura, U.; Del-Corso, A. The Use of Dimethylsulfoxide as a Solvent in Enzyme Inhibition Studies: The Case of Aldose Reductase. *J. Enzym. Inhib. Med. Chem.* **2017**, *32*, 1152–1158. [CrossRef] [PubMed]

46. Morrison, J.F. Kinetics of the Reversible Inhibition of Enzyme-Catalysed Reactions by Tight-Binding Inhibitors. *Biochim. Biophys. Acta BBA—Enzymol.* **1969**, *185*, 269–286. [CrossRef] [PubMed]

47. Ottanà, R.; Paoli, P.; Cappiello, M.; Nguyen, T.N.; Adornato, I.; del Corso, A.; Genovese, M.; Nesi, I.; Moschini, R.; Naß, A.; et al. In Search for Multi-Target Ligands as Potential Agents for Diabetes Mellitus and Its Complications—A Structure-Activity Relationship Study on Inhibitors of Aldose Reductase and Protein Tyrosine Phosphatase 1B. *Molecules* **2021**, *26*, 330. [CrossRef]

48. Trott, O.; Olson, A.J. AutoDock Vina: Improving the Speed and Accuracy of Docking with a New Scoring Function, Efficient Optimization, and Multithreading. *J. Comput. Chem.* **2009**, *31*, 455–461. [CrossRef]

49. Šali, A.; Blundell, T.L. Comparative Protein Modelling by Satisfaction of Spatial Restraints. *J. Mol. Biol.* **1993**, *234*, 779–815. [CrossRef]

50. Pettersen, E.F.; Goddard, T.D.; Huang, C.C.; Couch, G.S.; Greenblatt, D.M.; Meng, E.C.; Ferrin, T.E. UCSF Chimera? A Visualization System for Exploratory Research and Analysis. *J. Comput. Chem.* **2004**, *25*, 1605–1612. [CrossRef]

51. Laskowski, R.A.; Swindells, M.B. LigPlot+: Multiple Ligand–Protein Interaction Diagrams for Drug Discovery. *J. Chem. Inf. Model.* **2011**, *51*, 2778–2786. [CrossRef]

52. *Dassault Systèmes*; BIOVIA, Discovery Studio: Vélizy-Villacoublay, France, 2021.

53. Pranzini, E.; Pardella, E.; Muccillo, L.; Leo, A.; Nesi, I.; Santi, A.; Parri, M.; Zhang, T.; Uribe, A.H.; Lottini, T.; et al. SHMT2-Mediated Mitochondrial Serine Metabolism Drives 5-FU Resistance by Fueling Nucleotide Biosynthesis. *Cell Rep.* **2022**, *40*, 111233. [CrossRef]

54. Maccari, R.; del Corso, A.; Paoli, P.; Adornato, I.; Lori, G.; Balestri, F.; Cappiello, M.; Naß, A.; Wolber, G.; Ottanà, R. An Investigation on 4-Thiazolidinone Derivatives as Dual Inhibitors of Aldose Reductase and Protein Tyrosine Phosphatase 1B, in the Search for Potential Agents for the Treatment of Type 2 Diabetes Mellitus and Its Complications. *Bioorganic Med. Chem. Lett.* **2018**, *28*, 3712–3720. [CrossRef]

55. Nandi, S.; Saxena, M. Potential Inhibitors of Protein Tyrosine Phosphatase (PTP1B) Enzyme: Promising Target for Type-II Diabetes Mellitus. *Curr. Top. Med. Chem.* **2020**, *20*, 2692–2707. [CrossRef] [PubMed]

56. Singh, J.P.; Lin, M.-J.; Hsu, S.-F.; Peti, W.; Lee, C.-C.; Meng, T.-C. Crystal Structure of TCPTP Unravels an Allosteric Regulatory Role of Helix A7 in Phosphatase Activity. *Biochemistry* **2021**, *60*, 3856–3867. [CrossRef] [PubMed]

57. Brewer, P.D.; Habtemichael, E.N.; Romenskaia, I.; Mastick, C.C.; Coster, A.C.F. Insulin-Regulated Glut4 Translocation. *J. Biol. Chem.* **2014**, *289*, 17280–17298. [CrossRef]

58. Ryder, J.W.; Kawano, Y.; Chibalin, A.V.; Rincón, J.; Tsao, T.S.; Stenbit, A.E.; Combatsiaris, T.; Yang, J.; Holman, G.D.; Charron, M.J.; et al. In Vitro Analysis of the Glu-Cose-Transport System in GLUT4-Null Skeletal Muscle. *Biochem. J.* **1999**, *342*, 321–328. [CrossRef]

59. Divakaruni, A.S.; Wiley, S.E.; Rogers, G.W.; Andreyev, A.Y.; Petrosyan, S.; Loviscach, M.; Wall, E.A.; Yadava, N.; Heuck, A.P.; Ferrick, D.A.; et al. Thiazolidinediones Are Acute, Specific Inhibitors of the Mitochondrial Pyruvate Carrier. *Proc. Natl. Acad. Sci. USA* **2013**, *110*, 5422–5427. [CrossRef] [PubMed]

60. Huneif, M.A.; Alshehri, D.B.; Alshaibari, K.S.; Dammaj, M.Z.; Mahnashi, M.H.; Majid, S.U.; Javed, M.A.; Ahmad, S.; Rashid, U.; Sadiq, A. Design, Synthesis and Bioevaluation of New Vanillin Hybrid as Multitarget Inhibitor of α-Glucosidase, α-Amylase, PTP-1B and DPP4 for the Treatment of Type-II Diabetes. *Biomed. Pharmacother.* **2022**, *150*, 113038. [CrossRef]

61. Zhang, Z.-Y. Drugging the Undruggable: Therapeutic Potential of Targeting Protein Tyrosine Phosphatases. *Acc. Chem. Res.* **2017**, *50*, 122–129. [CrossRef]

62. Posner, B.I. Insulin Signalling: The Inside Story. *Can. J. Diabetes* **2017**, *41*, 108–113. [CrossRef]

63. Teimouri, M.; Hosseini, H.; ArabSadeghabadi, Z.; Babaei-Khorzoughi, R.; Gorgani-Firuzjaee, S.; Meshkani, R. The Role of Protein Tyrosine Phosphatase 1B (PTP1B) in the Pathogenesis of Type 2 Diabetes Mellitus and Its Complications. *J. Physiol. Biochem.* **2022**, *78*, 307–322. [CrossRef]

64. Koren, S.; Fantus, I.G. Inhibition of the Protein Tyrosine Phosphatase PTP1B: Potential Therapy for Obesity, Insulin Resistance and Type-2 Diabetes Mellitus. *Best Pract. Res. Clin. Endocrinol. Metab.* **2007**, *21*, 621–640. [CrossRef] [PubMed]

65. SOMWAR, R.; KIM, D.Y.; SWEENEY, G.; HUANG, C.; NIU, W.; LADOR, C.; RAMLAL, T.; KLIP, A. GLUT4 Translocation Precedes the Stimulation of Glucose Uptake by Insulin in Muscle Cells: Potential Activation of GLUT4 via P38 Mitogen-Activated Protein Kinase. *Biochem. J.* **2001**, *359*, 639. [CrossRef]

66. Nicholls, D.G. Spare Respiratory Capacity, Oxidative Stress and Excitotoxicity. *Biochem. Soc. Trans.* **2009**, *37*, 1385–1388. [CrossRef] [PubMed]

67. Marchetti, P.; Fovez, Q.; Germain, N.; Khamari, R.; Kluza, J. Mitochondrial Spare Respiratory Capacity: Mechanisms, Regulation, and Significance in Non-transformed and Cancer Cells. *FASEB J.* **2020**, *34*, 13106–13124. [CrossRef] [PubMed]

68. Kelley, D.E.; He, J.; Menshikova, E.V.; Ritov, V.B. Dysfunction of Mitochondria in Human Skeletal Muscle in Type 2 Diabetes. *Diabetes* **2002**, *51*, 2944–2950. [CrossRef]

69. Lee, A.Y.; Chung, S.K.; Chung, S.S. Demonstration That Polyol Accumulation Is Responsible for Diabetic Cataract by the Use of Transgenic Mice Expressing the Aldose Reductase Gene in the Lens. *Proc. Natl. Acad. Sci. USA* **1995**, *92*, 2780–2784. [CrossRef]

70. Hwang, Y.C.; Kaneko, M.; Bakr, S.; Liao, H.; Lu, Y.; Lewis, E.R.; Yan, S.; Ii, S.; Itakura, M.; Rui, L.; et al. Central Role for Aldose Reductase Pathway in Myocardial Ischemic Injury. *FASEB J.* **2004**, *18*, 1192–1199. [CrossRef]
71. Kador, P.F.; Wyman, M.; Oates, P.J. Aldose Reductase, Ocular Diabetic Complications and the Development of Topical Kinostat®. *Prog. Retin. Eye Res.* **2016**, *54*, 1–29. [CrossRef]

 *pharmaceutics*

Article

An Exploratory Study of the Safety and Efficacy of a *Trigonella foenum-graecum* Seed Extract in Early Glucose Dysregulation: A Double-Blind Randomized Placebo-Controlled Trial

Emily Pickering [1,2], Elizabeth Steels [1,2,*], Amanda Rao [3] and Kathryn J. Steadman [1]

[1] School of Pharmacy, University of Queensland, Brisbane, QLD 4102, Australia
[2] Evidence Sciences Pty, Ltd., Brisbane, QLD 4005, Australia
[3] School of Human Movement and Nutrition Sciences, University of Queensland, Brisbane, QLD 4102, Australia
* Correspondence: e.steels@uq.edu.au

Abstract: Background: This was an exploratory study to assess the safety and efficacy of a specialized *Trigonella foenum-graceum* L. seed extract for supporting healthy blood glucose metabolism in a pre-diabetic cohort. Methods: Fifty-four participants were randomised to receive 500 mg/day of *T. foenum-graecum* seed extract or matching placebo daily for 12 weeks. Fasting blood glucose (FBG), post-prandial glucose (PPBG), HbA1c, fasting insulin (FI), post-prandial insulin (PPI) and C-peptide were assessed at baseline, week 6 and week 12. Lipid levels, liver enzymes and C-reactive protein (CRP), along with safety markers and tolerability were also assessed at baseline and week 12. Results: By week 12 there was a significant difference in FBG ($p < 0.001$), PPBG ($p = 0.007$) and triglycerides ($p = 0.030$) between treatment groups, with no changes in HbA1c ($p = 0.41$), FI ($p = 0.12$), PPI ($p = 0.50$) or C-peptide ($p = 0.80$). There was no difference in total cholesterol ($p = 0.99$), high-density lipoprotein ($p = 0.35$), low density lipoprotein ($p = 0.60$) or CRP ($p = 0.79$). There was no change in safety markers and the treatment was well tolerated. Conclusions: The results of the study indicated that *T. foenum-graecum* seed extract may influence blood glucose metabolism and larger studies are warranted to evaluate efficacy and potential mechanisms of action.

Keywords: *Trigonella foenum-graecum*; fenugreek; prediabetes; type 2 diabetes; fasting blood glucose; post-prandial glucose; insulin

Citation: Pickering, E.; Steels, E.; Rao, A.; Steadman, K.J. An Exploratory Study of the Safety and Efficacy of a *Trigonella foenum-graecum* Seed Extract in Early Glucose Dysregulation: A Double-Blind Randomized Placebo-Controlled Trial. *Pharmaceutics* **2022**, *14*, 2453. https://doi.org/10.3390/pharmaceutics14112453

Academic Editors: Diana Marcela Aragon Novoa and Fátima Regina Mena Barreto Silva

Received: 5 October 2022
Accepted: 9 November 2022
Published: 14 November 2022

Publisher's Note: MDPI stays neutral with regard to jurisdictional claims in published maps and institutional affiliations.

1. Introduction

Type 2 diabetes mellitus (T2DM) is a growing epidemic across the world and currently accounts for over 90% of all world-wide cases of diabetes, equating to approximately 483 million people in 2021 and predicted to rise to 705 million by 2045 [1]. The preceding condition, termed prediabetes, is a major risk factor for developing T2DM and is characterized as either or both impaired fasting blood glucose (IFG) and/or impaired blood glucose tolerance (IGT). Over 319 million people (6.2% of the world population) are recorded to have IFG, with predictions to rise to 440 million (6.9%) by 2045, while the impact of IGT is higher again, with 541 million (10.6%) currently affected and predicted to rise to 730 million (11.4%) by the year 2045 [1]. In addition, the prediabetic condition poses its own potential health risks as IGT is associated with metabolic syndrome as well as being a strong predictor of atherosclerotic cardiovascular disease, potentially increasing the development cardiovascular disease by approximately 15% [2].

Insulin resistance and uncontrolled calorie consumption are recognized as key contributors to the conversion of prediabetes to T2DM [1]. Treatments for prediabetes include dietary and lifestyle changes as first-line strategies. Interventions that reduce insulin resistance and the need for insulin secretion by pancreatic beta-cells, such as glucose-lowering medications (i.e., metformin) may also reduce the likelihood of prediabetes progressing to

T2DM [3]. Other interventions such as functional foods and herbal medicines are also an ongoing area of research interest for managing early glucose dysregulation.

Herbal remedies, such as *Trigonella foenum-graceum*, *Gymnema sylvestre*, and *Momordica charantia* have been used as preventative measures and as treatments for managing glucose dysregulation [4]. The seeds of some plants, such as *Linum usitatissimum* L. (flax), *Salvia hispanica* (chia) and *T. foenum-graceum* (fenugreek) have been used as ingredients to help lower glycemic loads of foods, thus reducing the glycemic impact of carbohydrates and simple sugars [5,6]. *T. foenum-graecum*, from the family Fabaceae (syn Leguminosae), is a culinary herb, the seeds of which have traditionally been used in the management of blood glucose levels [7]. There is support from clinical studies of *T. foenum-graecum* seeds in treating early and established T2DM as a food ingredient [6,8,9], raw powdered seeds added to water [10–12] and seed extracts [13–16]. In a recent clinical trial, 10 g/day of a *T. foenum-graecum* powder was found to have a protective effect against progression from prediabetes to T2DM when used long term [17]. Despite the large body of previous studies, including a meta-analysis [18], showing *T. foenum-graecum* seeds can reduce blood glucose levels in those type T2DM, less research exists regarding *T. foenum-graecum* seed extracts as a treatment for prediabetes. It is important to distinguish the effects of *T. foenum-graceum* seed extracts between those with T2DM and prediabetes as efficacy with the lower blood glucose levels featured in prediabetes may differ from that of T2DM. The safety of *T. foeneum-graceum* seed has previously been assessed in those with T2DM as both raw seed [19] and seed extracts [20] and is generally considered to not cause any adverse or intolerable effects. This is yet to be established in those with prediabetes.

The aim of this study is to assess the efficacy and safety of a specialised *T. foenum-graecum* seed extract in supporting healthy glucose metabolism in adults diagnosed with prediabetes (IFG and/or IGT) as a stand-alone treatment, over a 12-week period.

2. Materials and Methods

2.1. Trial Design

This clinical trial was a double-blind, placebo-controlled randomized study of participants who were diagnosed with pre-diabetes, and not currently prescribed antidiabetic medications, to assess the effectiveness of *T. foenum-graecum* extract on reducing fasting blood glucose (FBG) and post-prandial blood glucose (PPBG) and associated metabolic parameters over 12 weeks. The study was conducted in Brisbane, Australia between March 2018 and March 2020 and was carried out according to the principles expressed in the Declaration of Helsinki. It was approved by the Bellberry Ethics Committee No: 2017-08-601 and registered with the Australian New Zealand Clinical Trials Registry (ANZCTR) No: ACTRN12618000031268.

2.2. Investigational Products

The investigational product was a dark green film-coated tablet containing 250 mg of *T. foenum-graecum* L. seed, hydro-alcoholic extract (marketed by Gencor Pacific Ltd., Lantau Island, Hong Kong, under the brand name Trigogen™), standardized to no less than 20% trigonelline and no less than 20% amino acids which was identified by high performance thin layer chromatography (HPTLC) by Indus Biotech Pty Ltd. The placebo product was an identical dark green film-coated tablet containing maltodextrin. The products were provided by Gencor Pacific Ltd., Unit 3, 1/F, Office Building Block 2, 96 Siena Avenue, Discovery Bay North, Lantau Island, N.T., Hong Kong. Participants took either the investigational product or the placebo tablets, twice a day (one in the morning and one in the evening) for 12 weeks.

2.3. Study Outcome Measures

Primary outcome: The primary outcome was evaluation of fasting blood glucose (FBG) as part of the standard 2 h oral glucose tolerance test (2hOGTT) at baseline, 6 weeks and 12 weeks. FBG was used to assess IFG.

Secondary outcomes: The secondary outcomes included post-prandial blood glucose (PPBG) using the standard 2hOGTT at baseline, 6 weeks and 12 weeks. PPBG was used to assess IGT. Fasting insulin (FI), and post-prandial insulin (PPI) were also assessed at baseline, 6 weeks and 12 weeks. The other secondary outcomes were measured at baseline and 12 weeks: glycated haemoglobin (HbA1c), C-peptide, C-reactive protein (CRP) and lipids (total cholesterol, triglycerides (TG), low density lipoprotein (LDL), high density lipoprotein (HDL)) and the safety markers; full blood count (FBC) and a comprehensive metabolic panel (including electrolytes, liver, and kidney function) at baseline and at 12 weeks. The Depression Anxiety Stress Scale (DASS-21) was used to assess stress, anxiety and depression levels of participants at baseline and 12 weeks [21]. Product tolerability was evaluated at clinic interviews by investigators at all clinic appointments. An adverse event (AE) was defined as any untoward medical occurrence and that does not necessarily have a causal relationship with this treatment, and safety testing procedures followed the NHMRC Safety Monitoring Guidelines and Reporting in Clinical Trials involving Therapeutic Goods, 2016 [22].

2.4. Inclusion, Exclusion, and Screening

Inclusion criteria: Participants were included if they had a FBG > 5.5 mmol/L or a PPBG of > 7.8 mmol/L and agreed to have a 2hOGTT on three (3) occasions; at baseline, 6 weeks and 12 weeks. The cut-off levels for FBG and PPBG matched the classification system used in Australia for diabetes screening and diagnosis at the time of the study [23].

Exclusion criteria: Those with a normal glucose profile or a previous diagnosis of T2DM, were outside healthy weight reference range of 18 to 35 BMI or were taking anti-obesity medications or oral blood glucose-lowering medications such as sulfonylureas, biguanides, alpha-glucosidase inhibitors, thiazolidinedione or taking any dietary supplements or herbal medicines specifically for management of blood glucose levels were excluded. Other exclusions included recent episodes of symptomatic coronary arterial disease, stroke, any cardiovascular events including infarction, treatment for cancers within last 5 years, those taking anticoagulants, those with known past reactions to *T. foenum-graecum* seeds, cucumber or maltodextrin, those with an admission of alcohol abuse or the use of illicit drugs, and pregnancy or breastfeeding or not having had a normal menstrual cycle. Potential participants were also excluded if they were considering commencing new lifestyle interventions, including changing diet or physical activity for the purpose of managing body weight or blood glucose levels, or if they had completed participation in any other clinical trial during past month.

Screening: The initial screening process (screening 1), conducted online and through telephone screening, identified potential participants that met all inclusion and exclusion criteria and were: (1) at risk of T2DM on the Australian Type 2 Diabetes Risk Assessment Tool (AUSDRISK) questionnaire, score > 6, or (2) were those that had been medically diagnosed as pre-type 2 diabetic (pre-diabetic) by their medical practitioner. These potential participants underwent further screening (screening 2) and were invited to undertake an in-clinic FBG test (taken at 8 am, after fasting the previous night from 10 pm), using handheld Accu-Chek glucose meters. They were deemed eligible to participate in the study if they had a FBG > 5.5 mmol/L at this time. Assessment of the inclusion criteria of PPBG of >7.8 mmol/L was only possible with the baseline 2hOGTT, and in a prediabetic state FBG levels are still variable, so enrolled/consented participants that were found in the assessments made at baseline to have either a normal FBG, or PPBG > 7.8 mmol/L were subsequently excluded and referred to their medical practitioner.

2.5. Randomisation, Blinding and Study Procedures

The randomisation of the products (1:1 ratio, in blocks of 30) was performed independently of the investigators using computer generated random allocation software. The products were blinded independently of the investigators. The trial products were delivered to the investigators in trial product containers that were identical in function and appearance, with the only difference being the trial number on the labels of the containers. The investigators allocated the next available numbered container in the sequence as each participant enrolled in the study, ensuring that all participants had an equal chance of receiving the investigational product or the placebo product. The investigators and participants were therefore blinded for the duration of the study.

Study procedures: At the baseline visit, study procedures were explained, and participants provided informed written consent. The participant provided health information including health history, current medications/supplements, and had body weight, height and blood pressure measured. Participants were provided with their allocated trial product and were instructed to maintain the same diet and level of physical activity for the duration of the study. Prior to commencing the product, participants were asked to attend an independent pathology clinic for the baseline pathology tests, which were 2hOGTT (for IFG, PPBG, FI and PPI), HbA1c, C-peptide, lipid studies, CRP, FBC and a comprehensive metabolic panel were completed by an independent laboratory and results provided to investigators via an online portal. For the 2hOGTT, participants attended the pathology clinic at 8 am in a fasted state (no food since 10 pm the previous night). A blood draw was taken at time 0 (baseline), followed by ingestion of a flavoured 70 g glucose drink (within 5 min) and then subsequent blood draws at 1 and 2 h post-prandial to measure both blood glucose and insulin levels. At week 6 the participants were assessed for compliance and for adverse events and completed a 2hOGTT. At week 12 (final interview), participants returned all containers and any unused products and had the third (final) pathology test which was a repeat of the baseline pathology tests. Compliance was monitored throughout the study by the trial investigators and participants were considered compliant if they had taken over 80% of the product.

2.6. Sample Size and Data Analysis

Pre-study, it was determined that a sample size of 58 participants would be required to detect a 0.37 mmol/L change in FBG between active treatment group and placebo group over 12 weeks, assuming a 0.4 mmol/L change score, at 80% power and a type I error of 5%. A total of 57 participants were enrolled, as recruitment was suspended in March 2020 due to the COVID-19 pandemic, resulting in 48 participants being included in the analysis. Post-study, a 0.37 mmol/L difference in FBG was observed with the smaller sample size. The effect size was calculated from the difference between the two means to provide a quantitative measure of the magnitude of the experimental effect and to be reflective of clinical significance. Analysis of the effect size for the primary outcome (FBG) was calculated using Cohen's D, and found to be d = 1.108, indicating a large effect size with > 79% of control group measuring higher than the mean of the experimental group, thus making the sample size sufficient for this study.

All data is expressed as mean ± standard deviation unless otherwise specified. A modified intent to treat (ITT) approach was used for data analysis of the primary outcome, where all participants who completed midpoint pathology were included in the analysis. The ITT single missing data points for pathology data was managed using the simple imputation method, with the data from the last observation carried forward. Normality between groups was assessed using the Kolmogorov–Smirnov and Sharpiro-Wilk tests. The glucose pathology parameters FBG, FI, PPBG and PPI, and DASS-21 were assessed as change scores from baseline to week 12 and analyzed using 2-sided student t-tests. All other pathology tests were assessed by difference between groups at baseline and 12 weeks using 2-sided t-tests. Cohen's d test was used to determine effect size and correlations were performed using Pearson's coefficient using the Statistical Packages for the Social Sciences (SPSS) software version 21 (statistical significance set at $p < 0.05$).

3. Results

3.1. Demographics

The study initially enrolled 57 participants (Figure 1, Table 1). Of the 30 participants allocated to active treatment group, 25 were included in analysis, with the five withdrawals due to being inducted but not completing baseline ($n = 2$), not pre-diabetic on baseline pathology assessment ($n = 1$), withdrew prior to week 6 ($n = 1$) and an adverse event ($n = 1$). There were 27 participants allocated to placebo group, with 23 participants included in analysis, and 4 withdrawals due to not being pre-diabetic on baseline pathology assessment ($n = 1$) or withdrew prior to 6 weeks ($n = 3$). Participants included these previously identified as pre-diabetic by their medical practitioner (active treatment group, 60%; placebo group, 35%), with the remainder being those who were found to be prediabetic upon screening.

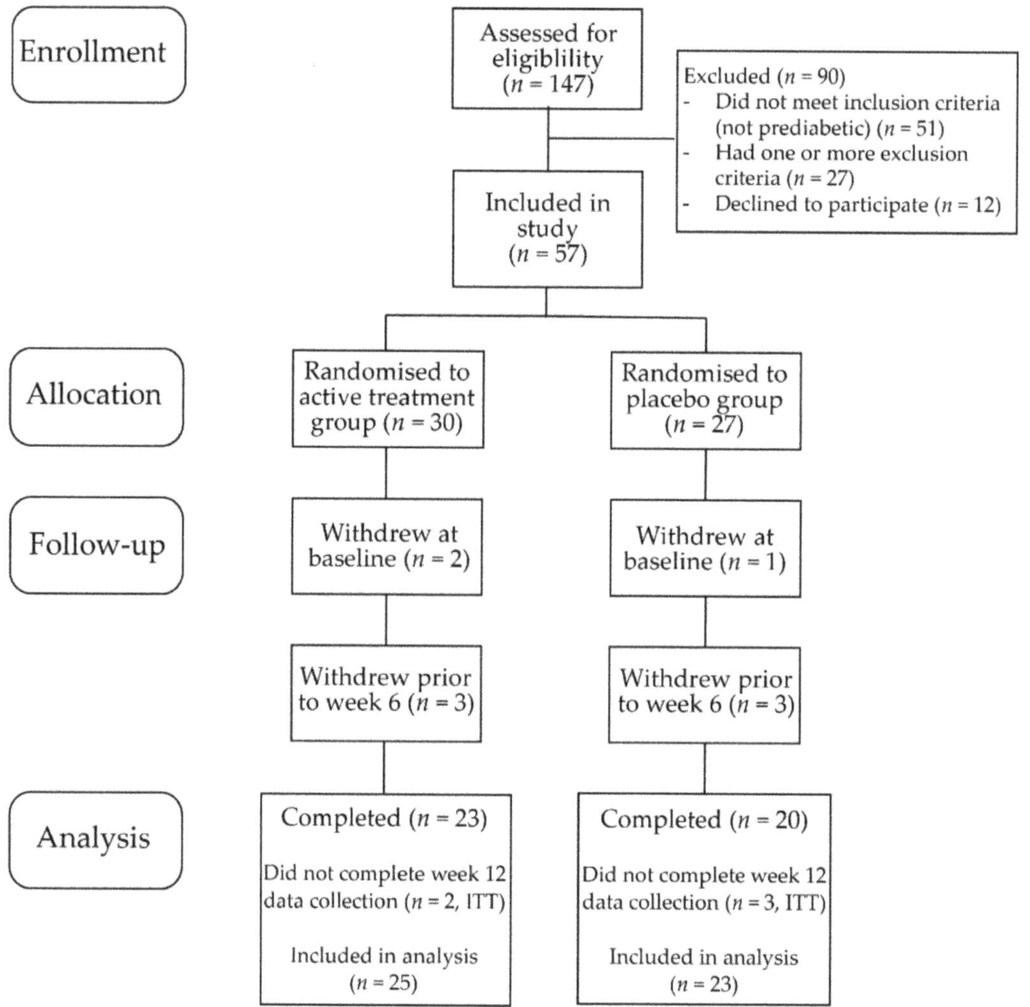

Figure 1. CONSORT participant flow chart.

Table 1. Demographics of participants in the Active treatment group and Placebo group at baseline.

	Active Treatment Group			Placebo Group		
	Total	Men	Women	Total	Men	Women
Participant No.	25	16	9	23	12	11
Age (average)	58.2	58.5	54.9	59.7	61.8	65.8
BMI (average)	30.9	31.2	30.4	32.5	33.8	31.2
Previously diagnosed as pre-diabetic (*n*)	14	10	4	8	4	4
Diagnosed as pre-diabetic at screening (*n*)	11	6	5	15	8	7
Taking medication for hypertension (*n*)	8	5	3	5	5	0
Blood pressure -Systolic/Diastolic	129/88	130/89	128/87	128/86	129/88	126/84
Known family history of T2DM (*n*)	15	9	6	16	8	8
Drink alcohol regularly (*n*)	14	11	3	11	6	5
Sedentary lifestyle (*n*)	18	11	7	15	7	8
Exercise regularly (*n*)	8	5	3	8	5	3

3.2. Effect of Treatment on Glucose Metabolism Markers

Fasting blood glucose (FBG): The baseline FBG levels appeared to be slightly higher for the active treatment group (7 mmol/L) than the placebo group (6 mmol/L) but this was borderline non-significant ($p = 0.052$) (Table 2). There was no difference in FBG between groups at 6 weeks ($p = 0.20$), however, by 12 weeks a statistically significant difference was observed ($p < 0.001$) (Figure 2A, Table 2).

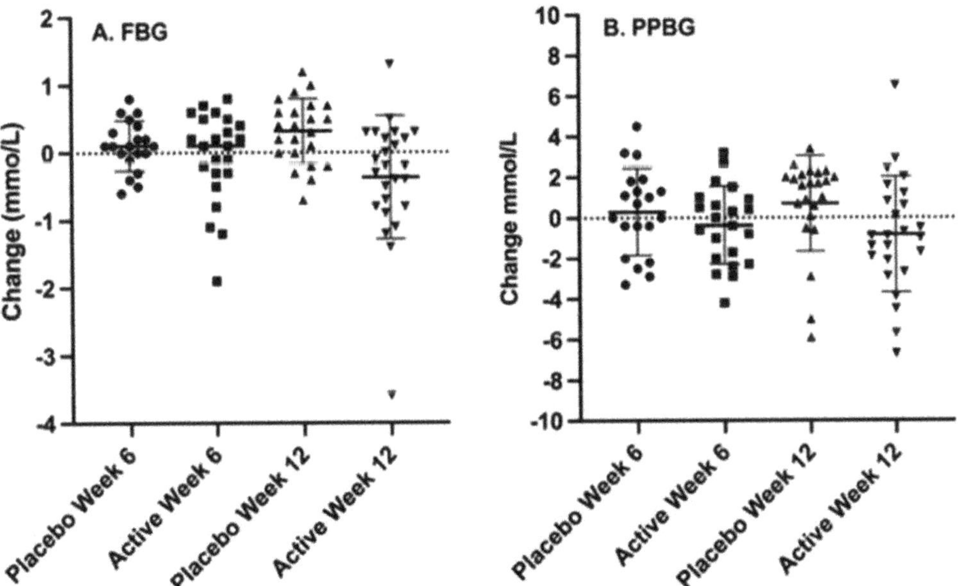

Figure 2. Change from baseline in fasting blood glucose levels ((**A**) FBG) and 2 h post prandial blood glucose levels ((**B**) PPBG) for Active treatment group and Placebo group at week 6 and week 12.

Post-prandial blood glucose (PPBG): Baseline PPBG levels, assessing blood glucose in the fed state, were similar for both groups at one and two hours post-prandial ($p = 0.31$, and $p = 0.21$, respectively) (Table 2). While the 1 h PPBG levels in the active treatment group showed a trend of reduction at 6 weeks and further reduction at 12 weeks, reflecting a total reduction of 0.73 mmol/L, the placebo group remained stable (Table 2). The 2 h PPBG in the active treatment group steadily reduced through the study and was significantly different from baseline. In contrast, in the placebo group, the 2 h PPBG levels increased steadily over the study period (Figure 2B), with the groups being significantly different by 12 weeks ($p = 0.007$).

HbA1c levels: The baseline HbA1c levels correlated with baseline FBG levels (r = 0.944). HbA1c levels were similar for both groups at baseline ($p = 0.15$) and were not statistically different at 12 weeks ($p = 0.41$) (Table 3).

Table 2. Blood glucose and insulin levels for Active treatment and Placebo groups at baseline, week 6 and week 12.

Test *	Time Point	Group	Mean ± SD	Change from Baseline	*p* Value #	Effect Size	95% CI
Fasting Glucose (<5.5 mmol/L)	Baseline	Active	6.95 ± 1.80	-	0.052	-	-
		Placebo	5.99 ± 0.91	-			
	Week 6	Active	6.81 ± 1.74	−0.10 ± 0.72	0.200	0.382	−0.225–0.984
		Placebo	6.13 ± 0.92	0.13 ± 0.38			
	Week 12	Active	6.52 ± 1.42	−0.43 ± 0.88	<0.001	1.108	0.493–1.713
		Placebo	6.34 ± 1.12	0.35 ± 0.46			
1 h post-prandial glucose (<11.1 mmol/L)	Baseline	Active	12.32 ± 3.52	-	0.310	-	-
		Placebo	11.86 ± 2.90	-			
	Week 6	Active	11.88 ± 3.54	−0.37 ± 1.59	0.540	0.193	−0.408–0.793
		Placebo	11.90 ± 3.32	−0.03 ± 1.97			
	Week 12	Active	11.59 ± 3.42	−0.74 ± 2.21	0.200	0.377	−0.197–0.946
		Placebo	11.97 ± 3.25	0.11 ± 2.30			
2 h post-prandial glucose (<7.8 mmol/L)	Baseline	Active	9.15 ± 4.78	-	0.210	-	-
		Placebo	8.11 ± 3.97	-			
	Week 6	Active	8.76 ± 4.35	−0.18 ± 1.84	0.220	0.388	−0.220–0.990
		Placebo	8.92 ± 3.27	0.59 ± 2.13			
	Week 12	Active	8.37 ± 3.83	−0.78 ± 2.74	0.007	0.806	0.212–1.392
		Placebo	9.32 ± 3.82	1.21 ± 2.12			
Fasting Insulin (6–22 mU/L)	Baseline	Active	18.60 ± 9.69	-	0.200	-	-
		Placebo	14.52 ± 6.41	-			
	Week 6	Active	15.80 ± 9.90	−2.52 ± 4.50	0.600	0.157	−0.444–0.756
		Placebo	12.55 ± 5.94	−1.90 ± 3.22			
	Week 12	Active	15.36 ± 7.68	−3.24 ± 7.38	0.140	0.421	−0.154–0.992
		Placebo	13.74 ± 6.61	−0.78 ± 3.42			
1 h post-prandial insulin (40–90 mU/L)	Baseline	Active	101.48 ± 55.49	-	0.270	-	-
		Placebo	113.00 ± 67.88	-			
	Week 6	Active	112.32 ± 75.45	5.95 ± 50.67	0.690	−0.131	−0.761–0.500
		Placebo	128.25 ± 127.04	−1.17 ± 58.02			
	Week 12	Active	122.32 ± 92.23	13.26 ± 59.48	0.690	−0.563	−1.164–0.044
		Placebo	117.41 ± 115.31	−16.00 ± 42.17			
2 h post-prandial insulin (15–65 mU/L)	Baseline	Active	101.76 ± 89.67	-	0.130	-	-
		Placebo	76.91 ± 57.67	-			
	Week 6	Active	93.12 ± 97.09	−7.43 ± 64.00	0.890	0.215	−0.563–0.652
		Placebo	75.10 ± 60.08	−4.42 ± 71.27			
	Week 12	Active	91.08 ± 87.91	−9.54 ± 58.87	0.470	0.215	−0.367–0.794
		Placebo	82.23 ± 49.76	3.09 ± 58.71			

$n = 25$ active group, $n = 23$ placebo group. * Numbers in brackets are healthy reference ranges. # Baseline *p* values are based on difference between group means. Week 6 and Week 12 *p* values are based on difference between group change from baseline.

C-peptide: C-peptide levels were similar between the active treatment and placebo groups at baseline ($p = 0.075$) and also stayed similar at week 12 ($p = 0.79$) (Table 3).

Insulin levels: The fasting insulin (FI) levels varied widely, as expected in a prediabetic state. The mean FI appeared to be slightly higher in the active treatment group though this was not significant ($p = 0.20$) and may reflect the corresponding slightly higher FBG also observed at baseline (Table 2). The FI levels in the active treatment group had a greater reduction by 12 weeks in contrast to the placebo group although this was not significantly different ($p = 0.14$). Both the 1 h and 2 h postprandial insulin (PPI) levels were similar between groups at baseline, 6 weeks and 12 weeks (Table 2).

Table 3. HbA1c, C-peptide, lipid profile, atherogenic potential and C-reactive protein for Active treatment and Placebo at baseline and week 12.

Test *	Group	Baseline Mean ± SD	p Value	Week 12 Mean ± SD	p Value	Effect Size	95% CI
HbA1c % (<6.0%)	Active	6.03 ± 0.75	0.150	5.93 ± 0.78	0.410	−0.250	−0.850–0.343
	Placebo	5.83 ± 0.58		5.75 ± 0.69			
HbA1c mmol/L (<42 mmol/mol)	Active	42.48 ± 8.13	0.130	41.70 ± 8.33	0.370	−0.260	−0.861–0.343
	Placebo	40.00 ± 6.35		39.60 ± 7.72			
C-peptide (0.8–1.9 µg/L)	Active	0.92 ± 0.36	0.075	0.78 ± 0.25	0.792	0.082	−0.518–0.681
	Placebo	0.79 ± 0.27		0.81 ± 0.30			
Cholesterol (3.6–6.9 mmol/L)	Active	5.31 ± 1.15	0.564	5.21 ± 1.02	0.875	−0.047	−0.613–0.520
	Placebo	5.51 ± 1.29		5.15 ± 1.53			
High Density Lipoprotein (HDL) (>0.9 mmol/L)	Active	1.31 ± 0.31	0.241	1.30 ± 0.36	0.172	−0.394	−0.964–0.180
	Placebo	1.21 ± 0.28		1.18 ± 0.36			
Low Density Lipoprotein (LDL) (<2.0 mmol/L)	Active	3.14 ± 1.17	0.643	3.08 ± 1.10	0.917	0.030	−0.536–0.597
	Placebo	3.31 ± 1.29		3.11 ± 1.15			
Total/HDL ratio	Active	4.28 ± 1.14	0.284	4.32 ± 1.13	0.633	0.138	−0.430–0.704
	Placebo	4.60 ± 0.90		4.46 ± 0.89			
Triglycerides (0.3–4.0 mmol/L)	Active	1.62 ± 0.88	0.238	1.58 ± 0.76	0.030	0.667	0.081–1.246
	Placebo	1.97 ± 1.15		2.28 ± 1.28			
C-reactive protein (CRP) (<5 mmol/L)	Active	5.56 ± 4.14	0.797	5.34 ± 4.11	0.789	−0.078	−0.644–0.489
	Placebo	5.24 ± 4.41		5.00 ± 4.59			

n = 25 active group, n = 23 placebo group. * Numbers in brackets are healthy reference ranges.

3.3. Effect of Treatment on Blood Lipids, C-Reactive Protein and DASS-21

The total cholesterol levels, HDL and LDL levels were similar in both groups at baseline and week 12 (Table 3). The triglyceride levels were similar at baseline (p = 0.23), with a significant difference observed between groups after 12 weeks (p = 0.030) due to an increase in levels for the placebo group and no change in those taking the active treatment. C-reactive protein, used as marker of inflammation, was not significantly different between groups at baseline or 12 weeks (Table 3). The assessment of mood (DASS-21) indicated that both groups reported similar normal to low levels of depression (active, 2.28 vs. placebo, 1.52, p = 0.20), anxiety (0.76 vs. 1.30, p = 0.31) and stress (2.68 vs. 1.83, p = 0.15) at baseline, with no significant changes at week 6 (depression: -0.44 vs. −0.44, p = 0.99; anxiety: −0.28 vs. −0.48, p = 0.51; stress: −0.12 vs. −0.09, p = 0.95) or week 12 (depression: −0.80 vs. −0.52, p = 0.56; anxiety: −0.24 vs. −0.70, p = 0.27; stress: 0.04 vs. −0.52, p = 0.28).

3.4. Safety, Adverse Events and Tolerability

Safety was assessed by the FBC and a comprehensive metabolic panel (including electrolytes, liver and kidney function). These markers were in healthy reference range for both groups at baseline and week 12 (Table 4). There were no reported changes in diet or physical activity during the study. There was a single report of feeling "light-headed", possibly hypoglycemia, resulting in withdrawal from the study, with no other adverse events reported.

Table 4. Safety profile and metabolic panel for Active treatment and Placebo at baseline and week 12.

Test *	Baseline			Week 12		
	Active #	Placebo #	p-Value	Active #	Placebo #	p-Value
Haemoglobin (115–160 g/L)	146 ± 11 (128–165)	145 ± 12 (124–167)	0.807	145 ± 13 (123–166)	141 ± 9 (130–175)	0.225
Red Cell Count (3.6–5.2 × 10^{12}/L)	4.9 ± 0.5 (4.4–6.2)	4.9 ± 0.5 (4.2–6.2)	0.717	4.9 ± 0.5 (3.8–5.6)	4.7 ± 0.4 (3.9–5.8)	0.204
Haematocrit (0.33–0.46)	0.45 ± 0.03 (0.39–0.51)	0.44 ± 0.04 (0.38–0.51)	0.645	0.44 ± 0.03 (0.36–0.51)	0.43 ± 0.03 (0.39–0.51)	0.251

Table 4. *Cont.*

Test *	Baseline			Week 12		
	Active #	**Placebo #**	*p*-**Value**	**Active #**	**Placebo #**	*p*-**Value**
Mean Cell Volume (80–98 fL)	91 ± 4.3 (82–100)	91 ± 3.4 (82–97)	0.889	91 ± 4 (86–100)	92 ± 4 (86–102)	0.430
Mean Cell Haemoglobin (27–35 pg)	29.9 ± 1.5 (26–32)	30.1 ± 1.5 (27–33)	0.699	30 ± 2 (28–34)	30.2 ± 1.7 (28–34)	0.515
Platelet Count (150–450×10^9/L)	266 ± 56 (171–385)	260 ± 48 (156–371)	0.697	271 ± 57 (131–326)	253 ± 48 (154–344)	0.264
White Blood Cells (4.0–11.0×10^9/L)	6.6 ± 1.4 (4.4–8.9)	7.2 ± 1.8 (4.1–10.8)	0.230	6.8 ± 1.9 (4.3–9.7)	7.1 ± 1.3 (3.8–12)	0.593
Neutrophils (2.0–7.5×10^9/L)	3.6 ± 0.9 (2.10–5.2)	4.1 ± 1.1 (2.3–8.5)	0.070	3.8 ± 1.7 (2.1–5.8)	4.0 ± 1.0 (1.7–5.6)	0.625
Lymphocytes (1.1–4.0×10^9/L)	2.2 ± 0.6 (1–3.4)	2.3 ± 0.9 (1.1–3.8)	0.674	2.2 ± 0.6 (1.1–3.9)	2.2 ± 0.7 (1.3–5.5)	0.893
Monocytes (0.2–1.0×10^9/L)	0.58 ± 0.14 (0.4–0.8)	0.58 ± 0.24 (0.3–1.4)	0.916	0.66 ± 0.27 (0.4–1.6)	0.59 ± 0.56 (0.3–1.4)	0.363
Eosinophils (0.04–0.40×10^9/L)	0.24 ± 0.17 (0.0–0.7)	0.20 ± 0.17 (0.0–0.7)	0.459	0.24 ± 0.17 (0.07–0.36)	0.18 ± 0.07 (0.0–0.8)	0.143
Basophils ($<0.21 \times 10^9$/L)	0.06 ± 0.02 (0.01–0.09)	0.11 ± 0.22 (0.0–0.08)	0.248	0.06 ± 0.03 (0.0–0.12)	0.06 ± 0.03 (0.0–0.8)	0.481
Sodium (137–$147 \times$ mmol/L)	140 ± 2 (135–145)	139 ± 2 (136–146)	0.253	140 ± 3 (135–146)	140 ± 2 (137–144)	0.643
Potassium (3.5–$5.0 \times$ mmol/L)	4.5 ± 0.3 (3.5–5.0)	4.3 ± 0.4 (3.6–5.4)	0.120	4.4 ± 0.3 (3.8–5.0)	4.4 ± 0.3 (4.0–4.8)	0.687
Chloride (96–$109 \times$ mmol/L)	105 ± 2 (100–110)	104 ± 2 (99–109)	0.439	105 ± 2 (101–108)	105 ± 3 (101–111)	0.785
Bicarbonate (25–$33 \times$ mmol/L)	28 ± 2 (24–32)	28 ± 2 (24–32)	0.489	28 ± 2 (25–31)	27 ± 2 (23–30)	0.420
Creatinine (40–$110 \times$ umol/L)	74 ± 11 (51–93)	74 ± 13 (52–93)	0.911	75 ± 12 (47–95)	71 ± 14 (44–95)	0.419
eGFR (>59 mL/min)	81 ± 5 (59–90)	76 ± 7 (65–90)	0.400	81 ± 7 (66–89)	78 ± 8 (63–90)	0.305
Urea (3.0–8.5 mmol/L)	5.92 ± 1.35 (4.5–9.4)	6.18 ± 1.73 (3.5–11.0)	0.560	5.85 ± 1.59 (4.2–9.0)	6.01 ± 1.37 (3.40–9.10)	0.721
Total Bilirubin (2–$20 \times$ umol/L)	11 ± 6 (4–28)	10 ± 4 (4–20)	0.415	12 ± 7 (5–36)	10 ± 4 (4–19)	0.125
Alk Phosphatase (ALP) (30–$115 \times$ U/L)	73 ± 18 (36–107)	78 ± 18 (43–115)	0.368	71 ± 18 (31–112)	75 ± 19 (37–108)	0.506
Gamma-glutamyl transferase (GGT) (0–45 U/L)	37 ± 23 (11–119)	49 ± 51 (10–262)	0.313	33 ± 16 (10–68)	43 ± 44 (14–208)	0.303
Alanine aminotransferase (ALT) (0–45 U/L)	40 ± 21 (17–110)	35 ± 16 (16–74)	0.437	36 ± 14 (13–68)	29 ± 12 (14–57)	0.083
Aspartate aminotransferase (AST) (0–41 U/L)	30 ± 9 (18–66)	26 ± 7 (15–40)	0.117	30 ± 8 (19–45)	23 ± 7 (12–40)	0.007 ^
Lactate Dehydrogenase (LD) (120–250 U/L)	199 ± 31 (159–269)	183 ± 32 (114–227)	0.081	195 ± 25 (161–254)	170 ± 23 (100–200)	0.001 ^
Calcium (2.25–$2.65 \times$ mmol/L)	2.37 ± 0.065 (2.26–2.53)	2.34 ± 0.09 (2.15–2.51)	0.140	2.37 ± 0.07 (2.18–2.48)	2.34 ± 0.07 (2.23–2.49)	0.218
Phosphate (0.8–$1.5 \times$ mmol/L)	1.14 ± 0.16 (0.7–1.4)	1.09 ± 0.20 (0.8–1.5)	0.358	1.11 ± 0.14 (0.8–1.3)	1.17 ± 0.22 (0.7–1.5)	0.252
Total Protein (60–$82 \times$ g/L)	71 ± 4 (64–79)	70 ± 4 (66–78)	0.632	70 ± 3 (62–76)	69 ± 3 (63–76)	0.128
Albumin (35–$50 \times$ g/L)	42 ± 2 (39–47)	42 ± 2 (38–47)	0.640	42 ± 2 (38–47)	42 ± 2 (37–45)	0.439
Globulin (20–$40 \times$ g/L)	28 ± 5 (21–36)	28 ± 3 (23–32)	0.634	28 ± 3 (20–34)	26 ± 5 (5–31)	0.144

$n = 25$ active group, $n = 23$ placebo group. * Numbers in brackets are healthy reference ranges. # Data written as mean $\pm$ SD with range (in brackets). ^ AST and LD levels, whilst significantly different, were all within Healthy Reference Range, so unlikely to be clinically significant.

4. Discussion

The results of this study demonstrated, in a group of pre-diabetic adults, that 500 mg per day of this hydro-alcoholic *T. foenum-graceum* seed extract was associated with a significant positive effect on fasting and post-prandial blood glucose levels against worsening FBG levels in the placebo group. In addition, those taking the *T. foenum-graceum* extract maintained triglyceride levels while the placebo group had an increase in triglyceride levels over the course of the 12-week study, indicating that the condition was progressing in those with untreated prediabetes.

The results of this study support a previous long-term study in pre-diabetics using 10 g of powdered *T. foenum-graceum* seeds taken daily for three years, showing a protective effect in the prevention of T2DM in those with pre-diabetes [17]. The beneficial reduction of FBG levels by *T. foenum-graceum* seed preparations has also been witnessed in adults with T2DM. In a study that added 10 g of *T. foenum-graceum* seeds soaked in hot water alongside anti-diabetic medication in 60 patients with T2DM, there was a greater reduction in fasting blood glucose levels than with the anti-diabetic medications alone [12]. *T. foenum-graceum* (8 g/day of powdered dried seeds for 8 weeks) was also advantageous as an add-on to exercise in men with T2DM with a reduction in FBG and 2 h PPBG at a level greater than exercise or *T. foenum-graceum* alone [24]. The mechanism of action in these whole seed studies is likely linked to the fiber content, as has been found with other seeds such as flaxseed [25] chia [5].

Extracts of *T. foenum-graceum* seeds differ from the abovementioned studies, in that they do not contain fiber. A hydro-alcoholic *T. foenum-graceum* seed extract at a dose of 1 g per day for 12 weeks significantly reduced FBG levels compared to standard care (dietary control and exercise) alongside placebo in 25 patients with newly diagnosed T2DM [13]. A saponin-rich *T. foenum-graceum* seed extract at 6.3 g per day as an add-on to oral sulphonylurea medication in 69 poorly controlled T2DM patients was reported to maintain blood glucose levels, and significantly decrease FBG and PPBG levels [14]. A more recent RCT study of 154 people with T2DM assessing 1 g of a patented furostanolic saponin-enriched *T. foenum-graceum* seed extract as an add-on medication to oral anti-diabetic prescription also reported a significant decrease in FBG and 2 h PPBG glucose after 12 weeks [20].

In this study, there was no reduction in HbA1c levels over the 12 weeks. There are conflicting results for the effect of *T. foenum-graceum* seeds on HbA1c levels in T2DM conducted over a similar time of 12 weeks, with positive results recorded in stand-alone studies using the whole seeds [12,24] and seed extracts [13–15], but with no effect observed in a study where the *T. foenum-graceum* seed extract was provided as an add-on medication to oral anti-diabetics [20]. Interestingly, reductions in HbA1c were observed in a longer-term study on T2DM [12], indicating that studies of longer duration may be required. This is especially important in a pre-diabetic population as HbA1c may take longer to change due to blood glucose being known to fluctuate between normal and elevated levels [26]. There was also a stabilization of triglyceride levels observed in this study, which has been reported in previous studies of *T. foenum-graceum* seeds in T2DM [11,27]. This indicates that *T. foenum-graceum* may also exert a cardio-protective role in those with glucose dysregulation, although it is unknown if this effect is directly or indirectly due to the insulin sensitizing effects of the *T. foenum-graceum* seed extract.

The effects of *T. foenum-graceum* seeds and seed extracts on glucose metabolism and insulin sensitivity have been attributed to several constituents, including fiber, trigonelline, 4-hydroxyisoleucine (4-HIL), and saponins [28]. It is noteworthy that the extract used in this study was standardized to 20% trigonelline. The use of isolated trigonelline has been shown in an animal model of T2DM to lower blood glucose, free fatty acids, TNF-α and IL-6 levels [29] as well as improve serum insulin, leptin, pancreatic antioxidant status. Additionally, trigonelline was shown to have an insulin-sensitizing effect as measured by homeostasis model of insulin resistance (HOMA-IR) and homeostasis model assessment of β-cell function (HOMA-B). In an in vitro study of HepG2 cells [30], 4-HIL was found to decrease the production of TNF-α and increase the actions of insulin receptor substrate-1 and glucose transporter type 4 (GLUT-4). Previous clinical studies have also found that 4-HIL exerts an insulin secretory and mimicry effect in response to glucose by stimulating pancreatic beta cells [31]. The saponins in *T. foenum-graceum* seeds have been demonstrated in animal studies to decrease FBG and PPBG, insulin and insulin resistance markers [32], IL-6, TNF-α [33], increase expression of GLUT-4 in skeletal muscle [34], and significantly increase pancreatic beta cell function [29]. It is interesting that these constituents of *T. foenum-graceum* seeds may support healthy blood glucose levels without inducing hypoglycemia in those with prediabetes, in addition to T2DM, as witnessed in this study.

In addition to effects on glucose, safety measurements for the effect of *T. foenum-graceum* seed extract on those with prediabetes. There were no adverse changes to kidney, liver, and metabolic markers. There was a statistical difference in AST and LD markers between groups at week 12, however this was due to decreased levels in the placebo group rather than an increase in the treatment group. To date, this is the first study to report comprehensive safety data of *T. foenum-graceum* extract use in those with prediabetes. There were no adverse events reported in the long-term *T. foenum-graceum* powdered seed and prediabetes study by Gaddam et al. [17], however no safety measurements were reported. Safety measurements have been reported in those taking *T. foenum-graceum* with T2DM. Whole *T. foenum-graceum* seed studies of 8 weeks duration have reported no adverse changes to kidney and liver markers in those with T2DM [15,19,35]. Benefits including improvements to liver enzyme markers ALT, ALP [19] and AST, as well as systolic blood pressure, blood urea nitrogen levels [15], creatinine and triglycerides [35] were also reported. Safety parameters have been measured in some *T. foenum-graceum* seed extract studies [16,20], with no reported liver or kidney toxicity, however the constituent standardization in these studies was different to the extract used in this study, so results are not fully comparable. Therefore, although there is no indication of any concern about safety, longer studies would be required to assess this extract for longer term use in prediabetics.

This study was conducted during the COVID-19 pandemic which affected recruitment and participant drop-out rate, resulting in a lower sample size than originally planned. Given the positive results of this small, 3-month study, further investigations with a larger cohort and longer duration are warranted. This study has also highlighted safety issues with the use of a placebo control in those with unmedicated prediabetes over long periods of time.

5. Conclusions

The results of this study indicate that this *T. foenum-graceum* seed extract may be an effective treatment for impaired fasting glucose and impaired glucose tolerance, both of which are associated with pre-diabetes. In addition, this *T. foenum-graceum* seed extract may provide cardioprotective support for those with prediabetes by stabilizing triglycerides.

Author Contributions: Conceptualization and methodology, E.S.; Data curation, E.S. and E.P.; Formal analysis, E.S., A.R. and E.P.; Investigation, E.S.; Methodology, E.S.; Supervision, K.J.S.; Visualization, A.R. and E.P.; Writing—original draft, E.P.; Writing—review and editing, E.P., E.S., A.R. and K.J.S. All authors have read and agreed to the published version of the manuscript.

Funding: The sponsor for this trial was Gencor Pacific Ltd., Hong Kong. The sponsor had no role in the design of the clinical trial and no influence over the analysis, reporting, and interpretation of data. E.P. was supported by the University of Queensland Graduate School scholarship.

Institutional Review Board Statement: The study was conducted in accordance with the Declaration of Helsinki and approved by the Bellberry Ethics Committee (No: 2017-08-601).

Informed Consent Statement: Informed consent was obtained from all subjects involved in the study.

Data Availability Statement: Date described in the manuscript will be made available pending application and approval.

Acknowledgments: E.P. was supported by a University of Queensland Graduate School scholarship. The authors would like to acknowledge and thank Rommy Castaneda for proof-reading and review of the manuscript. E.P. would like to thank Christopher Bailey for statistical analysis support.

Conflicts of Interest: The authors declare no conflict of interest. The funders had no role in the design of the study; in the collection, analyses, or interpretation of data; in the writing of the manuscript; or in the decision to publish the results.

References

1. International Diabetes Federation. *IDF Diabetes Atlas*, 10th ed.; International Diabetes Federation: Brussels, Belgium, 2021.
2. Cai, X.; Zhang, Y.; Li, M.; Wu, J.; Mai, L.; Li, J.; Yang, Y.; Hu, Y.; Huang, Y. Association between prediabetes and risk of all cause mortality and cardiovascular disease: Updated meta-analysis. *BMJ* **2020**, *370*, m2297. [CrossRef] [PubMed]
3. Kanat, M.; DeFronzo, R.A.; Abdul-Ghani, M.A. Treatment of prediabetes. *World J. Diabetes* **2015**, *6*, 1207–1222. [CrossRef] [PubMed]
4. Salehi, B.; Ata, A.; Kumar, N.V.A.; Sharopov, F.; Ramírez-Alarcón, L.; Ruiz-Ortega, A.; Ayatollahi, S.; Fokou, P.V.T.; Kobarfard, F.; Zakaria, Z.A.; et al. Antidiabetic potential of medicinal plants and their active components. *Biomolecules* **2019**, *9*, 551. [CrossRef] [PubMed]
5. Vuksan, V.; Choleva, L.; Jovanovski, E.; Jenkins, A.L.; Au-Yeung, F.; Dias, A.G.; Ho, H.V.T.; Zurbau, A.; Duvnjak, L. Comparison of flax (*Linum usitatissimum*) and Salba-chia (*Salvia hispanica* L.) seeds on postprandial glycemia and satiety in healthy individuals: A randomized, controlled, crossover study. *Eur. J. Clin. Nutr.* **2017**, *71*, 234–238. [CrossRef] [PubMed]
6. Robert, S.D.; Ismail, A.A.; Rosli, W.I. Reduction of postprandial blood glucose in healthy subjects by buns and flatbreads incorporated with fenugreek seed powder. *Eur. J. Nutr.* **2016**, *55*, 2275–2280. [CrossRef]
7. Yadav, U.C.; Baquer, N.Z. Pharmacological effects of *Trigonella foenum-graecum* L. in health and disease. *Pharm. Biol.* **2014**, *52*, 243–254. [CrossRef]
8. Kassaian, N.; Azadbakht, L.; Forghani, B.; Amini, M. Effect of fenugreek seeds on blood glucose and lipid profiles in type 2 diabetic patients. *Int. J. Vitam. Nutr. Res.* **2009**, *79*, 34–39. [CrossRef]
9. Losso, J.N.; Holliday, D.L.; Finley, J.W.; Martin, R.J.; Rood, J.C.; Yu, Y.; Greenway, F.L. Fenugreek bread: A treatment for diabetes mellitus. *J. Med. Food* **2009**, *12*, 1046–1049. [CrossRef]
10. Bhaktha, G.; Nayak, S.; Shantaram, M. Management of newly diagnosed diabetes by *Trigonella foenum-graecum*. *Int. J. Res. Ayu. Pharm.* **2011**, *2*, 1231–1234.
11. Rafraf, M.; Malekiyan, M.; Asghari-Jafarabadi, M.; Aliasgarzadeh, A. Effect of fenugreek seeds on serum metabolic factors and adiponectin levels in type 2 diabetic patients. *Int. J. Vitam. Nutr. Res.* **2014**, *84*, 196–205. [CrossRef]
12. Kanade, M.; Mudgalkar, N. A simple dietary addition of fenugreek seed leads to the reduction in blood glucose levels: A parallel group, randomized single-blind trial. *Ayu* **2017**, *38*, 24–27. [CrossRef]
13. Gupta, A.; Gupta, R.; Lal, B. Effect of *Trigonella foenum-graecum* (fenugreek) seeds on glycaemic control and insulin resistance in type 2 diabetes mellitus: A double blind placebo controlled study. *J. Assoc. Physicians India* **2001**, *49*, 1057–1061.
14. Lu, F.R.; Shen, L.; Qin, Y.; Gao, L.; Li, H.; Dai, Y. Clinical observation on *Trigonella foenum-graecum* L. total saponins in combination with sulfonylureas in the treatment of type 2 diabetes mellitus. *Chin. J. Integr. Med.* **2008**, *14*, 56–60. [CrossRef]
15. Najdi, R.A.; Hagras, M.M.; Kamel, F.O.; Magadmi, R.M. A randomized controlled clinical trial evaluating the effect of *Trigonella foenum-graecum* (fenugreek) versus glibenclamide in patients with diabetes. *Afr. Health Sci.* **2019**, *19*, 1594–1601. [CrossRef]
16. Verma, N.; Usman, K.; Awasthi, V.; Goel, P.K.; Lamgora, G. Clinical evaluation fenugreek seed extract in patients with type-2 diabetes. An add-on study in 154 patients. *World J. Pharm. Res.* **2015**, *4*, 2266–2279.
17. Gaddam, A.; Galla, C.; Thummisetti, S.; Marikanty, R.K.; Palanisamy, U.D.; Rao, P.V. Role of fenugreek in the prevention of type 2 diabetes mellitus in prediabetes. *J. Diabetes Metab. Disord.* **2015**, *14*, 74. [CrossRef]
18. Gong, J.; Fang, K.; Dong, H.; Wang, D.; Hu, M.; Lu, F. Effect of fenugreek on hyperglycaemia and hyperlipidemia in diabetes and prediabetes: A meta-analysis. *J. Ethnopharmacol.* **2016**, *194*, 260–268. [CrossRef]
19. Hadi, A.; Arab, A.; Hajianfar, H.; Talaei, B.; Miraghajani, M.; Babajafari, S.; Marx, W.; Tavakoly, R. The effect of fenugreek seed supplementation on serum irisin levels, blood pressure, and liver and kidney function in patients with type 2 diabetes mellitus: A parallel randomized clinical trial. *Complement. Ther. Med.* **2020**, *49*, 102315. [CrossRef]
20. Verma, N.; Usman, K.; Patel, N.; Jain, A.; Dhakre, S.; Swaroop, A.; Bagchi, M.; Kumar, P.; Preuss, H.G.; Bagchi, D. A multicenter clinical study to determine the efficacy of a novel fenugreek seed (*Trigonella foenum-graecum*) extract (Fenfuro) in patients with type 2 diabetes. *Food Nutr. Res.* **2016**, *60*, 32382. [CrossRef]
21. Ng, F.; Trauer, T.; Dodd, S.; Callaly, T.; Campbell, S.; Berk, M. The validity of the 21-item version of the Depression Anxiety Stress Scales as a routine clinical outcome measure. *Acta Neuropsychiatr.* **2007**, *19*, 304–310. [CrossRef]
22. National Health and Medical Research Council. *Guidance: Safety Monitoring and Reporting in Clinical Trials Involving Therapeutic Goods*; National Health and Medical Research Council: Canberra, Australia, 2016.
23. Royal Australian College of General Practitioners and Diabetes Australia. *Management of Type 2 Diabetes: A Handbook for General Practice*; The Royal Australian College of General Practitioners: Melbourne, Australia, 2021; pp. 1–198.

24. Karim, A.; Siroosce, C.; Jabrayel, P.J. Antidiabetic effects of exercise and fenugreek supplementation in males with NIDDM. *Med. Dello Sport* **2009**, *62*, 315–324.

25. Villarreal-Renteria, A.I.; Herrera-Echauri, D.D.; Rodríguez-Rocha, N.P.; Zuñiga, L.Y.; Muñoz-Valle, J.F.; García-Arellano, S.; Bernal-Orozco, M.F.; Macedo-Ojeda, G. Effect of flaxseed (*Linum usitatissimum*) supplementation on glycemic control and insulin resistance in prediabetes and type 2 diabetes: A systematic review and meta-analysis of randomized controlled trials. *Complement. Ther. Med.* **2022**, *70*, 102852. [CrossRef] [PubMed]

26. Chakarova, N.; Dimova, R.; Grozeva, G.; Tankova, T. Assessment of glucose variability in subjects with prediabetes. *Diabetes Res. Clin. Pract.* **2019**, *151*, 56–64. [CrossRef] [PubMed]

27. Geberemeskel, G.A.; Debebe, Y.G.; Nguse, N.A. Antidiabetic effect of fenugreek seed powder solution (*Trigonella foenum-graecum* L.) on hyperlipidemia in diabetic patients. *J. Diabetes Res.* **2019**, *2019*, 8507453. [CrossRef] [PubMed]

28. Gupta, R.C.; Doss, R.B.; Garg, R.C.; Srivastava, A.; Lall, R.; Sinha, A. Fenugreek: Multiple health benefits. In *Nutraceuticals*, 2nd ed.; Gupta, R.C., Lall, R., Srivastava, A., Eds.; Academic Press: Cambridge, MA, USA, 2021; pp. 585–602. [CrossRef]

29. Tharaheswari, M.; Jayachandra Reddy, N.; Kumar, R.; Varshney, K.C.; Kannan, M.; Sudha Rani, S. Trigonelline and diosgenin attenuate ER stress, oxidative stress-mediated damage in pancreas and enhance adipose tissue PPARγ activity in type 2 diabetic rats. *Mol. Cell Biochem.* **2014**, *396*, 161–174. [CrossRef]

30. Gao, F.; Jian, L.; Zafar, M.I.; Du, W.; Cai, Q.; Shafqat, R.A.; Lu, F. 4-Hydroxyisoleucine improves insulin resistance in HepG2 cells by decreasing TNF-alpha and regulating the expression of insulin signal transduction proteins. *Mol. Med. Rep.* **2015**, *12*, 6555–6560. [CrossRef]

31. Broca, C.; Gross, R.; Petit, P.; Sauvaire, Y.; Manteghetti, M.; Tournier, M.; Masiello, P.; Gomis, R.; Ribes, G. 4-Hydroxyisoleucine: Experimental evidence of its insulinotropic and antidiabetic properties. *Am. J. Physiol.* **1999**, *277*, E617–E623. [CrossRef]

32. Naidu, P.B.; Ponmurugan, P.; Begum, M.S.; Mohan, K.; Meriga, B.; RavindarNaik, R.; Saravanan, G. Diosgenin reorganises hyperglycaemia and distorted tissue lipid profile in high-fat diet–streptozotocin-induced diabetic rats. *J. Sci. Food Agri.* **2015**, *95*, 3177–3182. [CrossRef]

33. Ghosh, S.; More, P.; Derle, A.; Patil, A.B.; Markad, P.; Asok, A.; Kumbhar, N.; Shaikh, M.L.; Ramanamurthy, B.; Shinde, V.S.; et al. Diosgenin from Dioscorea bulbifera: Novel hit for treatment of type II diabetes mellitus with inhibitory activity against α-amylase and α-glucosidase. *PLoS ONE* **2014**, *9*, e106039. [CrossRef]

34. Fang, K.; Dong, H.; Jiang, S.; Li, F.; Wang, D.; Yang, D.; Gong, J.; Huang, W.; Lu, F. Diosgenin and 5-methoxypsoralen ameliorate insulin resistance through ER-α/PI3K/Akt-signaling pathways in HepG2 cells. *Evid-Based Comp. Alt. Med.* **2016**, *2016*, 7493694. [CrossRef]

35. Elsaadany, M.A.; AlTwejry, H.M.; Zabran, R.A.; AlShuraim, S.A.; AlShaia, W.A.; Abuzaid, O.I.; AlBaker, W.I. Antihyperglycemic effect of fenugreek and ginger in patients with type 2 diabetes: A double-blind, placebo-controlled study. *Curr. Nutr. Food Sci.* **2022**, *18*, 231–237. [CrossRef]

Article

Gut Microbiota and Bile Acids Mediate the Clinical Benefits of YH1 in Male Patients with Type 2 Diabetes Mellitus: A Pilot Observational Study

Yueh-Hsiang Huang [1,2,3,†], Yi-Hong Wu [1,2,3,†], Hsiang-Yu Tang [4], Szu-Tah Chen [5], Chih-Ching Wang [5], Wan-Jing Ho [6], Yi-Hsuan Lin [5], Geng-Hao Liu [2,3,7], Pei-Yeh Lin [8], Chi-Jen Lo [4], Yuan-Ming Yeh [9,10,*] and Mei-Ling Cheng [4,11,12,*]

[1] Department of Traditional Chinese Medicine, Chang Gung Memorial Hospital, Taipei 105, Taiwan
[2] Graduate Institute of Clinical Medical Sciences, Chang Gung University, Taoyuan 333, Taiwan
[3] School of Traditional Chinese Medicine, College of Medicine, Chang Gung University, Taoyuan 333, Taiwan
[4] Metabolomics Core Laboratory, Healthy Aging Research Center, Chang Gung University, Taoyuan 333, Taiwan
[5] Division of Endocrinology and Metabolism, Department of Internal Medicine, Chang Gung Memorial Hospital, Linkou Branch, Taoyuan 333, Taiwan
[6] Division of Cardiology, Department of Internal Medicine, Chang Gung Memorial Hospital, Linkou Branch, Taoyuan 333, Taiwan
[7] Department of Traditional Chinese Medicine, Chang Gung Memorial Hospital, Linkou Branch, Taoyuan 333, Taiwan
[8] Department of Medical Nutrition Therapy, Chang Gung Memorial Hospital, Taoyuan 333, Taiwan
[9] Genomic Medicine Core Laboratory, Chang Gung Memorial Hospital, Linkou Branch, Taoyuan 333, Taiwan
[10] Graduate Institute of Health Industry Technology, Chang Gung University of Science and Technology, Taoyuan 333, Taiwan
[11] Clinical Metabolomics Core Laboratory, Chang Gung Memorial Hospital, Taoyuan 333, Taiwan
[12] Department of Biomedical Sciences, College of Medicine, Chang Gung University, Taoyuan 333, Taiwan
*   Correspondence: ymyeh@cgmh.org.tw (Y.-M.Y.); chengm@mail.cgu.edu.tw (M.-L.C.)
†   These authors contributed equally to this work.

Citation: Huang, Y.-H.; Wu, Y.-H.; Tang, H.-Y.; Chen, S.-T.; Wang, C.-C.; Ho, W.-J.; Lin, Y.-H.; Liu, G.-H.; Lin, P.-Y.; Lo, C.-J.; et al. Gut Microbiota and Bile Acids Mediate the Clinical Benefits of YH1 in Male Patients with Type 2 Diabetes Mellitus: A Pilot Observational Study. *Pharmaceutics* **2022**, *14*, 1857. https://doi.org/10.3390/pharmaceutics14091857

Academic Editor: Crispin R. Dass

Received: 16 August 2022
Accepted: 31 August 2022
Published: 2 September 2022

Publisher's Note: MDPI stays neutral with regard to jurisdictional claims in published maps and institutional affiliations.

Abstract: Our previous clinical trial showed that a novel concentrated herbal extract formula, YH1 (*Rhizoma coptidis* and Shen-Ling-Bai-Zhu-San), improved blood glucose and lipid control. This pilot observational study investigated whether YH1 affects microbiota, plasma, and fecal bile acid (BA) compositions in ten untreated male patients with type 2 diabetes (T2D), hyperlipidemia, and a body mass index ≥ 23 kg/m^2. Stool and plasma samples were collected for microbiome, BA, and biochemical analyses before and after 4 weeks of YH1 therapy. As previous studies found, the glycated albumin, 2-h postprandial glucose, triglycerides, total cholesterol, and low-density lipoprotein cholesterol levels were significantly improved after YH1 treatment. Gut microbiota revealed an increased abundance of the short-chain fatty acid-producing bacteria *Anaerostipes* and *Escherichia/Shigella*. Furthermore, YH1 inhibited specific phylotypes of bile salt hydrolase-expressing bacteria, including *Parabacteroides*, *Bifidobacterium*, and *Bacteroides caccae*. Stool tauro-conjugated BA levels increased after YH1 treatment. Plasma total BAs and 7α-hydroxy-4-cholesten-3-one (C4), a BA synthesis indicator, were elevated. The reduced deconjugation of BAs and increased plasma conjugated BAs, especially tauro-conjugated BAs, led to a decreased glyco-to tauro-conjugated BA ratio and reduced unconjugated secondary BAs. These results suggest that YH1 ameliorates T2D and hyperlipidemia by modulating microbiota constituents that alter fecal and plasma BA compositions and promote liver cholesterol-to-BA conversion and glucose homeostasis.

Keywords: type 2 diabetes; Chinese herbal medicine; YH1; gut microbiota; bile acids

1. Introduction

Traditional Chinese medicine (TCM) has been an excellent resource for developing new medications for metabolic disorders, such as type 2 diabetes (T2D), since ancient times [1]. Increasing evidence indicates that TCM can treat metabolic disorders by regulating the composition of gut microbiota, such as bacteria with bile salt hydrolase (BSH) activity and short-chain fatty acid (SCFA)-producing bacteria [2]. Our previous randomized double-blind placebo-controlled pilot study discovered that YH1, an innovative antidiabetic medication, could improve hypoglycemic action, β-cell function, and lipid metabolism in overweight/obese patients with poorly controlled T2D who had taken three or more classes of oral hypoglycemic agents (OHAs) before enrollment [3]. YH1 was designed as a concentrated herbal granule containing *Rhizoma coptidis* (RC) and Shen-Ling-Bai-Zhu-San (SLBZS). Modern pharmacological research has identified that the major chemical constituents of RC are alkaloids, and berberine serves as its main bioactive alkaloid with antidiabetic, antidyslipidemic, antiobesity, antibiotic, antioxidant, and anti-inflammatory activities [4]. However, the bioavailability of berberine is far below 1% in plasma, due to poor intestinal absorption and efflux via the action of P-glycoprotein [5]. Previous studies in mice showed that RC alkaloids induced hypolipidemic effects through the modulation of gut microbiota and hepatic lipid metabolism [2,6]. In addition, recent experimental and clinical studies have reported that berberine can reduce inflammation and regulate blood sugar and lipid levels by changing the composition of gut microbes and bile acids (BAs) [2,6–9]. As shown in both in vivo and in vitro studies [7], berberine exposure changed bacterial community composition and function, particularly reducing BSH-expressing bacteria such as *Clostridium spp.* Metagenomic and metabolomic studies demonstrated that the antidiabetic effect of berberine was mediated by the inhibition of deoxycholic acid (DCA) biotransformation by *Ruminococcus bromii* [9], suggesting a potential therapeutic approach for patients with T2D through the modulation of microbiome dysbiosis. Therefore, it is critical to understand whether the composition of gut microbes and BAs of patients with T2D and hyperlipidemia are modulated by YH1 treatment.

There are bidirectional interactions between gut microbiota and BA synthesis [10]. Gut microbiota affects the BA pool size, composition, and metabolism, including deconjugation and 7α-dehydroxylation of conjugated and primary BAs, as well as the formation of secondary BAs, whereas BAs can control intestinal bacterial overgrowth and prevent inflammation [11]. In addition, BAs play important roles in the homeostasis of lipid, glucose, and energy metabolism by regulating two major BA receptors: the farnesoid X receptor (FXR) and Takeda G protein-coupled receptor 5 (TGR5) [12]. FXR is the major sensor and regulator of BA metabolism, which regulates BA pool size via a negative feedback loop involving intestinal and hepatic FXR signaling [13]. Deconjugation of glyco (G)- and tauro (T)-conjugated BAs is performed by microbial BSH. A recent study reclassified BSHs into eight phylotypes and demonstrated the variation in different microbial deconjugation abilities of glyco- and tauro-conjugated BAs [14]. Therefore, changes in gut microbiota composition affect BA metabolism and are thus associated with BA pool size alteration, the conjugated to unconjugated BA ratio, G/T-BAs ratio, and primary to secondary BA ratio. The most potent endogenous agonists for TGR5 are tauro-conjugated BAs, followed by unconjugated BAs and glyco-conjugated BAs [15]. Furthermore, the activation of TGR5 by secondary BAs is greater than that of primary BAs, so taurolithocholic acid (TLCA) is the most potent endogenous activator of TGR5 [16]. However, conjugated BAs are reportedly inactive to nuclear receptors (FXR) in the absence of a specific transporter, and the effective dose of BAs for TGR5 was lower than that for FXR [17]. Activation of TGR5 in intestinal L cells by conjugated BAs enhanced the secretion of glucagon-like peptide-1 (GLP-1), and FXR exhibited crosstalk with TGR5 to control GLP-1 secretion [11]. Overall, in addition to BA synthesis and excretion playing a critical role in affecting lipid catabolism, alteration of BA composition directly affects their signaling due to different affinities to FXR and TGR5 in the host, thus regulating lipid, glucose, and energy homeostasis [12].

SCFAs, including acetate, propionate, and butyrate, are well-known to induce GLP-1 secretion from gut L cells, and thus indirectly regulate glucose homeostasis by promoting insulin secretion and reducing pancreatic glucagon secretion [18]. SLBZS was reported to modulate the gut microbiota during the treatment of functional dyspepsia or inflammatory bowel disease by enriching SCFA-producing bacteria [19,20]. Moreover, berberine can enrich SCFA-producing bacteria, especially butyrate-producing bacteria, such as *Enterobacter* and *Escherichia/Shigella*, resulting in reduced plasma glucose and lipid levels [21]. A daily dose of YH1 contains 360.9 mg of berberine. Although the YH1 treatment used in our previous study had a four-fold lower berberine dosage than the dose of berberine used alone in another study [22], YH1 treatment achieved an 11.1% better reduction in glycated hemoglobin (HbA1c) levels, along with larger, favorable decreases in serum triglycerides (TG), total cholesterol, and low-density lipoprotein cholesterol (LDL-C) levels [3]. This suggests that YH1 treatment has superior antidiabetic and antidyslipidemic activities compared with using the single pure compound alone. To evaluate the mechanisms of YH1-mediated hypoglycemic and hypolipidemic effects, we investigated alterations in the stool microbiome, as well as in BA composition of plasma and fecal samples from patients with T2D and hyperlipidemia following 4 weeks of YH1 treatment.

2. Materials and Methods

2.1. Study Design and Participants

We conducted this pilot observational study in the TCM clinics of the Taoyuan branches of Chang Gung Memorial Hospital from January 2020 to December 2021 (ClinicalTrials.gov number, NCT04194515). This study was approved by the Committee on Research Ethics of the Chang Gung Memorial Hospital in Taiwan (No: 201901022B0A3). All participants provided written informed consent prior to participation. Eligible participants were male outpatients with T2D and hyperlipidemia who were not receiving treatment and consented to taking 6 g of YH1 [3] three times daily for 4 weeks. Patients were eligible to be enrolled in this study only if they met all of the following criteria: (1) Male patients with T2D who did not take hypoglycemic agents in the past month; (2) Aged 20–65 years; (3) Body mass index (BMI) $\geq$ 23 kg/m^2; (4) Received dietary and exercise education from a nutritionist for at least one month; (5) HbA1c $\geq$ 6.5%; (6) LDL-C $\geq$ 130 mg/dl; (7) Agreed to take 6 g of YH1 three times per day for 4 weeks.

Patients were excluded if they met any of the following criteria: (1) Type 1 or other specific type of diabetes; (2) Female sex; (3) Use of oral hypoglycemic agents or insulin within the past month; (4) Use of lipid-lowering agents within the past month; (5) Serious gastrointestinal (GI) tract diseases, such as peptic ulcers or GI tract bleeding; (6) Hepatic insufficiency with alanine aminotransferase (ALT) > 108 U/L or renal insufficiency with estimated glomerular filtration rate (eGFR) < 60 mL/min/1.73 m^2; (7) stressful situations, including diabetic ketoacidosis, nonketotic hyperosmolar diabetic coma, severe infection, or surgery in the previous month; (8) Mental illness; (9) Addiction to tobacco, alcohol or other drugs; (10) Hemoglobin-related disease or chronic anemia; (11) Underlying conditions that could lead to poor compliance; (12) Severe organ disease, including cancer, coronary artery disease, myocardial infarction, or cerebrovascular diseases; (13) Consecutive use of antibiotics, probiotics, or weight loss drugs for more than 3 days within the 3 months prior to participation in this study; (14) Uncontrolled hypertension (blood pressure $\geq$ 160/100 mmHg); (15) Chinese medicine treatment in the past month.

All participants were instructed to maintain their eating habits and lifestyles during the study period, and dietary recalls were obtained to document usual dietary intake by the same nutritionist before and 4 weeks after YH1 treatment. Patients were evaluated for medication history, Bristol Stool Form Scale (BSFS), Constipation Assessment Scale (CAS), prescription adherence to YH1, and adverse events at 0 and 4 weeks. Additionally, physical examination parameters, including body weight, BMI, waist circumference, blood pressure, and heart rate, were measured. Fecal specimens were collected at home, with a clean stool collection container with a spoon attached to the lid, prior to and upon conclusion of YH1 treatment. Stool samples were stored in the freezer immediately after collection, transferred to the laboratory in the frozen state within 3 days, and stored at −80 °C before analysis. In addition, blood samples were collected while fasting and 2 h post-breakfast for assessment of metabolic and biochemical parameters before and 4 weeks after YH1 treatment. A standardized breakfast containing a Chinese omelet with pork chop and a slice of pan-fried radish cake (total 440 calories, including 45 g carbohydrates, 20 g protein, and 20 g fat) was provided at 0 and 4 weeks to avoid the effect of different foods on BA profiles.

2.2. Microbiome, Metabolomic, and Biochemical Measures

For stool microbiome analysis, we conducted DNA extraction, polymerase chain reaction (PCR) amplification, and 16S rRNA sequencing, as previously described [23]. Total bacterial genomic DNA was isolated from fecal specimens using a QIAamp PowerFecal® DNA Kit (Qiagen, USA) in accordance with the manufacturer's protocol. Genomic DNA was amplified by PCR with primers targeting the 16S rRNA V3-V4 region, with a product size of 460 bp. The detailed protocol for PCR and sequencing has been described in a previous study [24]. Briefly, adapter overhang nucleotide sequences (Illumina Inc) were added to the gene-specific sequences (forward primer sequence, 5'-TCGTCGGCAGCGTCAGATGTGT ATAAGAGACAGCCTACGGGNGGCWGCAG-3', and reverse primer sequence, 5'-GTCTC GTGGGCTCGGAGATGTGTATAAGAGACAGGACTACHVGGGTATCTAATCC-3'). Subsequently, amplicon products were purified using Agencourt AMPure XP beads (Beckman Coulter), followed by indexing and barcoding PCR. The final amplicon libraries were approximately 630 bp, and the Agilent 2100 Bioanalyzer with the Agilent HS DNA Kit was used for size and quality validation. Multiplexed, pooled library sequencing was conducted on the MiSeq System with MiSeq Reagent Kit v3 (600-cycle) (Illumina, San Diego, CA, USA). Microbiome analysis was executed by USEARCH (v11), following the methods described in a previous study [24]. Subsequent analysis was performed only for sequence tags ≥400 bp in length. The operational taxonomic units (OTUs) were clustered by the UPARSE [25] algorithm. In addition, the RDP training set (v16) was used as a database of reference species, and the SINTAX algorithm was performed for final taxonomic classification [26]. Measurements of α-diversity (Chao1 and Shannon index) and β-diversity (Bray–Curtis dissimilarity) were assessed by QIIME (version 1.9.1) [27] via DESeq2 [28] normalization.

Fasting plasma glucose (FPG), 2-h postprandial glucose (2hPG), and glycated albumin were measured at weeks 0 and 4. Homeostatic model assessment of insulin resistance (HOMA-IR), β-cell function index (HOMA-β), lipid profiles, C-reactive protein (CRP), as well as hepatic and renal function were assessed at the beginning and end of the study. Furthermore, 2-h postprandial blood samples were collected for measurement of 2hPG and stored at $-80\,^{\circ}\mathrm{C}$ until analyses of plasma BA profiles, 7α-hydroxy-4-cholesten-3-one (C4), and fibroblast growth factor 19 (FGF19), before and after YH1 therapy. C4 and FGF19 serum biomarkers were indicators of BA synthesis and an index of the negative feedback loop involved in BA metabolism, respectively.

Plasma and fecal sample preparation and analysis of BA by a UPLC–MS/MS system (Waters, Milford, CT, USA) covering 15 major BA species are described in Protocol S1. The full term of each BA and its abbreviation are presented in the table of Protocol S1. We assessed the total and individual concentrations of the 15 BAs, as well as the ratios of classified BAs, including ratios of 12α-hydroxylated BAs (CA, GCA, TCA, DCA, GDCA, and TDCA) to non-12α-hydroxylated BAs (CDCA, GCDCA, TCDCA, LCA, GLCA, TLCA, UDCA, GUDCA, and TUDCA); primary (CA, GCA, TCA, CDCA, GCDCA, TCDCA) and secondary BAs (DCA, GDCA, TDCA, LCA, GLCA, TLCA, UDCA, GDCA, TUDCA); and glyco- plus tauro-conjugated (GCA, GCDCA, GDCA, GLCA, GUDCA, TCA, TCDCA, TDCA, TLCA, TUDCA) to unconjugated BAs (CA, CDCA, DCA, LCA, UDCA).

For plasma C4 analysis, 300 µL methanol was added to 100 µL plasma samples for liquid–liquid extraction. Samples of the mixture were incubated on ice for 30 min and then centrifuged to precipitate protein at 12,000 rpm for 30 min at 4 °C. The supernatant was transferred to a sample vial and analyzed in an LC–MS system (UPLC with Xevo TQS MS, Waters, Manchester, UK) in negative atmospheric pressure chemical ionization (APCI) mode with multiple reaction monitoring. The chromatographic separation was achieved on an Acquity HSS pentafluorophenyl (PFP) column (2.1 × 100 mm, particle size of 1.8 µm, Waters Corp., Milford, CT, USA) at 25 °C with mobile phase A (25% acetonitrile with 0.1% formic acid) and mobile phase B (methanol). The flow rate was set to 0.3 mL/min. The gradient profile was as follows: 70% B for 1 min; linear gradient 70–75.7% B, 8 min; 75.7–100% B, 1 min; and 100% B, 2.2 min. The column was then re-equilibrated for 3 min. The parameters of MS were as follows: corona was 1 µA, desolvation gas flow was 800 L/h at 600°C, and cone gas flow was 150 L/h at 150 °C. System operation and data acquisition were controlled using Mass Lynx software and targeted metabolic data were analyzed by TargetLynx (Waters, Milford, CT, USA). In addition, serum FGF19 was measured by using commercially available enzyme-linked immunosorbent assay kits (R&D Systems, Minneapolis, MN, USA) according to the manufacturer's instructions.

2.3. Statistical Analysis

IBM SPSS statistical software (Version 21, Armonk, NY, USA) was used for nonparametric analysis and verification. Measurement data are presented as the median (minimum, maximum), and categorical data are described by numbers. The Wilcoxon signed-rank test was used to detect differences in gut microbial features (richness, diversity, and relative abundances of species), plasma and fecal BA levels, and clinical outcomes between baseline and post-treatment measurements. Fisher's exact test was performed to assess categorical variables. Spearman's test was used for correlation analyses, not only between the percentage change in C4 and FGF19, but also between the gut microbiome and host metabolome or clinical parameters. All statistics were performed using a two-tailed test. Values of $p < 0.05$ (*) were considered statistically significant, and $p < 0.01$ (**) was considered highly significant.

3. Results

3.1. Patient Characteristics

Ten male patients with hyperglycemia and hyperlipidemia who agreed to take 6 g of YH1 three times per day for 4 weeks were recruited for the study, and no participants dropped out. Prescription adherence of all patients was above 80%, and no participant was excluded from the data analysis. Participants were aged 38 to 57 years, with a median age of 48 years. The median duration of diabetes was 2.0 years, with a range from 0.5 to 14.0 years. Previous medications prescribed to treat chronic diseases included hypouricemic agents (sulfinpyrazone and benzbromarone) taken by two patients for a long time, four antihypertensive drugs taken by three participants (one individual received Exforge, one used olmesartan, and one took both Sevikar and lercanidipine), antihistamines (levocetirizine and denosin) used by two people, and silymarin used by two individuals. Participants continued to take the above medication without dosage or medication changes during the study.

3.2. Microbiota Profiles

Stool samples were collected from the participants before and after 4 weeks of YH1 treatment. Microbiota profiles were investigated using 16S rRNA sequencing. Microbial community variation, as indicated by Chao1 and Shannon indices (α-diversity), was slightly decreased after YH1 treatment without a significant difference (Figure 1a). Principal coordinate analysis (PCoA) revealed that the β-diversity of the gut microbiota at baseline and after treatment did not show significant changes (Figure 1b). Taxonomic profiles of the bacterial community at the genus level revealed elevated abundances of *Anaerostipes* (0.09 to 0.56%, $p = 0.015$) and *Escherichia/Shigella* (2.22 to 10.18%, $p < 0.005$), but reduced abundances of *Parabacteroides* (1.88 to 0.52%, $p < 0.005$), *Bifidobacterium* (1.31 to 0.09%, $p = 0.036$), and *Romboutsia* (0.25 to 0.01%, $p < 0.005$), after YH1 therapy (Figure 1c). YH1 reduced the abundance of bacteria species (Figure 1d) including *Parabacteroides distasonis* (0.92 to 0.32%, $p = 0.043$), *Parabacteroides merdae* (0.84 to 0.14%, $p = 0.031$), *Oscillibacter ruminantium* (0.39 to 0.13%, $p = 0.031$), and *Bacteroides caccae* (0.91 to 0.30%, $p = 0.011$). Collectively, our microbiome analysis indicated that, although YH1 did not significantly alter the richness and diversity of gut microbiota, specific gut bacteria showed prominently different abundances at the genus and species levels after 4 weeks of YH1 treatment. YH1 enriched SCFA-producing bacteria, including *Anaerostipes* and *Escherichia/Shigella*, and inhibited BSH-active clones, such as *Parabacteroides*, *Bifidobacterium*, and *Bacteroides caccae*.

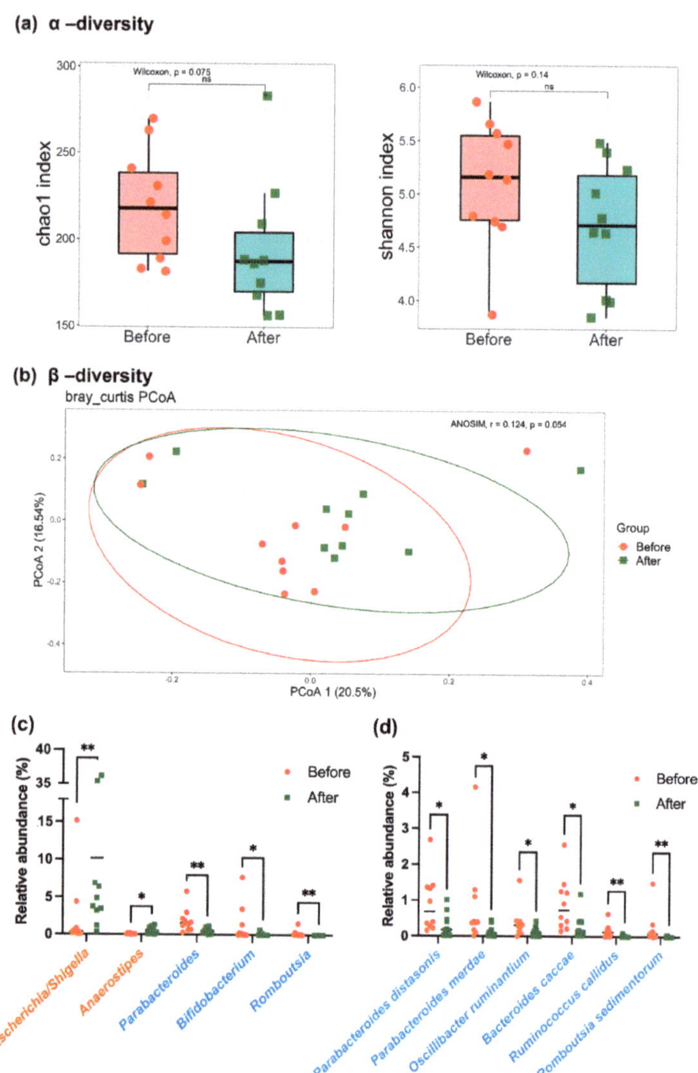

Figure 1. Oral YH1 modulated the composition of gut microbiota in men with type 2 diabetes and hyperlipidemia. (**a**) Chao1 and Shannon indices were used to analyze the α-diversity of fecal microbiota before (red circles) and after (green squares) YH1 treatment in ten patients. (**b**) The β-diversity of fecal microbiota at baseline and after YH1 treatment was determined by principal coordinate analysis (PCoA) based on the Bray-Curtis dissimilarity index. Significant changes in the fecal microbiota at the (**c**) genus and (**d**) species level before and after YH1 treatment are presented. Genus and species names in brown or blue font represent significant enrichment or depletion of bacteria after YH1 treatments, respectively. Statistical significance (* $p < 0.05$; ** $p < 0.01$) were detected by the Wilcoxon signed-rank test.

3.3. Changes in Plasma and Fecal Bile Acid Profiles, Plasma C4, and FGF19 Levels

In this study, plasma and stool samples were collected from ten patients at baseline and 4 weeks after oral administration of YH1. The changes in 15 types of plasma BAs are shown in Figure 2. Plasma levels of glyco- and tauro-conjugated primary BAs (GCA, TCA, GCDCA, TCDCA) all significantly increased. The concentrations of three other tauro-conjugated secondary BAs (TDCA, TLCA, TUDCA) were also elevated. Regarding glyco-conjugated secondary BAs, only levels of GDCA significantly increased. However, two unconjugated secondary BAs (DCA and LCA) showed significantly reduced concentrations. Regarding the change in the percentage composition of total BAs after YH1 treatment (Figure S1), the median percentage of CAs (sum of CA, GCA, and TCA) significantly increased from 10.3 to 15.9% ($p = 0.013$). The median percentage of CDCAs (sum of CDCA, GCDCA, and TCDCA) also increased prominently from 41.9 to 56.7% ($p = 0.007$). However, the median percentage of secondary BAs, including DCAs (sum of DCA, GDCA, and TDCA), LCAs (sum of LCA, GLCA, and TLCA), and UDCAs (sum of UDCA, GUDCA, and TUDCA), significantly decreased from 33.7 to 24.1% ($p = 0.007$), 2.3 to 1.5% ($p = 0.037$), and 6.9 to 3.2% ($p = 0.005$), respectively. Therefore, the ratios of CAs/DCAs and CDCAs/DCAs both increased significantly after YH1 treatment, indicating that YH1 reduced the production of secondary BAs.

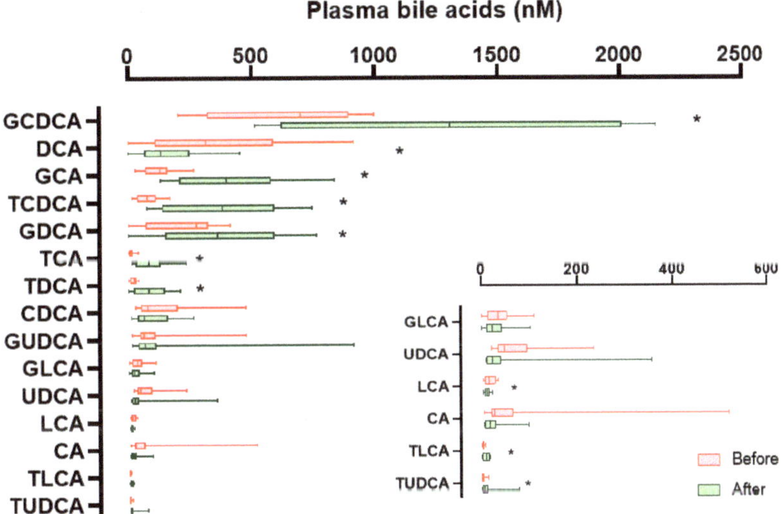

Figure 2. YH1 altered plasma bile acid profiles in men with type 2 diabetes and hyperlipidemia. The concentration of each bile acid in the plasma of ten patients before (red) and after (green) YH1 treatment is represented by a box-and-whisker plot. The line inside the box represents the median value, and the box is between the first and third quartiles. The ends of the whiskers represent the minimum and maximum values. Wilcoxon signed-rank test was used, and * $p < 0.05$ was considered statistically significant.

Regarding the change in BA classification after YH1 treatment from baseline (Figure 3), the levels of 12α-hydroxylated BAs increased after YH1 treatment ($p = 0.047$), with the median concentration increasing from 827.1 to 1073.2 nM. Similarly, the median concentration of non-12α-hydroxylated BAs significantly increased from 1036.9 to 2037.8 nM at the end of YH1 treatment ($p = 0.005$). Levels of primary BAs in plasma increased significantly ($p = 0.005$), with the median concentration increasing from 1118.4 to 2354.2 nM, whereas levels of secondary BAs did not change. Conjugated BA levels were significantly elevated ($p = 0.005$), with a median increase from 1261.4 to 2872.0 nM. Interestingly, levels of both glyco- and tauro-conjugated BAs were strikingly elevated, and the increase in tauro-conjugated BA levels was significantly greater than that of glyco-conjugated BA levels after YH1 treatment. Therefore, the median ratio of G/T-BAs decreased from 10.7 to 4.2 (p = 0.005), and a decline in the G/T ratio was observed in both primary and secondary BAs. Furthermore, the median unconjugated BA level decreased from 545.0 to 289.2 after treatment (p = 0.047). The ratio of different classifications of plasma BAs before and after YH1 treatment are presented in Figure 4. The ratio of 12α-hydroxylated to non-12α-hydroxylated BAs showed no significant change. The ratio of primary to secondary BAs increased significantly, with a median increase from 1.4- to 2.5-fold (p = 0.005). The ratio of conjugated to unconjugated BAs also increased prominently, with a median increase from 2.0- to 7.0-fold (p = 0.007), suggesting that YH1 inhibited the deconjugation of conjugated BAs by modulating the gut microbiome.

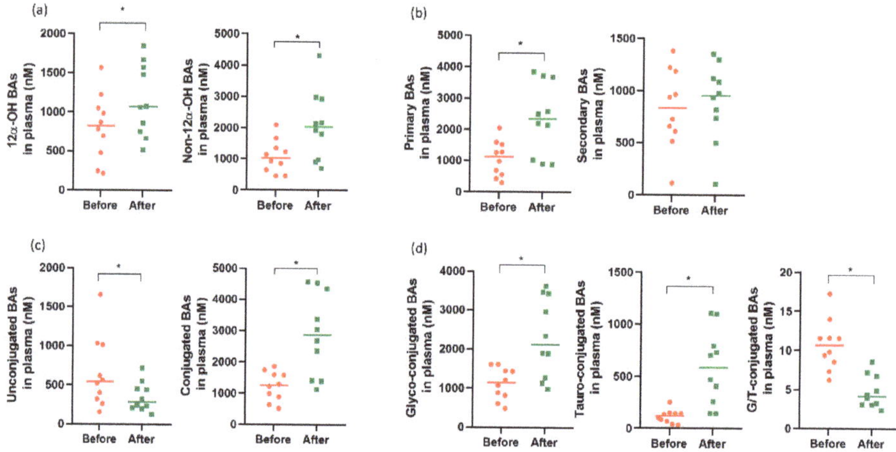

Figure 3. Changes in the composition of bile acids (BAs) of various categories in plasma before and after treatment with YH1. The concentrations of classified BAs in plasma before (red circles) and after (green squares) YH1 treatment in the ten patients are shown as scatter plots, with lines representing median values. Statistical differences (* p < 0.05) were detected by the Wilcoxon signed-rank test. After four weeks of YH1 treatment, the levels of (**a**) 12α-hydroxylated BAs and non-12α-hydroxylated BAs, (**b**) primary BAs, (**c**) conjugated BAs, and (**d**) glyco (G)- and tauro (T)-conjugated BAs in plasma were all significantly elevated. (**c**) Unconjugated BAs and (**d**) G/T ratio of conjugated BAs were significantly reduced. 12α-OH BAs, 12α-hydroxylated bile acids; Non-12α-OH BAs, non-12α-hydroxylated bile acids.

Figure 4 shows that the total levels of plasma BAs increased significantly after YH1 treatment (p = 0.017), with the median concentration increasing from 1914.5 to 3200.5 nM. C4 levels were also markedly elevated (p = 0.037), but plasma FGF19 levels showed no significant change before and after YH1 treatment (Figure S2). However, there was a significant negative correlation between the percentage change in C4 and FGF19 (r = −0.7; p = 0.025). Intestinal FXR regulated the production of FGF19, a hormone that travels via enterohepatic circulation with a negative feedback effect of hepatic CYP7A1 activity on BA and lipid metabolism. A lower FGF19 concentration suggested less activation of intestinal FXR signaling, thus leading to increases in BA biosynthesis and plasma C4 levels after YH1 treatment.

The analysis of BAs in fecal samples indicated that concentrations of stool TCDCA, TDCA, and TLCA were significantly increased after YH1 therapy (Figure S3). The total levels of tauro-conjugated BAs in feces increased significantly, with the median concentration rising from 363.2 to 747.3 nmol/mg after treatment (p = 0.047). However, there were no significant differences in total stool BAs or the ratios of different classifications in feces (Figure 4). Therefore, clinical YH1 treatment was found to be related to prominent modulation of plasma BA profiles and elevated levels of the fecal tauro-conjugated BAs in male individuals with T2D and hyperlipidemia.

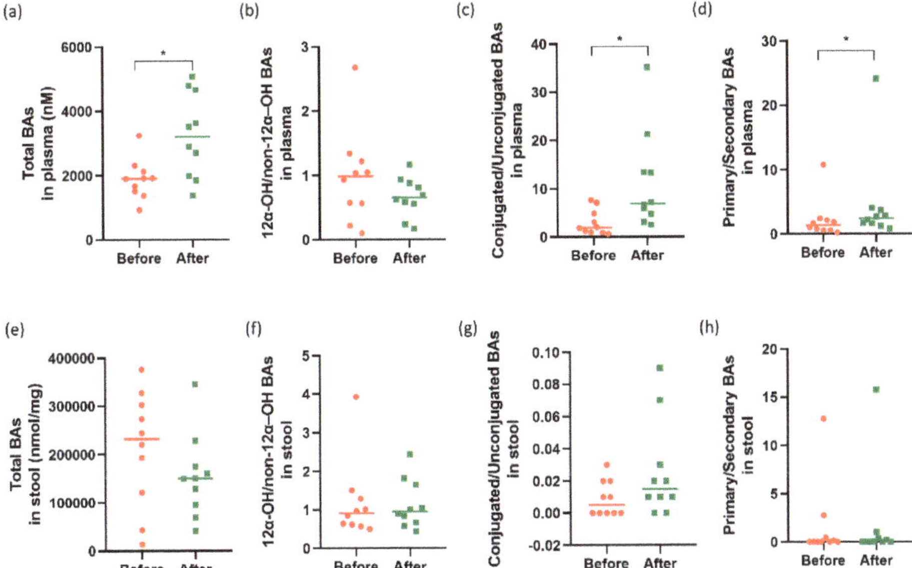

Figure 4. Changes in total and classification ratios of bile acids (BAs) in plasma and feces after YH1 treatment. Total BAs and classification ratios of BAs in plasma and feces before (red circles) and after (green squares) YH1 treatment in the ten patients are shown as scatter plots. The line in the graph represents the median value. The Wilcoxon signed-rank test was used to determine whether there was a statistical difference before and after treatment (* $p < 0.05$). (**a**) Total plasma BAs increased significantly after YH1 treatment. (**b**) The ratio of 12α-hydroxylated to non-12α-hydroxylated BAs didn't show significant change. Changes in the ratio of BA classification were only significantly elevated in plasma (**c**) conjugated/unconjugated BAs and (**d**) primary/secondary BAs. (**e**) Total stool BAs had no significant change before and after YH1 treatment. In addition, no significant difference was found in the analysis of fecal BA profiles (**f**–**h**) before and after treatment. 12α-OH/non-12α-OH BAs, 12α-hydroxylated/non-12α-hydroxylated bile acids.

3.4. Changes in Clinical Parameters and the Correlation between YH1-Responsive Microbiota and BA or Clinical Outcomes

Changes in various clinical parameters before and after four weeks of YH1 treatment are shown in Table 1. Regarding anthropometric characteristics, there were no significant differences in body weight, waist circumference, blood pressure, or heart rate. After YH1 treatment, the median glycated albumin level significantly decreased from 16.8 to 15.3% ($p = 0.005$), without hypoglycemic episodes. The median 2hPG level also declined from 158.0 to 117.5 mg/d ($p = 0.007$). However, no significant changes were found in FPG, HOMA-IR, or HOMA-β scores at the end of this study. Regarding the parameters of lipid metabolism, the 4-week YH1 treatment resulted in significant reductions ($p < 0.01$) in total cholesterol, LDL-C, and TG levels compared with the baseline values (Table 1). Although high-density lipoprotein cholesterol (HDL-C) levels decreased synchronously, the TG/HDL-C value, a predictor of cardiovascular disease, also decreased significantly after the treatment, with a median reduction from 4.4 to 2.7. Levels of the liver enzyme alanine aminotransferase (ALT) showed a decreasing trend after YH1 treatment ($p = 0.07$), with a median decrease from 49.5 to 39.5 U/L. However, neither inflammatory marker (CRP) levels nor renal function was significantly different after YH1 treatment in this study. Regarding the assessment of bowel movements, there were no significant changes in CAS or BSFS scores during this study. Adverse GI events were found once in two patients, experiencing mild nausea and bloating, respectively. The remaining eight patients did not experience any side effects during this trial. The severity of the above GI events was below grade two, as measured by the Common Terminology Criteria for Adverse Events (CTCAE, Version 4.0) grading system. YH1 treatment was temporarily paused for one day when the two participants reported discomfort, and treatment was resumed once symptoms were resolved without other medical intervention. No patients experienced hypoglycemia during this study.

Table 1. Clinical parameters before and after YH1 treatment.

	Pre-Treatment ($n = 10$)	Post-Treatment ($n = 10$)	p Value
	Median (Min, Max)	Median (Min, Max)	
Demographic characteristics			
Age (years)	48 (38, 57)		
Gender, Male/Female	10/0		
Duration of DM (years)	2.0 (0.5, 14.0)		
Anthropometric characteristics and vital signs			
Weight (kg)	78.8 (68.4, 102.6)	79.5 (68.2, 101.8)	0.76
BMI (kg/m^2)	27.2 (24.8, 34.2)	27.4 (24.7, 33.3)	0.77
SBP (mmHg)	133.0 (104.0, 153.0)	128.5 (103.0, 150.0)	0.42
DBP (mmHg)	86.0 (51.0, 101.0)	77.0 (62.0, 104.0)	0.17
HR (beat/min)	76.0 (60.0, 89.0)	73.0 (56.0, 80.0)	0.18
Laboratory data			
Glycated albumin (%)	16.8 (14.0, 32.1)	15.3 (12.8, 29.8)	0.005 **
FPG (mg/dL)	129.5 (81.0, 277.0)	130.0 (86.0, 255.0)	0.21
2hPG (mg/dL)	158.0 (111.0, 335.0)	117.5 (98.0, 304.0)	0.007 **
ALT (U/L)	49.5 (21.0, 69.0)	39.5 (12.0, 74.0)	0.07
Cr (mg/dL)	0.9 (0.5, 0.9)	0.7 (0.6, 1.0)	0.10
Total cholesterol(mg/dL)	231.5 (212.0, 262.0)	193.0 (141.0, 223.0)	0.007 **
HDL-C (mg/dL)	43.5 (34.0, 61.0)	40.0 (29.0, 50.0)	0.02 *
LDL-C (mg/dL)	150.0 (139.0, 196.0)	134.5 (88.0, 150.0)	0.007 **
Triglycerides (mg/dL)	194.0 (74.0, 497.0)	103.5 (58.0, 211.0)	0.005 **
Triglycerides/HDL-C	4.4 (1.2, 11.8)	2.7 (1.2, 4.9)	0.009 **
Fasting insulin (μU/mL)	10.2 (5.7, 20.1)	11.0 (5.3, 26.6)	0.14
HOMA-IR	3.8 (2.1, 7.7)	3.5 (2.0, 10.1)	0.24
HOMA-β	55.3 (13.0, 402.0)	81.9 (16.9, 275.1)	0.14
CRP (mg/L)	2.5 (0.4, 4.8) [#]	1.9 (0.6, 13.5) [#]	0.95

[#] Only nine subjects were analyzed because of missing data. The data are presented as medians (min, max). Statistical significance (* $p < 0.05$; ** $p < 0.01$) was obtained by the Wilcoxon signed-rank test. After 4 weeks of YH1 treatment, there were significant reductions in glycated albumin, 2hPG, total cholesterol, HDL-C, LDL-C, triglycerides, and triglycerides/HDL-C. There was a trend of decreased ALT level after YH1 treatment, but the difference was not significant. ALT, alanine aminotransferase; BMI, body mass index; Cr, creatinine; DBP, diastolic blood pressure; DM, diabetes mellitus; FPG, fasting plasma glucose; HDL-C, high-density lipoprotein cholesterol; HOMA-IR, homeostatic model assessment of insulin resistance; HOMA-β, homeostatic model assessment of β cell function; HR, heart rate; LDL-C, low-density lipoprotein cholesterol; 2hPG, 2-h postprandial glucose; SBP, systolic blood pressure.

There were some correlations between YH1-responsive microbiota and BAs or clinical outcomes. At the genus level, the increase in *Escherichia/Shigella* abundance represented a positive correlation with elevated TLCA (r = 0.648; p = 0.043) in plasma. *Parabacteroides* was negatively correlated with plasma TCDCA (r = −0.685; p = 0.029) and tauro-conjugated BAs (r = −0.661; p = 0.038), but positively correlated with LDL-C levels (r = 0.685; p = 0.029). In addition, *Bifidobacterium* was negatively correlated with TDCA (r = −0.697; p = 0.025) and tauro-conjugated secondary BAs (r = −0.648; p = 0.043) in plasma. Therefore, a decrease in *Parabacteroides* and *Bifidobacterium* abundances led to an increase in plasma levels of tauro-conjugated BAs and a decrease in LDL-C levels. At the species level, the reduction in *Parabacteroides distasonis* abundance also showed a positive correlation with decreased LDL-C levels (r = −0.782; *p* = 0.008), and the decrease in *Parabacteroides merdae* abundance was positively correlated with decreased plasma unconjugated BA levels (r = 0.745; *p* = 0.013). The decrease in *Oscillibacter ruminantium* abundance was positively correlated with reduced FPG levels (r = 0.867; *p* = 0.001), and negatively correlated with both elevated HOMA-β scores (r = −0.818; *p* = 0.004) and stool TLCA levels (r = −0.770; *p* = 0.009). In brief, YH1 significantly inhibited some microbiota in the gut, especially *Parabacteroides* and *Bifidobacterium*, which were related to the changes in BA composition, as well as in the clinical parameters.

4. Discussion

This pilot observational clinical study is the first to report that 4 weeks of oral YH1 treatment alone significantly alters the stool microbiome and increases levels of conjugated BAs, especially tauro-conjugated BAs, in plasma and stool. These changes could effectively improve glycemic levels and lipid profiles for middle-aged, T2D male patients with LDL-C ≥ 130 mg/dl and BMI ≥ 23 kg/m^2. YH1 increased the abundances of SCFA-producing bacteria *Anaerostipes* and *Escherichia/Shigella* and reduced the abundances of *Parabacteroides, Bifidobacterium, Romboutsia, Oscillibacter ruminantium,* and *Bacteroides caccae.* Similarly, metformin improved T2D by modulating the microbiota, such as increasing the abundance of *Escherichia/Shigella* and decreasing that of *Romboutsia, Oscillibacter, Bacteroides,* and *Parabacteroides* [29]. Other antidiabetic drugs, such as DPP-4 inhibitors, reduce *Oscillibacter* abundance, and GLP-1 receptor agonist increases *Anaerostipes* abundance [29]. Therefore, YH1 could serve as an add-on medication to enhance the hypoglycemic effects of OHAs to achieve better glycemic control [3], by potentially modulating gut microbiota in patients with T2D.

Although YH1 contained berberine, the changes in microbiota and plasma BAs after YH1 treatment were partially different from those observed after a single berberine compound treatment in previous studies [7,9,30]. Previous experimental research indicated that oral administration of berberine could alter the composition of BAs by affecting gut microbes via a reduction in or elimination of *Clostridium* spp. Therefore, BSH activity was decreased, leading to an accumulation of TCA in the intestine, thus acting on FXR receptors and affecting lipid metabolism [7,30]. A clinical study reported that the hypoglycemic effect of berberine was mediated by inhibiting *Ruminococcus bromii* to attenuate DCA biotransformation [9]. In addition, berberine also depleted species including *Parabacteroides merdae* and *Bifidobacterium* spp., but enriched species such as *Parabacteroides distasonis.* Our study proved that YH1 decreased BSH-expressing microbiota, such as *Bifidobacterium, Parabacteroides merdae, Parabacteroides distasonis,* and *Bacteroides caccae,* elevating the ratio of conjugated to unconjugated BAs. Interestingly, we found that YH1 treatment decreased the G/T ratio of BAs, which was not reported in a previous clinical trial [9]. Gut microbiota with different phylotypes of BSH display some degree of selectivity for conjugated BA substrates. The phylotypes of BSH-T5 and BSH-T6 mainly comprised *Bacteroides* and *Parabacteroides,* exhibiting different deconjugating activity between tauro- and glyco-conjugated BAs [14]. Similarly, other studies suggested that the deconjugation activities of some bacteria, such as *Bacteroidetes,* was more selective for tauro-conjugated BAs than glyco-conjugated BAs [31,32]. Therefore, YH1 increased the levels of conjugated BAs, especially tauro-conjugated BAs, probably by inhibiting specific BSH-phylotype bacteria, including *Parabacteroides* and *Bacteroides caccae.* Furthermore, YH1 treatment increased the ratio of primary to secondary BAs in plasma, but the levels of primary unconjugated BAs (CA and CDCA) were not elevated. As the abundances of 7α-dehydroxylated bacteria, such as *Clostridium* (clusters XIVa and XI) and *Eubacterium,* did not show significant changes in this study and these bacteria can metabolize only unconjugated-BAs [11], the reduced production of secondary BAs (DCA and LCA) by YH1 might be related to the decrease in unconjugated BAs. Interestingly, secondary BAs and SCFAs are two major types of gut bacterial metabolites and have opposing effects on colonic inflammation. Secondary BAs, including DCA and LCA, are risk factors for gut inflammation and cancer, whereas an increase in SCFA production is associated with anti-inflammation and anticancer effects [33]. Our study demonstrated that YH1 treatment not only significantly decreased the levels of unconjugated secondary BAs, but also considerably elevated the abundances of SCFA-producing bacteria, which may play a role in protection from gut inflammation. Overall, YH1 differed from berberine in modulating gut microbiota and BA composition, which may help explain the more potent hypoglycemic and hypolipidemic effects of this herbal formula than that of the pure compound.

BAs regulate glucose, lipid, and energy homeostasis mainly through nuclear FXR and TGR5, and have pathophysiologic roles in obesity, T2D, dyslipidemia, and nonalcoholic steatohepatitis [34,35]. In addition, tauro- or glyco-conjugated forms of BAs showed agonistic activity on TGR5, with taurine conjugates being more potent than glycine conjugates [17]. In this study, YH1 treatment increased the stool levels of tauro-conjugated BAs, which can activate TGR5 in gut L cells and lead to GLP-1 secretion and glucose homeostasis. A previous research showed that berberine-enriched SCFA-producing bacteria, especially butyrate-producing bacteria, could have beneficial effects on the host [21]. Our study shows that YH1 increases the abundance of *Anaerostipes*, as well as *Escherichia/Shigella*, and both are butyrate-producing strains in the intestine. Through SCFA receptors, gut microbiota-derived SCFAs have roles in GLP-1 release, insulin secretion, and regulation of energy expenditure [18]. Collectively, the findings of this study suggest that YH1 increases not only stool tauro-conjugated BA levels, but also *Anaerostipes* and *Escherichia/Shigella*-derived SCFA levels that target receptors in gut L cells. This results in improvements in GLP-1 secretion and hyperglycemia, without any side effects of hypoglycemia. Conjugated BAs were associated with less activation of intestinal FXR signaling, so less negative feedback regulation promoted CYP7A1 activity with elevation of plasma C4 levels in our study. This led to liver cholesterol-to-BA conversion and hyperlipidemia reduction. Hence, this clinical research on YH1 could support the therapeutic mechanisms of TCM in improving metabolic diseases by modulating the gut microbiota and BAs found in animal studies [2].

Comparing our previous clinical trial [3] with the current study, YH1 not only was effective in patients with treatment-resistant T2D, but also had beneficial outcomes for drug-naïve T2D patients. Similarly, the incidence of adverse GI events, including nausea or bloating, in the YH1 group of previous trial and current study were 21.7 and 20%, respectively. In contrast to the hypoglycemic effects of OHAs alone, hyperglycemia and hyperlipidemia were significantly improved after YH1 treatment in both studies. There was no significant difference in the change of body weight or waist circumference after 4 weeks of oral YH1 treatment, possibly due to the shorter treatment period of this study. Although HDL-C was slightly reduced at week 4, the TG/HDL-C ratio decreased significantly after YH1 treatment in this study ($p = 0.009$), with a median decrease from 4.4 to 2.7. According to the previous literature, a TG/HDL-C ratio > 3.5 is linked with higher risks of cardiovascular disease and coronary heart disease-related death. It is anticipated that the future annual incidence of diabetes will increase two-fold for individuals with a TG/HDL-C ratio > 3.5 [36,37]. Nine out of ten YH1 subjects recruited for this trial met the diagnostic criteria for metabolic syndrome, which was associated with a two-fold increase in the risk of stroke and heart disease in the future compared to that in the normal population [38,39]. This study suggests that YH1 treatment improves glycemic and lipid profiles, which could result in the reduced risk of patients developing cardiovascular disease.

This study has the following limitations. First, the sample size was small, and participation in our study was limited to males only. Females were excluded to avoid blood contamination of fecal samples during the menstrual period and to avoid sex differences in the microbiota and BA compositions. The standardized breakfast was provided in this study to prevent different diet-related BA secretion, and dietary records of patients were monitored by the same nutritionist during the study. Therefore, we controlled and minimized variables that could affect gut microbiota and BA composition to make our findings more reliable. Larger randomized controlled trials are warranted to confirm these pilot results. Future research that enrolls women together with men could determine if the influences of YH1 on gut microbiota and BA are universal or gender specific for patients with T2D and hyperlipidemia. Second, this study enrolled middle-aged, overweight or obese patients with treatment-naïve T2D and hyperlipidemia in Taiwan. The results may not be applicable to individuals in different stages of T2D or other racial/ethnic populations. Finally, the 4-week study period was relatively short. Longer treatment periods are required to evaluate the long-term effects of YH1 on microbiota and BA metabolism. Notwithstanding these limitations, our study still provides important evidence that YH1 can manage T2D and hyperlipidemia by regulating the microbiome and BA composition.

5. Conclusions

We found that YH1 improved T2D and hyperlipidemia, by potentially inhibiting some gut bacteria with specific phenotypes of BSH, resulting in reduced deconjugation abilities. This led to elevated levels of conjugated BAs, especially tauro-conjugated BAs in plasma and stool. Consequently, YH1 potentially inhibited the activity of intestinal FXR, thereby decreasing hyperlipidemia by reducing negative feedback regulation and promoting the conversion of hepatic cholesterol to BAs. In addition, increased levels of conjugated BAs plus *Anaerostipes* and *Escherichia/Shigella*-derived SCFAs might further improve hyperglycemia by regulating the release of GLP-1 in gut L cells. The detailed and complicated mechanism of YH1 requires more clinical and basic research in the future.

Supplementary Materials: The following supporting information can be downloaded at: https://www.mdpi.com/article/10.3390/pharmaceutics14091857/s1, Protocol S1: Analysis of stool and plasma bile acids as well as plasma C4; Figure S1: Changes in the percentage composition of total bile acids after YH1 treatment; Figure S2: Plasma C4 and FGF19 levels before and after YH1 treatment; Figure S3: Stool bile acid profiles before and after YH1 treatment.

Author Contributions: Conceptualization, Y.-H-H. and Y.-H.W.; methodology, Y.-H.H., Y.-M.Y. and M.-L.C.; software, G.-H.L. and Y.-M.Y.; formal analysis, H.-Y.T. and Y.-M.Y.; investigation and project administration, Y.-H.H., S.-T.C., C.-C.W., W.-J.H., Y.-H.L. and P.-Y.L.; resources, Y.-M.Y. and M.-L.C.; data curation, Y.-H.H. and G.-H.L.; writing—original draft preparation, Y.-H.H.; writing—review and editing, Y.-H.W., H.-Y.T., C.-J.L., Y.-M.Y. and M.-L.C.; supervision, Y.-H.W., Y.-M.Y. and M.-L.C. All authors have read and agreed to the published version of the manuscript.

Funding: This study was supported by research grants from Chang Gung Memorial Hospital Research Program (CMRPG1K0011) to Y.-H.H. and Ministry of Science and Technology (MOST 110-2320-B-182A-004) to Y.-M.Y. The study sponsor had no role in study design, data collection, data analysis, results interpretation, preparation of the manuscript, or decision to publish.

Institutional Review Board Statement: The study was conducted in accordance with the Declaration of Helsinki and was approved by the Committee on Research Ethics of the Chang Gung Memorial Hospital in Taiwan (No. 201901022B0A3; date 13 August 2019).

Informed Consent Statement: Informed consent was obtained from all subjects involved in the study. Trial registration: ClinicalTrials.gov NCT04194515.

Data Availability Statement: The data presented in this study are available on request from the corresponding author.

Acknowledgments: We are grateful to the nurses and research assistants, especially Shu Chuan Shih, for their practical work during the study.

Conflicts of Interest: The authors declare no conflict of interest.

References

1. Tong, X.L.; Dong, L.; Chen, L.; Zhen, Z. Treatment of diabetes using traditional Chinese medicine: Past, present and future. *Am. J. Chin. Med.* **2012**, *40*, 877–886. [CrossRef]
2. Zhang, H.Y.; Tian, J.X.; Lian, F.M.; Li, M.; Liu, W.K.; Zhen, Z.; Liao, J.Q.; Tong, X.L. Therapeutic mechanisms of traditional Chinese medicine to improve metabolic diseases via the gut microbiota. *Biomed. Pharmacother.* **2021**, *133*, 110857. [CrossRef] [PubMed]

3. Huang, Y.H.; Chen, S.T.; Liu, F.H.; Hsieh, S.H.; Lin, C.H.; Liou, M.J.; Wang, C.C.; Huang, C.H.; Liu, G.H.; Lin, J.R.; et al. The efficacy and safety of concentrated herbal extract granules, YH1, as an add-on medication in poorly controlled type 2 diabetes: A randomized, double-blind, placebo-controlled pilot trial. *PLoS ONE* **2019**, *14*, e0221199. [CrossRef] [PubMed]

4. Yin, J.; Ye, J.; Jia, W. Effects and mechanisms of berberine in diabetes treatment. *Acta Pharm. Sin. B* **2012**, *2*, 327–334. [CrossRef]

5. Habtemariam, S. The Quest to Enhance the Efficacy of Berberine for Type-2 Diabetes and Associated Diseases: Physicochemical Modification Approaches. *Biomedicines* **2020**, *8*, 90. [CrossRef]

6. He, K.; Hu, Y.; Ma, H.; Zou, Z.; Xiao, Y.; Yang, Y.; Feng, M.; Li, X.; Ye, X. *Rhizoma coptidis* alkaloids alleviate hyperlipidemia in B6 mice by modulating gut microbiota and bile acid pathways. *Biochim. Biophys. Acta* **2016**, *1862*, 1696–1709. [CrossRef]

7. Tian, Y.; Cai, J.; Gui, W.; Nichols, R.G.; Koo, I.; Zhang, J.; Anitha, M.; Patterson, A.D. Berberine Directly Affects the Gut Microbiota to Promote Intestinal Farnesoid X Receptor Activation. *Drug Metab. Dispos.* **2019**, *47*, 86–93. [CrossRef]

8. Wolf, P.G.; Devendran, S.; Doden, H.L.; Ly, L.K.; Moore, T.; Takei, H.; Nittono, H.; Murai, T.; Kurosawa, T.; Chlipala, G.E.; et al. Berberine alters gut microbial function through modulation of bile acids. *BMC Microbiol.* **2021**, *21*, 24. [CrossRef]

9. Zhang, Y.; Gu, Y.; Ren, H.; Wang, S.; Zhong, H.; Zhao, X.; Ma, J.; Gu, X.; Xue, Y.; Huang, S.; et al. Gut microbiome-related effects of berberine and probiotics on type 2 diabetes (the PREMOTE study). *Nat. Commun.* **2020**, *11*, 5015. [CrossRef]

10. Ramírez-Pérez, O.; Cruz-Ramón, V.; Chinchilla-López, P.; Méndez-Sánchez, N. The Role of the Gut Microbiota in Bile Acid Metabolism. *Ann. Hepatol.* **2017**, *16* (Suppl. S1), S21–S26. [CrossRef]

11. Wahlström, A.; Sayin, S.I.; Marschall, H.U.; Bäckhed, F. Intestinal Crosstalk between Bile Acids and Microbiota and Its Impact on Host Metabolism. *Cell Metab.* **2016**, *24*, 41–50. [CrossRef] [PubMed]

12. Ferrell, J.M.; Chiang, J.Y.L. Understanding Bile Acid Signaling in Diabetes: From Pathophysiology to Therapeutic Targets. *Diabetes Metab. J.* **2019**, *43*, 257–272. [CrossRef] [PubMed]

13. Panzitt, K.; Zollner, G.; Marschall, H.U.; Wagner, M. Recent advances on FXR-targeting therapeutics. *Mol. Cell. Endocrinol.* **2022**, *552*, 111678. [CrossRef]

14. Song, Z.; Cai, Y.; Lao, X.; Wang, X.; Lin, X.; Cui, Y.; Kalavagunta, P.K.; Liao, J.; Jin, L.; Shang, J.; et al. Taxonomic profiling and populational patterns of bacterial bile salt hydrolase (BSH) genes based on worldwide human gut microbiome. *Microbiome* **2019**, *7*, 9. [CrossRef]

15. Duboc, H.; Taché, Y.; Hofmann, A.F. The bile acid TGR5 membrane receptor: From basic research to clinical application. *Dig. Liver Dis.* **2014**, *46*, 302–312. [CrossRef] [PubMed]

16. Sepe, V.; Festa, C.; Renga, B.; Carino, A.; Cipriani, S.; Finamore, C.; Masullo, D.; Del Gaudio, F.; Monti, M.C.; Fiorucci, S.; et al. Insights on FXR selective modulation. Speculation on bile acid chemical space in the discovery of potent and selective agonists. *Sci. Rep.* **2016**, *6*, 19008. [CrossRef]

17. Kawamata, Y.; Fujii, R.; Hosoya, M.; Harada, M.; Yoshida, H.; Miwa, M.; Fukusumi, S.; Habata, Y.; Itoh, T.; Shintani, Y.; et al. A G protein-coupled receptor responsive to bile acids. *J. Biol. Chem.* **2003**, *278*, 9435–9440. [CrossRef]

18. Priyadarshini, M.; Wicksteed, B.; Schiltz, G.E.; Gilchrist, A.; Layden, B.T. SCFA Receptors in Pancreatic Beta Cells: Novel Diabetes Targets? *Trends Endocrinol. Metab.* **2016**, *27*, 653–664. [CrossRef]

19. Zhang, S.; Lin, L.; Liu, W.; Zou, B.; Cai, Y.; Liu, D.; Xiao, D.; Chen, J.; Li, P.; Zhong, Y.; et al. Shen-Ling-Bai-Zhu-San alleviates functional dyspepsia in rats and modulates the composition of the gut microbiota. *Nutr. Res.* **2019**, *71*, 89–99. [CrossRef]

20. Lv, W.J.; Liu, C.; Li, Y.F.; Chen, W.Q.; Li, Z.Q.; Li, Y.; Xiong, Y.; Chao, L.M.; Xu, X.L.; Guo, S.N. Systems Pharmacology and Microbiome Dissection of Shen Ling Bai Zhu San Reveal Multiscale Treatment Strategy for IBD. *Oxid. Med. Cell. Longev.* **2019**, *2019*, 8194804. [CrossRef]

21. Wang, Y.; Shou, J.W.; Li, X.Y.; Zhao, Z.X.; Fu, J.; He, C.Y.; Feng, R.; Ma, C.; Wen, B.Y.; Guo, F.; et al. Berberine-induced bioactive metabolites of the gut microbiota improve energy metabolism. *Metabolism* **2017**, *70*, 72–84. [CrossRef]

22. Yin, J.; Xing, H.; Ye, J. Efficacy of berberine in patients with type 2 diabetes mellitus. *Metabolism* **2008**, *57*, 712–717. [CrossRef]

23. Yang, C.Y.; Yeh, Y.M.; Yu, H.Y.; Chin, C.Y.; Hsu, C.W.; Liu, H.; Huang, P.J.; Hu, S.N.; Liao, C.T.; Chang, K.P.; et al. Oral Microbiota Community Dynamics Associated with Oral Squamous Cell Carcinoma Staging. *Front. Microbiol.* **2018**, *9*, 862. [CrossRef]

24. Chiu, S.F.; Huang, P.J.; Cheng, W.H.; Huang, C.Y.; Chu, L.J.; Lee, C.C.; Lin, H.C.; Chen, L.C.; Lin, W.N.; Tsao, C.H.; et al. Vaginal Microbiota of the Sexually Transmitted Infections Caused by Chlamydia trachomatis and Trichomonas vaginalis in Women with Vaginitis in Taiwan. *Microorganisms* **2021**, *9*, 1864. [CrossRef]

25. Edgar, R.C. UPARSE: Highly accurate OTU sequences from microbial amplicon reads. *Nat. Methods* **2013**, *10*, 996–998. [CrossRef]

26. Edgar, R.C. SINTAX: A simple non-Bayesian taxonomy classifier for 16S and ITS sequences. *bioRxiv* **2016**. bioRxiv:074161. [CrossRef]

27. Caporaso, J.G.; Kuczynski, J.; Stombaugh, J.; Bittinger, K.; Bushman, F.D.; Costello, E.K.; Fierer, N.; Peña, A.G.; Goodrich, J.K.; Gordon, J.I.; et al. QIIME allows analysis of high-throughput community sequencing data. *Nat. Methods* **2010**, *7*, 335–336. [CrossRef]

28. Love, M.I.; Huber, W.; Anders, S. Moderated estimation of fold change and dispersion for RNA-seq data with DESeq2. *Genome Biol.* **2014**, *15*, 550. [CrossRef]

29. Gurung, M.; Li, Z.; You, H.; Rodrigues, R.; Jump, D.B.; Morgun, A.; Shulzhenko, N. Role of gut microbiota in type 2 diabetes pathophysiology. *eBioMedicine* **2020**, *51*, 102590. [CrossRef]

30. Sun, R.; Yang, N.; Kong, B.; Cao, B.; Feng, D.; Yu, X.; Ge, C.; Huang, J.; Shen, J.; Wang, P.; et al. Orally Administered Berberine Modulates Hepatic Lipid Metabolism by Altering Microbial Bile Acid Metabolism and the Intestinal FXR Signaling Pathway. *Mol. Pharmacol.* **2017**, *91*, 110–122. [CrossRef]

31. Staley, C.; Weingarden, A.R.; Khoruts, A.; Sadowsky, M.J. Interaction of gut microbiota with bile acid metabolism and its influence on disease states. *Appl. Microbiol. Biotechnol.* **2017**, *101*, 47–64. [CrossRef]

32. Yao, L.; Seaton, S.C.; Ndousse-Fetter, S.; Adhikari, A.A.; DiBenedetto, N.; Mina, A.I.; Banks, A.S.; Bry, L.; Devlin, A.S. A selective gut bacterial bile salt hydrolase alters host metabolism. *eLife* **2018**, *7*, e37182. [CrossRef]

33. Zeng, H.; Umar, S.; Rust, B.; Lazarova, D.; Bordonaro, M. Secondary Bile Acids and Short Chain Fatty Acids in the Colon: A Focus on Colonic Microbiome, Cell Proliferation, Inflammation, and Cancer. *Int. J. Mol. Sci.* **2019**, *20*, 1214. [CrossRef]

34. Porez, G.; Prawitt, J.; Gross, B.; Staels, B. Bile acid receptors as targets for the treatment of dyslipidemia and cardiovascular disease. *J. Lipid Res.* **2012**, *53*, 1723–1737. [CrossRef]

35. Chávez-Talavera, O.; Tailleux, A.; Lefebvre, P.; Staels, B. Bile Acid Control of Metabolism and Inflammation in Obesity, Type 2 Diabetes, Dyslipidemia, and Nonalcoholic Fatty Liver Disease. *Gastroenterology* **2017**, *152*, 1679–1694.e1673. [CrossRef]

36. Hajian-Tilaki, K.; Heidari, B.; Bakhtiari, A. Triglyceride to high-density lipoprotein cholesterol and low-density lipoprotein cholestrol to high-density lipoprotein cholesterol ratios are predictors of cardiovascular risk in Iranian adults: Evidence from a population-based cross-sectional study. *Casp. J. Intern. Med.* **2020**, *11*, 53–61. [CrossRef]

37. Vega, G.L.; Barlow, C.E.; Grundy, S.M.; Leonard, D.; DeFina, L.F. Triglyceride-to-high-density-lipoprotein-cholesterol ratio is an index of heart disease mortality and of incidence of type 2 diabetes mellitus in men. *J. Investig. Med.* **2014**, *62*, 345–349. [CrossRef]

38. Mottillo, S.; Filion, K.B.; Genest, J.; Joseph, L.; Pilote, L.; Poirier, P.; Rinfret, S.; Schiffrin, E.L.; Eisenberg, M.J. The metabolic syndrome and cardiovascular risk a systematic review and meta-analysis. *J. Am. Coll. Cardiol.* **2010**, *56*, 1113–1132. [CrossRef]

39. Kurl, S.; Laukkanen, J.A.; Niskanen, L.; Laaksonen, D.; Sivenius, J.; Nyyssönen, K.; Salonen, J.T. Metabolic syndrome and the risk of stroke in middle-aged men. *Stroke* **2006**, *37*, 806–811. [CrossRef]

MDPI

St. Alban-Anlage 66

4052 Basel

Switzerland

www.mdpi.com

Pharmaceutics Editorial Office

E-mail: pharmaceutics@mdpi.com

www.mdpi.com/journal/pharmaceutics